Chiral
Photochemistry

MOLECULAR AND SUPRAMOLECULAR PHOTOCHEMISTRY

Series Editors
V. RAMAMURTHY
Professor
Department of Chemistry
Tulane University
New Orleans, Louisiana

KIRK S. SCHANZE
Professor
Department of Chemistry
University of Florida
Gainesville, Florida

ADDITIONAL VOLUMES IN PREPARATION

Chiral
Photochemistry

edited by
Yoshihisa Inoue
Osaka University
Suita, and
Japan Science and Technology Agency
Kawaguchi, Japan

V. Ramamurthy
Tulane University
New Orleans, Louisiana, U.S.A.

MARCEL DEKKER

NEW YORK

SEP/AE

CHEM

Library of Congress Cataloging-in-Publication Data
A catalog record for this book is available from the Library of Congress.

ISBN: 0-8247-5710-6

This book is printed on acid-free paper.

Headquarters
Marcel Dekker, 270 Madison Avenue, New York, NY 10016, U.S.A.
tel: 212-696-9000; fax: 212-685-4540

Distribution and Customer Service
Marcel Dekker, Cimarron Road, Monticello, New York 12701, U.S.A.
tel: 800-228-1160; fax: 845-796-1772

World Wide Web
http://www.dekker.com

The publisher offers discounts on this book when ordered in bulk quantities. For more information, write to Special Sales/Professional Marketing at the headquarters address above.

Current printing (last digit):

10   9   8   7   6   5   4   3   2   1

PRINTED IN THE UNITED STATES OF AMERICA

Nov. 16, 2004
ac

ac
12/16/04

Preface

Control of molecular chirality is central to contemporary chemistry- and biology-related areas, such as pharmaceutical, medicinal, agricultural, environmental, and materials science and technology. Thus, a wide variety of sophisticated chiral reagents, auxiliary, catalysts, hosts, and methodologies have been developed particularly in recent years as tools for controlling chiral reactions, equilibria, and recognition in both chemistry and biology.

Traditionally, chirality control has been achieved predominantly through the ground-state interactions of substrate/guest with chiral reagent/auxiliary/catalyst/host/receptor/enzyme, which are structurally and energetically well defined and characterized in considerable detail. In sharp contrast, the possibility of chirality control in the electronically excited state has not been extensively explored experimentally or theoretically, until recently. This is simply because of the long held belief that in the excited states chiral interactions, particularly the intermolecular ones, are too weak and short-lived in general to achieve high stereochemical recognition and differentiation. However, this has turned out not to be true, and several recent works have clearly demonstrated that chirality control in the excited state is not a difficult but rather a fascinating subject to study, displaying various novel phenomena which are unexpected and unprecedented in the conventional *thermal* chiral chemistry occurring in the ground electronic state.

The origin of chiral photochemistry, or photochirogenesis, dates back to the late 19th Century, when le Bel (1875) and van't Hoff (1894) suggested the use of circularly polarized light for so-called "absolute asymmetric synthesis (AAS)." The first successful AAS was achieved by Kuhn in 1929, immediately after the discovery of circular dichroism by Cotton. Modern chiral organic photochemistry may be traced back to the work of Hammond and Cole reported in 1965; they demonstrated for the first time that chiral information could be transferred from an optically active sensitizer to a substrate upon photosensitization. In the 1970s, considerable effort was devoted to the AAS of helicene precursors by Kagan's and Calvin's groups, while more recently, and particularly in the last decade, molecular and supramolecular chiral photochemistry has attracted growing interest as a unique methodology for inducing molecular chirality in the photoproducts. Consequently, an increasing number of studies have been reported at a variety of conferences and in journals of different disciplines of science and technology. This rejuvenation has even expanded to fields other than chemistry such as biology, pharmacology, physics, and even astronomy. Under such circumstances, the First International Symposium on Asymmetric Photochemistry was held in Osaka, Japan, on September 4-6, 2001, and the Second Symposium in Nara, Japan, on July 30-31, 2003, which clearly demonstrates the growing activity and interest in this interdisciplinary field of science and technology.

In this exclusive volume on chiral photochemistry, leading scientists in the field vividly depict the most recent achievements in this interdisciplinary field of photochemistry and chiral chemistry, which can be applied to a wide variety of science and technology. The chapters herein are up-to-date, comprehensive, and authoritative. This book is a valuable resource for advanced courses in chemistry, biochemistry, and pharmacology.

Yoshihisa Inoue
V. Ramamurthy

Contents

Contributors

Thorsten Bach   Department of Chemistry, Technische Universität München, München, Germany

Paul Brumer   Department of Chemistry, University of Toronto, Toronto, Ontario, Canada

Benjamin Grosch   Department of Chemistry, Technische Universität München, München, Germany

Taisuke Hamada   Department of Bioresources Engineering, Okinawa National College of Technology, Okinawa, Japan

Norbert Hoffmann   Université de Reims Champagne-Ardenne, Reims, France

Yoshihisa Inoue   Department of Molecular Chemistry, Osaka University, Suita, and ICORP Entropy Control Project, Japan Science and Technology Agency, Kawaguchi, Japan

Abraham Joy   Department of Chemistry, Georgia Institute of Technology, Atlanta, Georgia, U.S.A.

Lakshmi S. Kaanumalle   Department of Chemistry, Tulane University, New Orleans, Louisiana, U.S.A.

S. Karthikeyan   Department of Chemistry, Tulane University, New Orleans, Louisiana, U.S.A.

Hideko Koshima   Department of Applied Chemistry, Ehime University, Matsuyama, Japan

Reiko Kuroda   Department of Life Sciences, The University of Tokyo, Tokyo, and Kuroda Chiromorphology Project, ERATO, Japan Science and Technology Agency, Kawaguchi, Japan

Arunkumar Natarajan   Department of Chemistry, Tulane University, New Orleans, Louisiana, U.S.A.

Yuji Ohashi   Department of Chemistry and Materials Science, Tokyo Institute of Technology, Tokyo, Japan

Jean-Pierre Pete   Université de Reims Champagne-Ardenne, Reims, France

V. Ramamurthy   Department of Chemistry, Tulane University, New Orleans, Louisiana, U.S.A.

Hermann Rau   Universität Hohenheim, Stuttgart, Germany

G. L. J. A. Rikken   Laboratoire National des Champs Magnétiques Pulsés, CNRS–INS–UPS, Toulouse, France

Masako Saito   Department of Advanced Materials Chemistry, Yokohama National University, Yokohama, Japan

Shigeyoshi Sakaki   Department of Molecular Engineering, Kyoto University, Kyoto, Japan

Masami Sakamoto   Department of Applied Chemistry and Biotechnology, Chiba University, Chiba, Japan

John R. Scheffer   Department of Chemistry, University of British Columbia, Vancouver, British Columbia, Canada

J. Shailaja   Department of Chemistry, University of Colorado, Boulder, Colorado, U.S.A.

Moshe Shapiro   The Weizmann Institute, Rehovot, Israel, and Departments of Chemistry and Physics, University of British Columbia, Vancouver, British Columbia, Canada

J. Sivaguru   Department of Chemistry, Columbia University, New York, New York, U.S.A.

Takehiko Wada   Department of Molecular Chemistry, Osaka University, Suita, and PRESTO, Japan Science and Technology Agency, Kawaguchi, Japan

Eiji Yashima   Department of Molecular Design and Engineering, Nagoya University, Nagoya, Japan

Yasushi Yokoyama   Department of Advanced Materials Chemistry, Yokohama National University, Yokohama, Japan

Chiral
Photochemistry

1

Direct Asymmetric Photochemistry with Circularly Polarized Light

Hermann Rau
Universität Hohenheim
Stuttgart, Germany

I. INTRODUCTION

Asymmetry in chemistry is realized by chiral molecules. *Chirality* (or handedness) is a symmetry property: chiral structures lack a center of inversion and a mirror plane. In group theory they belong to the nonaxial point group C_1 or the purely rotational groups C_n and D_n. Chiral molecules always exist in two forms: the structures of the two *enantiomers* (the old term *optical antipodes* is considered obsolete [1]) are like left and right hands or differ by a left or right screw sense in their structure. The labeling of the individual enantiomers follows different nomenclatures. The enantiomers of most organic molecules, especially those with an asymmetric C atom, are identified by the affixes R and S [2], helicenes by P and M, and octahedral bidentate complexes by Δ and Λ.

The bond lengths and bond angles of enantiomers have the same magnitude, but at least some bond angles have identical positive values in one and negative values in the other enantiomer. Thus the nonvectorial properties, as e.g. energies of formation and activation, reaction rates, or excited state lifetimes, are identical for both enantiomers. Therefore in nonasymmetric chemistry both enantiomers are produced in equal quantities, and such a mixture is called a *racemic mixture*, or more loosely a *racemate*. The individual enantiomer of a chiral molecule can be identified only by its interaction with a chiral probe. The interactions between one enantiomer of the object and the two of the chiral probing system leads to states or complexes of different energies, reaction rates, or lifetimes, which can be discriminated by normal analytical tools. This is the important principle of

1

diastereomerism. In all asymmetric reactions, an element of diastereomerism, i.e., a combination of, for example, R/R and R/S, or S/Λ and S/Δ, is apparent or hidden. Important probing systems are circularly polarized light (cpl) or chromatography at chiral phases.

A. Circularly Polarized Light, ORD and CD

Chirality is the origin of the spectroscopic property *optical activity.* The interaction of light and matter is characterized by the refractive index and the absorption coefficient. For chiral molecules, both the refractive index and the absorbance coefficient of one enantiomer differ for right and left circularly polarized light (r-cpl and l-cpl).

Light is a transversal electromagnetic wave, i.e., the electric and magnetic field vectors (*E* and *H*, respectively) are perpendicular to the direction of propagation *z* (and perpendicular to one another). In polarized radiation there are spatial and temporal restrictions for these vectors. Linear polarization requires a constant direction of the vectors, the point of the electric field vector *E* moves up and down on a line if a spectator observes it in the direction of propagation. If two coherent perpendicularly linearly polarized light beams are superimposed, the *E* vector point leaves the line and may in general create an elliptic figure. With equal amplitude and a $+\pi$ and $-\pi$ phase shift, the vector point rotates on a circle in a right or left handed sense. Light is addressed as l-cpl if the point of the *E* vector rotates counterclockwise when the light *approaches* the observer (Fig.1). On the other hand, the superposition of r-cpl and l-cpl of equal phase and amplitude realizes linearly polarized light. Thus r-cpl and l-cpl on one side and two perpendicularly linearly polarized beams are both bases to describe polarization phenomena. If the amplitudes are not equal, the degree of circular polarization γ is less than unity.

Experimentally optical activity becomes manifest in the phenomena of *optical rotatory dispersion* (ORD) and of *circular dichroism* (CD), both

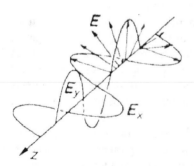

Figure 1   Left circularly polarized light.

wavelength dependent. ORD is related to the differential refractive indices $\Delta n_\lambda = (n_{\text{l-cpl}} - n_{\text{r-cpl}})_\lambda$, CD to the differential absorption coefficients $\Delta\varepsilon_\lambda = (\varepsilon_{\text{l-cpl}} - \varepsilon_{\text{r-cpl}})_\lambda$; Δn and $\Delta\varepsilon$ generally are small differences of large numbers. However, ORD is measured by the rotation of the polarization plane of linearly polarized light that is tied directly to the difference $(n_{\text{l-cpl}} - n_{\text{r-cpl}})_\lambda$ and does not require the subtraction of two separate measurements of $n_{\text{l-cpl}}$ and $n_{\text{r-cpl}}$. So high precision is possible, and very small effects can be detected. For the determination of CD, the two absorbances of one enantiomer versus l-cpl and r-cpl have to be measured separately. [*] By symmetry the different absorption of one kind of light by the two enantiomers is also $\Delta\varepsilon_\lambda = \varepsilon_{S,\lambda} - \varepsilon_{R,\lambda}$. The same holds for the refractive indices. CD is mostly quantified by the *molar ellipticity* $[\theta]_\lambda$. The units of the chiroptical properties are, owing to the historic background, somewhat odd:

$$[\theta]_\lambda = \frac{100\theta_\lambda}{dc} = 3300\Delta\varepsilon_\lambda \tag{1}$$

is expressed in deg cm^2/dmol, where θ_λ is the ellipticity instrument reading at wavelength λ in degrees, d the path length in cm, and c the concentration in mol/L. The factor 3300 adjusts also the units.

The magnitude of CD should be seen in relation to the magnitude of absorption. Kuhn [3] has introduced a wavelength-dependent anisotropy factor g_λ defined by

$$g = \frac{\varepsilon_l - \varepsilon_r}{\varepsilon} = \frac{\Delta\varepsilon}{\varepsilon} = \frac{\Delta\varepsilon}{(\varepsilon_r + \varepsilon_l)/2} = \frac{\varepsilon_S - \varepsilon_R}{\varepsilon} = \frac{\Delta\varepsilon}{(\varepsilon_R + \varepsilon_S)/2} \tag{2}$$

which ranges between 0 and 2. For symmetry reasons, $\varepsilon_S = \varepsilon + \Delta\varepsilon/2$ and $\varepsilon_R = \varepsilon - \Delta\varepsilon/2$.

Figure 1 may help to understand the different interactions of l-cpl with the two enantiomers, which allow us to discriminate the two enantiomers. Consider the pair of enantiomers of the spiropyrazoline **1**. If the light beam travels through the molecule along the long axis, it has to sense whether the transition moment of the second unit is rotated clockwise or counterclockwise compared to that of the first unit. It is obvious that this would be easy if the *E* vector of the light would exhibit a considerable rotation within the molecule. However, the size of the molecule is about 0.7 nm, and the wavelength of the probing radiation in the order of 200 to 700 nm, and thus the difference in interaction energy is small. Therefore CD is a small effect compared to absorption. The same holds for ORD.

---

[*] This is the reason that in the early days optical activity meant only the rotation of the polarization plane. The first commercial CD instruments did not appear before the 1960s.

This is evident in a comparison of the expression for the magnitudes of optical activity, the rotator strength R_{ik}, which is for an electronic transition $0 \to k$,

$$R_{0k} = Im\{<0|\mu|k> \cdot <k|\mathbf{m}|0>\} \tag{3}$$

and of absorption, the oscillator strength, which is (essentially)

$$S_{0k} = |<0|\mu|\mathrm{k}>|^2 \tag{4}$$

where $\langle 0|\mu|k \rangle$ is the electric transition dipole moment and $\langle k|\mathbf{m}|0 \rangle$ the magnetic transition dipole moment. Note that in Eq. (3) the two moments must have parallel components (vector point product). The fact that R_{0k} is purely imaginary will not be discussed further, but the magnitude of the magnetic dipole moment is governed by r/λ, the ratio of molecular and wavelength dimensions. Thus optical activity is a second-order, i.e. a small, effect.

This holds for the realm of linear optics. There is, however, new theoretical [4,5] and some experimental [4,6] work that seems to indicate that for 2-photon excitation, the existence of a CD effect can be due to pure electronic-dipole allowed transitions. Damping of the virtual intermediate state seems to be a requirement [5]. This phenomenon waits for broader investigation.

As the transition moments are coupling the ground and the excited state, emission is also circularly polarized. Circularly polarized luminescence was reviewed by Riehl [7] and by Shippers and Dekkers [8]. It is also characterized by a g factor (excited state racemization excluded) of

$$g_{em} = \frac{I_1 - I_r}{I_1 + I_r} \tag{5}$$

for excitation with, say, r-cpl; the total emission g factor contains the absorption and the emission anisotropy [9]

$$g_{em}(\lambda_{em}) = \frac{g^R_{abs}(\lambda_{abs}) \cdot g^R_{em}(\lambda_{em})}{2} \tag{6}$$

Circularly polarized emission is, however, not of importance in the scope of this book.

B. Direct Asymmetric Photochemistry

Photochemical reactions—asymmetric as well as nonasymmetric ones—take their outset from electronically excited states of the reactant. In this chapter, direct cpl excitation is covered, sensitized reactions are discussed in Chap. 4. Asymmetric photochemistry produces new chirality in a reaction system. The cpl-induced reactions are often called *absolute asymmetric*, as there is no net chirality in the reactants. This discriminates asymmetric photochemistry from the photochemistry of chiral molecules, which can be induced by nonpolarized light. The newly created chirality in cpl-induced reactions becomes apparent in an excess of the amount of one reactant or product enantiomer over the other. This *enantiomeric excess* (*ee*) in a mixture of R and S enantiomers is defined as

$$ee = \frac{c_S - c_R}{c_S + c_R} \tag{7}$$

ee is often also called optical or enantiomeric purity or (sometimes) optical yield.

Direct asymmetric cpl-induced photoreactions are only observed if there are two enantiomeric ground state reactants present that absorb different amounts of light. Thus the asymmetry of the cpl source is transformed into a different concentration of excited-state enantiomeric species, which becomes obvious in emitting systems in the circular polarization of luminescence [7]. These in turn react in a nonchiral environment with the same rate constants for the different deactivation channels. Thus asymmetric photoreactions are dependent or independent parallel reactions of the enantiomers with different net rates.

The expectation of the magnitude of the *ee*s that can be achieved by cpl photochemistry is governed by the information of Sec. I.A: we have to expect small differences in the absorption according to small $\Delta\varepsilon$ values. Without autocatalytic or other amplification, this leads to only partial resolution of the racemates with small *ee*s. Cpl-induced photoreactions are, however, often very clean and allow rigid kinetic treatment. Cpl is also useful in mechanistic investigations of unresolved enantiomers [10], and cpl effects are considered important in the creation of homochirality of biomolecules and the origin of life [11] (see also Sec. IV).

Some attention has to be given to the degree of circular polarization γ and thus to the creation of cpl. Two experimental principles are common: the use of a quarter wavelength plate and the use of a Fresnel rhomb or a Solcil-Babinet compensator; in a few experiments, radiation from a cyclotron after passage through a polarizing undulator was used. The degree of circular polarization is easily jeopardized when a λ/4 plate is employed, as this is an interference device and has a very narrow wavelength domain of good polarization performance. This device should only be used in laser experiments, and care should be taken for exact matching of laser wavelength and plate thickness. For broader bandwidth irradiation, like that by different lamps, a Fresnel rhomb or a Solcil-Babinet compensator should be used, as their performance is based on the refractive index; this is a slowly changing function of wavelength. Still, good parallel light beams are required in order to get high degrees of circular polarization. We obtained a $\gamma = 96 \pm 5\%$ circular polarization with a beam made parallel by two pinholes at a distance of 4 m; with care, 90% can be reached in a normal irradiation setup of lamp/interference filter/lens system/linear polarizer/Fresnel rhomb. Poor irradiation geometries may reduce the effect of cpl considerably [12]. Thus quantitative results should take reference to the degree of circular polarization; if this is not controlled, too low asymmetry effects may be the consequence. An example is shown in Fig. 2.

One big caveat should be kept in mind: in some cases, the results in asymmetric chemistry have not proved to be reproducible, and the same holds for asymmetric photochemistry as well, especially when small effects are observed, and even more so if ORD is used as an analytical tool at long wavelengths. In this case, every optically active species in the sample contributes to ORD. Therefore CD should be preferred where the spectrum can be used to identify the species.

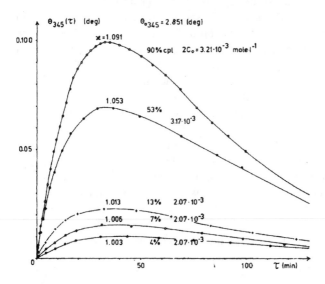

Figure 2 Influence of the degree of cpl-polarization γ on the magnitude of ellipticity during cpl-induced photodestruction of **1**. (Reprinted from Ref. 12. Copyright American Chemical Society.)

Examples for the many cases where the observed development of optical activity in a reaction could not be reproduced are the photoaddition of H_2O_2 to diethylfumarate [13] or the thermal decarboxylation of 2-phenyl-2-carboxylbutyric acid in cholesteric liquid crystals [14]. On the other hand, spurious optically active impurities may, especially in autocatalytic systems, cause considerable asymmetry effects. This exceptional case was demonstrated by Singleton and Vo [15] in the thermal enantioselective synthesis discovered by Soai and coworkers [16].

C. Some Remarks on the Development of the Field

Both the pioneers in understanding chirality, van't Hoff and Le Bel, pointed out that cpl might be used to induce asymmetry in chemical reactions [17,18]. In the following decades there were many vain efforts to realize this idea [19,20–22,24–27]. The failure, most probably, was due to the small effects caused by the small differences in $\Delta\varepsilon$. The first successful asymmetric photoreactions were found by W. Kuhn, who published a series of papers in 1929 and 1930 [3,28–28] on the photolysis of racemic α-azido propionic dimethylamide **2** and α-bromopropionic ethyl ester **3** derivatives. Mitchell [33] reported in 1930 on humulene nitrosite (probably **4***) and Tsuchida et al in 1935 on $[\mathrm{Co(ox)_3}]^{3-}$ Ca-malogue to 5) [98].

| 2 | 3 | 4 |

The interest in cpl-induced photochemistry peaked again in the 1970s, when reliable spectropolarimeters for CD measurements became commercially available. Stevenson and Verdieck [34] published their first papers on photoderacemization, the groups of Calvin and of Kagan produced a series of papers on (absolute) asymmetric synthesis, and Kagan and his group, on the basis of Kuhn's results, deepened our knowledge of asymmetric destruction. Above all they laid the basis for the kinetics of asymmetric photoreactions.

Most recently, the range of asymmetric photochemistry was extended to shorter wavelengths by the use of synchrotron radiation and two-photon excitation. New developments are visible with high-intensity and ultrashort pulse lasers (see Sec. D of this chapter and Chap. 2 of this book).

---

* In Ref. 33 only the formula $C_{15}H_{24}N_2O_3$ is given, no structure. See also footnote h of Table 1

Cpl-induced asymmetric photochemistry was reviewed at regular intervals: in 1978 by Kagan and Fiaud [35], in 1983 by Rau [36], in 1988 by Bonner [11], in 1992 by Inoue [37], in 1999 by Everitt and Inoue [38], and in 2002 by Griesbeck and Meierhenrich [39], usually in the general context of asymmetric photochemistry.

The kinetics of asymmetric photoreactions were developed by several groups [12,34,40]. General reviews of the kinetics of asymmetric thermal reactions, which can be adapted to photoreactions, were written by Straathoff and Jongejan in 1977 [41] and Kagan and Fiaud in 1998 [42]. In the following sections of this chapter the specialties of photoreaction kinetics are given in detail.

II. ASYMMETRIC PHOTOREACTIONS WITH CPL

This chapter covers photoreactions where the asymmetric induction is introduced by cpl, i.e., in the absorption process. This requires that two differently cpl-absorbing ground state species, i.e., enantiomers or diastereomers, are present. Treatment of the latters' photoreactions is not covered, as for diastereomers non-polarized light is sufficient to modify the optical activity of a sample. Starting from this a general reaction scheme is presented in Scheme 1, from which the different types of cpl-induced reactions will be deducted.

In Scheme 1 the respective rate constants of thermal interconversion of R and S(k), R* and S*(k*), and of deactivation of R* and S*(B) are equal. If this were not so, spontaneous or nonpolarized-light induced deracemization would occur, which never has been observed. The interconversion rates of P_R and P_S(k_3 and k_{-3}) may be different, if the products are diastereomers. If the ground state enantiomerization is slow compared to the other reactions then the kinetics of independent parallel reactions apply; if not, those of dependent parallel reactions.

Buchardt [43] introduced the tri-fold classification of asymmetric photoreactions:

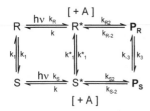

Scheme 1   General scheme for asymmetric photoreactions.

1. *Photoderacemization** is observed if the product formation is slow compared to the enantiomerization in the excited state (Sec. II.A).
2. *Asymmetric photodestruction* is observed if the product formation is fast compared to deactivation and enantiomerization in the excited state and if the reactants are considered (Sec. II.B).
3. *Asymmetric synthesis* is observed if the interest focuses on the product side of the reaction (Sec. II.C). The nonracemic products may be created by asymmetric destruction or stem from reactants kept racemic by fast ground state enantiomerization.

Unequivocal classification is sometimes not possible, but the type of kinetic development is taken as a guideline.

In Scheme 1, bimolecular reactions of R* and S* with a different molecule A are not excluded. However, if A is a nonchiral reaction partner, then the concentration ratio of the R* and S* excited species created by cpl is not influenced by A, nor are the following reactions asymmetric. The *ee* of the reaction is thus not changed. If A is a chiral molecule in racemic concentration, then the rate equations cannot be integrated in closed form in most cases. Second-order asymmetry effects may appear, but the amplification, if existent, is estimated to be very small. Experimental investigation of this question is lacking, so that amplification of the *ee*, which is created by cpl excitation, in a bimolecular reaction of the excited species, is not a known procedure. In the following schemes, the possibility of bimolecular reactions is kept as an option, but the kinetics are developed only for first-order reactions. Also the following kinetic schemes are set up on the basis of linear kinetic relations. Nonlinear effects, i.e., cases where a reaction product might influence the stereoselectivity [44] or where the chemical reactivity is dependent on its enantiomeric composition [45], have not yet appeared important in asymmetric photochemistry.

A. Photoderacemization

Photoderacemization is the simplest case of direct asymmetric photoreactions induced by cpl. The enantiomers are interconverted, and the mixture becomes optically active. Reaction scheme 2 is a modification of Scheme 1; ground state racemization is excluded. The enantiomerization step R* $\leftrightarrows$ S* was observed directly by Metcalf et al. [9] by means of the time-resolved circularly polarized luminescence of europium-tris(bipicolinate). By means of a cpl laser pulse, a difference in the excited state population is created, and the decay of circular

---

* We call the interconversion of enantiomers *enantiomerization*, the disappearance of optical activity by enantiomerization *racemization*, and the induction of optical activity by enantiomerization *deracemization*.

polarization in the emission is monitored. It is characterized by $\tau = 1/k + 1/k^*_1$, and emission anisotropies of $g_{em} = 0.08$ have been found, which are temperature dependent with $E_a \approx 45$–50 kJ mol^{-1}.

The reaction kinetics of the even more reduced scheme R $\rightleftarrows$ S were first developed by Stevenson and Verdieck [34]. It is evident that the total concentration c_0 in reactions following Scheme 2 will remain constant, and so does the absorbance. With l- or r-cpl, the ratio of the concentrations of R and S forms becomes larger or smaller than 1, depending on which enantiomer absorbs more strongly, and for long irradiation times, a photostationary state with a certain CD value is reached. Of course, photoenantiomerization occurs also with nonpolarized light (npl), but for an optically inactive educt mixture it does not become apparent, because of the symmetry of the reaction system. If, however, pure enantiomers or an unequal mixture of the enantiomers is irradiated with npl, this photoenantiomerization leads to a racemic mixture of R and S.

Scheme 2 is the simplest but not the only reaction scheme conceivable. The deactivation of, say, S* may proceed via an excimer (SS)* or a diastereomeric exciplex (SR)*. The same deactivation pattern holds, of course, for R*, so that the deactivation *in toto* is nonasymmetric. If there were exciplex emission, circularly polarized fluorescence might be observed. Inoue discusses also a "hot ground state" path of deactivation [37].

The rate equation of photoderacemization of R and S according to Scheme 2 is determined using the steady-state approximation for the short-lived excited enantiomers:

$$\frac{dc_R}{dt} = -\frac{dc_S}{dt} = -\frac{k_1^*}{k + 2k_1^*}\left(k_R\, c_R - k_S\, c_S\right)$$

$$(8)$$

The asymmetry of the reaction is determined by the two rate constants k_S and k_R. With the excitation rate constant,

$$k_R = 1000\, I_0'\, \frac{\left(1 - 10^{-A'(t)}\right)}{A'(t)}\, \varepsilon_R'$$

$$(9)$$

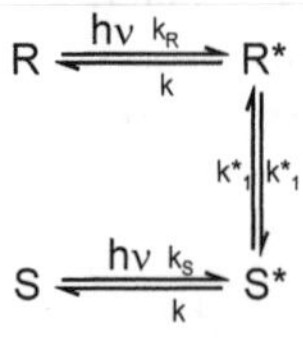

Scheme 2   Photoderacemization by photoenantiomerization.

and with the corresponding k_S we define the asymmetry of a reaction

$$\kappa = \frac{k_S}{k_R} = \frac{\varepsilon'_S}{\varepsilon'_R} \tag{10}$$

The rate equations are

$$\frac{dc_R}{dt} = -\frac{dc_S}{dt} = \frac{k_1^*}{k+2k_1^*} \; 1000 \; I'_0 \frac{\left(1-10^{-A'(t)}\right)}{A'(t)} \left(\varepsilon'_R \; c_R - \varepsilon'_S \; c_S\right) \tag{11}$$

Here some conventions are assumed:

    1. All photokinetic equations are for a cell length of 1 cm.

    2. The primed, wavelength-dependent quantities are those taken at the excitation wavelength; unprimed ones refer to an arbitrary analysis wavelength.

    3. The expression

$$F = \frac{1-10^{-A'(t)}}{A'(t)} \tag{12}$$

the so-called photokinetic factor, accounts for the change of the absorption of the solution with time due to the photoreaction and also for absorption of nonreacting species. For photoderacemization, F is practically time independent, as $\varepsilon_S \approx \varepsilon_R$ and the absorbance is virtually constant.

    4. The factor $k_1^*/(k+2k_1^*) = \phi$ is the photochemical quantum yield of photoenantiomerization. This number is the same for R and S for symmetry reasons (otherwise a CD effect would be observed when npl excites the photoenantiomerization system).

    Equation 11 indicates a photostationary state (pss) when $dc_R/dt = dc_S/dt = 0$. Then

$$\varepsilon'_R \; c_R = \varepsilon'_S \; c_S \tag{13}$$

with $\varepsilon'_R = \varepsilon' - \Delta\varepsilon'/2$ and $\varepsilon'_S = \varepsilon' + \Delta\varepsilon'/2$; with $c_0 = c_R + c_S$ this gives

$$(c_S - c_R)_{pss} = \frac{\Delta\varepsilon'}{2\varepsilon'} (c_S + c_R) = \frac{\Delta\varepsilon'}{2\varepsilon'} c_0 = \frac{g'}{2} c_0 \tag{14}$$

According to Scheme 2, R and S are interconverted, and the total concentration is constant. Then Eq. (8) can be transformed into

$$\frac{dc_R}{dt} = -1000 \; I'_0 \; F\phi \left[\left(\varepsilon'_R + \varepsilon'_S\right)c_R - \varepsilon'_S \; c_0\right]$$

and

$$\frac{dc_S}{dt} = -1000 \; I'_0 \; F\phi \left[\left(\varepsilon'_R + \varepsilon'_S\right)c_S - \varepsilon'_R \; c_0\right] \tag{15}$$

The difference of the rate constants is the rate constant of the difference

$$\frac{dc_S}{dt} - \frac{dc_R}{dt} = \frac{d(c_S - c_R)}{dt} = 1000 \, I_0' \, F\phi \left[\left(\varepsilon_S' - \varepsilon_R' \right) c_0 - \left(\varepsilon_S' + \varepsilon_R' \right) \left(c_S - c_R \right) \right] \qquad (16)$$

Under the condition that F is time independent, Eq. 16 can be integrated and, after some rearrangements, gives

$$\begin{aligned} c_S - c_R &= -\frac{\varepsilon_S' - \varepsilon_R'}{\varepsilon_S' + \varepsilon_R'} c_0 \left(1 - \exp\left[-1000 \, I_0' \, F\phi (\varepsilon_S' + \varepsilon_R') \cdot t \right] \right) \\ &= -\frac{\Delta\varepsilon'}{2\varepsilon'} c_0 \left(1 - \exp\left[-1000 \, I_0' \, F\phi 2\varepsilon' \cdot t \right] \right) \\ &= -\frac{g'}{2} c_0 \left(1 - \exp\left[-2000 \, I_0' \, F\phi \, \varepsilon' \cdot t \right] \right) \end{aligned} \qquad (17)$$

As $\varepsilon = (\varepsilon_S + \varepsilon_R)/2$ and $\varepsilon_S \approx \varepsilon_R$, the equation in terms of absorbance is

$$\Delta A = -\frac{g'}{2} A_0 \left(1 - \exp\left[-2000 \, I_0' \, F\phi \, \varepsilon' \cdot t \right] \right) \qquad (18)$$

Note: the unprimed symbol A means that an arbitrary analysis wavelength λ can be chosen, which may well be the irradiation wavelength λ'. According to Eq. (1), ΔA transforms into $(1/33)\theta$.

The same steady state is approached in a monoexponential way, regardless whether the starting solution contains pure R form, pure S form, or is racemic (Fig. 3A). In Fig. 3B this is demonstrated for a 541 nm irradiation of racemic Cr(acac)$_3$ with the data of Stevenson [46] ($\Delta\varepsilon' = -4.2$ mol^{-1} cm^{-1}, $\varepsilon' = 66$ mol^{-1} cm^{-1}, $g' = 0.064$, $\phi = 0.006$, and $c_0 = 0.02$ mol L^{-1}) and an assumed intensity of $I_0' = 10^{-8}$ Einst. cm^{-2} s^{-1}. The absorbance is constant, $A = 1.32$. The same steady state of $\Delta A = [(4.2/66)/2] \, 1.32 = 0.042$ is reached also starting from pure Δ or Λ forms monoexponentially, however, with $\Delta\varepsilon = 4.2$ mol^{-1} cm^{-1} the starting point and most of the plot would be far out of the scale of Fig. 3B.

Only few examples of cpl-induced photoderacemization are known, and most of them are not free of side reactions. This is easily controlled by monitoring the absorbance of the reaction mixtures, which should stay constant. When very long irradiation times are necessary, this may require special precautions to avoid, for example, solvent evaporation.

One group of examples comprises isomerizations of inorganic bidentate octahedral complexes [Me(L-L)$_3$]^x. An example for the structure of the enantiomers is the chromium-trisoxalato complex [Cr(ox)$_3$]$^{3-}$ **5** on which the first paper by Stevenson and Verdieck appeared in 1968 [47]. They took this work up again and expanded it to tris(malonato)-, tris(dithiooxalato)-, tris(tartrato)-, and ethylenediamidodioxalato chromate and gave the kinetic analysis [34]. They stated that photoderacemization is exponentially related to the quantum efficiency of photoracemization [34]. In the case of [Cr(acac)$_3$]$^{3+}$ there is negligible thermal racemi-

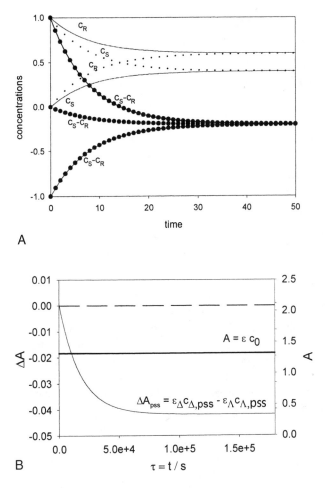

Figure 3 Kinetics of photoderacemization. (A) Calculated traces with $\kappa = 1.5$ (k_R and = 0.75) starting from pure R form (——) and pure S form ($\cdots$). ●●●●● c_S-c_R starting from pure S form, pure R form and racemic mixture. (B) Experimental results of Cr(acac)$_3$ by means of 541 nm cpl with the data of Ref. 46.

zation, but the photoderacemization and thus the excited state interconversion has an activation energy of only 9.5 kJ mol^{-1} [46]. Laser excitation was applied to the same complex [48,49], and Kane-Maguire and Langford [49] report steady state *ee*s of 1–2% in agreement with expectations in partial photoresolution by deracemization of Cr[(C$_2$O$_4$)phen$_2$]$^+$, Cr[(C$_2$O$_4$)$_2$phen]$^-$, and Cr[(bipy)$_3$]$^{3+}$ on irradiation with circularly polarized light from an argon ion laser at 496.5 or 514.5 nm. Nordén [50] found that $[Cr(ox)_3]^{3-}$ was racemized thermally by the catalytic activity of strychnine and partially photoresolved by r-cpl and l-cpl to equal *ee*s but of opposite sign. Anzai et al. used the cpl-induced enantiomerization of chromium tris(3-butyl-2,4-pentanedione) to switch phases of a liquid crystal [51].

$$\Delta - 5 \qquad\qquad\qquad \Lambda - 5$$

Photoenantiomerization of different classes of organic molecules was investigated. However, the expectation that if a pure enantiomer can be racemized by light then photoderacemization of the racemate with cpl should lead to an optically active photostationary state, does not hold in all cases. Even when enantiomerization is fast, the $\Delta\varepsilon$ values are often too small. Molecules like **6** and **7** were extensively studied by Schuster and his group (e.g., [52,53]) in search of a molecule suited for an optical switch for mesogenic phase transitions.

6   **7**

An especially interesting case [53] is the photochromic pair of enantiomerically stable 1,1′-binaphthylpyran **8** and 2-hydroxy- 2′-hydroxymethyl-1,1′-binaphthylene **10**. The enantiomers of the pyran **8** can be separated by chromatography

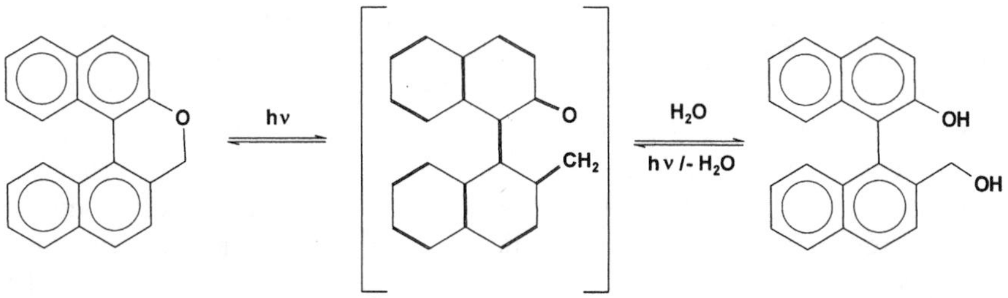

Scheme 3 (Adapted with permission from Ref. 53. Copyright American Chemical Society).

and racemize on irradiation via the 1,1′-binaphthyl quinone methide **9**. However, the enantiomers of the pyran could not be resolved even partly by cpl.

Still, in many cases photoderacemization of racemic mixtures has been successful. Irie et al. announced [54] the detection of the effect, but a detailed study of this system was published only as a patent [55]: in CH_2Cl_2 solutions of 1,1′-binaphthyl they found values of $[\alpha]_{345} = .0.75$ and $+ 0.7$ for l- and r-cpl with a 270–300 nm band of a 500 W Xe source at room temperature in ca. 5 min. Here two aromatic units are connected by a single bond. These systems are realizations of Kuhn's coupled oscillator model of optical activity [56]. The steric overlap provides chirality, ground state enantiomerization has $\Delta H^{\neq} \approx 92$ kJ mol^{-1} [57], and the triplet state enantiomerizes fast, with 8 kJ mol^{-1}.

Photoderacemization is improved if the single bond between the aromatic moieties is replaced by a double bond (Scheme 4): with sufficient steric crowding, chiral ethenes are the result, and here the enantiomerization is at the same time

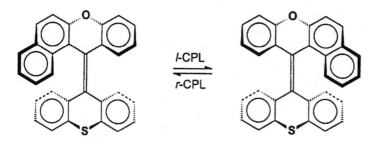

Scheme 4 (Reprinted with permission from Ref. 61. Copyright Wiley-VcH.)

a cis–trans isomerization. Feringa [58] reported the cpl-induced photoderacemization of such systems, among others that of **11**. The cpl photostationary state of **11** reaches only an *ee* of $+0.07\%$ or -0.07%, but this type of molecule has potential for optical switches and data storage devices [58], as it is photochemically and thermally stable, has a high isomerization quantum yield of 0.4, and has a sufficient *g* factor of 0.006. Feringa has used the molecular enantiomerization as a trigger for amplification, e.g., by addition of ca. 20% of **11** to a liquid crystalline mesophase of 4'-(pentyloxy)-4-biphenylcarbonitrile (5-CB). The formation of the cholesteric phase can be induced by cpl, the nematic one restored by lpl. The cholesteric phases with right- or left-hand pitch can be interconverted by l- or r-cpl [59,60]. For extensive information consult Feringa and Van Delden's review article [61] and Feringa's contribution to this book.

The same principle, where photoisomerization is tied to photoenantiomerization, appears even clearer in the bicylcic ketone **12** of Zhang and Schuster [62]. **12** has a high *g* factor ($g = 0.0502$) in the absorption band of the carbonyl group. On irradiation of this group, intersystem crossing to the triplet is fast and so is energy transfer to the ethenic triplet, where the geometrical isomerization takes place. With 313 nm irradiation, a photostationary state with 1.6% *ee* was reached. This unit was used to switch the backbone helicity of liquid crystals like polyisocyanates by cpl [63,64]; the pitch of the cholesteric phase is proportional to the *ee* [65] of the photostationary state.

12   **14**

Photo induced interconversion of the configuration at an "asymmetric" C atom, has not been reported. For tyrosine it has been excluded even under high-intensity laser irradiation [66]. This chiral unit seems to be very stable. A bicyclic ketone system with (formal) double epimerization is described by Zandomeneghi et al. [67] (Eq. 33). This reaction proceeds, however, by a complex rearrangement mechanism and will be discussed in Sec. II.B.

Promising systems for photoderacemization are solutions of 1,3-dimethylindene or 1,3-diphenylindene where a photoreaction is "narcissistic" because of a photoinduced 1,3 hydrogen shift. Unfortunately we were not able to induce detectable optical activity in such solutions, probably because of too-small *g* factors [68].

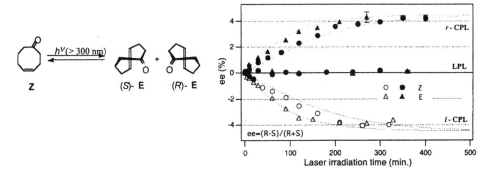

Figure 4 The 4-cyclo-octenone system. (Reprinted from Ref. 69.)

There are several types of systems where the photoenantiomerization proceeds through a defined intermediate in a photoinduced stationary state. All molecules must absorb; therefore not too many types are known. Z-E isomerizations are examples, e.g., 4-cyclooctenone **13** (Fig. 1.4), where an *ee* of up to 4% is reached. The highly distorted **E-13** has a high *g* factor of 0.15 [69]. In photochromic systems [70] where the photochromic properties are due to cyclo/seco reactions, as in spiro compounds, fulgides, or dihydroindazolidines, right- or left-handed screws are formed from an open form. Therefore these systems belong to Sec. II.D. Chap. 9 of this book is dedicated to this group.

An interesting approach is a method that uses the modulation of the degree of circular polarization (e.g., between pure l-cpl and linearly polarized light) to determine fast isomerization kinetics [71]. However, preliminary results on [Co(ox)(phen)$_2$]$^-$ and rhodamin 6G **14** were not followed by the announced full papers, and the arcticle is rarely cited. In principle the concept may be used to determine interconversion and the CD spectra of rotamers (like **14**) which are the reactants in asymmetric synthesis (see Sec. II.C.).

B. Asymmetric Photodestruction

In asymmetric photodestruction the interest concentrates on the reactants. They, again, absorb at different rates, and different concentrations of R* and S* are created. Product formation is treated in Sec. II.C, but this is not of interest in this context. The reactions of R* and S* may be mono- or bimolecular, k_{S2} and k_{D2} of Scheme may be identical or different, the products P$_R$ and P$_S$ may or may not be optically active or even be identical molecules. The rates of product formation just influence the magnitude of the quantum yield ϕ. However, the

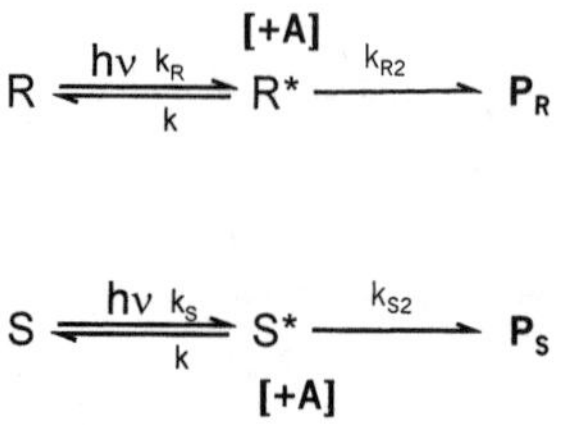

$$R \xrightleftharpoons[k]{hv\ k_R} R^* \xrightarrow{\substack{[+A] \\ k_{R2}}} P_R$$

$$S \xrightleftharpoons[k]{hv\ k_S} S^* \xrightarrow{k_{S2}} P_S$$
$$[+A]$$

Scheme 5 Independent parallel reactions.

ground state and excited state thermal interconversion must be slow, i.e., k_1 and k_1^* are small compared to k_{R2} and k_{S2}. Thus Scheme 1 is modified to Scheme 5.

This scheme represents two independent parallel reactions of the enantiomers. A general treatment for thermal kinetic resolution was given by Kagan and Fiaud [42]. For unimolecular photoreactions first-order equations seem to be appropriate [40]. Accordingly the rates are

$$\frac{dc_R}{dt} = -k_R \phi c_R = -1000\ I_0' \frac{\left(1-10^{-A'(t)}\right)}{A'(t)} \varepsilon_R' \phi\ c_R$$

$$\frac{dc_S}{dt} = -k_S\ \phi\ c_S = -1000\ I_0' \frac{\left(1-10^{-A'(t)}\right)}{A'(t)} \varepsilon_S' \phi\ c_S \tag{19}$$

Here $\phi = k_{R2}/(k + k_{R2})$ is the photochemical product formation yield. Note that the absorbance and ε values are those at the exciting wavelength. In contrast to the case of photo deracemization, in photodestruction reactions the absorbance [and consequently the photokinetic factor F, Eq. (12)] is not constant; therefore the integral

$$\int_0^t \frac{\left(1-10^{-A'(t)}\right)}{A'(t)} dt \equiv \tau(t) \tag{20}$$

represents a transformed time axis [72], which reflects the number of absorbed photons, and it is only in this time axis that the destruction of R and S is of first order [12]:

$$c_R = \frac{c_0}{2} \cdot e^{-1000\ I_0'\ \varepsilon_R'\ \phi\ \tau}$$

$$c_S = \frac{c_0}{2} \cdot e^{-1000\ I_0'\ \varepsilon_S'\ \phi\ \tau} \tag{21}$$

The absorbance of the reaction mixture at an arbitrary analysis wavelength is $A = \varepsilon_S c_S + \varepsilon_R c_R \approx \varepsilon\ (c_S + c_R)$, and $\varepsilon_S \approx \varepsilon_R \approx \varepsilon$, so

$$A(\tau) = A_0 \cdot e^{-1000 I_0' \varepsilon' \phi \tau} \tag{22}$$

The deviation from the monoexponential relation due to the small $\Delta\varepsilon$ is usually buried in the error margin.

The ellipticity is proportional to the difference of the concentrations of S and R educt.

$$\theta(\tau) = \frac{[\theta]}{100}(c_S - c_R) = \frac{[\theta]}{200} c_0 \left[e^{-1000 I_0' \varepsilon_S' \phi \tau} - e^{-1000 I_0' \varepsilon_S' \phi \tau} \right] \tag{23}$$

These relations of Eqs. (21) to (23) are demonstrated in Fig. 5 for the 332 nm cpl-induced photodestruction of trans-3,5-diphenylpyrazoline **15** [73]. For this molecule $\varepsilon_{332} = 304$ L mol^{-1} cm^{-1}, $\Delta\varepsilon_{332} = 17.5$ L mol^{-1} cm^{-1}, $g_{332} = 0.058$. Figure 5B also demonstrates the difference in the t and τ irradiation times.

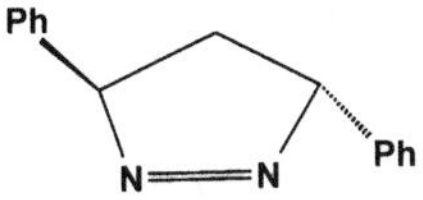

15

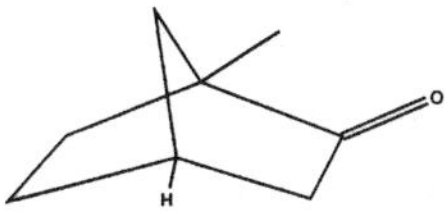

16

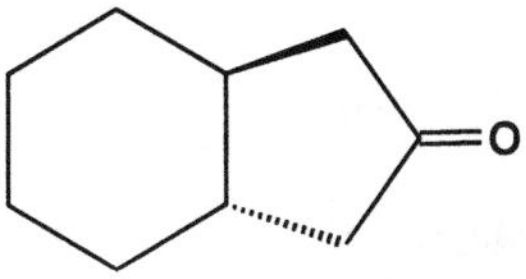

17

Bimolecular reactions with a nonchiral or chiral partner A do not change the type of relations; they may change the yield ϕ in Eq. (19).

Equation 23 and Fig. 5 show that during the photodestruction of a racemic mixture the CD (and ORD) first increase and then decrease again. This maximum is due to the superposition of increasing relative concentration difference and decreasing overall concentration. This curve has some interesting features [12,74]: the maximum is at

$$\tau_{max} = \frac{\ln \kappa}{k_S - k_R} = \frac{\ln(\varepsilon_S' / \varepsilon_R')}{1000 I_0' \phi (\varepsilon_S' - \varepsilon_R')} \tag{24}$$

$$\frac{c_R(\tau_{max})}{c_S(\tau_{max})} = \kappa \tag{25}$$

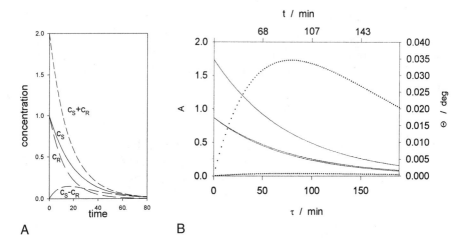

Figure 5   cpl-induced photodestruction. (A) Simulation with $k_S = 0.01$ and $k_R = 0.015$.
(B) Of trans-3,5-diphenylpyrazoline **15**. Absorbance of R and S enantiomers and of the
mixture ellipticity of the mixture, ($\cdots$) (upper curve $\times$ 50, right hand scale). Note the
difference between irradiation time t (upper labels) and τ (lower labels, linear).

$$\theta(\tau_{max}) = \frac{[\theta]^{\cdot}}{100} \frac{c_0}{2} \left(1 - \frac{1}{\kappa}\right) \cdot \exp\left(-\frac{\ln \kappa}{1-\kappa}\right) \tag{26}$$

Additional relations are available for the inflection point, which occurs at $t = 2\tau_{max}$. These relationships are valid for arbitrary analysis wavelengths including, of course, the excitation wavelength. The time-axis transformation allows rigid kinetic treatment of the independent parallel reactions; it is, however, rarely applied.

Partial photoresolution by asymmetric photodestruction can be used to determine the chiroptical constants of the pure enantiomer of the reactant, without their separation. A hint to this is found in the paper of Balavoine et al. [40], a detailed development of this concept has been published by the Rau group [12,73]. By the combination of ε' and CD$'$ data *at the irradiation wavelength*, with the extent of reaction expressed as $\xi = [(A(\tau_{max}) - A(\tau \to \infty)]/[(A(\tau = 0) - A(\tau \to \infty)]$, one obtains

$$\left[\theta'\right]^2 = \frac{66}{\gamma \cdot (\varepsilon' c_0)} 10^4 \varepsilon'^2 \left(1/\frac{A(\tau_{max}) - A(\tau \to \infty)}{A(\tau = 0) - A(\tau \to \infty)}\right) \cdot \theta'(\tau_{max}) \tag{27}$$

If $\kappa < 1.5$ (which usually is the case), then the CD maximum occurs at about 63% photolysis, i.e., the extent of the reaction term in Eq. 27 is $1/e$ within experimental precision. If at the start of the reaction the reactant is the only absorbing species in the solution, Eq. (27) reduces to

$$\left[\theta'\right]^2 = \frac{66}{\gamma \cdot A_0} 10^4 \varepsilon'^2 \cdot e \cdot \theta'(\tau_{max})$$

(28)

Additional relations are obtained from the initial and final slopes of θ vs. ξ [74], which are useful as internal criteria of the method.

The transformation of the time axis is avoided in a different type of graphic representation. Balavoine et al. [40] introduced a plot of ee vs. the extent of reaction ξ with the g factor as a parameter according to Eq. (29) (Fig. 6), which underlines the development of enantiomeric excess during the photoreaction:

$$\xi = 1 - \frac{1}{2}\left[\left(\frac{1+ee}{1-ee}\right)^{1/2-1/g'} + \left(\frac{1+ee}{1-ee}\right)^{-1/2-1/g'}\right]$$

(29)

This representation is quite popular in the literature, a detailed derivation of Eq. (29) is found in Nakamura et al. [75]. The general relation between ξ, ee, and κ is [35]

$$\kappa = \frac{\ln\left[(1-\xi)(1-ee)\right]}{\ln\left[(1-\xi(1+ee))\right]}$$

(30)

and this relation is also valid for bimolecular reactions of the second order according to Scheme 5.

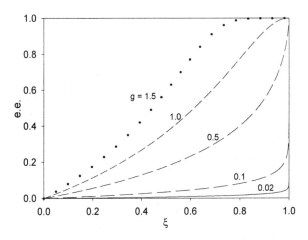

Figure 6 Enantiomeric excess ee as a function of extent of reaction ξ and anisotropy factor g. (Adapted from Ref. 40. Copyright American Chemical Society).

Asymmetric photodestruction can be used as a method to achieve high enantiomeric excess values. As the g factors are generally small, this is only reached at a large ξ where, according to Fig. 5, most of the starting material has disappeared. So Fig. 6 in some way is deceptive. The optical yield in reference to the starting material can be defined by

$$P_{rel} = 100 \cdot \frac{c_R - c_S}{c_0} = 100 \cdot \frac{\theta \, 100}{[\theta]c_0} \% \tag{31}$$

This curve mimics the $\theta(\tau)$ curve (e.g., Fig. 5B). The optical yield is a maximum at the maximum of the $\theta - \tau$ curve, i.e., at 63% photodestruction for $\kappa < 1.5$. Examples are trans-3,5-diphenyl-pyrazoline **15** with $P_{rel}(\tau_{max}) = 1.1\%$ [$g_{332} = 0.058$, $ee(\tau_{max}) = 3\%$] or the spiropyrazoline **1** with $P_{rel}(\tau_{max}) = 1.7\%$ [$g = 0.096$, $ee(\tau_{max}) = 4.3\%$]. Indeed, most papers published on cpl photochemistry for these reasons deal with asymmetric photodestruction. Balavoine et al. [40] reached 20% ee at 99% destruction for camphor **16**; even higher ee of 30% was reached for *trans*-bicylco[4.3.0]nonane-8-one **17** [76,77]. Emeis et al. [78] reported that **17** also shows circularly polarized luminescence with $g_{em} = 0.035$ [79].

In order to summarize the papers on asymmetric destruction, the format of Inoue [37], who used a table, seems to be appropriate. His listing is completed in Table 1.

In many investigations photolysis was not conducted to more that 63% completion (see Table 1). Then the CD or ORD vs. time plots give the deceptive impression of a continuous rise or approaching a limit, and this impression is increased by the dilatation of the t axis compared to the τ axis (see Fig. 5B).

The photoelimination reactions seem to present the cleanest progress; the products mostly do not absorb at the irradiation wavelength. These reactions are the prime candidates for verification of the kinetic relations [12,40]. In many other cases, photoracemization, side reactions or, if the primary products absorb at the irradiation wavelength, secondary photolysis occur and obscure the kinetics of CD development. The isomerization of E-cyclooctene **42** also is a well-defined reaction. The α vs. time plots show a maximum [103], but a photostationary state will be reached at long times as the $Z \rightarrow E$ isomerization is also active under irradiation.

$$\text{(-) (R) - 42E} \qquad \text{(+) (S)- 42E} \qquad \text{42 Z} \tag{32}$$

Table 1 Direct Asymmetric Destruction with Circularly Polarized Light in Solution[a]

Substrate (g factor)[b]	Irradiation λ, nm (cpl)	Conversion[c], %	α_{max}^{d} or θ_{max}^{e} deg	ee[f]%	Ref(s)
Elimination reactions (Hal, NO, N₂)					
ethyl 2-bromopropionate (**3**)	280 (r)	50	α_D +0.05	g	28
	280 (l)	50	α_D —0.05	g	
2-azido-N,N-dimethylpropionamide (**2**)	ca. 300 (r)	40	$[\alpha]_D$ +0.78	g	3,32
	ca. 300 (l)	40	$[\alpha]_D$ −1.04	g	
2 (0.02)	313 (r)	50	g	0.6	40
"humulene nitrosite" (**4**)[h]	ca. 700 (r)	g	α_{546} −0.21	g	33
	ca. 700 (l)	g	α_{546} +0.21	g	
4,4,8,8-tetramethyl-2,3,6,7-tetraaza-spiro[4.4] nona-2,6-diene (**1**) (0.096)	345 (r or l)[g]	59	θ^{345} +0.099	4.3	12,80
		83	θ_{345} +0.088	8.4	
trans-3,5-diphenylpyrazoline (**15**) (0.058)	332 (r or l)[g]	63	θ_{332} +0.036	2.8	73,80
		90	θ_{332} +0.019	6.5	
2-chloro-2-nitroso-1,4-diphenylbutane (**18**)	>600 (r)	90	$[\alpha]_{530}$ −2.50	g	81
	>600 (l)	90	$[\alpha]_{530}$ +2.75	g	
Ketone reactions					
trans-tricyclo[5.3.0.0^{2,6}]decane-3,10-dione (**19**)	330 (l)	g	$[\alpha]_{385}$ −0.13	g	109
camphor (**16**) (0.09)	313 (r)	99	g	20	43
20 (0.15)	313 (r)	60	$[\alpha]_D$ +8.20	3.0	76
21	365 (r)	40	$[\alpha]_{546}$ −0.004	g	82
	365 (l)	46	$[\alpha]_{546}$ +0.004	g	
bicyclo[3.2.0]hepta-3,6-dien-2-one (**22**)[i]	351, 363 (l)	g	g	1.5	67,83,84,85
tricyclo[5.4.0.0^{1,5}]undecan-8-one (**23**)	334 (l)	54	$[\theta]_{334}$ −1.82	g	86

(Continued)

Table 1 Continued

Substrate (g factor)[b]	Irradiation λ, nm (cpl)	Conversion[c], %	α_{max}^{d} or θ_{max}^{e} deg	ee[f]%	Ref(s)
1-(1-hydroxy-1-methylethyl)tricyclo[4.1.0.0$^{2.7}$]-hepta-4-en-3-one (24)[k]	351, 363 (r)	35	$[\theta]_{353}$ +43.6	g	87
	351, 363 (l)	38	$[\theta]_{353}$ +18.5	g	
4-acetyl-2-cyclopentenone (25)	351, 363 (l)	30	$[\theta]_{578}$ +0.92	g	88
trans-bicyclo[4.3.0]nonan-8-one (17) (0.24)	313 (r)	99	g	30	76,77
2-phenylcyclohexanone (26) (0.030)	308 l/r	78–95	g	2.5/3	89
2-phnylcycloheptanone (27) (0.020)	308 l/r	78–95	g	2.5/−0.1	89
2-phenylcyclooctantone (28) (0.017)	308 l/r	78–95	g	1.2/−1.2	89
Amino acid photolysis					
alanine (29) (0.007)	>200 (r/l)	>20	θ_{210} +9.10^{-6}	0.06	90
(pH 1) (0.029)	213	24	θ_{215} 4.3·10^{-4}	0.2–0.27	91
glutamic acid (30) (0.008)	>200 (r/l)	52	θ_{210} +36.10^{-6}	0.22	90
leucine (32) (pH 2)(0.0244)	213 (r)	59	g	2.0	92
	213 (l)	75	g	2.5	
(pH 1) (0.028)	215	17.4	θ_{215} 3.5·10^{-4}	0.2,	91,93
solid (cyclotron)	183	33	g	2.6	94
tryptophan (34)(10K, 10^{12} photons/cm^{-1}s^{-1})	254	g	g	3	95
tyrosine (35) (8.8·10^{-5}, 0.024^{l})	266 (r, 23 ps)^{m}	51	g	3.6	96,66
	266 (l, 23 ps)^{m}	45	g	2.4	
	266 (r, 10 ns)^{m}	48	g	≈0	
	266 (l, 10 ns)^{m}	48	g	≈0	

other reactions

2,6-dimethyl-3-ethylcarboxy-4-(2¹-nitrophenyl)-5-acetyl-1,4-dihydropyridine(43)	366 (r or l)[g]	40	α_D −0.056	g	97
potassium trioxalatocobaltate(III)	589 (r)	g	α_{499} −0.16	g	98
	589 (l)	g	α_{499} +0.11	g	
Tris(acetylacetonato)chromium(III) High intensity laser	1064 (30 ps)			0	
	800 (50 fs)			0.73	99
	400			0.26	110
hexahelicene (37)	313 (r)	25	$[\alpha]_D$ +2.6	g	100
	313 (l)	25	$[\alpha]_D$ −1.1	g	
1,2-dithiane (38)[n] (0.02)	>290 (r)	32	θ_{300} +0.0025	g	90
	>200 (r/l)	>20	θ_{210} +42·10⁻⁶	0.11	
tartaric acid (39) (0.02)	351 (2 photon, l)			0.11	101
	351 (2 photon, r)			0.07	101
tropone-iron tricarbonyl (40)	380–500 (r)	3	α_D +0.012	g	102
	380–500 (l)	3	α_D −0.010	g	
7-dehydrocholesterol (41)[o] (1.5·10⁻⁴, 0.044[p])	305 (r, 100 fs)[m]	3–18		g	96
	305 (l, 100 fs)[m]	3–16		g	
	308 (r, 6 ns)[m]	23–31		g	
	308 (l, 6 ns)[m]	24–29		g	
E-cyclooctene (42)	190	75	$[\alpha]_D$ 0.5	0.12	103

[a] Irradiation performed with racemic substrate at room temperature, unless noted otherwise. [b]Anisotropy (g) factor at or around irradiation wavelength, if reported or estimated. [c]Extent of destruction. [d]Maximum observed rotation α of irradiated solution, or specific rotation [α] of isolated sample or of residue obtained upon evaporation. [e]Maximum observed ellipticity of irradiated solution or molar ellipticity of isolated sample. [f]Enantiomeric excess of isolated sample. [g]Not reported. [h]Compound (mp 113 °C) of unknown structure, obtained in a reaction of humulene with sodium nitrite, according to the reported procedure; Chapman, AC. *J. Chem. Soc.* **1895**;67:780. [i]A mixed case of asymmetric destruction and photoderacemization; irradiation performed at 0° C. [j]Enantiomerically enriched sample used. [k]Estimated g factor enhanced by two-quantum excitation with high intensity picosecond laser pulse. [m]High-intensity laser of indicated pulse duration used. [n]Irradiation performed at 77 **K** in a hydrocarbon glass matrix. [o]Optically pure sample photolyzed only to evaluate the enhanced g factor. [p]Estimated g factor enhanced by two—quantum excitation with high-intensity femtosecond laser pulse.

Source: adapted with permission from Ref. 37. Copyright American Chemical Society.

In a series of papers, Zandomeneghi, Cavazza and coworkers [67,83] investigate the photo induced ''double'' (oxa)-di-n-methane rearrangement of a bicylcic ketone structure **22**.

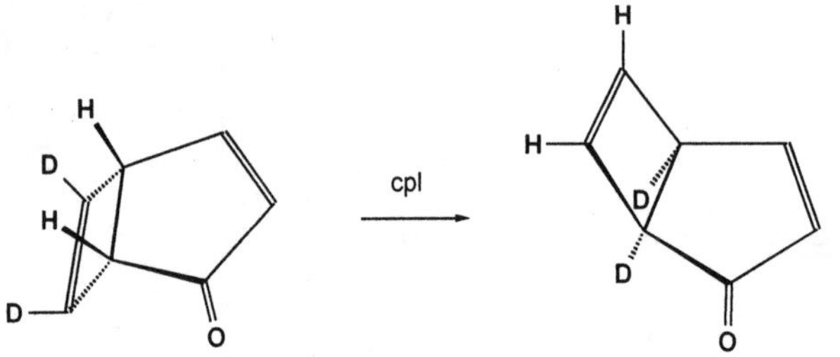

(-)(1S,5S) - 22a   **(+)(1R,5R) - 22b**   (33)

The development of optical activity by cpl is reported for the D_2 molecule **22a** [67]. Although both stereogenic centers are reverted, **22a** and **22b** are not enantiomers. One of the four hydrogen atoms drawn in Eq. (33) may be replaced by other groups like -OCH_3 [84] or -$NHCOCH_3$ [85], and the results of the irradiation of these molecules exclude a simple epimerization at the two bridgehead carbon atoms.

An interesting case is that of 2,6-dimethyl-3-ethylcarboxyl-4-(2'-nitrophenyl)-5-acetyl-1.4-dihydropyridine **43**. Here the chiral element ''asymmetric carbon atom'' is replaced by atropisomerism as the chiral principle.

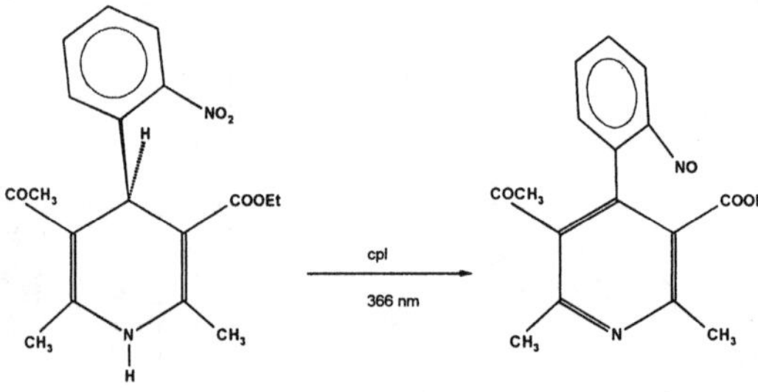

(34)

43

The photolysis of hydroxy- and amino acids requires short wavelength irradiation sources. Flores et al. used the fifth harmonic of a Nd/YAG laser at 212.8 nm [92]. Shimizu [101] used a focused 351 nm XeF laser as a source and photolyzed racemic tartaric acid by two-photon excitation. In this paper highly selective mineralization (CO_2, CO, H_2O) of L-tartaric acid **39** by r-cpl and the enantiomer by l-cpl is found. No racemization or formation of a meso form is found. The *ee* in a focused laser is up to 11%! Inoue and his group [91,93] used 200 and/or 215 nm cpl-radiation from a polarizing undulator installed in an electron storage ring and conducted a thorough investigation on the npl- and cpl-photodecomposition of amino acids (glycine, alanine **29**, valine **31**, leucine **32**, and isoleucine **33**). Glycine and alanine show slow photodeamination/hydroxylation, whereas valine, leucine, and isoleucine decompose in a Norrish type II reaction [91,104]. The spectral and chiroptical properties and the types of photodecomposition reactions are pH (ionic state) dependent. At pH 1 the acid form develops the *ee* according to Scheme 5 and the formulas derived therefrom while at pH 7 the zwitterionic form gives enantiospecifically the α-hydroxycarbonic acid of opposite chiroptical effects. Thus in this case the temporal CD development is a canceling superposition of the contributions of reactant amino acid; and product hydroxycarbonic acid.

An interesting increase of the *g* factor for tyrosin **35** in nanocrystals of up to 0.1 has been detected by Paul and Siegmann [105]. They attribute this to ionized nanoparticles. They did not, however, investigate the chemistry of the process.

C. Asymmetric Synthesis

As stated in Sec. B, asymmetric synthesis is the other side of the coin of asymmetric destruction: the products are in the focus of interest. However, there are at least two types of reactions to be considered: independent and dependent parallel reactions. The latter, starting from a prochiral racemic mixture of reactants, are often considered the real representatives of asymmetric synthesis.

For *independent parallel reactions*, Scheme 5 is appropriate. The rate equations should be expressed in the time coordinate τ and are

$$\frac{dc_{P_R}}{d\tau} = -\frac{dc_R}{d\tau} = \frac{k_R \, k_{R2}}{k + k_{R2}} c_R = k_R \phi c_R \quad \text{and} \quad \frac{dc_{P_S}}{d\tau} = -\frac{dc_S}{d\tau} = k_S \, \phi \, c_S \quad (35)$$

and further,

$$c_{P_R} = \frac{c_0}{2} - c_R \quad \text{and} \quad c_{psR} = \frac{c_0}{2} c_s \quad (36)$$

where from

$$c_{P_R} = \frac{c_0}{2}(1 - e^{-1000 I_0' \, \varepsilon_R' \, \phi \, \tau}) \quad \text{and} \quad c_{P_S} = \frac{c_0}{2}(1 - e^{-1000 I_0' \, \varepsilon_S' \, \phi \, \tau}) \quad (37)$$

The optical purity of the product in solution thus is

$$ee \text{ (product)} = \frac{c_{P_S} - c_{P_R}}{c_{P_S} + c_{P_R}} = \frac{c_S - c_R}{c_S + c_R - c_0} \quad (38)$$

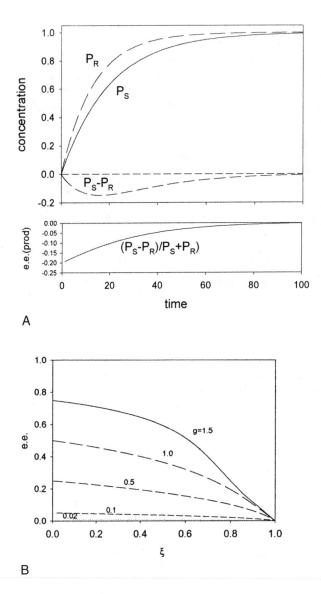

Figure 7 Calculated development of (A) the concentrations and *ee* of products with time ($\kappa = 1.5$) and of (B) the *ee* of the products with extent of reaction [106]. Note that the absolute value of *ee* in both (A) and (B) decreases.

The time and extent of reaction dependencies are given in Fig. 7. According to Eq. (38), the enantiomer produced from the preferentially destroyed reactant enantiomer is enriched in the product solution. The *ee* of the product vs. the extent of reaction was given by Nishino, et al. [106] as

$$y_{Pr} = -y_{Ed} \frac{1-\xi}{\xi} \tag{39}$$

For *dependent parallel reactions*, Scheme 6 is representative. Here ground state interconversion is fast, i.e., the rate constant k_1 is large. This means that at any time $c_S = c_R$. and $dc_S/dt = dc_R/dt$. Then elementary textbook arithmetic for parallel reactions can be used for the pseudospecies $S + R$, and the concentrations of P_R and P_S are, for Scheme 6,

$$\frac{dc_{P_R}}{dt} = k_R \frac{k_{R2}}{k + k_{R2}} c_R = k_R \phi \frac{c_S + c_R}{2} \tag{40}$$

$$\frac{d(c_S + c_R)}{dt} = -k_S \phi c_S - k_R \phi c_R = -(k_S + k_R)\phi c_R$$
$$= -\frac{1}{2}(k_S + k_R)\phi(c_S + c_R) \tag{41}$$

$$(c_S + c_R) = c_0 \, e^{-\frac{1}{2} 1000 \, I_0' \, \phi \, (\epsilon_S' + \epsilon_R') \tau} \tag{42}$$

With Eq. (42) in Eq. (40) we get under consideration of k_R and k_S in Eq. (19) and $\tau = f(t)$ in Eq. (20),

$$c_{P_R} = -\frac{k_R}{k_R + k_S} c_0 \left(1 - e^{-\frac{1}{2}\phi(k_S + k_R)\tau} \right)$$
$$= -\frac{\epsilon_R'}{\epsilon_S' + \epsilon_R'} c_0 \left(1 - e^{-\frac{1}{2} 1000 \, I_0' \, \phi \, (\epsilon_S' + \epsilon_R') \tau} \right)$$
$$c_{P_S} = -\frac{k_S}{k_R + k_S} c_0 \left(1 - e^{-\frac{1}{2}\phi(k_S + k_R)\tau} \right)$$
$$= -\frac{\epsilon_S'}{\epsilon_S' + \epsilon_R'} c_0 \left(1 - e^{-\frac{1}{2} 1000 \, I_0' \phi(\epsilon_S' + \epsilon_R') \tau} \right) \tag{43}$$

Both product enantiomers are created with the same time dependence; the absorption coefficients at the irradiation wavelength determine the ratio of the concentrations. In this case the enantiomeric excess is constant over the whole reaction period. Figure 8 shows these exponential dependencies in the corrected time axis τ.

Soon after Kuhn's photodestruction experiments around 1930, Karagunis and Drikos tried to perform an asymmetric synthesis with an asymmetrically substituted triphenylmethyl radical, but there was no optical activity created [107]. In 1935 Davis and Heggie [108] reported a total asymmetric synthesis by addition of bromine to 2,4,6-trinitrostilbene (they found, however, a maximum curve of

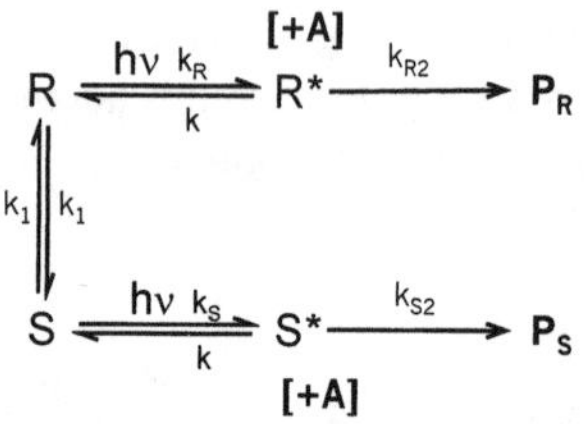

R $\underset{k}{\overset{h\nu\ k_R}{\rightleftharpoons}}$ R* $\xrightarrow{\overset{[+A]}{k_{R2}}}$ P$_R$

k$_1$ | k$_1$

S $\underset{k}{\overset{h\nu\ k_S}{\rightleftharpoons}}$ S* $\xrightarrow{k_{S2}}$ P$_S$

[+A]

Scheme 6 Dependent parallel reactions.

rotation vs. irradiation time, which indicates asymmetric photodestruction rather than asymmetric photosynthesis). Later Davis and Ackerman reported the asymmetric cpl-induced addition of H_2O_2 to diethyl fumarate [13]. At that time the analysis was only possible by measuring the rotation outside of the absorption regions, and every optically active species in the solution contributed to that, so

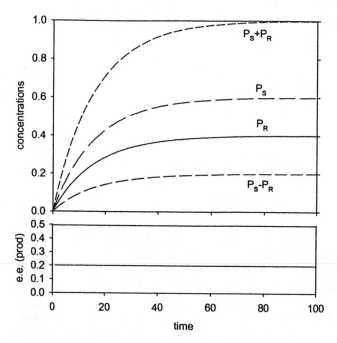

Figure 8 Calculation of the time development of the product enantiomer concentrations (rate constants 0.05 and 0.075, $\kappa = 1.5$, time τ).

it is very difficult to identify the relevant species. To this author it seems thinkable, if the effect is real at all (and reproducibility has not been confirmed [43,109]), that the ethenes were photoisomerized and that the cis compounds provide differential absorption and thus a nonracemic stationary enantiomeric mixture, which then disappears by the addition reaction. These examples were discussed here in order to exemplify the difficulties and questions often met with in reading the literature. Not only major components of a solution but impurities e.g., in solvents (in extreme cases down to parts per trillion [15]) can fake asymmetry in a reaction.

Several of the papers dealing with asymmetric destruction also report information about the enantiomeric excess of the products in the sense of independent asymmetric reactions. If the data of both reactants and products can be determined independently, the degree of transfer of asymmetry may be used for mechanistic questions [10] according to Eq. (38).

Most of the known asymmetric synthesis reactions equilibrating from prochiral racemic mixtures are photocyclizations where ring systems that include one or two "asymmetric" C atoms are created as a first or the final product.

One example is the formation of helicenes by dehydrocyclization of diaryl ethenes (Scheme 7), which was well investigated by the groups of Calvin and Kagan [110–114].

As in Calvin's group, ORD was the analytical tool the nature of the optically active species could not be determined directly as is possible with CD measurements. Buchardt [43] gives the arguments for the exclusion of an asymmetric destruction of the newly formed helicene and of partial photodestruction of the dihydrohelicene intermediate. The reacting diarylethene is screw-like already, but the two enantiomers can easily be interconverted by rotation around a single bond. So dependent parallel reactions are expected (Scheme **6**). If, however, the *p*-position in the phenyl ring is substituted by larger groups, the *ee* in the helicenes

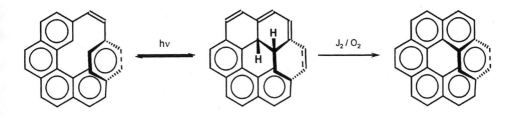

36   **37**

Scheme 7   (reprinted from ref. 38)

increases because the ground state enantiomerization is impeded, and the two parallel reactions become more and more independent.

Scheme 7 shows the ''1–4'' addition starting from 1-(5,6-benzophenanthtr-3-yl)-2-phenylethene; an alternative is the ''2–3'' addition starting from 1-(3-phenanthryl)-2(2-naphthyl)ethene. They have different enantiomeric excesses: 1–4 reaches $ee = 0.06\%$, while 2–3 gives $ee = 0.3\%$ [43]. Obviously the ees are quite low. The difference is explained by thermal rotation in the excited Z-ethene around the single ethene-arene bonds which in the 1–4 case leads to some excited state racemization, decreasing the majority conformation for ring closure, whereas the rotation in the 2–3 case is more difficult. By the same token, substitution in one o-position of the phenyl ring increases the ee of the hexahelicene formed up to 0.33% in the case of o-chloro substitution [114].

Other examples are the photocyclization of suitably substituted N-methyl-N-aryl-enamins (Scheme 8), which may undergo rearomatization [115] (**44** gives 0.2% ee in the product **45**) or the internal cyclization of 2-methoxytropone **46** which results in the bicyclic ketone 1-methoxy-bicyclo[3.2.0]hepta-3,6-dien-2-one (1-CH$_3$O-**22**) [116].

This type of photocyclization is also found in the unimolecular open form/closed form isomerism of photochromic systems [70] as, for example, spiropyrans and spirodihydroindolazines, fulgides. The chirality in such photochromic reactions will be covered by Chap. 9 of this book; an example is **47** in Scheme 9.

D. Asymmetric Photoisomerization

This very interesting reaction scheme was studied in detail by Nishino et al. [117]. They used Scheme 10, which represents the independent reversible photo-isomerization of the enantiomers. This scheme leads to a photostationary state. The kinetics are quite complicated. The solution of the differential rate equations is only possible by numerical integration with definite numbers for six parameters,

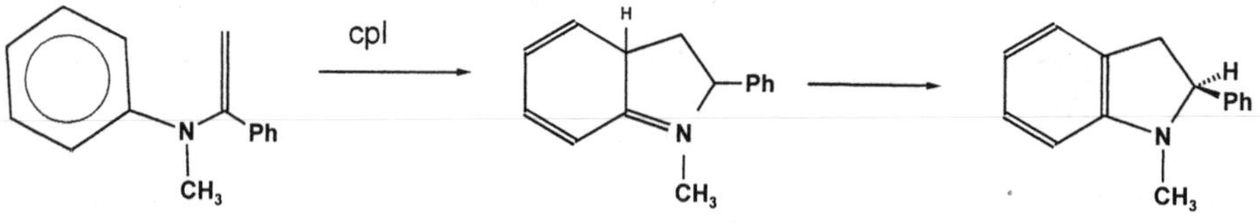

44

45

Scheme 8

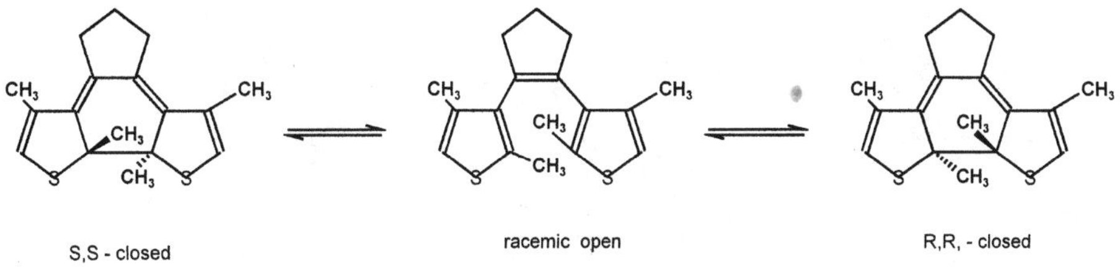

S,S - closed   racemic open   R,R, - closed

47

Scheme 9

the four absorption coefficients, and the two quantum yields. The result of simulations [117] is presented in Fig. 9.

In Figure 9 the *ee* of reactant A and product B is shown as a function of the extent of reaction ξ expressed as $C_B/(C_A + C_B)$. The scale ends at $\xi = 0.1$, as the composition of the photostationary state is a parameter chosen by $K = (\varepsilon_B \phi_B)/(\varepsilon_A \phi_A) = 0.1$. The influence of identical or opposite signs of CD at A and B is characterized by the parameters $g_A' = g_B'$ or $g_A' = -g_B'$, which are arbitrarily chosen to be $|g'| = 1$ (this does not, however, mean that the molar ellipticities $[\theta']$ of A and B were equal). Figure 9 shows that when the signs of CD of both products and educts are identical there is an increase and following decrease of the *ee* or ellipticity, which in the special case of $g' = 1$ leads to the disappearance of CD at the photostationary state. In the case of opposite signs of CD of reactant and product, the *ee* increases with the progress of the reaction and reaches more than 50% in the product at the photostationary state (again for $g' = +$ or -1).

However, the anisotropy factors are always smaller than $g' = 1$. In Fig. 10 the corresponding plot for asymmetric photoisomerization of quadricyclane to norbornadiene is shown. The lines are calculated with $g'(\mathbf{48}) = -0.0074$ and $g'(\mathbf{47}) = +0.005$ for r-cpl and the g factors of opposite sign for l-cpl [117].

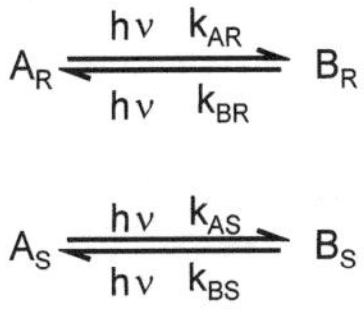

Scheme 10   Independent reversible photoreactions.

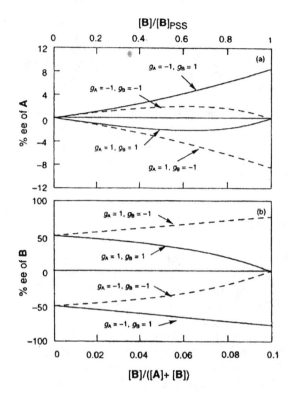

Figure 9   Simulation of reaction of Scheme 10 with the photostationary state at 10%.
(From Ref. 117. Copyright The Royal Society of Chemistry.)

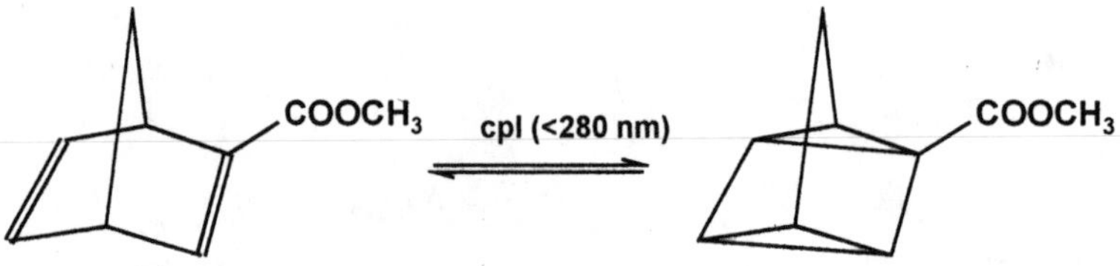

48   **49**

Scheme 11

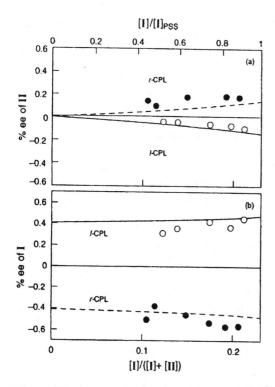

Figure 10 Asymmetric photoisomerization of **48** ($\leftarrow$ **II**) and **49** ($\leftarrow$ **I**) (see Scheme 11) with r-cpl and l-cpl. Points: experimental; lines: calculated with $g'(49) = -0.0074$ and $g'(48) = +0.005$ for r-cpl and g factors of opposite sign for l-cpl. (From Ref. 117. Copyright The Royal Society of Chemistry.)

The quadricyclane-1-carboxylate /norbornadiene-2-carboxylate **49/48** seems to be the only asymmetric isomerization studied successfully [106,117]. We tried the isomerization of 3-(1-methylpropyl)azobenzene with cpl, but no CD effect appeared.

III. THE COMBINATION OF RADIATION AND MAGNETIC FIELDS

Barron defined "true" and "false" chirality in systems and pointed out that only true chirality in a system allows absolute asymmetric synthesis when the system is near thermodynamic equilibrium, but that false chirality may also be effective in systems far from equilibrium [118], Feringa and van Delden reviewed some consequences [61]. Cpl is a true chiral system, and so is the combination of a constant magnetic field in the direction of the propagation of a linearly polarized light beam. The latter arrangement induces magnetic optical activity (MCD, MORD) in nonchiral molecules (Faraday effect) [119], or magnetochiral optical activity (MchOA) and magnetochiral dichroism (MChD) in chiral molecules.

After detecting magnetochiral anisotropy in the emission of the tris-(trifluoroacetylcamphor)-europium(3 +) ion, Rikken and Raupach used the MChD effect arrangement and found photoderacemization of $[Cr(C_2O_4)_3]^{3-}$ [120]. On a perpendicular arrangement of light and the magnetic field vector, no *ee* appeared [121,122]. The same group gives a detailed theory of magnetochiral anisotropy [122]. A deeply founded general discussion of potentially asymmetric reactions under physical fields is given by Avalos et al. [123]. Chapter 3 of this book is dedicated to magnetochiral photochemistry.

A different way, how cpl-induced photoreactions might be influenced by a magnetic field, is to exploit the magnetic field effect on the spin dynamics of intersystem crossing, which is important in the competition between cage recombination and cage escape of radicals [124]. Kohtani et al. [89] investigated the reaction of biradicals X excited from R and S forms according to the kinetic scheme, Scheme 12.

The rate constants are

$$\frac{dc_R}{dt} = -1000 \, I'_0 \left[\frac{1-10^{-A'}}{A'} \right] \left[\varepsilon'_R c_R - \frac{k_c}{2k_c + k_z} \cdot \frac{k_{isc}}{k_{isc} + k_y} A' \right]$$

$$\frac{dc_S}{dt} = -1000 \, I'_0 \left[\frac{1-10^{-A'}}{A'} \right] \left[\varepsilon'_S c_S - \frac{k_c}{2k_c + k_z} \cdot \frac{k_{isc}}{k_{isc} + k_y} A' \right] \qquad (44)$$

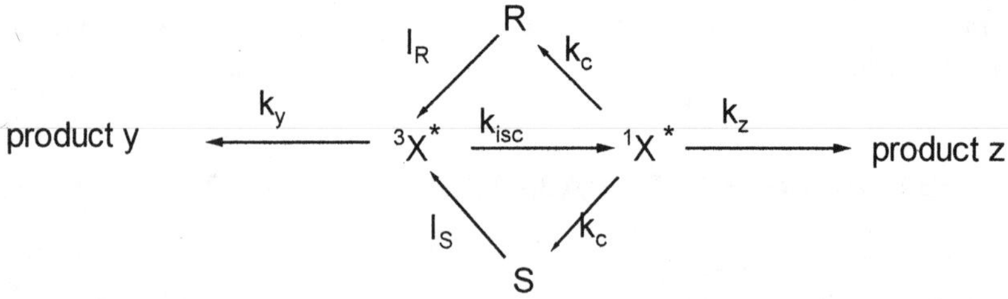

Scheme 12 (Adapted from Ref 89. Copyright Chemical Society of Japan.)

By subtraction of the equal term from $\varepsilon'_R c_R$ and $\varepsilon'_S c_S$ the two rates become smaller, but their ratio becomes larger and there should be a larger κ. The authors searched for an enhancement of *ee* in the cpl-photoreaction of 2-phenylcyclohexanone **27**, 2-phenylcycloheptanone **28** and 2-phenylcyclooctanone **29**, but found no magnetic field effect.

IV. PHOTOLYSIS BY MEANS OF HIGH-INTENSITY AND OF ULTRA-SHORT LASER PULSES

A Russian group [66] investigated the results of high-power laser pulse irradiation of tyrosine (see Table 1), which according to the authors caused a stepwise two-photon absorption to higher excited states. Off-hand, one would expect that the product of the small g factors of the $S_0 \rightarrow S_1$ and $S_1 \rightarrow S_n$ transitions governed the results. However, what was observed was a considerable increase in *ee* when ps pulses were used rather than ns pulses. The photon density of the ps pulses was 3 orders of magnitude higher, which may lead to a direct two-photon absorption via virtual states to states that carry magnetic dipole strength with high g factor or to coherence phenomena like absorption of photon 1, enantiomeric dephasing in state S_1, absorption of photon 2, reaction of state S_n. This will be covered in Chap. 2 of this book.

Taniguchi, et al. [99] have tested the one- and two-photon excitation for $[Cr(acac)_3]^{3-}$ for several irradiation wavelengths but found no indication of a superlinear dependence of photoracemization or photodecomposition. This seems to be in line with the experiments of Gunde and Richardson [6] on $Na_3[Gd(O_2C-CH_2-O-CH_2-CO_2)_3] \cdot 2NaClO_4 \cdot 6H_2O$ crystals. The possibility discussed in several papers [4–6] that in high-intensity radiation fields a pure electric dipole 2-photon absorption may lead to CD effects adds a new possibility of rationalization of the high-power laser results.

The advent of fs pulse lasers recently opened new perspectives for asymmetric photochemistry. The elaboration of this field still is in the theoretical realm. Pulse sequence [125,126] and coherence [127] scenarios are set up for chiral molecular products from achiral precursors. If, for example, phosphinothiotic acid $H_2PO(SH)$ molecules are preoriented, which can be effected by laser action, and a special sequence of cpl pulses is used, then the theoretical prediction is that the L enantiomer is transformed to the R enantiomer, but the reverse process is suppressed and vice versa for a different pulse sequence [125]. Chapter 2 of this book is dedicated to these coherent phenomena controlling asymmetric photoreactions.

V. AMPLIFICATION OF SMALL CPL-INDUCED ENANTIOMERIC EXCESSES AND CPL AS A SOURCE OF HOMOCHIRALITY IN NATURE

The cpl-induced asymmetry in photoreactions as described in Sec. B. of this chapter is not very pronounced. In order to obtain *ee*s in excess of a few percent, photodestruction must be chosen and most of the reactant material must be sacrificed. Therefore amplification mechanisms for all types of cpl-induced asymmetric photoreactions would be highly desirable. Autocatalysis, i.e., an asymmetric synthesis where a chiral product acts as a catalyst for its own production [128], and autoinduction, i.e., the stimulation of a chiral catalyst by a chiral product [44,129], are options. Autocatalytic systems that will tilt to one enantiomeric side were introduced by Frank [130] and Seelig [131].

Knowledge on these kinds of reactions is being accumulated for thermal reactions, which is also relevant for photoreactions. The topic is thoroughly discussed by Todd [132]. Soai and his group have presented a very efficient amplification system, the addition of dialkyl zinc to pyridine-3-carbaldehyde **49** [16,133], in which the chiral alcohol produced catalyzes the addition.

This reaction can be realized in a cascade (Fig. 11); the *ee* formed in the alcohol of a foregoing reaction acts as a catalyst in the next generation so the *ee*s of the alcohol could be bought up to more than 80% in the fifth generation [15]. Very recently Soai et al. were able to reach air *ee* of the alcohol 50% by addition of a tiny $5.10^{-5}\%ee$ in the starting mixture [134].

The Soai system is sensitive to any chiral additive. $NaClO_3$ single crystals [135], inorganic complexes [136], penta- and hexahelicenes [137], or *d*- and *l*-quartz [138] are examples; *ee* values of up to $> 95\%$ of the secondary alcohol can be obtained. The system responds to the sense of chirality of the dopant enantiomers also with opposite signs of *ee* and is ready for stimulation by cpl via the induced *ee*s of reactants or products [139]: the cpl-induced hexahelicene formation with 2% *ee* amplifies to an *ee* of $> 90\%$ in a Soai system [140]. A review is found in Ref. 141. Unfortunately, the system does not amplify the *ee* of the dopant.

Whereas the Soai system does not develop intrinsic spontaneous asymmetry (the *ee* in an undoped Soai system, which pretends to do so, is not statistical, and therefore the appearance of asymmetry is attributed to the action of a chiral impurity [15]), Asakura et al. report random variations of *ee* values (up to $+$ or -25–30%) in the creation of asymmetry by chiral autocatalysis in the reaction of a trinuclear Co complex with ammonium bromide [142]. The authors propose a stochastic model. The reactions quickly reach a state of supersaturation of the racemic chiral product, and enantiopure autocatalytic clusters of 10 and more product molecules are formed, which tips the reaction to one enantiomeric side.

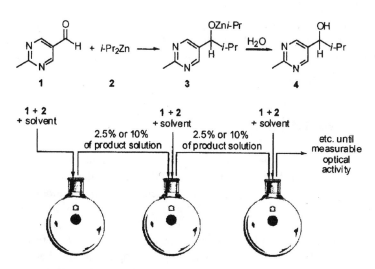

Figure 11 A process for replicative asymmetric amplification. (Reprinted from Ref. 15. Copyright American Chemical Society.)

So far no system has been found that e.g. amplifies the *ee*s of amino acids created in cpl-photolysis that might explain the homochirality in nature.

The question of the (nearly) exclusive use of L amino acids and D sugars in living organisms is one of the unsolved enigmas of the chemical sciences. As there are nonracemic amino acids found in meteorites [143,144] and as cpl occurs in outer space [145] and can be produced under natural terrestrial environmental conditions, cpl photolysis is discussed as a source of homochirality (up to scenarios like the synchronization of terrestrial cpl day cycles with the periodicity of tides [146]). The question of amino acid homochirality being due to asymmetric photolysis in space is discussed by Cerf and Jorisson [147], who summarize, ''In conclusion, we hope to have convinced the reader that the role of extraterrestrial asymmetric photolysis in the origin of the homochirality of natural amino acids on Earth, if at all involved, is far more complicated than is usually apprehended in the astronomical literature.'' The same authors more recently gave a critical review of the ample literature on asymmetric photochemistry as a possible source of homochirality [148].

If direct cpl-induced one-photon photochemistry should be the source of homochirality, the assumption of the activity of amplification mechanisms is necessary. But cpl photolysis is only one of the mechanisms discussed for symmetry breaking and creation of homochirality. Bonner [11] and Feringa and van Delden [61] reviewed this topic and give sources for further information. On the other hand, some deny the necessity of specific symmetry-breaking reasons: Siegel [149], for example, stresses that molecular evolution should by necessity lead to homochirality. However, this issue is not in the scope of this review on asymmetric photochemistry with cpl.

VI. CONCLUDING REMARKS

Cpl-induced asymmetric photochemistry is a way to conduct absolute asymmetric synthesis. The field is well developed. The method depends on the differential absorption of left and right circularly polarized radiation by the enantiomeric species of the reactants and thus on their g factor. The kinetic schemes are well developed, and for mechanistic questions as well as for the determination of molecular chiroptical constants cpl irradiation can be used with success. Unfortunately, the smallness of the enantiodifferentiating g factor prevents the method from being a match to the methods exploiting diastereomeric intermolecular interactions for thermal synthesis purposes.

New and promising developments can be seen in high-intensity (2-photon) laser photolysis, in the combination of fields and radiation (see Chap. 3), in exploitation of the coherence properties (see Chap. 2), and in the detection of more autocatalytic or autoinductive reaction systems.

Asymmetric photochemistry is intensively discussed as one of the possibilities for the origin of the observed homochirality of amino acids and saccharides in biological systems. There is no doubt that also in natural environments cpl exists and may create an enantiomeric excess in a photoreaction, but according to all knowledge, there must be an amplification mechanism, which has not yet been safely detected.

REFERENCES

1. Mislow K. Chirality 2002; 14:126–134.
2. Cahn RS, Ingold CK, Prelog V. Angew Chem Int Ed 1966; 5:335–415.
3. Kuhn W, Knopf E. Z Phys Chem B 1930; 7:292–310.
4. Gunde KE, Burdick GW, Richardson FS. Chem Phys 1996; 208:195–219.
5. Naguleswaran S, Reid MF, Stedman GE. Chem Phys 2000; 256:207–212.
6. Gunde KE, Richardson FS. Chem Phys 1995; 194:195–205.
7. Riehl JP, Richardson FS. Chem Rev 1986; 86:1–16.
8. Shippers PH, Dekkers MPJM. Tetrahedron 1982; 38:2089–2096.
9. Metcalf DH, Snyder SW, Demas JN, Richardson FS. J Am Chem Soc 1990; 112: 469–479.
10. Hörmann M, Ufermann D, Schneider MP, Rau H. J Photochem 1981; 15:259–262.

11. Bonner WA. Top Stereochem 1988; 18:1–96.
12. Blume R, Rau H, Schuster O. J Am Chem Soc 1976; 98:6583–6586.
13. Davis TL, Ackerman Jr. J Am Chem Soc 1945; 67:486–489.
14. Eskenazi C, Nicaud JF, Kagan HB. J Org Chem 1979; 44:995–999.
15. Singleton DA, Vo LK. J Am Chem Soc 2002; 124:10010–10011.
16. Soai K, Niwa S, Hori H. J Chem Soc, Chem Commun 1990:982–983.
17. Le Bel JA. Bull Soc Chim (Paris) 1874; 22:337–347.
18. van't Hoff JH. Die Lagerung der Atome im Raum. 2d ed.. Braunschweig. Vieweg, 1894:30.
19. Cotton A, Hebd Séanc C R. Acad Sci Paris 1895; 120:989–991.
20. Cotton A. Ann Chim Phys. 1896; 8:347–350.
21. Freundler P. Bull Soc Chim Fr 1907; 1:657–659.
22. Henle F, Haakh H. Ber Deut Chem Ges 1908; 41:4261–4264.
23. Freundler P. Ber Deut Chem Ges 1909; 42:233–234.
24. Guye P, Drouginine G. J Chim Phys 1909; 7:96–100.
25. Padoa M. Gazz Chim Ital 1911; 41:469–472.
26. Pirak J. Biochem Z 1922; 130:76–79.
27. Bredig G. Z Angew Chem 1923; 36:456–458.
28. Kuhn W, Braun E. Naturwiss 1929; 17:227–228.
29. Kuhn W. Z Angew Chem 1929; 42:296–296.
30. Kuhn W. Z Angew Chem 1929; 42:828–828.
31. Kuhn W. Trans Faraday Soc 1930; 26:293–308.
32. Kuhn W. Naturwiss 1930; 18:183–183.
33. Mitchell S. J Chem Soc 1930:1829–1834.
34. Stevenson KL, Verdieck JF. Mol Photochem 1969; 1:271–278.
35. Kagan HB, Fiaud JC. Top Stereochem 1978; 10:175–285.
36. Rau H. Chem Rev 1983; 83:535–547.
37. Inoue Y. Chem Rev 1992; 92:741–770.
38. Everitt SRL, Inoue Y. In: Ramamurthy V , Schanze KS, Eds. Organic Molecular Photochemistry. New York: Marcel Dekker, 1999:71–130.
39. Griesbeck AG, Meierhenrich UJ. Angew Chem Int Ed 2002; 41:3147–3152.
40. Balavoine G, Moradpour A, Kagan HB. J Am Chem Soc 1974; 96:5152–5158.
41. Straathof AJJ, Jongejan JA. Enzyme Microb Technol 1997; 21:559–571.
42. Kagan HB, Fiaud JC. Top Stereochem 1998; 18:249–330.
43. Buchardt O. Angew Chem Int Ed 1974; 13:179–185.
44. Alberts AH, Wynberg H. J Am Chem Soc 1989; 111:7265–7266.
45. Wynberg H, Feringa BL. Tetrahedron 1976; 32:2831–2834.
46. Stevenson KL. J Am Chem Soc 1972; 94:6652–6654.
47. Stevenson KL, Verdieck JF. J Am Chem Soc 1968; 90:2974–2975.
48. Yoneda H, Yakashima Y, Sakaguchi U. Chem Lett 1973:1343–1346.
49. Kane-Maguire NAP, Langford CH. Can J Chem 1972; 50:3381–3383.
50. Nordén B. Acta Chem Scand 1970; 24:349–351.
51. Anzai N, Machida S, Horie K. Chem Lett 2001:888–889.
52. Lemieux RP, Schuster GB. J Org Chem 1993; 58:100–110.
53. Burnham KS, Schuster GB. J Am Chem Soc 1998; 120:12619–12626.

54. Irie M, Yoshidy K, Hayashi K. J Phys Chem 1977; 81:969–972.
55. Haysashi K, Irie M. Chem Abstr 1978; 89:6151j.
56. Kuhn W. In: Freudenberg K, Ed. Stereochemie. Wien. Deudicke, 1933:317–423.
57. Colter AK, Clemens LM. J Phys Chem 1974; 68:651–654.
58. Feringa BL, van Delden RA, Koumora N, Geertsema EM. Chem Rev 2000; 100: 1789–1816.
59. Huck NPM, Jager WF, de Lange B, Feringa BL. Science 1996; 273:1686–1688.
60. Feringa BL, Huck NPM, Schoevaars AM. Adv Mater 1996; 8:681–684.
61. Feringa BL, van Delden RA. Angew Chem Int Ed 1999; 38:3418–3438.
62. Zhang M, Schuster GB. J Org Chem 1995; 60:7192–7197.
63. Li J, Schuster GB, Cheon KS, Green MM, Selinger JV. J Am Chem Soc 2000; 122:2603–2612.
64. Li J, Schuster GB, Green MM. Mol Cryst Liq Cryst 2000; 344:7–13.
65. Burnham KS, Schuster GB. J Am Chem Soc 1999; 121:10245–10246.
66. Nikogosyan DN, Repeyev YA, Khoroshilova EV, Kryukov IV, Khoroshilov EV, Sharkov AV. Chem Phys 1990; 147:437–445.
67. Zandomeneghi M, Cavazza M, Pietra F. J Am Chem Soc 1984; 106:7261–7262.
68. Rau H. unpublished.
69. Fukui K, Naito Y, Taniguchi S, Inoue Y. Book of Abstracts. 1st. Osaka: Intl Symp Asymm Photochem, 2001:L 106.
70. Dürr H, Bouas-Laurent H. Photochromism. Molecules and Materials. Amsterdam: Elsevier, 1990.
71. Rubalcava H, Fitzmaurice DJ. J Chem Phys 1990; 92:5975–5987.
72. Mauser H. Formale Kinetik. Düsseldorf: Bertelsmann Universitätsverlag, 1974:166 ff.
73. Blume R, Rau H, Schneider M, Schuster O. Ber Bunsenges Phys Chem 1977; 81: 33–39.
74. Rau H, Hörmann M. J Photochem 1981; 16:231–247.
75. Nakamura A, Nishino H, Inoue Y. J Chem Soc, Perkin Trans 2 2001:1701–1705.
76. Nicoud JF, Eskenazi C, Kagan HB. J Org Chem 1977; 42:4270–4272.
77. Kagan HB, Fiaud JC. Top Stereochem 1988; 18(249–330):321.
78. Emeis CA, Oosterhoff L, deVries G. J Chem Phys 1971; 54:4809–4819.
79. Emeis CA, Oosterhoff L. Chem Phys Lett 1967; 1:129–132.
80. Schneider M, Schuster O, Rau H. Chem Ber 1977; 110:2180–2188.
81. Mitchell S, Dawson IM. J Chem Soc 1944:452–454.
82. Quinkert G, Schmieder KR, Dürner G, Hache K, Stegk A, Barton DHR. Chem Ber 1977; 110:3582–3614.
83. Cavazza M, Zandomeneghi M, Festa C, Lupi E, Sammuri M, Pietra F. Tetrahedron Lett 1982; 23:1387–1390.
84. Zandomeneghi M, Cavazza M, Festa C, Pietera F. J Am Chem Soc 1983; 105: 1839–1843.
85. Cavazza M, Morganti G, Zandomeneghi M. J Chem Soc, Perkin Trans 2 1984: 891–895.
86. Cavazza M, Zandomeneghi M, Ciacchini G, Pietra F. Tetrahedron 1985; 41: 1989–1990.

87. Zandomeneghi M, Cavazza M, Festa C, Fissi A. Gazz Chim Ital 1987; 117:255–258.
88. Cavazza M, Zandomeneghi M. Gazz Chim Ital 1987; 117:17–21.
89. Kohtani S, Sugiyama M, Fujiwara Y, Tanimoto Y, Nakagaki R. Bull Chem Soc Jpn 2002; 75:1223–1233.
90. Nordén B. Nature 1977; 266:567–568.
91. Nishino H, Kosaka A, Hembury GA, Aoki M, Miyauchi K, Shitomi H, Onuki H, Inoue Y. J Am Chem Soc 2002; 124:11618–11627.
92. Flores JJ, Bonner WA, Massey GA. J Am Chem Soc 1977; 99:3622–3624.
93. Nishino H, Kosaka A, Hembury GA, Shitomi H, Onuki H, Inoue Y. Org Lett 2001; 3:921–924.
94. Meierhenrich UJ, Jaquet R, Chabin A, Alcaraz C, Brack A, Nahon L, Barbier B. Book of Abstracts, 1st Intl Symp on Asymm Photochem, Osaka, 2001:L 104.
94. Meierhenrich UJ, Barbier B, Jacquet R, Chabin A, Alcaraz C, Nahon L, Brack A. Europ Space Agency 2000; SP-496:167–170.
95. Greenberg JM, Kouchi A, Niessen W, Irth H, van Paradijs J, de Groot M, Hermsen W. J Biol Phys 1994; 20:61–70.
96. Nikogosyan DN, Repeyev YA, Khoroshilova EV, Kryukov IV, Khoroshilov EV, Sharkov AV. Chem Phys 1990; 147:437–445.
97. Berson JA, Brown E. J Am Chem Soc 1955; 77:450–453.
98. Tsuchida R, Nakamura A, Kobayashi M. Nihon Kagakukaishi (J Chem Soc Jpn). Chem Abstr 1935; 30:7452.
99. Taniguchi S, Naitoh Y, Inoue Y. Book of Abstracts. 1st: Intl Symp on Asymm Photochem: Osaka, 2001:L 106.
100. Nelander B, Nordén B. Chem Phys Lett 1974; 28:384–386.
101. Shimizu Y. J Chem Soc, Perkin Trans 1 1997:1275–1278.
102. Litman S, Gedanken A, Goldschmidt Z, Bakal Y. J Chem Soc, Chem Commun 1978:983–984.
103. Inoue Y, Tsuneishi H, Hakushi T, Yagi K, Awazu K, Onuki H. J Chem Soc, Chem Commun 1996:2627–2628.
104. Wolff G, Ourisson G. Tetrahedron Lett 1981; 22:1441–1442.
105. Paul J, Siegmann K. Chem Phys Lett 1999; 304:23–27.
106. Nishino H, Nakamura A, Inoue Y. J Chem Soc, Perkin Trans 2 2001:1693–1700.
107. Karagunis G, Drikos G. Naturwiss 1933; 21:607–607.
108. Davis TL, Heggie R. J Am Chem Soc 1935; 57:377–379.
109. Boldt P, Thielecke W, Luthe H. Chem Ber 1971; 104:353–359.
110. Moradpour A, Nicoud JF, Balavoine G, Kagan HB, Tsoucaris G. J Am Chem Soc 1971; 93:2353–2354.
111. Bernstein WJ, Calvin M, Buchardt O. J Am Chem Soc 1972; 94:494–498.
112. Bernstein WJ, Calvin M, Buchardt O. Tetrahedron Lett 1972:2195–2198.
113. Kagan HB, Moradpour A, Nicoud JF, Balavoine G, Martin RH, Cosyn JP. Tetrahedron Lett 1971:2479–2482.
114. Bernstein WJ, Calvin M, Buchardt O. J Am Chem Soc 1973; 95:527–532.
115. Nicaud JF, Kagan HB. Israel J Chem 1976/77; 15:78–81.
116. Zandomeneghi M, Cavazza M, Gozzini A, Alzetta G, Lupi E, Samurri M, Pietra F. Lett Nuovo Cimento Soc Ital Fis 1981; 30:189 (cited after Ref. 37).

117. Nishino H, Nakamura A, Shitomi H, Onuki H, Inoue Y. J Chem Soc, Perkin Trans 2001; 2:1706–1713.
118. Barron LD. J Am Chem Soc 1986; 108:5539–5542.
119. E.g., Caldwell DJ, Eyring H. The Theory of Optical Activity. New York: Wiley-Interscience, 1971, Chap. 6.
120. Rikken GLJA, Raupach E. Nature 1977; 390:493–494.
121. Rikken GLJA, Raupach E, Roth T. Physica B 2001; 194:1–4.
122. Rikken GLJA, Raupach E, Krstic V, Roth S. Mol Phys 2002; 100:1155–1160.
123. Avalos M, Babiano R, Cintas P, Jimenez JL, Palacios JC. Chem Rev 1998; 98: 2391–2404.
124. Steiner U. Ber Bunsenges Phys Chem 1981; 85:228–233.
125. Hoki K, Ohtsuki Y, Fujimura Y. J Chem Phys 2001; 114:1575–1581.
126. Fujimura Y, González L, Hoki K, Kröner D, Manz J, Ohtsuki Y. Angew Chem Int Ed 2000; 24:4586–4588.
127. Shapiro M, Brumer P. J Chem Phys 1991; 95:8658–8661.
128. Soai K, Sato I. Acc Chem Res 2000; 33:382–390.
129. Soai K, Inoue Y, Takahashi T, Shibata T. Tetrahedron 1996; 52:13355–13362.
130. Frank FC. Biochim Biophys Acta 1953; 11:459–463.
131. Seelig FF. J Theoret Biol 1971; 31:355–361.
132. Todd MH. Chem Soc Rev 2002; 31:211–222.
133. Soai K, Niwa S, Hori H. J Chem Soc, Chem Commun 1990:982–983.
134. Soai K, Sato I, Uvabe H, Ishigura S, Shibata T. Angew Chem Int Ed 2003; 42: 315–317.
135. Sato I, Kadowaki K, Soai K. Angew Chem Int Ed 2000; 39:1510–1512.
136. Sato I, Kadowaki K, Ohgo Y, Soai K, Ogino H, Thede R, Heller D. J Chem Soc, Chem Commun 2001:1022–1023.
137. Sato I, Yamashima R, Kodawaki K, Yamamoto J, Shibata T, Soai K. Angew Chem Int Ed 2001; 40:1096–1098.
138. Soai K, Osanai S, Kadowaki K, Yonekubo S, Shibata T, Sato I. J Am Chem Soc 1999; 121:11235–11236.
139. Shibata T, Yamamoto J, Matsumoto N, Yonekubo S, Osanai S, Soai K. J Am Chem Soc 1998; 120:12157–12158.
140. Soai K, Sato I, Shibata T. Yuki Gosei Kagaku Kyokaishi. 2002; 60:668–678. Chem Abstr. AN 2002 531585.
141. Soai K, Sato I, Shibata T. Chem Rec 2001; 1:321–332.
142. Asakura K, Ikumo A, Kurihara K, Osanai S, Kondepudi DK. J Phys Chem A 2000; 104:2689–2694.
143. Engel HM, Macko SA, Silfer JA. Nature 1990; 348:47–49.
144. Cronin JR, Pizzarello S. Science 1997; 275:951–955.
145. Bailey J, Chrisostomou A, Hough JH, Geldhill TM, McCall A, Clark S, Menard F, Tamura M. Science 1998; 281:672–674.
146. Popa R. J Mol Evol 1997; 44:121–127.
147. Cerf C, Jorisson A. Space Sci Rev 2000; 92:603–612.
148. Jorissen A, Cerf C. Orig Life Evol Biosp 2002; 32:129–142.
149. Siegel J. Chirality 1998; 10:24–27.

2
Coherent Laser Control of the Handedness of Chiral Molecules

Paul Brumer
University of Toronto
Toronto, Ontario, Canada

Moshe Shapiro
The Weizmann Institute, Rehovot, Israel, and
University of British Columbia
Vancouver, British Columbia, Canada

I. INTRODUCTION

Selecting a desired enantiomer from a racemic mixture (enantioselectivity), and, more ambitiously, converting a racemic mixture to the desired enantiomer (enantiopurification) using optical means have been the foci of great fascination over the years [1–17]. Earlier suggestions of enantioselective optical methods centered on the use of circularly polarized light [2–6], which relies upon the very weak magnetic dipole interaction with the field. Recently, owing to the emergence of the field of quantum control [18–68], we [7–13] and others [14–17] have shown that it is possible to utilize the far stronger electric dipole interaction with the electric field (i.e., $\mathbf{d} \cdot E$) to achieve enantioselectivity and even purification. A common feature to all the methods outlined below is that the handedness of the molecular system is determined by the phase of the light fields involved. Which handedness is formed is determined by the combined *phase* of the optical fields and molecular dipole matrix elements. In this sense phase information is equivalent to handedness.

We begin this review by discussing, in Sec. II, a general theorem [13] that shows that enantioselectivity and purification can be achieved in a laser–molecule scenario that is reliant upon the $\mathbf{d} \cdot E$ interaction only if the time evolution resulting from the laser–molecule interaction depends on the sign of the electric field vector. We then consider specific realizations of this general theorem. In Sec. III we introduce the relatively simple case of the dissociation of a prochiral molecule. Specifically we look at the two-photon dissociation from a single initial state of a $B - A - B'$ molecule to yield $BA + B'$ and $B + AB'$, where B and B' are chiral enantiomers. We demonstrate two results: (1) as expected, one-photon dissociation of the prochiral BAB' molecule with linearly polarized light yields identical cross sections for the production of the right-handed (B) and the left-handed (B') fragment, and (2) enantioselectivity of the photofragments can be induced by a coherent two-pulse dissociation process, in which one (excitation) pulse forms an intermediate wave packet and a second pulse is used to dissociate it. The nature of the enantiomer formed is controlled by appropriately varying the delay time (i.e., the molecular phase) between the excitation pulse and the dissociation pulse.

In Secs. IV to VIII we address the more difficult task of purifying a *racemic* mixture. In Sec. IV we discuss the so-called laser distillation scheme in which purification is achieved via the use of many excitation–relaxation cycles induced by three lasers of parallel polarization. In that Section we consider the case of a magnetically (M) polarized sample, where M is the projection of the total angular momentum J along the polarization direction of the incident light. We show that this scheme, specialized to a four-level model, can lead to a very substantial degree of enantioselectivity, allowing use to convert, in some cases, a racemic mixture to the pure enantiomeric form of choice. Extensive control is also shown even in the presence of decoherence. The application of this scheme to the chiral purification of the 1,3 dimethylallens [9] using *ab-initio* surfaces [10] is then discussed. We computationally demonstrate enantiomeric excesses of over 90%, using lasers of moderate intensities.

In Sec. V we extend the treatment of Sec. IV to the purification of *unpolarized* racemic mixtures. We show that when the three laser polarization directions are mutually perpendicular (thus forming a system of axes with definite handedness) one can purify ordinary (unpolarized) racemic mixtures. In accordance with the general theorem of Sec. II, a change in the handedness of the three polarization vectors results in a change in the handedness of the enantiomer to which the racemic mixture is being converted. It then follows that, as discussed in Sec. VI, enantioselectivity can also be achieved by a two-pulse process performed on a sample in which the *axis* of the molecules (rather than their M projection) has been preoriented.

In order to avoid the need to cycle repeatedly the excitation–relaxation process, we present in Sec. VII an alternative approach to the laser distillation scheme of Sec. IV in which one can affect enantioselectivity of the sample by a *single* laser pulse. The method exploits the coexistence, owing to the lack of an inversion center, of one- and two-photon transitions between the *same* chiral molecules' quantum state. This scheme is further improved in Sec. VIII, in which we show how to use the scheme of Sec. VII to achieve full enantiopurification, using a short sequence of laser pulses.

II. PRINCIPLES OF ELECTRIC DIPOLE–ALLOWED ENANTIOMERIC CONTROL

In this section we establish the general conditions under which the electric dipole electromagnetic field interaction may be used to attain selective control over the population of a desired enantiomer. Consider a molecule, described by the *total* Hamiltonian (including electrons and nuclei) H_{MT}, which has eigenstates describing the L and D enantiomers, denoted $|L_i \rangle$ and $|D_i \rangle$ ($i = 1,2,3, \ldots$) that satisfy

$$I|L_i\rangle = -|D_i\rangle \quad I|D_i\rangle = -|L_i\rangle \tag{1}$$

where I is the operator that inverts all space fixed coordinates through the origin. Note that neither $|L_i\rangle$ nor $|D_i\rangle$ has well-defined parity, since they are not eigenstates of I. Strictly speaking, neither $|L_i\rangle$ nor $|D_i\rangle$ is a true eigenstate of H_{MT}. We treat them as such whenever the tunneling splittings between the true (symmetric and antisymmetric) eigenstates are so small that the $|L_i\rangle$ and $|D_i\rangle$ states are stable over periods much larger than our measurement times.

The dipole interaction of this molecule with an incident time dependent electric field $\mathbf{E}(t)$ is described by the Hamiltonian

$$H(\mathbf{E}) = H_{MT} - \mathbf{d} \cdot \mathbf{E} \tag{2}$$

Here $\mathbf{d}$ is the total dipole operator, including both electron and nuclear contributions. Consider now the effect of inversion on H. Noting that I operates on the coordinates of the molecule, that $I^\dagger = I$ and that $[H_{MT}, I] = 0$, we have [69] that

$$IH(\mathbf{E})I = H(-\mathbf{E}) \tag{3}$$

where $H(-\mathbf{E}) = H_{MT} + \mathbf{d} \cdot \mathbf{E}$. Further, if we define $U(\mathbf{E})$ and $U(-\mathbf{E})$ as the propagators corresponding to dynamics under $H(\mathbf{E})$ and $H(-\mathbf{E})$, respectively, then

$$U(\mathbf{E})I = IU(-\mathbf{E}) \tag{4}$$

We now examine whether an enantiomeric excess δ, i.e., the difference between the number of D and L molecules, can result from the irradiation of a 50 : 50 racemic mixture by any combination of transform limited light pulses in the absence of collisions. We have that

$$\delta = \sum_i P_i \sum_j \left[\left| \langle D_j | U(\mathbf{E}) | D_i \rangle \right|^2 + \left| \langle D_j | U(\mathbf{E}) | L_i \rangle \right|^2 \right]$$
$$- \left[\left| \langle L_j | U(\mathbf{E}) | D_i \rangle \right|^2 + \left| \langle L_j | U(\mathbf{E}) | L_i \rangle \right|^2 \right] \quad (5)$$

where P_i is the probability of observing state $|L_i\rangle$ (also equal to that of state $|D_i\rangle$) in the initial racemic mixture.

To determine the conditions under which δ is nonzero, we rewrite Eq. (5) as

$$\delta = \sum_i P_i \sum_j \left[\left| \langle D_j | U(\mathbf{E}) | D_i \rangle \right|^2 - \left| \langle L_j | U(\mathbf{E}) | L_i \rangle \right|^2 \right]$$
$$+ \left[\left| \langle D_j | U(\mathbf{E}) | L_i \rangle \right|^2 - \left| \langle L_j | U(\mathbf{E}) | D_i \rangle \right|^2 \right] \quad (6)$$

and recast the second and third terms using

$$\left| \langle L_j | U(\mathbf{E}) | L_i \rangle \right|^2 = \left| \langle D_j | I^\dagger U(\mathbf{E}) I | D_i \rangle \right|^2 = \left| \langle D_j | U(-\mathbf{E}) | D_i \rangle \right|^2$$
$$\left| \langle D_j | U(\mathbf{E}) | L_i \rangle \right|^2 = \left| \langle D_j | U(\mathbf{E}) I | D_i \rangle \right|^2 = \left| \langle D_j | I U(-\mathbf{E}) | D_i \rangle \right|^2$$
$$= \left| \langle L_j | U(-\mathbf{E}) | D_i \rangle \right|^2 \quad (7)$$

giving

$$\delta = \sum_i P_i \sum_k \left[\left| \langle D_k | U(\mathbf{E}) | D_i \rangle \right|^2 - \left| \langle D_k | U(-\mathbf{E}) | D_i \rangle \right|^2 \right]$$
$$+ \left[\left| \langle L_k | U(-\mathbf{E}) | D_i \rangle \right|^2 - \left| \langle L_k | U(\mathbf{E}) | D_i \rangle \right|^2 \right] \quad (8)$$

Equation (8), the essential result of this section, provides the general condition under which electric fields, assuming a dipole interaction, can break the right/left symmetry of the initial state and result in enhanced production of a desired enantiomer. Specifically, the difference between the amount of D and L formed is seen to depend entirely on the difference between the molecular dynamics when irradiated by $\mathbf{E}$ and by $-\mathbf{E}$. Hence we can state that *a necessary condition for creating nonzero handedness, and the breaking of the left–right symmetry, is that the dynamics depend on the sign of the electric field.* Note that the fact that molecular dynamics can depend on the phase of the incident electric field is well substantiated [72,73], but its utility for asymmetric synthesis is only evident from this result. Finally, note that the result is completely consistent with symmetry-based arguments that can usefully provide conditions under which δ must equal zero. For example, a racemic mixture of thermally equilibrated molecules is rotationally invariant. Hence any rotation that converts $\mathbf{E}$ to $-\mathbf{E}$ could not, in this case, result in enantiomeric control. In particular, in this case, as shown in Sec. V, the sum over M (where M is the component of the total angular momentum along the direction of laser polarization) implicit in the sum over P_i

in Eq. (8) would result in $\delta = 0$. By contrast, as discussed below, a racemic mixture of M polarized molecules irradiated with linearly polarized light, gives nonzero δ, as does a racemic mixture of unpolarized molecules when irradiated with three perpendicular laser polarizations [74].

Both qualitative and quantitative applications of Eq. (8) are possible. Qualitatively, for example, a traditional scheme where the ground electronic state of L and D are incoherently excited to bound levels of an excited state gives $\delta = 0$. This is because all processes connecting the initial and final $|L_i\rangle$ and $|D_i\rangle$ states, i.e., contributions to the matrix elements in Eq. (8), are even in the power of the electric field. Hence propagation under $\mathbf{E}$ and $-\mathbf{E}$ are identical. By contrast, consider the four-level model scheme in Fig. 1, discussed in detail in Sec. IV. When $\varepsilon_0(t) \neq 0$, there exist processes connecting the initial and final $|L\rangle$ and $|D\rangle$ states that are of the form $|L\rangle \rightarrow |1\rangle \rightarrow |2\rangle \rightarrow |D\rangle$, and hence there are terms in Eq. (8) that are odd in the power of the electric field. One therefore anticipates the possibility of altering the handedness using this combination of pulses, providing the basis for the control results reported below. Further, if $\varepsilon_0 = 0$, then the situation reverts to the case discussed above, where only processes even in the electric field contribute to transitions between the initial $|D\rangle$, $|L\rangle$ and final $|D\rangle$, $|L\rangle$ transitions, and hence control over the handedness is lost. For this reason, the

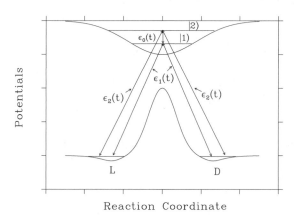

Reaction Coordinate

Figure 1 The "laser distillation" control scenario discussed in detail in Sec. IV. Two lasers, with pulse envelopes $\varepsilon_1(t)$ and $\varepsilon_2(t)$ couple, by virtue of the dipole operator, the states of the D and L enantiomers to two vib-rotational states $|\,1\,\rangle$ and $|\,2\,\rangle$ (denoted $|E_1\rangle$ and $|E_2\rangle$ in the text) in the excited electronic manifold. A third laser pulse with envelope $\varepsilon_0(t)$ couples the excited $|E_1\rangle$ and $|E_2\rangle$ states to one another. The system is allowed to absorb a photon and relax back to the ground state. After many such excitation–relaxation cycles, a significant degree of handedness is obtained, as explained in Sec. IV.

$\varepsilon_0(t)$ coupling laser is crucial to enantiomeric control. This qualitative picture is substantiated quantitatively in Sec. IV.

What is required experimentally to achieve this kind of control is the ability to manipulate the phase of the electric field. The ability to perform such manipulations, which are at the heart of coherent control, are by now well documented [18–68,75,76].

III. SYMMETRY BREAKING IN THE TWO-PHOTON DISSOCIATION OF PURE STATES

Postponing the discussion of the purification of racemic mixtures to later sections, we consider in this section the simpler case of controlling the handedness of photo-fragments resulting from the dissociation of a 'prochiral' molecule of the type BAB' where B and B' are enantiomers.

The prochiral molecule is not chiral because it possesses a hyperplane of symmetry, denoted as σ, which is the set of points of equal $B - A$ and $A - B'$ distances. The operator corresponding to reflection in this plane is denoted as σ_h. In order to control coherently the dissociation of this system, we take advantage of the existence of degenerate continuum states, which do not possess this reflection symmetry. That is, these molecules possess degenerate continuum states $|E, \mathbf{n}, D^-\rangle$ and $|E, \mathbf{n}, L^-\rangle$ that correlate asymptotically with the dissociation of the right B' group and left B group, respectively. The collective quantum index $\mathbf{n}$ in the states $|E, \mathbf{n}, D^-\rangle$ and $|E, \mathbf{n}, L^-\rangle$ includes m, the magnetic quantum number of the B or B' fragment. These states are neither symmetric nor antisymmetric with respect to the reflection operator σ_h, although linear combinations of these states might possess this symmetry.

We consider a BAB' molecule irradiated by two pulses, termed the pump and dump pulses. These pulses are assumed to be temporally separated by a time delay Δ_d. Our aim is to control the probabilities to yield of two products $P_{q,\mathbf{n}}(E)$, with q labeling either the right ($q = D$) or left ($q = L$) handed product. The application of this scenario to the chiral synthesis case is depicted schematically in Fig. 2.

The BAB' molecule is assumed initially ($t = 0$) to be in an eigenstate $|E_g\rangle$ of the molecular Hamiltonian H_M. It is subjected to two pulses whose electric field is

$$\mathbf{E}(\tau) = \mathbf{E}_x(\tau) + \mathbf{E}_d(\tau) \tag{9}$$

where $\mathbf{E}_x(\tau)$ is the pump pulse and $\mathbf{E}_d(\tau)$ is the dump (or dissociation pulse), with τ being the retarded time ($t - z/c$). For both pulses, the electric field is of the form $\mathbf{E}(\tau) = 2\hat{\varepsilon}(\tau) \cos(\omega_o \tau)$. Here ω_o is the carrier frequency and $\varepsilon(\tau)$ de-

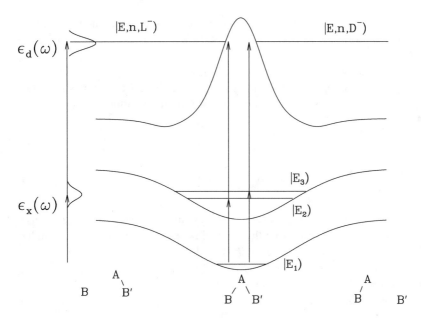

Figure 2 A schematic showing the controlled dissociation of the molecule $B-A-B'$ to yield the $B - A + B'$ or the $B + A - B'$ products, where B and B' are two enantiomers. The molecule is excited from an initial state $|E_1\rangle$ ($= |E_g\rangle$ of the text) to a superposition of antisymmetric $|E_2\rangle$ ($= |E_1\rangle$ of the text) and symmetric, $|E_3\rangle$ ($= |E_2\rangle$ of the text) vibrational states belonging to an excited electronic state, by an excitation pulse $\varepsilon_x(\omega)$. After an appropriate delay time, the molecule is dissociated by a second pulse $\varepsilon_d(\omega)$ to the $|E, \mathbf{n}, D^-\rangle$ or $|E, \mathbf{n}, L^-\rangle$ continuum state.

scribes the pulse envelopes, assumed for convenience to be Gaussians peaking at $t = t_\chi$ and t_d, respectively,

$$\mathbf{E}_x(\tau) = \hat{\varepsilon}_x \varepsilon_x \exp[-i(\omega_x \tau + \delta_x)] \frac{\exp\left[-(\tau - t_x)^2 / \tau_x^2\right]}{2} \tag{10}$$

Each Gaussian pulse is assumed to carry an overall phase δ_χ. The associated frequency profile is given by the Fourier transform of Eq. (10):

$$\overline{\varepsilon}_x(\omega) \equiv \left|\overline{\varepsilon}_x(\omega)\right| \exp[i\phi(\omega)]$$
$$= \sqrt{\pi/2}\,\varepsilon_x \gamma_x \exp[-\gamma_x^2(\omega_x - \omega)^2 / 4]\exp[-i(\omega_x - \omega)t_x - i\delta_x] \tag{11}$$

The analogous quantities for the dissociation laser, $\mathbf{E}_d(\tau)$, $\varepsilon_d(\omega)$, and $\overline{\varepsilon}_d(\omega)$, are defined similarly, with the parameters t_d and ω_d replacing t_x and ω_x, etc.

The bandwidth of the pump pulse $\sim 2/\tau_x$ is chosen sufficiently narrow to excite only two excited intermediate states, $|E_1\rangle$ and $|E_2\rangle$. The dump pulse

$E_d(\tau)$ dissociates the molecule by further exciting it to the continuous part of the spectrum. Both fields are chosen sufficiently weak for perturbation theory to be valid.

Since the two pulses are temporally distinct, it is convenient to deal with their effects consecutively. After the first pulse is over, the superposition state prepared by the $E_x(\tau)$ pulse is given in first-order perturbation theory as

$$|\phi(t)\rangle = |E_g\rangle e^{-iE_g t/\hbar} + b_1 |E_1\rangle e^{-iE_1 t/\hbar} + b_2 |E_2\rangle e^{-iE_2 t/\hbar} \qquad (12)$$

where

$$b_k = \frac{2\pi i}{\hbar} \langle E_k | \hat{\varepsilon} \cdot \mathbf{d} | E_g \rangle \bar{\varepsilon}_x(\omega_{k,g}) \qquad k = 1,2, \qquad (13)$$

with $\omega_{k,g} \equiv (E_k - E_g)/\hbar$.

After a delay time of $\Delta_d \equiv t_d - t_x$ the system is subjected to the $E_d(\tau)$ pulse. It follows from Eq. (12) that after this delay time, each preparation coefficient has picked up an extra phase of $-iE_k\Delta_d/\hbar$, $k = 1, 2$. Hence the phase of b_1 relative to b_2 at that time increases by $[-(E_1 - E_2)\Delta_d/\hbar = \omega_{2,1}\Delta_d]$. Thus the natural two-state time evolution controls the relative phase of the two terms.

After the action and subsequent decay of the $E_d(\tau)$ pulse, the system wave function is

$$|\psi(t)\rangle = |\phi(t)\rangle + \sum_{m,q} \int dE \, b_{E,m,q}(t) |E,m,q^-\rangle e^{-iEt/\hbar} \qquad (14)$$

The probability of observing the $q\,(= D, L)$ product at total energy E in the remote future is given as the square of the asymptotic form of the $b_{E,m,q}(t)$ coefficients,

$$p_q(E) = \sum_m |b_{E,m,q}(t = \infty)|^2 \qquad (15)$$

where $b_{E,m,q}(t = \infty)$ is given by firstorder perturbation theory as

$$b_{E,m,q}(t = \infty) = \frac{\sqrt{2\pi}}{\hbar} \sum_{k=1,2} b_k \langle E,m,q^- | d_{e,g} | E_k \rangle \bar{\varepsilon}_d(\omega_{EE_k}) \qquad (16)$$

where $\omega_{EE_k} = (E - E_k)/\hbar$, b_k is given by Eq. (13), and $\bar{\varepsilon}_d(\omega)$ is given via an expression analogous to Eq. (11).

Expanding the square in Eq. (15) and using the Gaussian pulse shape [Eqs. (10) and (11)] gives

$$P_q(E) = \frac{2\pi}{\hbar^2} \Big[|b_1|^2 \, d_q(11)\bar{\varepsilon}_1^2 + |b_2|^2 \, d_q(22)\bar{\varepsilon}_2^2 \\ + 2|b_1 b_2^* \bar{\varepsilon}_1 \bar{\varepsilon}_2^* d_q(21)|\cos(\omega_{2,1}\Delta_d + \alpha_q(21) + \chi) \Big] \qquad (17)$$

where $\bar{\varepsilon}_i = |\bar{\varepsilon}_d(\omega_{EE_i})|$, $\omega_{21} = (E_2 - E_1)/\hbar$, an and the phases χ, $\alpha_q(21)$ are defined via

$$\langle E_1 |d_{e,g}| E_g \rangle \langle E_g |d_{g,e}| E_2 \rangle \equiv \left| \langle E_1 |d_{e,g}| E_g \rangle \langle E_g |d_{g,e}| E_2 \rangle \right| e^{i\chi}$$

$$d_q(ki) \equiv \left| d_q(ki) \right| e^{i\alpha_q(ki)} = \sum_m \langle E_k |d_{g,e}| E, \mathbf{m}, q^- \rangle \langle E, \mathbf{m}, q^- |d_{e,g}| E_i \rangle \tag{18}$$

Equation (17) needs to be averaged over the full width of the dump pulse to obtain P_q, the probabilities of forming channels $q = D, L$.

We see that P_D and P_L are functions of the delay time $\Delta_d = (t_d - t_x)$ between pulses and the ratio $x = |b_1/b_2|$; the latter is controlled by varying the carrier frequency of the initial excitation pulse. Active control over the products $B + AB'$ vs. $B' + AB$, i.e., a variation of $R_{DL} \equiv P_D/P_L$ with Δ_d and x, and hence control over left vs. right handed products, will result only if P_D and P_L have different functional dependences on the control parameters x and Δ_d.

To show that this is the case we simplify the discussion of the optical excitation of the $B - A - B'$ molecule by focusing upon transitions between electronic states of the same representations, e.g., A' to A' or A'' to A'' (where A' denotes the symmetric representation and A'' the antisymmetric representation of the C_s group). We further assume that the ground vibronic state belongs to the A' representation. To obtain control we choose the intermediate state $| E_2 \rangle$ to be *symmetric*, and the intermediate state $| E_1 \rangle$ to be *antisymmetric*, with respect to reflection in the σ hyperplane. Hence we must first demonstrate that it is possible optically to excite, simultaneously, both the symmetric $|E_2\rangle$ and antisymmetric $|E_1\rangle$ states from the ground state $|E_g\rangle$. This requires the existence of both a symmetric dipole component, denoted $\mathbf{d}_s$, and an antisymmetric component, denoted $\mathbf{d}_a$, with respect to reflection in the σ hyperplane, because, by the symmetry properties of $|E_2\rangle$ and $|E_1\rangle$,

$$\langle E_2 |\mathbf{d} \cdot \hat{\varepsilon}| E_g \rangle = \langle E_2 |\mathbf{d}_s \cdot \hat{\varepsilon}| E_g \rangle \quad \langle E_1 |\mathbf{d} \cdot \hat{\varepsilon}| E_g \rangle = \langle E_1 |\mathbf{d}_a \cdot \hat{\varepsilon}| E_g \rangle \tag{19}$$

We note that the coexistence of symmetric and antisymmetric components of the dipole moment is with respect to σ_h. Since the σ plane rotates with the molecule, the σ_h operation is said to be body fixed (or molecule fixed). Both the body fixed symmetric $\mathbf{d}_s$ and the body fixed antisymmetric $\mathbf{d}_a$ dipole moment components do occur in $A' \rightarrow A'$ electronic transitions whenever the geometry of a bent $B' - A - B$ molecule deviates considerably from the points on the σ hyperplane, characterized by the points of equidistance ($C_{2\theta}$) geometries (where $\mathbf{d}_a = 0$). See Fig. 3. The deviation of $\mathbf{d}_a$ from zero on the σ plane necessitates going beyond the Franck–Condon approximation, which assumes that the electronic dipole moment does not change as the molecule vibrates. (In the terminology of the theory of vibronic transitions, both symmetric and antisymmetric components can be non-zero owing to a Herzberg–Teller intensity borrowing [77] mechanism.)

Note also that the dipole moment operator, being a vector, must reverse its sign under inversion I. Hence, with respect to I, the dipole moment is always *antisymmetric*. Thus for the integrals in Eq. (19) to be nonzero also requires that

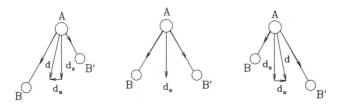

Figure 3 The emergence of an antisymmetric dipole component d_a in addition to the symmetric component d_s in a bent BAB′ triatomic molecule as a result of an asymmetric stretching vibration, assuming that the dipole is a vectorial sum of bond dipoles that are proportional to the bond lengths.

$|E_2\rangle$ and $|E_1\rangle$ be of opposite symmetry with respect to inversion. Given the extant conditions on the behavior of $|E_2\rangle$ and $|E_1\rangle$ with respect to the reflection σ_h, the symmetry requirements with respect to I are most easily accommodated through the rotational components of the $|E_2\rangle$ and $|E_1\rangle$ states.

Thus the excitation pulse can create a superposition of $|E_1\rangle$, $|E_2\rangle$ consisting of two states of different reflection symmetry. The resultant superposition possesses no symmetry properties with respect to reflection [78]. We now show that the broken symmetry created by this excitation of *nondegenerate* bound states translates into a nonsymmetry in the probability of populating the *degenerate* $|E, \mathbf{n}, D^-\rangle$, $|E, \mathbf{n}, L^-\rangle$ continuum states upon subsequent excitation. To do so we examine the properties of the bound–free transition matrix elements $\langle E, \mathbf{n}, q^- \mid d_{e,g}|E_k\rangle$ that enter into the probability of dissociation. Note first that although the continuum states $|E, \mathbf{n}, q^-\rangle$ are nonsymmetric with respect to reflection, we can define symmetric and antisymmetric continuum eigenfunctions $|E, \mathbf{n}, s^-\rangle$ and $|E, \mathbf{n}, a^-\rangle$ via the relations

$$|E,\mathbf{n},D^-\rangle \equiv \frac{|E,\mathbf{n},s^-\rangle + |E,\mathbf{n},a^-\rangle}{\sqrt{2}} \tag{20}$$

$$|E,\mathbf{n},L^-\rangle \equiv \frac{|E,\mathbf{n},s^-\rangle - |E,\mathbf{n},a^-\rangle}{\sqrt{2}} \tag{21}$$

using the fact that $\sigma_h |E, \mathbf{n}, D^-\rangle = |E, \mathbf{n}, L^-\rangle$.

Consider first the nature of the $d_q(ij)$ that enter Eq. (18), prior to averaging over product scattering angles, and denoted $d_q(ij; \hat{\mathbf{k}})$, where $\hat{\mathbf{k}}$ is the scattering direction. Since $|E_2\rangle$ is symmetric and $|E_1\rangle$ is antisymmetric, and adopting the notation $A_{s1} \equiv \langle E, \mathbf{n}, s^- |d_a|E_1\rangle$, $S_{a2} \equiv \langle E, \mathbf{n}, a^- |d_s| E_2\rangle$, etc. we have [see Eq. (18)].

$$d_q(22; \hat{\mathbf{k}}) = \sum{}'' \left[|S_{s2}|^2 + |A_{a2}|^2 \pm 2R_e(A_{a2}S_{s2}^*) \right]$$

$$d_q(11; \hat{\mathbf{k}}) = \sum{}'' \left[|A_{s1}|^2 + |S_{a1}|^2 \pm 2R_e(A_{s1}S_{a1}^*) \right]$$

$$d_q(21; \hat{\mathbf{k}}) = \sum{}'' \left[S_{s2}A_{s1}^* + A_{a2}S_{a1}^* \pm S_{s2}S_{a1}^* \pm A_{a2}A_{s1}^* \right] \tag{22}$$

where the plus sign applies for $q = D$, the minus sign applies for $q = L$, and $d_q(12; \hat{\mathbf{k}}) = d_q^*(21; \hat{\mathbf{k}})$. The double prime on the sum denotes a summation over all q, $\mathbf{n}$ other than the scattering angles and the product m, where m denotes the projection of the product angular momentum along the axis of laser polarization.

Equation (22) takes on a simpler form after angular averaging. The reason for this is that the overall parity of a state with respect to the inversion operation, I, *must* change upon photon absorption since a photon has odd parity. As a result, if we have a single photon absorption process in which the parity of a vibrational state is unchanged, then the parity of the rotational states must change, and vice versa. Close examination of Eq. (22) reveals that the S^*_{s2} term does not involve a change in the parity of the vibrational state, whereas the A_{a2} term does. As a result, the rotational wavefunctions associated with each term must have opposing parities, and the angular integral of the product must vanish. The same goes for the $A_{s1}S^*_{a1}$ term. In a similar manner the $S_{s2}A^*_{s1} + A_{a2}S^*_{a1}$ term vanishes in the $d_q(21)$ interference term. By contrast, the $\pm S_{s2}S^*_{a1} \pm A_{a2}A^*_{s1}$ terms do not vanish upon angular integration since they correspond to final rotational states that have the same parity.

As a consequence, the net result is that, after angular averaging, Eq. (22) becomes

$$d_q(22) = \sum{}' \left[|S_{s2}|^2 + |A_{a2}|^2 \right]$$

$$d_q(11) = \sum{}' \left[|A_{s1}|^2 + |S_{a1}|^2 \right]$$

$$d_q(21) = \sum{}' \pm \left[S_{s2}S_{a1}^* + A_{a2}A_{s1}^* \right] \tag{23}$$

where single primes on the sums indicate that the sum over product m is not carried out.

These equations display two noteworthy features:

1 $d_L(jj) = d_D(jj)$, $j = 1, 2$, i.e., lacking interference, no discrimination between the left-handed and right-handed products is possible.

2 $d_L(12) \neq d_D(12)$, i.e., laser controlled symmetry breaking, which depends upon $d_q(12)$ in accordance with Eq. (17), is possible. As noted below, this type of discrimination is possible only if we select the direction of the angular momentum of the products (m polarization).

To demonstrate the extent of expected control, as well as the effect of m summation, we considered a model of enantiomer selectivity, i.e., HOH photodis-

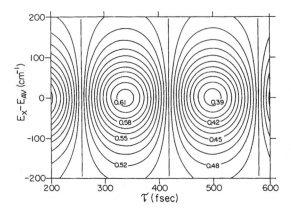

Figure 4   Contour plot of percent HO + H (as distinct from H + OH) in HOH photodissociation. The ordinate is the detuning from $E_{av} = (E_2 - E_1)/2$, and the abscissa is the
time delay between the pulses. (Taken from Fig. 9, Ref. 20.)

sociation in three dimensions, where the two hydrogens are assumed distinguishable,

$$H_a O + H_b \leftarrow H_a OH_b \rightarrow H_a + OH_b \tag{23a}$$

The computations of the ratio R_{DL} of the HO + H (as distinct from the H +
OH) product in a fixed m state were done using the formulation and computational methodology of Refs. 79 and 80. Figure 4 shows the result of first
exciting the superposition of symmetric plus asymmetric vibrational modes
$[(1, 0, 0) + (0, 0, 1)]$ with $J_i = J_k = 0$ in the ground electronic state, followed
by dissociation at 70,700 cm^{-1} to the B state using a pulse width of 200 cm^{-1}.
The results show that varying the time delay between pulses allows for controlled
variation of P_D from 61% to 39%. This variation is significant since it reveals
the symmetry breaking arising within this scenario.

IV.   PURIFICATION OF RACEMIC MIXTURES BY LASER
DISTILLATION

In Sec. III we showed that coherent control techniques can be used to direct the
photodissociation of a BAB' molecule to yield an excess of a desired B or B'
enantiomer. Throughout the treatment we assumed that prior to dissociation the
BAB' molecule exists in single quantum state $|E_i, J_i, p_i\rangle$. While the above process
is of great scientific interest, practically speaking, we usually want to separate a

racemic mixture of the B and B' enantiomers. If we were to use the scenario of Sec. III to accomplish this task, we would have first to prepare the BAB' prochiral molecule in a pure state. Since the preparation of BAB', and especially its separation from the BAB and $B'AB'$ molecules that would inevitably accompany it, is not a trivial task, it is preferable to find control methods that could separate the B and B' racemic mixture directly. In this section we outline a method that can achieve this much more ambitious task.

Consider then a molecular system composed of a pair of stable nuclear configurations, denoted L and D, with L being the (distinguishable) mirror image of D. Since L and D are assumed stable, it follows that the ground potential energy surface must possess a sufficiently high barrier at nuclear coordinates separating L and D so that the rate of interconversion between them by tunneling is negligible. By contrast, L and D need not be stable on an excited potential energy surface. To this end, we assume that there is at least one excited potential surface, denoted G, which possesses a potential well midway between the L and the D geometries. (See Fig. 1.) (A number of molecules expected to be of this type are tabulated in Ref. 5, and a number of examples are discussed below.) Hence the interconversion between L and D on the excited surface G is expected to be very facile. A direct consequence of the potential well midway between the L and the D geometries is the existence of stable vibrational eigenstates. Because of the symmetry of G, the vibrational eigenstates must be either symmetric or antisymmetric with respect to σ_h.

The procedure that we propose in order to enhance the concentration of a particular enantiomer when starting with a racemic mixture, i.e., to "purify" the mixture, is as follows. The statistical (racemic) mixture of L and D is irradiated with a specific sequence of three coherent laser pulses, as described below. These pulses excite a coherent superposition of symmetric and antisymmetric vibrational states of G. After each pulse the excited system is allowed to relax back to the ground electronic state by spontaneous emission or any other nonradiative process. By allowing the system to go through many irradiation and relaxation cycles, we show below that the concentration of the selected enantiomer L or D can be enhanced, depending on the laser characteristics. We call this scenario laser distillation of chiral enantiomers.

We note at the outset that detailed angular momentum considerations [74] show that if the three incident lasers are of the same polarization then control results only if we do not average over M, the projection of the total angular momentum of the reactant along the z-axis (chosen as the direction of laser polarization). In particular, enantiomeric enhancement of one enantiomer from molecules in state M is exactly counterbalanced by enantiomeric enhancement of the other enantiomer by molecules in state $-|M|$. Hence enantiomeric control in this scenario requires prior M selection of the molecules. This scenario is discussed below. Results are also provided for the case of three lasers of perpendicular polarization, where M averaging is nondestructive.

Consider then a molecule with Hamiltonian H_M, in the presence of a series of laser pulses. The interaction between the molecule and radiation is given by

$$H_{MR}(t) = -\mathbf{d} \cdot \mathbf{E}(t) = -2\mathbf{d} \cdot \sum_k R_e[\hat{\varepsilon}_k \varepsilon_k(t) \exp(-i\omega_k t)] \tag{24}$$

Here $\varepsilon_k(t)$ is the pulse envelope, ω_k is the carrier laser frequency, and $\hat{\varepsilon}_k$ is the polarization direction. Expanding $|\Psi(t)\rangle$ in eigenstates $|E_j\rangle$ of the molecular Hamiltonian (i.e., $H_M |E_j\rangle = E_j |E_j\rangle$), we find

$$|\psi(t)\rangle = \sum_j b_j \exp\left(\frac{-iE_j t}{\hbar}\right) |E_j\rangle \tag{25}$$

and substituting Eq. (25) into the time-dependent Schrödinger equation gives the standard set of coupled equations:

$$\dot{b}_i = \frac{-i}{\hbar} \sum_{jk} b_j \exp(-i\omega_{ji}t)\langle E_i | H_{MR}(t) | E_j \rangle \tag{26}$$

where $\omega_{ji} = (E_j - E_i)/\hbar$.

As an example of an effective control scenario, consider the molecules D and L in their ground electronic states and in vib-rotational states $|E_D\rangle$ and $|E_L\rangle$, of energy $E_D = E_L$. We choose $\mathbf{E}(t)$ so as to excite the system to two eigenstates $|E_1\rangle$ and $|E_2\rangle$ of the electronically excited potential surface G. The states $|E_1\rangle$ and $|E_2\rangle$ are also coupled by an additional laser field (see Fig. 1).

Specifically, we choose $\mathbf{E}(t)$ to be composed of three linearly polarized light pulses (all of the same polarization),

$$\mathbf{E}(t) = \sum_{k=0,1,2} 2\mathrm{Re}[\varepsilon_k(t)\exp(-i\omega_k t)] \hat{\varepsilon}_k \tag{27}$$

with ω_0 in near resonance with $\omega_{2,1} \equiv (E_2 - E_1)/\hbar$, ω_1 chosen to be near resonant with $\omega_{1,D} \equiv (E_1 - E_D)/\hbar$, and ω_2 near resonant with $\omega_{2,D} \equiv (E_2 - E_D)/\hbar$ (see Fig. 1). In this case, only four molecular states are relevant, and Eq. (25) becomes

$$|\psi\rangle = b_D(t)\exp\left(\frac{-iE_D t}{\hbar}\right)|E_D\rangle + b_L(t)\exp\left(\frac{-iE_L t}{\hbar}\right)| E_L\rangle$$
$$+ b_1(t)\exp\left(\frac{-iE_1 t}{\hbar}\right)| E_1\rangle + b_2(t)\exp\left(\frac{-iE_2 t}{\hbar}\right)| E_2\rangle \tag{28}$$

Equation (26), in the rotating wave approximation, is then given by

$$\dot{b}_1 = i\exp(i\Delta_1 t)[\Omega_{D,1}^* b_D + \Omega_{L,1}^* b_L] + i\exp(-i\Delta_0 t)\Omega_0^* b_2$$
$$\dot{b}_2 = i\exp(i\Delta_2 t)[\Omega_{D,2}^* b_D + \Omega_{L,2}^* b_L] + i\exp(-i\Delta_0 t)\Omega_0^* b$$
$$\dot{b}_D = i\exp(-i\Delta_1 t) \Omega_{D,1} b_1 + i\exp(-i\Delta_2 t)\Omega_{D,2} b_2$$
$$\dot{b}_L = i\exp(-i\Delta_1 t) \Omega_{L,1} b_1 + i\exp(-i\Delta_2 t)\Omega_{L,2} b_2 \tag{29}$$

where $\Omega_{i,j}(t) \equiv d_{i,j}^{(j)} \varepsilon_1(t)/\hbar$, $\Omega_0 \equiv d_{2,1}^{(0)} \varepsilon_0(t)/\hbar$, $\Delta_j \equiv \omega_{j,D} - \omega_1$, $\Delta_0 \equiv \omega_{2,1} - \omega_0$, and $d_{i,j}^{(k)} \equiv \langle E_i | \mathbf{d} \cdot \hat{\varepsilon}_k | E_j \rangle$, with $i = D, L; k = 0, 1, 2$, and $j = 1, 2$.

The essence of the laser distillation process lies in choosing the laser of carrier frequency ω_1 so that it excites the system to a state $|E_1\rangle$ which is *symmetric* with respect to inversion I, and to a state $|E_2\rangle$ which is *antisymmetric* with respect to I. By contrast, $|E_D\rangle$ and $|E_L\rangle$ do not share these symmetries but are related to one another, [i.e., $I| E_D\rangle = |E_L\rangle$ and $I| E_L\rangle, = |E_D\rangle$, whereas $I| E_1\rangle = |E_1\rangle$ and $I| E_2\rangle = -|E_2\rangle$.

To consider the nature of the Rabi frequencies Σ in Eq. (29), we rewrite $|E_D\rangle$ and $|E_L\rangle$ in terms of a symmetric state $|S\rangle$ and an antisymmetric state $|A\rangle$:

$$\begin{aligned} |E_D\rangle &= |A\rangle + |S\rangle \\ |E_L\rangle &= |A\rangle - |S\rangle \end{aligned} \tag{30}$$

Given that the dipole operator must be antisymmetric with respect to I, the relevant matrix elements satisfy the following relations:

$$\begin{aligned} \langle 1| d^{(1)} | D\rangle &= \langle 1| d^{(1)} | A+S\rangle = \langle 1| d^{(1)} | A\rangle \\ \langle 1| d^{(1)} | L\rangle &= \langle 1| d^{(1)} | A-S\rangle = \langle 1| d^{(1)} | A\rangle \\ \langle 2| d^{(2)} | D\rangle &= \langle 2| d^{(2)} | A+S\rangle = \langle 2| d^{(2)} | S\rangle \\ \langle 2| d^{(2)} | L\rangle &= \langle 2| d^{(2)} | A-S\rangle = -\langle 2| d^{(2)} | S\rangle \end{aligned} \tag{31}$$

That is,

$$\Omega_{D,1} = \Omega_{L,1} \quad \Omega_{D,2} = -\Omega_{L,2} \tag{32}$$

Given Eq. (32), Eq. (29) becomes

$$\begin{aligned} \dot{b}_1 &= i\exp(i\Delta_1 t)\Omega_{D,1}^*[b_D + b_L] + i\exp(-i\Delta_0 t)\Omega_0^* b_2 \\ \dot{b}_2 &= i\exp(i\Delta_2 t)\Omega_{D,2}^*[b_D - b_L] + i\exp(i\Delta_0 t)\Omega_0 b_1 \\ \dot{b}_D &= i\exp(-i\Delta_1 t)\Omega_{D,1}^* b_1 + i\exp(-i\Delta_2 t)\Omega_{D,2} b_2 \\ \dot{b}_L &= i\exp(-i\Delta_1 t)\Omega_{D,1} b_1 - i\exp(-i\Delta_2 t)\Omega_{D,2} b_2 \end{aligned} \tag{33}$$

The essence of optically controlled enantioselectivity in this scenario lies in Eq. (32) and the effect of these relationships on the dynamical equations for the level amplitudes [Eq. (33)]. Note specifically that the equation for $\dot{b}_D(t)$ is different from the equation for $\dot{b}_L(t)$, owing to the sign difference in the last term in Eq. (33). Although not sufficient to ensure enantiomeric selectivity, the ultimate consequence of this difference is that populations of $|E_D\rangle$ and $|E_L\rangle$ after laser excitation are different when there is radiative coupling between levels $|E_1\rangle$ and $|E_2\rangle$.

Note, in accord with Sec. II, the behavior of Eq. (33) under the transformation $\mathbf{E} \rightarrow -\mathbf{E}$. Specifically, changing $\mathbf{E}$ to $-\mathbf{E}$ means changing all $\varepsilon_j(t)$ to $-\varepsilon_j(t)$. Doing so, and defining $b'_1 = -b_1$ and $b'_2 = -b_2$, converts Eq. (33) into

$$\dot{b}_1' = i\exp(i\Delta_1 t)\Omega_{D,1}^*[b_D + b_L] - i\exp(i\Delta_0 t)\Omega_0^* b_2'$$
$$\dot{b}_2' = i\exp(i\Delta_2 t)\Omega_{D,2}^*[b_D - b_L] - i\exp(-i\Delta_0 t)\Omega_0 b_1'$$
$$\dot{b}_D = i\exp(-i\Delta_1 t)\Omega_{D,1}b_1' + i\exp(-i\Delta_2 t)\Omega_{D,2}b_2'$$
$$\dot{b}_L = i\exp(-i\Delta_1 t)\Omega_{D,1}b_1' - i\exp(-i\Delta_2 t)\Omega_{D,2}b_2' \tag{34}$$

Clearly, Eq. (34) is the same as Eq. (33) barring the change of sign in the Ω_0 terms. Thus the solution to Eq. (33) depends on the sign of $\mathbf{E}$ when $\varepsilon_0 \neq 0$. Hence, by the argument in Sec. II, this scenario allows for handedness control when $\varepsilon_0(t) \neq 0$. For $\varepsilon_0(t) = 0$, Eq. (34) is the same as Eq. (33), so that enantiomer control is not possible.

To obtain quantitative estimates for the extent of obtainable control, we consider results for model cases assuming Gaussian pulses

$$\varepsilon_k(t) = \varepsilon_k \exp\left[-\left(\frac{t-t_k}{\alpha_k}\right)^2\right] \quad (k = 0,1,2) \tag{35}$$

and system parameters $\langle 1|d^{(1)}|D\rangle = \langle 1|d^{(1)}|L\rangle = \langle 2|d^{(2)}|L\rangle = -\langle 2|d^{(2)}|D\rangle = 1$ a.u., $\langle 1|d^{(0)}|2\rangle = 1$ a.u., $\omega_{2,1} = 100$ cm^{-1}, and $\Delta_0 = 0$. Figure 5 displays the final probabilities $P_D = |b_D(\infty)|^2$, $P_L = |b_L(\infty)|^2$ of population in $|E_D\rangle$ and $|E_L\rangle$, after a single pulse, for a variety of pulse parameters. Results are shown for various values of Δ_1 at various different pulse powers, assuming that one starts solely with D, solely with L, or with a racemic mixture of both enantiomers. Clearly, for particular parameters, one can significantly enhance the population of one chiral enantiomer over the other. For example, for $\Delta_1 = -115$ cm^{-1}, $\varepsilon_0 = \varepsilon_1 = 4.5 \times 10^{-4}$, a racemic mixture of D and L can be converted, after a single pulse, to a enantiomerically enriched mixture with predominantly D.

Control is strongly affected by the relative phase θ of the ε_1 and ε_0 fields, as shown in Fig. 6. Here it is clear that changing θ by π interchanges the dynamical evolution of the L and D enantiomers.

Although not immediately obvious, this control scenario relies entirely upon quantum interference effects. To see this note that in the absence of an $\varepsilon_0(t)$ pulse, excitation from $|D\rangle$ or $|L\rangle$ to level $|E_i\rangle$, for example, occurs via one photon excitation with $\varepsilon_i(t)$, $i = 1, 2$. In this case, as noted above, there is no chiral control. By contrast, with nonzero $\varepsilon_0(t)$, there is an additional (interfering) route to $|E_i\rangle$, i.e., a two-photon route using $\varepsilon_j(t)$ excitation to level $|E_j\rangle$, $j \neq i$, followed by an $\varepsilon_0(t)$-induced transition from $|E_j\rangle$ to $|E_i\rangle$. The one- and two-photon routes interfere, thus causing symmetry breaking transitions.

The computation with results in Fig. 5, which gives the result of a single pulse, provides input into a calculation of the overall result. In the overall process we begin with an incoherent mixture of N_D molecules of type D and N_L molecules of type L. In the first step the system is excited, as above, with a laser pulse sequence. In the second step, the system collisionally and radiatively relaxes so that all the population returns to the ground state to produce an incoherent mixture

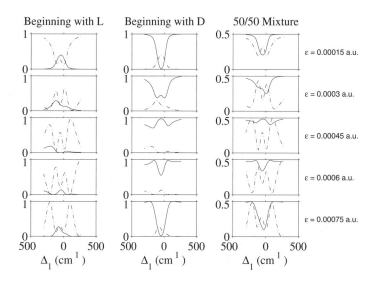

Figure 5 Probabilities of populating the $|E_D\rangle$ (solid lines) and $|E_L\rangle$ (dot-dash lines) after laser excitation, but prior to relaxation, as a function of the detuning Δ_1. Three different cases are shown, corresponding to three different initial conditions: (1) only state $|E_L\rangle$ is initially populated; (2) only state $|E_D\rangle$ is initially populated; (3) a statistical mixture made up of equal shares of the $|E_D\rangle$ and $|E_L\rangle$ states is initially populated. Results are shown for five different $\varepsilon_1 = \varepsilon_0 \equiv \varepsilon$ laser peak electric fields, where Gaussian pulses are assumed with $\alpha_0 = \alpha_1 = 0.15$ ps, and $t_0 = t_1$.

of $|E_L\rangle$ and $|E_D\rangle$. This pair of steps is then repeated until the populations of $|E_L\rangle$ and $|E_D\rangle$ reach convergence.

To obtain the result computationally note that the population after laser excitation, but before relaxation, consists of the weighted sum of the results of two computations: N_D times the results of laser excitation starting solely with molecules in $|E_D\rangle$, plus N_L times the results of laser excitation starting solely with molecules in $|E_L\rangle$. If $P_{D \leftarrow D}$ and $P_{L \leftarrow D}$ denote the probabilities of $|E_D\rangle$ and $|E_L\rangle$ resulting from laser excitation assuming the first of these initial conditions, and $P_{D \leftarrow L}$ and $P_{L \leftarrow L}$ for the results of excitation following from the second of these initial conditions, then the populations of $|E_D\rangle$ and $|E_L\rangle$ after laser excitation of the mixture are $N_D P_{D \leftarrow D} + N_L P_{D \leftarrow L}$ and $N_D P_{L \leftarrow D} + N_L P_{L \leftarrow L}$, respectively. The remainder of the population, $N_D [1 - P_{D \leftarrow D} - P_{L \leftarrow D}] + N_L [1 - P_{D \leftarrow L} - P_{L \leftarrow L}]$, is in the upper two levels $|E_1\rangle$ and $|E_2\rangle$. Relaxation from levels $|E_1\rangle$ and $|E_2\rangle$ then follows, with the excited population dividing itself equally between $|E_D\rangle$

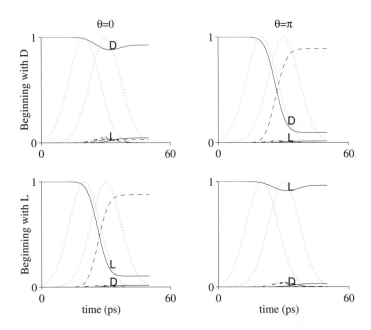

Figure 6 The time evolution of the enantiomeric populations for two different relative phases θ between the ε_1 and ε_0 beams. (___) gives the population in the D or L enantiomer; (....) shows the ε_1 and ε_0 laser pulses; (---) denotes the excited state population in levels $|E_1\rangle + |E_2\rangle$.

and $|E_L\rangle$. The resultant populations N_D and $\mathcal{N}_L$ in ground state $|E_D\rangle$ and $|E_L\rangle$ is then

$$\mathcal{N}_D = 0.5 N_D [1 + P_{D\leftarrow D} - P_{L\leftarrow D}] + 0.5 N_L [1 + P_{D\leftarrow L} - P_{L\leftarrow L}]$$
$$\mathcal{N}_L = 0.5 N_D [1 + P_{L\leftarrow D} - P_{D\leftarrow D}] + 0.5 N_L [1 + P_{L\leftarrow L} - P_{D\leftarrow L}] \tag{36}$$

The sequence of laser excitation followed by collisional relaxation and radiative emission is then iterated to convergence. In the second step, for example, the populations in Eq. (36) are taken as the initial populations for two independent computations, one assuming a population of $\mathcal{N}_D$ in $|E_D\rangle$, with $|E_L\rangle$ unpopulated, and the second assuming a population of $\mathcal{N}_L$ in $|E_L\rangle$, with $|E_D\rangle$ unpopulated.

Clearly, convergence is obtained when the populations, postrelaxation, are the same as those prior to laser excitation, i.e., when $\mathcal{N}_D = N_D$, and $\mathcal{N}_L = N_L$. These conditions reduce to

$$N_D(1 - P_{D\leftarrow D} + P_{L\leftarrow D}) = N_L(1 + P_{D\leftarrow L} - P_{L\leftarrow L}) \tag{37}$$

If the total population is chosen to be normalized ($N_D + N_L = 1$), then the final probabilities P_D, P_L of populating states $|E_D\rangle$ and $|E_L\rangle$ are

$$P_D = \frac{1 + P_{D \leftarrow L} - P_{L \leftarrow L}}{2 - P_{D \leftarrow D} + P_{L \leftarrow D} + P_{D \leftarrow L} - P_{L \leftarrow L}}$$

$$P_L = \frac{1 - P_{D \leftarrow D} + P_{L \leftarrow D}}{2 - P_{D \leftarrow D} + P_{L \leftarrow D} + P'_{D \leftarrow L} - P_{L \leftarrow L}} \tag{38}$$

and the equilibrium enantiomeric branching ratio is simply

$$R_{D,L} \equiv \frac{P_D}{P_L} = \frac{1 + P_{D \leftarrow L} - P_{L \leftarrow L}}{1 - P_{D \leftarrow D} + P_{L \leftarrow D}} \tag{39}$$

Results for the converged probabilities for the cases depicted in Fig. 5, are shown in Fig. 7. The results clearly show substantially enhanced enantiomeric ratios at various choices of control parameters. For example, at $\varepsilon_0 = \varepsilon_1 = 1.5 \times 10^{-3}$, tuning Δ_1 to 50 cm^{-1} gives a preponderance of L, whereas tuning to the $\Delta_1 = -125$ cm^{-1} gives more D.

Numerous other parameters in this system, such as the pulse shape, time delay between pulses, pulse frequencies, and pulse powers, etc., can be varied to affect the final L to D ratio [70] resulting in a very versatile approach to asymmetric synthesis.

Finally, note that although we have only included two ground state levels, the method applies equally well when a large number of ground state levels are

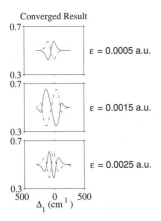

Converged Result

$\varepsilon = 0.0005$ a.u.

$\varepsilon = 0.0015$ a.u.

$\varepsilon = 0.0025$ a.u.

Δ_1 (cm^{-1})

Figure 7 Results for laser distillation after a convergent series of steps including radiative excitation followed by collisional and radiative relaxation. Shown are the results at three different field strengths.

included. In this case, relaxation will be among all of these ground state levels, but the proposed scenario, tuned to the above set of transitions, will bleed population from one M level of the desired enantiomer. As relaxation refills this level it will continue to be pumped over to the other enantiomer, with the overall effect that the major amount of the population will be transferred from one enantiomer to the other.

As a realization of the above scheme we now examine [10] the case of enantiomer control in dimethylallene, a molecule shown in Fig. 8. Note that, at equilibrium in the ground state, the H—C—CH$_3$ groups at both ends of the molecule lie on planes that are perpendicular to one another, resulting a molecule that is chiral. By contrast, in the excited state, the C=C double bond breaks, allowing for rotation of one plane relative to the other. Cuts through the ground and first two excited state potential energy surfaces for this molecule along the α and θ coordinates (see Fig. 8) are shown in Fig. 9. The potentials show the features required for control in this scenario, i.e., a minimum in the excited state potential surface at the geometry corresponding to the potential energy maximum on the ground state potential.

The results of a computation [71] on the control of L vs. D 1,3-dimethylallene are shown in Fig. 10. Outstanding enantiomeric control over the dimethylallene enantiomers is evident for a wide variety of powers. For example, a most impressive result is achieved for $\Delta_1 = 0.0986$ cm^{-1} and $\varepsilon_0 = 1.5 \times 10^{-4}$ a.u., $\varepsilon_1 = \varepsilon_2 = 4.31 \times 10^{-5}$ a.u., corresponding to laser powers of 7.90×10^8 W/cm^2 and 6.52×10^7 W/cm^2, respectively. Here a racemic mixture of dimethylallene in

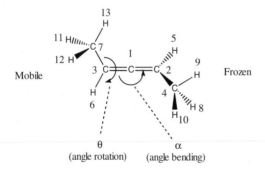

Figure 8   The geometry of the 1,3 dimethylallene and the two angles θ and α that were varied to scan the potential energy surface. Here θ is the dihedral angle between the H$_3$C···C—C and the C—C···CH$_3$ planes, and α is the C—C—C bending angle, here shown by an arrow that brings the H$_3$C···C···H out of the plane of the paper. (From Fig. 2, Ref. 10.)

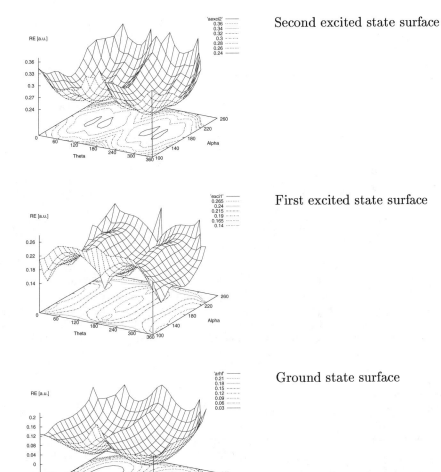

Figure 9 Potential energy surfaces for 1,3-dimethylallene. Here we show in-plane surfaces for the ground and first two excited electronic states. (From Fig. 4, Ref. 10.)

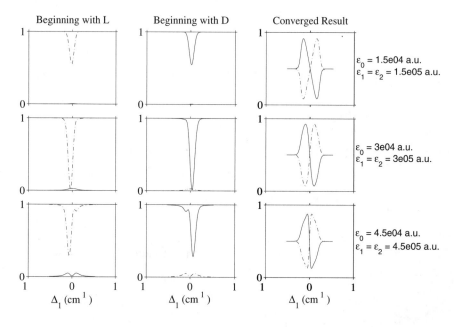

Figure 10 Control over dimethylallene enantiomer populations as a function of the detuning Δ_1 for various laser powers. The first column corresponds to probabilities of L (dot-dash curves) and D (solid curves) after a single laser pulse, assuming that the initial state is all L. The second column is similar, but for an initial state that is all D. The rightmost column corresponds to the probabilities L and D after repeated excitation–relaxation cycles, as described in the text. (This is a corrected version of Fig. 2, Ref. 71.)

a specific J, M, λ state can be converted, after a series of pulses, to a mixture of dimethylallene, containing 92.7% of the D-dimethylallene in this state. (Here λ is the projection of the total angular momentum J along an axis fixed in the molecule). Similarly, detuning to $\Delta_1 = -0.0986$ cm^{-1} results in a similar enhancement of L-dimethylallene. Slightly lower extremes of control are seen to be achievable for the two other laser powers shown. Further, control was achievable to field strengths down to 10^4 W/cm^2. Note, however, that this computation neglects the competitive process of internal conversion, discussed later below.

It is of some interest to note the character of the eigenstates $|E_1\rangle$ and $|E_2\rangle$ that contribute to these results; they are shown in Fig. 11. Clearly they are states with considerable vibrational energy, so that they are broad enough in configuration space to overlap the ground electronic state, ground vibrational state wave-

Eigenvector |2)

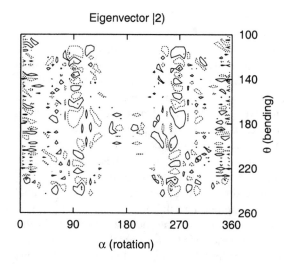

Eigenvector |1)

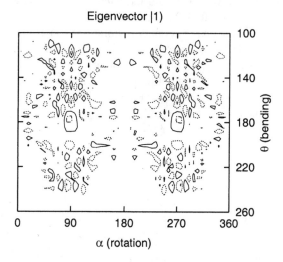

Figure 11 Contour plots of $|E_1\rangle$ and $|E_2\rangle$ where dash-dash lines = 0.012 a.u., dot-dot lines = 0.0004 a.u., solid lines = −0.004 a.u. and dot-dash lines = −0.012 a.u. Note that $|E_1\rangle$ is symmetric with respect to reflection and $|E_2\rangle$ is antisymmetric. Reflection here corresponds to changing ($\alpha - 180°$) to ($\alpha + 180°$). (From Fig. 1, Ref. 71.)

functions. If this is not the case then the dipole matrix elements are too small to allow control at reasonable laser intensities.

The primary experimental difficulty associated with this scenario is the requirement to isolate a particular subset of M levels, in order to avoid cancellation of M and $-|M|$ control. That is, from the viewpoint of the M structure, this scenario is associated with the level structure shown in Fig. 12.

To remove this restriction we introduced another scenario [74] in which all of the three laser polarizations, $\hat{\varepsilon}_0$, $\hat{\varepsilon}_1$, and $\hat{\varepsilon}_2$, are perpendicular to one another. This laser arrangement now allows for transitions between different M levels. The first few of these levels is shown in Fig. 13. Under these circumstances, control survives averaging over M levels [74], as discussed in Sec. V.F.

Enantiomeric control is more difficult if the excited molecular potential energy surfaces do not possess an appropriate minimum at the σ_h hyperplane configurations (see Figs. 1 and 2). In this case the method introduced in this section is not applicable. One may however be able to apply the laser distillation procedure by adding a molecule B to the initial L, D mixture to form weakly bound $L - B$ and $B - D$, which are themselves right- and left-handed enantiomeric pairs [83]. The molecule B is chosen so that electronic excitation of $B - D$ and $L - B$ forms an excited species G, which has stationary rovibrational states that are either symmetric or antisymmetric with respect to reflection through σ_h. The species $L - B$ and $B - D$ now serve as the L and D enantiomers in the general scenario above, and the laser distillation procedure described above then applies. Further, the molecule B serves as a catalyst that may be removed from the final product by traditional chemical means.

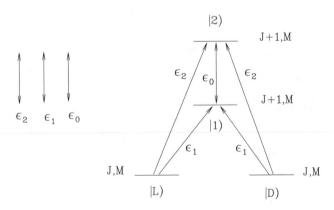

Figure 12 Schematic level diagram for three lasers of parallel polarization for the four-level scheme of Fig. 1.

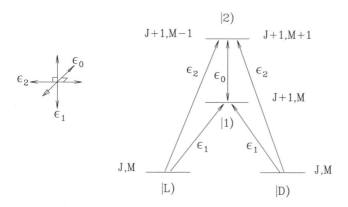

Figure 13 Schematic level diagram when three lasers of perpendicular polarizations irradiate the D and L enantiomers. Only the first five levels that are coupled by these lasers are shown.

For example, L and D might be the left- and right-handed enantiomers of a chiral alcohol, and B is the ketone derived from this alcohol (see Fig. 14). In this case, studies [83] of the electronic structure of the alcohol–ketone system show that there are weakly bound chiral alcohol–ketone minima in the ground electronic state, as desired. The particular advantage of using the ketone–alcohol complex is that the ketone, which is "recycled" after the conversion of one enantiomer to another, serves as a catalyst for the process.

The results in this section make clear that a chiral outcome, the enhancement of a particular enantiomer, can arise by coherently encoding quantum interference information in the excitation of a racemic mixture. The fact that the initial state displays a broken symmetry and that the excited state has states that are either symmetric or antisymmetric with respect to σ_h allows for the creation of a superposition state that does not have these transformation properties. Radiatively coupling the states in the superposition then allows for the transition probabilities from L and D to differ, allowing for depletion of the desired enantiomer.

V. ANGULAR MOMENTUM CONSIDERATIONS

In Sec. IV we examined enantiopurification in a racemic mixture based on four levels ($|D\rangle$, $|L\rangle$, $|1\rangle$, $|2\rangle$) using three parallel linearly polarized laser fields (see

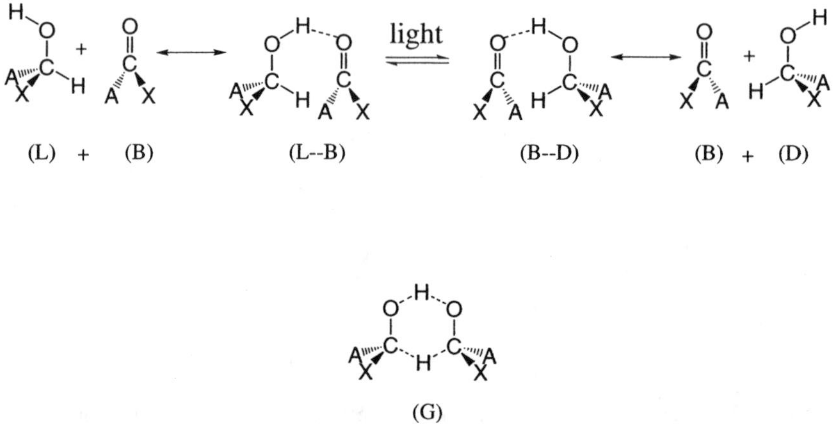

Figure 14 Sample scenario for enhanced enantiomeric selectivity in a racemic mixture of two chiral alcohols related by inversion. An alcohol and a ketone exchange two hydrogen atoms so as to produce the ketone, but with an alcohol of the reverse handedness. Here A and X are distinct organic groups, and dashes denote, in the upper panel, hydrogen bonds. The electronically excited species G, which is formed upon excitation with light, is postulated to be given by the structure at the bottom of the figure. In this case the topmost and bottommost hydrogens are attached to the oxygens and carbons, respectively, by "half-bonds." (From Fig. 4, Ref. 70.)

Fig. 12), which we denote here as configuration **A**. We show below that in this configuration we can only affect enantioselectivity in *polarized* systems, i.e., systems with a single well-defined M, where M is the projection of the total angular momentum J along the polarization direction of the incident light.

The polarization requirement adds an experimental complexity we would like to avoid. Luckily this can be done by working in a configuration we term configuration **B**, in which the racemic mixture is irradiated with three beams of mutually perpendicular linear polarizations. We show below that in this configuration no M polarization is necessary.

In both configurations we first consider control for one value of M, limiting attention to the ladder of states accessed by weak to moderately strong laser fields. In the case of configuration **A** this corresponds to the scheme in Fig. 12. For configuration **B**, more levels are involved, as shown in Fig. 13. Subsequently, we consider averaging over M, which in both configurations can be done analytically. We show that in configuration **A** the handedness generated by states with quantum numbers J, M is minus that generated with quantum numbers $J, -M$. Hence control is lost upon M averaging. This is not the case in configuration **B**, where the handedness generated by the J, M molecules equals exactly that of the J, $-M$ molecules. Thus the effect survives M averaging, and configuration **B** therefore provides a general method for enantiomeric control, via the $\mathbf{d} \cdot \mathbf{E}$ interaction, in traditional racemic mixtures.

A. Changing the Sign of the Fields

The time evolution of the amplitude in levels $|D\rangle$, $|L\rangle$, $|1\rangle$, and $|2\rangle$ in configuration **A** is described by Eqs. (33). As noted in Sec. II, enantiomeric control is only possible when the dynamics depend on the sign of the overall electric field [13]. Consider then the effect of changing $\mathbf{E} \rightarrow -\mathbf{E}$, i.e., $\varepsilon_0 \rightarrow -\varepsilon_0$, $\varepsilon_1 \rightarrow -\varepsilon_1$, and $\varepsilon_2 \rightarrow -\varepsilon_2$. Under this transformation, Eq. (33) become

$$\dot{b}_1 = -i\exp(i\Delta_1 t)[\Omega^*_{D,1}b_D + \Omega^*_{D,1}b_L] - i\exp(-i\Delta_0 t)\Omega^*_0 b_2$$

$$\dot{b}_2 = -i\exp(i\Delta_2 t)[\Omega^*_{D,2}b_D - \Omega^*_{D,2}b_L] - i\exp(i\Delta_0 t)\Omega_0 b_1$$

$$\dot{b}_D = -i\exp(-i\Delta_1 t)\Omega_{D,1}b_1 - i\exp(-i\Delta_2 t)\Omega_{D,2}b_2$$

$$\dot{b}_L = -i\exp(-i\Delta_1 t)\Omega_{D,1}b_1 + i\exp(-i\Delta_2 t)\Omega_{D,2}b_2 \tag{40}$$

By defining $b'_1 = -b_1$ Eq. (40) can be rewritten as

$$\dot{b}'_1 = i\exp(i\Delta_1 t)[\Omega^*_{D,1}b_D + \Omega^*_{D,1}b_L] + i\exp(-i\Delta_0 t)\Omega^*_0 b_2$$

$$\dot{b}_2 = i\exp(i\Delta_2 t)[-\Omega^*_{D,2}b_D + \Omega^*_{D,2}b_L] + i\exp(i\Delta_0 t)\Omega_0 b'_1$$

$$\dot{b}_D = i\exp(-i\Delta_1 t)\Omega_{D,1}b'_1 - i\exp(-i\Delta_2 t)\Omega_{D,2}b_2$$

$$\dot{b}_L = i\exp(-i\Delta_1 t)\Omega_{D,1}b'_1 + i\exp(-i\Delta_2 t)\Omega_{D,2}b_2 \tag{41}$$

These equations are identical to those in Eq. (33) except that $b_D \Leftrightarrow b_L$. Hence, $\mathbf{E} \rightarrow -\mathbf{E}$ affects the dynamics, and enantiomeric control is possible, consistent with previous studies [8]. Of additional interest is that the effect is precisely that which one would expect from the inversion of the total system, i.e., of both the molecule and the electric field, in which the dynamics of the D enantiomer and the L enantiomer interchange with one another.

   Note, for interest below, that in this scenario changing the sign of any one of the three electric field constituents ε_i has the same effect as changing the sign of total electric field insofar as it reverses the roles of enantiomeric selectivity. Specifically, $L \Leftrightarrow D$ reversal occurs when we change the sign of ε_1, ε_2, or ε_0. This can easily be verified by defining new (primed) probability amplitudes such as $b'_2 = -b_2$ for $\varepsilon_0 \rightarrow -\varepsilon_0$ and $b'_D = -b_D$ and $b'_L = -b_L$ for $\varepsilon_1 \rightarrow -\varepsilon_1$. Similarly, changing the sign of ε_2 in Eq. (33) directly gives Eq. (41) without further algebra. Note also that if we change the sign of any two electric field constituents ε_i then the enantiomeric selectivity remains unchanged. For example, consider sign reversal of both ε_0 and ε_1. In this case the original equations [Eq. (33)] are recovered by defining $b'_2 = -b_2$, $b'_D = -b_D$, and $b'_L = -b_L$.

B. Introducing Rotations

The above discussion does not consider the role of angular momentum, shown here to be crucial to overall control. Specifically, to consider the effect of averaging over M [84], the situation that would exist in a nonpolarized medium, we associate with each of the $|D\rangle$, $|L\rangle$, $|1\rangle$, and $|2\rangle$ states a parity-adapted symmetric-top wave function of the type

$$D_{\lambda,M}^{J,p}(\phi,\theta,\chi) \equiv t_\lambda \left(\frac{2J+1}{8\pi^2}\right)^{\frac{1}{2}} \left\{ D_{\lambda,M}^{J}(\phi,\theta,\chi) + p(-1)^\lambda D_{-\lambda,M}^{J}(\phi,\theta,\chi) \right\} \tag{42}$$

where (ϕ, θ, χ) are three Euler angles of rotation in a space-fixed coordinate system, $D_{\lambda,M}^{J}(\phi, \theta, \chi)$ are the rotational matrices in Edmonds' notation [85], $p = \pm 1$, and $t_\lambda = 2^{-1/2}$ for $\lambda > 0$ and $t_\lambda = 1/2$ for $\lambda = 0$. Here J is the total angular momentum, M is its projection on a laboratory-fixed z axis, and λ is its projection on a molecular-fixed axis. It is seen from Eq. (42) that every vib-rotational level is doubly degenerate in p (corresponding to the two signs of λ) except for $\lambda = 0$. The extension to the asymmetric top is straightforward insofar as the p degeneracy is removed and the wave functions maintain the same M dependence as the symmetric top.

The ''parity-adapted'' $D_{\lambda,M}^{J,p}(\phi, \theta, \chi)$ functions are eigenfunctions of the inversion operator I,

$$I D_{\lambda,M}^{J,p}(\phi,\theta,\chi) = p(-1)^J D_{\lambda,M}^{J,p}(\phi,\theta,\chi) \tag{43}$$

a property that follows from the basic relation $I D_{\lambda,M}^{J}(\phi, \theta, \chi) = (-1)^{J+\lambda} D_{-\lambda, M}^{J}(\phi, \theta, \chi)$.

Given the above rotational wave functions, the wave functions in the ground electronic state are written as

$$\left|L_p\right\rangle = \left|v_L\right\rangle D_{\lambda,M}^{J,p} \qquad \left|D_p\right\rangle = \left|v_D\right\rangle D_{\lambda,M}^{J,p} \tag{44}$$

and the excited state wavefunctions are written (see Fig. 12) as

$$\left|L_{p'}\right\rangle = \left|v_1\right\rangle D_{\lambda',M'}^{J',p'} \qquad \left|D_{p''}\right\rangle = \left|v_2\right\rangle D_{\lambda'',M''}^{J'',p''} \tag{45}$$

where the $|v_i\rangle$ are the vibrational components of the wave function. Since the eigenvalues of the symmetric top Hamiltonian depend upon J and λ, we can excite between selected levels J, λ, J', λ', and J'', λ'' by choosing the appropriate laser frequency. The energy does not, however, depend upon the values of p, p', p''. Hence the system comprises eight levels, i.e., four energetically degenerate sets (as shown in Fig. 12 but where each level is twofold degenerate in p).

We assume, with no loss of generality, that $|1_{p'}\rangle$ is symmetric with respect to inversion, and therefore that $|1_{-p'}\rangle$ is antisymmetric. Similarly, $|2_{p''}\rangle$ is assumed symmetric and hence $|2_{-p''}\rangle$ is antisymmetric. It is then notationally convenient to relabel the states as

$$|1\rangle \equiv |1_{-p'}\rangle \quad |2\rangle \equiv |1_{p'}\rangle \quad |3\rangle \equiv |2_{-p''}\rangle \quad |4\rangle \equiv |2_{p''}\rangle \tag{46}$$

With this notation, $|i\rangle$ with even values of i are symmetric with respect to inversion and $|i\rangle$ with odd values of i are antisymmetric with respect to inversion.

The state wave function can then be expanded in terms of the above set of vib-rotational wave functions as

$$\begin{aligned}
|\Psi\rangle = {} & b_{D_p} \exp(-iE_D t/\hbar)|D_p\rangle + b_{D_{-p}} \exp(-iE_D t/\hbar)|D_{-p}\rangle \\
& + b_{L_p} \exp(-iE_L t/\hbar)|L_p\rangle + b_{L_{-p}} \exp(-iE_L t/\hbar)|L_{-p}\rangle \\
& + b_1 \exp(-iE_1 t/\hbar)|1\rangle + b_2 \exp(-iE_2 t/\hbar)|2\rangle \\
& + b_3 \exp(-iE_3 t/\hbar)|3\rangle + b_4 \exp(-iE_4 t/\hbar)|4\rangle
\end{aligned} \tag{47}$$

giving the following set of coupled equations in the rotating wave approximation:

$$\begin{aligned}
\dot{b}_{Lp} &= i\exp(-i\Delta_1 t)(b_1\Omega^*_{1,Lp} + b_2\Omega^*_{2,Lp}) + i\exp(-i\Delta_2 t)(b_3\Omega^*_{3,Lp} + b_4\Omega^*_{4,Lp}) \\
\dot{b}_{L-p} &= i\exp(-i\Delta_1 t)(b_1\Omega^*_{1,L-p} + b_2\Omega^*_{2,L-p}) + i\exp(-i\Delta_2 t)(b_3\Omega^*_{3,L-p} + b_4\Omega^*_{4,L-p}) \\
\dot{b}_{Dp} &= i\exp(-i\Delta_1 t)(b_1\Omega^*_{1,Dp} + b_2\Omega^*_{2,Dp}) + i\exp(-i\Delta_2 t)(b_3\Omega^*_{3,Dp} + b_4\Omega^*_{4,Dp}) \\
\dot{b}_{D-p} &= i\exp(-i\Delta_1 t)(b_1\Omega^*_{1,D-p} + b_2\Omega^*_{2,D-p}) + i\exp(-i\Delta_2 t)(b_3\Omega^*_{3,D-p} + b_4\Omega^*_{4,D-p}) \\
\dot{b}_1 &= i\exp(i\Delta_1 t)(b_{L_p}\Omega_{1,L_p} + b_{L_{-p}}\Omega_{1,L-p} + b_{D_p}\Omega_{1,D_p} + b_{D_{-p}}\Omega_{1,D-p}) + ib_4\exp(-i\Delta_o t)\Omega^*_{4,1} \\
\dot{b}_2 &= i\exp(i\Delta_1 t)(b_{L_p}\Omega_{2,L_p} + b_{L_{-p}}\Omega_{2,L-p} + b_{D_p}\Omega_{2,D_p} + b_{D_{-p}}\Omega_{2,D-p}) + ib_3\exp(-i\Delta_o t)\Omega^*_{3,2} \\
\dot{b}_3 &= i\exp(i\Delta_2 t)(b_{L_p}\Omega_{3,L_p} + b_{L_{-p}}\Omega_{3,L-p} + b_{D_p}\Omega_{3,D_p} + b_{D_{-p}}\Omega_{3,D-p}) + ib_2\exp(i\Delta_o t)\Omega^*_{3,2} \\
\dot{b}_4 &= i\exp(i\Delta_2 t)(b_{L_p}\Omega_{4,L_p} + b_{L_{-p}}\Omega_{4,L-p} + b_{D_p}\Omega_{4,D_p} + b_{D_{-p}}\Omega_{4,D-p}) + ib_1\exp(i\Delta_o t)\Omega^*_{4,1}
\end{aligned} \tag{48}$$

Here $\Omega_{i,j} \equiv \langle i|\,\mu_k\,|j\rangle\,\varepsilon_k/\hbar$, $\Delta_k \equiv \omega_{i,j} - \omega_k$, $i = 1,2,3,4$, $j = L_p, L_{-p}, D_p, D_{-p}$, $k = 0,1,2$. In a fashion similar to that of the four: level scheme above [Eq. (30)], states $|L_p\rangle$ and $|D_p\rangle$ can be rewritten in terms of definite parity states $|S_p\rangle$ and $|A_p\rangle$, and again without loss of generality we can assign states $|S_p\rangle$ and $|A_p\rangle$ to be overall symmetric and antisymmetric, respectively. Their counterparts $|L_{-p}\rangle$ and $|D_{-p}\rangle$, obtained by changing p to $-p$, are of symmetry opposite to $|L_p\rangle$ and $|D_p\rangle$. Therefore one can show that

$$\begin{aligned}
\Omega_{1,L_p} &= -\Omega_{1,D_p} = \Omega_{1,S_p} & \Omega_{3,L_p} &= -\Omega_{3,D_p} = \Omega_{3,S_p} \\
\Omega_{1,L_{-p}} &= \Omega_{1,D_{-p}} = \Omega_{1,A_{-p}} & \Omega_{3,L_{-p}} &= \Omega_{3,D_{-p}} = \Omega_{3,A_{-p}} \\
\Omega_{2,L_p} &= \Omega_{2,D_p} = \Omega_{2,A_p} & \Omega_{4,L_p} &= \Omega_{4,D_p} = \Omega_{4,A_p} \\
\Omega_{2,L_{-p}} &= -\Omega_{2,D_{-p}} = \Omega_{2,S_{-p}} & \Omega_{4,L_{-p}} &= -\Omega_{4,D_{-p}} = \Omega_{4,S_{-p}}
\end{aligned} \tag{49}$$

Hence Eq. (48) can be rewritten as

$$\dot{b}_{L_p} = i\exp(-i\Delta_1 t)(b_2\Omega^*_{2,L_p} - b_1\Omega^*_{1,L_p}) + i\exp(-i\Delta_2 t)(b_4\Omega^*_{4,L_p} - b_3\Omega^*_{3,L_p})$$

$$\dot{b}_{D_p} = i\exp(-i\Delta_1 t)(b_2\Omega^*_{2,L_p} + b_1\Omega^*_{1,L_p}) + i\exp(-i\Delta_2 t)(b_4\Omega^*_{4,L_p} + b_3\Omega^*_{3,L_p})$$

$$\dot{b}_{L_{-p}} = i\exp(-i\Delta_1 t)(b_1\Omega^*_{1,L_{-p}} - b_2\Omega^*_{2,L_{-p}}) + i\exp(-i\Delta_2 t)(b_3\Omega^*_{3,L_{-p}} - b_4\Omega^*_{4,L_{-p}})$$

$$\dot{b}_{D_{-p}} = i\exp(-i\Delta_1 t)(b_1\Omega^*_{1,L_{-p}} + b_2\Omega^*_{2,L_{-p}}) + i\exp(-i\Delta_2 t)(b_3\Omega^*_{3,L_{-p}} + b_4\Omega^*_{4,L_{-p}})$$

$$\dot{b}_1 = i\exp(i\Delta_1 t)\left[\Omega_{1,L_p}(b_{D_p} - b_{L_p}) + \Omega_{1,L_{-p}}(b_{D_{-p}} + b_{L_{-p}})\right] + ib_4\exp(-i\Delta_0 t)\Omega^*_{4,1}$$

$$\dot{b}_2 = i\exp(i\Delta_1 t)\left[\Omega_{2,L_p}(b_{D_p} + b_{L_p}) + \Omega_{2,L_{-p}}(b_{D_{-p}} - b_{L_{-p}})\right] + ib_3\exp(-i\Delta_0 t)\Omega^*_{3,2}$$

$$\dot{b}_3 = i\exp(i\Delta_2 t)\left[\Omega_{3,L_p}(b_{D_p} - b_{L_p}) + \Omega_{3,L_{-p}}(b_{D_{-p}} + b_{L_{-p}})\right] + ib_2\exp(i\Delta_0 t)\Omega_{3,2}$$

$$\dot{b}_4 = i\exp(i\Delta_2 t)\left[\Omega_{4,L_p}(b_{D_p} + b_{L_p}) + \Omega_{4,L_{-p}}(b_{D_{-p}} - b_{L_{-p}})\right] + ib_1\exp(i\Delta_0 t)\Omega_{4,1} \qquad (50)$$

We can readily show that the dynamics in Eq. (50) are sensitive to the sign of the total electric field and hence [13] that enantiomeric control is possible. Further, for use below, we note that changing the sign of any one component of the electric field, i.e., $\varepsilon_0 \to -\varepsilon_0$, or $\varepsilon_1 \to -\varepsilon_1$, or $\varepsilon_2 \to -\varepsilon_2$, results in a reversal of the role of the enantiomeric selectivity; specifically, $L_{-p} \Leftrightarrow D_{-p}$ and $L_p \Leftrightarrow D_p$. Similar conclusions as those reached following Eq. (41) can be made when only one or two electric fields change sign. That is, the roles of L and D are reversed when the sign of only one field component is changed but remain the same if two field components are changed. As an example, consider the change $\varepsilon_0 \to -\varepsilon_0$. By defining $b'_1 = -b_1, b'_3 = -b_3, b'_{L_{-p}} = -b_{L_{-p}}$ and $b'_{D_{-p}} = -b_{D_{-p}}$, it can be shown that the roles of enantiomeric selectivity are reversed [86].

C. *M* Averaging in Configuration A

To see the effect of averaging over the M quantum number we examine the relationship between enantiomeric control with a given value of M and with the corresponding value of $-M$. Consider then the M dependence of the dipole transition matrix elements. The projection of the dl ($l = 0, 1, 2$) dipole operators on the polarization directions $\hat{\varepsilon}_l$ can be related (see Ref. 87) to the (x,y,z) body-fixed components of the dipole operators as

$$\mathrm{d}_l \equiv \vec{\mathrm{d}}_l \cdot \hat{\varepsilon}_l = f_l\, t_{q_l} \sum_{k=-1}^{K=1} \mathrm{d}_l^{(K)}\left\{D^1_{K,q_l}(\phi,\theta,\chi) + p_l(-1)^{q_l}D^1_{K,-q_l}(\phi,\theta,\chi)\right\} \qquad (51)$$

where

$$\mathrm{d}_l^{(0)} \equiv \mathrm{d}_l^z, \quad \mathrm{d}_l^{(-1)} \equiv \frac{(\mathrm{d}_l^x \to i\mathrm{d}_l^y)}{\sqrt{2}} \quad \mathrm{d}_l^{(1)} \equiv \frac{-(\mathrm{d}_l^x + i\mathrm{d}_l^y)}{\sqrt{2}} \qquad (52)$$

Here $f_l = 1$, $q_l = 0$, $p_l = 1$ for $\hat{\varepsilon}_l$ along the Z axis, $f_l = -1$, $q_l = 1$, $p_l = 1$ for $\hat{\varepsilon}_l$ along the X axis, and $f_l = i$, $q_l = 1$, $p_l = -1$ for $\hat{\varepsilon}_l$ along the Y axis.

Given Eqs. (44), (45) and (51), consideration of dipole transition matrix elements requires the following integrals [85], Eq. (4.6.2)]:

$$\left(D_{\lambda',M'}^{J',p'} \left| D_{K,q}^1 \right| D_{\lambda,M}^{J,p} \right) \equiv \int_0^{2\pi} d\phi \int_0^\pi \sin\theta \, d\theta \int_0^{2\pi} d\chi \, D_{\lambda',M'}^{J',p'*}(\phi,\theta,\chi) D_{K,q}^1(\phi,\theta,\chi) D_{\lambda,M}^{J,p}(\phi,\theta,\chi)$$

$$= (-1)^{-M'} t_{\lambda'} t_\lambda [(2J'+1)(2J+1)]^{\frac{1}{2}} \begin{pmatrix} J' & 1 & J \\ -M & q & M \end{pmatrix} F(J',\lambda',p';J,\lambda,p;K)$$

$$\equiv (-1)^{M'} \begin{pmatrix} J' & 1 & J \\ -M' & q & M \end{pmatrix} B(J',\lambda',p';j,\lambda,p;K)$$

$$(53)$$

where nonzero Clebsch–Gordan coefficients require $M + q - M' = 0$ and where

$$F(J',\lambda',p';J,\lambda,p;K) = (-1)^{\lambda'} \begin{pmatrix} J' & 1 & J \\ -\lambda' & K & \lambda \end{pmatrix} + p(-1)^{\lambda+\lambda'} \begin{pmatrix} J' & 1 & J \\ -\lambda' & K & \lambda \end{pmatrix}$$

$$+ p' \begin{pmatrix} J' & 1 & J \\ \lambda' & K & \lambda \end{pmatrix} + p'p(-1)^{l} \begin{pmatrix} J' & 1 & J \\ -\lambda' & K & -\lambda \end{pmatrix}$$

$$(54)$$

$B(J', \lambda', p'; j, \lambda, p; K)$ has been defined in Eq. (53) in order to expose the M and M' dependence of the matrix element.

Consider now the average over the M quantum number where all components ε_i are Z polarized. The relevant integrals in Eq. (53) would then have $q = 0$. In accord with this equation the entire M dependence, within the transition dipole moments, is contained in a single 3-j symbol. The property of the 3-j coefficients with respect to $M \rightarrow -M$ transformation is given by [85]

$$\begin{pmatrix} J_2 & 1 & J_1 \\ M' & 0 & -M \end{pmatrix} = (-1)^{J_1+1+J_2} \begin{pmatrix} J_2 & 1 & J_1 \\ -M' & 0 & M \end{pmatrix}$$

$$(55)$$

Here nonzero Clebsch–Gordan coefficients require $M = M'$. As a consequence of Eq. (55), some or all of the transition matrix elements will change signs when $M \rightarrow -M$, depending on which J states are involved in the excitation scheme. Since these matrix elements enter into the dynamics via the $\Omega's$, such a change in sign is equivalent to changing the sign of the associated electric field component.

The selection rule $\Delta J = 0, \pm 1$, and the requirement that dipole transitions induced by all of the ε_i be nonzero, allow for only two possible angular momentum scenarios: the first, where all J states have the same value, and the second, where two states have the same J and the other state has angular momentum $J \pm 1$. As an example, consider excitation from the ground $| L_p \rangle$ and $| D_p \rangle$ states with angular momentum J to excited state levels $|i\rangle$, $i = 1 \ldots 4$, with angular momentum $J + 1$. In this case, dipole coupling matrix elements associated with ε_1 and ε_2 are unaffected by $M \rightarrow -M$, but couplings associated with ε_0 will change sign. Hence the $-M$ case behaves in the same way as that of M with field component $-\varepsilon_0$. In accord with the discussion following Eq. (50), in this case the enhancement of D over L (or vice-versa) resulting from the M case is countered by the enhancement of L over D from $-M$. Further, for $M = 0$, the coupling associated with at least one of the ε_i components will be zero. This is analogous to turning off the associated ε_i field component, so that the $M = 0$ case shows no enantiomeric control. Hence, control is lost upon M averaging.

In the other possible scenario, where all J quantum numbers are the same, all transition dipole moments change sign with $M \rightarrow -M$. The effect is similar to changing the sign of all three electric field components. As noted in the discussion following Eq. (50), this interchanges the role of D and L, hence ensuring loss of control upon M averaging.

We note that this argument is independent of the p values of the energy levels. As such the arguments apply equally well to the asymmetric top where the p degeneracy does not exist [88].

D. Enantioselectivity with Three Mutually Perpendicular Polarizations: Laser Configuration B

Consider now laser configuration B, where the three field polarizations are mutually orthogonal to one another. Here we show that it is possible to generate handedness in a non-M-polarized medium, i.e., enantiomeric control is not lost upon M averaging. Specifically, consider the case where $\hat{\varepsilon}_1$ lies along the Z direction, $\hat{\varepsilon}_2$ lies along the X direction, and $\hat{\varepsilon}_0$ lies along the Y direction. All fields are linearly polarized. The $|D\rangle$, $|L\rangle$, $|1\rangle$, and $|2\rangle$ states in this case are identical to those of configuration **A**. However, the selection rule for the coupling to levels $|3\rangle$ and $|4\rangle$ now allows for $\Delta M = \pm 1$, doubling the number of coupled excited states. To retain the previous notation we use $|3^+\rangle$ and $|4^+\rangle$ to designate $+M$ and $|3^-\rangle$ and $|4^-\rangle$ to designate $-M$ components of the indicated states. Note that the value of the M quantum number does not alter the parity of the states. Note also that additional M states, coupled in higher order, are not included in this section but are treated computationally in subsection V.F.

In this case, ten molecular states are relevant in lowest order, and Eq. (25) becomes

$$|\Psi\rangle = b_{D_p} \exp\left(\frac{-iE_D t}{\hbar}\right)|D_p\rangle + b_{D_{-p}} \exp\left(\frac{-iE_D t}{\hbar}\right)|D_{-p}\rangle$$

$$+ b_{L_p} \exp\left(\frac{-iE_L}{\hbar}\right)|L_p\rangle + b_{L_{-p}} \exp\left(\frac{-iE_L t}{\hbar}\right)|L_{-p}\rangle$$

$$+ b_1 \exp\left(\frac{-iE_1 t}{\hbar}\right)|1\rangle + b_2 \exp\left(\frac{-iE_2 t}{\hbar}\right)|2\rangle$$

$$+ \exp\left(\frac{-iE_3 t}{\hbar}\right)(b_{3^+}|3^+\rangle + b_{3^-}|3^-\rangle) + \exp\left(\frac{-iE_4 t}{\hbar}\right)|(b_{4^+}|4^+\rangle + b_{4^-}|4^-\rangle) \quad (56)$$

Substituting into the Schrödinger equation gives, in the rotating wave approximation,

$$
\begin{aligned}
\dot{b}_{L_p} &= i\exp(-i\Delta_1 t)\,(b_2\Omega^*_{2,L_p} - b_1\Omega^*_{1,L_p}) \\
&\quad + i\exp(-i\Delta_2 t)\Big[(b_{4^+}\Omega^*_{4^+,L_p} + b_{4^-}\Omega^*_{4^-,L_p}) - (b_{3^+}\Omega^*_{3^+,L_p} + b_{3^-}\Omega^*_{3^-,L_p})\Big] \\
\dot{b}_{D_p} &= i\exp(-i\Delta_1 t)\,(b_2\Omega^*_{2,L_p} + b_1\Omega^*_{1,L_p}) \\
&\quad + i\exp(-i\Delta_2 t)\Big[(b_{4^+}\Omega^*_{4^+,L_p} + b_{4^-}\Omega^*_{4^-,L_p}) + (b_{3^+}\Omega^*_{3^+,L_p} + b_{3^-}\Omega^*_{3^-,L_p})\Big] \\
\dot{b}_{L_{-p}} &= i\exp(-i\Delta_1 t)\,(b_1\Omega^*_{1,L_{-p}} - b_2\Omega^*_{2,L_{-p}}) \\
&\quad + i\exp(-i\Delta_2 t)\Big[(b_{3^+}\Omega^*_{3^+,L_{-p}} + b_{3^-}\Omega^*_{3^-,L_{-p}}) - (b_{4^+}\Omega^*_{4^+,L_{-p}} + b_{4^-}\Omega^*_{4^-,L_{-p}})\Big] \\
\dot{b}_{D_{-p}} &= i\exp(-i\Delta_1 t)\,(b_1\Omega^*_{1,L_{-p}} + b_2\Omega^*_{2,L_{-p}}) \\
&\quad + i\exp(-i\Delta_2 t)\Big[(b_{3^+}\Omega^*_{3^+,L_{-p}} + b_{3^-}\Omega^*_{3^-,L_{-p}}) + (b_{4^+}\Omega^*_{4^+,L_{-p}} + b_{4^-}\Omega^*_{4^-,L_{-p}})\Big] \\
\dot{b}_1 &= i\exp(i\Delta_1 t)\Omega_{1,L_p}(b_{D_p} - b_{L_p}) + i\exp(i\Delta_1 t)\Omega_{1,L_{-p}}(b_{D_{-p}} + b_{L_{-p}}) \\
&\quad + i\exp(-i\Delta_0 t)(b_{4^+}\Omega^*_{4^+,1} + b_{4^-}\Omega^*_{4^-,1}) \\
\dot{b}_2 &= i\exp(i\Delta_1 t)\Omega_{2,L_p}(b_{D_p} + b_{L_p}) + i\exp(i\Delta_1 t)\Omega_{2,L_{-p}}(b_{D_{-p}} - b_{L_{-p}}) \\
&\quad + i\exp(-i\Delta_0 t)(b_{3^+}\Omega^*_{3^+,2} + b_{3^-}\Omega^*_{3^-,2}) \\
\dot{b}_{3^+} &= i\exp(i\Delta_2 t)\Omega_{3^+,L_p}(b_{D_p} - b_{L_p}) + i\exp(i\Delta_2 t)\Omega_{3^+,L_{-p}}(b_{D_{-p}} + b_{L_{-p}}) \\
&\quad + i\exp(i\Delta_0 t)b_2\Omega_{3^+,2} \\
\dot{b}_{3^-} &= i\exp(i\Delta_2 t)\Omega_{3^-,L_p}(b_{D_p} - b_{L_p}) + i\exp(i\Delta_2 t)\Omega_{3^-,L_{-p}}(b_{D_{-p}} + b_{L_{-p}}) \\
&\quad + i\exp(i\Delta_0 t)b_2\Omega_{3^-,2} \\
\dot{b}_{4^+} &= i\exp(i\Delta_2 t)\Omega_{4^+,L_p}(b_{D_p} + b_{L_p}) + i\exp(i\Delta_2 t)\Omega_{4^+,L_{-p}}(b_{D_{-p}} + b_{L_{-p}}) \\
&\quad + i\exp(i\Delta_0 t)b_1\Omega_{4^+,1} \\
\dot{b}_{4^-} &= i\exp(i\Delta_2 t)\Omega_{4^-,L_p}(b_{D_p} + b_{L_p}) + i\exp(i\Delta_2 t)\Omega_{4^-,L_{-p}}(b_{D_{-p}} - b_{L_{-p}}) \\
&\quad + i\exp(i\Delta_0 t)b_1\Omega_{4^-,1}
\end{aligned}
\tag{57}
$$

where we have made use of Eq. (49) and the notation is a natural extension of that previously introduced [Eq. (48)].

Once again, changing the sign of the electric fields results in a change in dynamics. Hence [13] this ten-level scheme allows for enantiomeric control. Further, as in configuration **A**, changing the sign of either one ε_i field component or of all three field components results in loss of enantioselectivity, whereas if two field components change sign then enantioselectivity is maintained. Qualitative motivation for this result resides in that changing the sign of either one field or three results from reflecting the field through either one plane or three perpendicular planes, respectively. Since each reflection equals inversion followed by rotation, this corresponds to an overall operation with an odd number of inversion steps. This being so, one expects that the effect of the inversions on the electric field would be similar to that of the inversions on the molecule—i.e., the interchange of L and D, which is indeed the case. By contrast, changing the sign of two field components implies reflecting the field through perpendicular planes.

This corresponds to two steps of rotation and inversion. However, since rotations and inversions commute, the result of changing the sign of two field components is equivalent to a rotation of the electric field vector. Such a rotation cannot affect the role of L and D and hence does not affect the enantiomeric selectivity.

E. *M* Averaging in Configuration B

As in configuration **A**, an examination of the angular momentum attributes of these differential equations provides insight into the effects of averaging over M. In this case, however, we can show that in the ten-level case **B** the M and $-M$ terms contribute equally, so that control is maintained after M averaging. To see this, consider how each M: dependent 3-j symbol changes when $M \to -M$. Given Eqs. (51) and (53), we have for the dipole transition matrix element associated with ε_1 that

$$\begin{pmatrix} J' & 1 & J \\ -M & 0 & M \end{pmatrix} \xrightarrow{M \to -M} (-1)^{J'+J+1} \begin{pmatrix} J' & 1 & J \\ -M & 0 & M \end{pmatrix} \tag{58}$$

Similarly, for the term associated with ε_2,

$$\left[\begin{pmatrix} J' & 1 & J \\ -(M+1) & 1 & M \end{pmatrix} - \begin{pmatrix} J' & 1 & J \\ -(M-1) & -1 & M \end{pmatrix} \right]$$

$$\xrightarrow{M \to -M} (-1)^{J'+J} \left[\begin{pmatrix} J' & 1 & J \\ -(M+1) & 1 & M \end{pmatrix} - \begin{pmatrix} J' & 1 & J \\ -(M-1) & -1 & M \end{pmatrix} \right] \tag{59}$$

and for the term associated with ε_0,

$$\left[\begin{pmatrix} J' & 1 & J \\ -(M+1) & 1 & M \end{pmatrix} + \begin{pmatrix} J' & 1 & J \\ -(M-1) & -1 & M \end{pmatrix} \right]$$

$$\xrightarrow{M \to -M} (-1)^{J'+J+1} \left[\begin{pmatrix} J' & 1 & J \\ -(M+1) & 1 & M \end{pmatrix} + \begin{pmatrix} J' & 1 & J \\ -(M-1) & -1 & M \end{pmatrix} \right] \tag{60}$$

As in configuration **A**, depending on the values of J, some or all of the transition dipole moments may or may not change sign. Consider then the excitation scheme where all J quantum numbers are the same. It is easily seen from Eqs. (58), (59), and (60) that $M \to -M$ has the same effect as $\varepsilon_1 \to -\varepsilon_1$, $\varepsilon_0 \to -\varepsilon_0$, and $\varepsilon_2 \to \varepsilon_2$. Therefore, from previous discussions, both M and $-M$ contribute equally, and control is maintained with M averaging. Similar results obtain for the case where two levels are of similar J and the third is of angular momentum $J \pm 1$. Hence enantiomeric control survives M averaging in configuration **B**. Note that the same M considerations are applicable to the asymmetric top.

F. Computational Tests: Laser Distillation

The results above indicate that M averaged control is possible for configuration **B**, assuming that the fields significantly couple only the ten indicated levels. A generalized, albeit computational, extension of these results to include all M levels is provided in this section. Specifically, we show that the full system of states that are coupled by three perpendicular linearly polarized lasers shows control, even upon M averaging. Note, however, that numerical studies do show that the ten-level approximation is, in most cases, an excellent approximation.

To outline the computation associated with laser distillation we describe, for notational simplicity, the case where there is no p degeneracy (e.g., the asymmetric top). Computations below are, however, for the symmetric top, where the p degeneracy is added as noted later below. Consider the system initiated in a racemic mixture of $|L, M\rangle$s and $|D, M\rangle$ of fixed J. Here we have explicitly indicated the M quantum number of the ground state. In the first step, the system is excited with a laser pulse sequence as described above. In the second step, the system collisionally and radiatively relaxes so that all the population returns to the ground state to produce an incoherent mixture of $|L, M\rangle$s and $|D, M\rangle$s. This pair of steps is then repeated until the populations of $|L, M\rangle$ and $|D, M\rangle$ reach convergence.

To obtain the equilibrium result computationally, we repeat the procedure of Eq. (36) to Eq. (39) while explicitly taking into account the M averaging. We note that the population after laser excitation, but before radiative relaxation, consists of the weighted sum of the results of $2 \times (2J + 1)$ computations: $N_D/(2J + 1)$ times the results of laser excitation starting solely with molecules in $|D, M\rangle$ for $M = -J, \ldots, J$, plus $N_L/(2J + 1)$ times the results of laser excitation starting solely with molecules in $|L, M\rangle$ for $M = -J, \ldots, J$.

Let $P_{D \leftarrow D}^{M,M_1}$ and $P_{L \leftarrow D}^{M,M_1}$ denote the probabilities of $|D, M\rangle$ and $|L, M\rangle$ resulting from laser exciltation, assuming that the initial system is composed solely of molecules in $|D, M_1\rangle$; and let $P_{D \leftarrow L}^{M,M_1}$ and $P_{L \leftarrow L}^{M,M_1}$ be the corresponding probabilities for excitation from an initial system composed solely of $|L, M_1\rangle$.

The populations of $|D, M\rangle$ and $|L, M\rangle$ after laser excitation of the mixture are

$$\frac{1}{2J+1} \sum_{M_1} \left[N_D P_{D \leftarrow D}^{M,M_1} + N_L P_{D \leftarrow L}^{M,M_1} \right] \tag{61}$$

and

$$\frac{1}{2J+1} \sum_{M_1} \left[N_D P_{L \leftarrow D}^{M,M_1} + N_L P_{L \leftarrow L}^{M,M_1} \right] \tag{62}$$

The total D and L populations are therefore $N_D P_{D \leftarrow D} + N_L P_{D \leftarrow L}$ and $N_L P_{D \leftarrow L} + N_L P_{L \leftarrow L}$, respectively, where we have denoted

$$P_{q \leftarrow q'} \equiv \frac{1}{2J+1} \sum_{M_1, M} P_{q \leftarrow q'}^{M, M_1} \quad q, q' = L, D \tag{63}$$

The remainder of the population,

$$N^{ex} = N_D [1 - P_{D \leftarrow D} - P_{L \leftarrow D}] + N_L [1 - P_{D \leftarrow L} - P_{L \leftarrow L}] \tag{64}$$

is in the upper levels $|1, M\rangle$ and $|2, M\rangle$, $M = -J - 1, \ldots, J + 1$.

Radiative emission, or collision relaxation, from these excited levels then follows, with the excited population dividing itself equally between the $2 \times (2J + 1) |D, M\rangle$ and $|L, M\rangle$ states. The effect of collisions is also such as to equilibrate the population among the M states. The resultant M summed populations N_D and N_L in ground states $|D\rangle$ and $|L\rangle$ can then be computed as in Eqs. (37) to (39).

Note that although we focus upon a single J level, the scenario extends to the entire set of equilibrated J levels. Specifically, as the laser-targeted J level is enantiomerically enriched, the other J levels feed population into the depleted J level during the collisional relaxation step, so that the entire system ultimately becomes enantiomerically enriched.

In the case of the symmetric top, explicitly computed below, the above argument is extended by introducing the p degeneracy. Thus, for example, the system is initially composed of a racemic mixture of states $|L, M, p\rangle$ $|D, M, p\rangle$ with $M = -J, -J + 1, \ldots J - 1, J, p = 1$ and $p = -1$. Independent computations are done starting with each of the $2 \times 2 \times (2J + 1)$ independent initial states, which serve as input into the total calculation.

We consider the following model for computational purposes: an initial state composed of an unpolarized racemic mixture of states is irradiated with pulses having Gaussian envelopes,

$$\varepsilon_l = \beta_l e^{i\theta l} e^{-[(t-t_l)/\alpha_l]^2} \tag{65}$$

The values of the vibrational part of the transition dipole (see appendix) were taken as, $d^x = 0.9$, $d^y = 1.0$, and $d^z = 1.1$ for all transitions. We have chosen $J = 1$ $\lambda = 0$, $\lambda' = 1$, and $\lambda'' = 1$. In this case the ground states only have $p = 1$ allowed.

Results subsequent to laser excitation were obtained by numerically solving the full set of coupled levels for all relevant M and p states. The excess after a single pulse is given as the differences between the D and L populations in an initial racemic mixture where each of the allowed states is equally populated. The converged results pertain to the results after iteratively exciting and relaxing the system as described above. The particular results shown below were obtained by optimizing the converged excess using a simulated annealing routine [90] with respect to the following free parameters of the system: the field amplitudes β_l, phases θ_l, pulse peak times t_l, durations α_l, and detunings Δ_l.

Figure 15 shows the populations of the D and L enantiomers after a single pulse (and before relaxation) and after a convergent sequence of pulses followed by relaxation for a number of different parameters. In the cases shown, all variables other than that explicitly varied along the abscissa are those for which the converged excess reached a maximal (or minimal) value. Note that many such maxima and minima with values close to 1 (or -1) occur in this multidimensional parameter space.

In Fig. 15a the phase θ_0 of the field ε_0 is varied, whereas in 15b the amplitude β_0 is varied. As can be seen, a maximal converged excess of close to unity

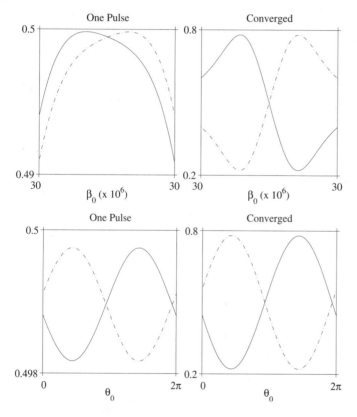

Figure 15 Left, enantiomeric excess after one pulse (and before relaxation); right, the converged degree of handedness when we vary (a) the phase θ_0 and (b) the amplitude β_0. The values of the other parameters are $J = 1$, $\lambda = 0$, $\lambda' = 1$, $\lambda'' = 1$, $\alpha_0 = 7.66$ ps, $\alpha_1 = 10.51$ ps, $\alpha_2 = 11.21$ ps, $\beta_1 = 3.98*$ 10^{-6} a.u., $\beta_2 = 5.00 * 10^{-6}$ a.u., $t_0 = 0.0$ ps, $t_1 = -23.24$ ps, $t_2 = -2.06$ ps, $\theta_1 = 3.99$, $\theta_2 = 5.95$, $\Delta_1 = 1.43$ cm^{-1}, $\Delta_2 = -2.11$ cm^{-1}, and (a) $\beta_0 = 1.28 \times 10^{-6}$, (b) $\theta_0 = 1.39$. Note the different ordinate scales.

can be attained, even though the single-pulse selectivity is very small. This is because at the end of the pulse one of the enantiomers is almost unaffected, while the other enantiomer is somewhat depleted into the excited states.

Computations using configuration **B** were also carried out with various models of dimethylallene. The simplest results for the three-laser case with perpendicular polarizations are shown in Fig. 16, first row, where extensive control is evident. Here, even with M averaging, one can choose to convert the racemic mixture to over 90% of the L enantiomer, or of the D enantiomer, depending on

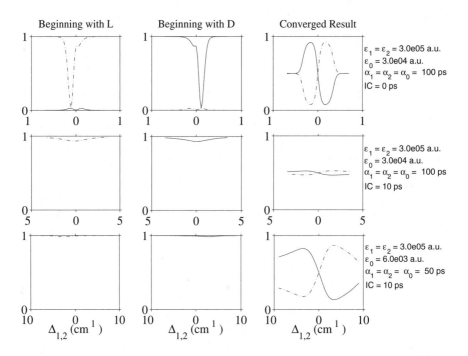

Figure 16 Control over dimethylallene enantiomer populations as a function of the detuning Δ_1 for various laser powers. The first column corresponds to probabilities of L (dot-dash curves) and D (solid curves) after a single laser pulse, assuming that the initial state is all L. The second column is similar, but for an initial state which is all D. The rightmost column corresponds to the probabilities L and D after repeated excitation–relaxation cycles, as described in the text. The first row corresponds to control using the laser parameters on the extreme right, in which there is no internal conversion; the second row uses the same laser parameters as does the first row, but with an internal conversion time of 10 ps; the bottom row shows results for an internal conversion time of 10 ps, but with the modified laser parameters shown.

the detuning. In this case the 1,3-dimethylallene was treated as an asymmetric top, and averaging was carried out over all M levels using $J = 1$.

A realistic model of dimethylallene control must also recognize the possibility of internal conversion to the ground state. In this process, the electronically excited molecule undergoes a radiationless transition to the ground electronic state, leaving a highly vibrationally excited species. Only a few estimates or measurements of the internal conversion time scales for molecules are available [81,82], and dimethylallene has not been explored. Further, after internal conversion one expects, in the dimethylallene case, that the excited molecule subsequently dissociates, leaving molecular fragments that no longer participate in the control scenario. Hence the process of internal conversion serves as a decoherence mechanism that can reduce control. Further decoherence effects, but on a slower time scale, would arise, for example, if the control was carried out in solution.

Results on a preliminary study are shown in Fig. 16. Specifically, the second row shows control with similar parameters as in the first row, but in the presence of a T_2 relaxation time associated with internal conversion chosen arbitrarily as 10 ps. Clearly, almost all of the control is lost. However, if the laser parameters are changed to those shown on the right hand side of the figure, bottommost row, then significant control is restored once again. In this case, however, the process occurs with the loss of considerable reactant population to dissociated dimethylallene. A more realistic picture of dimethylallene control [88] recognizes that collisional effects can *aid* control insofar as they stabilize (by inducing transitions to low-lying vibrational states of the ground electronic state) molecules which have undergone internal conversion from the excited electronic state to highly vibrationally excited states of the ground electronic state. Hence, collisions can deactivate these excited molecules prior to dissociation. In doing so, these excited molecules are returned to the pool of ground state molecules that can be enantiomerically purified.

Results of a computation that optimizes collision rates are shown in Fig. 17. Here we show populations of D and L after a converged set of laser distillation steps, assuming an internal conversion time of 10 ps. The abscissa consists of the detuning of ω_1 from the transition between $|E_1\rangle$ and $|D\rangle$ and of ω_2 from the transition between $|E_2\rangle$ and $|D\rangle$, denoted $\Delta_{A,S}$. In the first case, Fig. 17A, excellent control is achieved for a collision rate that is $\approx 10^3$ times faster than that of dissociation, leading to a small population loss of 0.5%. For example, a positive detuning of $\Delta_{A,S} = 3.478$ cm^{-1} will result in a 75.9% enhancement of the D enantiomer; similarly, a negative detuning of $\Delta_{A,S} = -3.112$ cm^{-1} will result in the increase of the L enantiomer to 72.3%. Alternatively, keeping the same laser parameters, a reduction in the collision rate will reduce the decoherence and consequently increase the control, as seen in Fig. 17B. Indeed, a detuning of $\Delta_{A,S} = 3.295$ cm^{-1} will result in a control of 82.06% of the D enantiomer, whereas a detuning of $\Delta_{A,S} = -2.929$ cm^{-1} will augment the L enantiomer to

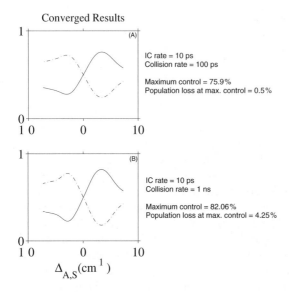

Converged Results

Figure 17 Control over dimethylallene enantiomer populations with maximum control and minimal population loss. The D population is represented by a solid curve and the L by a dot-dash curve. In both cases, the laser parameters are the same with pulse widths of $\alpha_0 = 75$ ps, $\alpha_1 = 31$ ps, and $\alpha_2 = 31.5$ ps, field strengths of $\beta_0 = 1.00 \times 10^{-3}$ a.u., $\beta_1 = 3.401 \times 10^{-6}$ a.u., and $\beta_2 = 8.08 \times 10^{-6}$ a.u., and central frequencies of $\omega_0 = 261.8$ cm^{-1}, $\omega_1 = 50{,}989$ cm^{-1}, and $\omega_2 = 50{,}727$ cm^{-1}. (From Ref. 88.)

77.5%. However, the cost of this extra control comes from an increase in population loss, which is now 4.25%.

VI. ENANTIOMERIC CONTROL VIA STRONG FIELD ORIENTATION OF MOLECULES

An alternative way to introduce handedness into the interaction of matter with light using linearly polarized light has been introduced by Manz, Fujimura et al. [14–17]. In this approach one first preorients the racemic mixture of D and L along some axis. Under these circumstances, there is a difference in the direction of the transition dipole moments of the left: and right-handed enantiomers. That is, matrix elements like $\langle E_i | \mathbf{d} \cdot \hat{\varepsilon}_k | E_j \rangle$ are different for the two enantiomers $i = D$ or $i = L$. This distinction between L and D suffices to allow for the possibility of control over enantiomers.

As a simple example, consider the model [15] shown in Fig. 18. The system is initially in a mixture of the ground vibrational state of D and L. The pump laser carries the system to a single excited vibrational state of the excited electronic state, and the dump laser returns the system to an excited vibrational state of the ground electronic state. This then is a pump–dump scenario, but transitions are solely between bound states.

To appreciate the essence of this control scenario, recall the results of applying a laser pulse $\varepsilon(t)$ to induce a transition between two bound states $|E_i\rangle$ and $|E_j\rangle$. We denote the dipole transition matrix element between these two states by $d_{i,j}$ and define $\kappa_{i,j} = 2d_{i,j}/\hbar$. Then it is well known [89] that complete population transfer between these levels can be accomplished by using a π pulse, i.e., a pulse of duration t satisfying

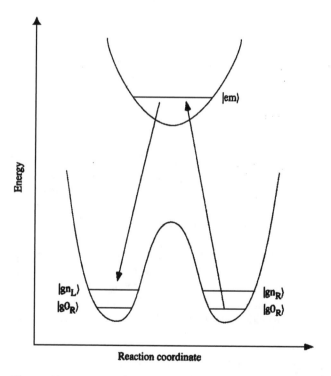

Figure 18 Model system for enantiomer control in accord with Ref. 15. The notation is such that $|gn_R\rangle$ denotes the D enantiomer on the ground electronic state in vibrational state n, $|gn_L\rangle$ is the analogous state of the L enantiomer, and $|em\rangle$ is the mth level of the excited electronic state. (From Fig. 1, Ref. 15.)

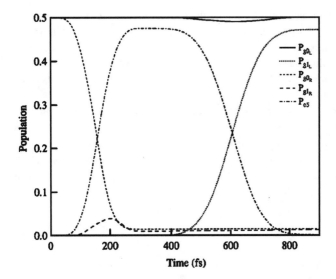

Figure 19 Sample computation of control in the pump–dump scenario for controlling handedness. Populations shown are defined in the upper right-hand corner of the figure. For example, P_{g0R} denotes the population of $|g0R\rangle$, etc. (From Fig. 4, Ref. 15.)

$$\int_{-\infty}^{t} \kappa_{i,j}\, \varepsilon(t')dt' = \pi \tag{66}$$

Consider then the scenario in Fig. 18, and suppose that we wish to transfer population from L to D. Then, since $\langle E_i\, |\mathbf{d} \cdot \hat{\varepsilon}_k| E_j\rangle$ differ for the L and D states, we can choose a laser polarization such that this matrix element is zero for excitation of the ground state of D but not L. Application of a π pulse at this polarization will then transfer the ground state L population to the excited state. Application of a second π pulse of different polarization can then transfer this population to the excited vibrational state of the ground electronic state of D, by now choosing a polarization that does not couple the excited state to L.

Sample results for the control over the oriented enantiomers of H$_2$POSH are shown in Fig. 19. Clearly, as proposed, the method is very effective. In this case the primary experimental challenge is to orient the system prior to irradiation.

VII. CHIRAL SEPARATION BY CYCLIC POPULATION TRANSFER

Rather than going through many excitation–relaxation cycles as in the laser distillation scenario, described in Sec. IV, it is possible to affect the enantio-selectivity by a single application of two or three laser pulses of different carrier frequencies. This can be done by inducing an adiabatic passage (AP) between levels, a phenomenon discussed extensively in the optics literature [91–94], according to which by using strong enough lasers that vary sufficiently slowly one can make the system "adiabatically" follow the field-dressed states and execute a complete population transfer between quantum states. In the particular realization of AP, called stimulated rapid adiabatic passage (STIRAP) [92,93], population in state $|1\rangle$ is transferred to state $|3\rangle$ by a sequence of two one-photon transitions, using as an intermediate state $|2\rangle$. The method has been applied to atomic and molecular systems [92,93], as well as to quantum dots [94].

Ordinary STIRAP is only sensitive to the energy levels and the *magnitudes* of transition-dipole coupling matrix elements between them. These quantities are identical for enantiomers. Its insensitivity to the *phase* of the transition-dipole matrix elements renders STIRAP incapable of selecting between enantiomers. Recently we have demonstrated [11] that precisely the *lack of inversion center*, which characterizes chiral molecules, allows us to combine the weak-field one- and two-photon interference control method [29,54,95,96] with, the strong-field STIRAP to render a phase-sensitive AP method. In this method, which we termed cyclic population transfer (CPT), one forms a STIRAP "loop" by supplementing the usual STIRAP $|1\rangle \Leftrightarrow |2\rangle \Leftrightarrow |3\rangle$ two-photon process by a one-photon process $|1\rangle \Leftrightarrow |3\rangle$. The lack of inversion center is essential, because one-photon and two-photon processes cannot connect the same states in the presence of an inversion center, where all states have a *well defined parity*, because a one-photon absorption (or emission) between states $|1\rangle$ and $|3\rangle$ requires that these states have opposite parities, whereas a two-photon process requires that these states have the same parity.

Contrary to systems possessing an inversion center in which the interference between a one-photon and a two-photon process can only lead to phase control of *differential* properties, e.g., current directionality [29,54,95,96], we have shown that the CPT process of broken symmetry systems allows us to control *integral* properties as well, a prime example of which is the control of the excited states population of two enantiomers.

The scheme is illustrated in Fig. 20. In addition to chiral molecules, one can apply this method to two asymmetric quantum wells, to two heteronuclear molecules aligned in an external DC electric field [97]. In the setup of Fig. 20 (lower plot), we consider operating on states $|i\rangle$ and their mirror images $|i\rangle_M$ by three pulses in a counterintuitive order [92,93], i.e., *two* "pump" pulses with Rabi frequencies $\Omega_{12}(t)$ and $\Omega_{13}(t)$, which follow a dump pulse $\Omega_{23}(t)$. The Rabi frequencies are defined as, $\Omega_{ij}(t) \equiv d_{ij}\,\varepsilon_{ij}(t)/\hbar = |\Omega_{ij}(t)|e^{i\phi_{ij}} = \Omega^*_{ji}(t)$, where d_{ij} and $\varepsilon_{ij}(t)$ are, respectively, the transition dipoles and the envelopes of electric fields, of carrier frequencies ω_{ij}, operating between states $i \neq j$ ($i, j = 1, 2, 3$). If we symmetrically detune the pulse center frequencies, as shown in Fig. 20,

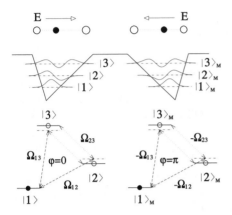

Figure 20 (upper plot) An asymmetric quantum well and its mirror image. Also shown are two field-oriented heteronuclear molecules. (lower plot) Illustration of the three pulses used in these CPT systems. The two systems can be discriminated by their different matter–radiation phases ϕ.

we satisfy the $|1\rangle \Leftrightarrow |2\rangle \Leftrightarrow |3\rangle$ and $|1\rangle \Leftrightarrow |3\rangle \Leftrightarrow |2\rangle$ two-photon resonance condition, while keeping the one-photon processes, $|1\rangle \Leftrightarrow |3\rangle$ and $|1\rangle \Leftrightarrow |2\rangle$, off-resonance. As a result, the loop formed by the three transitions is not resonantly closed. Therefore the one and two-photon processes only interfere at isolated points in time when the pulses are on.

We now consider this scenario in detail by first writing the CPT (radiation + matter) Hamiltonian in the rotating wave approximation as

$$H = \sum_{j=1}^{3} \omega_j |j\rangle \langle j| + \sum_{i>j=1}^{3} (\Omega_{ij}(t)e^{-iw_{ij}t} |i\rangle \langle j| + h.c.)$$

(67)

where ω_j are the energies of the states $|j\rangle$; atomic units ($\hbar = 1$) are used throughout. The system wave function can be written as

$$|\psi(t)\rangle = \sum_{n=1}^{3} c_n(t)e^{-i\omega_n t} |n\rangle$$

(68)

where $\mathbf{c}(t)$, the column vector of the (slow varying) coefficients $\mathbf{c} = (c_1, c_2, c_3)$ can be evaluated from the Schrödinger equation

$$\dot{\mathbf{c}}(t) = -i\mathrm{H}(t) \cdot \mathbf{c}(t)$$

(69)

with $\mathrm{H}(t)$, the effective Hamiltonian matrix, given as

$$H = \begin{bmatrix} 0 & \Omega_{12}^* e^{i\Delta_{12}t} & \Omega_{13}^* e^{i\Delta_{13}t} \\ \Omega_{12} e^{-i\Delta_{12}t} & 0 & \Omega_{23}^* e^{i\Delta_{23}t} \\ \Omega_{13} e^{-i\Delta_{13}t} & \Omega_{23} e^{-i\Delta_{23}t} & 0 \end{bmatrix} \tag{70}$$

where the detunings are defined as $\Delta_{ij} = \omega_i - \omega_j + \omega_{ij} = -\Delta_{ji}$. For brevity, we have omitted writing explicitly the time-dependence of $\Omega_{ij}(t)$.

In contrast to ordinary STIRAP, unless $\Omega \equiv \Delta_{12} + \Delta_{23} + \Delta_{31} = 0$, it is not possible to transform away the rapidly oscillating $e^{-i\Delta_{ij}t}$ components from the CPT Hamiltonian [Eq. (70)]. As a result, the system phase factor varies as $(e^{-i\Sigma t})$ during the time when the three pulses overlap. As a result, in CPT, unless $\Omega = 0$, null states (i.e., states with zero eigenvalue) cease to be so when the pulses overlap. Moreover, owing to nonadiabatic couplings, the population does not follow the adiabatic states during the entire time evolution, migrating at the near crossing region from the initially occupied null state.

We can quantify the above statements by examining the eigenvalues of the Hamiltonian [Eq. (70)], given by

$$E_2 = \frac{2^{1/3} a}{3c} + \frac{c}{3 \, 2^{1/3}}$$

$$E_{1,3} = \frac{-(1 \pm i\sqrt{3})a}{3 \, 2^{2/3} c} - \frac{(1 \mp i\sqrt{3})c}{6 \, 2^{1/3}} \tag{71}$$

where $a = 3 \, (|\Omega_{12}|^2 + |\Omega_{23}|^2 + |\Omega_{31}|^2)$, $b = 3^3 \, \mathrm{Det}(H) = 3^3 \, 2 \, \mathrm{Re} \, \mathbb{O}$ and $c = (b + \sqrt{b^2 + 4 \, (-a)^3})^{1/3}$, with $\mathbb{O} = \Omega_{12} \Omega_{23} \Omega_{31} \, e^{-i\Sigma t}$.

We see that the three eigenvalues depend only on the *overall phase* of $\mathbb{O}$. This phase is composed of a *time-independent* part $\varphi \equiv \phi_{12} + \phi_{23} + \phi_{31}$, of the product of the Rabi frequencies, and a *time-dependent* part Σt. In particular, it follows from Eq. (71) that when $\varphi = \pm \pi/2$ and $\Sigma = 0$, $b = 0$; hence $c = i2^{1/3}a^{1/2}$ and $E_2 = 0$.

In Fig. 21 we present the time dependence of the eigenvalues $E_i(t)$ ($i = 1,2,3$) for three Gaussian pulses parameterized as $|\Sigma_{23}(t)| = \Omega_{max} \exp[-t^2/\tau^2]$, $|\Omega_{12}(t)| = 0.7 \, \Omega_{max} \exp[-(t - t_2)^2/\tau^2]$ and $|\Omega_{13}(t)| = 0.7 \, \Omega_{max} \exp[-(t - t_3)^2/\tau^2]$, with $\Omega_{max} = 30/\tau$, where τ is the pulse duration. The pulse delays are $t_3 = t_2 = 2\tau$, and the detunings, chosen to give maximal selectivity, are $\Delta_{12} = -\Delta_{13} = -\Delta_{23} = 0.08/\tau$. The eigenvalues are presented for the phases $\varphi = 0.235\pi$ (see Fig. 23) and $\varphi = (0.235 + 0.5) \, \pi$.

For the problem defined by the parameters of Fig. 21, $(|c_1|, |c_2|, |c_3|)$, the vector of magnitudes of the expansion coefficients of the $|E_i\rangle$, eigenvectors in the "bare" basis starts in the remote past:

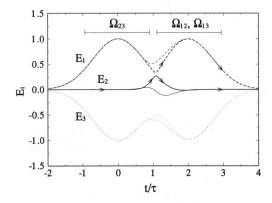

Figure 21 The three dressed eigenvalues E_i (t) at two different phases. The solution for $\phi = 0.235\,\pi$ and $\phi = (0.235 + 0.5)\pi$ is plotted by thick and thin lines, respectively. An initial population at state $|1\rangle$ stays on the null state $|E_2(t)\rangle$ with $E_2(t) \approx 0$ up to the avoided crossing region where the population becomes shared with the eigenstate $|E_1(t)\rangle$ or $|E_3(t)\rangle$, depending on the phase ϕ. The horizontal short lines denote the approximate times of action of the Rabi frequencies Ω_{ij}.

$$\left(|c_1|, |c_2|, |c_3|\right) \xrightarrow{t \to -\infty} (1, 0, 0) \text{ for } |E_2\rangle$$
$$(0, 1, 1)/\sqrt{2} \text{ for } |E_1\rangle \text{ and } |E_3\rangle$$

At the end of the process, we have that

$$\left(|c_1|, |c_2|, |c_3|\right) \xrightarrow{t \to \infty} (0, 1, 1)/\sqrt{2} \text{ for } |E_2\rangle$$
$$(\sqrt{2}, 1, 1)/2 \text{ for } |E_1\rangle \text{ and } |E_3\rangle$$

Since the evolution starts with bare state $|1\rangle$, only the $|E_2\rangle$ eigenstate gets initially populated.

Figure 21 clearly shows that the system evolution is governed by the interference between the "clockwise" ($|1\rangle \rightarrow |3\rangle \rightarrow |2\rangle$) and the "counter clockwise" ($|1\rangle \rightarrow |2\rangle \rightarrow |3\rangle$) two-photon processes depicted in Fig. 20. This interference results in the appearance of an avoided crossing between the $E_2(t)$ eigenvalue and (depending on the phase φ) either the $E_1(t)$ or the $E_3(t)$ eigenvalue. In the crossing region, the adiabatic description ceases to be valid, and the system populates a *superposition state* $\alpha_2 |E_2\rangle + \alpha_i |E_i\rangle$ ($i = 1$ or $i = 3$).

Figure 22 displays the evolution of the populations S_i $(t) \equiv |\langle E_i(t) |\psi(t)\rangle|^2$ of the field dressed states, having started with $|\psi(t = 0)\rangle = |1\rangle$. The parameters are as in Fig. 21, with φ being confined to the 0.235 π value. We see that the eigenstate $|E_2\rangle$ is populated exclusively until the avoided crossing region, where

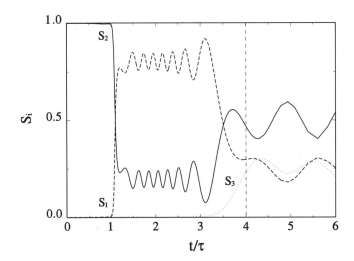

Figure 22   The populations, given as $S_i(t) \equiv |\langle E_i(t)|\psi(t)\rangle|^2$, of the field dressed states, given that $|\psi(t=0)\rangle = |1\rangle$ and $\varphi = 0.235\,\pi$. All other parameters are as in Fig. 2. The thin vertical ($-\cdot-\cdot-\cdot$) line points at the time after which the populations in the bare states $|i\rangle$ roughly cease to vary.

the system goes to the state $\alpha_2|E_2\rangle + \alpha_1|E_1\rangle$. As the pulses wane and all $\Omega_{ij}(t) \to 0$, *nonadiabatic* processes populate also the $|E_3\rangle$ state. The populations of the $|E_1\rangle$ and $|E_3\rangle$ states have roughly the same magnitudes $S_1 \approx S_3$ at the end of the process, as expected from the roughly equal final values of the $|c_1|$, $|c_2|$, $|c_3|$ coefficients shown above. Hence, by varying φ and Σ we can adjust the α_i coefficients so that $\Sigma_i\,\alpha_i|E_i\rangle \xrightarrow{t\to\infty} |2\rangle$ or $|3\rangle$.

An example of the degree of control attainable in this manner is given in Fig. 23, where we display the phase dependence of the final populations p_i of the bare states $|i\rangle$, using the parameters of Figs. 21 and 22. The main feature of Fig. 23 is that the roles of state $|2\rangle$ and state $|3\rangle$ are *reversed* as we translate the phase φ by π. This features serves, as discussed below, to establish the discrimination between left-handed and right-handed chiral systems.

The calculations of Fig. 23 show enhanced sensitivity of the final populations p_i on φ at small detunings Δ_{ij}. The population transfer can be made essentially complete by choosing $\varphi \approx 0.235\,\pi$ (denoted by a small arrow at the bottom of Fig. 23). In that case, 99% of the population is transferred from state $|1\rangle$ to state $|3\rangle$. As the phase φ is shifted by π, the system switches over, with the same efficiency, to the $|1\rangle \to |2\rangle$ population transfer process.

The phase dependence of CPT can be used to discriminate between left- and right-handed chiral systems. Denoting by $|i^+\rangle$ (formerly $|i\rangle$) a given

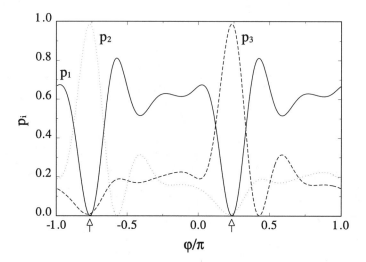

Figure 23 Dependence of p_i, the final populations of the bare states $|i\rangle$ on the phase φ. The two vertical arrows show the phases for the best separation of the chiral systems, where the population is transferred from state $|1\rangle$ to state $|2\rangle$ or $|3\rangle$.

symmetry-broken state and by $|i^-\rangle$ (formerly $|i\rangle_M$) its mirror image, we can write these states in terms of symmetric $|S_i\rangle$ and antisymmetric $|A_i\rangle$ states of the two systems as [5,1]

$$\left|i^\pm\right\rangle = |S_i\rangle \pm |A_i\rangle$$

Because dipole moments can only connect states of opposite parity, we obtain that the Rabi frequencies for transition between different symmetry-broken states $|i^\pm\rangle$ and $|j^\pm\rangle$ are given as

$$\Omega_{ij}^\pm = \pm[\langle S_i|d|A_j\rangle + \langle A_i|d|S_j\rangle]\varepsilon_{ij}$$

We see that the Rabi frequencies between any pair of left and right-handed states differ by a sign, i.e., a phase factor of π. Since in the CPT processes the two enantiomers are influenced by the phase $\varphi^\pm$ of the products $\Omega_{12}^\pm \, \Omega_{23}^\pm \, \Omega_{31}^\pm$, we always have that $\varphi^- - \varphi^+ = \pi$. This property is *invariant* to any arbitrary phase change in the individual wave functions of the states $|i^\pm\rangle$.

It therefore follows from Fig. 23, where a change in π of the phase φ is seen to switch the population transfer process from $|1\rangle \rightarrow |2\rangle$ to $|1\rangle \rightarrow |3\rangle$, and vice versa, that we can affect the transfer of population in one chiral system relative to its mirror image. Because the overall *material phase* $\varphi_s^\pm$ of the product of the dipole matrix elements $d_{12}^\pm \, d_{23}^\pm \, d_{31}^\pm$ is a fixed quantity ($\varphi_s^- - \varphi_s^+ = \pi$), and $\varphi^{\pm} = \varphi_s^{\pm} + \varphi_f$, it is the overall phase φ_f of the *three laser fields* ε_{ij} that acts as the laboratory knob allowing us to determine which population transfer process is experienced by each of the two enantiomers.

The ability of CPT to separate two enantiomers also depends on the individual detuning parameters Δ_{ij} and on the related dynamical phase $2 \Sigma \tau$. At resonance $\Delta_{ij} = 0$ and $\varphi = \pm \pi/2$, the exact null eigenstate $|E_2(t)\rangle$ gives a complete population transfer from state $|1\rangle$ to a combination of states $|2\rangle$ and $|3\rangle$. In that case, the p_2/p_3 branching ratio of the final populations is given, as in the double STIRAP case [98,99], by the $|\Omega_{12}/\Omega_{13}|^2$ ratio, and no enantiomeric selectivity is then possible.

Once each enantiomer has been excited to a different state ($|2\rangle$ or $|3\rangle$), the pair can be physically separated using a variety of energy-dependent processes, such as ionization, followed by ions extraction by an electric field. If we execute the excitation in the IR range and ionize the chosen excited enantiomer after only a few ns delay, losses from fluorescence, whose typical lifetimes in that regime are in the ms range, are expected to be minimal.

VIII. AN OPTICAL "SWITCH" FOR CHIRAL CONVERSION AND PURIFICATION

In Sec. VII we introduced the cyclic population transfer (CPT) [11] method for achieving enantioselectivity. The great advantage of the method is that it leads to the separation of a racemic mixture to two enantiomers by the application of only three laser pulses. Another advantage is that all the optical transitions can take place on the ground electronic surface, thereby avoiding the disruptive competing processes, such as dissociation and internal conversion, discussed in Sec. V.F, that abound in the UV-excitation-dependent laser distillation scheme [70]. The disadvantage of the CPT scheme is its great sensitivity to small variations of the laser parameters. Moreover, in contrast to the laser distillation of Sec. IV, CPT is a separation, not a *purification* scheme: it does not convert one enantiomer to another. Thus a mixture of two enantiomers in state $|1\rangle$ can be separated into a mixture in which one enantiomer is excited to state $|2\rangle$, while its mirror image is transferred to state $|3\rangle$.

A scheme that effectively overcomes both these drawbacks has been recently suggested [12]. It is composed of two steps, one called enantiodiscriminator, which is a robust variant of the CPT process in which we excite just one enantiomer, while leaving the other in its ground state. In the second step, called enantioconverter, consisting of a phase-sensitive population transfer process [102], the particular enantiomer excited by the enantiodiscriminator is converted to its mirror image form. Thus, in principle, in just two steps, an entire racemic mixture of (nonoriented) chiral molecules can be turned into a single preselected enantiomer.

The general scenario described above has been computationally demonstrated [12] for the realistic case of the (transiently chiral) D_2S_2 molecule, shown schematically in Fig. 24. The molecule has six vibrational degrees of freedom, the large amplitude torsional motion of the D atoms about the S – S bond being the most relevant for our purposes. Based on a thorough *ab initio* electronic structure calculation for this molecule [12], it was determined that there are two energetic barriers for enantiomutation, one a cis barrier, roughly 2800 cm^{-1} in height, the other a trans barrier, 1900 cm^{-1} in height.

A one-dimensional cut of the ground electronic potential energy surface along the enantiomutative path involving the trans barrier is shown in Fig. 24. The tunneling splitting of the lowest torsional states ($v_t = 0, 1$), of which there is no experimental information, were calculated [12,103] to be very small, 10^{-9} cm^{-1} and $2 \cdot 10^{-7}$ cm^{-1}, respectively. These values correspond to enantiomeric lifetimes of 33 and 0.165 ms, respectively. Thus, although D_2S_2 is not a chiral molecule in the conventional sense, molecular configurations described by superpositions of the lowest torsional states, localized in one minimum of the double well potential, stay chiral for sufficiently long times to be detectable.

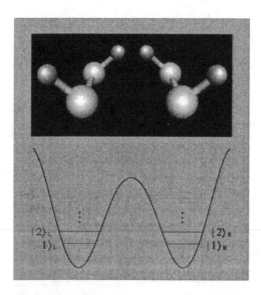

Figure 24   The double well potential energy for the torsional motion of the D_2S_2 molecule. The two enantiomers connected by this enantiomutation are shown above the respective wells.

The same scenario has also been applied to the chiral purification of dimethylallene [104], as indicated later below.

A. Description of the System

Using the same formulation of the Hamiltonian as in Sec. VII [specifically Eqs. (67)–(70)], the two-step process makes use of five pairs of rovibrational states (specified explicitly below). The vibrational eigenstates correspond to the combined torsional and S–D asymmetric stretching modes. The rotational eigenfunctions are the parity-adapted symmetric top wave functions. Each eigenstate has additionally an S/A label denoting its symmetry with respect to inversion. Within the pairs used, the observable chiral states are composed as

$$|k\rangle_{L,D} = \frac{1}{\sqrt{2}}(|k\rangle_S \pm |k\rangle_A) \quad k = 1,\ldots,4 \tag{72}$$

The $|k\rangle_L \Leftrightarrow |k\rangle_D$ interconversion accompanying the tunneling occurs within $\tau_s \approx$ 33, 3.3, 0.165 ms for $k = 1, 2, 3$, respectively, and $\tau_s \approx 0.05$ μs for $k = 4$. The duration of the pulses τ is typically much less than the interconversion times for the four cases, i.e., $\tau \ll \tau_s$, thus justifying the L and D assignments. The higher lying $|5\rangle_S$ and $|5\rangle_A$ states are separated by $\Delta E_{S,A}^5 = 0.38$ cm^{-1}, for which $\tau_s \approx$ 0.1 ns. A 1 ns pulse will therefore address either the $|5\rangle_S$ or the $|5\rangle_A$ states, which are now the physically meaningful ones. Thus pulse durations $\tau \sim 1$ ns and Rabi frequencies $\Omega \approx 20/\tau$ satisfy all the requirements for the scheme described below.

B. The Enantiodiscriminator

A three-level enantiodiscriminator is shown in the upper panel of Fig. 25. Assuming that we start with a mixture of the $|1\rangle_D$ and $|1\rangle_L$ states, the task of the discriminator is to transfer selectively one enantiomer to the $|3\rangle$ state and to keep the other in the $|1\rangle$ state. This is not an easy task, because owing to the near degeneracy of the $|i\rangle_L$ and $|i\rangle_D$ levels ($i = 1, 2, 3$), the driving light field $\mathbf{E}(t)$ is expected to excite simultaneously the resonant $|i\rangle_{L,D} \Leftrightarrow |j\rangle_{L,D}$, $i \neq j = 1, 2, 3$, transitions of *both* enantiomers [105].

The effective Hamiltonian matrix of the enantiodiscriminator is given by

$$H(t) = \begin{bmatrix} 0 & \Omega_{12}^*(t) & \Omega_{13}^*(t) \\ \Omega_{12}(t) & 0 & \Omega_{23}^*(t) \\ \Omega_{13}(t) & \Omega_{23}(t) & 0 \end{bmatrix} \tag{73}$$

As in the above, ϕ_{ij}, the phases of the Rabi frequencies $\Omega_{ij}(t)$, are given as $\phi_{ij} = \phi_{ij}^d + \phi_{ij}^E$, where ϕ_{ij}^d are the phases of the dipole matrix elements d_{ij}, and ϕ_{ij}^E are the phases of the the electric field components ε_{ij}. The total phase φ of the

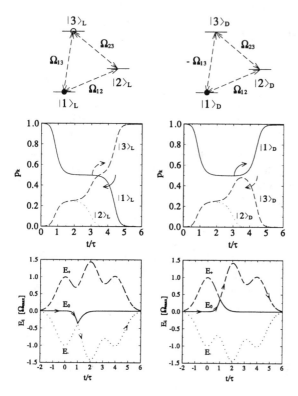

Figure 25 (upper plot) A schematic plot of the enantiodiscriminator. The three levels of each enantiomer are resonantly coupled by three fields. The dipole moments of the two enantiomers have opposite signs. (middle plot) The time evolution of the population of the three levels. The D and L enantiomers start in the $|1\rangle$ state. At the end of the process one enantiomer is found in the $|3\rangle$ state and the other in the $|1\rangle$ state. (lower plot) The time-dependence of the eigenvalues of the Hamiltonian of Eq. (73). The population initially follows the $|E_0\rangle$ dark state. At $t \approx \tau$ the population crosses over diabatically to $|E_-\rangle$ for one enantiomer and to $|E_+\rangle$ for the other.

coupling terms is defined as $\varphi \equiv \phi_{12} + \phi_{23} + \phi_{31}$. It turns out the outcome of the evolution depends mainly on this phase.

We now outline the particular implementation of CPT that forms the basis of the enantiodiscriminator. We denote by $|E_-\rangle$, $|E_0\rangle$, and $|E_+\rangle$ the (adiabatic) eigenstates of the Hamiltonian of Eq. (73). It is easy to show that if we choose $\varphi = 0$ or $\varphi = \pi$, two of these eigenstates become degenerate, at a specific time point $t = \tau$ for which $|\Omega_{12}| = |\Omega_{13}| = |\Omega_{23}|$. Specifically, at $t = \tau$, for $\varphi = 0$, $E_+ = 2\Omega$ and $E_- = E_0 = 2\Omega \cos\left(\dfrac{2\pi}{3}\right)$

for $\varphi = \pi$, $E_- = -2\Omega$ and $E_+ = E_0 = -2\Omega \cos\left(\dfrac{2\pi}{3}\right)$

Thus (as shown in the lower panel of Fig. 25), at $t = \tau$ there is an *exact* crossing between two eigenvalues, the identity of which depending on the overall phase, hence on the enantiomer.

We utilize this fact in the following way: As in the CPT scheme (Sec. VII), the excitation process is envisioned to start with a dump pulse $\varepsilon_{23}(t) = \varepsilon_{23}^{max} f(t)$ that couples the $|2\rangle$ and $|3\rangle$ states. In the present application we have chosen $f(t) = \exp[-t^2/\tau^2]$ and $\varepsilon_{23}^{max} \equiv \Omega^{max}/d_{23}$, with $\Omega^{max} = 30/\tau$. At this stage of the process all the population resides in the $|E_0\rangle$ adiabatic state. In the second stage of the CPT process we *simultaneously* add two pump pulses $\Omega_{12}(t) = \Omega_{13}(t)$ that couple the $|1\rangle \Leftrightarrow |2\rangle$ and the $|1\rangle \Leftrightarrow |3\rangle$ states. In contrast to the CPT scheme of Sec. VII, all three pulses are chosen now to be in exact resonance with the three transitions of interest. We also choose the phases of the optical fields such that either $\varphi = 0$ or $\varphi = \pi$. The degeneracy between the two states noted above at $t = \tau$, when all three Rabi frequencies assume the same magnitude, results in an exact crossing of the initial adiabatic level $|E_0\rangle$ with either $|E_-\rangle$ or the $|E_+\rangle$, adiabatic states. The situation is depicted in the lower panel of Fig. 25. When an exact crossing occurs, the system goes smoothly through the crossing region and the population gets transferred *diabatically* from $|E_0\rangle$ to either $|E_-\rangle$ or $|E_+\rangle$, depending on whether $\varphi = 0$ or $\varphi = \pi$, i.e., on the identity of the enantiomer (since φ differs by π for the two enantiomers [11]). Therefore by changing the overall phase of the light fields by π we can control which enantiomer is to be excited.

After the crossing is complete (at $t > \tau$), the process becomes adiabatic again, with the population remaining fully confined to either the $|E_-\rangle$ or the $|E_+\rangle$ adiabatic states, depending on the enantiomer. At this stage we slowly switch off the $\varepsilon_{12}(t)$ pulse while making sure that the $\varepsilon_{13}(t)$ field is on. We do so by choosing $\varepsilon_{13}(t)$ to be of the form $\varepsilon_{13}(t) = \varepsilon_{13}^{max} (f(t - 2\tau) + f(t - 4\tau) \exp[-it\Omega^{max} f(t - 6\tau))$. As a result, the zero adiabatic eigenstate $|E_0\rangle$ correlates adiabatically with state $|2\rangle$, while the $|E_\pm\rangle$ states correlate as $|E_\pm\rangle \rightarrow (|1\rangle \pm |3\rangle)/\sqrt{2}$.

The *chirp* in the second term of $\varepsilon_{13}(t)$, i.e., $(\varepsilon_{13}^{max} f(t - 4\tau) \exp[-i t \Omega^{max} f(t - 6\tau)])$, is meant to cause a $\pi/2$ rotation in the $[|1\rangle, |3\rangle]$ subspace at $t \approx 5\tau$. Specifically, by introducing the chirp into the effective Hamiltonian of Eq. (73) in this subspace,

$$H_{rot}(t) = \Omega^{max} \begin{bmatrix} \dfrac{f(t-6\tau)}{2} & f(t-4\tau) \\ \dfrac{}{2} & -f(t-6\tau) \\ f(t-4\tau) & \dfrac{}{2} \end{bmatrix} \tag{74}$$

we can cause the $|E_+\rangle$ state to go to state $|3\rangle$ and the $|E_-\rangle$ state to go to state $|1\rangle$, or vice versa, depending on φ. Thus, depending on φ, one enantiomer is made to return to its initial $|1\rangle$ state and the other to switch over to the $|3\rangle$ state. As shown in the middle panel of Fig. 25, the enantiodiscriminator is very robust, with all the population transfer processes occurring in a smooth nonoscillatory fashion.

C. The Enantioconverter

In the next step (called enantioconverter) we selectively convert the enantiomer excited to level $|3\rangle_L$ to its mirror-imaged form. This is achieved using the scheme [102] depicted in Fig. 26 composed of two adiabatic passages (AP),

$$|3\rangle_L \rightarrow |4\rangle_L \rightarrow |5\rangle, \quad \rightarrow |5\rangle \rightarrow |4\rangle_D \rightarrow |3\rangle_D \tag{75}$$

induced by four separated pulses. In the first AP, the population in the $|3\rangle_L$ state is transferred to a superposition state $|5\rangle = \alpha\, e^{-i\omega_S^5 t}\, |5\rangle_S + \beta\, e^{-i\omega_A^5 t}\, |5\rangle_A$ by simultaneously using two $\Omega_{45S}(t)$, $\Omega_{45A}(t)$ (dump) pulses of duration $\tau \gg (\omega_S^5 - \omega_A^5)^{-1}$, followed by a single $\Omega_{34}(t)$ (pump) pulse. In the second step population is transferred from the superposition state to the $|3\rangle_D$ state, by reversing the order of the pulses and shifting the phase of $\Omega_{45S(A)}(t)$ by π relative to the $\Omega_{45S(A)}(t)$ pulse used in the first AP process. In this manner we guarantee that the system that had started in the $|3\rangle_L$ state ends up in the $|3\rangle_D$ state.

In greater detail, the six-level structure of the converter is described by a coupled STIRAP[106] Hamiltonian matrix,

$$|H(t) = \begin{bmatrix} 0 & \Omega_{34} & 0 & 0 & 0 & 0 \\ \Omega_{34}^* & 0 & \Omega_{45S} & \Omega_{45A} & 0 & 0 \\ 0 & \Omega_{45S}^* & 0 & 0 & -\Omega_{45S} & 0 \\ 0 & \Omega_{45A}^* & 0 & 0 & \Omega_{45A} & 0 \\ 0 & 0 & -\Omega_{45S}^* & \Omega_{45A}^* & 0 & -\Omega_{34}^* \\ 0 & 0 & 0 & 0 & \Omega_{34} & 0 \end{bmatrix}$$

$$\tag{76}$$

The process starts in the $|3\rangle_L$ state corresponding to the initial vector $\mathbf{c}(0) = (c_{3L}, c_{4L}, c_{5S}, c_{5A}, c_{4D}, c_{3D}) = (1,0,0,0,0,0)$, and ends in the $|3\rangle_D$ state with the vector $\mathbf{c}(t_{end}) = (0,0,0,0,0,1)$. The Rabi frequencies $\Omega_{45S}(t) = \Omega^{max}\, (f(t) - f(t - 6\tau))$, $\Omega_{45A}(t) = 0.5\, \Omega^{max}\, (f(t) + f(t - 6\tau))$, $\Omega_{34}(t) = \Omega^{max}\, (f(t - 2\tau) + f(t - 4\tau))$, with $f(t)$ and Ω^{max} the same as in the discriminator, correspond to the above described sequence of two two-photon processes. Notice that in the second process we shift by π the phase of the (pump) pulse in $\Omega_{45S}(t)$, which is centered

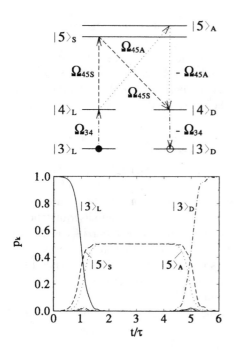

Figure 26 (upper plot) Scheme of the enantioconverter. In the first transfer the population passes from the $|3\rangle_L$, state to the $|5\rangle_{S,A}$ states. In the second transfer, it passes to the $|3\rangle_D$ state. (lower plot) The time-dependence of these two transitions.

at $t = 6\tau$. All transitions can be induced by z linearly polarized light with $d_{34}^L = -d_{34}^R$, $d_{45}^{LS} = -d_{45}^{RS}$, and $d_{45}^{LA} = d_{45}^{RA}$[107].

The evoluation of the p_i populations is shown in the lower plot of Fig. 26. Interestingly, the populations of the $|5\rangle_{S,A}$ states are nearly the same, independently on the ratio $\Omega_{45S}/\Omega_{45A}$ of the Rabi frequencies. This is due to the special symmetry of the six-level scheme [106], with two pairs of degencrate L, D states. More surprising is that the complete population transfer from L to D can occur, in the sequence of two two-photon transitions, although the $|5\rangle_S$, $|5\rangle_A$ states are simultaneously coupled to *both* the $|4\rangle_L$, $|4\rangle_D$ states, in each of these transitions. The reason for this is a destructive interference of the fluxes to the $|4\rangle_D$ and $|4\rangle_L$ states in the first and second process, respectively, due to the opposite *structural* phases of the dipole moments (the Rabi frequencies Ω_{45A}) in the two (L/D) channels. We note that we must complete the selector and converter processes before collisional relaxation of the $|i\rangle$ ($i = 1$–5) states can occur, it being the only relevant relaxation process within the pulse durations chosen.

In addition to D_2S_2, this method was computationally tested for the chiral purification of $J = 1\ 1, 3$ dimethylallene [104], described in Sec. V.F. As in D_2S_2 we use three perpendicular laser polarizations so as to be able to work with unpolarized samples. Here, as distinct from Eq. (75), levels $|3\rangle$ and $|5\rangle$ are coupled as well as $|4\rangle$ and $|5\rangle$. In this case there is no population in level $|5\rangle$ during the time evolution so that level $|5\rangle$, a state chosen as one that would undergo internal conversion, does not suffer such losses.

Sample results are shown in Fig. 27. In this example, the enantiodiscriminator results in the excitation of only 90% of the D enantiomer to the state $|3_D\rangle$ with the remaining population residing in the $|1_D\rangle$, $|2_D\rangle$, and (8%) $|3_L\rangle$ states. The discrimination is not perfect because of M averaging. Since the Clebsch–Gordan coefficients depend upon M, the fields that result in optimal discrimination differ with M, thus leading upon M averaging to an imperfect enantiodiscrimination.

Figure 27 shows the populations of three levels of the ground L enantiomer (denoted P_{iL} for $i = 1,3$) as a function of time in the enantioconverter step. Essentially all of the initial $|3_L\rangle$ is successfully converted to the $|4_D\rangle$, as desired.

In summary, we have presented in this section a two-step optical enantiopurification method, composed of a discriminator and a converter. We have illustrated its action on racemic mixtures of the D_2S_2 and 1,3 dimethylallene mole-

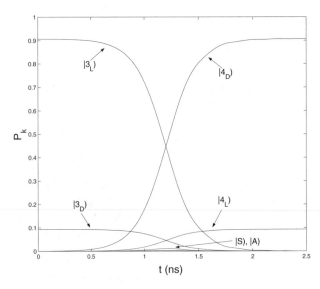

Figure 27 Population of states in dimethylallene as a function of time in the enantioconverter step.

cules. In the discriminator step, the population of *only* one enantiomer of choice is transferred to an excited state. In the converter, the enantiomer excited in the first step is converted to the form of opposite handedness. Thus in just two steps one achieves a transfer of essentially all the population of an enantiomer of a chosen handedness contained in a racemic mixture to its mirror-image form.

ACKNOWLEDGMENTS

We thank E. Frishman, D. Gerbasi, P. Král, and I. Thanopulos for discussions. Support by the U.S. Office of Naval Research is gratefully acknowledged.

REFERENCES

1. See, e.g. Morrison RT, Boyd RN. Organic Chemistry. 6th ed.: Benjamin Cummings, 1992.
2. Barron LD. Molecular Light Scattering and Optical Activity. Cambridge: Cambridge Univ. Press, 1982.
3. Walker DC, ed. Origins of Optical Activity in Nature. Amsterdam: Elsevier, 1979.
4. Barron L.D. Chem. Soc. Rev. 1986, 15, 189. For historical perspectives, see Bel, J.A. Bull. Soc. Chim. Fr. 1874, 22, 337; Van't Hoff, J.H. Die Lagerung der Atome im Raume, 2d ed. Vieweg: Braunschweig, 1894, p. 30.
5. Quack M. Angew Chem. Int. Ed. 1989; 28:571.
6. Salam A, Meath WJ. Chem. Phys. 1998; 228:115.
7. Shapiro M, Brumer P. J. Chem. Phys. 1991; 95:8658.
8. Shapiro M, Frishman E, Brumer P. Phys. Rev. Lett. 2000; 84:1669.
9. Gerbasi D, Shapiro M, Brumer P. J. Chem. Phys. 2001; 115:5349.
10. Deretey E, Shapiro M, Brumer P. J. Phys. Chem. 2001; A105:9509.
11. Král P, Shapiro M. Phys. Rev. Lett. 2001; 87:183002.
12. Král P, Thanopulos I, Shapiro M, Cohen D. Phys. Rev. Lett. 2003; 90:033001.
13. Brumer P, Frishman E, Shapiro M. Phys. Rev. A 2001; 65:015401.
14. See Fujimura Y, Gonalez L, Hoki K, Manz J, Ohtsuki Y. Chem. Phys. Lett. 1999, 306, 1. errata: Chem. Phys. Lett. 1999, 310, 578, for control over an initial superposition state and the following papers for control over a racemic mixture: Fujimura Y, Gonzalez L, Hoki K, Kroener D, Manz J, Ohtsuki Y. Angew Chem. Int. Ed. 2000, 39, 4586. Hoki K, Oht suki Y, Fujimura Y. J. Chem. Phys. 2001, 114, 1575. Hoki K, Kroener D, Manz J. Chem. Phys. 2001, 267, 59. Gonzalez L, Kroner D, Sola IR, J. Chem. Phys. 2001, 115, 2519. Umeda H, Takagi M, Yameda S, Koseki S, Fujimura Y. J. Am. Chem. Soc. 2002, 124, 9265.
15. Hoki K, Gonzalez L, Fujimura Y. J. Chem. Phys. 2002; 116:2433.
16. Ohta Y, Hoki K, Fujimura Y. J. Chem. Phys. 2002; 116:7509.
17. Hoki K, Gonzalez L, Fujimura Y. J. Chem. Phys. 2002; 116:8799.

18. Shapiro M, Brumer P. Principles of the Quantum Control of Molecular Processes. New York: John Wiley, 2003.
19. Rice SA, Zhao M. Optical Control of Molecular Dynamics. New York: John Wiley, 2000.
20. Shapiro M, Brumer P. In. Advances in Atomic, Molecular and Optical Physics Bederson B. , Walther H., Eds, Academic Press. 1999; Vol. 42:287–343.
21. Gordon RJ, Rice SA. Ann. Rev. Phys. Chem. 1997; 48:595.
22. Tannor DJ, Rice SA. J. Chem. Phys. 1985; 83:5013.
23. Brumer P, Shapiro M. Chem. Phys. Lett. 1986; 126:541.
24. Tannor DJ, Kosloff R, Rice SA. J. Chem. Phys. 1986; 85:5805.
25. Shapiro M, Brumer P. Faraday Disc. Chem. Soc. 1987; 82:177.
26. Shi S, Woody A, Rabitz H. J. Chem. Phys. 1988; 88:6870.
26a. Peirce AP, Dahleh MA, Rabitz H. Phys. Rev. A 1988; 37:4950.
27. Shapiro M, Hepburn JW, Brumer P. Chem. Phys. Lett. 1988; 149:451.
28. Tannor DJ, Rice SA. Adv. Chem. Phys. 1988; 70:441.
29. Kurizki G, Shapiro M, Brumer P. Phys. Rev. B 1989; 39:3435.
30. Shi S, Rabitz H. Chem. Phys. 1989; 139:185.
31. Kosloff R, Rice SA, Gaspard P, Tersigni S, Tannor DJ. Chem. Phys. 1989; 139: 201.
32. Baranova BA, Chudinov AN, Zel'dovitch B, Ya B. Optics Comm. 1990; 79:116.
33. Peirce AP, Dahleh MA, Rabitz H. Phys. Rev. A 1990; 42:1065.
34. Jakubetz W, Manz J, Schreier H-J. Chem. Phys. Lett. 1990; 165:100.
35. Muller HG, Bucksbaum PH, Schumacher DW, Zavriyev A. J. Phys. B. 1990; 23: 2761.
36. Chen C, Yin Y-Y, Elliott DS. Phys. Rev. Lett. 1990; 64:507.
36. Phys. Rev. Lett. 1990; 65:1737.
37. Park SM, Lu S-P, Gordon RJ. J. Chem. Phys. 1991; 94:8622.
38. Shi S, Rabitz H. Comp. Phys. Comm. 1991; 63:71.
39. Potvliege RM, Smith PHG. J. Phys. B. 1992; 25:2501.
40. Schafer KJ, Kulander KC. Phys. Rev. A 1992; 45:8026.
41. Lu S-P, Park SM, Xie Y, Gordon RJ. J. Chem. Phys. 1992; 96:6613.
42. Brumer P, Shapiro M. Ann. Rev. Phys. Chem. 1992; 43:257.
43. Bandrauk AD, Gauthier J-M, McCann JF. Chem. Phys. Lett. 1992; 200:399.
44. Judson RS, Rabitz H. Phys. Rev. Lett. 1992; 68:1500.
45. Shapiro M, Brumer P. Chem. Phys. Lett. 1993; 208:193.
46. Ivanov MYu, Corkum PB, Dietrich P. Las. Phys. 1993; 3:375.
47. Charron E, Guisti-Suzor A, Mies FH. Phys. Rev. Lett. 1993; 71:692.
48. Warren WS, Rabitz H, Dahleh M. Science 1993; 259:1581.
49. Yan Y, Gillian RE, Whitnell RM, Wilson KR. J. Phys. Chem. 1993; 97:2320.
49. Krause JL, Whitnell RM, Wilson KR, Yan Y, Mukamel S. J. Chem. Phys. 1993; 99:6562.
50. Shapiro M, Brumer P. Int. Rev. Phys. Chem. 1994; 13:187.
51. Kleiman VD, Zhu L, Li X, Gordon RJ. J. Chem. Phys. 1995; 102:5863.
52. Zhu L, Kleiman VD, Li X, Lu S, Trentelman K, Gordon RJ. Science 1995; 270: 77.

53. Kohler B, Krause JL, Raski F, Wilson KR, Yakovlev VV, Whitnell RM, Yan Y. Acct. Chem. Res. 1995; 28:133.

54. Dupont E, Corkum PB, Liu HC, Buchanan M, Wasilewski ZR. Phys. Rev. Lett. 1995; 74:3596.

55. Sheehy B, Walker B, DiMauro LF. Phys. Rev. Lett. 1995; 74:4799.

56. Papanikolas JM, Williams RM, Kleiber PD, Hart JL, Brink C, Price SD, Leone SR. J. Chem. Phys. 1995; 103:7269.

57. Shapiro M, Brumer P. Trans. Faraday Soc. 1997; 93:1263.

58. Zhu L, Suto K, Fiss JA, Wada R, Seideman T, Gordon RJ. Phys. Rev. Lett. 1997; 79:4108.

59. Wang X, Bersohn R, Takahashi K, Kawasaki M, Kim HL. J. Chem. Phys. 1996; 105:2992.

60. Bardeen CJ, Yakovlev VV, Wilson KR, Carpenter SD, Weber PM, Warren WS. Chem. Phys. Lett. 1997; 280:151.

61. Yelin D, Meshulach D, Silberberg Y. Optics Lett. 1997; 22:1793.

62. Assion T, Baumert T, Bergt M, Brixner T, Kiefer B, Seyfried V, Strehle V, Gerber G. Science 1998; 282:919.

63. Gordon RJ, Zhu LC, Scideman T. Acct. Chem. Res. 1999; 32:1007.

64. Shapiro M. Adv. Chem. Phys. 2000; 114:123.

65. Uberna R, Amitay Z, Qian CXW, Leone SR. J. Chem. Phys. 2001; 114:10311.

66. Shapiro M, Brumer P. J. Phys. Chem. 2001; 105:2897.

67. Levis RJ, Menkir GM, Rabitz H. Science 2001; 292:709.

68. Herek JL, Wohlleben W, Cogdell RJ, Zeidler D, Motzkus M. Nature 2002; 417: 533.

69. Maierle CS, Harris RA. J. Chem. Phys. 1998; 109:3713.

70. Shapiro M, Frishman E, Brumer P. Phys. Rev. Lett. 2000; 84:1669.

71. Gerbasi D, Shapiro M, Brumer P. J. Chem. Phys. 2001; 115:5349.

72. Shirley JH. Phys. Rev. B 1965; 138:979.

73. Brown A, Meath WJ. J. Chem. Phys. 1998; 109:9351.

74. Shapiro M, Frishman E, Brumer P. Enantiomeric Purification of Non-polarized Racemic Mixtures with Coherent Light. J. Chem. Phys.. submitted.

75. Apolonski A, Poppe A, Tempea G, Spielmann C, Udem T, Holzwarth R, Hansch TW, Krausz E. Phys. Rev. Lett. 2000; 85:740.

76. Raman C, Weinacht TC, Bucksbaum PH. Phys. Rev. A 1997; 55:R3995.

77. Hollas JM. High Resolution Spectroscopy. London: Butterworths, 1982.

78. On the preparation and measurement of a superposition of chiral states, see also Harris RA, Shi Y, Cina JA, J. Chem. Phys. 1994, 101, 3459. Cina JA, Harris RA. J. Chem. Phys. 1994, 100, 2531>.

79. Shapiro M. J. Chem. Phys. 1972; 56:2582.

80. Segev E, Shapiro M. J. Chem. Phys. 1980, 73, 2001; J. Chem. Phys, 1982.

81. Hayashi M, Mebel AM, Liang KK, Lin SH. J. Chem. Phys. 1993; 108:2044.

82. Lochbrunner S, Schults T, Schmitt M, Shaffer JP, Zgierski MZ, Stolow A. J. Chem. Phys. 2001; 114:2519.

83. Brumer Y, Shapiro M, Brumer P, Balderidge K. J. Phys. Chem. 2002; 106:9512.

84. Frishman E, Shapiro M, Gerbasi D, Brumer P. Enantiomeric Purification of Non-polarized Racemic Mixtures with Coherent Light. J. Chem. Phys. submitted.

85. Edmonds AR. Angular Momentum in Quantum Mechanics. 2d ed.. Princeton: Princeton University Press, 1960.

86. Results for the other possible cases are obtained using the following transformations: case 1 $\varepsilon_1 \rightarrow -\varepsilon_1$, $b'_2 = -b_2$, $b'_3 = -b_3$, $b'_{L_{-p}} = b_{L_{-p}}$, and $b'_{D_{-p}} = -b_{D_{-p}}$; case 2: $\varepsilon_2 \rightarrow -\varepsilon_2$, $b'_2 = -b_2$, $b'_4 = -b_4$, $b'_{L_{-p}} = b_{L_{-p}}$, and $b'_{D_{-p}} = -b_{D_{-p}}$; case 3: $\varepsilon_1 \rightarrow -\varepsilon_1$ and $\varepsilon_2 \rightarrow -\varepsilon_2$, $b'_1 = -b_1$, $b'_2 = -b_2$, $b'_3 = -b_3$, $b'_4 = -b_4$; case 4: $\varepsilon_1 \rightarrow -\varepsilon_1$ and $\varepsilon_0 \rightarrow -\varepsilon_0$, $b'_1 = -b_1$, $b'_2 = -b_2$; case 5: $\varepsilon_2 \rightarrow -\varepsilon_2$ and $\varepsilon_0 \rightarrow -\varepsilon_0$, $b'_3 = -b_3$, $b'_4 = -b_4$.

87. Balint-Kurti GG, Shapiro M. Chem. Phys. 1981; 61:137. Our Eq. (51) extends Eq. (14) of this reference to arbitrary direction of polarization, and corrects a typographical error regarding the complex conjugate.

88. Gerbasi D, Shapiro M, Brumer P. (in progress).

89. Allen L, Eberly JH. Optical Resonance and Two-Level Atoms. New York: John Wiley, 1975.

90a. Corana A, Marchesi M, Martini C, Ridella S. ACM Transactions on Mathematical Software 1987; 13:262.

90b. Goffe WL, Ferrier GD, Rogers J. Journal of Econometrics 1994; 60:65. Fortran source code by Goffe, W.L.

91. Grischkowsky D, Loy MMT. Phys. Rev. A. 1975, 12, 1117; Phys. Rev. A. 1975; 12:2514.

92. For a review see Bergmann K, Theuer H, Shore BW. Rev. Mod. Phys., 1998.

93. Gaubatz U, Rudecki P, Schiemann S, Bergmann K. J. Chem. Phys. 1990; 92:5363.

93. Theuer H, Unanyan RG, Habscheid C, Klein K, Bergmann K. Optics Express 1999; 4:77.

94. Hohenester U, Troiani F, Molinari E, Panzarini G, Macchiavello C. Appl. Phys. Lett. 2000; 77:1864.

95. Atanasov R, Haché A, Hughes LP, van Driel HM, Sipe JE. Phys. Rev. Lett. 1996; 76:1703.

96. Král P, Tománek D. Phys. Rev. Lett. 1999; 82:5373.

97. Ortigoso J, Fraser GT, Pate BH. Phys. Rev. Lett. 1999; 82:2856.

98. Unayan R, Fleischhauer M, Shore BW, Bergmann K. Optics Comm. 1998; 155: 144.

99. Kobrak MN, Rice SA. Phys. Rev. A 1998; 57:2885.

100. Bodenha;aufer K, Hierlemann A, Seemann J, Gauglitz G, Koppenhoefer B, Ga; aupel W. Nature 1997; 387.

101. McKendry R, Theoclitou M-E, Rayment T, Abell C. Nature 1998; 391.

102. Král P, Amitay Z, Shapiro M. Phys. Rev. Lett. 2002; 89:063002.

103. Gottselig M, Luckhaus D, Quack M, Stohner J, Willeke M. Helv. Chim. Acta 2000; 84:1846.

104. Gerbasi D, Brumer P, Thanopulos I, Král P, Shapiro M. 'Theory of the Two Step Enantiomeric Purification of Dimethylallene'. J. Chem. Phys.. submitted.

105. The electric fields have mutually orthogonal polarizations so as to overcome the deleterious effect of the projection of the angular momentum directions (M) averag-

ing [84]. The respective L/D dipole-matrix elements differ by a sign, $d_{ij}^L = -d_{ij}^D$, for transitions along the z direction, while there is no sign difference in case of x, y. Hence $\phi^L = \phi^D \pm \pi$, giving rise to different excitation paths for each enantiomer, as demonstrated in Fig. 25. The pulses of $\tau \approx 20$ ns are used and the dipole moments for the x, y, and z transitions are given in the Methods. The intensities for the $|2\rangle \leftrightarrow |3\rangle$, $|1\rangle \leftrightarrow |2\rangle$, and $|1\rangle \leftrightarrow |3\rangle$ transitions used are ≈ 30, 1.2, 0.01 GW cm^{-2}, respectively.

106.   Král P, Fiurásek J, Shapiro M. Phys. Rev. A 2001; 64:023414.
107.   The corresponding transition moments are given in the Methods. We use pulses with $\tau \approx 2$ ns and intensity ≈ 50 GW cm^{-2} and ≈ 15 GW cm^{-2} for the $|3\rangle \leftrightarrow |4\rangle$ and $|4\rangle \leftrightarrow |5\rangle$ transitions, respectively.

3

Magnetochiral Anisotropy in Asymmetric Photochemistry

G. L. J. A. Rikken
CNRS–INS–UPS
Toulouse, France

I. INTRODUCTION

Natural optical activity (NOA), which occurs exclusively in chiral media, and magnetic optical activity (MOA), which is induced by a longitudinal magnetic field, show a strong phenomenological resemblance. In both cases, the direction of polarization of linearly polarized light is rotated during propagation through the medium. Interpreting magnetic optical activity as a sign of magnetically induced chirality, Pasteur was the first to search—in vain—for an enantioselective effect of magnetic fields [1]. These searches were partly motivated by the hope of finding an explanation for the homochirality of life [2]. Although several positive results were reported, all were revoked or could not be confirmed [3,4]. The symmetry requirements for any process to yield a chiral result were formulated by Barron, who pointed out that no enantioselectivity of magnetic fields per se is allowed [5,6].

In 1962 the first implicit prediction appeared of a cross-effect between natural and magnetic optical activity, which discriminates between the two enantiomers of chiral molecules [7]. This was followed independently by a prediction of magnetospatial dispersion in noncentrosymmetrical crystalline materials [8]. This cross-effect has been called magnetochiral anisotropy and has since been predicted independently several times more [9–12]. Its existence can be appreciated by expanding the dielectric tensor of a chiral medium subject to a magnetic field to first order in the wave vector k and magnetic field B [8]:

$$\varepsilon_{ij}\ (\omega,\ k,\ B)\ =\ \varepsilon_{ij}\ (\omega)\ +\ \alpha_{ijl}\ (\omega)k_l\ +\ \beta_{ijl}\ (\omega)B_l\ +\ \gamma_{ijlm}\ (\omega)k_lB_m \qquad (1)$$

For high-symmetry media like gases, liquids, cubic crystals, or uniaxial crystals with their optical axis parallel to B, and with the propagation direction of the light parallel to B, the optical eigenmodes are right- and left-handed circularly polarized waves, denoted by $+$ and $-$. For such media, Eq. (1) can be simplified to [8,9]

$$\varepsilon_{\pm}(\omega,\ k,\ B)\ =\ \varepsilon(\omega)\ \pm\ \alpha^{DIL}\ (\omega)k\ \pm\ \beta(\omega)B\ +\ \gamma^{DIL}\ (\omega)k \cdot B \qquad (2)$$

where $x^D(\omega) = -x^L\ (\omega)$ refers to right (D)- and left (L)-handed media, α describes natural optical activity and β describes magnetic optical activity. The material parameters ε, α, β, and γ are all generally complex-valued, and we will denote their real and imaginary parts by x' and x'', respectively. The essential features of MChA are (1) the dependence on the relative orientation of k and B, (2) the dependence on the handedness of the chiral medium (enantioselectivity), and (3) the independence of the polarization state of the light. The simple physical picture behind MChA is that any spinning particle, which also moves parallel to its rotation axis, is a chiral object [5]. An electron in a magnetic field is spinning because of the induction of a net spin or orbital angular momentum by the field. When it moves parallel to the magnetic field, e.g., because it has absorbed linear momentum from a photon or an electric field, it is chiral, and its interaction with the chiral geometry of the molecule depends on the relative handedness of the electronic motion and the molecular framework.

Baranova et al. were the first to present a simple microscopic model [10] that can give an order-of-magnitude estimate for MChA. It is an extension of the classical Becquerel model for MOA [13], and it interprets MChA as a result of the Larmor precession in NOA. Baranova et al. find (CGS units)

$$\gamma(\omega) = \frac{e}{2mc}\frac{\partial\alpha}{\partial\omega} \qquad (3)$$

where e and m are the electron charge and mass. When studying natural circular dichroism (NCD) or magnetic circular dichroism (MCD), it is convenient to normalize the dichroism by the normal absorption in dimensionless dissymmetry factors $g \equiv 2(A_+ - A_-)/(A_+ + A_-)$, where $A_{\pm}$ is the optical extinction coefficient for right/left circularly polarized light (see Appendix A). If $\varepsilon'_{\pm} \gg \varepsilon''_{\pm}$, the dissymmetry factors for NCD and MCD can be simply expressed in the imaginary parts of the terms of Eq. (2):

$$g_{NCD} = \frac{2\alpha''k}{\varepsilon''} \qquad (4)$$

$$g_{MCD} = \frac{2\,\beta''B}{\varepsilon'} \tag{5}$$

Here we define the magnetochiral anisotropy factor similarly:

$$g_{MChA} \equiv 2\,\frac{A(B\uparrow\uparrow k) - A(B\uparrow\downarrow k)}{A(B\uparrow\uparrow k) + A(B\uparrow\downarrow k)} = \frac{2\gamma''kB}{\varepsilon''} \tag{6}$$

The g_{NCD} is usually quite constant across one given optical transition. Then we can write $\alpha''(\omega) \equiv C \cdot \varepsilon''(\omega)$, where C is a constant. Using the Becquerel result $\beta(\omega) = e/2mc \cdot \partial\varepsilon/\partial\omega$ leads to

$$g_{MChA} = \frac{eCkB}{mc\varepsilon''}\frac{\partial\varepsilon''}{\partial\omega} = \frac{g_{NCD}g_{MCD}}{2} \tag{7}$$

This simple model therefore gives as an estimate for the relative strength of MChA in absorption, the product of the relative strengths of NCD and MCD, a result that seems in line with physical intuition for a cross-effect. A detailed molecular theory for MChA in molecular liquids and gases has been formulated by Barron and Vrbancich [14]. It requires complete knowledge of all molecular transition moments involved and therefore cannot be easily used to obtain quantitative predictions. Very recently, the first *ab initio* calculations of MChA in simple molecules have appeared [15], and a prediction of very strong MChA in optically pumped atomic systems [16] was made.

II. MChA IN LUMINESCENCE

Although there always seemed to be theoretical unanimity on the existence of magnetochiral anisotropy, it was not observed experimentally until a few years ago. The most strongly chiral optical transitions reported in the literature are the $^5D_0 \rightarrow {}^7F_{1,2}$ luminescent transitions in tris(3-trifluoroacetyl-$\pm$-camphorato) europium(III) complexes (Eu(($\pm$)tfc)$_3$). These transitions also have a considerable MCD. Such complexes are therefore likely candidates to show a significant magnetochiral effect. The experiment performed by us to observe MChA measures the difference in luminescence intensity in the directions parallel and antiparallel to B [17,18]. In order to increase sensitivity, the magnetic field is alternated and the intensity difference $I_{\hat{B}\uparrow\uparrow\hat{k}} - I_{\hat{B}\uparrow\downarrow\hat{k}}$ is phase-sensitively detected by a lock-in. Factors related to excitation intensity, complex concentration, and sample geometry are eliminated by dividing the lock-in output by the total static luminescence signal. The magnetochiral luminescence anisotropy factor g_{MChA} can then be expressed as

$$g_{MChA} = \frac{\left(\frac{\partial}{\partial B}\right)\left(I_{\hat{B}\uparrow\uparrow\hat{k}} - I_{\hat{B}\uparrow\downarrow\hat{k}}\right)}{I_{\hat{B}\uparrow\uparrow\hat{k}} + I_{\hat{B}\uparrow\downarrow\hat{k}}}B \tag{8}$$

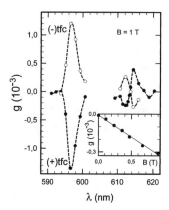

Figure 1   MChA in luminescence spectra of the enantiomers of Eu(([±])tfc) (5% wt/wt in dimethylsulfoxide), excitation at 350 nm wavelength. Inset shows MChA of Eu(([−])tfc) as a function of magnetic field, at a wavelength of 615.8 nm. (From Ref. 17.)

Figure 1 shows the experimental results for g_{MChA} for the two enantiomers of the complex, showing significant MChA at the two transitions $^5D_0 \rightarrow {}^7F_1$ and $^5D_0 \rightarrow {}^7F_2$. An essential characteristic of MChA is that g_{MChA} should be of opposite sign for the two enantiomers, as observed. The inset shows the expected linear magnetic field dependence of g_{MChA}. Figure 1 therefore constitutes the complete proof for the existence of MChA in luminescence and all of its characteristics. The Baranova model of Eq. (7) predicts the anisotropy factor to be $g_{MChA}/B \approx 5.10^{-3}\ T^{-1}$ for the $^5D_0 \rightarrow {}^7F_1$ transition and $g_{MChA}/B \approx 4.10^{-4}\ T^{-1}$ for the $^5D_0 \rightarrow {}^7F^2$ transition, in reasonable agreement with our observations. This simple model therefore seems to be useful to find order-of-magnitude estimates of MChA. More recently, MChA has also been observed in refraction [19,20], and in absorption, as discussed below.

III.   MChA IN ABSORPTION

The absorption of unpolarized light by a chiral medium with a dielectric constant as given by Eq. (2) is most readily calculated by considering linearly polarized light, which can be decomposed into two circularly polarized waves of the same amplitude and opposite handedness. If $\varepsilon'_\pm \gg \varepsilon''_\pm$, $|\alpha k|, |\beta B|, |\gamma kB|$ and neglecting reflection, the transmission coefficient T for linearly polarized light of such a medium with thickness L is found to be

$$T(\omega, k, B) = \exp\left\{-kL\left(2n'' + \frac{\gamma''k \cdot B}{n'}\right)\right\}\cosh\left\{\frac{kL}{n'}(\alpha''k + \beta''B)\right\} \tag{9}$$

where $n' + in'' \equiv \sqrt{\varepsilon}$. This result holds for an arbitrary linear polarization and therefore also for unpolarized light. When alternating the magnetic field at a frequency Ω and using Eq. (9), one finds for the ratio between the modulated transmitted intensity I_Ω and the static transmitted intensity I_0, for $I_\Omega \ll I_0$,

$$\frac{I_\Omega}{I_0} = -\frac{kBL}{\sqrt{2}n'}\left(\gamma''k - \beta'' \tanh\frac{\alpha''k^2 L}{n'}\right) \tag{10}$$

The first term on the right-hand side represents the pure magnetochiral anisotropy in absorption. The second term stems from a cascading of natural and magnetic circular dichroism. Cascading occurs because NCD creates an excess of one circularly polarized component in the initially unpolarized light. Because of this excess, the MCD then leads to an intensity modulation at Ω. This cascaded MChA shows all the essential features of MChA given above but can be discriminated from the pure effect by its dependence on the sample thickness and on the concentration of the active species.

We have studied the chiral uniaxial crystal $\alpha - \text{NiSO}_4 \cdot 6\text{H}_2\text{O}$ because of its very large g_{NCD} and its reasonably large g_{MCD} [21]. Because of the weakly allowed character of the visible and near-infrared optical transitions of the $[\text{Ni(OH}_2)_6]^{2+}$ complex, $\alpha' k$, $\beta' B$, $\gamma' kB \ll \varepsilon'$, and the analysis outlined above should apply quantitatively. The handedness of each crystal was determined by measuring its NCD, and MCD was found to be the same for the two crystal enantiomers. The transmission of the crystals, with their optical axes and the propagation direction of the light parallel to B, is measured with unpolarized light, with the magnetic field alternating and the transmitted intensity modulation being phase-sensitively detected. The ratio between the modulated intensity and the static intensity is equal to $\Delta A_{\text{MChA}} \cdot L \cdot B$, where

$$\Delta A_{MChA} = \frac{\gamma''k^2 - \beta''k \tanh\dfrac{(\alpha''k^2 L)}{n'}}{\sqrt{2}n'} \tag{11}$$

Figure 2a shows the experimental spectra for ΔA_{MChA} for two crystals of opposite handedness, proving the essential characteristic of MChA that ΔA_{MChA} should be of opposite sign for the two enantiomers. Also shown is the cascaded MChA calculated from the second term in Eqs. (10) and (11), using the observed NCD and MCD. From both the magnitude and the line shape of this calculated cascaded contribution to the MChA, it is clear that $\alpha - \text{NiSO}_4 \cdot 6\text{H}_2\text{O}$ shows predominantly the pure effect. The prediction for MChA on the basis of the Baranova model is also shown in Fig. 2a, although the MCD of $\alpha - \text{NiSO}_4 \cdot 6\text{H}_2\text{O}$ does not fulfil the validity requirements for this model. The line shape is evidently not correct,

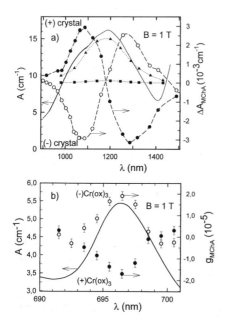

Figure 2 (a) MChA in absorption of both enantiomers of $\alpha - NiSO_4 \bullet 6H_2O$ crystals at the $^3A_{2g} \rightarrow {}^3T_{2g}$ transition. Solid line is the absorption spectrum. Squares are the calculated cascaded contribution, triangles the Baranova model (see text). (From Ref. 21.) (b) MChA spectra of both Cr(III)tris-oxalato enantiomers, dissolved in dimethylsulfoxide, around the $^4A_{2g} \rightarrow {}^2E_g$ transition. Solid line is the absorption spectrum. (From Ref. 18.)

but the predicted magnitude agrees well with the experimental results, again confirming the usefulness of this model.

We have also studied the MChA in absorption for the Cr(III)tris-oxalato complex in solution [22]. Both the NCD and the MCD spectrum of this complex are known. In particular, the spin-forbidden transition from the ground state $^4A_{2g}$ to the excited ^2E_g state shows fairly large values for both g_{NCD} and g_{MCD}, so we can expect a substantial g_{MChA} at this transition. The samples consisted of solutions of the pure enantiomers, resolved according to a literature method [23], in dimethylsulfoxide (DMSO). This solvent was used because in DMSO we have found the thermal recemization rate of the tris(oxalato)Cr(III) ion to be very low. No measurable decay of NCD occurred in such solutions over periods of days at room temperature, whereas in aqueous solution the NCD started to decrease after several hours. Concentrations were varied up to 0.3 molar. Figure 2b shows our result for the MChA spectrum of the two enantiomers. These spectra are clearly of opposite sign. It was verified that g_{MChA} varies linearly in B, again

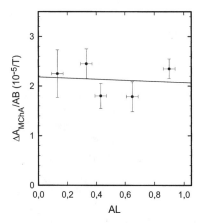

Figure 3 $\Delta A_{MChA}/AB$ as a function of concentration of the I-tris(oxalato)Cr(III) ion in DMSO, expressed as optical density, at a wavelength of 696 nm. ($AL = 1$ corresponds to a concentration of 0.3 molar.) Solid line is linear fit to the data. (From Ref. 18.)

proving the existence of MChA in absorption. The dispersive-type line shape is indicative of so-called A terms, implying that the magnetic field influence on the optical properties is through the Zeeman effect. As discussed above, the order of magnitude of the MChA is estimated to be approximately, $g_{NCD} \cdot g_{MCD}/2$. At $\lambda = 701$ nm we have found $g_{MCD}/B = 1.2 10^{-3}$ T^{-1} and $g_{NCD} = 2.2 10^{-3}$, which yields an estimate $g_{MChA} = 1.3\ 10^{-6}$ T^{-1}, close to the observed value. Figure 3 shows the MChA of this complex at a given wavelength as a function of the concentration of the complex, expressed as optical extinction for the given sample length. No significant concentration dependence of the experimental MChA is observed. This is to be expected for the true MChA, as it is a purely molecular property. The cascaded effect however should change with concentration [Eq. (10)]. Figure 3 therefore shows that under these conditions, the true MChA dominates the cascaded effect. We estimate that the cascaded effect contributes less than 10% to the observed MChA.

Very recently, the observation of MChA for the absorption of x-rays in Cr$_2$O$_3$ crystals was reported [24].

IV. ASYMMETRIC MChA PHOTOCHEMISTRY

Asymmetric photochemistry with circularly polarized light (CPL), based on NCD, is well-established and has been extensively reviewed [25,26]. The Cr(III)tris-oxalato complex discussed above also has been studied in this context, in a process called photoresolution [27]. This complex is unstable in aqueous solution and spontaneously dissociates and reassociates. In equilibrium, one has a racemic mixture (equal concentrations of right- and left-handed complex). The dissociation is accelerated by the absorption of light, so under CPL irradiation, the more absorbing enantiomer will dissociate more often, whereas the subsequent random reassociation yields equal amounts of both enantiomers. This leads to an excess of the less absorbing enantiomer, the handedness of which depends on the handedness of the CPL. If this photoresolution is much faster than the thermal racemization, the size of the enantiomeric excess ee in dynamic equilibrium can be shown to be given by $ee_{CPL} = g_{NCD}/2$ [27]. As soon as the irradiation stops, the system will return to the racemic state owing to thermal dissociation and random reassociation of the complexes.

Following the same reasoning, one should expect to obtain an enantiomeric excess through MChA when one-sidedly irradiating a racemic solution of this complex with unpolarized light in a magnetic field parallel to the irradiation direction. The handedness of the excess will be determined by the relative orientation of the light and the magnetic field. The situation is however more complex than for photoresolution by CPL, because, as with the absorption case discussed above, unpolarized incident light may become circularly polarized by MCD or NCD. This CPL component will then through the classical photoresolution process also give rise to an enantiomeric excess. Below we will detail our model to obtain a quantitative description of MChA in photoresolution, taking into account all possible causes for an enatiomeric excess. This model also provides all the elements necessary to the quantitatively model MChA equivalents of other asymmetric photochemical processes with CPL that are well established and understood, like photodecomposition and direct photosynthesis [25,26].

A. Mathematical Description

We assume a racemic mixture of equal concentrations of dextro- (N_D) and levo-enantiomers (N_L) of the complex, together with partly dissociated achiral complexes (N_X). The incident unpolarized light beam is described as equal intensities of left- and right-circularly polarized light, I_l and I_r, respectively, propagating along the z direction. Because of optical absorption, molecular diffusion, and chemical reactions, I_l, I_r, N_D, and N_L will depend on the spatial coordinate z and time t. This can be expressed by the following differential equations:

$$\frac{\partial N_D(z,t)}{\partial t} = -D\frac{\partial^2 N_D(z,t)}{\partial z^2} - YN_D(z,t)[I_r(z,t)\alpha_{D,r}$$
$$+ I_l(z,t)\alpha_{D,l}] + KN_X(z,t) - pN_D(z,t)$$

$$\frac{\partial N_L(z,t)}{\partial t} = -D\frac{\partial^2 N_L(z,t)}{\partial z^2} - YN_L(z,t)[I_r(z,t)\alpha_{L,r}$$
$$+ I_l(z,t)\alpha_{L,l}] + KN_X(z,t) - pN_L(z,t) \qquad (12)$$

$$\frac{\partial I_r}{\partial z}(z,t) = -I_r(z,t)\left[N_D(z,t)\alpha_{D,r} + N_L(z,t)\alpha_{L,r} + N_X(z,t)\alpha_{x,r}\right]$$

$$\frac{\partial I_l}{\partial z}(z,t) = -I_l(z,t)\left[N_D(z,t)\alpha_{D,l} + N_L(z,t)\alpha_{L,l} + N_X(z,t)\alpha_{x,l}\right] \tag{13}$$

where D describes molecular diffusion of chiral complexes, Y the photodissociation process, K the random reassociation into chiral complexes, and p the thermal dissociation of chiral complexes. The different molar absorption coefficients for levo- and dextrocomplexes and left- and right-circularly polarized light are given by (see Appendix A)

$$\alpha_{D,r} = \frac{k}{n'}(\tilde{\varepsilon}'' + \alpha''k + \beta''B + \gamma''\mathbf{k}\cdot\mathbf{B})$$

$$\alpha_{D,l} = \frac{k}{n'}(\tilde{\varepsilon}'' - \alpha''k - \beta''B + \gamma''\mathbf{k}\cdot\mathbf{B})$$

$$\alpha_{L,r} = \frac{k}{n'}(\tilde{\varepsilon}'' - \alpha''k + \beta''B - \gamma''\mathbf{k}\cdot\mathbf{B})$$

$$\alpha_{L,l} = \frac{k}{n'}(\tilde{\varepsilon}'' + \alpha''k - \beta''B - \gamma''\mathbf{k}\cdot\mathbf{B}) \tag{14}$$

where α'' describes NCD, β'' describes MCD, and γ'' describes MChD. Eqs. (12) and (13) have to be solved self-consistently, with the following boundary conditions:

1. Racemic mixture before irradiation:

$$N_D(z,t = 0) = N_L(z,t = 0) = \text{const} \equiv \frac{N}{2} \tag{15}$$

2. Cr(III) concentration independent of z and t:

$$N_D(z,t) + N_L(z,t) + N_X(z,t) = \text{const} = N_{\text{total}} \tag{16}$$

3. Irradiation with unpolarized light:

$$I_r(z = 0,t) = I_l(z = 0,t) = \text{const} \equiv \frac{I_0}{2} \tag{17}$$

For the moment we neglect diffusion, i.e., we put $D = 0$ in Eq. (12). In Appendix B, we will show that even infinitely rapid diffusion does not affect the dominant terms in the results for the enantiomeric excess. The essential features of the model can already be obtained from Eq. (12) in the photostationary state. This yields, upon rearrangement,

$$\frac{\Delta N(z)}{N(z)} \equiv \frac{N_D(z) - N_L(z)}{N_D(z) + N_L(z)} = \frac{I_r(z)(\alpha_{L,r} - \alpha_{D,r}) + I_l(z)(\alpha_{L,l} - \alpha_{D,l})}{I_r(z)(\alpha_{L,r} + \alpha_{D,r}) + I_l(z)(\alpha_{L,l} + \alpha_{D,l}) + \frac{2p}{Y}} \tag{18}$$

By insertion of Eqs. (13) and (14) into (18) and defining $\Delta I \equiv I_l - I_r$ and $I \equiv I_l + I_r$, we obtain

$$\frac{\Delta N(z)}{N(z)} = \frac{\alpha'' k \Delta I(z)}{\beta'' B \Delta I(z) + \varepsilon'' I(z) + \dfrac{n'p}{kY}} + \frac{\gamma'' \mathbf{k} \cdot \mathbf{B} I \ (z)}{\beta'' B \Delta I(z) + \varepsilon'' I(z) + \dfrac{n'p}{kY}} \tag{19}$$

We see that there are two terms that can lead to an enantiomeric excess ($\Delta N \neq 0$). The first term on the right-hand side is proportional to α'' (i.e., NCD) and to ΔI (i.e., a net circularly polarized component in the light). This is the classical photoresolution mechanism studied by Stevenson and Verdieck [27]. It will be clear that if a net circular component is generated upon propagation of an initially unpolarized light beam, this term will also lead to an enantiomeric excess. This is exactly what MCD does, as will be shown below. The second term on the right-hand side of Eq. (19), proportional to the magnetochiral parameter γ'', describes the true magnetochiral asymmetric photochemistry, which only depends on the total intensity I, independent of the state of polarization. Equation (19) can be simplified because for all cases $\beta'' B[I_r \ (z) - I_l(z)] \ll \varepsilon'' \ [I_r \ (z) + I_l(z)]$, and for our experimental conditions (see below), thermal dissociation is negligible ($p = 0$). This yields

$$\frac{\Delta N(z)}{N(z)} \approx \frac{\alpha'' k \Delta I(z)}{\varepsilon'' I(z)} + \frac{\gamma'' \mathbf{k} \cdot \mathbf{B}}{\varepsilon''} = \frac{g_{\mathrm{NCD}}}{2} \frac{\Delta I(z)}{I(z)} + \frac{g_{\mathrm{MChA}}}{2} \tag{20}$$

(For the definition of the asymmetry factors g_x, see Appendix A.) The first term is the Stevenson and Verdieck result [27] (for circularly polarized light, $\Delta I/I = 1$). The second term describes the enantiomeric excess due to the true magnetochiral anisotropy.

From the equations above, it will be clear, that as far as the light is concerned, the only important quantity is $\Delta I/I$. It can be easily shown that in the photostationary state, Eqs. (13) and (14) lead to

$$\frac{\Delta I(z)}{I(z)} = -\tanh\left[\frac{k}{n'} \int_0^z \alpha'' k (N_D - N_L)(z') + \beta'' B (N_D + N_L)(z') dz' \right] \tag{21}$$

where we have used $\alpha_{Xr} = \alpha_{Xl}$ as the (partly) dissociated ion is no longer chiral. Equation (21) only contains the parameters α'' and β'', combined with ΔN and N, respectively, indicating that the appearing circular polarization is exclusively due to MCD and NCD in accordance with our understanding of MChD as a polarization-independent phenomenon. Equation (21) can be simplified by inserting the extinction coefficient $A \approx N \ (k''/\sqrt{\varepsilon'}$ and the refractive index $n' \approx \sqrt{\varepsilon'}$, which yields

$$\frac{\Delta I(z)}{I(z)} = -\tanh\left(\frac{A}{2}\int_0^z\left(g_{NCD}\left(\frac{\Delta N(z')}{N(z')}\right) + g_{MCD}\right)dz'\right)$$ (22)

The differential equations (12) and (13) have now been reduced to the integro equations (20) and (22), which can be solved iteratively.

B. Iterative Solution Method

The cascaded magnetochiral photochemistry can be described as follows. In the incident unpolarized light the amounts of right and left circularly polarized components are equal ($\Delta I = 0$). Upon propagation through the initially racemic sample ($\Delta N = 0$) in the presence of a magnetic field, I_r and I_l become different as result of MCD (preferential absorption of one handed component of the incident light beam, independent of the handedness of the medium), i.e., the light beam inside the sample acquires a larger and larger circularly polarized component ($\Delta I \neq 0$). This creates an excess of one enantiomer owing to NCD by preferential decomposition of the other, more absorbing, enantiomer owing to the classical photoresolution process [27]. This excess ($\Delta N \neq 0$) again results in an additional change of the polarization ΔI because of NCD, which in turn changes ΔN and so on (cascading effect). Simultaneously, the true MChD also generates an enantiomeric excess that is independent of the polarization state and intensity of the light, as long as the latter is high enough so that photoresolution is much faster than thermal racemization.

The zero-th order approximation is $(\Delta N/N)_0 = 0$. We calculate the first-order circular polarization $(\Delta I(z)/I(z))_1$ by use of Eq. (22). Then we insert the result for $(\Delta I(z)/I(z))_1$ into Eq. (20), which yields the first-order approximation z-dependent enantiomeric excess $(\Delta N(z)/N(z))_1$. This result is then again inserted into (4.11), which yields the second-order approximation for the circular polarization $(\Delta I(z)/I(z))_2$. We insert this into Eq. (20) to calculate to second approximation the enantiomeric excess $(\Delta N(z)/N(z))_2$, and so on. The general iteration formalism, neglecting diffusion and thermal dissociation is therefore

$$\left(\frac{\Delta I(z)}{I(z)}\right)_n = -\tanh\left(\frac{A}{2}\int_0^z\left(g_{NCD}\left(\frac{\Delta N(z')}{N(z')}\right)_{n-1} + g_{MCD}\right)dz'\right)$$ (23)

and

$$\left(\frac{\Delta N(z)}{N(z)}\right)_n = \frac{g_{NCD}}{2}\left(\frac{\Delta I(z)}{I(z)}\right)_n + \frac{g_{MChA}}{2}$$ (24)

with the starting condition

$$\left(\frac{\Delta N}{N}\right)_0 = 0$$ (25)

The enantiomeric excess after a certain irradiation period was experimentally obtained by means of an NCD measurement after its z-dependence had been averaged away by stirring. To compare calculated and measured results, we average the enantiomeric excess over the sample length L:

$$ee_n \equiv \frac{1}{L} \int_0^L \left(\frac{\Delta N(z)}{N(z)} \right)_n dz \tag{26}$$

C. Execution of Iteration

Insertion of (25) into (23) gives, with $\tanh(x) = x - (x^3/3) + O(x^5)$,

$$\left(\frac{\Delta I(z)}{I(z)} \right)_1 \equiv -\tanh\left(\frac{g_{MCD}}{2} Az \right) \approx \frac{g_{MCD}}{2} Az + O(g^3) \tag{27}$$

By use of (24) we obtain

$$\left(\frac{\Delta N(z)}{N(z)} \right)_1 = -\frac{g_{NCD}}{2} \tanh\left(\frac{g_{MCD}Az}{2} \right) + \frac{g_{MChA}}{2}$$

$$\approx \frac{g_{MChA}}{2} - \frac{g_{NCD}g_{MCD}}{4} Az + O(g^4) \tag{28}$$

and with (26),

$$ee_1 = \frac{g_{MChA}}{2} - \frac{g_{NCD}g_{MCD}AL}{8} + O(g^4) \tag{29}$$

We now calculate the second order, by inserting the obtained value for $(\Delta N/N)_1$ from Eqs. (28) into (23):

$$\left(\frac{\Delta I(z)}{I(z)} \right)_2 = -\tanh\left(\frac{g_{MCD}}{2} Az - \frac{g^2{}_{NCD}g_{MCD}}{16} A^2 z^2 \right.$$

$$\left. + \frac{g_{NCD}g_{MChA}}{4} Az + O(g^4) \right)$$

$$\approx -\frac{g_{MCD}}{2} Az + \frac{g^2{}_{NCD}g_{MCD}}{16} A^2 z^2 + \frac{g^3{}_{MCD}}{24} A^3 z^3$$

$$- \frac{g_{NCD}g_{MChA}}{4} Az + O(g^4) \tag{30}$$

Insertion into Eq. (24) gives

$$\left(\frac{\Delta N(z)}{N(z)} \right)_2 \approx \frac{g_{MChA}}{2} - \frac{g_{NCD}g_{MCD}}{4} Az + \frac{g^3{}_{MCD}g_{NCD}}{48} A^3 z^3$$

$$+ \frac{g^3{}_{NCD}g_{MCD}}{32} A^2 z^2 - \frac{g^2{}_{NCD}g_{MChA}}{8} Az + O(g^5) \tag{31}$$

leading to

$$
\begin{aligned}
ee_2 = \frac{g_{MChA}}{2} &- \frac{g_{NCD}\, g_{MCD}AL}{8} - \frac{g^2{}_{NCD}\, g_{MChA}AL}{8} \\
&+ \frac{g^3{}_{MCD}\, g_{MCD}(AL)^3}{192} + \frac{g^3{}_{NCD}\, g_{MCD}(AL)^2}{96} + O\!\left(g^5\right)
\end{aligned}
\tag{32}
$$

It is easy to see that in each higher order n, additional terms of the order g^{2n} are obtained. The first term on the right-hand side of Eq. (32) describes the true enantioselective magnetochiral photochemistry. All other subsequent terms are due to cascading. For our experiment, $AL \approx 1$, $g_{MCD} \approx 10^{-2}$, and $g_{NCD} \approx 10^{-3}$, so we can conclude that among the cascading terms, the term $g_{MCD}g_{NCD}AL/8$ is dominant and that Eq. (29) is already a good estimate of the enantiomeric excess. This term describes the first cascading step, i.e., the enhancement of one enantiomer over the other owing to the effect of a net circularly polarized light component that results from MCD. As MCD only depends on $N_D + N_L$, the diffusion of the enantiomers of the complex does not in first order affect the cascading mechanism. This justifies the neglect of diffusion. In Appendix B, we show that even infinitely rapid diffusion does not change the first order result.

D. Comparison with Experiment

Our experimental results for the enantioselective magnetochiral photochemistry of the Cr(III) tris-oxalate complex are described in Ref. 28. These experiments were done under such conditions that the true magnetochiral effect should dominate the cascaded one. Therefore the first term on the right-hand side of Eq. (29) is expected to dominate. Here we will experimentally verify the validity of this expression.

The MChA spectrum of the resolved Cr(III) tris-oxalate complex in DMSO was shown in Fig. 2b. Figure 4 shows the *ee* obtained by photoresolution in aqeous solution with unpolarized light, as a function of magnetic field [28], confirming the linear dependence on **k·B**. In Fig. 5, we compare our results for the photochemical enantiomeric excess spectrum, obtained in Ref. 28 by photoresolution on racemic mixtures, with our MChA spectrum of the resolved ion. The MChA spectrum was obtained at a concentration that is a factor 2 lower than the photochemical excess spectrum, and with a sample length that is also a factor 2 smaller. Therefore AL is a factor of 4 smaller in the MChD spectrum than in the photochemical spectrum. The good agreement between the two spectra again confirms that the cascading term [the second term on the right-hand side of Eq. (29), which depends on AL] does not contribute significantly and that the true magnetochiral effect dominates the cascaded one.

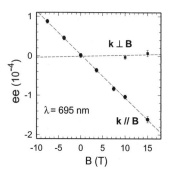

Figure 4 Enantiomeric excess (*ee*) obtained after irradiation of a racemic Cr(III)tris-oxalato solution with unpolarized light at $\lambda = 695.5$ nm, as a function of magnetic field, with an irradiation direction **k** either parallel or perpendicular to the magnetic field **B**. Each point was obtained with a fresh racemic starting solution. (From Ref. 28.)

E. Discussion

The good agreement between our model calculations and our experimental results shows that we have a quantitative understanding of the mechanism of MChA in photoresolution. For the case considered here, the cascading mechanism does not contribute significantly. However, in other systems, in particular in those consist-

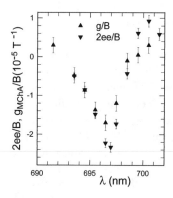

Figure 5 Comparison of the photochemical enantiomeric excess spectrum, obtained in Ref. 28 in water (triangles down), with the MChA spectrum of the (+)-tris(oxalato)Cr(III) ion in DMSO (triangles up). The water spectrum is corrected for a small spectral blue shift. (From Ref. 22.)

ing of two components, one of which has a large MCD and the other a large NCD, the cascading mechanism may be dominant. Based on existing literature values for g_{MCD} and g_{NCD}, quite large values for the ee can be envisaged.

The description that we have developed also allows to describe the effect of MChA in other asymmetric photochemical processes. The case of MChA in photodecomposition can easily be related to Eqs. (12) and (13) by a proper choice of the parameters.

MChA has been suggested as an explanation for the homochirality of life [29]. So far, the two possible causes for this homochirality considered most likely are photochemistry with CPL and the electroweak interaction [2]. We can only wonder if the magnetic fields observed in nature, ranging from the Earth's magnetic field of 10^{-4} T up to the field of neutron stars (10^8 T on the surface), might through MChA lead to sufficiently large ee. Furthermore, issues of spectral, spatial, and temporal averaging have to be addressed, as with the case of photochemistry with CPL. Our results only show that MChA merits consideration in the discussion on the homochirality of life.

V. MChA IN DIFFUSION

Chirality is a very general symmetry concept, and its consequences are not limited to the optical properties of systems. An electrical conductor for instance may be chiral because of several reasons. The material may crystallize in a chiral space group, like tellurium or β-manganese [30], or be composed of chiral subunits like chiral conducting polymers [31] and Langmuir–Blodgett films [32] or vapors [33] of chiral molecules. Even if the material itself is nonchiral, it may still be formed into a chiral shape, like a helix. In all these cases, the conductor can exist in two enantiomeric forms.

As the existence of MChA can be deduced by very general symmetry arguments and the effect does not depend on the presence of a particular polarization, one may wonder if something like MChA can also exist outside optical phenomena, e.g. in electrical conduction or molecular diffusion. Time-reversal symmetry arguments cannot be applied directly to the case of diffusive transport, as diffusion inherently breaks this symmetry. Instead, one has to use the Onsager relation. (For a discussion see, e.g., Refs. 34 and 35.) For any generalized transport coefficient σ_{ij} (e.g., the electrical conductivity or molecular diffusion tensor) close to thermodynamic equilibrium, Onsager has shown that one can write

$$\sigma_{ij} = \sigma_{ji}^{\dagger} \tag{33}$$

where $\dagger$ denotes time reversal. In a magnetic field, one therefore finds

$$\sigma_{ij}(\mathbf{B}) = \sigma_{ij}(-\mathbf{B}) \tag{34}$$

This is equivalent to the statement that any two-terminal transport phenomenon can only have an even magnetic field dependence. In chiral systems, symmetry allows all microscopic properties to have in principle an odd dependence on the wave vector **k** of the moving particles. As the wave vector is also odd under time-reversal, from Eq. (33) it follows that

$$\sigma_{ij}(\mathbf{k} \cdot \mathbf{B}) = \sigma_{ji}(-\mathbf{k} \cdot -\mathbf{B}) = \sigma_{ji}(\mathbf{k} \cdot \mathbf{B}) \tag{35}$$

and so there are no time-reversal symmetry objections against a linear dependence of σ_{ij}, and therefore of any two-terminal diffusion phenomenon, on **k·B**. Parity-reversal symmetry tell us that

$$\sigma_{ii}(\mathbf{k,B}) = \sigma_{ii}^*(-\mathbf{k,B}) \tag{36}$$

where* denotes the parity reversed system, i.e. the opposite enantiomer. One can therefore conjecture that, e.g., the electrical resistance R of any chiral conductor subject to a magnetic field **B** is of the form

$$R(\mathbf{B,I})^{D/L} = R_0(1 + \beta B^2 + \chi^{D/L}\mathbf{B} \cdot \mathbf{I}) \tag{37}$$

where the current $\mathbf{I} \propto <\mathbf{k}>$ and **k** is the wave vector of the charge carriers and $\chi^D = -\chi^L$. The parameter β describes the normal magnetoresistance that is allowed in all diffusive conductors. In analogy to the optical case, we call the effect corresponding to the last term on the right-hand side electrical magnetochiral anisotropy. The spatial averaging takes into account the diffusive aspect. As averaging is a linear operation, it does not affect the symmetry properties of the problem. We have shown experimentally that electrical MChA exists for metallic helices and for wires with helical deformations [36], confirming Eq. (37). In order to study the existence of MChA in electrical conduction on a molecular scale, we have performed magnetoresistance measurements on single-walled carbon nanotubes (SWNT). Principally, an SWNT can be thought of as a graphene sheet some microns in length and a few nanometers in width that has been rolled up seamlessly [37]. The way the graphene sheet is cut is classified by a pair of indices (n, m). These indices determine the diameter and whether a SWNT is metallic or semiconducting. The cutting can occur in two directions, leading in general to left-(L) and right-(D)-handed chiral nanotubes. Our measurements have been carried out on two-terminal contacted metallic SWNT bundles with **B** along the tube axis [38]. The resistance anisotropy $\Delta R(B,I) \equiv \delta R(B,I) - \delta R(-B,I)$, where $\delta R(B,I) \equiv R(B,I) - R(B, -I)$, is determined by standard lock-in techniques. In Fig. 6a the resistance anisotropy of a nanotube is plotted as a function of B, and indeed a linear dependence, characteristic for MChA, is observed. Electrical MChA is also predicted to be linear in the current traversing the nanotube and this is indeed what we observe, as shown in Fig. 6b. At this moment, no experimental control over the handedness of SWNT exists, and the handedness

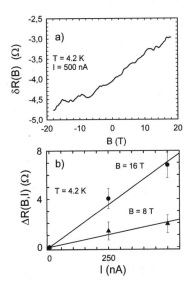

Figure 6   (a) δR of a SWNT as a function of magnetic field. (b) ΔR as a function of the current for another SWNT. (From Ref. 38.)

of each SWNT is arbitrary. The observed relative occurrences of both signs of MChA are consistent with that.

Our results prove that MChA also exists in electrical transport on a molecular scale, which implies that electrochemistry and electrophoresis in a magnetic field can in principle be asymmetric. Such experiments have been performed, but an initial positive result [39] could not be confirmed [40]. The photochemical results above show that MChA in general leads to quite small *ee*, even in very high magnetic fields, so the electrochemical experiments should be repeated with a carefully chosen model system, and with a very high sensitivity for the resulting *ee*.

As Onsager's aproach to diffusive transport is quite general, also other diffusive transport phenomena, like electric discharges, electrophoresis, thermal conductivity, and molecular diffusion, should show MChA. Again, only small effects should be expected, but they are important from a fundamental point of view.

VI. CONCLUSION

In conclusion, we have described our observations of magnetochiral anisotropy in luminescence, absorption, asymmetric photochemistry, and electrical resistivity. Our results suggest that MChA is a very general phenomenon by which a flux of particles parallel to a magnetic field can can have asymmetric interactions with the two enantiomers of chiral systems.

ACKNOWLEDGMENTS

This work was performed in collaboration with E. Raupach, V. Krstić, J. Fölling, C. Train, and B. Malezieux. I gratefully acknowledge H. Krath for technical assistance, and P. Wyder, G. Martinez, B. van Tiggelen, L. Barron, and G. Wagnière for stimulating discussions.

APPENDIX A. OPTICAL MATERIAL PARAMETERS

The unpolarized incident light beam is described as an electromagnetic wave consisting of two circularly polarized components of equal amplitude, propagating in the z-direction. The medium is subject to a static magnetic field in the same direction. We can write the electric field of the light as

$$\vec{E} = E_r \hat{e}_r \exp i(n_r kz - \omega t) + E_l \hat{e}_l \exp i(n_l kz - \omega t),$$

where $\quad \hat{e}_r = (\hat{x} + i\hat{y}) \quad$ and $\quad \hat{e}_l = (\hat{x} - i\hat{y})$ $\hfill$ (A.1)

The complex refractive index for right and left circularly polarized light is given by

$$\varepsilon_{rll}(\omega, \vec{B}, \vec{k}) = (n'_{rll} + i\, n''_{rll})^2 = \tilde{\varepsilon}(\omega) \pm a(\omega)k \pm b(\omega)B + c(w)\mathbf{k} \cdot \mathbf{B} + \ldots$$

where ε, a, b, and c are complex (see Ref., [21]) and reference therein). This gives

$$n_{rll} \approx \sqrt{\tilde{\varepsilon}}\left(1 \pm \frac{ak}{2\varepsilon} \pm \frac{bB}{2\varepsilon} + \frac{c\mathbf{k} \cdot \mathbf{B}}{2\varepsilon}\right) = \tilde{n} \pm \frac{ak}{2n} \pm \frac{bB}{2n} + \frac{c\mathbf{k} \cdot \mathbf{B}}{2n}$$ (A.2)

So with $n \approx n'$ we obtain

$$\vec{E}_r = \hat{e}_r \exp i(\tilde{n}kz - \omega t)\hat{e}_r \exp i\left(\frac{kz}{2n'}(ak + bB + c\mathbf{k} \cdot \mathbf{B})\right)$$

$$\vec{E}_l = \hat{e}_l \exp i(\tilde{n}kz - \omega t)\hat{e}_l \exp i\left(\frac{kz}{2n'}(-ak - bB + c\mathbf{k} \cdot \mathbf{B})\right)$$

$\hfill$ (A.3)

$$I_r = 2\hat{E}_r^2 \exp(-2\tilde{n}''kz)\exp\left(\frac{kz}{n'}(-a''k - b''B + c''\mathbf{k}\cdot\mathbf{B})\right)$$

$$I_l = 2\hat{E}_l^2 \exp(-2\tilde{n}''kz)\exp\left(\frac{kz}{n'}(a''k + b''B - c''\mathbf{k}\cdot\mathbf{B})\right) \qquad \text{(A.4)}$$

Comparing Eq. (A.4) with the Lambert–Beer absorption law $I = I_0 e^{-AL}$, where A is the extinction coefficient and L the sample length, and considering that a'' and c'' have opposite signs for two enantiomers of a chiral substance, and that a'' and b'' have opposite sign for opposite handedness of light, we find that the extinction coefficients are given by $A_{x,y} = N_X \alpha_{X,y}$, where

$$\alpha_{D,r} = \frac{k}{n'}(\tilde{\varepsilon}'' + \alpha''k + \beta''B + \gamma''\mathbf{k}\cdot\mathbf{B})$$

$$\alpha_{D,l} = \frac{k}{n'}(\tilde{\varepsilon}'' - \alpha''k - \beta''B + \gamma''\mathbf{k}\cdot\mathbf{B})$$

$$\alpha_{L,r} = \frac{k}{n'}(\tilde{\varepsilon}'' - \alpha''k + \beta''B - \gamma''\mathbf{k}\cdot\mathbf{B})$$

$$\alpha_{L,l} = \frac{k}{n'}(\tilde{\varepsilon}'' + \alpha''k - \beta''B - \gamma''\mathbf{k}\cdot\mathbf{B}) \qquad \text{(A.5)}$$

where the $\alpha_{D/L,\ r/l}$ are now the molar absorption coefficients of the dextro/levo forms of the molecules, for right/left circularly polarized light. It is often convenient to express the molecular parameters α'', β'', γ'' in dimensionless, so-called asymmetry factors. These are defined as

$$g_x \equiv 2\frac{A^{x+} - A^{x-}}{A^{x+} + A^{x-}},$$

where $x \pm$ refers to the properly chosen symmetry condition of the optical process. This yields

$$g_{MCD} \approx \frac{2\beta''B}{\varepsilon''} \qquad g_{NCD} \approx \frac{2\alpha''k}{\varepsilon''} \qquad g_{MChA} \approx \frac{2\gamma''\vec{B}\vec{k}}{\varepsilon''} \qquad \text{(A.6)}$$

Note that all asymmetry factors are much smaller than unity.

APPENDIX B: THE INFLUENCE OF DIFFUSION ON MChA IN PHOTORESOLUTION

In solving the equations for ΔN and ΔI, we can take diffusion of the enantiomers into account by averaging the concentrations over the sample length after every iteration step, so that the incident unpolarized light always feels a completely homogenous solution. In this way we mimic the extreme case of infinitely fast diffusion. We take the first-order result without diffusion and then assume instantaneous diffusion such that the z-dependence of $\Delta N/N$ is averaged out:

$$\left(\frac{\Delta N}{N}\right)_{1,D} \equiv \frac{1}{L}\int_0^L \left(\frac{\Delta N(z)}{N(z)}\right)_1 dz = \frac{g_{MChA}}{2} - \frac{g_{NCD}g_{MCD}AL}{8} + O\left(g^4\right) \tag{B.1}$$

We obtain the second order approximation, by inserting the value for $(\Delta N/N)_{1D}$ obtained from Eq. (B.1) into Eq. (23)

$$\left(\frac{\Delta I(z)}{I(z)}\right)_{2,D} \approx -\frac{g_{MCD}}{2} Az + \frac{g_{NCD}^2\, g_{MCD}}{16} A^2 Lz$$
$$+ \frac{g_{MCD}^3}{24} A^3 z^3 - \frac{g_{NCD}g_{MChA}}{4} Az + O\left(g^4\right) \tag{B.2}$$

Insertion into Eq. (23) gives

$$\left(\frac{\Delta N(z)}{N(z)}\right)_2 \approx \frac{g_{MChA}}{2} - \frac{g_{NCD}g_{MCD}}{4} Az + \frac{g_{MCD}^3\, g_{NCD}}{48} A^3 z^3$$
$$+ \frac{g_{NCD}^3\, g_{MCD}}{32} A^2 Lz - \frac{g_{NCD}^2\, g_{MChA}}{8} Az + O\left(g^5\right) \tag{B.3}$$

We again apply spatial averaging as in (B.1), assuming infinitely fast diffusion, which gives

$$\left(\frac{\Delta N}{N}\right)_{2D} \approx \frac{g_{MChA}}{2} - \frac{g_{NCD}g_{MCD}AL}{8} - \frac{g_{NCD}^2\, g_{MChA}AL}{16}$$
$$+ \frac{g_{NCD}^3\, g_{MCD}(AL)^2}{64} + \frac{g_{MCD}^2\, g_{NCD}(AL)^3}{192} + O\left(g^5\right) \tag{B.4}$$

which by definition equals ee_{2D}. Comparing Eq. (B.4) with Eq. (32), we directly see that the true magnetochiral term and the dominant cascading term in the presence of infinitely fast diffusion are the same as without any diffusion, thereby justifying the neglect of diffusion.

REFERENCES

1. Mason SF. In: Circular Dichroism. K Nakanishi, Berova N, RW Woody, Eds. New York: VCH, 1994.
2. Cline DB, ed. Physical Origin of Homochirality in Life. Am. Inst. Physics. New York, 1996.
3. Avalos M, Babiano R, Cintas P, Jimenez JL, Palacios JC, Barron LD. Chem. Rev. 1998; 98:2391–2404.
4. Bonner WA. Origins of Life and Evolution of the Biosphere 1995; 25:175–190.
5. Barron LD. J. Am. Chem. Soc. 1986; 108:5539–5542.
6. Barron LD. Science 1994; 266:1491–1492.
7. Groenewege MP. Mol. Phys. 1962; 5:541–553.
8. Portigal DL, Burstein E. J. Phys. Chem. Solids 1971; 32:603–608.
9. Baranova NB, Bogdanov Y, Zeldovich BY. Opt. Commun. 1977; 22:243–247.

10. Baranova NB, Zeldovich BY. Mol. Phys. 1979; 38:1085–1098.
11. G Wagnière, A Meier. Chem. Phys. Lett. 1982; 93:78–82.
12. G Wagnière. Chem. Phys. Lett. 1984; 110:546–549.
13. Landau LD, Lifshitz EM, Pitaevski LP. Electrodynamics of Continuous Media. Oxford: Pergamon, 1984.
14. Barron LD, Vrbancich J. Mol. Phys. 1984; 51:715–730.
15. Coriani S, Pecul M, Rizzo A, Jorgensen P, M. Jaszunski. J. Chem. Phys 2002; 117: 6417–6428.
16. Agarwal GS, Dasgupta S. accepted for publication in Phys. Rev A.
17. GLJA Rikken, E Raupach. Nature 1997; 390:493–494.
18. Raupach E. The Magnetochiral AnisotropyUniversity of Konstanz. Hartung Gorre Verlag, Konstanz, 2002.
19. Kleindienst P, G Wagnière. Chem. Phys. Let. 1998; 288:89–97.
20. Vallet M, Ghosh R, A Le Floch, Ruchon T, Bretenaker F, JY Thépot. Phys. Rev. Lett. 2001; 87:183003-1–183003-4.
21. GLJA Rikken, E Raupach. Phys. Rev. E 1998; 58:5081–5084.
22. Raupach E, GLJA Rikken, Train C, B Malézieux. Chem. Phys. 2000; 261:373–380.
23. Kaufmann GB, Takahashi LT, Sugisaka N. Inorg. Synth. 1966; 8:207–209.
24. Goulon J, Rogalev A, Wilhelm F, C Goulon-Ginet, Carra P, Cabaret D, C Brouder. Phys. Rev. Lett. 2002; 88:237401-1–237401-4.
25. Rau H. Chem. Rev. 1983; 83:535–547.
26. Inoue Y. Chem. Rev. 1992; 92:741–770.
27. Stevenson KL, Verdieck JF. Mol. Photochem. 1969; 1:271–288.
28. GLJA Rikken, E Raupach. Nature 2000; 405:932–935.
29. G Wagnière, A Meier. Experientia 1983; 39:1090–1091.
30. Lide DR, ed. CRC Handbook of Chemistry and Physics. 77th ed.. Boca Raton: CRC Press, 1997.
31. Akagi K, Piao G, Kaneko S, Sakamaki K, Shirakawa H, Kyotani M. Science 1998; 282:1683–1686.
32. Ray K, Ananthavel SP, Waldeck DH, Naaman R. Science 1999; 283:814–816.
33. Mayer S, Kessler J. Phys. Rev. Lett. 1995; 74:4803–4806.
34. Landau LD, Lifshitz EM. Statistical Physics, Part 1. 3d ed.. Oxford: Pergamon, 1980: 359ff.
35. Reif F. Fundamentals of Statistical and Thermal Physics: Int. Student Ed. Tokyo McGraw Hill kogakusha, 1965:598 ff.
36. GLJA Rikken, J Fölling, P Wyder. Phys. Rev. Lett. 2001; 87:236602-1–236602-4.
37. Saito R, Dresselhaus G, Dresselhaus MS. Physical Properties of Carbon Nanotubes. London: Imperial College Press, 1998.
38. V. Krstić, S. Roth, M. Burghard, K. Kern, G.L.J.A. Rikken. J. Chem. Phys. 2002; 117:11315–11319.
39. Takahashi F, Tomii K, Takahashi H. Electrochim. Acta 1986; 31:127–130.
40. Bonner WA. Origins of Life and Evolution of the Biosphere. 1990; 20:1–13.

4

Enantiodifferentiating Photosensitized Reactions

Yoshihisa Inoue
Osaka University, Suita, and
Japan Science and Technology Agency
Kawaguchi, Japan

I. INTRODUCTION

Enantiodifferentiating photosensitization is a unique photochemical invention, in which chiral information of an optically active sensitizer is transferred to a prochiral or racemic substrate through noncovalent interactions in the excited state, allowing chirality amplification [1]. Like the conventional catalytic and enzymatic asymmetric syntheses in the ground state, the enantiodifferentiating photosensitization is an essentially chirogen-efficient methodology, which necessitates only a catalytic amount of optically active compound as chiral sensitizer/catalyst. This approach can be distinguished from other asymmetric photochemical strategies, which employ circularly polarized light, chiral complexing agents, chiral auxiliaries, chiral supramolecular hosts, and chiral crystal lattices, since the chiral interaction occurs exclusively in the excited state, often involving an exciplex intermediate. Hence chiral sensitizers can influence the stereochemical outcome of prochiral or racemic substrates through the excited-state sensitizer–substrate interactions, including quenching, exciplex formation, relaxation, and/or decay processes. It should be noted that the sensitization may proceed through a singlet, triplet, or electron transfer mechanism, depending on the nature of sensitizer, substrate, and solvent (e.g., energy relationship, spin multiplicity, and solvent polarity). These features do not mean complication but are rather advantages of photosensitized asymmetric synthesis, providing us with many accesses to the dynamic control of photochemical enantiodifferentiation processes.

Since the very first example of enantiodifferentiating photosensitization reported by Hammond and Cole in 1965 [2], studies of enantiodifferentiating photosensitization have focused mostly on unimolecular reactions such as decomposition and isomerization. These studies have revealed the unique features of enantiodifferentiating photosensitization, providing us with a variety of intriguing mechanistic and synthetic insights as well as high optical yields, the details of which will be described below. Recent studies have also dealt with bimolecular photoreactions, such as cyclodimerization and polar addition. For these reactions, the stereocontrol of termolecular interactions (of chiral sensitizer, substrate, and reagent in the excited state) and the elucidation of the enantiodifferentiation mechanism appear to be more difficult to achieve. Nevertheless, recent endeavors to explicate the detailed enantiodifferentiation mechanism have led to several general strategies for more precise stereocontrol and also for obtaining greatly improved optical yields in photosensitized enantiodifferentiating bimolecular reactions. Traditionally, these uni- and bimolecular enantiodifferentiating photosensitized reactions are executed under conventional conditions using chiral sensitizers in isotropic fluid solutions. More recently, novel supramolecular approaches, exploiting chiral sensitizing host molecules, molecular assemblies, and organized media have also been developed. In this strategy, allowing the control of a wide variety of noncovalent interactions in both ground and excited states, the scope of enantiodifferentiating photosensitization will be further expanded; see Chap. 8 for details.

The title topic has been reviewed frequently in the general context of chiral photochemistry [1,3], and hence we will only briefly refer to the earlier works and put more emphasis on the latest results in this chapter devoted solely to enantiodifferentiating photosensitized reactions.

II. ENANTIODIFFERENTIATING UNIMOLECULAR PHOTOREACTIONS

Historically, this is the first and most investigated area of enantiodifferentiating photosensitization, in which the chirality transfer occurs upon energy or electron transfer from chiral sensitizer to reactant, often via an exciplex intermediate, with the excited species produced undergoing a unimolecular process to give an optically active reactant and/or product. This category of photoreaction may be divided into three subcategories [1b,c]: (1) photodestruction/production, (2) photoderacemization, and (3) geometrical photoisomerization. Since the previous general reviews [1a–c] were published, several important reactions and phenomena have been reported in particular for the last subcategory.

A. Photodestruction/production

This class of photosensitized enantiodifferentiations includes a variety of irreversible photodecomposition and photorearrangement reactions induced by energy or electron transfer from chiral sensitizers to racemic substrates.

The first enantiodifferentiating electron transfer and subsequent decomposition of *rac*-Co(acac)$_3$ **1** to achiral Co(acac)$_2$ was sensitized by Δ-Ru(bpy)$_3$ **2** to exhibit circular dichroism spectral growth corresponding to Δ-**1** of 4% *ee* (Scheme 1, top) [4]. The relative rate of enantiodifferentiating quenching by Δ- and Λ-**1** is a modest $k_\Lambda/k_\Delta = 1.08$. Phosphorescence quenching of Δ-**3** by (+)- and (−)-Co(edta)$^-$ was demonstrated to occur with a similar order of enantiodifferentiation, $k_+/k_- = 1.1$–1.2 [5]. Ruthenium complexes with chirally modified bipyridine ligands **3** (Scheme 1, R = (−)-menthyloxycarbonyl) and **4** (R = (−)-1-phenylethylaminocarbonyl) were also used for the same reaction to give better enantioselective quenching and decomposition of **1**, $k_\Lambda/k_\Delta = 1.33$ [6]. This system has further been explored in recent years by the same group to give higher enantiodifferentiation [7], details of which will be described in Chap. 7. A higher enantiodifferentiation was achieved upon phosphorescence quenching of Δ- and Λ-**3** by an optically active methylviologen derivative **5**$^{2+}$, affording $k_\Lambda/k_\Delta = 1.95$, although the relative rate of enantiodifferentiating reduction, affording **5**$^{+\cdot}$, was reduced to $k_\Lambda/k_\Delta = 1.32$ due to the competing back electron and the dissociation of the geminate ion pair (Scheme 1) [8].

A rather rare example of a successive enantiodifferentiating photodestruction/production process was reported for 3*H*-2-oxepinone **6** [9a].

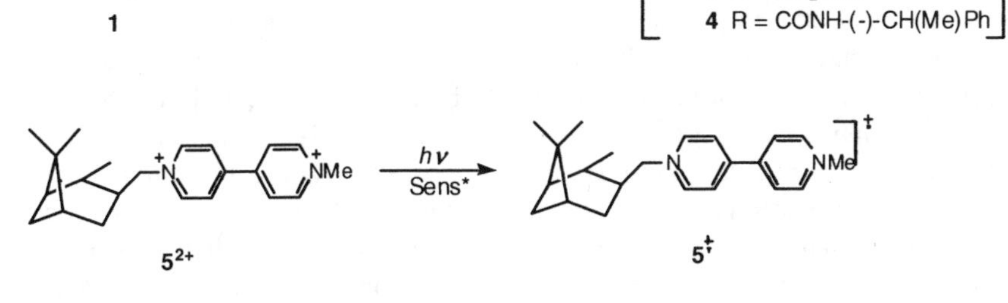

Scheme 1

Upon photosensitization with optically active naphthalenecarboxamide **9**, naphthalenecarboxylate **10**, and naphthyl amide **11**, *rac*-**6** first rearranges to the labile 7*H*-2-oxepinone **7**, which in turn undergoes di-π-methane rearrangement to give oxabicylco[4.1.0]heptenone **8** as the final product (Scheme 2, top). The enantioselectivity upon excitation, as determined from the recovered substrate **6** at 50% conversion, was low, with the highest k_R/k_S of 1.04 obtained with the chiral amide sensitizer **9**. Unfortunately, the specific rotation of optically pure **8** was not known at that time, and hence the enantioselectivity upon the second enantiodifferentiating sensitization step was not determined; only relatively small optical rotations were reported for product **8**, showing a tendency similar to that of the k_R/k_S of recovered **6**. A photosensitized enantiodifferentiating oxa-di-π-methane rearrangement was also reported for bicyclo[2.2.2]oct-5-en-2-one **12** (Scheme 2, middle) [9b]. Upon triplet sensitization with chiral indanone **14**, bicyclooctenone **12** gave tricyclooctanone **13** in 4.5% *ee* at room temperature and in 10% *ee* at −78°C after 7–44% conversion. No *ee* was reported for the remaining substrate **12**.

More recently, an interesting solid-state approach was employed for the enantiodifferentiating di-π-methane rearrangement of benzonorbornadienecarboxylic acid **15** to tricyclic **16** (Scheme 2, bottom) [10]. In this strategy, differing from the usual ionic chiral auxiliary method [11], enantiopure amino acid esters of aromatic ketones **17** and **18**, introduced as countercations of the prochiral substrate **15** in the crystalline salt, played the dual role of ionic chiral auxiliary and triplet sensitizer upon solid-state photoirradiation. By using the acetylbenzyl ester of L-phenylalanine **17** as a chiral sensitizer/countercation, moderate-to-good *ee*s of up to 70% were obtained upon irradiation at −20°C, while the benzoylphenyl ester of L-valine **18** gave **16** in 84% *ee* at room temperature and 91% *ee* at −20°C. The effects of temperature and irradiation period upon the product *ee* are small to moderate, giving more or less lower *ee*s at higher temperatures and conversions. Although an equimolar amount of chiral sensitizer is inevitably required in this strategy, the enantiomeric excess of 91% obtained at the 100% conversion is remarkable.

Inter- and intramolecularly sensitized enantiodifferentiating photodecompositions of pyrazoline derivatives **19t** and **23** were also examined [12]. The triplet sensitized photodecomposition of *trans*-3,5-diphenylpyrazoline **19t** with (−)-rotenone **21** and (+)-testosterone **22** afforded chiral *trans*- and achiral *cis*-1,2-diphenylcyclopropanes **20t** and **20c** (Scheme 3, top). The enantioselectivity, determined by monitoring the circular dichroism (CD) spectral changes upon photodecomposition of **19t**, was not very high, giving k_R/k_S values of 1.080 and 1.016 upon sensitization with **21** and **22**, respectively. For the intramolecular photosensitized decomposition (although this is a diastereodifferentiating process in the strict sense), (+)-camphor-3-spiropyrazoline **23** was chosen as a chiral substrate with a built-in sensitizer, where the camphor ketone moiety was used as a chiral antenna chromophore and intramolecular singlet/triplet sensitizer

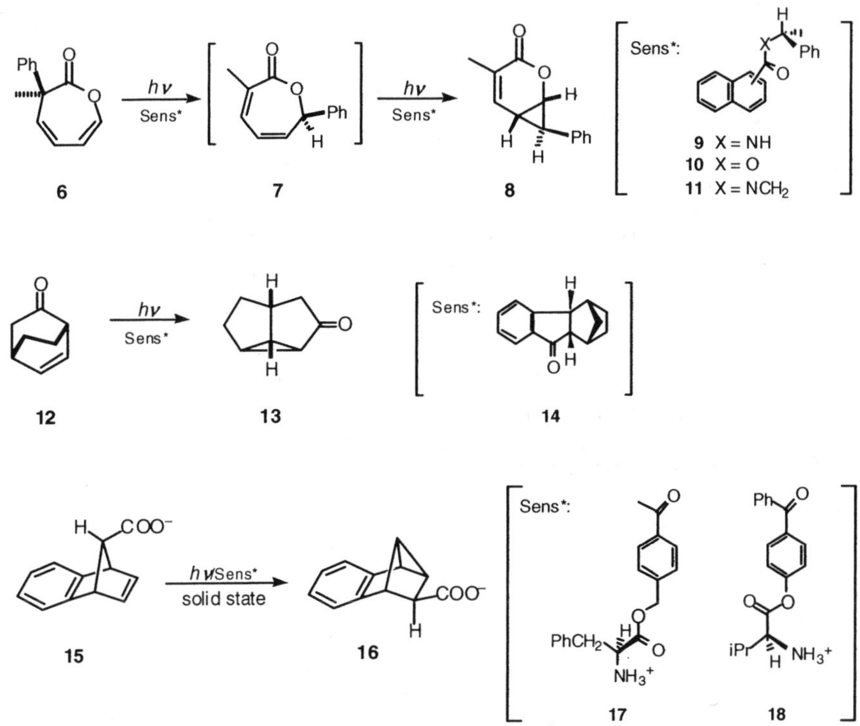

Scheme 2

(Scheme 3, bottom). If the energy is transferred to the enantiomeric pyrazoline moiety at different rates, appreciable CD signal may evolve in the course of the photolysis. However, the energy transfer was found to be too fast and/or too efficient to discriminate the enantiomeric configuration of the substrate moiety.

B. Photoderacemization

Photosensitized deracemization, or enantiomerization, is a method for shifting the equilibrium between enantiomers through the excited-state interaction with a chiral sensitizer. This is unique to photochemistry, as the ground-state thermodynamics do not allow the deviation of the equilibrium constant from unity. However, only a limited amount of effort has hitherto been devoted to this unique methodology. Thus practically only two types of substrate, i.e., sulfoxide and allene, have been subjected to photosensitized deracemization (Scheme 4), and the reported examples do not appear to be very successful.

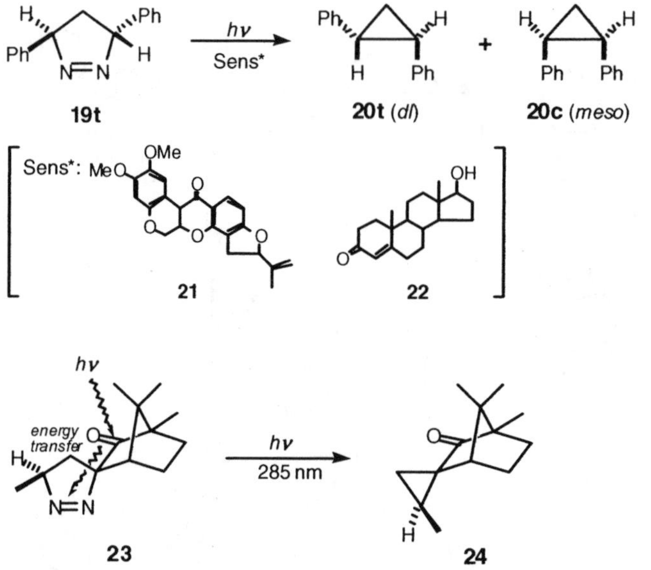

Scheme 3

It was known that photoracemization of optically active sulfoxides can be induced upon sensitization with naphthalene [13]. Hence, by using (R)-N-acetyl-1-naphthylethylamine **26** as a chiral sensitizer, the first photoderacemization of methyl p-tolyl sulfoxide **25a** was performed to give an *ee* of 2.25% at the apparent photostationary state (pss) after prolonged irradiation [14]. However, more detailed examination, starting with **25a** of 3.2% *ee* and 5.1% *ee*, led to the conclusion that the ultimate value at the pss is 4.1% *ee*. With more a bulky substrate **25b**, a higher *ee* of 12% was obtained upon photosensitization with (+)-N-(trifluoromethyl)-1-naphthylethylamine **27** [15].

Chiral allenes can also be deracemized by photosensitization. Thus 2,3-pentadiene **28** was photoderacemized upon sensitization with the chiral aromatic steroid, 21,22-dihydroneoergosterol **29** to give 3.4% *ee* at the pss [16]. Cyclic allene **30** was also subjected to photoderacemization sensitized by (−)-menthyl or (−)-bornyl benzene(di)carboxylate, but the *ee*s obtained were low (< 5%) [17].

C. Geometrical Photoisomerization

1. Cyclopropanes and Oxiranes

The first successful enantiodifferentiating photosensitization was achieved in 1965 for the geometrical photoisomerization of 1,2-diphenylcyclopropane **20**, in

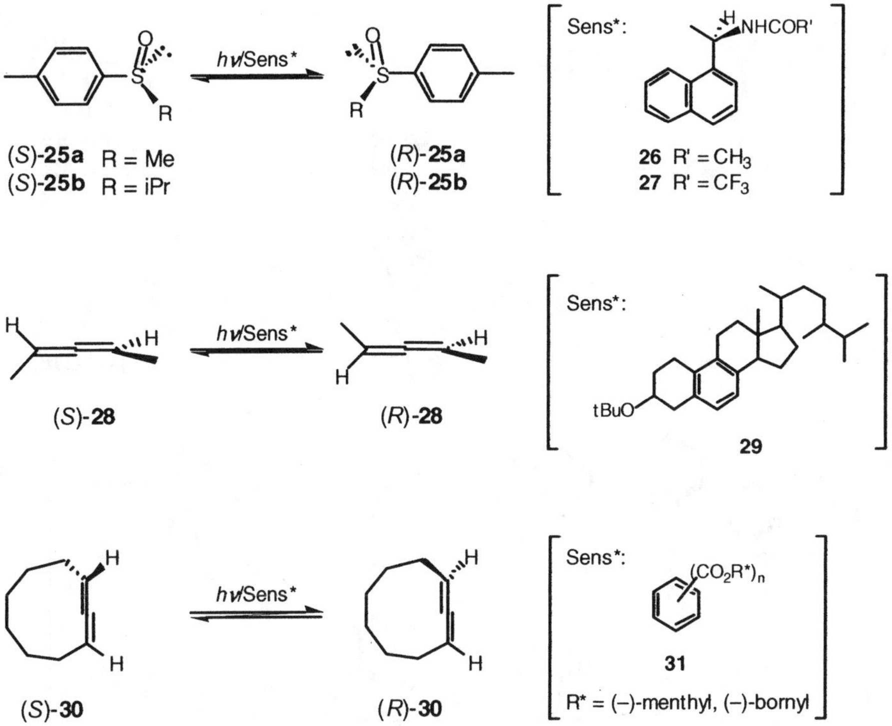

Scheme 4

which racemic *trans*-1,2-diphenylcyclopropane **20t** was employed as the starting material and (*R*)-*N*-acetyl-1-naphthylethylamine **26** as the chiral sensitizer to give optically active **20t** after prolonged irradiation (Scheme 5) [2]; afterward, the singlet mechanism was established for this sensitization [18]. This enantiodifferentiation process, starting with a racemic mixture of **20t**, may be called deracemizing photoisomerization, as the enantiomeric enrichment is realized not through direct photoderacemization but via the cis–trans isomerization of the substrate. Although the obtained *ee* was not very high (6.7% *ee*) [19], this pioneering work by Hammond and Cole opened a new era in asymmetric photochemistry, unambiguously demonstrating that chiral discrimination can occur through excited-state interactions between chiral sensitizer and prochiral substrate.

Several research groups further examined this particular reaction by using a variety of chiral singlet, triplet, and electron transfer sensitizers, illustrated in Scheme 5. Triplet sensitization of **20** with chiral aromatic ketones **34–39** [20,21],

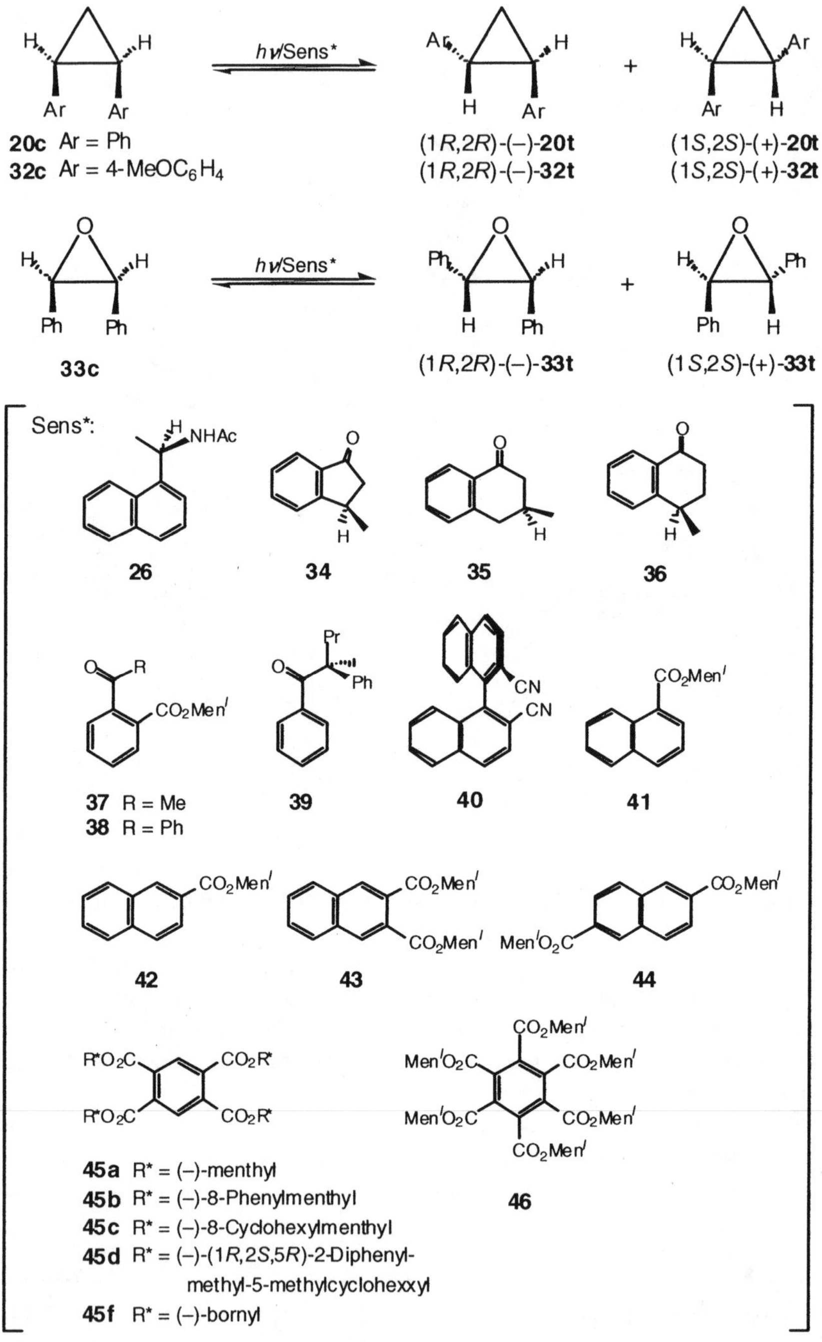
hν/Sens*
20c Ar = Ph
32c Ar = 4-MeOC6H4
(1R,2R)-(−)-20t
(1R,2R)-(−)-32t
(1S,2S)-(+)-20t
(1S,2S)-(+)-32t
hν/Sens*
33c
(1R,2R)-(−)-33t
(1S,2S)-(+)-33t
Sens*:
26
34
35
36
37 R = Me
38 R = Ph
39
40
41
42
43
44
45a R* = (−)-menthyl
45b R* = (−)-8-Phenylmenthyl
45c R* = (−)-8-Cyclohexylmenthyl
45d R* = (−)-(1R,2S,5R)-2-Diphenyl-
methyl-5-methylcyclohexxyl
45f R* = (−)-bornyl
46

Scheme 5

and singlet sensitization with chiral naphthoate **42** [22], afforded poor *ee*s of 1–3%, while the photosensitization of **20c** with chiral naphthylethyl amide bound to a silica surface gave the practically racemic product (1% *ee*) [23]. It was also shown that chiral 1,1'-bis(2-cyanonaphthyl) **40** can sensitize the cis–trans isomerization of **20** through a photoinduced electron transfer mechanism to afford **20t** in low *ee*s: i.e., 0.1% *ee* in acetonitrile at room temperature and 4% *ee* in toluene at $-30°C$ [24]. The extremely low *ee* obtained in acetonitrile is attributable to the intervention of a free or solvent-separated radical ion pair, which collapses the intimate sensitizer–substrate interactions indispensable for enantiomeric photoisomerization.

However, photosensitization of **20** with optically active (poly)alkyl benzene(poly)carboxylates **45a–d,f** and **46** (Men^l = (−)-menthyl) led to improved *ee*s of up to 10% in pentane, 8–9% in ether, and 3–4% in acetonitrile (all at 25°C), whereas the use of naphthalene(di)carboxylates **41–44** as chiral sensitizer resulted in low *ee*s of 0–2% even in pentane at 25°C [25]. Judging from the rapidly growing *ee* obtained upon sensitization of the pure *cis*-isomer **20c**, it is plausible that the enantiodifferentiating step is not the decay from the intervening exciplex or radical–ion pair but the quenching of the chiral sensitizer by the trans isomer **20t** [25]. Interestingly, the antipodal photoproducts were obtained by changing the solvent polarity, affording (+)-**20t** in 4.7% *ee* in pentane and (−)-**20t** in 4.1% *ee* in acetonitrile at 0°C. Although the product *ee*s reported are not particularly high, this is the first case that clearly demonstrates the switching of product chirality by solvent. A 1,3-biradical mechanism involving singlet species in a nonpolar solvent, or a triplet species produced from the back electron transfer of the initially formed radical ion pair, was proposed for the sensitized photoisomerization of diphenylcyclopropane **20** on the basis of the less pronounced effect of solvent polarity and the appreciable *ee*s obtained even in acetonitrile [25].

In contrast, the more electron-rich 1,2-dianisylcyclopropane **32** and 1,2-diphenyloxirane **33** gave merely racemic trans isomers **32t** and **33t** (*ee* < 1%) upon sensitization with chiral benzenepolycarboxylates **45** and **46** in pentane or acetonitrile [25]. This was attributed to the intervention of a radical cationic or zwitterionic ring-opened intermediate, which freely racemizes in the absence of a chiral sensitizer in its vicinity, particularly in polar solvents.

2. Cyclooctene

Constrained (*E*)-isomers of medium-sized cycloalkenes and cycloalkadienes have long been known to be chiral and indeed the enantiomerically pure or enriched (*E*)-cyclooctene/cyclooctadienes were resolved optically and/or synthesized via thermal reactions in the 1960s [26]; photochemical preparation of (*E*)-cyclooctene from the (*Z*)-isomer was also reported in the late 1960s [27]. Nevertheless, the photochemical asymmetrical synthesis of (*E*)-cycloalkenes was not reported until 1978, when the enantiodifferentiating geometrical photoisomerization of (*Z*)-cyclooctene **47Z** (Scheme 6) was examined for the first time by using chiral benzenecarboxylate sensitizers to give optically active (*E*)-isomer **47E** in low *ee* (≤ 4%) [28]. A subsequent study of the enantiodifferentiating photosensitizations of **47Z** [29] did not lead to an improvement of the product *ee*, affording a best *ee* of 4% for **47E**. Such rather disappointing results would often have discouraged further exploration of the enantiodifferentiating photosensitization of cycloalkene isomerization.

Temperature Effects.   A breakthrough was made in 1989 for the enantio-differentiating photoisomerization of cyclooctene **47** by using chiral *ortho*-ben-zenepolycarboxylates **49a,f** and **45a,f** as singlet sensitizers and also by changing the irradiation temperature [30]. Aside from the greatly improved product *ee*s of up to 41% obtained for **47E** upon sensitization with (−)-bornyl pyromellitate **45f** at −88°C, a more intriguing phenomenon observed is the unprecedented switching of product chirality caused by altering the irradiation temperature (Fig. 1). From a detailed mechanistic study, it was revealed that the rotational relaxation around the olefinic C=C bond in the exciplex of chiral sensitizer with prochiral substrate **47Z** is the critical enantiodifferentiating process ($k_S \neq k_R$), while the quenching of chiral sensitizer by (*S*)- and (*R*)-**47E** is essentially nonenantiodiffer-entiating ($k_{qS} = k_{qR}$); see Scheme 7.

This apparently unexpected inversion of product chirality by temperature (*T*) is reasonably accounted for in terms of the pronounced contribution of the entropy term in the enantiodifferentiation step, as can be recognized from the differential Eyring Eq. (1):

$$\ln(k_S/k_R) = -\Delta\Delta H_{S\text{-}R}^{\ddagger}/RT + \Delta\Delta S_{S\text{-}R}^{\ddagger}/R \qquad (1)$$

where k_S and k_R represent the formation rate constants for (*S*)- and (*R*)-**47E**, and $\Delta\Delta H_{S\text{-}R}^{\ddagger}$ and $\Delta\Delta S_{S\text{-}R}^{\ddagger}$ the differential activation enthalpy and entropy for enantiodifferentiation, respectively. As can be seen from Eq. (1), the enthalpy term governs the product chirality at low *T*, while the entropic contribution be-comes dominant as *T* increases, leading to the inversion of the sign of $\ln(k_S/k_R)$ at the critical, or equipodal, temperature (T_0). This is exactly the case with the photosensitization by *ortho*-benzenepolycarboxylates, as exemplified for **45a** in Fig. 1.

Since then, similar temperature-switching behavior has been frequently ob-served upon sensitization with *ortho*-benzenepolycarboxylates **45**, **46**, and **49** with a variety of chiral auxiliaries **a–n** and **p–u**, illustrated in Scheme 6 [31,32], and the best *ee*s of product **47E** obtained at ambient and low temperatures were further raised up to 49% (at 25°C) and 64% (at −89°C) by using a highly bulky chiral sensitizer **45c** [32].

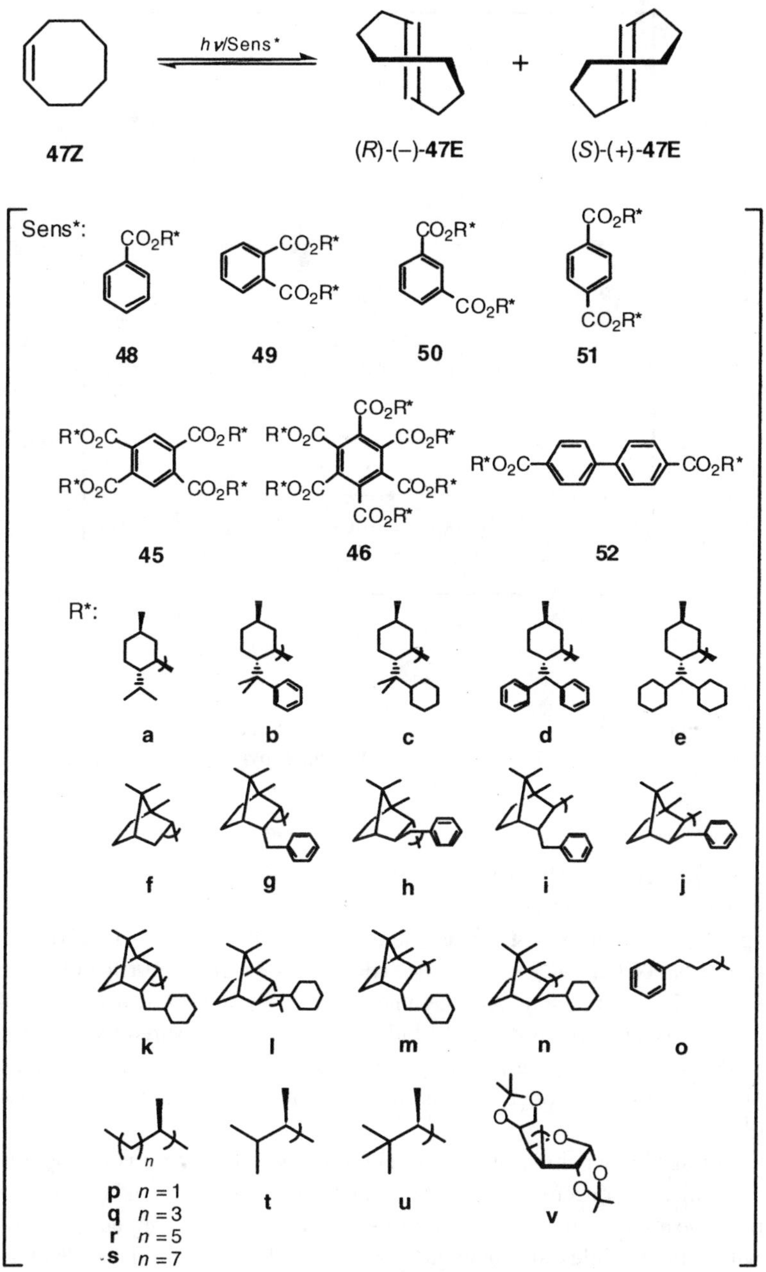

Scheme 6

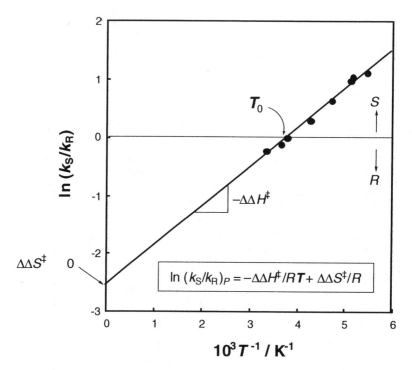

Figure 1 Temperature switching of product chirality in enantiodifferentiating photo-isomerization of cyclooctene **47** sensitized by (−)-menthyl pyromellitate **45a** in pentane.

Steric effects of the chiral auxiliary upon the product *ee* were systematically investigated by employing two series of 1-methylalkyls (R*) of varying chain length (**p–s**) and bulkiness (**p, t**, and **u**) (Scheme 6); (1) R* = $CH_3CH(CH_2)_nCH_3$ (n = 1, 3, 5, and 7) and (2) R* = CH_3CHR' (R' = Et, iPr, and tBu) [31]. As expected, the differential activation parameters $\Delta\Delta H^{\ddagger}_{S-R}$ and $\Delta\Delta S^{\ddagger}_{S-R}$ are critical functions of the chain length n and the bulkiness of R' as measured by Eliel's A value. However, the changing profile is significantly different for each variant, showing a saturation for longer chains of $n \geq 3$ or a steady increase even for R' = tBu.

The dramatic switching of product chirality by temperature is entropic in origin, for which the different degrees of conformational changes induced by the rotational relaxation to the enantiomeric twisted cyclooctene singlets (R)- and (S)-^1p within the exciplex are thought to be responsible [30,31]. This idea was supported experimentally by using chiral pyromellitate sensitizers **45g–j**. Carrying chiral auxiliaries with an electron-rich aromatic substituent as a donor moiety

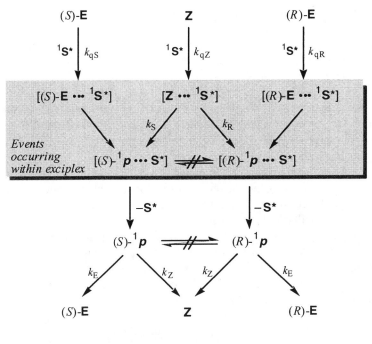

Scheme 7

(D), the benzenepolycarboxylate core behaves as the acceptor moiety (A) to form an A-D-D′-type triplex with substrate cyclooctene (D′) in the excited state. The triplex formation is expected to affect the enantiodifferentiating step, and therefore the activation parameters, through the control of conformational flexibility. Indeed, striking differences were observed in the photosensitization behavior of *endo,endo-* and *exo,exo*-3-benzyl-2-bornyl pyromellitates **45g,j** and *endo,exo-* and *exo,endo*-3-benzyl-2-bornyl pyromellitates **45h,i**. Thus the *endo,endo-* and *exo,exo*-sensitizers **45g,j** consistently give low *ee*s over a wide range of temperature with very small $\Delta\Delta H^{\ddagger}_{S\text{-}R}$ and $\Delta\Delta S^{\ddagger}_{S\text{-}R}$ values. In sharp contrast, the *endo,exo-* and *exo,endo*-sensitizers **45h,i** led to highly temperature-dependent *ee*s and large activation parameters. This contrasting behavior is reasonable, since the former sensitizers (**45g,j**), carrying a benzyl group *syn* to the pyromellitate moiety, can readily form a triplex upon interaction with cyclooctene in the excited state, which in turn reduces the conformational freedom resulting in the temperature-insensitive *ee*s. For the latter, sensitizers with *anti*-benzyl group (**45h,i**) and the formation of intramolecular exciplex and of triplex with **47** are not sterically feasible, and the conformationally flexible chiral moiety leads to dynamic dependence of product *ee* upon temperature [32].

Pressure Effects.   Since hydrostatic pressure can alter the freedom and conformation of molecules in the system through the volume change, one can use pressure as an entropy-related tool for controlling the reaction rates and equilibria in the ground and excited states [33]. Indeed, the regio- and diastereoselectivities of photocycloadditions are known to be significantly affected by applying pressure [34], while the pressure effect upon enantiodifferentiating photoreaction has only recently been examined. The enantiodifferentiating photoisomerization of cyclooctene **47Z**, sensitized by chiral benzene(poly)carboxylates **45**, **46**, and **48–53** with chiral auxiliaries **a**, **f**, and **p–r** (Scheme 6), was investigated under pressures ranging from 0.1 to 400 MPa (1–4000 bar) [35]. Upon sensitization in particular with chiral *ortho*-benzenepolycarboxylates **45**, **46**, and **49**, the *ee* of the photoproduct **47E** displayed a critical dependence on pressure. In an extreme case using sensitizer **45a**, the product chirality was inverted at a critical pressure (P_0) of 120 MPa (Fig. 2), as was the case with the temperature variation described above.

The pressure dependence of relative rate constant k_S/k_R at a constant temperature is expressed by Eq. 2:

$$\ln(k_S/k_R)_T = -(\Delta\Delta V^{\ddagger}_{S\text{-}R}/RT)\, P + C \tag{2}$$

where $\Delta\Delta V^{\ddagger}_{S\text{-}R}$ is the activation volume difference between the transition states of (*S*)- and (*R*)-$^1\boldsymbol{p}$ in Scheme 7, while the constant C is equal to $\ln(k_S/k_R)$ at $P = 0$. Indeed, the plot of $\ln(k_S/k_R)$ as a function of P gives a good straight line, as exemplified for **45a** in Fig. 2. The $\Delta\Delta V^{\ddagger}_{S\text{-}R}$ obtained from the slope varies widely from 0.6 to 5.5 mL mol^{-1} at 25°C, some of which are exceptionally large as *differential* values (if compared with the regular $\Delta V^{\ddagger}$ values reported for thermal isomerizations [33]), demonstrating the substantial volume difference between the diastereomeric transition states. In this light, one might expect some correlation between the $\Delta\Delta S^{\ddagger}_{S\text{-}R}$ and $\Delta\Delta V^{\ddagger}_{S\text{-}R}$ values, which is however completely absent, as the temperature and pressure are mutually independent parameters in both a thermodynamic and a kinetic sense.

The pressure effect upon this enantiodifferentiating photoisomerization was further investigated at low temperatures down to -10°C, giving larger (for **45a** and **46r**) or smaller slopes or $\Delta\Delta V^{\ddagger}_{S\text{-}R}$ values (for **46a,f**). From these experiments, three-dimensional $\ln(k_S/k_R)$-*versus*-P-and-T^{-1} diagrams, illustrating the three categories of the *pressure-and-temperature* dependence of product *ee* and chirality were elucidated for the first time [35]. Very recently, this system was examined at yet higher pressures of up to 750 MPa, and the pressure effect on product *ee*

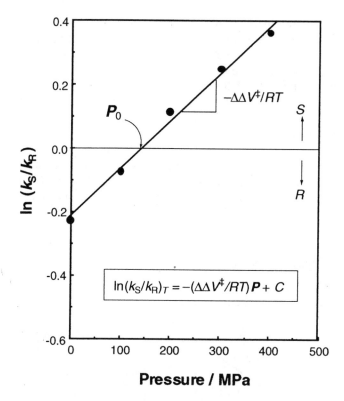

Figure 2 Pressure switching of product chirality in enantiodifferentiating photoisomerization of cyclooctene **47** sensitized by (−)-menthyl pyromellitate **45a** in pentane at 25°C.

was demonstrated not to be uniform but discontinuous over a wider pressure range [36]. Thus the plots of $\ln(k_S/k_R)$ against P gave two bends at 200 and 400 MPa, indicating the involvement of different excited states or conformations of the sensitizer. The latter seems more plausible since the pressures that give the bends are common to all of the four sensitizers employed.

Solvent Effects. From an entropic point of view, solvation is one of the most readily maneuverable environmental factors that directly manipulates the freedom of the relevant transition state. However, essentially no solvent effect upon product *ee* was observed in the photoisomerization of **47** sensitized by benzenepolycarboxylates **45**, **46**, and **48–53** with a variety of chiral (ar)alkyl groups **a–n** and **p–u** (Scheme 6).

Intriguingly, pyromellitate **45v**, possessing a protected saccharide auxiliary **v**, exhibited amazingly contrasting behavior in polar and nonpolar solvents [37].

At room temperature, this sensitizer affords (R)-$(-)$-**47E** in low *ee*s of around 5% in both pentane and diethyl ether. However, as the temperature decreases, the product *ee* starts to deviate from the original value to the *opposite* direction to give the antipodal products in these two solvents, as shown in Fig. 3. Then (R)-$(-)$-**47E** was produced in 40% *ee* in pentane at $-78°C$, while (S)-$(+)$-**47E** was obtained in ether in 50% *ee* at $-78°C$ and in a record-high *ee* of 73% at $-110°C$.

The excellent enthalpy–entropy compensation plot, obtained by using the $\Delta\Delta H^{\ddagger}_{S-R}$ and $\Delta\Delta S^{\ddagger}_{S-R}$ values, confirms that the enantiodifferentiation mechanism is not altered by a change in solvent [37,38]. Judging from the clearly segregated activation parameters obtained in polar solvents (positive $\Delta\Delta H^{\ddagger}_{S-R}$ and $\Delta\Delta S^{\ddagger}_{S-R}$) and nonpolar solvents (negative $\Delta\Delta H^{\ddagger}_{S-R}$ and $\Delta\Delta S^{\ddagger}_{S-R}$), this unprecedented sol-

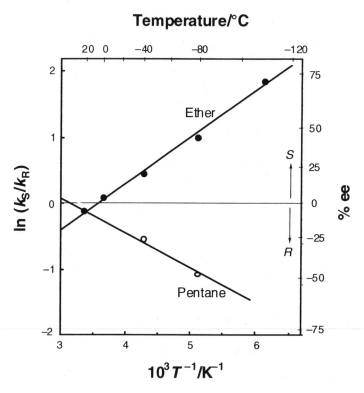

Figure 3 Solvent switching of product chirality in enantiodifferentiating photoisomerization of cyclooctene **47** sensitized by diacetone glucose (DAG) pyromellitate **45v** in pentane and ether.

vent switching of product chirality, specific to the saccharide sensitizer, is attributable to the solvation of the ether groups of the sensitizer's protected saccharide moieties. The existence of the selective solvation of the ether, probably through dipole–dipole interaction, is further supported by the nonlinear dependence of the product *ee* against the ether content in pentane, as illustrated in Fig. 4 [37]. In this case, the product chirality was switched at 8% ether, which is regarded as the antipodal concentration (C_0).

The above result unequivocally demonstrates that the seemingly weak solvation effect can play a decisive role, which controls and even switches the stereochemical outcome of the enantiodifferentiating photoisomerization. Temperature, pressure, and solvation, which function as environmental factors to control the enantiodifferentiation in the excited state, are all entropic in nature. Probably the key is the critical control of the weak interactions involved in the exciplex intermediate, as with the biological and supramolecular interactions in

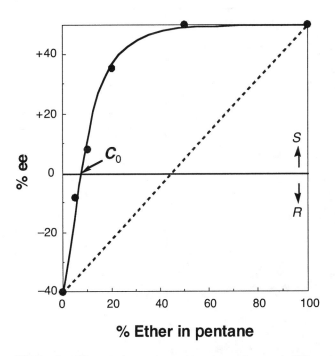

Figure 4   Effect of ether cotent upon product *ee* in enantiodifferentiating photoisomerization of cyclooctene **47** sensitized by diacetone glucose (DAG) pyromellitate **45v** in pentane–ether mixture at $-78°C$.

the ground state. It may be concluded therefore that we can rely on the entropy-related factors as long as weak interactions are involved in the essential step.

Supercritical Fluids. In view of the vital contribution of the entropy-related factors in the enantiodifferentiating photosensitizations described above, supercritical fluids (SCF) are of particular interest, since the use of an SCF as a reaction medium provides us with the rare opportunity to control enantioselectivity [39] through the dynamic changes of the solvent property and clustering behavior in relatively narrow ranges of pressure and/or temperature [40]. Indeed, the pronounced effects of SCF on the product chirality and enantioselectivity were reported recently for the enantiodifferentiating photoisomerization of cyclooctene **47** sensitized by chiral sensitizers (**45a,b,v**) in supercritical carbon dioxide (scCO$_2$) [41]. As can be seen from Fig. 5, the *ee* of product **47E** in scCO$_2$ is not

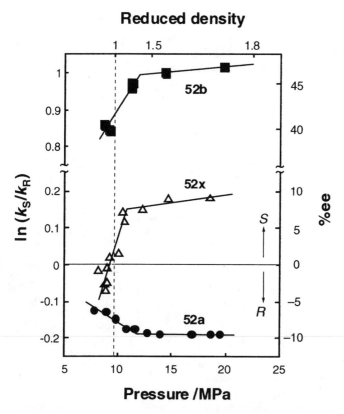

Figure 5 Pressure dependence of the product *ee*, in enantiodifferentiating photoisomerization of **47Z** sensitized by **45a**, **45b**, and **45x** in scCO$_2$ at 45°C.

a simple function of pressure but is undoubtedly discontinuous around the critical density, affording a bent plot for all the sensitizers. As a consequence of the very steep slope of the $\ln(k_S/k_R)$-*versus-P* plot, the differential activation volumes ($\Delta\Delta V^{\ddagger}$) obtained are extraordinarily large in the near-critical density region ($P = 8$–12 MPa), reaching 45–270 mL mol^{-1}, while scCO$_2$ at higher pressures ($P \geq 13$ MPa) gives much reduced, but more reasonable, $\Delta\Delta V^{\ddagger}$ values of 1.0–12.1 mL mol^{-1}, which are however still larger than the corresponding values obtained in pentane, i.e., $\Delta\Delta V^{\ddagger} = 0.2$–$3.7$ mL mol^{-1}. The unusually high sensitivity of *ee* to pressure in scCO$_2$, particularly in the near-critical density region, is ascribed to *solvent clustering*, which makes the local density of CO$_2$ around the solute molecules much higher than the bulk density [42], endowing considerably different microenvironmental cluster structures around the two diastereomeric exciplexes shown in Scheme 7. It is also interesting to note that, although the pressure dependence of *ee* gives a bent plot with two slopes (or two $\Delta\Delta V^{\ddagger}$ values) as illustrated in Fig. 5, the sign of the slope does not change before and after the bend in each case. This may indicate that the pressure effect on the *ee* is merely exaggerated by highly fluctuating CO$_2$ clustering near the critical density and that the exciplex structure is not essentially altered by clustering. Thus it is concluded that the entropy control of product chirality is more effective and critical in scCO$_2$ near the critical density than in liquid phase or highly compressed scCO$_2$.

Sensitization with Chiral Aromatic Amides, Phosphoryl Esters. Besides the above-mentioned studies employing aromatic carboxylic esters as sensitizers and naturally occurring alcohols as chiral auxiliaries, some attempts have been made to use other types of sensitizers, such as aromatic amides [43] and phosphoryl esters [44] with ($-$)-menthyl, as well as synthetic C_2-symmetric chiral auxiliaries, shown in Scheme 8. These chiral compounds can efficiently sensitize the Z–E photoisomerization of cyclooctene **47** to give moderate E/Z ratios of up to 0.17 and 028 and low-to-moderate *ee*s of up to 5% and 14% for the aromatic amides and phosphoryl esters, respectively [43,44].

Supramolecular Approaches. As can be seen from the foregoing discussion, most of the enantiodifferentiating photosensitizations of cyclooctene **47** have been conducted in conventional homogeneous solutions at varying temperatures and pressures. Recently a couple of supramolecular approaches have been attempted, using inherently chiral or chirally modified host-sensitizer systems, such as α- to γ-cyclodextrins modified with aromatic esters [45], nucleosides, DNA, and RNA [46], and NaY zeolite modified with chiral aromatic esters [47]. Although this topic is detailed in Chap. 8, it is still worth noting that the enantiodifferentiating photosensitization behavior in a supramolecular environment is not a simple extension of the homogeneous counterpart but significantly differs in the role of entropy factors. In this context, it is interesting that the product *ee* obtained in the photoisomerization of cyclooctene **47** included and sensitized by

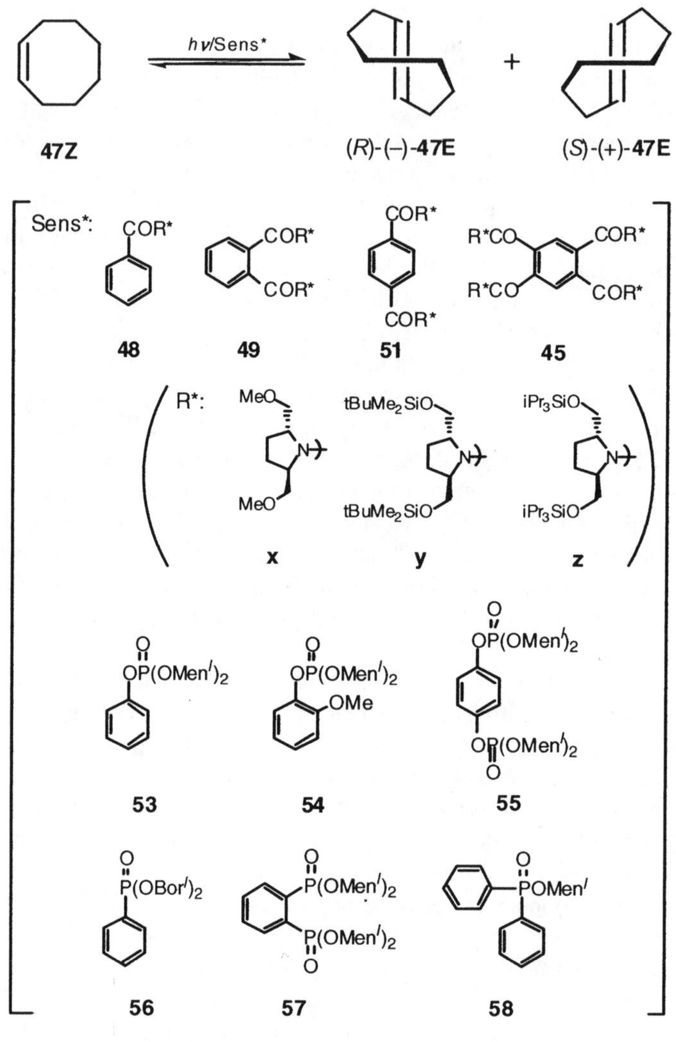
hν/Sens*
47Z
(R)-(–)-47E
+
(S)-(+)-47E
Sens*:
COR*
COR*
COR*
COR*
COR*
COR*
R*CO
R*CO
COR*
COR*
48
49
51
45
R*:
MeO
MeO
tBuMe2SiO
tBuMe2SiO
iPr3SiO
iPr3SiO
x
y
z
OP(OMen^l)2
OP(OMen^l)2
OMe
OP(OMen^l)2
OP(OMen^l)2
53
54
55
P(OBor^l)2
P(OMen^l)2
P(OMen^l)2
POMen^l
56
57
58

Scheme 8

β-cyclodextrin 6-O-phthalate does not immediately depend on the irradiation temperature but rather on the occupancy of the CD cavity, probably as a result of a nearly static quenching of the excited sensitizer moiety by **47** included in the cavity [45].

In the modified zeolite sensitization study, a significant enhancement of product *ee* was reported to occur by the introduction of a chiral sensitizer into the zeolite supercages that otherwise give an almost racemic product [47]. Further supramolecular approaches to the enantiodifferentiating photosensitization will lead to more solid conclusions on the mechanisms and factors that control supramolecular photochirogenesis and open a new channel to asymmetric supramolecular photochemistry in the near future; see Chap. 9.

Singlet Versus Triplet Sensitization. It is interesting to compare the effect of spin multiplicity upon the efficiency of chirality transfer in the excited state. For this purpose, two series of structurally related chiral sensitizers, i.e., (−)-menthyl benzene(di)carboxylates **48a**, **49a**, and **51a** as singlet sensitizers, and (−)-menthyl benzyl ether **59a** and xylylene ethers **60a** and **61a** as triplet sensitizers, were synthesized and used for the enantiodifferentiating photoisomerization of **47** (Scheme 9) [48]. It was found that the aromatic esters act as singlet sensitizers over a wide range of substrate concentrations up to 0.5 M of **47Z**, while the aromatic ethers exhibit mixed behavior, functioning as singlet sensitizers at high substrate concentrations (> 0.1 M) but as triplet sensitizers at several mM or lower concentrations of **47Z**, as judged from the photostationary state *E*/*Z* ratio and fluorescence quenching experiments. Although the absolute *ee* values obtained were low in general (≤ 10% *ee*), it was clearly revealed that the singlet sensitization gives much higher *ee*s than the triplet sensitization. This is probably due to the lack of a structurally well-defined "exciplex" intermediate in triplet sensitization. Hence the use of a chiral triplet sensitizer does not appear to be particularly beneficial and a conformationally well-structured singlet exciplex with a moderate charge-transfer character is more suitable for the excited-state chirality transfer through photosensitization [48].

3. Substituted and Tethered Cyclooctenes

Substituent effects on the enantiodifferentiating photosensitization of cyclooctene were examined by introducing a methyl group to cyclooctene. Thus the photoisomerization of (Z)-1-methylcyclooctene **62Z** was sensitized by (−)-menthyl benzoate **48a**, phthalate **49a**, isophthalate **50a**, terephthalate **51a**, trimesate, and pyromellitate **52a** to give moderate to extremely low *E*/*Z* ratios, which rapidly decrease with increasing number of ester groups (particularly at ortho position) in the sensitizer from *E*/*Z* = 0.21 for **48a** to 0.01 for **49a** and then to < 0.0005 for **52a** [49]. This result is entirely different from that obtained with **47**, which gives much higher *E*/*Z* ratios of up to 0.66 upon sensitization with **52a**. This

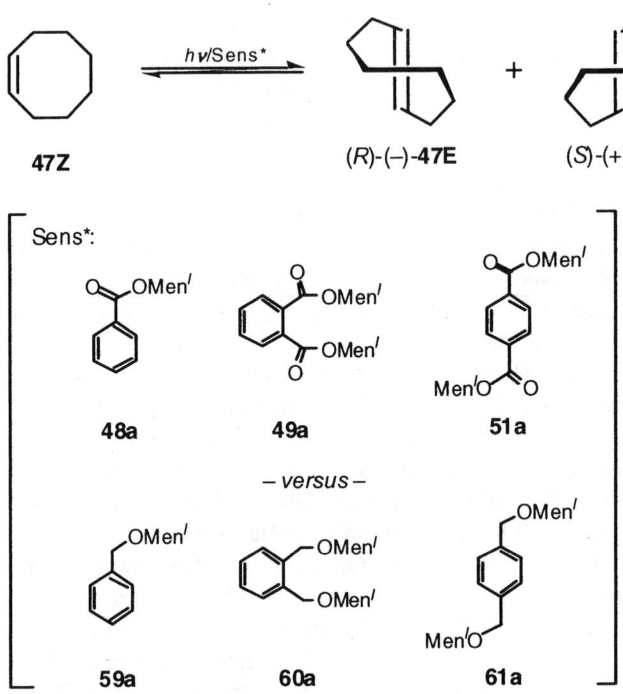

Scheme 9

contrasting behavior of the E/Z ratio for **62** is attributed to the steric hindrance of 1-methyl upon energy transfer through the exciplex formation with benzenepolycarboxylate sensitizers. As a consequence of the hindered intimate interactions of the exciplex, the ee of **62E** was generally low even at low temperatures, affording only 5–7% ee upon sensitization with **48a**, **50a**, **51a**, and trimesate.

In contrast, the intramolecular energy transfer in cyclooctene derivatives **63E–65E**, which are tethered to benzoate, isophthalate, and terephthalate moieties through a (R,R)-2,4-pentadiol linker, does not appear to be seriously affected by the steric bulk of 1-alkylation (Scheme 10) [50]. Upon irradiation at $-65°C$, **63**, **64**, and **65** afforded Z/E ratios of 0.8, 0.15, and 0.06, with diastereomeric excesses (des) of 33, 37, and 44%, respectively.

In relation to this, it is interesting to refer to the intramolecularly photosensitized diastereodifferentiating isomerization of 3-, 4-, and 5-benzoyloxycyclooctetenes **66Z–68Z** (Scheme 10) [51]. In particular, the de of photoproduct **66E** is a critical function of substrate concentration (Fig. 6), leading to a switching of the diastereoselectivity at 10 mM. This apparently unusual phenomenon is reasonably

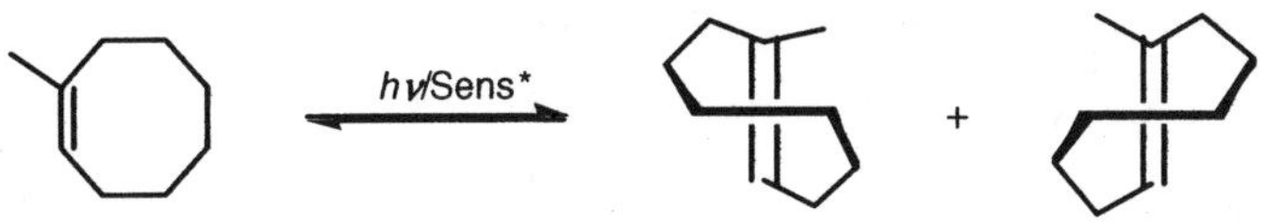

62Z (R)-62E (S)-62E

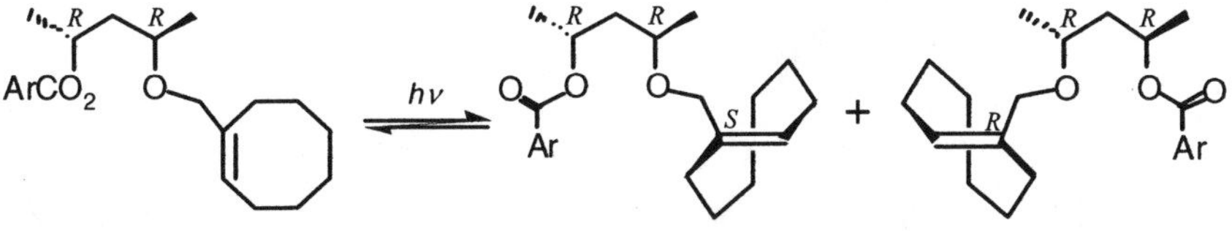

63E Ar = Ph (R,R,S)-63Z (R,R,R)-63Z
64E Ar = m-MeOCOC$_6$H$_4$ 64Z 64Z
65E Ar = p-MeOCOC$_6$H$_4$ 65Z 65Z

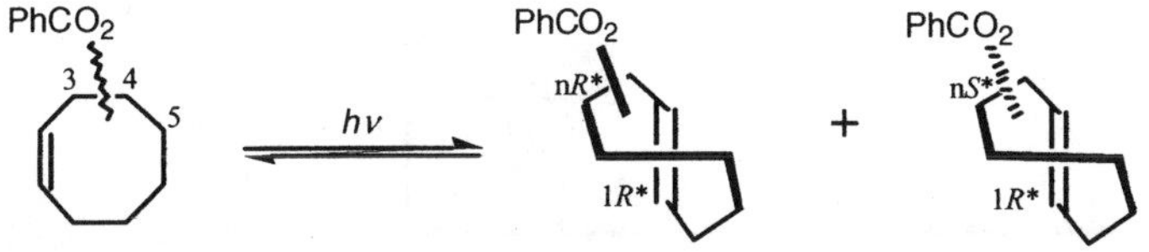

rac-66Z (3-PhCO2) (1R*,3R*)-66Z (1R*,3S*)-66Z
rac-67Z (4-PhCO2) (1R*,4R*)-67Z (1R*,4S*)-67Z
rac-68Z (5-PhCO2) (1R*,5R*)-68Z (1R*,5S*)-68Z

Scheme 10

accounted for in terms of the different sensitization mechanisms of opposite diastereoface-selectivities operating at low and high substrate concentrations. Thus only the intramolecular sensitization occurs at low concentrations to give (1*R**,3*R**)-**66E**, while the intermolecular process becomes dominant at high concentrations to give (1*R**,3*S**)-**66E** as the major product. On the other hand, such a dramatic diastereoselectivity switching was not observed for the positional isomers **67Z** and **68Z**, since the intra- and intermolecular sensitizations of **67Z** lead to the same diastereoselectivity, while only the intermolecular sensitization can take place in the case of **68Z** owing to the hindered intramolecular approach of the sensitizer moiety to the olefinic part.

4. 1,3- and 1,5-Cyclooctadienes

Possessing an extra double bond in a conjugating or transannular position, (*Z,Z*)-1,3-cyclooctadiene **69ZZ** and (*Z,Z*)-1,5-cyclooctadiene **70ZZ** are readily sensi-

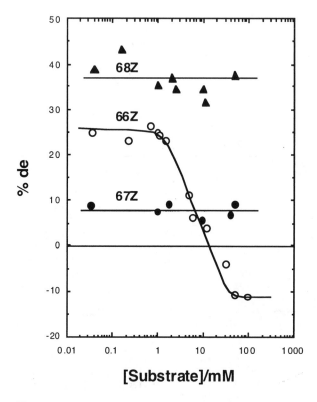

Figure 6 Diastereomeric excess of **66E–68E** produced upon self-sensitized diastereo-differentiating photoisomerization of (Z)-3-, 4-, and 5-benzoyloxycyclooctenes **66Z–68Z** at different concentrations.

tized by such aromatic compounds that can sensitize cyclooctene **47**, and hence their enantiodifferentiating photosensitization behavior has been investigated under analogous conditions.

Somewhat incredibly, optically active (E,Z)-1,3-cyclooctadiene **69EZ** were not known or prepared until rather recently, although enantiopure (E)-cyclooctene **47E** and (E,Z)-1,5-cyclooctadiene **70EZ** were synthesized by thermal reactions in the 1960s. The first synthesis and characterization of optically active (E,Z)-1,3-cyclooctadiene **69EZ** was prepared in 1997 by using the enantiodifferentiating photosensitization with chiral benzene-, naphthalene-, and anthracene(di)carboxylates **41–45** and **71–73** shown in Scheme 11 [52]. It is noted that not only pyromellitates **45a–c** and mellitates **46a–c** but also naphthalene- and anthracene(di)carboxylates **41a–43a** and **71a–73a** can sensitize the photoisomerization to give

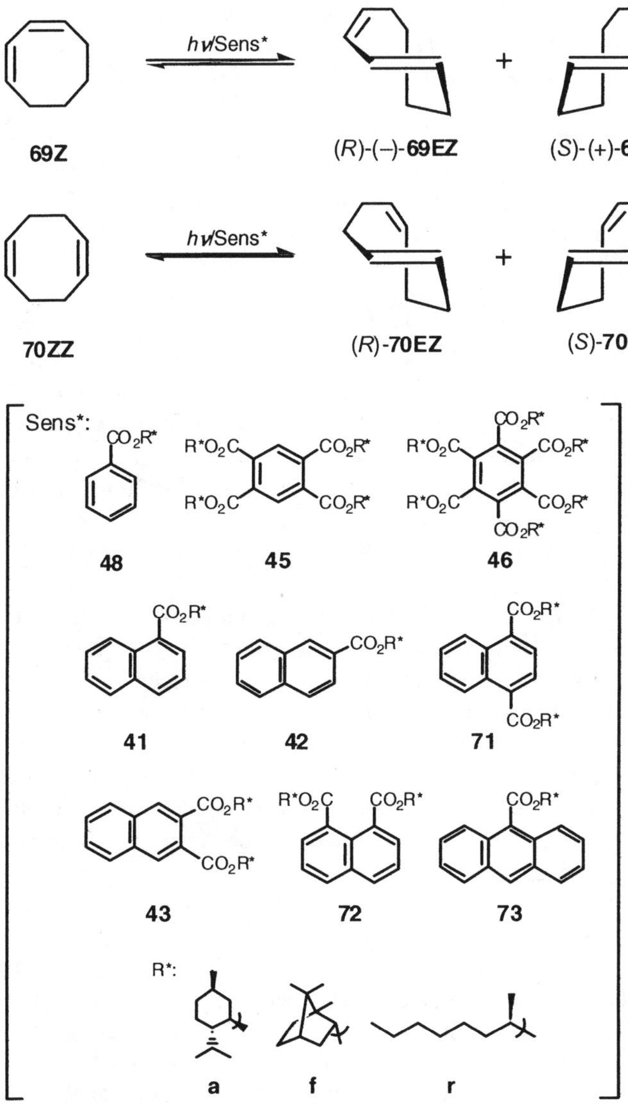

Scheme 11

appreciable *E*/*Z* ratios of 0.05–0.27, which is in sharp contrast to the photobehavior of cyclooctene **47** upon sensitization with the latter fused-aromatic sensitizers [53]. However, the ee's of **69EZ** produced are poor (0.3–2.5% at 25°C and 0.3–5.8% at −40 or −78°C) for most chiral sensitizers examined; the only exception is (−)-menthyl mellitate **46a**, which affords (*R*)-(−)-**69EZ** in 10% at 25°C and 18% at − 40°C in pentane, but in polar solvents the *ee* dramatically decreases to 0.2–1.3%, indicating the charge-transfer nature of the exciplex intermediate. Although the chemical and optical yields are only moderate to low, the optically active (*E,Z*)-1,3-cyclooctadiene **69EZ** shows unusual chiroptical properties, including the exceptionally large specific rotation of $[\alpha]_D = 1380°$ [after a correction for the optical purity (10.1% *ee*) of the photoproduced sample], which far exceeds the corresponding values reported for the structurally related (*E*)-cyclooctene **47E** ($[\alpha]_D = 426°$) [26b], 1-methylcyclooctene ($[\alpha]_D = 106°$) [49], and (*E,Z*)-1,5-cyclooctadiene **70EZ** ($[\alpha]_D = 152°$) [26d]. The other observed (chir)optical properties, such as oscillator and rotatory strengths of the first and second $\pi–\pi^*$ transitions, of optically active **69EZ** are in good agreement with those obtained by theoretical calculations [54].

This enantiodifferentiating photoisomerization of **69ZZ** was further studied by using chiral phthalamide **49y** and pyromellitamide **45y** (see Scheme 8 for structure) in pentane at − 67°C affording **69EZ** in 1.2% and 14.3% ee, respectively.

1,5-Cyclooctadiene **70ZZ** was also subjected to enantiodifferentiating photosensitization by (−)-menthyl benzoate in an early study, but the product *ee* was only 1.6% in pentane at room temperature [55]. Very recently, a further attempt was made to raise the product *ee* and also to examine the pressure effect on the *ee* by using (−)-1-methylheptyl pyromellitate **45r** as a sensitizer, to give (−)-**70EZ** in 4.4% *ee* at atmospheric pressure (0.1 MPa). Interestingly, by applying pressure, the product chirality was switched from levorotatory to dextrorotatory at ca. 120 MPa, and the antipodal (+)-**70EZ** was obtained in 4.2% at 300 MPa [36].

5. Cycloheptene

(*E*)-Cycloheptene **74E** (Scheme 12) was first generated and trapped as a reactive intermediate in a thermal elimination reaction in 1965 [26c]. Its existence as a relatively stable ground-state species (with lifetimes of 9.7 min at 1°C and 68 min at − 15°C), and its physical, spectral (UV and NMR), and chemical properties were revealed unequivocally with photochemically prepared **74E** in the 1980s [56,57]. In light of these foregoing studies, (*Z*)-cycloheptene **74Z** was irradiated at low temperatures, ranging from − 40 to − 70°C, in the presence of optically active alkyl pyromellitates **45a,b,f** [58]. To the irradiated solution was added a trapping agent, such as 1,3-diphenylisobenzofuran **75** or osmiumtetraoxide, at

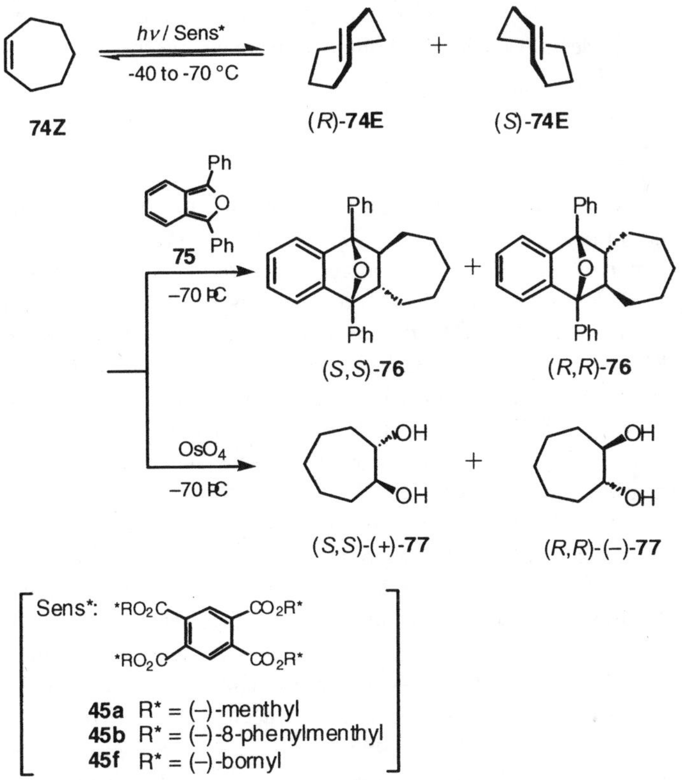

Scheme 12

$-70°C$ in the dark, to give the Diels–Alder adduct **76** or *trans*-1,2-diol **77**, respectively.

The subsequent thermal cycloaddition and oxidation processes were believed to proceed with stereoretention, and the absolute configuration and *ee* of **74E** produced photochemically were determined or evaluated from those of adduct **76** and/or diol **77**. The *ee*s of adduct **76**, and hence **74E**, are greater in general than those obtained for cyclooctene **47** under the comparable conditions, and reach 77% *ee* upon sensitization with (−)-menthyl pyromellitate **45a** in hexane at $-80°C$, although the sensitization in more polar solvents tends to give lower *ee*s. Similarly, the differential activation parameters obtained for **74** are much greater than the corresponding values for **47**, and the switching of product chirality was also observed in several cases. Fluorescence quenching experiments with C_5–C_8 cycloalkenes revealed that the higher *ee*s obtained for **74**, rather than

47, are attributable to the closer sensitizer–substrate contact due to the reduced steric hindrance, which secures the intimate interactions in the intervening diastereomeric exciplexes [58].

6. Cyclohexene

Both direct and sensitized irradiations of cyclohexene **78Z** to give *trans-anti-trans-*, *cis-trans-*, and *cis-anti-cis-*[2 + 2]-cyclodimers **80–82** in different ratios [59]. This photodimerization is believed to proceed through the initial formation of the highly strained (*E*)-isomer **78E**, which is followed by the thermal concerted and/or stepwise cyclodimerization with **78Z**, although no direct evidence for the intervention of **78E** has been obtained and the cyclization mechanism(s) involved are not very clear. In this photocyclodimerization, the enantiodifferentiation occurs not in the cyclodimerization but in the initial photoisomerization step; hence we classify this formally bimolecular reaction as a unimolecular enantiodifferentiating photosensitization.

The enantiodifferentiating photoisomerization/cyclodimerization of **78Z** was performed over a range of temperatures in the presence of chiral benzene(poly)carboxylates with a variety of terpenoid and saccharide auxiliaries, illustrated in Scheme 13 [60]. Benzoate, isophthalate, and terephthalate sensitizers **48**, **50**, and **51** afforded the cyclodimers in relatively low combined chemical and quantum yields ($\Phi = 0.02$–0.05), whereas the substituted benzoates **48** [X = 2-, 3-, 4-CF$_3$, 3,5-(CF$_3$)$_2$, 4-CN, and 2-OH] gave much reduced yields and the other *ortho*-benzenepolycarboxylates **45**, **46**, and **49** no cyclodimers. This sensitization behavior is very different from that observed with cyclooctene [30–32] or cycloheptene [58], for which the more strict energetic requirement (i.e., higher singlet energy for sensitization) would be responsible, as cyclohexene is conformationally more rigid and difficult to rotate around the double bond in the exciplex intermediate than the higher homologs.

Interestingly, of the two chiral products **79** and **80**, only the former was obtained optically active, while the latter was always racemic. This was reasonably accounted for in terms of mixed mechanisms, involving the concerted stereospecific [2π$_s$ + 2π$_a$] cyclodimerization to **79**, and the stepwise dimerization to a mixture of **79–81** via biradical **82** with a loss of the optical activity of **78E** induced photochemically [60].

In contrast to the cyclooctene [30–32] and cycloheptene cases [58], the plot of the logarithm of the relative rate constant did not give a straight line, particularly at high temperatures, as a result of the major contribution of the radical path (Fig. 7). Good linear plots were obtained, however, at lower temperatures, and the *ee* of **79** reached the highest value of 68% *ee* at $-78°C$ [60]. The differential activation parameters obtained are much larger than those for cyclooctene [30–32] and cycloheptene [58].

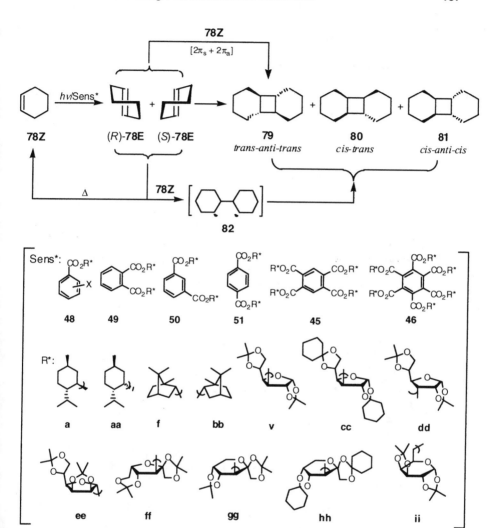

Scheme 13

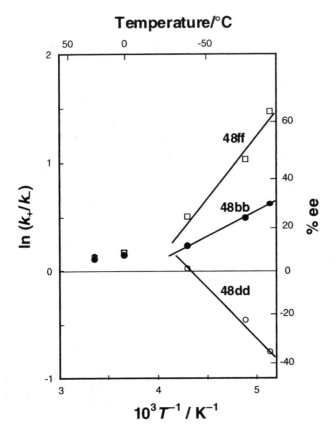

Figure 7 Temperature dependence of *ee* of cyclodimer **79** obtained upon enantiodiffer-entiating photocyclodimerization of cyclohexene **78Z** sensitized by chiral benzoates **48bb**, **48dd**, and **48ff** in pentane.

D. Photocyclization

5-Methyl-1-hexenoic acid **83** was known to cyclize photochemically to γ-lactones **84** and **85** upon sensitization with cyanoaromatics via an electron-transfer mecha-nism [61]. By using chirally modified 2,4,6-triphenylpyridinium (pyrilium) tetra-fluoroborates **86jj–86ll** as an electron-accepting triplet sensitizer, the enantiodif-ferentiating photocyclization of **83** was examined in dichloromethane to give the optically active photocyclization product **84** in 3–7% *ee* (Scheme 14) [62]. As judged from the efficient quenching of the excited pyrilium triplet by 1,4-cyclo-hexadiene, the excited state involved was believed to be triplet in nature [62], which would be responsible for the low *ee*s.

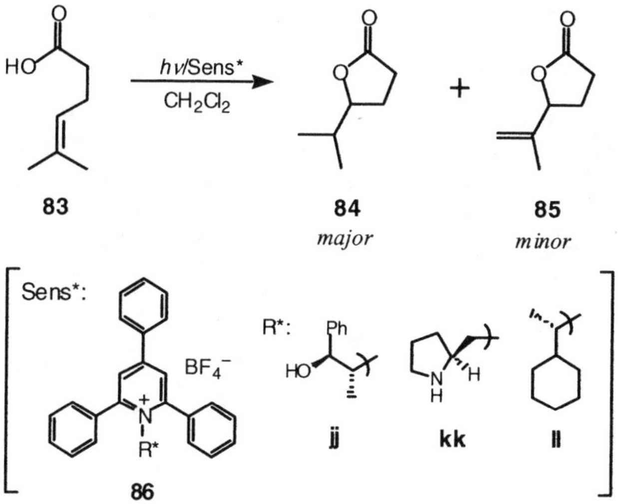

Scheme 14

III. ENANTIODIFFERENTIATING BIMOLECULAR PHOTOREACTIONS

In contrast to the enantiodifferentiating unimolecular photoreactions described above, the bimolecular counterpart has not been extensively explored in view of the vast variety of potential chiral photoaddition reactions hitherto reported [63]. This is partly because a more sophisticated control of the termolecular interaction in the triplex intermediate, or in the attack of a reagent to an exciplex intermediate, is required for obtaining an appreciable *ee* upon photosensitized enantiodifferentiating bimolecular reactions. Thus only the [2 + 2] and [4 + 2] photocycloadditions of alkenes and dienes, and the polar photoaddition of alcohols to aromatic olefins, have been subjected to enantiodifferentiating photoreactions in the presence of optically active sensitizers. Most of these bimolecular photoreactions take advantage of the donor–acceptor interaction in the excited state between the chiral sensitizer and substrate, and therefore involve a highly polarized exciplex or radical–ionic intermediate as the precursor to the chiral end product. However, the use of highly polar solvents inevitably leads to the complete transfer of the electron, generating a solvent-separated or free radical ion pair, for which no chiral sensitizer–substrate interaction is expected to occur. Hence the control of the degree of electron transfer by tuning the nature of the sensitizer, substrate, and solvent becomes a crucial issue in gaining a high *ee* in enantiodifferentiating bimolecular photoreactions.

A. Photocycloaddition

1. [2 + 2] Photocycloaddition

Upon photosensitization with (−)-menthyl salicylate **48a** (X = 2-OH), pyromel-litate **49a**, 1- and 2-naphthalenecarboxates **41a** and **42a**, 2,3- and 2,6-naphthalene-dicarboxylates **43a** and **91a**, and/or 9-anthracenecarboxylate **73a** in acetonitrile at 25 or − 40°C, aryl vinyl ethers **87** (X = H or Cl) and 4-methoxystyrene **89** cyclodimerize to the corresponding cyclobutane derivatives **88** and **90** (only the trans isomers are chiral) in low to good yields (Scheme 15) [64]. However, the

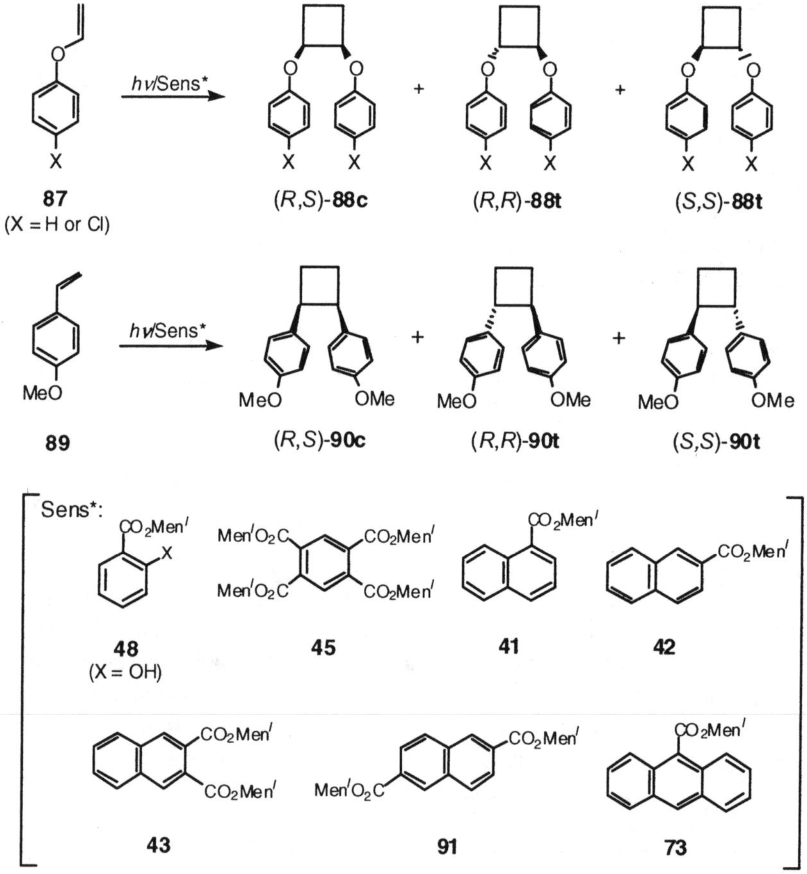

Scheme 15

ee of **88t** and the optical rotation of **90t** were extremely low (< 0.5% *ee* or 0.2°, respectively), for which the use of polar acetonitrile as a solvent is obviously responsible (attempted irradiations in less-polar pentane, benzene, or ether gave no products).

A supramolecular approach led to a better enantioselectivity in the photo-sensitized intramolecular [2 + 2] cycloaddition of butenylquinolonyl ether **92** to **93** (Scheme 16) [65]. Substoichiometric amounts (as low as 25% of substrate **92**) of "sensitizing receptor" **94**, which carries an amide/lactam receptor site and a triplet-sensitizing benzophenone, can bind the complementary lactam **92** to form the supramolecular complex **95**, which upon irradiation undergoes the intracomplex sensitization and cyclization of **92** to give photoadduct **93**. Although the product obtained at 30°C was racemic (even in the presence of up to two equivalents of receptor **94**), the *ee* was enhanced by decreasing the temperature, ultimately giving 19–22% *ee* at −70°C [65].

2. [4 + 2] Photocycloaddition

Radical cationic species of dienes, generated by photosensitized electron transfer, are known to exhibit an enhanced reactivity for [4 + 2] cycloaddition with

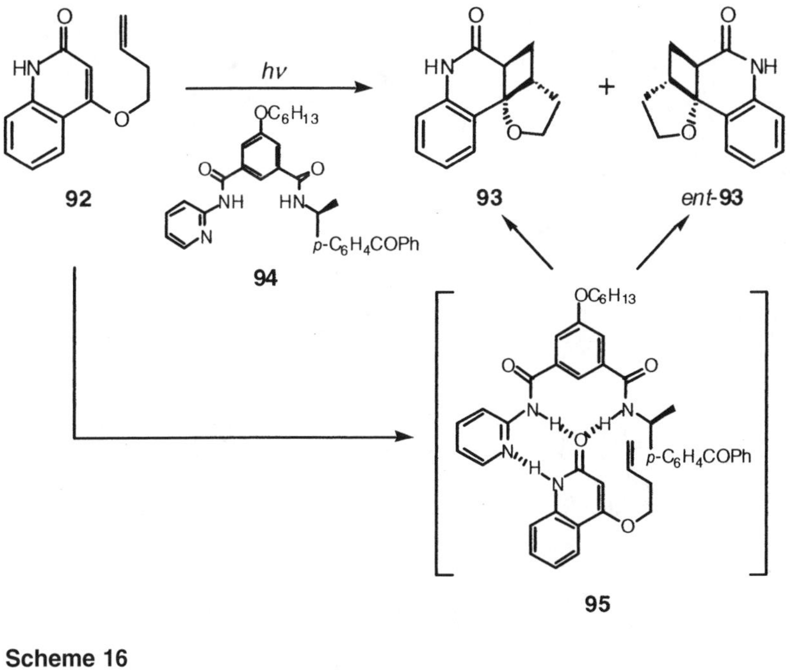

Scheme 16

nondienophilic alkenes that are inactive under typical Diels–Alder conditions [66]. However, only two such systems have been examined for enantiodifferentiating photosensitization, i.e., the [4 + 2] and [2 + 2] photocyclodimerizations of 1,3-cyclohexadiene **96** and [4 + 2] photocycloaddition of **96** to β-methylstyrene.

Although the use of axially chiral 1,1'-binaphthyl derivatives was proposed as chiral sensitizers for enantiodifferentiating [4 + 2] and [2 + 2] photocyclodimerizations of 1,3-cyclohexadiene **96** in 1990 [67], such a possibility was investigated only recently by using a variety of optically active benzene- and naphthalenepolycarboxylates shown in Scheme 17 [68]. The combined chemical yield of cycloadducts **97a,b** and **98a,b** varied from several to 80% depending on the solvent used, with their relative ratio also depending on the solvent used, giving more **97a** in acetonitrile but more **97b** and **98a** in less polar toluene, ether, and

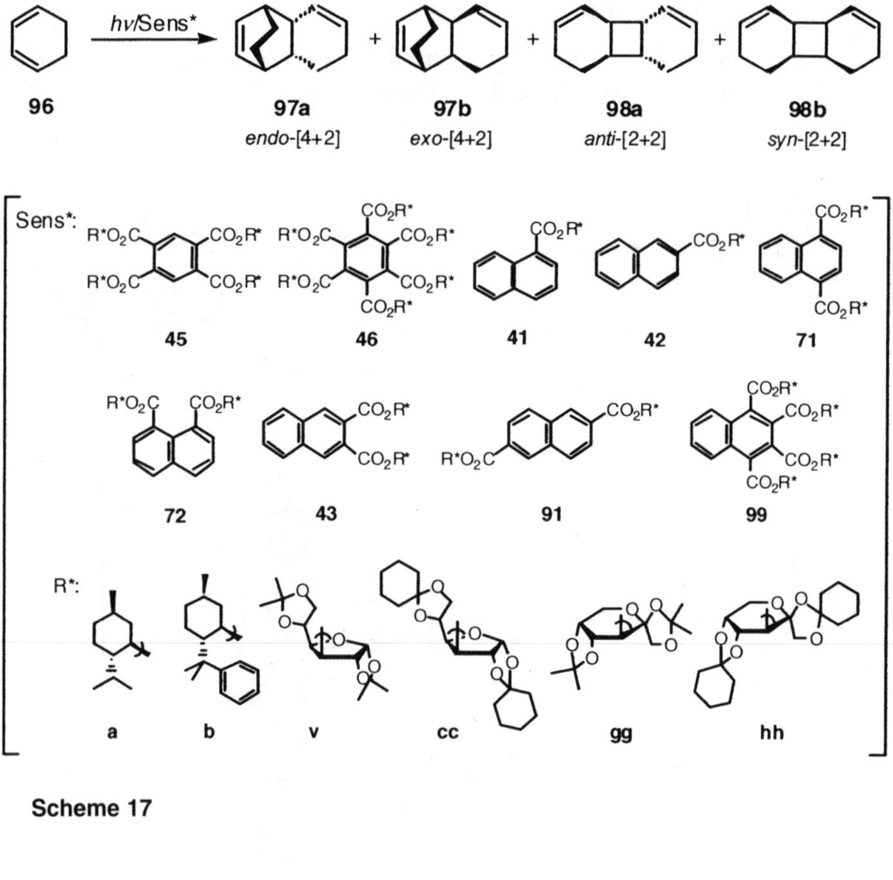

Scheme 17

pentane. Among the three chiral products (**97a,b** and **98a**), only **97b** was obtained as an appreciably optically active product with an *ee* of up to 8.2% upon sensitization with **71hh** in ether at −41°C. A complicated reaction mechanism was proposed, in which both the concerted [4 + 2] cyclodimerization to **97b** via a contact ion pair, and the stepwise path via a biradical and/or free radical ion, are involved. Since the observed *ee*s are likely to be diluted by the racemic product from the stepwise path, the "net" *ee* of **97b** produced via the concerted path is evaluated as 70% [68].

Enantiodifferentiating photo-Diels–Alder reactions of cyclohexadiene **96** with *p*-substituted β-methylstyrenes **100–103** sensitized by axially chiral tetracyanobinaphthyl **105** and tetracyanobianthryl **106** gave the *endo*-[4 + 2] cycloadduct **104** in moderate *ee*s (Scheme 18) [69].

The best *ee* of 23% was obtained for **104** upon sensitization of **100** with bianthryl **106** in toluene at −65°C, while *p*-substituted substrates **101–103** gave only comparable (with **105**) or lower *ee*s (with **106**) under the identical conditions. Irradiations at higher temperatures led to reduced *ee*s, and in acetonitrile only

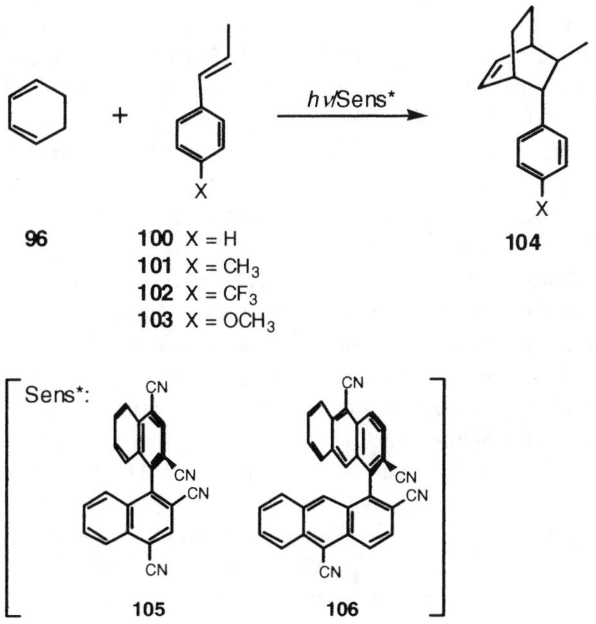

96   100 X = H   104
            101 X = CH₃
            102 X = CF₃
            103 X = OCH₃

Sens*:   105   106

Scheme 18

the racemic product was obtained even at $-30°C$. The existence of discrete diastereomeric exciplexes of sensitizer **105** with dienophile **100** was proved by a time-resolved fluorescence study. Thus the fluorescence decay kinetics of exciplex [**100**···**105**] in toluene, being single exponential at room temperature ($\tau = 26$ ns), became double exponential at $-65°C$ ($\tau = 37$ and 22 ns), indicating the existence of two diastereomeric exciplexes. It was further elucidated from the inverse dependence of product ee on the concentration of **96** that this lifetime difference is the origin of the observed enantiodifferentiation in this triplex Diels–Alder reaction, where the quenching of the excited sensitizer by methylstyrene is almost diffusion-limited and the exciplex formation is irreversible [69].

B. Polar Photoaddition

Enantiodifferentiating anti-Markovnikov polar photoadditions of alcohols to 1,1-diphenyl-1-alkenes **107** and **108** sensitized by optically active naphthalene(di)carboxylates **41–43**, **71**, **72**, and **91** were investigated in detail (Scheme 19) [70]. Since this photoaddition involves the attack of alcohol to a radical cationic species of the substrate alkene [71], the use of polar solvents is desirable for obtaining the adduct in a high yield. However, in polar solvents, the radical ionic sensitizer–substrate pair produced upon photoexcitation is immediately dissociated by solvation, and no chirality transfer is expected to occur. Thus the optical and chemical yields are often conflicting issues, and therefore the critical control of solvent polarity is essential for obtaining the optically active product with an appreciable ee in reasonable chemical yield. In fact, the initial attempts on **107**, employing naphthalenecarboxylate sensitizers with chiral terpenoid auxiliaries (**a–c** and **f**) and a pentane solvent afforded a best ee of 27% for adduct **110** ($R^2 = $ Me), but in $< 2\%$ yield [70a].

The use of naphthalenedicarboxylates **71** and **91**, however, which possess low reduction potentials and therefore give highly negative free-energy changes for electron transfer (ΔG_{et}), improved both chemical and optical yields of **110** [70b]. Further improvements in ee were achieved by using a series of saccharide esters of naphthalenedicarboxylic acids and more the bulky alcohols **109** (R = Et, iPr) [70b,c]. The best ees for methanol, ethanol, and isopropanol additions to **107** are 35%, 34%, and 58%, respectively, all of which are obtained upon sensitization with **91gg** in ether (containing 0.5 M alcohol) at 0°C. More bulky tert-butanol does not give any adduct, and diphenylbutene **108** affords adducts with methanol and ethanol in lower chemical and optical yields ($\leq 45\%$ ee) [70c].

From mechanistic studies, it was revealed that the exciplex formation is reversible and the product ee is determined by both the relative stability of the diastereomeric exciplexes and the relative rate of subsequent alcohol attack [70b,c]. It should be noted that the introduction of saccharide auxiliaries to the

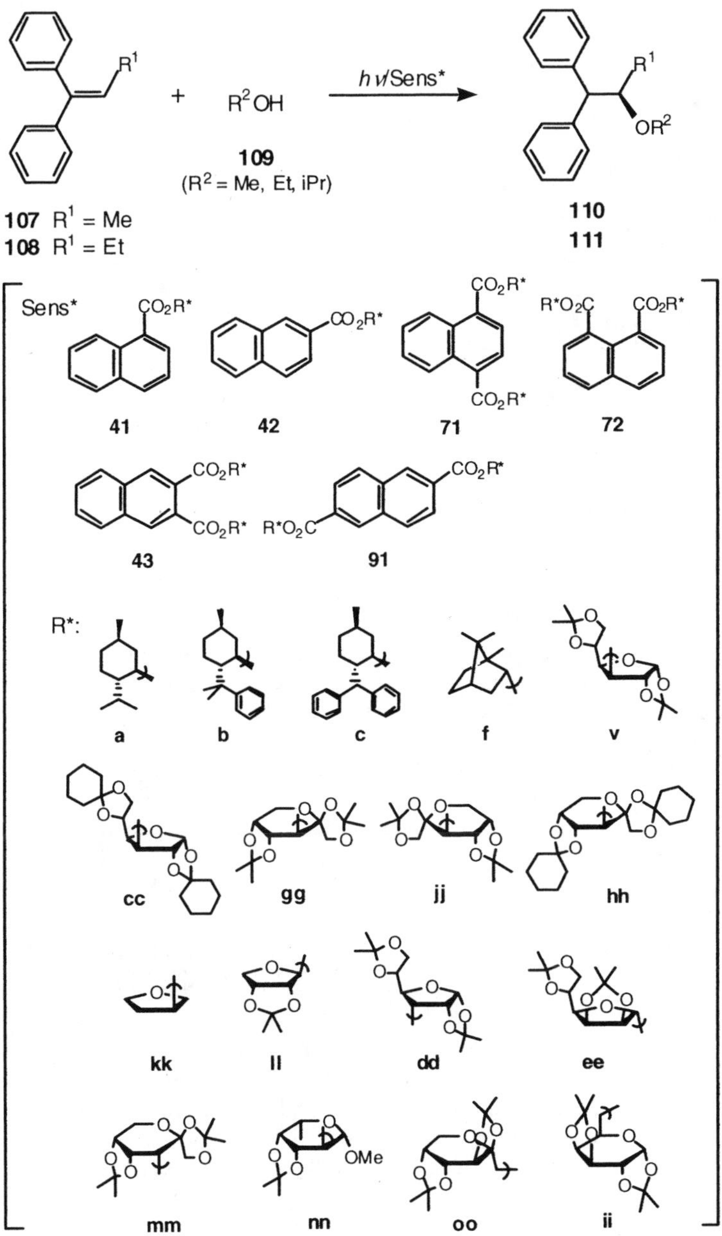
R1
+
R2OH
hν/Sens*
R1
OR2
109
(R2 = Me, Et, iPr)
107 R1 = Me
108 R1 = Et
110
111
Sens*
CO2R*
CO2R*
CO2R*
CO2R*
R*O2C
CO2R*
41
42
71
72
CO2R*
CO2R*
R*O2C
CO2R*
43
91
R*:
a
b
c
f
v
cc
gg
jj
hh
kk
ll
dd
ee
OMe
mm
nn
oo
ii

Scheme 19

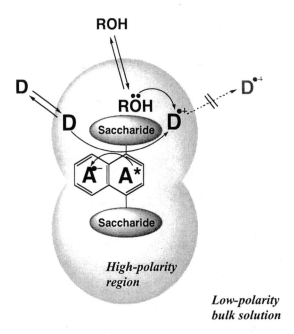

Figure 8 Microenvironmental polarity control upon enantiodifferentiating polar photo-addition of alcohol (ROH) to aromatic olefin (D) sensitized by naphthalenedicarboxylate with saccharide auxiliaries (A*); the local polarity is enhanced around the saccharide moieties, facilitating electron transfer from exited sensitizer (A*) to substrate olefin (D) to produce a radical cation (D* +). The radical cation produced cannot escape from the high polarity region around the saccharide to the low-polarity bulk solution and is accordingly attacked by ROH in the chiral environment of saccharide to produce the adduct in high *ee*.

sensitizer and the fine tuning of the bulk solvent polarity are highly crucial in terms of the microenvironmental polarity control around the sensitizer, which is crucial in order to accelerate the electron transfer from substrate to sensitizer, yet keeping the radical cationic substrate in the reach of the geminate radical anionic chiral sensitizer (Fig. 8). This novel strategy for enhancing the microenvironmental polarity by introducing a highly polar chiral auxiliary, such as protected saccharide, can overcome the normally accepted tradeoff between the chemical and optical yields, and is thought to be widely applicable to most enantiodifferentiating photosensitizations that involve an electron transfer process.

IV. CONCLUSION

In the last decade, great progress has been made in enatiodifferentiating photosensitized reactions; the best results hitherto obtained are summarized in Table 1.

As a consequence of the extensive efforts devoted to this attractive and intriguing area of asymmetrical photochemistry, the chirality transfer mechanisms operating in both uni- and bimolecular enantiodifferentiating photosensitizations have been understood in considerable detail, which in turn enabled us not only to obtain optical yields much higher than those achieved in earlier studies but also to utilize a variety of internal and external, or electronic, structural, and environmental, factors in the critical control of enantioselectivity in the excited state. From a wider chemical viewpoint, it should be emphasized that the entropy-related environmental factors, such as temperature, pressure, and solvent, play much more important roles than previously expected, and in typical cases even the product chirality may be switched by these apparently supplementary factors.

In the history of chemistry, the enthalpy factor has long been taken as the major factor controlling most chemical processes that involve covalent bond formation and cleavage, whereas the entropy factor does not appear to have been seriously utilized, in a positive sense, in the control of the rate and selectivity of chemical reactions. In contrast, a variety of weak interactions, such as dipole–dipole, van der Waals, hydrophobic, hydrogen bonding, charge transfer, and π–π interactions, play crucial roles in chemical and biological molecular recognition, where the entropy factor becomes more important [72]. This theory, being quite reasonable, has not been demonstrated experimentally or utilized explicitly as an effective tool for controlling the outcome, since thermal and biological reactions do not easily allow us to carry out experiments over a wide range of temperatures while maintaining the same reaction mechanism. However, photochemistry requires much less thermal energy to promote reactions and has the inherent advantage of allowing us to examine the contribution of the entropy factor. In the recent work on enantiodifferentiating photosensitization, it was revealed that the weak interactions that occur in the exciplex intermediate can be controlled by entropy-related factors. It is likely that the combined use of these factors will lead to a new era of asymmetrical photochemistry, enabling us directly and critically to maneuver the excited-state enantio- and diastereodifferentiating interactions under more readily accessible conditions. Finally, we should note that the concept of "entropy control" should not be restricted to photochemical processes but should be a key concept when discussing chemical and biological supramolecular interactions in which weak interactions are dominant.

ACKNOWLEDGMENT

I would like to thank Dr. Guy A. Hembury for assistance in the preparation of this manuscript.

Table 1 Highest Enantiomeric Excesses Reported for Each Type of Uni- and Bimolecular Enantiodifferentiating Photosensitizationr

Photoreaction	Sensitizer	% ee	Temperature	Ref.
Photodestruction/production				
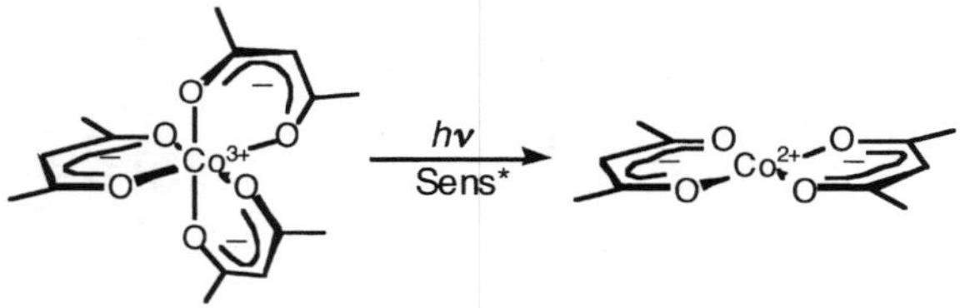 **1** $\xrightarrow[\text{Sens*}]{h\nu}$ (Co^{2+} complex)	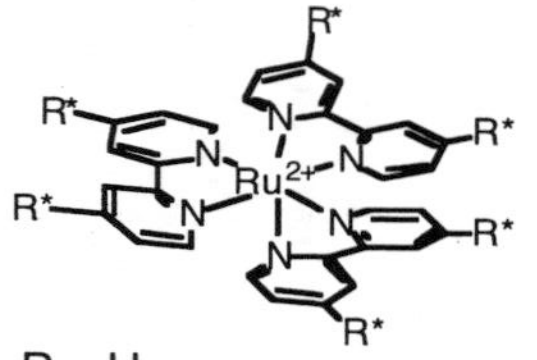 Ru^{2+}(R*) complex; **2** R = H; **3** R = CO$_2$Men^l; **4** R = CONH-(-)-CH(Me)Ph **3**	15	rt	6
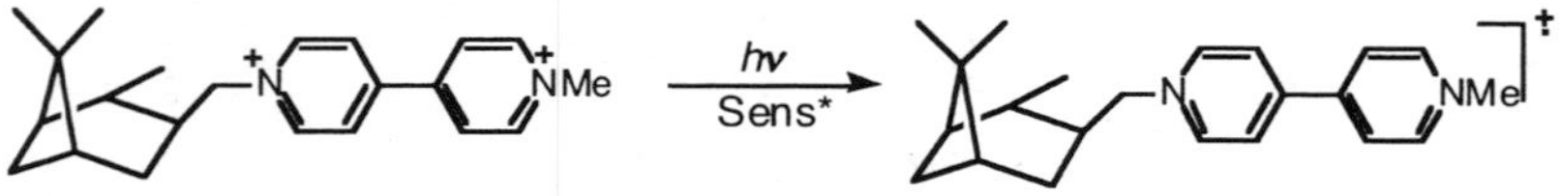 **5^{2+}** $\xrightarrow[\text{Sens*}]{h\nu}$ **5$^{•+}$**		14	rt	8
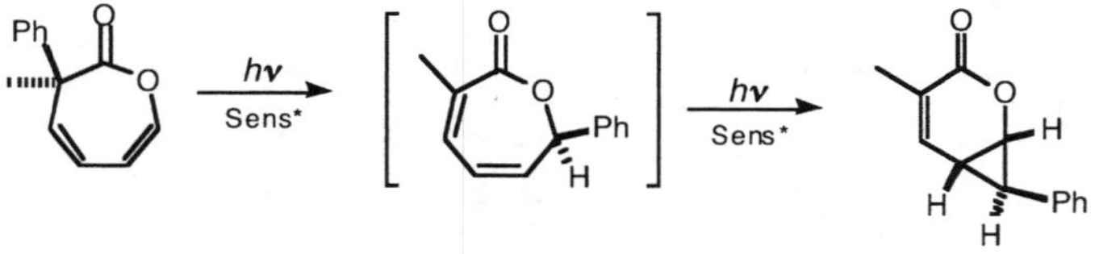 **12** (Ph lactone) $\xrightarrow[\text{Sens*}]{h\nu}$ [intermediate] → **8**	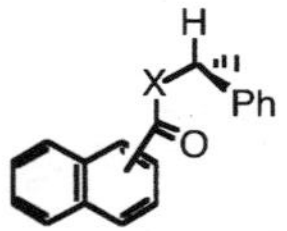 **9** X = NH; **10** X = O; **11** X = NCH$_2$	2	rt	9a
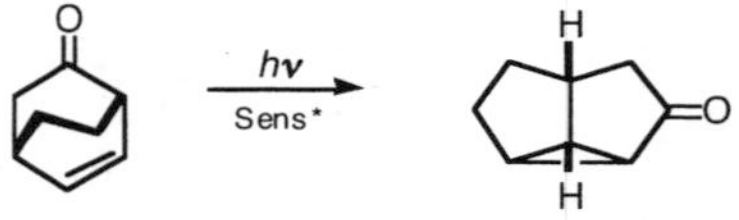 **12** $\xrightarrow[\text{Sens*}]{h\nu}$ **13**	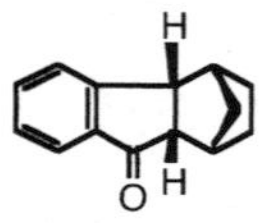 **14**	4.5 10	rt −78°C	9b

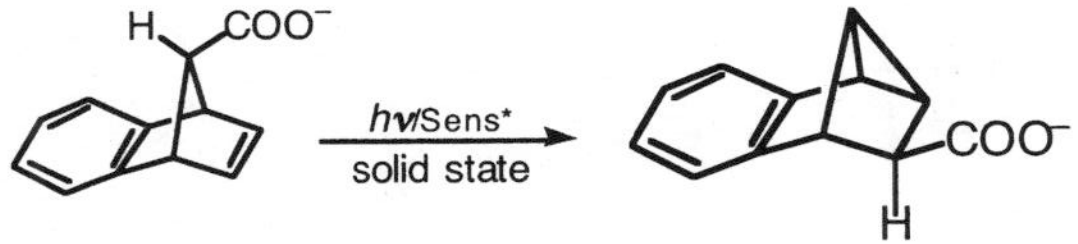 15 → 16	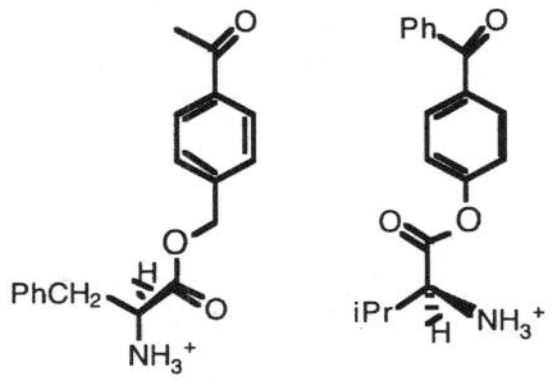 17 18	84 91	rt −20°C (both in solid state)	10
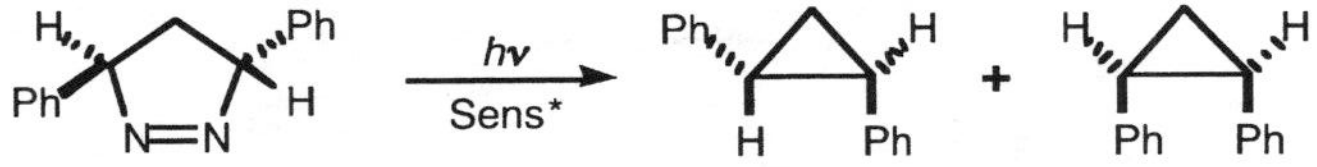 19t → 20t (*dl*) + 20c (*meso*)	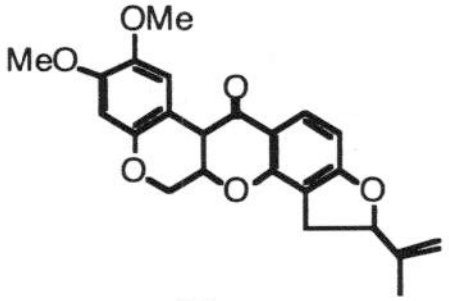 21	3.8	rt	12
Photoderacemization				
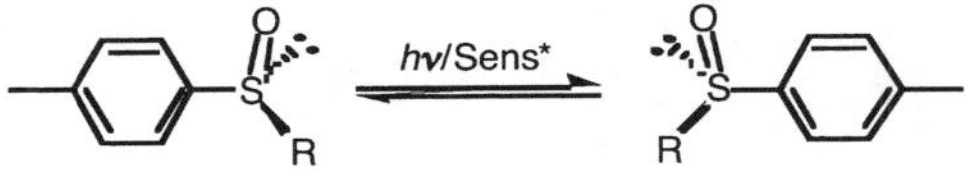 (*S*)-25a R = Me ⇌ (*R*)-25a (*S*)-25b R = iPr ⇌ (*R*)-25b	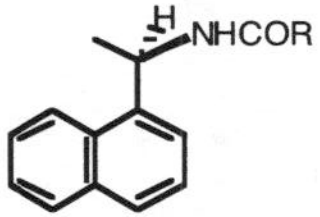 26 R′ = CH_3 27 R′ = CF_3	12	rt	15
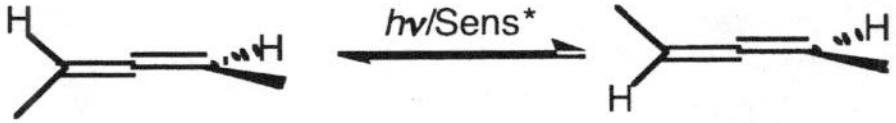 (*S*)-28 ⇌ (*R*)-28	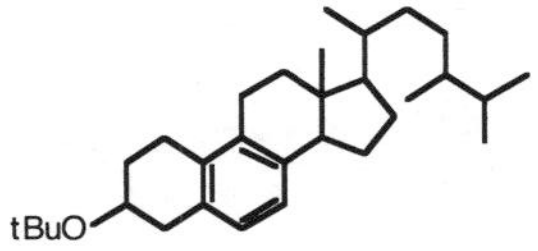 29	3.4	rt	16
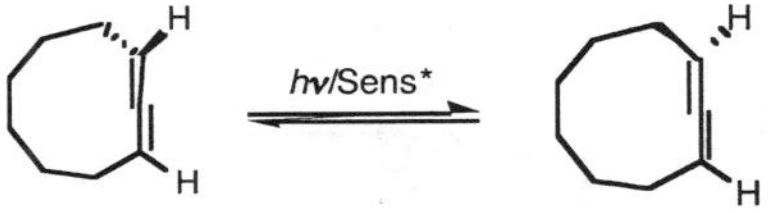 (*S*)-30 ⇌ (*R*)-30	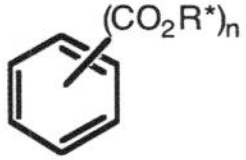 31 R* = (−)-menthyl, (−)-bornyl	<5	rt	17

(*Continued*)

Table 1 Continued

Photoreaction	Sensitizer	% ee	Temperature	Ref.
Geometrical photoisomerization				
20c Ar = Ph **32c** Ar = 4-MeOC$_6$H$_4$ (1R,2R)-(−)-**20t** 1R,2R)-(−)-**32t** + (1S,2S)-(+)-**20t** (1S,2S)-(+)-**32t**	R*O$_2$C—CO$_2$R* (cyclohexadiene diester) **45a** R* = (−)-menthyl	10.4	25°C	25b
	26 R*NHAc (naphthyl)	1.1	25°C	25b
(1R,2R)-(−)-**33t** + (1S,2S)-(+)-**33t** **33c**	R*O$_2$C—CO$_2$R* **45f** R* = (−)-bornyl	1.1	25°C	25b
47Z (R)-(−)-**47E** + (S)-(+)-**47E**	R*O$_2$C—CO$_2$R* (tetraester) R*: **45c**	49 64	25°C −89°C	32
	R*O$_2$C—CO$_2$R* R*: **45v**	73	−110°C (ether)	37
62Z (R)-**62E** + (S)-**62E**	OMen / O Ph (benzoate)	7.3	−78°C	49

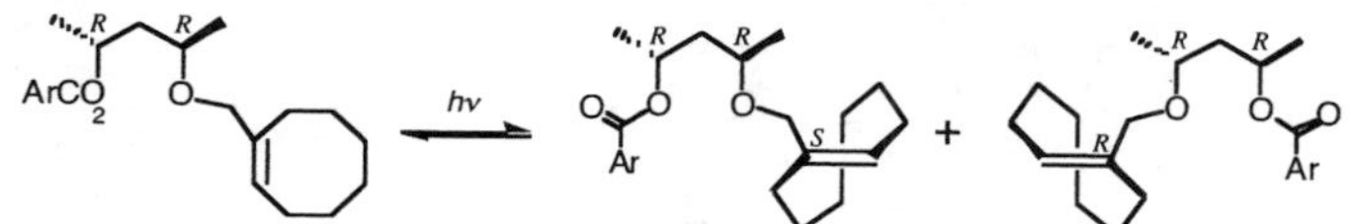

(44 de) −65°C 50

(43 de) 25°C 51b

17.6 −40°C 52

4.4 25°C 36

77 −80°C 58

(Continued)

(built-in)

(built-in)

CO_2R^*

46a R* = (−)-menthyl

45r R - (R)-1-methylheptyl

45a R* = (−)-menthyl

(R,R,R)-**63Z**
64Z
65Z

(R,R,S)-**63Z**
64Z
65Z

hv

63E Ar = Ph
64E Ar = m-MeOCOC$_6$H$_4$
65E Ar = p-MeOCOC$_6$H$_4$

$(1R^*,3S^*)$-**66Z**
$(1R^*,4S^*)$-**67Z**
$(1R^*,5S^*)$-**68Z**

$(1R^*,3R^*)$-**66Z**
$(1R^*,4R^*)$-**67Z**
$(1R^*,5R^*)$-**68Z**

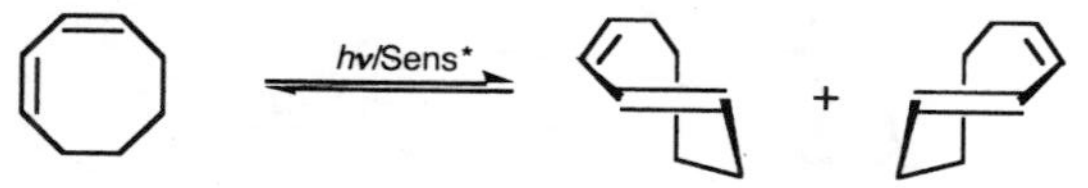

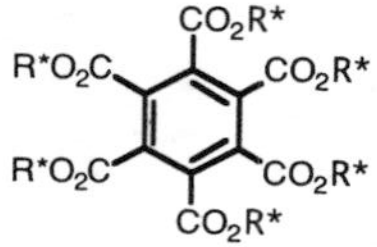

hv

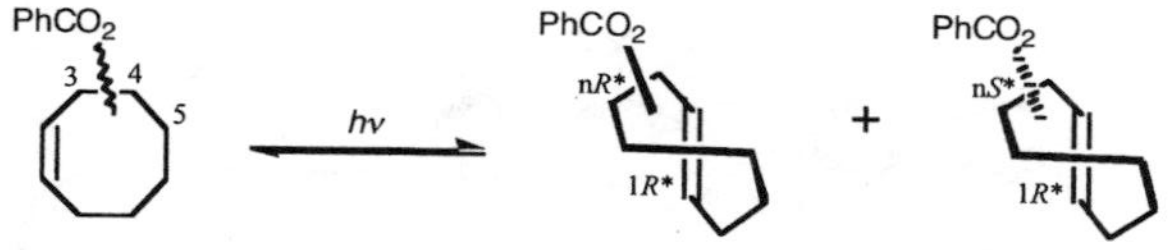

rac-**66Z** (3-PhCO2)
rac-**67Z** (4-PhCO2)
rac-**68Z** (5-PhCO2)

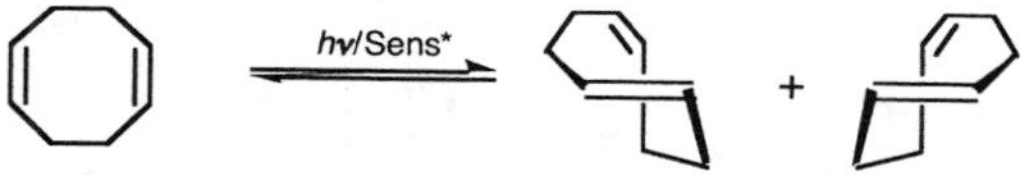

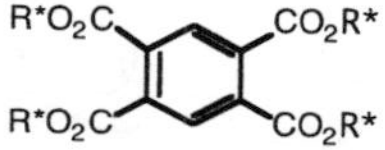

(R)-(−)-**69EZ**

(S)-(+)-**69EZ**

hv/Sens*

69Z

(R)-**70EZ**

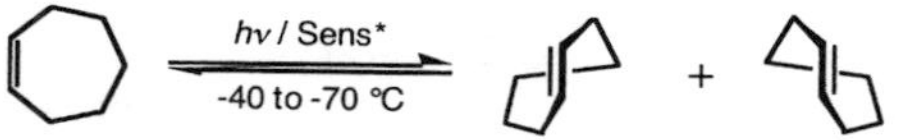

(S)-**70EZ**

hv/Sens*

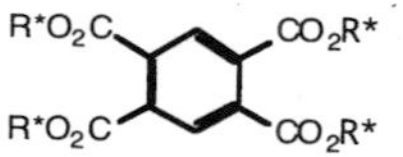

70ZZ

(R)-**74E**

(S)-**74E**

hv / Sens*
−40 to −70 °C

74Z

Table 1 Continued

Photoreaction	Sensitizer	% ee	Temperature	Ref.
78Z *Photocyclization*	**79** + **80** + **81**	68 for **79**	−78°C	60
	48ff **R***, **86**	7	25°C	62
83	**84** + **85**	0.5	25°C	64
Photocycloaddition	**45b** R*: **ii**, **jj** R*:, **kk**			
87 (X = H or Cl)	(R,R)-**88t** + (S,S)-**88t** + (R,S)-**88c**			
89	(R,R)-**90t** + (S,S)-**90t** + (R,S)-**90c**	n.d. ([α]$_D$ ≤ 0.2°)	25°C	64
	45a R* = (−)-menthyl			

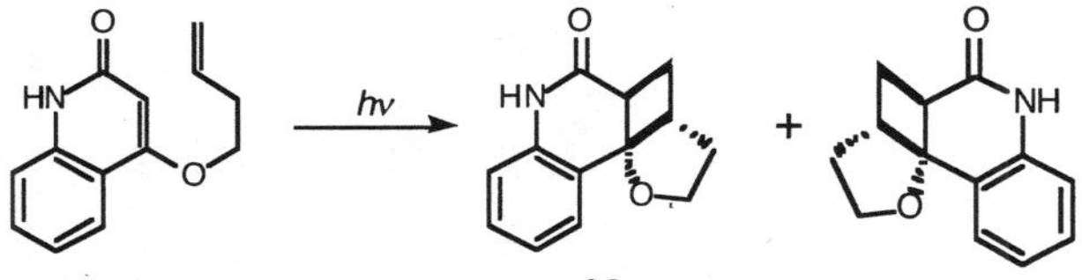

−70°C 65

22

−41°C 68

8.2
(70 net *ee*)
for **97B**

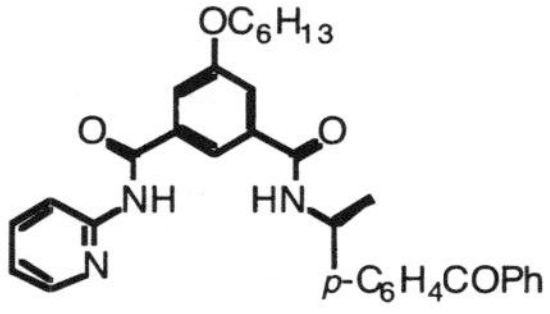

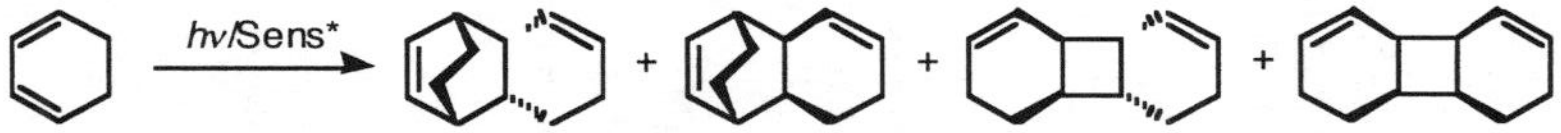

−65°C 69

23 with
106

0°C 71

58
(R¹ = Me,
R² = iPr)

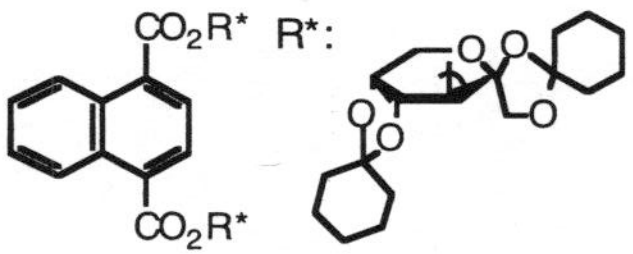

OC₆H₁₃

p-C₆H₄COPh

94

(host)

CO₂R* R*:

CO₂R*

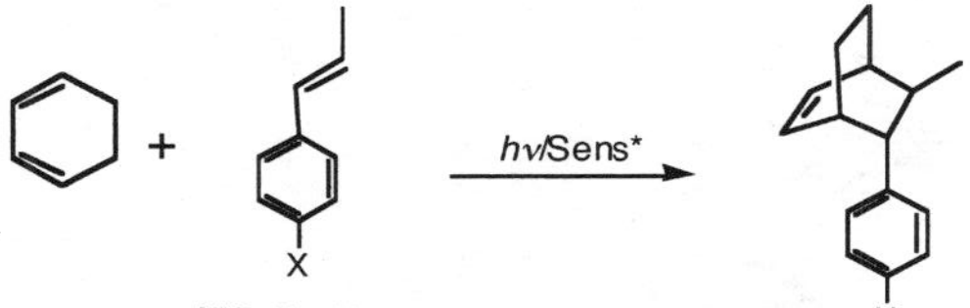

71hh

98b

97a **97b** **98a**

CN CN

CN

CN

106

CO₂R* R*:

91gg

R*O₂C

R²O₂C

ent-**93**

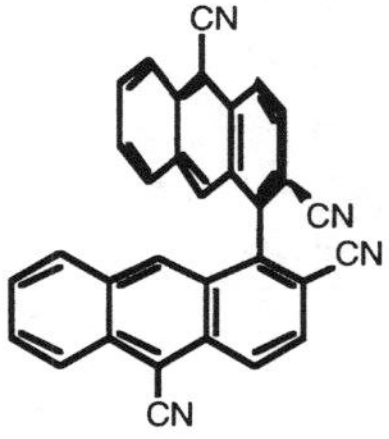

93

92

hv

hv/Sens*

96

104

100 X = H
101 X = CH₃
102 X = CF₃
103 X = OCH₃

96

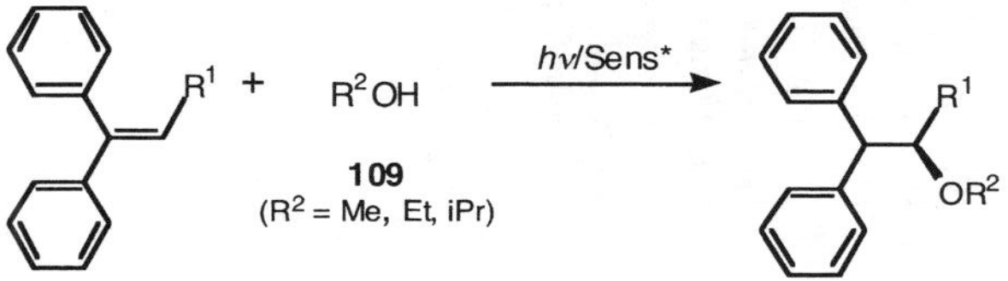

hv/Sens*

+

Polar photoaddition

R¹ + R²OH

109
(R² = Me, Et, iPr)

hv/Sens*

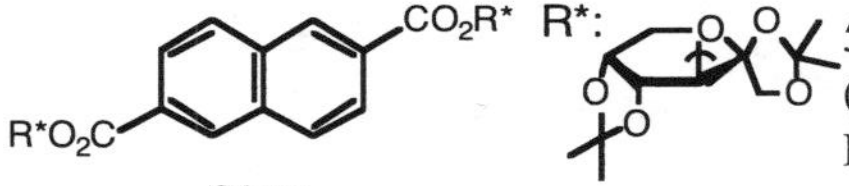

R¹ OR²

110
111

107 R¹ = Me
108 R¹ = Et

REFERENCES

1a.   Rau H. Chem Rev 1983; 83:535–547.
1b.   Inoue Y. Chem Rev 1992; 92:741–770.
1c.   Everitt SRL, Inoue Y. In: Ramamurthy V , Schanze KS, Eds. Organic Molecular Photochemistry (Molecular and Supramolecular Photochemistry, Vol. 3). New York: Marcel Dekker, 1999:71–130.
1d.   Griesbeck AG, Meierhenrich UJ. Angew Chem Int Ed 2002; 41:3147–3154.
 2.   Hammond GS, Cole RS. J Am Chem Soc 1965; 87:3256–3257.
3a.   Inoue Y, Wada T, Asaoka S, Sato H, Pete J-P. Chem Commun 2000:251–259.
3b.   Inoue Y, Sugahara N, Wada T. Pure Appl Chem 2001; 73:475–480.
4a.   Porter GB, Sparks RH. J Chem Soc, Chem Commun 1979:1094.
4b.   Porter GB, Sparks RH. J Photochem 1980; 13:123–131.
 5.   Kaizu Y, Mori T, Kobayashi H. J Phys Chem 1985; 89:332–335.
6a.   Ohkubo K, Hamada T, Inaoka T, Ishida H. Inorg Chem 1989; 28:2021–2022.
6b.   Ohkubo K, Ishida H, Hamada T, Inaoka T. Chem Lett 1989:1545–1548.
7a.   Ohkubo K, Fukushima M, Ohta H, Usui S. J Photochem Photobiol A 1996; 98: 137–140.
7b.   Ohkubo K, Yamashita K, Ishida H, Haramaki H, Sakamoto Y. J Chem Soc, Perkin Trans 2 1991:1833–1838.
7c.   Ohkubo K, Hamada T, Ishida H. J Chem Soc, Chem Commun 1993:1423–1425.
7d.   Ohkubo K, Hamada T, Watanabe M, Fukushima M. Chem Lett 1993:1651–1654.
7e.   Ohkubo K, Hamada T, Ishida H, Fukushima M, Watanabe M. J Mol Catal 1994; 89:L5–L10.
 8.   Rau H, Ratz R. Angew Chem Int Ed 1983; 22:550–551.
9a.   Hoshi N, Furukawa Y, Hagiwara H, Uda H, Sato K. Chem Lett 1980:47–51.
9b.   Demuth M, Raghavan PR, Carter C, Nakano K, Schaffner K. Helv Chim Acta 1980; 63:2434.
10.   Janz KM, Scheffer JR. Tetrahedron Lett 1999; 40:8725–8728.
11a. Scheffer JR, Scott C. Science 2001; 291:1712–1713.
11b. Scheffer JR. Can J Chem 2001; 79:349–357.
12a. Rau H, Hörmann M. J Photochem 1981; 16:231–247.
12b. Becker E, Weiland R, Rau H. J Photochem Photobiol A: Chem 1988; 41:311–330.
13a. Mislow K, Axelrod M, Rayner DR, Gotthardt H, Coyne LM, Hammond GS. J Am Chem Soc 1965; 87:4958–4959.
13b. Cooke RS, Hammond GS. J Am Chem Soc 1968; 90:2958–2959.
13c. Cooke RS, Hammond GS. J Am Chem Soc 1970; 92:2739–2745.
14.   Balavoine G, Jugè S, Kagan HB. Tetrahedron Lett 1973:4159–4162.
15.   Kagan HB, Fiaud JC. Top Stereochem 1978; 10:175–285.
16.   Drucker CS, Toscano VG, Weiss RG. J Am Chem Soc 1973; 95:6482–6484.
17.   Inoue Y. To be published.
18.   Murov SL, Cole RS, Hammond GS. J Am Chem Soc 1968; 90:2957–2958.
19.   This *ee* was calculated from the reported optical rotation of the photoproduct by using the specific rotation of enantiomerically pure **32t**, which was determined in later studies, e.g., Aratani T, Nakanisi Y, Nozaki H. Tetrahedron Lett 1809–1810,

1969; Aratani T, Nakanisi Y, Nozaki H. Tetrahedron 26: 1675–1684, 1970. However, our own experiments under comparable conditions gave lower *ees* of around 3–4% (see Note 285 in Ref. 1b).

20. Ouannès C, Beugelmans R, Roussi G. J Am Chem Soc 1973; 95:8472.
21. Kagan HB, Fiaud JC. Top Stereochem. (see Note 9) 1988; 18:249–331.
22. Horner L, Klaus J. Liebigs Ann Chem 1979:1232–1257.
23. Horner L, Klaus J. Liebigs Ann Chem 1981:792–810.
24. Vondenhof M, Mattay J. Chem Ber 1990; 123:2457.
25a. Inoue Y, Shimoyama H, Yamasaki N, Tai A. Chem Lett 1991:593–596.
25b. Inoue Y, Yamasaki N, Shimoyama H, Tai A. J Org Chem 1993; 58:1785–1793.
25c. For a more detailed examination of the involvement of 1,3-biradical species in electron transfer photoreaction, see Roth HD. J Phys Chem A. 2003; 107:3432–3437.
26a. Cope AC, Ganellin CR, Johnson HWJ, Winkler HJS. J Am Chem Soc 1963; 85: 3276–3279.
26b. Cope AC, Mehta AS. J Am Chem Soc 1964; 86:5626–5630.
26c. Corey EJ, Carey FA, Winter RAE. J Am Chem Soc 1965; 87:934–935.
26d. Cope AC, Hecht JK, Johnson JHW, Keller H, Winkler HJS. J Am Chem Soc 1966; 88:761–763.
26e. Cope AC, Howell CF, Bowers J, Lord RC, Whitesides GM. J Am Chem Soc 1967; 89:4024–4027.
26f. Hines JN, Peagram MJ, Whitham GH, Wright M. Chem Commun 1968:1593–1594.
27. Swenton JS. J Org Chem 1969; 34:3217–3218.
28. Inoue Y, Kunitomi Y, Takamuku S, Sakurai H. J Chem Soc, Chem Commun 1978: 1024–1025.
29. Inoue Y, Takamuku S, Kunitomi Y, Sakurai H. J Chem Soc, Perkin Trans 1980; 2: 1672–1677.
30a. Inoue Y, Yokoyama T, Yamasaki N, Tai A. J Am Chem Soc 1989; 111:6480–6482.
30b. Inoue Y, Yokoyama T, Yamasaki N, Tai A. Nature 1989; 341:225–226.
31. Inoue Y, Yamasaki N, Yokoyama T, Tai A. J Org Chem 1992; 57:1332–1345.
32. Inoue Y, Yamasaki N, Yokoyama T, Tai A. J Org Chem 1993; 58:1011–1018.
33a. Asano T, le Noble WJ. Chem Rev 1978; 78:407–489.
33b. van Eldik R, Asano T, le Noble WJ. Chem Rev 1989; 89:549–688.
34a. Chung WS, Turro NJ, Mertes J, Mattay J. J Org Chem 1989; 54:4881–4887.
34b. Buback M, Bünger J, Tietze LF. Chem Ber 1992; 125:2577–2582.
34c. Iwaoka T, Katagiri N, Sato M, Kaneko C. Chem Pharm Bull 1992; 40:2319–2324.
35. Inoue Y, Matsushima E, Wada T. J Am Chem Soc 1998; 120:10687–10696.
36. Kaneda M, Asaoka S, Ikeda H, Mori T, Wada T, Inoue Y. Chem Commun 2002: 1272–1273.
37. Inoue Y, Ikeda H, Kaneda M, Sumimura T, Everitt SRL, Wada T. J Am Chem Soc 2000; 122:406–407.
38. For a full account of the origin and analyses of the kinetic and thermodynamic enthalpy–entropy compensation effect, see Y Inoue, T Wada. In: GW Gokel, ed. Advances in Supramolecular Chemistry. Greenwich, CT: JAI Press, 1997, 55–96.
39a. Brennecke JF, Chateauneuf JE. Chem Rev 1999; 99:433–452.
39b. Baiker A. Chem Rev 1999; 99:453–474.

40a. Johnston KP, Pennenger ML, eds. Supercritical Fluid Science and Technology. Washington. D.C.: American Chemical Society, 1989.

40b. Bruno TJ, Ely JF, eds. Supercritical Fluid Technology. Reviews in Modern Theory and Application. Boca Raton. FL: CRC Press, 1991.

40c. Clifford T. Fundamentals of Supercritical Fluids, Oxford University Press. 1999.

41. Saito R, Kaneda M, Wada T, Katoh A, Inoue Y. Chem Lett 2002:860–861.

42. Tucker SC. Chem Rev 1999; 99:391–418.

43. Shi M, Inoue Y. J Chem Soc, Perkin Trans 1998; 2:1725–1729.

44. Shi M, Inoue Y. J Chem Soc, Perkin Trans 1998; 2:2421–2427.

45a. Inoue Y, Dong F, Yamamoto K, Tong L-H, Tsuneishi H, Hakushi T, Tai A. J Am Chem Soc 1995; 117:11033–11034.

45b. Inoue Y, Wada T, Sugahara N, Yamamoto K, Kimura K, Tong L-H, Gao X-M, Hou Z-J, Liu Y. J Org Chem 2000; 65:8041–8050.

46a. Sugahara N, Kawano M, Wada T, Inoue Y. Nucleic Acids Symp Ser 2000; 44: 115–116.

46b. Wada Y, Sugahara N, Kawano M, Inoue Y. Chem Lett 2000:1174–1175.

47. Wada T, Shikimi M, Inoue Y, Lem G, Turro NJ. Chem Commun 2001:1864–1865.

48. Tsuneishi H, Hakushi T, Inoue Y. J Chem Soc, Perkin Trans 1996; 2:1601–1605.

49. Tsuneishi H, Hakushi T, Tai A, Inoue Y. J Chem Soc, Perkin Trans 1995; 2: 2057–2062.

50. Sugimura T, Shimizu H, Umemoto Si, Tsuneishi H, Hakushi T, Inoue Y, Tai A. Chem Lett 1998:323–324.

51a. Inoue T, Matsuyama K, Inoue Y. J Am Chem Soc 1999; 121:9877–9878.

51b. Matsuyama K, Inoue T, Inoue Y. Synthesis 2001:1167–1174.

52. Inoue Y, Tsuneishi H, Hakushi T, Tai A. J Am Chem Soc 1997; 119:472–478.

53. Inoue Y, Nishida K, Ishibe H, Hakushi T, Turro NJ. Chem Lett 1982:471–474.

54. Bouman TD, Hansen AE. Croatia Chem Acta 1989; 62:227–243.

55. Goto S, Takamuku S, Sakurai H, Inoue Y, Hakushi T. J Chem Soc, Perkin Trans 1980; 2:1678–1682.

56a. Inoue Y, Ueoka T, Kuroda T, Hakushi T. J Chem Soc, Perkin Trans 1983; 2:983–988.

56b. Inoue Y, Ueoka T, Hakushi T. J Chem Soc, Perkin Trans 1984; 2:2053–2056.

56c. Inoue Y, Hagiwara S, Daino Y, Hakushi T. J Chem Soc, Chem Commun 1985: 1307–1309.

57. Squillacote M, Bergman A, De Felippis J. Tetrahedron Lett 1989; 30:6805–6808.

58. Hoffmann R, Inoue Y. J Am Chem Soc 1999; 121:10702–10710.

59. Kropp PJ, Krauss HJ. J Am Chem Soc 1967; 89:5199–5208.

59. Kropp PJ. J Am Chem Soc 1969; 91:5783–5791.

59. Kropp PJ, Snyder JJ, Rawlings PC, Fravel HG. J Org Chem 1980; 45:4471–4474.

59. Salomon RG, Folting K, Streib WE, Kochi JK. J Am Chem Soc 1974; 96:1145–1152.

60. Asaoka S, Horiguchi H, Wada T, Inoue Y. J Chem Soc, Perkin Trans 2000; 2: 737–747.

61a. Gassman PG, Boltorff KJ. J Am Chem Soc 1987; 109:7547–7548.

61b. Gassman PG, de Silva SA. J Am Chem Soc 1991; 111:9870–9872.

62. Alvaro M, Formentin P, Garcia H, Palomares E, Sabater MJ. J Org Chem 2002; 67: 5184–5189.

63. For example, see A Padwa, ed. 'Organic Photochemistry' Series, New York: Marcel Dekker. (b) V Ramamurthy, KS Schanze eds. 'Organic Molecular Photochemistry' Series, Marcel Dekker, New York, CRC Press.

64. Inoue Y, Okano T, Yamasaki N, Tai A. J Photochem Photobiol A: Chem 1992; 66: 61–68.

65. Cauble DF, Lynch V, Krische MJ. J Org Chem 2003; 68:15–21.

66. Mattay J. Angew Chem, Int Ed Engl 1987; 99:849.

67a. Vondenhof M, Mattay J. Tetrahedron Lett 1990; 31:985–988.

67b. Vondenhof M, Mattay J. Chem Ber 1990; 123:2457–2459.

68. Asaoka S, Ooi M, Jiang P, Wada T, Inoue Y. J Chem Soc, Perkin Trans 2000; 2: 77–84.

69a. Kim J-I, Schuster GB. J Am Chem Soc 1990; 112:9635–9637.

69b. Kim J-I, Schuster GB. J Am Chem Soc 1992; 114:9309–9317.

70a. Inoue Y, Okano T, Yamasaki N, Tai A. J Chem Soc, Chem Commun 1993:718–720.

70b. Asaoka S, Kitazawa T, Wada T, Inoue Y. J Am Chem Soc 1999; 121:8486–8498.

70c. Takehara Y, Ohta N, Shiraishi S, Asaoka S, Wada T, Inoue Y. J Photochem Photobiol, A: Chem 2001; 145:53–60.

70d. Asaoka S, Wada T, Inoue Y. J Am Chem Soc 2003; 125:3008–3027.

71. Mizuno K, Nakanishi I, Ichinose N, Otsuji Y. Chem Lett 1989:1095–1098.

72a. Inoue Y, Gokel GW, eds. Cation Binding by Macrocycles. New York: Marcel Dekker, 1990 Chapter 1.

72b. Whitesides GM, Simanek EE, Mathias JP, Seto CT, Chin D, Mammen M, Gordon DM. Acc Chem Res 1995; 28:37–44.

72c. Inoue Y, Wada T. In: Gokel GW, Ed. Advances in Supramolecular Chemistry. Greenwich. CT: JAI Press, 1997:55–96.

5
Diastereodifferentiating Photoreactions

Norbert Hoffmann and Jean-Pierre Pete
Université de Reims Champagne-Ardenne
Reims, France

I. INTRODUCTION

During the past twenty years, asymmetric synthesis has become so efficient that many reactions can be performed with an almost complete stereoselectivity. With the development of highly effective chiral catalysts, only reactions carried out with diastereoisomeric excess (*de*) or enantiomeric excess (*ee*) higher than 95% are considered to be useful for synthetic applications. In comparison with the high asymmetric inductions that can be obtained in the ground state, the synthetic applications of asymmetric photochemistry remained almost unexplored until recently. However, in a very important review by Rau in 1983, it was recognized that "photoreactions made asymmetric by chiral substituents of one reactant have the potential of satisfactory or even good optical yields" [1]. In agreement with this statement, photochemists have developed more and more highly stereoselective photoreactions and excellent reviews have appeared in the literature [2].

Numerous photoreactions are known to produce new stereogenic centers by either isomerization or addition processes. In asymmetric photoreactions, usually one of the possible isomers, which corresponds to the energetically more favorable transition state, predominates. In the presence of stereogenic centers in the starting substrates or reagents, an asymmetric induction occurs, and the resulting diastereoisomers can be easily isolated in pure form. For synthetic applications, the stereogenic centers of the reactants can be maintained as part of final elaborated structures or removed from the photoproducts. In the last case, a major achievement can be obtained when a removable chiral auxiliary is introduced on the starting substrate. Even if the asymmetric induction is quite low, the diastereoisomers can be easily separated. After the removal of the chiral auxiliary, the enantiomerically pure product can be considered as the result of an indirect and enantioselective photoreaction of prochiral starting materials.

The purpose of this review is to examine the recent progress in the field of asymmetric photoreactions involving a chiral reactant, and to discuss the factors that control the diastereodifferentiation. In the excited state as in the ground state, the level of asymmetric induction is strongly related to the conformational flexibility of the reactants. When carried out in the solid state, in a confined cavity of zeolites or in supramolecular scaffoldings, restrictions of the mobility occur, and photoreactions can become highly stereoselective. These aspects of asymmetric induction will be covered in separate chapters, and this review will consider only diastereoselective photoreactions carried out in solution.

Among these reactions, the photochemical cycloadditions of C=C bonds, which can create up to four asymmetric carbons during the photochemical step, are particularly interesting, and numerous synthetic applications of this reaction have been reported. Advances in the understanding of the origin of asymmetric induction, during addition of alkenes with carbonyl derivatives, cyclic enones, and aromatic compounds, will be discussed in detail.

Photoisomerizations involving electrocyclic or photoreductive cyclization processes, which are also able to create new stereogenic centers, are well known reactions. However, research on the asymmetric versions of these reactions have appeared only recently, and interesting features such as the transfer of chirality, also called the memory effect, have been discovered. The recent advances in the field seem to indicate that many photorearrangements of chiral substrates may be good candidates for highly diastereoselective syntheses. The factors that seem to control the diastereoselection in these photorearrangements will also be examined. Finally, unstable species such as dienols or radical species produced during the photochemical step can be transformed in the ground state with high *de*. The development of such a strategy as a useful tool for the asymmetric creation of new stereogenic centers will finally be described.

II. DIASTEREODIFFERENTIATION IN [2 + 2] PHOTOCYCLOADDITIONS INVOLVING ALKENES

A. The Paternò–Büchi Reaction

Oxetanes are versatile intermediates for organic synthesis. In a photochemical way, they are accessible by the Paternò–Büchi reaction of carbonyl compounds and almost all kinds of olefines (Scheme 1) [3]. However, the reaction is most efficient when it is carried out with electron-rich alkenes. In these cases, T_1 states of the carbonyl partner possessing $^3n,\pi^*$ character are involved in the reaction [4]. The polarity of carbonyl functions is reversed in this electronically excited

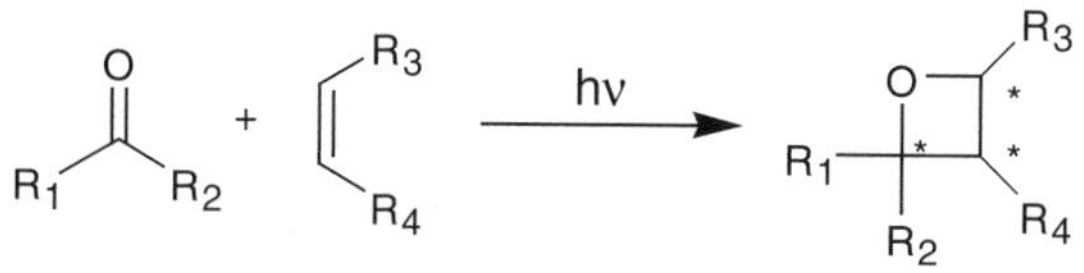

Scheme 1

state, and a positive partial charge is localized at the oxygen atom. They react easily with olefines by formation of a C–O σ-bond to yield 1,4-biradical intermediates of triplet multiplicity. More detailed mechanistic considerations include further, less stable intermediates for this step. In this context, exciplexes and radical ions have been proposed. However, the biradical intermediates play a dominant role as far as stereochemical factors are concerned. During the reaction, up to three stereogenic centers are created, and many efforts have been made to control their absolute configuration. Enantiopure products can be obtained from the transformation of enantiopure substrates. In this context, the ketone **1** derived from glucose added with furan **2** to yield four diastereoisomers **3**, **4**, **5**, and **6** (Scheme 2) [5].

The facial diastereoselectivity derived from the ratio $(3+4)/(5+6)$ was 50%, while the exo/endo selectivity derived from the product ratio $(3+5)/(4+6)$ was 40%. Oxetanes **9a,b** were obtained with a low diastereoselectivity from the reaction of (R)-isopropylideneglyceraldehyde **7** with 3,4-dimethylfuran **8** [6]. Oxetanes **9a,b** have been used for the synthesis of asteltoxin. Enantiopure acyl cyanides were used in the same way as chiral carbonyl reaction partners [7] and camphor for the addition with electron-poor alkenes like dicyanoethylene [8]. In the latter case the reaction occurs in the S_1 state of the carbonyl compound.

Several reactions were carried out with chiral olefines. For example, only one stereoisomer **11** was isolated from the Paternò–Büchi reaction of D-glucal triacetate **10** with acetone (Scheme 3) [9].

The malic acid derivative **12** reacted with benzaldehyde to yield the oxetanes **13a,b** with a diastereomeric excess of 80% (Scheme 4) [10]. It should also be mentioned that the regioselectivity and the exo/endo selectivity are complete. The favored formation of **13a** is explained by the dominant conformation depicted in **A**, **B**. The syn approach of benzaldehyde excited in the $^3n,\pi^*$ state with respect to the alkoxy substituent (transition state **A**) is hindered by electrostatic repulsion between the substituent and the carbonyl group having a reversed polarity in the excited state. The addition of benzophenone to the furan derivative **14** was stereospecific [11]. In this case, however, the attack of the $^3n,\pi^*$ excited ketone occurred in a syn manner with respect to the hydroxy function to yield **15**. The conformation indicated in the transition state **C** was supported by calculations.

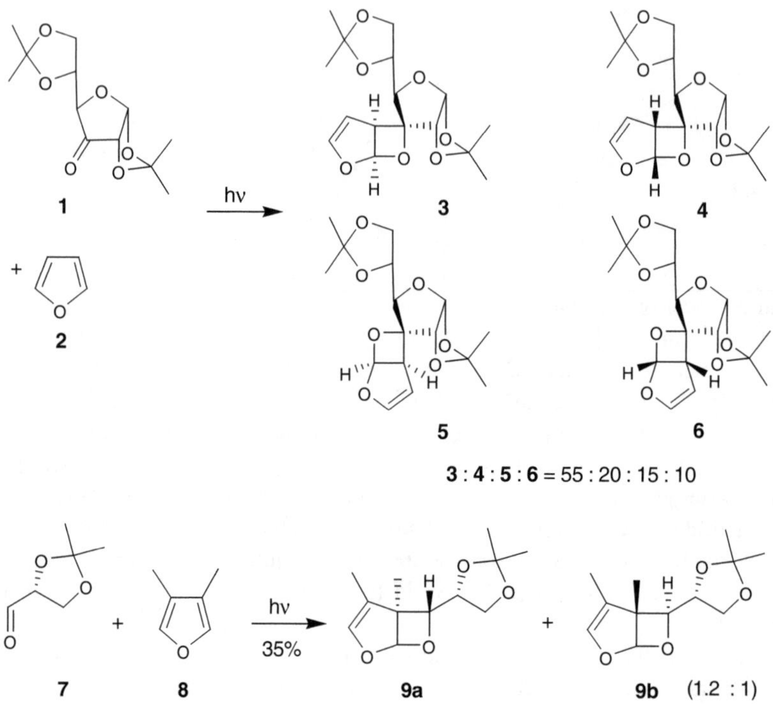

3 : 4 : 5 : 6 = 55 : 20 : 15 : 10

Scheme 2

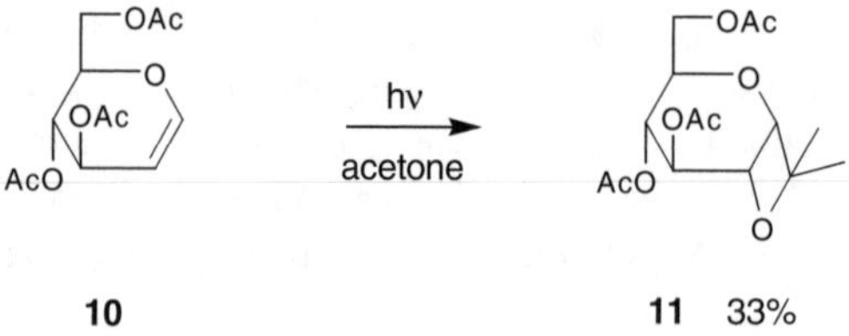

Scheme 3

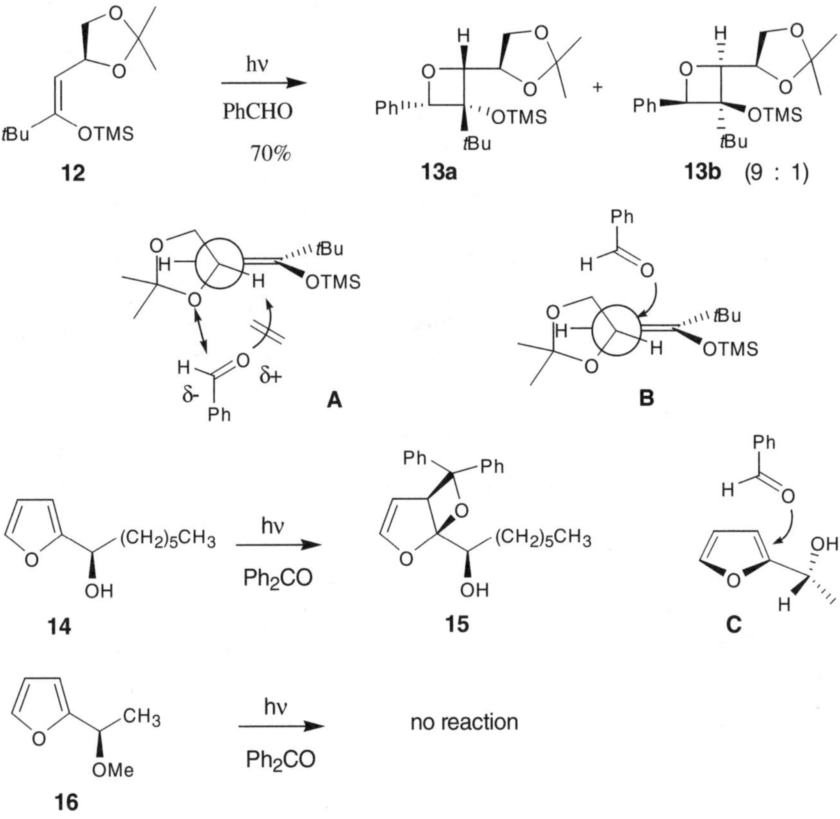

Scheme 4

It was proposed that either a hydrogen bond or polar interactions favor this approach. When the O-methylated furan derivative **16** was treated under the same conditions, no conversion was observed.

Chiral enamine derivatives have also been used as electron-rich alkenes. The oxazoline derivative **17** reacted with benzaldehyde to yield the two stereoisomeric oxetanes **18a** and **18b** with a diastereomeric excess of 67% (Scheme 5) [12]. A significantly higher diastereoselectivity was observed in the case of the reaction of the pyrrolidine derivatives **19** where the enamine function is localized inside the five membered ring [13]. Then the oxetane **20a** (R = n-C_9H_{19}) was used in an asymmetric synthesis of the antifungal alkaloid (+)-preussin. The approach of the ^{3}n,π* excited ketone preferentially occurred syn with respect to

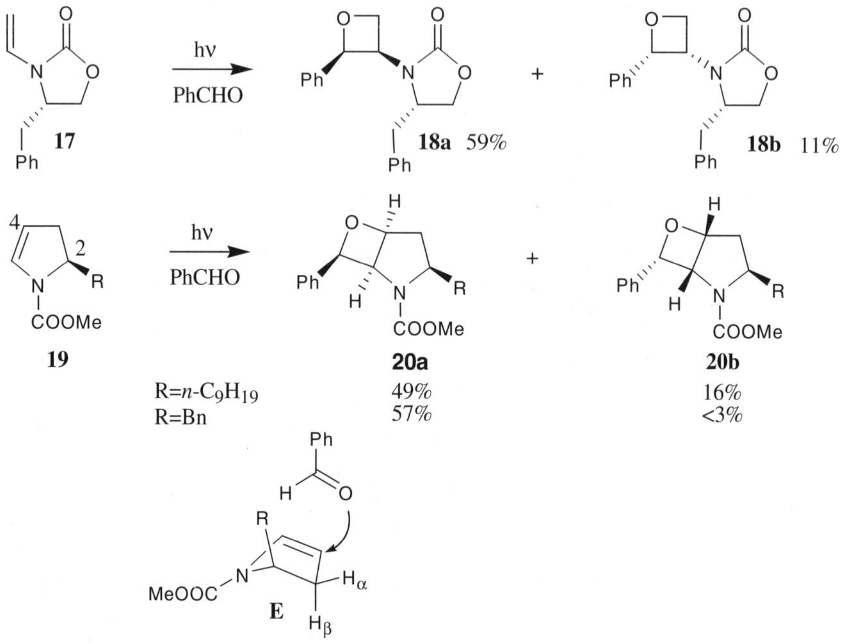

Scheme 5

the substituent R. Owing to a 1,3-allylic strain analog, R is directed in a pseudoaxial position (transition state **E**). The nitrogen atom is considered to be planar. Consequently, one of the hydrogen atoms of C-3 (H_β) is also in an axial position. The attack of the carbonyl compound at position 4 occurs antiperiplanar to this hydrogen to yield directly a staggered product.

Considerable work was done to induce chirality via chiral auxiliaries. Reactions with aromatic α-ketoesters like phenylglyoxylates **21** and electron-rich alkenes like dioxoles **22** and furan **23** were particularly efficient (Scheme 6). Yields up to 99% and diastereoselectivities higher than 96% have been observed when 8-phenylmenthol **21a** or 2-t-butylcyclohexanol **21b** were used as chiral auxiliaries [14–18]. It should be noted that only the exoisomers **24** and **25** were obtained from the reaction of dioxoles **22**. Furthermore, the reaction with furan **23** was regioselective. **24** were suitable intermediates in the synthesis of rare carbohydrate derivatives like branched chain sugars [16]. Other heterocyclic compounds like oxazole **28** [19] and imidazole **29** [20] derivatives as well as acyclic alkenes **30**, **31**, and **32** [14,15,21,22] were used as olefinic partners. Numerous cyclohexane derived alcohols [18,21–24] and carbohydrate derivatives [25] were used as chiral

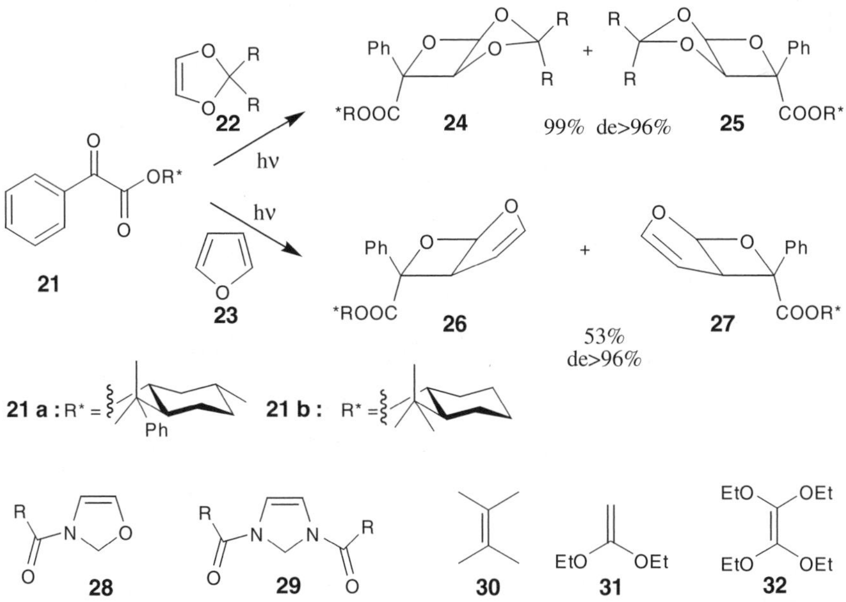

Scheme 6

auxiliaries. Acyclic chiral alcohols were less efficient auxiliaries [26]. Several heteroaromatic and substituted phenylglyoxylic acid derivatives as well as nonaromatic α-ketoesters have also been used as carbonyl reaction partners [26,27]. Recently, chiral auxiliaries able to complex the olefinic reaction partner very efficiently have been used successfully, and high diastereoselectivities (> 90%) could be obtained [28]. This strategy is discussed in a separate chapter.

The high diastereoselectivity of the reactions of the α-ketoesters was explained by the mechanism depicted in Scheme 7 [23]. After photochemical excitation of the carbonyl compound **21** and intersystem crossing, the substrate attains its $^3n,\pi^*$ state. The T_1 excited ketoester adds to the olefine **22** to yield the diastereomeric 1,4-biradicals **F** and **F'**. In this reaction step (the first step of the stereoselection), a C–O σ-bond is formed and chirality is induced. The amount of induced chirality can be increased in the following step (the second step of stereoselection). The intermediates **F** and **F'** have indeed two possibilities for reaction. They can form the final products **24** and **25** via the formation of a C–C σ-bond or fragment into the starting materials in their ground state. The disfavored intermediate **F'** preferentially undergoes fragmentation while the favored intermediate **F** yields the major diastereomeric oxetane **24**. The two steps of chiral induc-

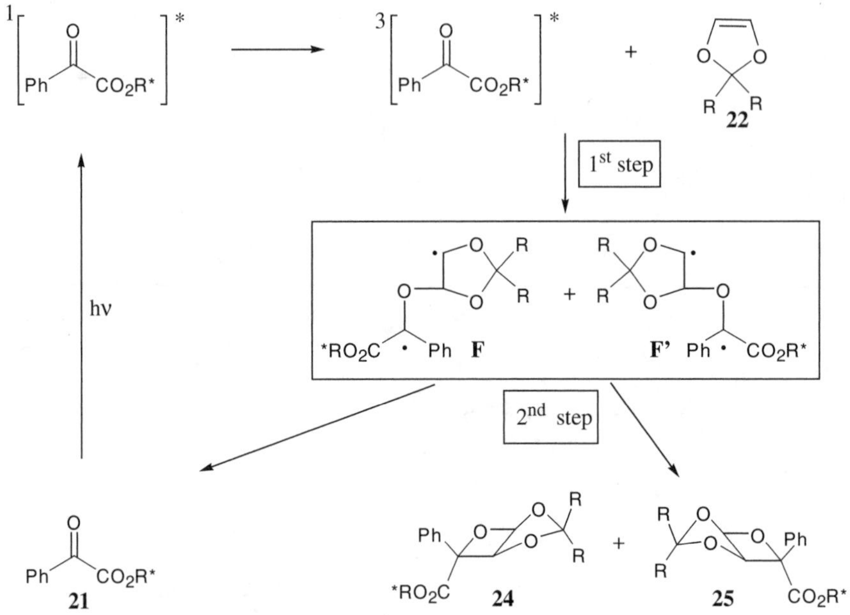

Scheme 7

tion were detected and characterized by a dependence of the diastereoselectivity with temperature. The enthalpy–entropy compensation that was described for the first time for the photochemical diastereodifferentiation seems to be of general importance [29,18]. Furthermore, it was found that the conformational equilibria of the cyclohexane auxiliaries could influence the competition between oxetane formation and fragmentation during the second selection step [24].

Recently, studies were carried out to explain the exo/endo selectivity of the Paternò–Büchi reaction [30]. These studies were carried out mostly with achiral or racemic substrates. Excited monocyclic aromatic aldehydes **33** react in their $^3n,\pi^*$ state with cyclic enol ether derivatives like 2,3-dihydrofuran **34** (Scheme 8) [31]. In these cases, the sterically disfavored endo isomer **35a** was obtained as major product. This result was explained by the fate of the triplet biradical intermediate **G**. In order to favor cyclization to the oxetanes **35a,b**, the radical p-orbitals have to approach in a perpendicular fashion to increase the spin-orbit coupling needed for the triplet to singlet intersystem crossing [32]. The sterically most favored arrangement of this intermediate is depicted as **G**. The encumbering Ar substituent is orientated upside and anti to the trihydrofuranyl moiety. Cyclization from this conformation yields the major isomer **35a**.

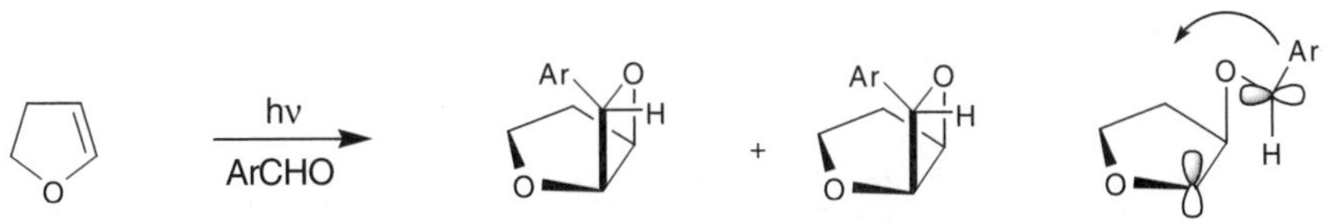

		35a		35b	
34	**33**	**35a**		**35b**	**G**
	Ar = Ph	88	:	12	
	o-Tol	92	:	8	
	Mes	>98	:	<2	
	2,4-di-tBu-6-Me-Ph	>98	:	<2	

Scheme 8

The asymmetric Paternò–Büchi reaction was also performed with aromatic thioketones and alkenes carrying an electron-withdrawing substituent. When xanthione **36** was irradiated in the presence of menthylmetacrylate, the corresponding thiethane **37** was isolated with a significantly higher diastereoselectivity when irradiated at $\lambda = 589$ nm rather than 400 nm (Scheme 9) [33]. Similar results were obtained with 4,4′-dimethoxythiobenzophenone. The dependence of the stereoselectivity on the wavelength of irradiation was explained by the formation of a low-energy triplet 1,4-biradical **H** in the first case and a more or less concerted reaction at a higher energy level in the second case. A more systematic investigation of the temperature dependence on the stereoselectivity of this reaction revealed that the 1,4-biradical intermediates play the same role as the corresponding intermediate **F,F′** (Scheme 7) in the classical Paternò–Büchi reaction [34]. When the irradiation was carried out at $\lambda = 589$ nm and at low temperature, the diastereoselectivity was significantly improved. The best *de* values were obtained in the reaction of phenylcyclohexyl tiglate or menthylmethacrylate **38** with xanthione **36**. It should also be mentioned that the reaction was regio and exo/endo selective.

B. The Reaction of Alkenes with Enones and Conjugated Esters

The mechanism of the [2 + 2] photocycloaddition of alkenes with cyclenones is a multistep process that involves the addition of alkenes to the $^3\pi,\pi^*$ triplet state of a cyclic enone and the formation of 1,4-biradical intermediates [35]. Except for the reaction of 1-alkenes, up to four new stereogenic centers are formed during the cycloaddition (Scheme 10).

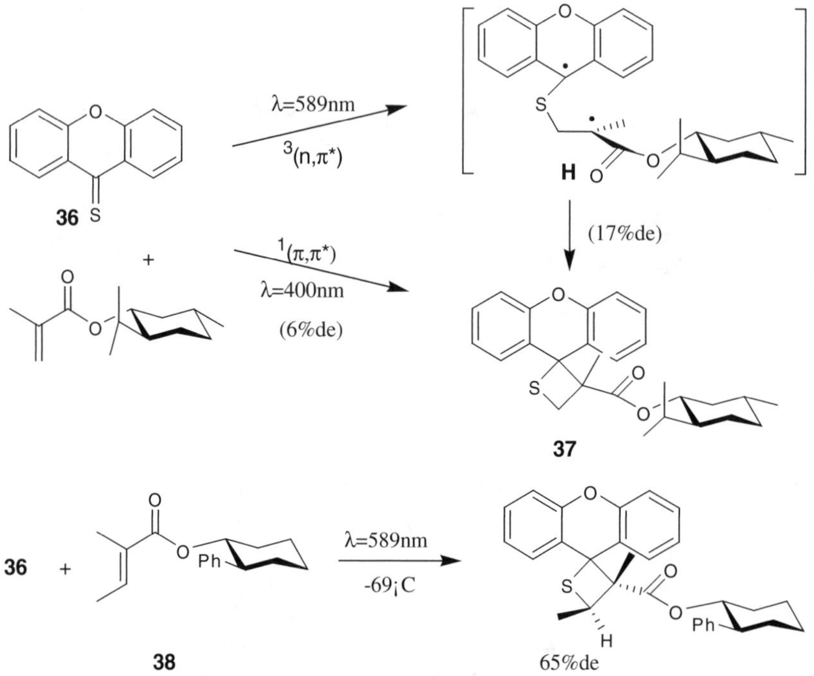

Scheme 9

1. Diastereodifferentiation in Intermolecular Processes

The formation of a cyclobutane introduces a large strain in the cycloadducts, and cis configurations of the bicyclic molecules are selectively obtained. When asymmetric centers are present in the starting material, the stereochemistry of the products is controlled by the relative rates of cyclization and of cleavage of these biradicals. For intermolecular processes, it is well established that steric

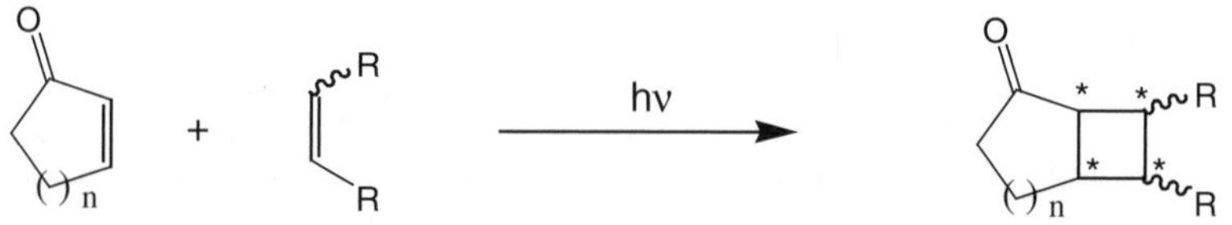

Scheme 10

hindrance of the reactants can influence strongly the stereochemistry of the new stereogenic centers [36]. Although the early examples of asymmetric induction in the [2 + 2] photocycloaddition will not be examined in detail, the main results will now be summarized.

With chiral cyclopentenes, which possess an almost planar structure, one face is selectively crowded by the largest substituent on the stereogenic center. For this reason, when cyclic enones such as **39–41** are irradiated in the presence of these chiral cyclopentenes, the cycloaddition occurs selectively from the less-hindered face, and usually only one stereoisomer can be isolated (Scheme 11) [37].

The reacting C=C bond of cyclic enones in its $^3\pi,\pi^*$ state is twisted and has a biradical character. As a consequence, the observed selectivity depends strongly on the conformational rigidity of the starting enone and the position of the stereogenic centers on the reactants **42–47**. The best facial selectivity is observed when alkyl substituents are introduced at allylic positions. When the stereogenic center is α- to the carbonyl group of **42** and **43**, the cycloaddition proceeds in high yields and with an almost exclusive approach of the alkene from the opposite side [38]. The diastereoselectivity is usually lower when the allylic substituent is located on the γ-carbon of a cyclohexenone [39] or of a furanone ring (Scheme 12) [40].

2. Diastereodifferentiation in Intramolecular Processes

With disymmetrical alkenes the intermolecular photocycloaddition reaction produces a mixture of regioisomers. In contrast, the intramolecular photocycloaddition process is highly regioselective, and only one regioisomer is usually observed. Furthermore, the conformational restrictions appearing in the corresponding transition states favor a highly diastereoselective process, even in the absence of stereogenic centers in the starting molecule. As observed for intermolecular photocycloadditions, a preferred approach from the less hindered side is observed for the corresponding intramolecular reactions of cyclohexenones or cyclopentenones substituted at Cα or Cβ by an unsaturated chain. With cyclohexenones **48** and **50** with a stereogenic center at C-4 and an ω-alkenyl chain at

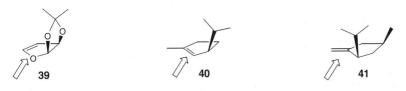

39   **40**   **41**

Scheme 11

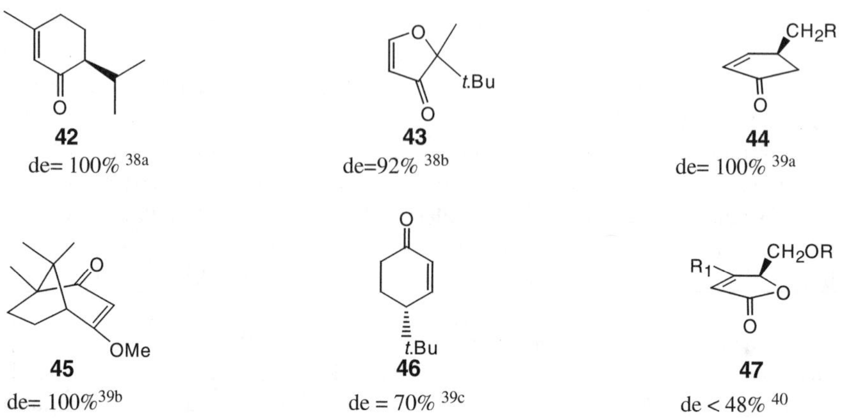

42
de= 100% [38a]

43
de=92% [38b]

44
de= 100% [39a]

45
de= 100%[39b]

46
de = 70% [39c]

47
de < 48% [40]

Scheme 12

either C-2 or C-3, an almost total control of the facial approach of the alkenyl group on the excited cyclohexenone ring is observed (Scheme 13) [41]. In the presence of ω-pentenyl substituents, formation of the biradical intermediates involves an initial formation of a cyclopentane ring according to the rule of five [42]. For steric reasons, the observed adduct results from a cyclization of the only 1,4-biradical that produces a *cis*-5, 4-fused system [43]. Interestingly, the facial selectivity is considerably lower for cyclopentenones like **53** substituted at C-3 by an ω-butenyl chain [44].

When the unsaturated side chain is fixed at the only stereogenic center of the cyclic enones **56–60**, bridged cycloadducts can be formed in high yields. An exclusive approach of the alkene from the same side of the enone as the alkenyl group on the stereogenic center is obtained [45]. This can be explained by the large steric strain present in the transition states leading to cycloadducts. Similar effects are observed with enones **61, 63,** or analogs such as maleimide **62, 64, 65** that are linked at the only stereogenic center of a cyclic alkene (Scheme 14) [46].

When stereogenic centers are introduced on the tether, a large asymmetric induction can also be observed, especially when the product results from an initial 1,5-ring closure. The relaxed $^3\pi,\pi^*$ excited state of the starting enone behaves as the 5-hexenyl radicals and prefers to cyclize in a *5-exotrig* process [47] according to Baldwin's rules [48]. The model used to determine the selectivity of the intramolecular addition of 5-hexenyl radicals can be applied to **66** and **68**. Then the stereochemistry of the major diastereoisomers **67** and **69** can be deduced from the transition states that place the largest substituent in equatorial position, as

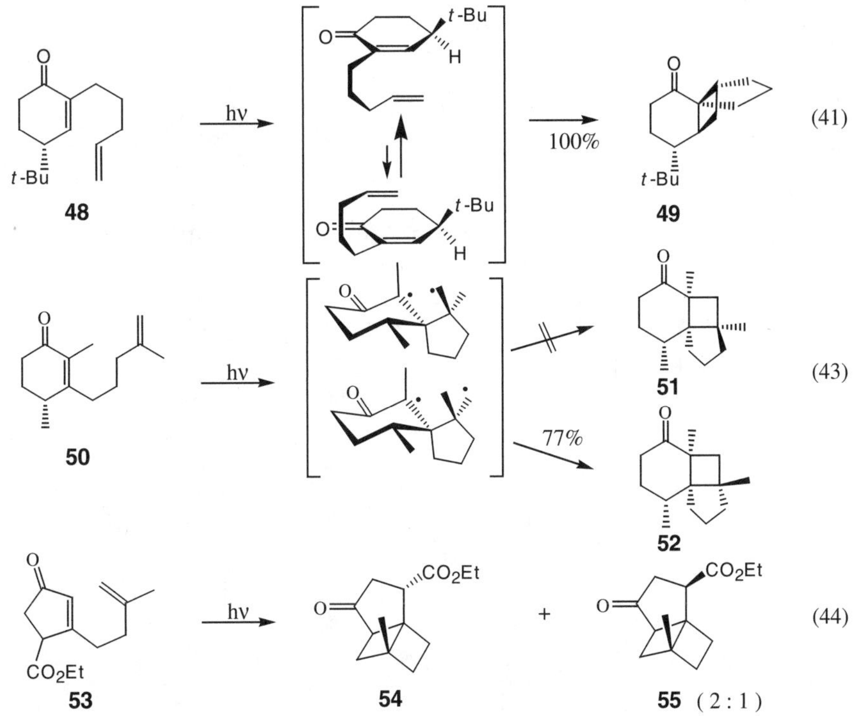

Scheme 13

shown in Scheme 15 [49]. The relative energy of the transition states based on MM2 calculations is in agreement with the observed selectivity. Similar conclusions can be obtained for the cycloaddition of cyclopentenones and cyclohexenones having a chiral ω-hexenyl chain [50].

The relative energy of diastereomeric transition states can be influenced not only by the position of substituents on the tether but also by their structure and the possibility of intramolecular hydrogen bonds. If an intramolecular hydrogen bond can be developed during the addition process, the preferred transition state can be modified and a reversal of the selectivity can result. This is illustrated in Scheme 16.

In the presence of silyl ethers in **70**, no hydrogen bonds can be created, and the large silyl group occupies an equatorial position in the chair transition state **TS A**. The cycloadduct **71** resulting from this approach is formed almost exclusively (*de* > 98%). When the hydroxyl group is not protected, an intramolecular hydrogen bond can be formed between the hydroxyl and carboethoxy groups,

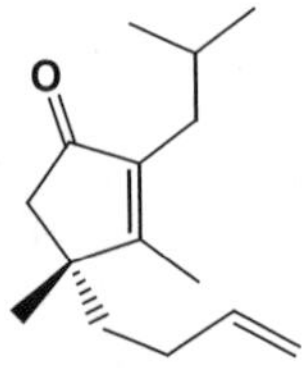

56

86% [45a]

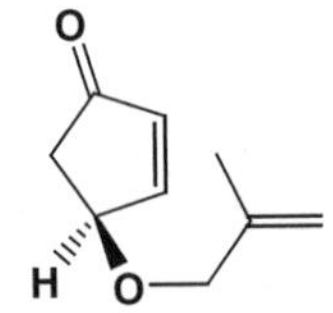

57

80% [45b]

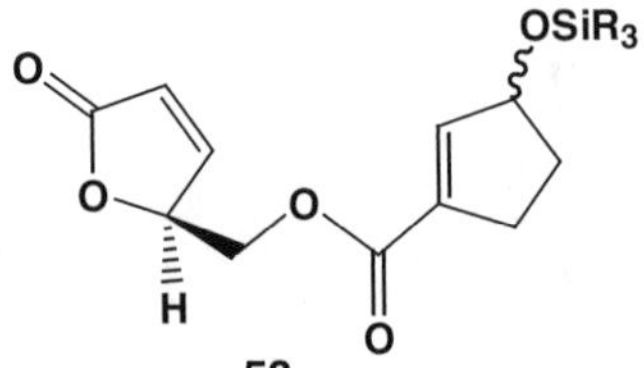

58

61% [45c]

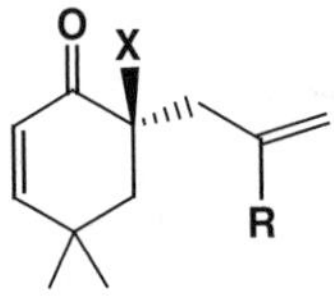

59

86% [45d]

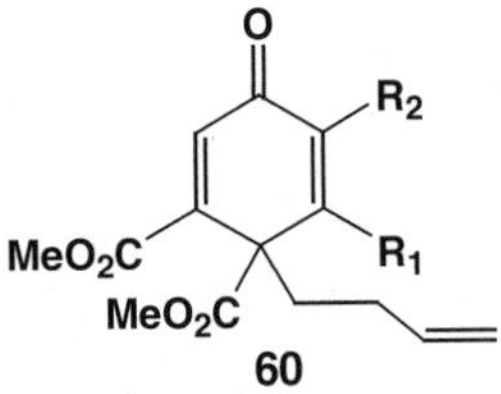

60

>90% [45e]

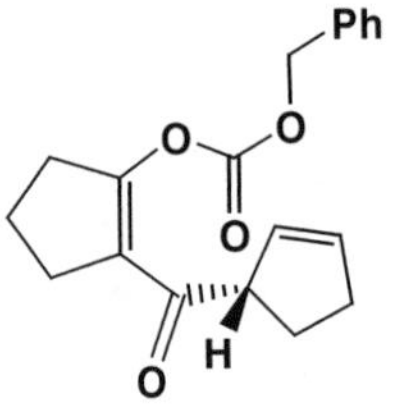

61

92% [46a,46b]

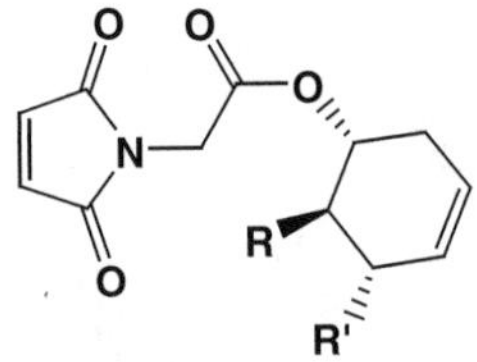

62

37-51% [46c]

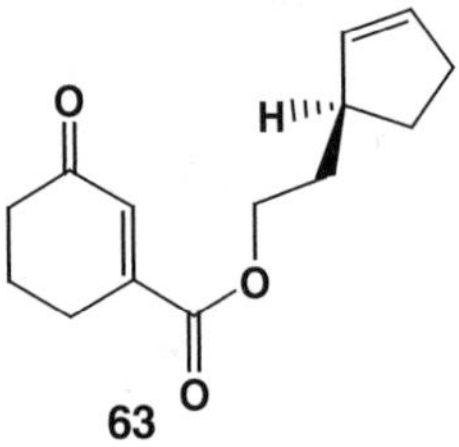

63

52% [46e]

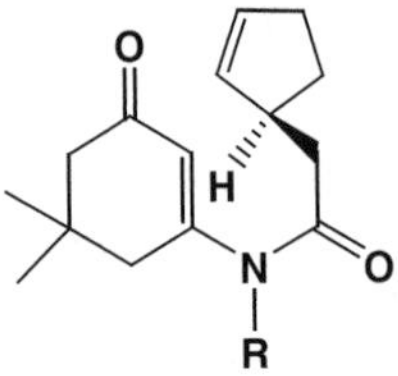

64 (R = PhCH$_2$, C$_6$H$_{13}$)

67-77% [46d]

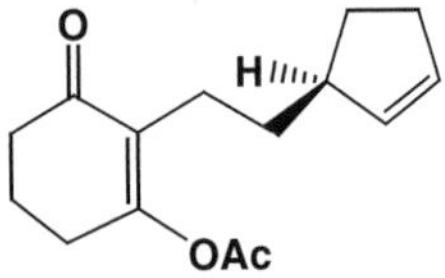

65

82% [46f]

Scheme 14

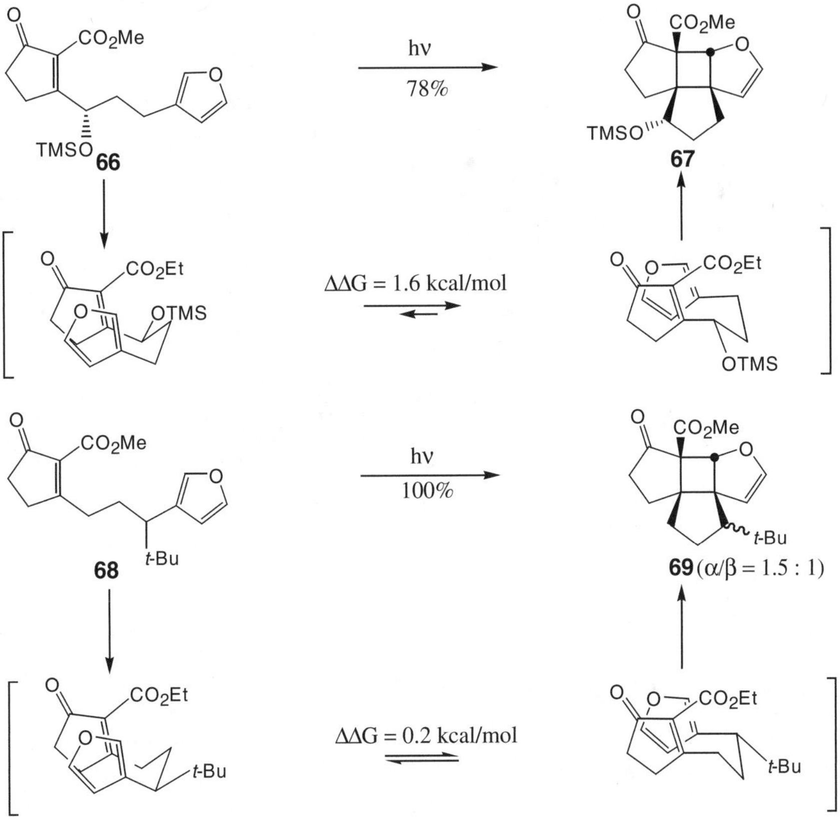

Scheme 15

and the transition state **TS B** becomes competitive in nonpolar solvents. As a consequence, a considerable lowering of the diastereoselectivity is observed. If the reaction is carried out in methanol, a competition between intra- and intermolecular hydrogen bonds is possible. This makes again transition state **TS A** more favorable, and an increase of the selectivity is obtained [51].

Photocycloaddition of allenes with cyclenones deserves some comments. The reaction is not only regioselective but highly diastereoselective, owing to the geometry of the allenyl group. In the presence of stereogenic centers on the enone, an important asymmetric induction is usually observed. In an attempt to explain the high selectivity, it was proposed that the preferred configuration of the major cycloadduct could be deduced from the geometry of the most favored

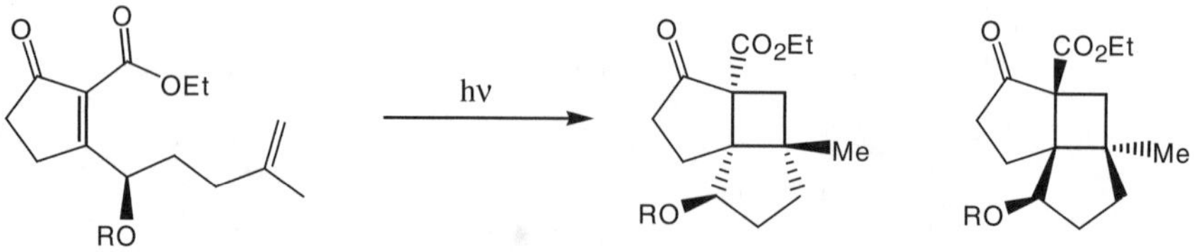

R = SiEt₃	hexanes	> 99	:	< 1
H		1.1	:	1
H	MeOH	5	:	1

$R = SiEt_3$

TS A :   **TS B :**

Scheme 16

biradical intermediate [52]. Interestingly, a similar approach based on minimum energy conformations of the 1,4-biradical intermediates, as determined by MM2 and MM3 calculations, was proposed recently to explain the diastereoselectivity of intramolecular photocycloadditions of enones [53]. This model considers that the lifetimes of the biradicals are long enough to allow a conformational relaxation, prior to cyclobutane formation or its cleavage to the starting materials. The factors governing the diastereodifferentiation of intramolecular cycloadditions of allenes or alkenes with cyclic enones remain similar [54].

Another illustration of the role of the conformational mobility on the level of the observed asymmetric induction is shown by the intramolecular photocycloaddition of cinnamyl derivatives of vinyl glycine **73** into a mixture of **74** and **75** (Scheme 17). In the sensitized reaction and for steric reasons, the attack of $^3\pi,\pi^*$ excited styrene on the vinyl oxazolidinone **76** gives mainly **77** [55].

It is well known that copper salts can catalyze the photocycloaddition of dienes [56]. The reaction can be very diastereoselective and useful as an access to polycyclic molecules, especially 1-vinyl-2-allylcyclanes [57]. The reaction of vinylglycine, cinnamyl derivatives **76**, and analogs is also catalyzed by Cu(I) produced in situ by the reduction of cupric triflate. The coordination of the double bonds to the copper ion during the [2 + 2]-photocycloaddition process increases

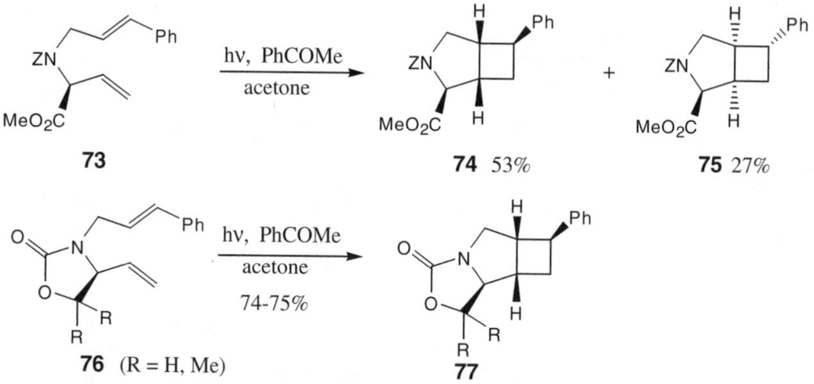

Scheme 17

the rigidity of the diastereoisomeric transition states. An increase of the chemical yield (> 81%) was observed and a single diastereoisomer was isolated [58].

3. Diastereodifferentiation by a Removable Chiral Auxiliary

In solution, the great conformational mobility of molecules does not favor tight transition states. For a good asymmetric induction, there is a need to decrease this mobility and to introduce the chiral information as close as possible from the prochiral centers. An efficient way to introduce the chiral environment in asymmetric [2 + 2]-photocycloadditions is to connect chiral auxiliaries to the reactants. When removable chiral auxiliaries are attached to the starting molecules before the photochemical step, photocycloadditions proceed diastereoselectively. After removal of the chiral auxiliary, the overall process can be considered as a byroad for an enantioselective reaction.

The first report of an enantioselective photocycloaddition of prochiral conjugated esters involved the introduction of (+)-2,3-di-O-methyl-erythritol as chiral auxiliary in a diastereoselective and intramolecular photocycloaddition of the bis cinnamate **78** (Scheme 18). Among the isolated isomers, the chiral δ-truxinate (+)-**79** could be obtained as the major product with a *de* up to 86% [59] (Scheme 18).

Despite the modest chemical yields, these results indicated that a high asymmetric induction was possible for bis cinnamates having a C-2 axis of symmetry. Furthermore, the structure of the preferred adduct could be deduced from the least strained transition state, assuming that the two cinnamyl groups approached in parallel planes.

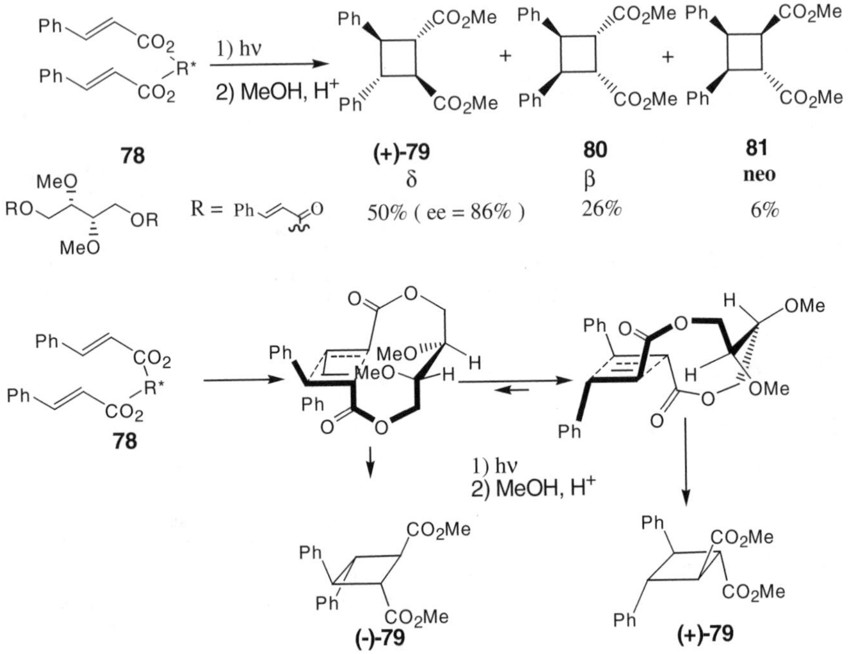

Scheme 18

Another breakthrough came several years later, when the photoadduct **84** of trans stilbene with chiral bornyl methyl fumarate **82** was obtained with a high diastereomeric excess [60]. Here again, a model involving an approach of the reagents in parallel planes was proposed to explain the observed stereoselectivity (Scheme 19). In an attempt to increase the observed *de*, the cycloaddition reaction of dibornyl fumarates was examined, but a far lower selectivity was observed. On this basis, a multistep process was proposed with control of the asymmetric induction by the rate of cyclization of the 1,4-biradical intermediates. The nature of the substituents, however, the complexity of the reaction mixture, and the low chemical yields of the chiral adducts are major limitations for synthetic applications [61].

Similarly, attempts to make enantioselective the intramolecular cycloaddition of dienes catalyzed by Cu(I), employing a removable chiral auxiliary, were described. The stereochemistry of the major diastereoisomer **86**, produced during the photochemical step of **85**, could be deduced from a comparison of the diastereoisomeric transition states as shown in Scheme 20 [62].

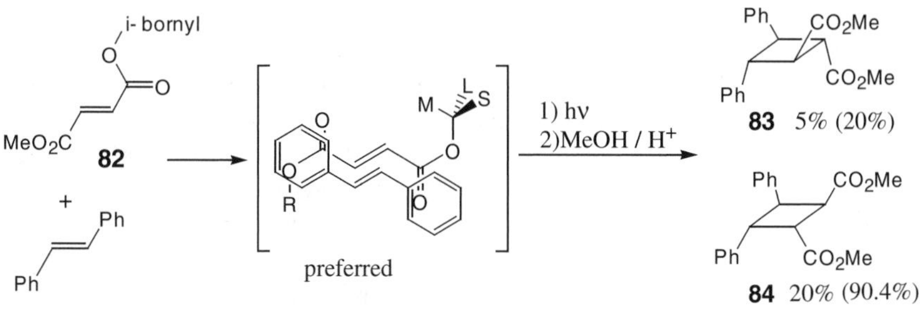

Scheme 19

Among all [2 + 2] asymmetric photocycloadditions that can be used in organic synthesis, most concern cyclic enones or related conjugated acid derivatives. As already indicated, the configuration of the new asymmetric centers is completely settled as soon as the first bond is formed. However, the retrocleavage of the C–C bond of 1,4-biradical intermediates is an important way of dissipation of the excitation energy, and the major stereoisomer does not necessarily come from the kinetically favored biradical. For this reason, attempts to create an asymmetric induction during the cycloaddition by using chiral sensitizers have little effect on the fate of biradicals and have not been successful.

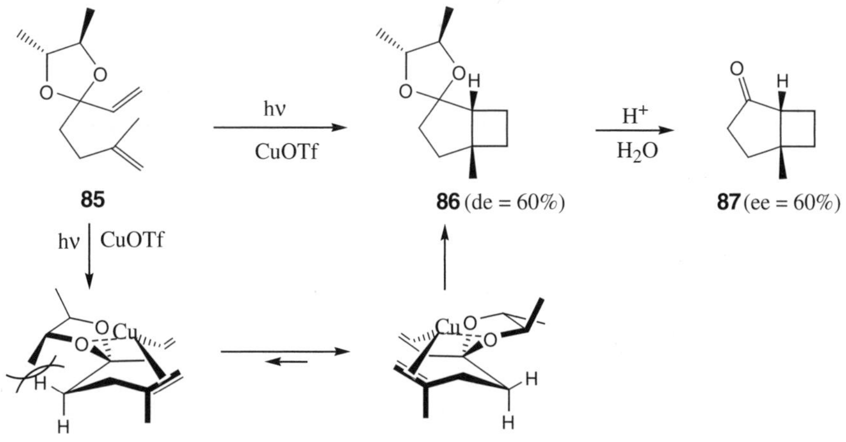

Scheme 20

Introduction of chiral auxiliaries in the starting materials is very attractive for applications to organic synthesis. However, to be of synthetic interest, the chiral auxiliaries have to be inexpensive, readily introduced on the starting material, inert in the conditions of irradiation, and readily removed from the photoadducts. Even if the first requirements can be easily satisfied with chiral ketals, esters, and amides, it is often difficult to avoid side reactions involving this auxiliary [63]. In order to control all the asymmetric centers created in the intermolecular photocycloadditions of cyclic enones with alkenes, esters of chiral alcohols were first considered. Although menthyl and bornyl derivatives gave only low *de*, 8-phenylmenthyl esters produced a far better asymmetric induction [64]. The facial selectivity was found to depend on the syn/anti nature of the cycloadducts and the structure and location of the chiral auxiliary on either the enone or the alkenyl moiety. More surprisingly the selectivity also depends strongly on the nature of the solvent (Scheme 21).

During the photocycloaddition of **88** with cyclopentene (Reaction 1), the *de* of the major isomer **89** increased from 30% in nonpolar solvents up to 68% in a mixture of methanol and acetic acid. When prochiral enone **91** was irradiated in the presence of a cyclopentene linked to the 8-phenylmenthol (Reaction 2), the best selectivity was now obtained in nonpolar solvents. To explain this effect, it was proposed that the facial selectivity is high in every case and that the diastereoselectivity depends on an *s*-cis: *s*-trans ratio of the conjugated esters influenced by hydrogen bonding [65]. Similar results were obtained with chiral

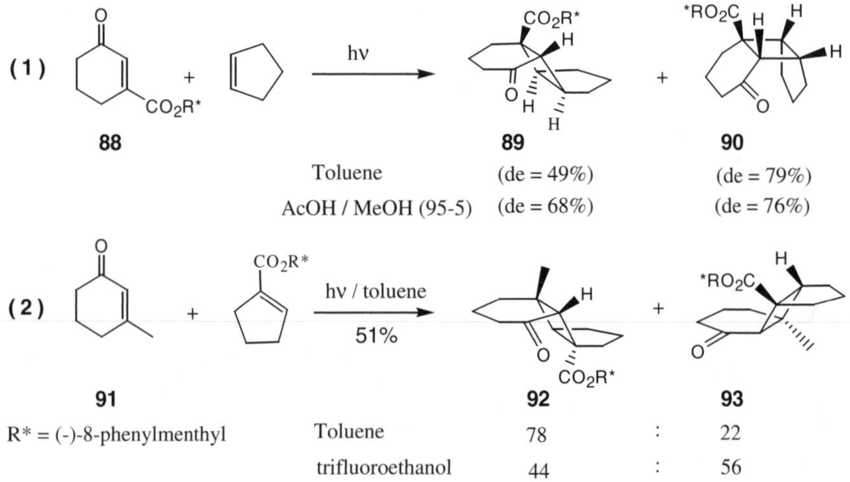

Scheme 21

cyclic enone-3-carboxylates and 1,1'-diethoxyethylene [66] or ethylene as the reacting alkenes [67].

Ketals have also been used as an alternative to link the chiral inductor to the starting reagents. An important asymmetric induction was observed during the cycloaddition of ketals **94** having a C_2 axis of symmetry, and a cyclopentenone or cyclohexenone derivative **95**. Unfortunately, the observed chemical yields of **96** remain low (Scheme 22) [68]. With ketals of aliphatic enones, the selectivity decreases, and complex mixtures of isomers were observed [69].

In connection with the total synthesis of grandisol, an asymmetric addition of ethylene on chiral heterocyclic aminals and ketals was examined (Scheme 23). The selectivity can be high, with a preferred approach of ethylene from the less hindered side, especially when chiral pyrrolidone **97** or furanones **100** were used in place of cyclic enones [70]. The diastereoisomeric excess of **101** or **102** remains modest with 5-menthyloxy furanone, even if the dark addition of nucleophiles or radicals on **100** occurs with a total facial selectivity. From a detailed analysis of the dependence of the product ratio with temperature and substituents, it was proposed that a pyramidalization of the β-carbon in the relaxed $^3\pi,\pi*$ of the excited furanones and a homoanomeric effect were responsible for the observed selectivity [71]. Excited cyclopentenones also possess a biradical character in their relaxed $^3\pi,\pi*$ state. However, no regio and no stereoselectivity could be detected when cyclopentenone was selectively excited in the presence of 5-men-thyloxyfuranone. An initial energy transfer, followed by a cycloaddition of the triplet excited furanone with cyclopentenone, explains these poor results. With more flexible or more electron deficient cyclenones the facial selectivity increases, but mixtures of regioisomers and syn/anti stereoisomers are obtained [72].

Chiral dioxenones constitute another alternative for synthetic applications. They are easily prepared from β-ketoacids and chiral ketones, and they give photocycloadducts in good yields. With menthone derivatives, the facial selectivity is high with a preference for an approach from the α-side of **103** and **106** to

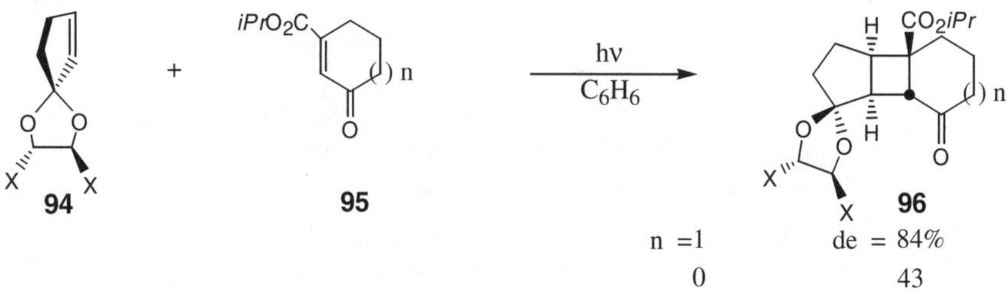

n =1   de = 84%
    0   43

Scheme 22

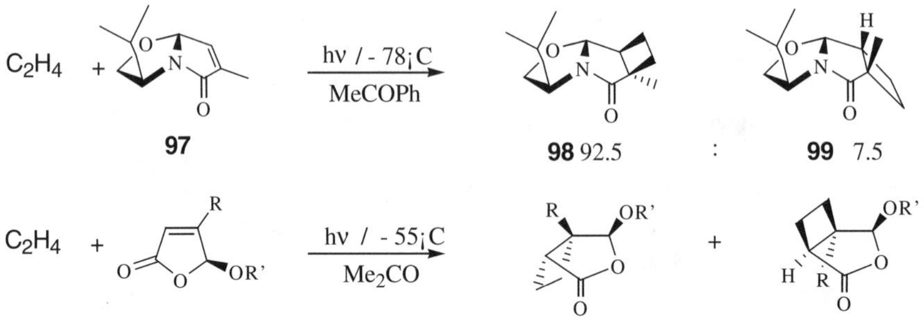

R	R'	101		102
H	Me, Et, iPr	73.5-76	:	26.5-24
Ph	Me	70.5	:	29.5
iPr	Me, Et, iPr	26.5-28.5	:	71.5-73.5

Scheme 23

give respectively **104**, **105** or **107**, and **108** as the major products (Scheme 24). Mixtures of at least four isomeric adducts were isolated with dissymmetric alkenes [73]. An intramolecular photocycloaddition of chiral and spirocyclic dioxinones was expected to control the regio- and stereoselectivity of the adducts. Unfortunately, the stereoselectivity observed with dioxinones, which are derived from menthone and substituted with ω-alkenyl groups, depends considerably on the chain length. With the ω-hexenyl derivative **109a**, a single cycloadduct **110** was isolated in 90% yield with a complete facial selectivity. In contrast, no selectivity could be detected with the ω-pentenyl derivative **109b**, and a complex mixture of products was obtained from the ω-butenyl derivative **109c** (Scheme 25) [74].

All these results indicate that enantioselective photocycloadditions of synthetic interest should be possible with the help of a removable chiral auxiliary, as soon as the right chiral auxiliaries could be defined. In order to test the limits of this strategy, functionalized cyclohexenones and cyclopentenones were selected to look for new chiral inductors. When ω-alkenyl substituents were attached to the cyclic enone through an enamide, a carboxamide, or an ester group (Scheme 26), photocycloadducts can also be efficiently produced.

Amides **113–115** are especially attractive, since the unsaturated chain, cyclic enone, and removable chiral auxiliary such as an α-phenethyl substituent are simultaneously attached to the nitrogen atom. However, the allylic or benzylic hydrogen atoms present in the starting molecule allow competitive hydrogen abstraction processes. In the presence of an α-phenethyl group, cycloadducts were indeed isolated but in low yields and poor diastereomeric excess, besides rearranged products [75].

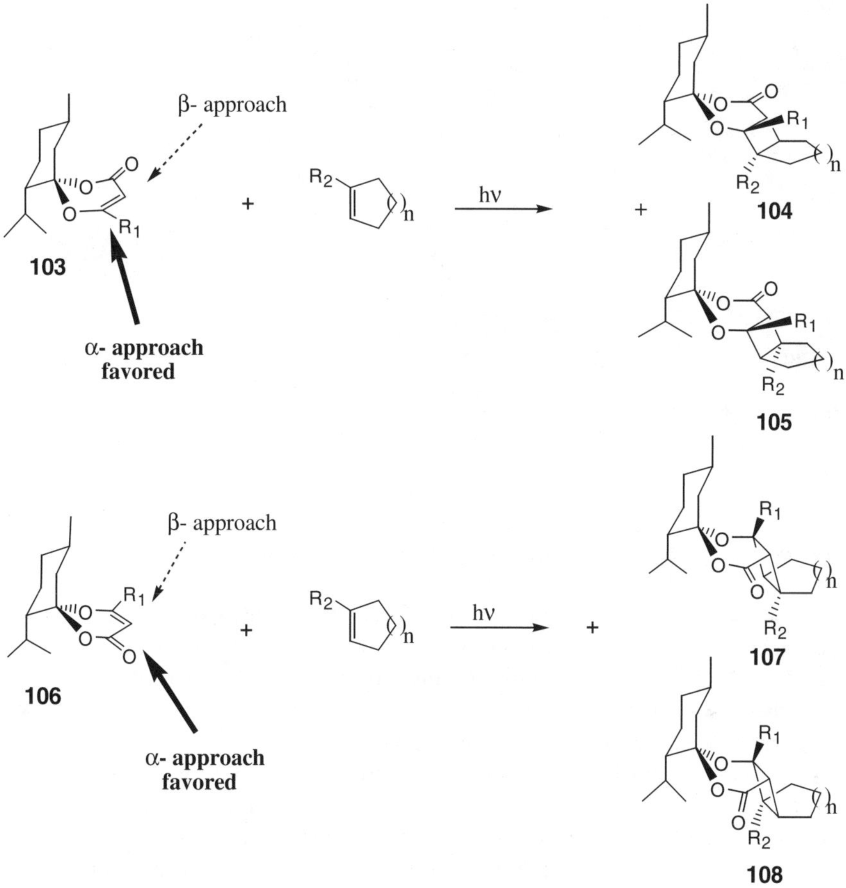

Scheme 24

The efficiency of intramolecular [2 + 2] photocycloadditions of 3-alkeny-loxycarbonyl cyclohexenones **116** is very sensitive to the chain length and to the steric restrictions introduced by the side chain. The reaction is clean and highly regio- and stereoselective with butenyl and pentenyl derivatives, and no intramolecular cycloaddition could be detected with heptenyl or longer chains. Interestingly, the reaction can proceed again in high yield when conformational restrictions are introduced in the tether. The conformational mobility of ester tethers can be considerably restricted as a consequence of the preferred s-trans conformation of the sigma CO–O bond. When two ester groups are introduced between

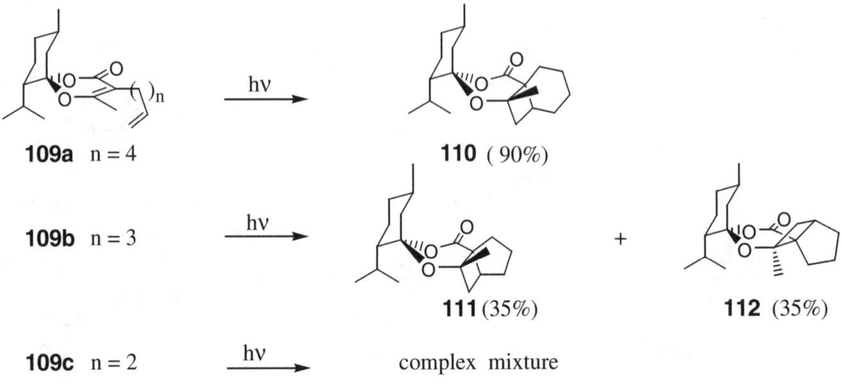

109a n = 4   hν →   **110** (90%)

109b n = 3   hν →   **111** (35%)   +   **112** (35%)

109c n = 2   hν →   complex mixture

Scheme 25

the reacting double bonds, the cycloaddition process occurs irrespective of the length of the tether [76].

With inexpensive hydroxyacids as chiral tethers, only two stereochemically pure head-to-head **118** and head-to-tail **119** cycloadducts were isolated from bis lactate **117**. With the corresponding dihydropyranone **120**, the facial selectivity decreased and head-to-head cycloadducts were obtained as a mixture of two diastereoisomers **121** and **122** (Scheme 27).

The observation of the head-to-tail regioisomer from bis lactate derivatives indicates that there is still a large conformational mobility in the starting material. Accordingly, with the corresponding butenyl lactate **124a** (R=CH$_3$), only two diastereoisomeric cycloadducts **125a, 126a** are formed selectively in high yield (Scheme 28).

Even if methyl groups are attached to the terminal double bond (**124f**), the cycloaddition proceeds with the same efficiency and diastereoselectivity [77].

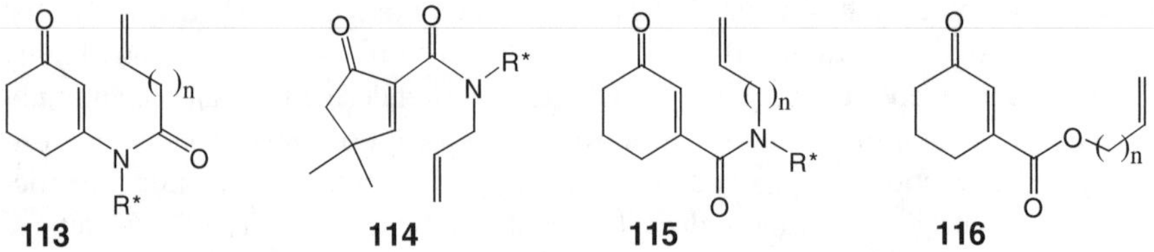

113   **114**   **115**   **116**

Scheme 26

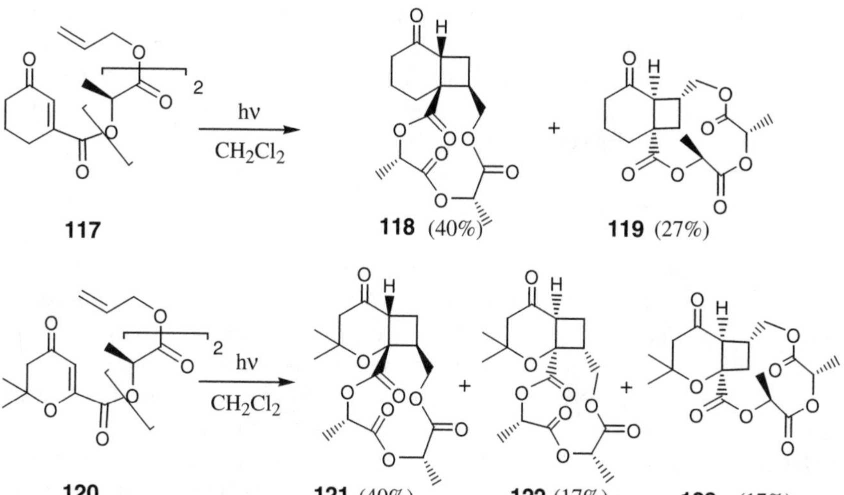

117   **118** (40%)   **119** (27%)

120   **121** (40%)   **122** (17%)   **123** (17%)

Scheme 27

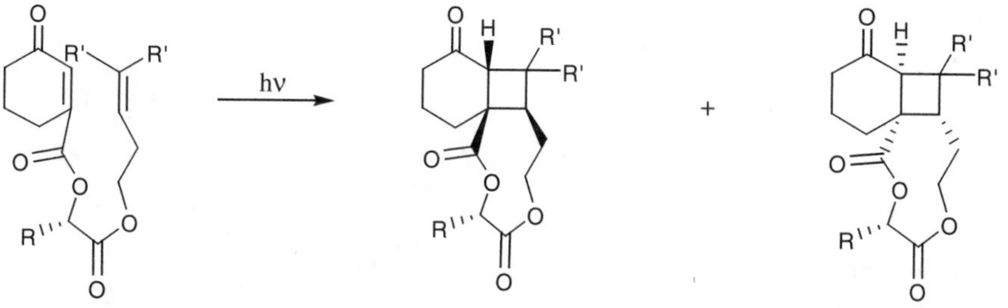

124   **125**   **126**

			125	126
a	R' = H	R = CH$_3$	89%	3%
b		i-Pr	83%	5%
c		Ph	88.5%	3.5%
d		CH$_2$Ph	64.5%	9.5%
e		CF$_3$(rac)	91%	3%
f	R' = CH$_3$	CH$_3$	78.5%	2.5%

Scheme 28

Furthermore, the size of the α substituents has little influence on the selectivity, and almost the same diastereomeric excess is observed with methyl, isopropyl, and phenyl substituents on the stereogenic center. This indicates that there is no direct interaction between the substituents of the chiral auxiliary and the reacting double bonds in the transition states [78].

With the same length of the tether, the replacement of a butenyl lactate by an allyl 3-hydroxy propanoate **127** led to comparable or lower diastereoselectivities (Scheme 29).

As expected, the position of the asymmetric center on the tether influences the selectivity, and a lower *de* is obtained with α- rather than with β-substituted butyrates [79]. Furthermore, for an identical chiral tether, the selectivity is usually higher with cyclohexenone than with cyclopentenone derivatives.

An interesting use of removable chiral auxiliaries in photocycloaddition reactions concerns imminium salts. With cyclic enones, the observed asymmetric induction does not result from an approach of the double bonds in parallel planes because of the triplet nature of the reactive excited state. In contrast, the corresponding imminium salts react through their singlet $^1\pi,\pi^*$ excited state, and an approach of the reactants in parallel planes is now required during the cycloaddition process. For chiral imminium salts **130** derived from a cyclohexenone and a pyrrolidine having a C_2 axis of symmetry, the intramolecular [2 + 2] photocycloaddition process occurs with a *de* up to 82%. As expected, the stereochemis-

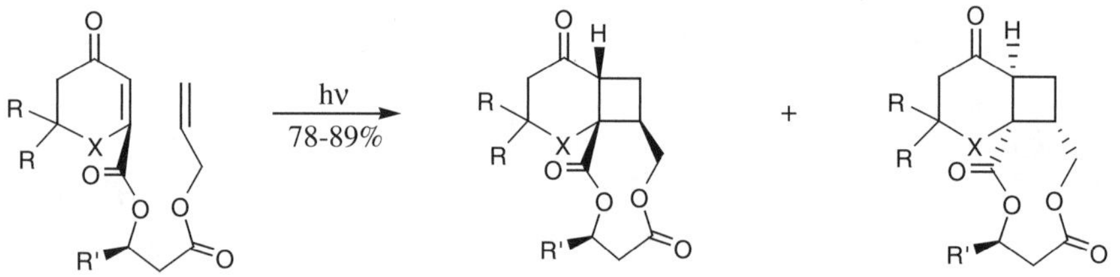

	X	R	R'			
a	CH$_2$	H	CH$_3$	82	:	18
b	—	H	CH$_3$	68	:	32
c	O	CH$_3$	CH$_3$	55	:	45
d	CH$_2$	H	i-Pr	77	:	23
e	O	CH$_3$	i-Pr	70	:	30

Scheme 29

try of the adducts **131** can be deduced from a comparison of the steric interactions involved in the diastereoisomeric transition states having their reactive double bonds in parallel planes (Scheme 30) [80]. Despite the high de, the modest chemical yields and the lack of generality of this reaction still constitute limitations for synthetic applications.

C. Photocycloaddition Between Aromatic Compounds and Alkenes

Three types of cycloaddition products are generally obtained from the photochemical reaction between aromatic compounds and alkenes (Scheme 31). While [2 + 2] (ortho) and [3 + 2] (meta) cycloaddition are frequently described, the [4 + 2] (para or photo-Diels–Alder reaction) pathway is rarely observed [81–83]. Starting from rather simple compounds, polycyclic products of high functionality are obtained in one step. With dissymmetric alkenes, several asymmetric carbons are created during the cycloaddition process. Since many of the resulting products are interesting intermediates for organic syntheses, it is particularly attractive to perform these reactions in a diastereoselective way.

[2 + 2] and [3 + 2] photocycloadditions are often competitive, and it has been proposed that the difference of the redox potentials of the reaction partners should play an important role for the control of the modes of cycloaddition [84]. However, treatments based on molecular orbital interactions could not differentiate between the two cycloaddition modes [85].

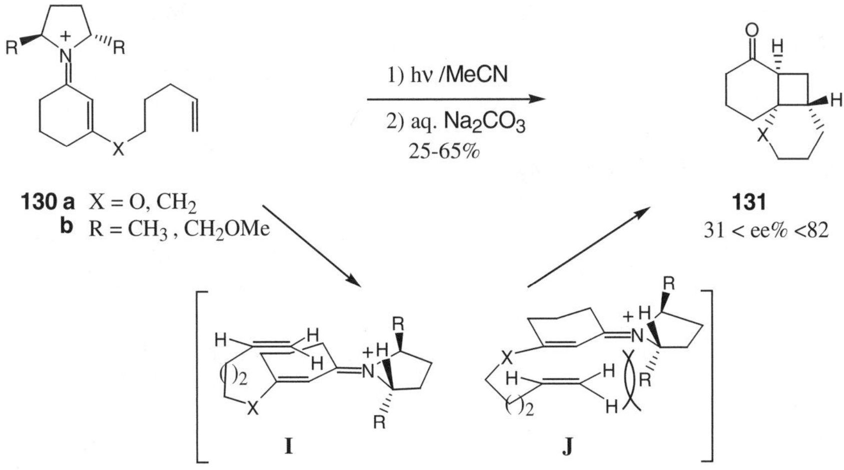

130 a X = O, CH$_2$
  b R = CH$_3$, CH$_2$OMe

1) hv /MeCN
2) aq. Na$_2$CO$_3$
25-65%

131
31 < ee% <82

Scheme 30

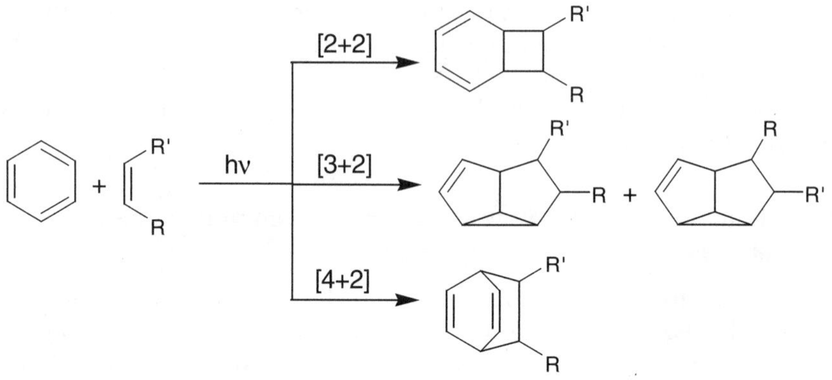

Scheme 31

The mechanism of the [3 + 2] cycloaddition is summarized in Scheme 32. The first intermediate results from charge transfer interaction between the electronically excited aromatic compound at its singlet state S1 with the alkene which leads to the formation of the exciplexes **K**. A more stable intermediate is then generated by the formation of two C–C bonds, leading to the intermediates **L**. These intermediates have still singlet multiplicity and therefore possess zwitterionic mesomeric structures mainly of type **M**. In most cases and especially in intramolecular reactions, chiral induction occurs during the formation of **L**. The final products are then obtained by cyclopropane formation in the last step.

Until now, only few efforts were undertaken to control the configuration of the new asymmetric centers and to obtain optical pure products from the [3 + 2] photocycloaddition. In one attempt, enantiomerically pure starting material **132** was used for the photochemical reaction (Scheme 33) [86]. The chiral information fixed in the γ position of the olefinic side chain permits a high facial (exo/endo) differentiation for the 1,3-attack (transition state **N**). The high diastereoselectivity

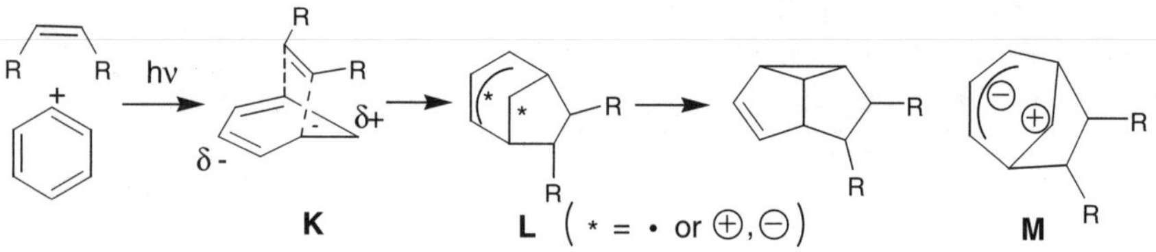

Scheme 32

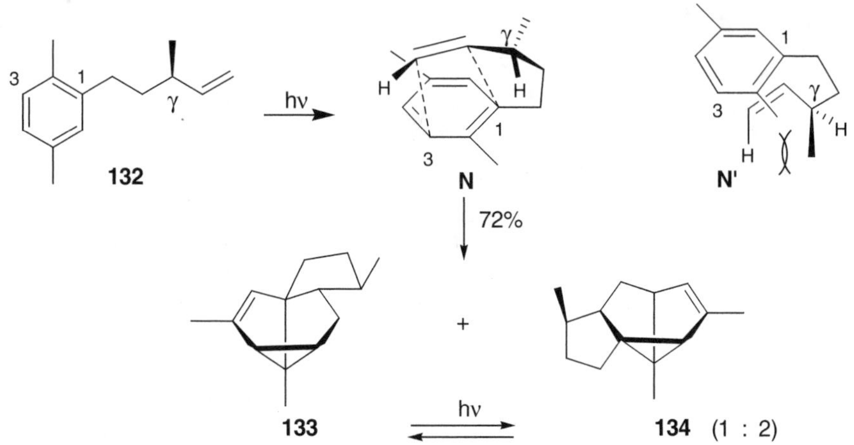

Scheme 33

can be explained by the presence of a 1,3-allylic strain effect on the olefinic side chain in the transition state **N′** involved for the alternative attack. The cyclopropanation leads to a 1 : 2 mixture of **133** and **134**, which are interconvertible by a vinylcyclopropane rearrangement. Then **134** can be transformed into the desired isomer **133** via photoequilibration in multigram quantities. The [3 + 2] photocycloaddition of **132** was used as one of the key steps of the asymmetric synthesis of retigeranic acid.

The facial diastereoselectivity of the reaction was also completely controlled, when a diether was used as chiral auxiliary (Scheme 34) [87]. Upon irradiation, both compounds **135** and **136** yield cycloadducts **137a,b** and **138** exclusively via intermediate **O**. In the case of **136**, possessing a side chain formally derived from the meso 2,4-pentanediol, the cyclopronation step occurred selectively. Only one of the two possible cyclopropanes **138** was observed, while the two isomers **137a** and **137b** were isolated from the reaction of **135**. Compounds **137a,b** and **138** were transformed into enantiomerically pure (−)-**139**.

Some efforts have also been done to induce chirality in the intramolecular [2 + 2] photocycloaddition of benzene derivatives using a chiral auxiliary [88]. **141b** was isolated with a diastereoselectivity of 17% from the intramolecular photocycloaddition of the salicylic acid derivative **140b** (Scheme 35) [89]. After an initial [2 + 2] cycloaddition, reversible thermal and photochemical rearrangements took place. These equilibria can be displaced by an acid catalyzed and irreversible addition of methanol to the intermediate **P**.

The derivatives carrying an acetyl group attached to the benzene ring **142a–d** exhibit a significantly enhanced photochemical reactivity (Scheme 36).

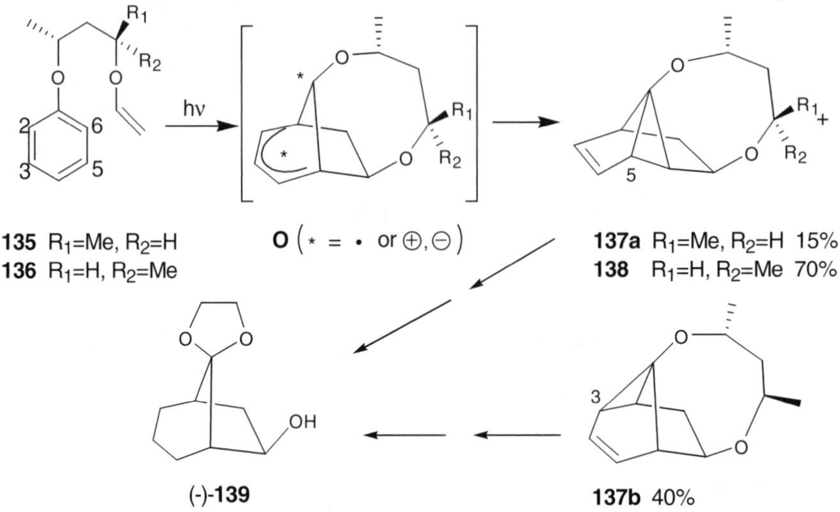

135 R₁=Me, R₂=H
136 R₁=H, R₂=Me

O (* = • or ⊕,⊖)

137a R₁=Me, R₂=H 15%
138 R₁=H, R₂=Me 70%

(-)-**139**

137b 40%

Scheme 34

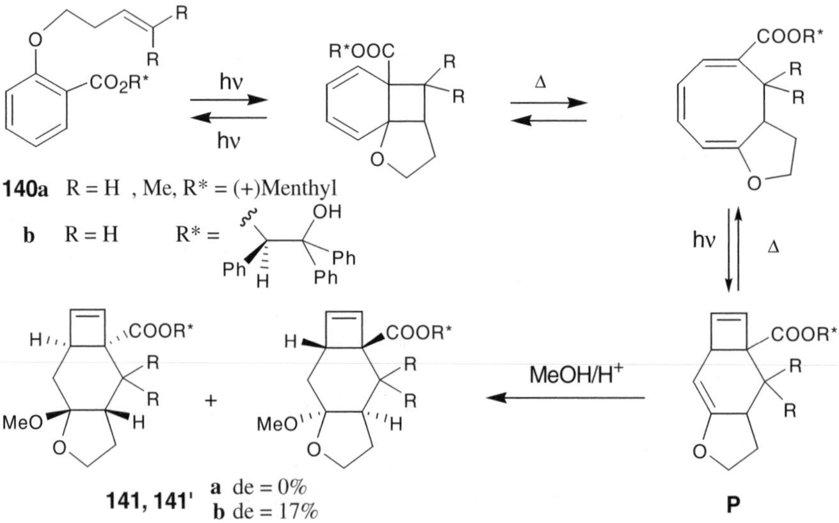

140a R = H , Me, R* = (+)Menthyl

b R = H   R* =

141, 141' **a** de = 0%
              b de = 17%

P

Scheme 35

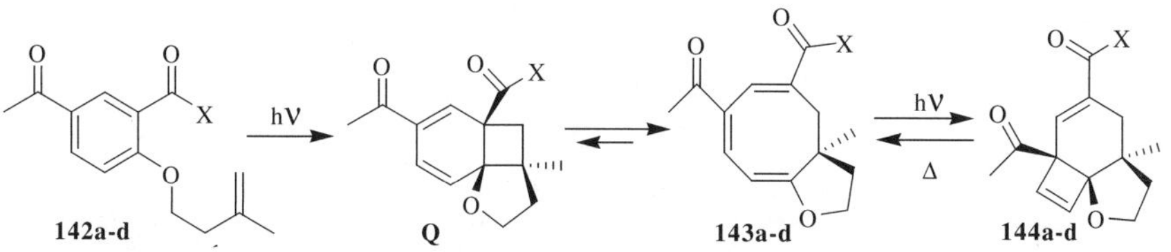

143	X-H	%de
a	(1R,2S,5R)-(-)-menthol	15
b	(S)-(-)-1-phenylethylamine	35
c	(2R,5R)-(-)-2,5-dimethylpyrrolidine	90
d	(7R)-(+)-camphorsultam	90

Scheme 36

Compounds **143a–d** could be isolated with a diastereomeric excess of up to 90% [90]. An initial [2 + 2] photocycloaddition involved in this reaction leads to intermediates **Q**. A consecutive thermal rearrangement produces **143a–d**. Finally, compounds **144a–d** are formed via further photochemical rearrangement. However, the thermal reversibility of the last step allows the formation of **143a–d** in yields of up to 90%.

The photo-Diels–Alder reaction of α-acetonaphthone **145** with the chiral α-enaminonitrile **146** yielded the cycloadduct **147** with almost complete diastereoselectivity (Scheme 37) [91,92]. The intermediately formed biradical **R** is particularly stable owing to delocalization of the radical on the aromatic moiety and to a captodative effect on the enamine moiety. In analogy to the Paternò–Büchi reaction (see Scheme 7), the chiral induction occurred in two steps. In the first step, a chiral center is created at the α-position of the acyl group. In the second step of the diastereoselection, one of the two diasteromeric intermediates undergoes preferential cyclization to yield the final product **147**, while the other one is decomposed to form the starting material [92].

Recently, many investigations have been carried out on the [4 + 4] photocycloaddition of 2-pyridones and its application to organic synthesis. Although the studied reactions involved only racemic materials, high diastereoselectivities have been obtained with chiral substrates. This makes this type of cycloaddition very promising for asymmetric synthesis [93].

III. DIASTEREODIFFERENTIATION IN PHOTOREARRANGEMENTS

A. Electrocyclic Processes

Electrocyclic reactions of conjugated polyenes create chiral molecules through stereospecific conrotatory or disrotatory processes. In solution, the two enantiom-

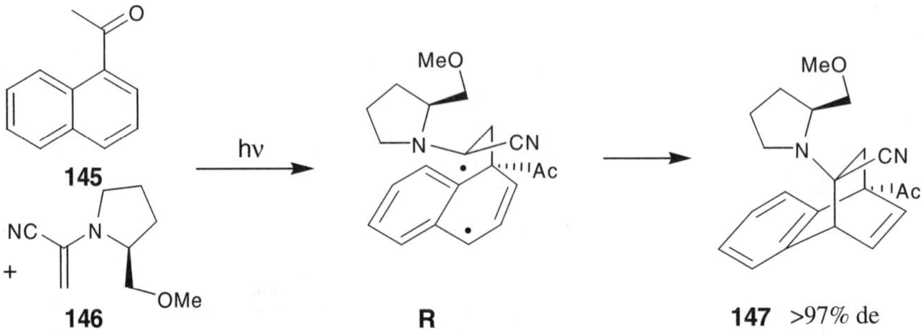

Scheme 37

ers are usually formed in equal amounts. When chiral substituents are introduced in the starting polyene, a diastereoselection is expected. However, the enrichment of one of the diastereoisomers is usually low under these conditions. In one early example, only a poor diastereodifferentiation was observed during the photocyclization of chiral diaryl ethylenes into helicenes [94]. In the field of photochromic compounds, the photochromism of chiral diaryl ethylene derivatives was recently investigated. Although the diastereoselectivity of the cyclization is usually very low in solution [95], *de's* of up to 86.6% have been observed in toluene at low temperature for **148** in the presence of a menthyl auxiliary [96]. A similar approach has been proposed for a diastereocontrolled synthesis of functionalized heterohelicenes [97]. The diastereodifferentiation can even be higher with binaphthol derivative **149** [98] (Scheme 38).

A modest diasterodifferentiation was reported during the photocyclization of chiral divinyl amine derivatives **150** into **151** [99]. Finally, photocyclization of the 4-(l-menthoxy)-pyridin-2(1H)-one-2 into the corresponding bicyclic β-lactams occurs only with a low diastereoselectivity [100].

B. Norrish–Yang and Related Reactions

When possible, a γ-hydrogen abstraction followed by a cleavage and a cyclization of the 1,4-biradical intermediate (Norrish–Yang reaction) is a very favorable process for phenyl ketones. Depending on the presence or absence of a heteroatom in the skeleton of the starting ketone, cyclobutanes, oxetanols, or azetidinols can be formed. When the γ-carbon is substituted, two new stereogenic centers are formed during the cyclization process. The efficiency of the cyclization process can be high when a heteroatom or a β-substituent is introduced on the side chain. In the absence of chiral centers in the starting ketone, cyclobutanols with the

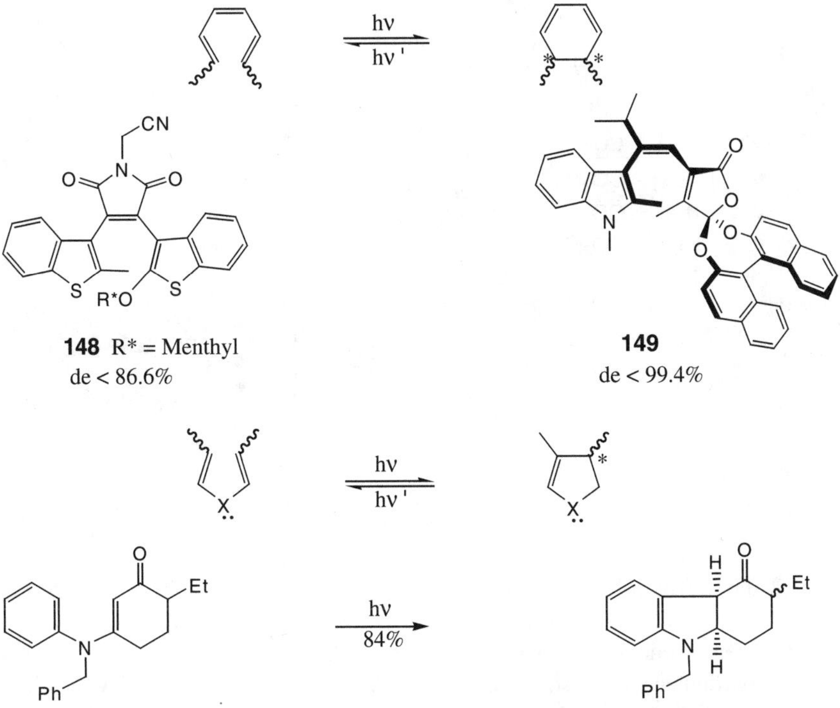

148 R* = Menthyl
de < 86.6%

149
de < 99.4%

150

151 de = 40%

Scheme 38

hydroxyl group and the substituent at C-2 in a cis relative configuration are formed preferentially [101]. When polar substituents are introduced on a stereogenic center of the starting ketone **152a–152c**, an asymmetric induction becomes possible, and one diastereoisomer **153** is selectively formed (Scheme 39) [102]. The asymmetric induction can be attributed to the stabilization of one conformer of the 1,4-biradical by an intramolecular hydrogen bond.

In the absence of γ-hydrogen, other intramolecular hydrogen abstraction processes can become competitive. For example, a highly diastereoselective photocyclization to *cis*-3-hydroxyproline derivatives is observed with substituted *N*-(2-benzoylethyl)-*N*-tosylglycine esters **155**. The products **156–157** involve a δ-hydrogen abstraction by the excited carbonyl and a cyclization of the 1,5-biradical intermediate. In the presence of a β-substituent, a high asymmetric induction is observed. The observed selectivity can be deduced from the preferred conforma-

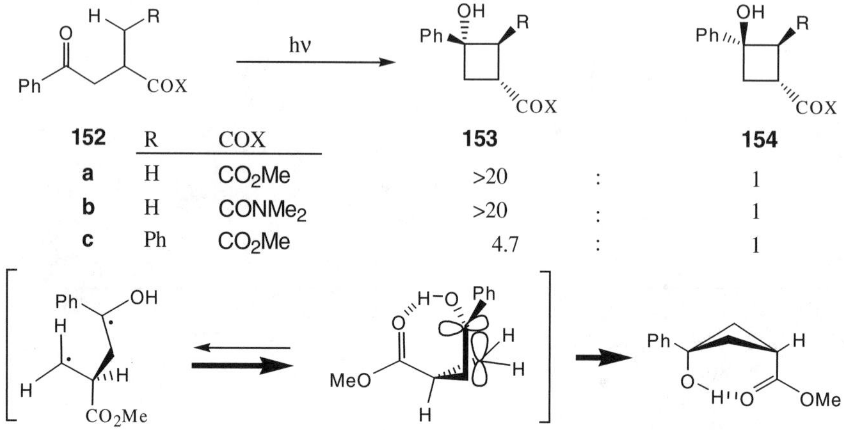

152	R	COX	153		154
a	H	CO$_2$Me	>20	:	1
b	H	CONMe$_2$	>20	:	1
c	Ph	CO$_2$Me	4.7	:	1

Scheme 39

tion of the 1,5-biradical. The asymmetric induction is especially high when a hydrogen bond is formed in the biradical between the hydroxyl group and the β-substituent (Scheme 40) [103].

A particularly interesting case of asymmetric induction occurred with chiral *N*-(2-benzoylethyl)-*N*-tosylglycinamides derived from pyrrolidines **158** having a C$_2$ axis of symmetry as the only source of chirality. A large chiral discrimination was indeed obtained between the diastereoisomeric transition states **S** and **T** involved during the cyclization of the 1,5-biradical intermediate into **159** and **160** (Scheme 41) [104]. Furthermore, a small temperature effect was detected on the selectivity and the *de* increases at low temperature (from 75.8% at 293K up to 86.4% at 213K) with 2*S*,5*S*-dicarbomethoxypyrrolidine as chiral auxiliary. With the more sterically hindered pyrrolidine derivative **161**, the corresponding cyclized product **162** was isolated in more than 72% yield.

Piperidones can also be formed through an ε-hydrogen abstraction process during the photocyclization of 4-oxo-4-phenylbutanoyl amines **163**. The products **164** and **165** are obtained with remarkable diastereoselectivities of up to 99% [105]. When 2-amino substituents are present in **166**, a large 1,4-asymmetric induction is observed during the photocyclization process, which leads to the corresponding lactames **167,168** (Scheme 42) [106].

C. Memory Effect

In the absence of γ- and δ-hydrogen atoms, the abstraction of an activated β-hydrogen by the n,π*-excited carbonyl group becomes possible, and cyclopropa-

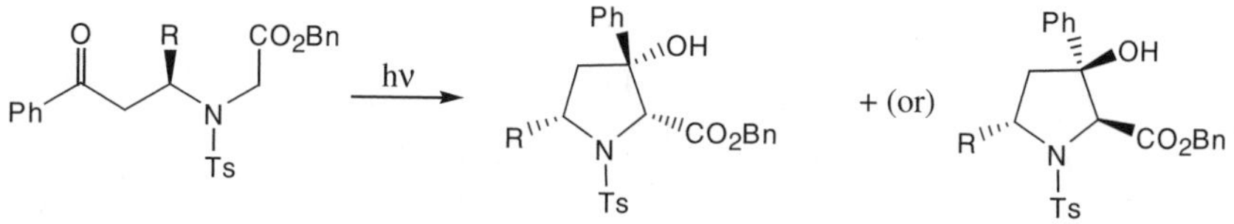

		155	156	157
a	R = CO_2Me		65% (de = 98%)	
b		CO_2(t-Bu)	57% (de = 98%)	
c		t-Bu		24% (de = 98%)

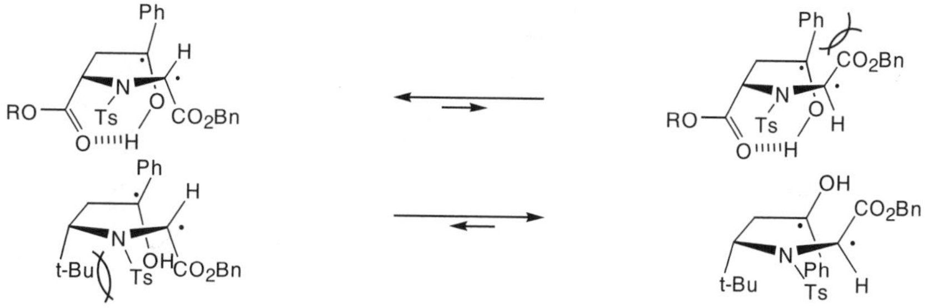

Scheme 40

nols can be isolated in good chemical yields. Interestingly, stereogenic centers at C_α and C_β induce a highly diastereoselective cyclization of the 1,3-biradical. Furthermore, when the only stereogenic center of the starting molecule **169a** is located at C_β the photocyclization is enantioselective. Although a planarization of the only stereogenic center occurs during the formation of the biradical intermediate, the final cyclopropanol **170** is obtained enantioselectively (Scheme 43) [107]. There is a memory of chirality and a conservation of the chirality of the starting molecule in the final cyclopropanol.

With phenylglyoxamides **171**, the Norrish–Yang reaction produces 3-hydroxy β-lactams **172,173**. The possibility of a memory effect was also examined, during the photoreaction of chiral phenyl glyoxamides of enantiomerically pure α-amino acid esters. As shown in Scheme 44, the chemical yields of the photocyclization are high. Unfortunately, the isomeric hydroxy-β-lactams were obtained as racemic mixtures, and similarly a 1|:|1 mixture of diastereoisomers **175** and **176** was obtained from amino acid derivatives **174** having two stereogenic centers. The absence of a memory effect, contrasting with the high retention of chirality previously described for the cyclopropanol and pyrrolidinol formation, may be due to the increased lifetime of the 1,4-delocalized biradical [108].

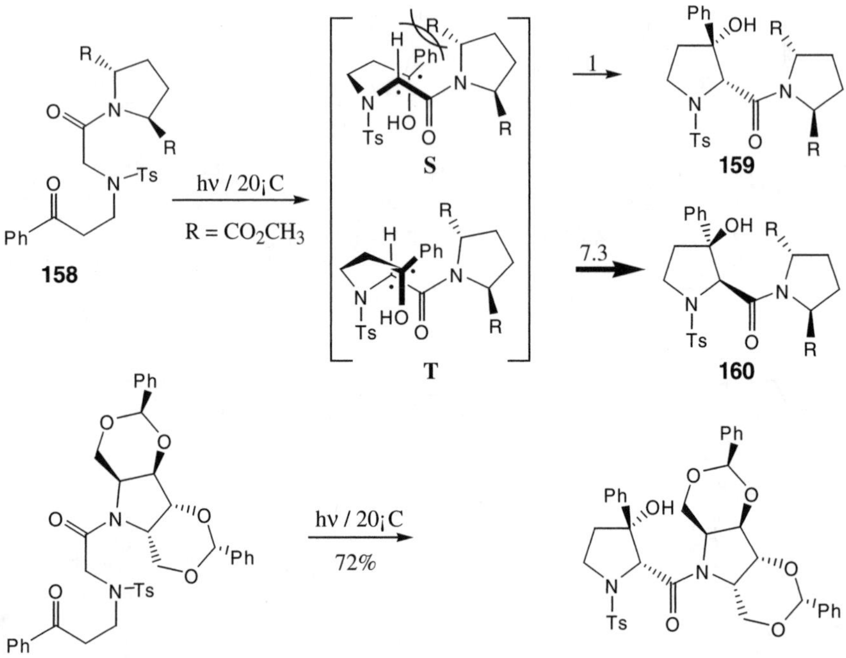

Scheme 41

When γ-hydrogen atoms are not available and in the presence of an activated δ-H atom at the only stereogenic center of **177**, the reaction goes through a 1,5-biradical sp_2-hybridized at the previous chiral center. However, a memory of chirality was observed, and proline derivatives **178** and **179** were formed enantioselectively with a preferred retention of configuration at the initial stereogenic center. From the observed selectivity, an activation energy of 2 kcal mol^{-1} was evaluated for the cyclization process. Such a value is far lower than the activation energy required for a racemization process by bond rotations around the single bonds of the biradical. With naphthalene as triplet quencher and singlet sensitizer the reaction was highly enantioselective. This indicated that the singlet biradical cyclized so fast that there was no time for rotations around the σ-bonds before the C–C bond formation. With benzophenone as triplet sensitizer, the triplet 1,5-biradical led to an almost complete racemized product. When the photoreaction was carried without benzophenone in the presence of oxygen to accelerate the triplet–singlet crossing and to decrease the lifetime of the biradical intermedi-

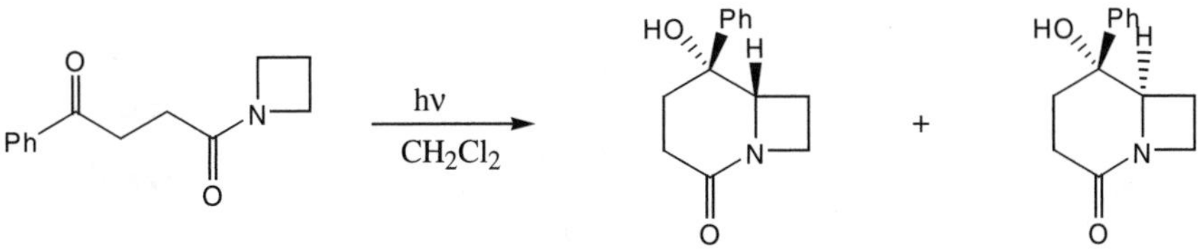

Scheme 42

| **163** | | **164** | : | **165** |
| | | >20 | | 1 |

| **166** | **a** | R = Tfa | **167** 52% | **168** 28% |
| | **b** | R = Cbz | 67% | 29% |

169	R^1	R^2	**170**
a	H	Ph	87%
b	Ph	H	98%
c	PhCH$_2$	H	75%

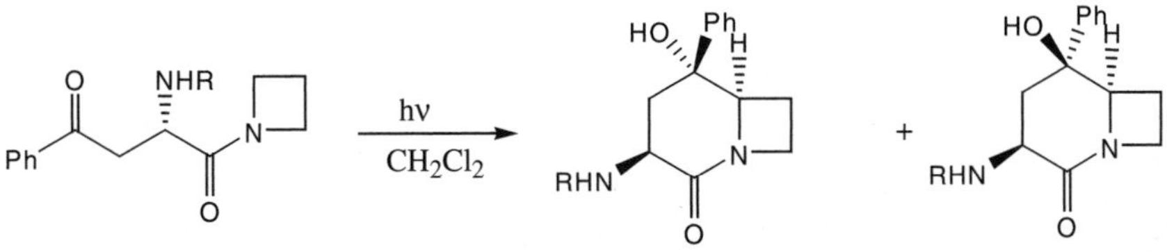

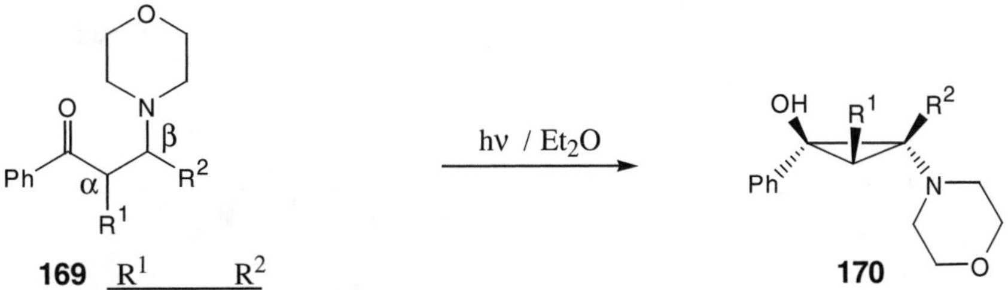

Scheme 43

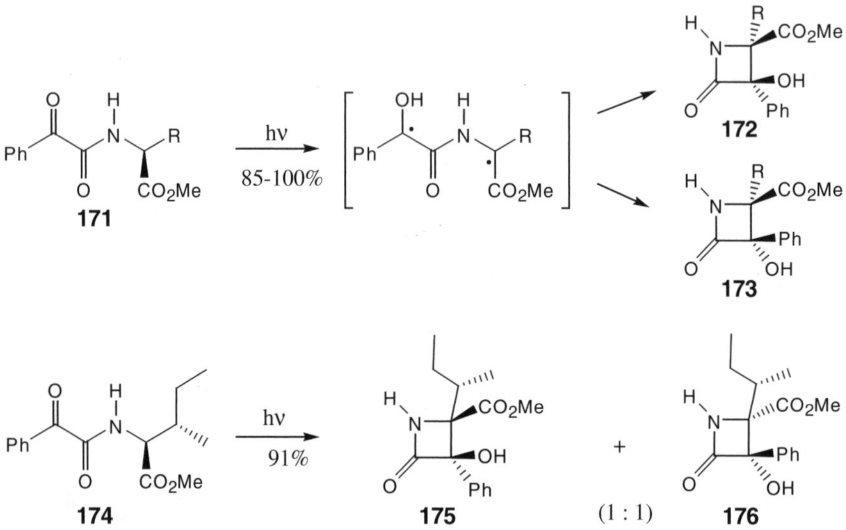

Scheme 44

ate, a high *ee* was again observed, in contrast to the results obtained under an argon atmosphere (Scheme 45) [109].

Interestingly, the same cyclic molecules can be obtained if a photodecarboxylation process is used to produce the 1,*n*-biradical intermediates. The decarboxylative photocyclization of chiral ω-phthalimidocarboxylates, derived from

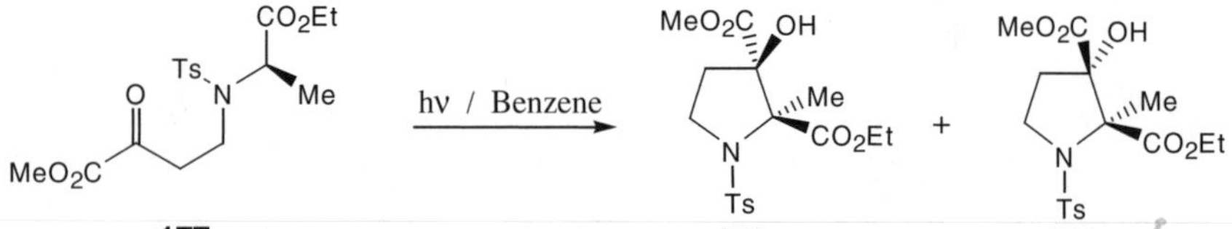

Additives	yield [%]	ee%	ee%
naphthalene (1M)	47	92	88
benzophenone (1M)	10	17	17
O$_2$	48	81	56.5

Scheme 45

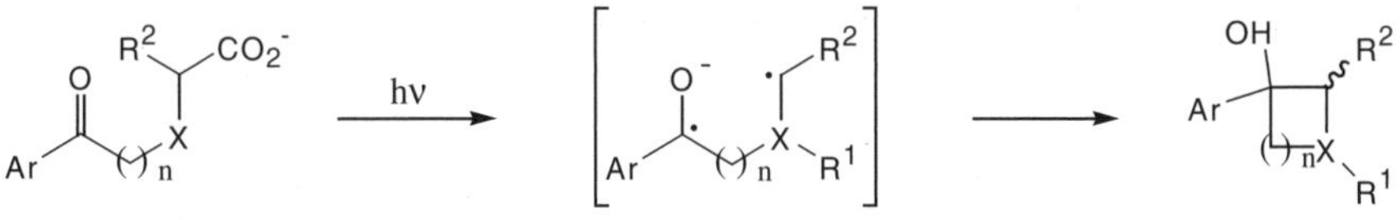

Scheme 46

amino acids and *N*-phthaloylanthranylic acid, also proceeds diastereoselectively with the formation of pure trans isomers (Scheme 46).

Surprisingly, the photocyclization is enantioselective with amino acid derivatives **180** of phthalimide. For example, an inversion of the starting stereogenic center on the pyrrolidine ring was observed from the corresponding proline derivatives **182** (Scheme 47) [110].

To explain the high degree of memory of chirality, it was proposed that there is a high activation barrier for the rotation around the central C–N single bond of the 1,7-biradical intermediate. Then the axial chirality can be preserved during the whole process.

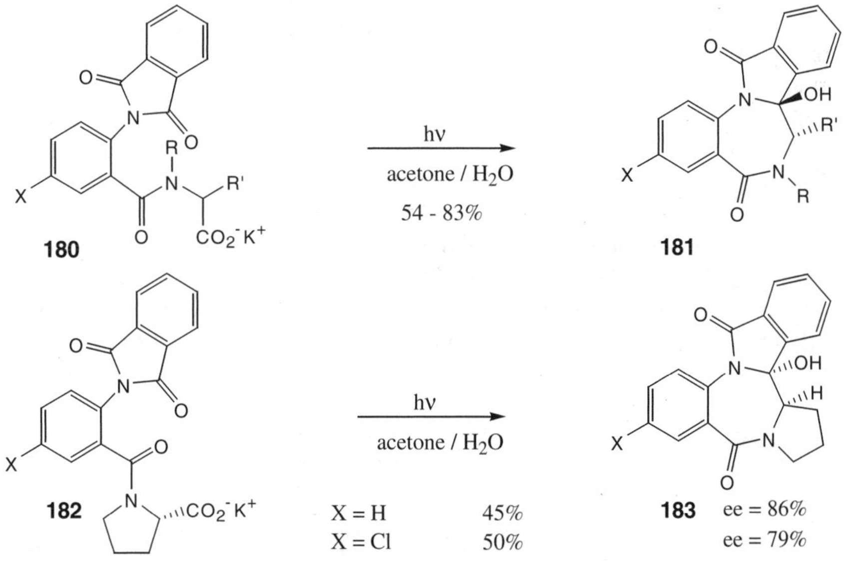

Scheme 47

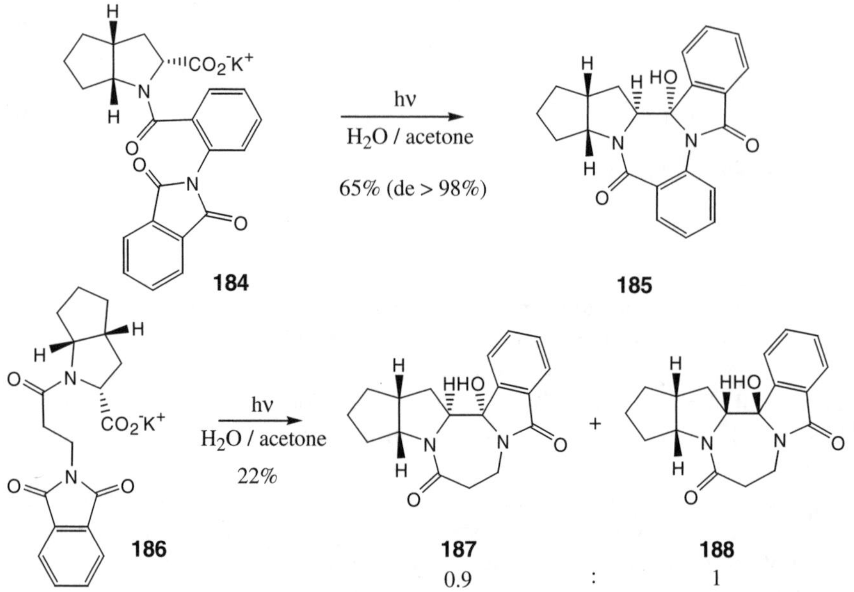

Scheme 48

Similarly, the photodecarboxylative cyclization of the 2-azabicyclo [3.3.0]octane-3-carboxylate derivative **184** produces only one stereoisomer **185** in high chemical yield. Here again, in the product there is an inversion of the configuration of the stereogenic center, formerly in α-position of the carboxyl group. However, when the conformational flexibility increased between the phthalimido and the amino acid unit in **186**, almost no diastereoselectivity could be detected in the cyclized products **187** and **188** (Scheme 48). This result is indicative of the importance of the conformational mobility on the memory effect, in reactions involving triplet 1,7-biradical intermediates. The observed relative endo configuration of the newly created stereogenic centers could be correlated with the endoselectivity observed in the Paterno–Büchi reaction [30,111].

IV. DIASTEREODIFFERENTIATION IN PHOTODECONJUGATION REACTIONS

It was shown previously that photodienols derived from α-substituted conjugated aliphatic esters could be protonated enantioselectively in aprotic solvents and in

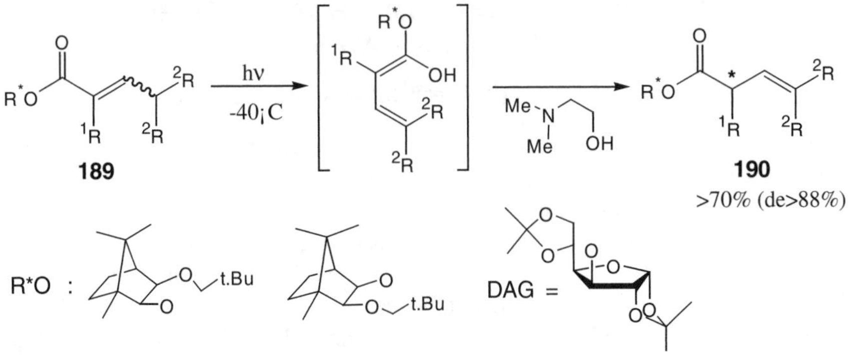

Scheme 49

the presence of small quantities of a chiral β-aminoalcohol [112]. The keys of the high asymmetric induction are the stereospecificity of the photoenolization step and the tight cyclic transition state required during the protonation step. An *ee* of up to 95% has been observed at low temperature and in the presence of catalytic amounts of aminoborneol derivatives [113]. However, the selectivity is strongly dependent on the nature of the substituents on the starting ester. For synthetic applications, a more general method almost independent of the substitution consists in the diastereoselective photodeconjugation of esters of chiral alcohols. Photodeconjugation of esters **189** of very crowded chiral alcohols, having a prochiral center at C_α and a γ-H, produces a mixture of diastereoisomers **190** with high diastereoselectivities (Scheme 49) [114].

For example, high *de* are obtained during the photodeconjugation of bornyl and isobornyl derivatives, in methylene chloride [115]. Although the presence of chiral aminoalcohols is not required for the diastereoselective protonation of the photoenol intermediate, the selectivity depends also on the reaction conditions and especially on the nature of additives in the reaction mixture. The observed diastereoselectivity can indeed be improved in the presence of small quantities of dimethylaminoethanol in the reaction mixture. With diacetonylglucose derivatives, the diastereodifferentiation process becomes very efficient and general. De higher than 95% are commonly observed for the photodeconjugation of numerous esters bearing a large variety of substituents on the acid skeleton [116]. The reaction has been applied to a diastereoselective synthesis of α-fluorocarboxylic derivatives [117] and to enantioselective syntheses of natural products such as (*R*)-lavandulol [118].

V. DIASTEREODIFFERENTIATION IN PHOTOSENSITIZED ADDITIONS ON CHIRAL SUBSTRATES

A. The Reaction of $^1O_2$ with Alkenes and Dienes

Singlet molecular oxygen ($^1O_2$) is a very small electrophilic reagent that can be produced by photosensitization and react easily with dienes and alkenes in [4 + 2] or [2 + 2] cycloadditions or ene-reactions. In order to develop synthetic applications of photooxygenations, numerous studies have recently been developed with chiral substrates and especially in the presence of chiral auxiliaries.

The [4 + 2] photooxygenation of dienes and benzene rings leads to endoperoxides. With chiral naphtyl alcohols **191**, **192**, and **193** were formed with a diastereomeric excess up to 90%, when the reaction was carried out at − 30°C in CDCl₃ (Scheme 50). In order to explain the selectivity in aprotic solvents, it was proposed that an intramolecular hydrogen bond can be developed between the hydroxyl group and $^1O_2$ during the approach of the dienophile from one or the other side of the naphthalene ring, favoring one of the two diastereoisomeric transition states.

When hydrogen bonds can be formed between $^1O_2$ and hydroxylic solvents such as CD₃OD, the selectivity dropped considerably, as expected [119]. The [4 + 2] cycloaddition of $^1O_2$ can also be highly diastereoselective with aliphatic dienes. With sorbic acid derivatives **194** having a removable chiral 2,2-dimethyloxazoline auxiliary, **195** was isolated with an almost perfect stereocontrol during the addition process (Scheme 51) [120].

A large diastereodifferentiation can also be observed in the [2 + 2] cycloaddition of $^1O_2$ with allylic alcohols. When adamantylene-substituted allylic alcohol **196** was irradiated in CDCl₃ in the presence of oxygen and a sensitizer, the favored dioxetane **197** was formed with a diastereoisomeric excess higher than 90%. The observed selectivity was attributed to a synergy between the directing effect of hydrogen bonding and the conformational restrictions due to 1,3-allylic strain. In the preferred transition state (**TS A**) an intramolecular hydrogen bonding

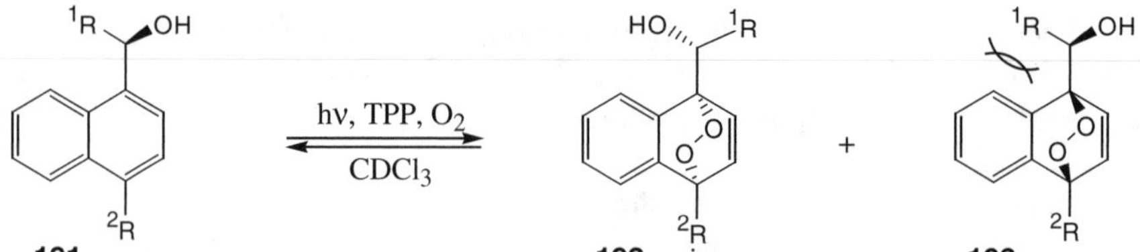

Scheme 50

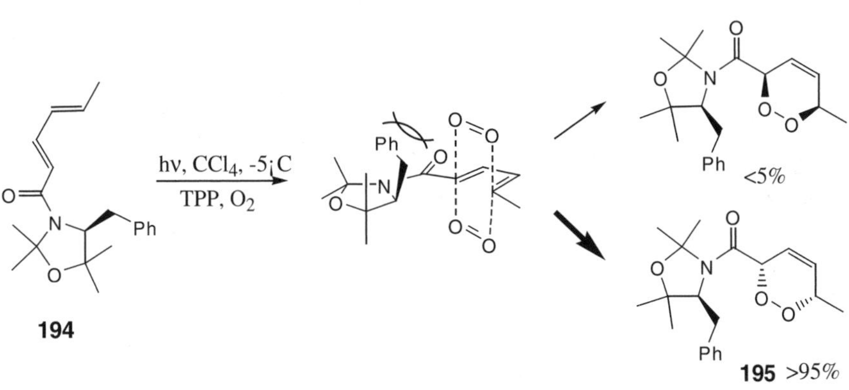

Scheme 51

was developed between the hydroxyl group and $^1O_2$, in a conformation minimizing the 1,3-allylic strain between the allylic methyl group and the adamantylene group (Scheme 52) [121]. As expected, the selectivity decreases significantly in a mixture of CCl_4-CD_3OD enabling some external hydrogen bonding.

Aliphatic enamine derivatives **198** linked to a chiral oxazolidone allow an almost totally diastereoselective [2 + 2] photocycloaddition of $^1O_2$ (Scheme 53). This efficiency was attributed to a selective shielding of the lower face by the R substituents in the preferred conformation of the chiral oxazilidone enecarbamate. Interestingly, the *de* of **199** did not depend on the configuration of the stereocenter on the exocyclic appendage [122].

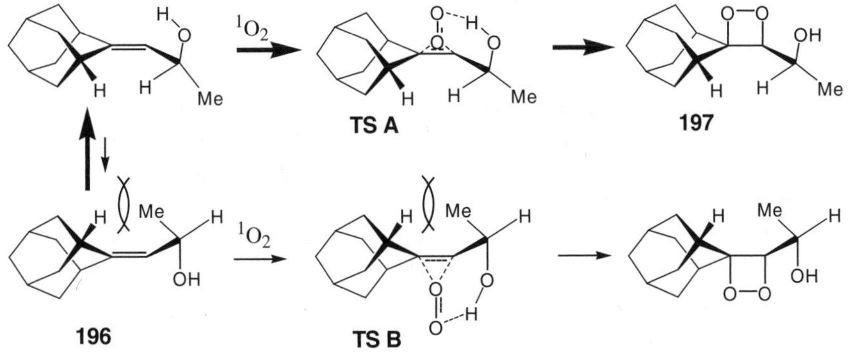

Scheme 52

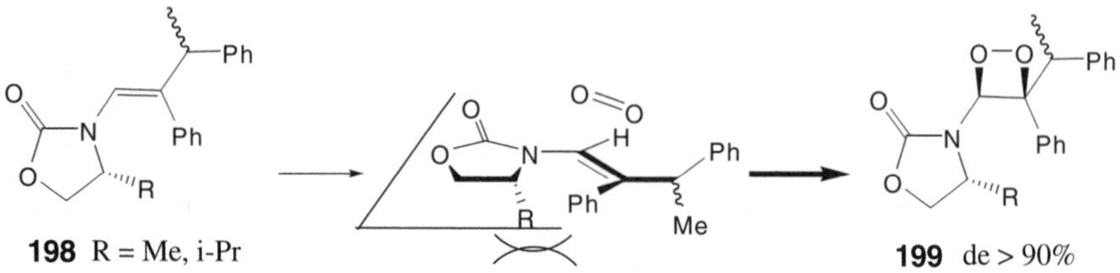

198 R = Me, i-Pr   **199** de > 90%

Scheme 53

The stereoselectivity of the allylic hydroperoxidation also depends on several factors. With chiral allylic alcohols or allylic amines **200**, a hydrogen bond is developed between $^1O_2$ and the vicinal hydroxyl or amino group. The facial differentiation results from an approach of $^1O_2$ in the transition state that minimizes 1,3-allylic strain. **201** and **202** can be obtained with a diastereoisomeric excess higher than 90% in CCl_4. As previously indicated for the formation of dioxetanes and 1,4-endoperoxides, the selectivity decreases considerably in the presence of hydroxylic solvents [123]. When hydrogen bonding is no more possible in **203**, the stereofacial differentiation is steered by steric and electronic repulsion effects at the level of the possible diastereoisomeric transition states and **204** is formed selectively (Scheme 54).

The control of the diastereoselectivity of the allylic ene reaction could be extended to alkenes **205** substituted by a removable chiral auxiliary. Here again the best *de* are obtained when hydrogen bonds can be developed between the substrate and $^1O_2$ (Scheme 55) [124].

B. The Addition of Alcohols, Acetals, and Amines to Alkenes

Although there exist numerous ground state reactions, photochemically induced asymmetric radical additions can be very efficient and even highly stereoselective [125]. Furthermore, no particular functionalization of the starting material is necessary prior to the formation of a C–C bond. In this context, the photosensitized addition of alcohols, cyclic acetals, and tertiary amines to electron-deficient alkenes has been particularly studied. This will be illustrated by a few examples.

First attempts to induce chirality in the photoinduced addition of ketyl radicals (e.g., **U**) involved α,β-usaturated carbonyl compounds such as **208** derived from carbohydrates (Scheme 56) [126]. With benzophenone as sensitizer, these radicals could be added stereoselectively, and similar reactions were carried out with dioxolane and α,β-usaturated nitropyranones [127].

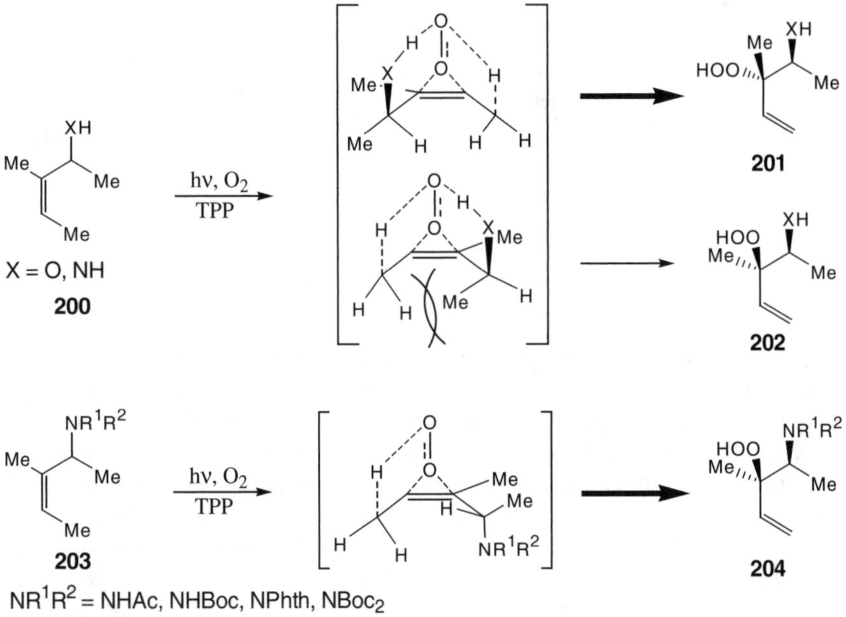

X = O, NH
200

NR^{1}R^2 = NHAc, NHBoc, NPhth, NBoc$_2$

Scheme 54

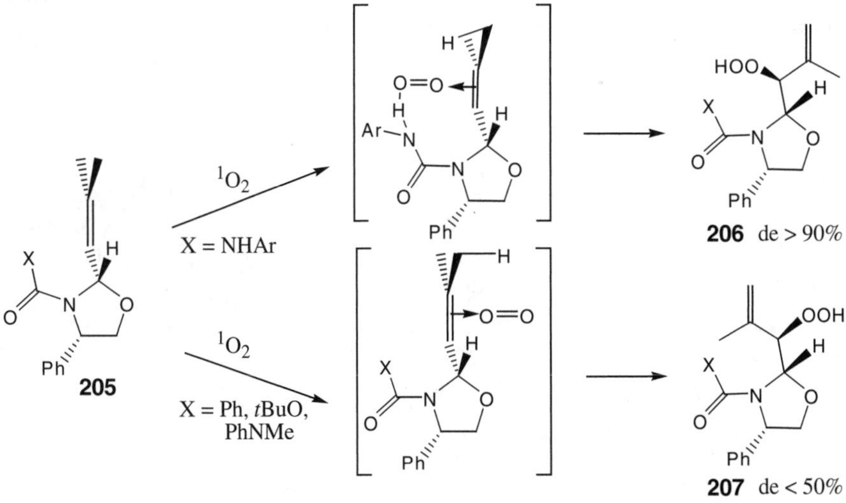

Scheme 55

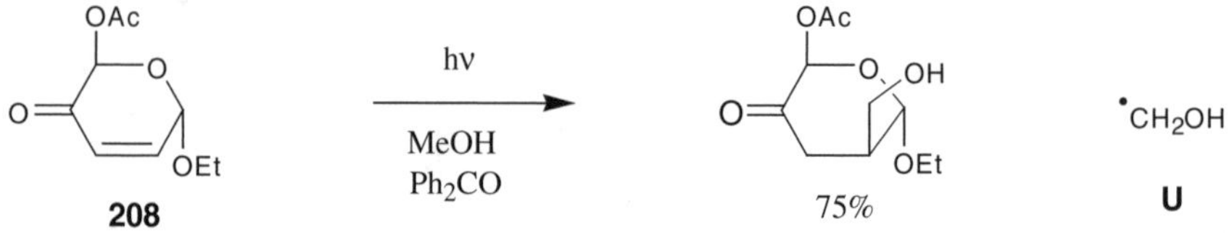

Scheme 56

In the same way isopropanol was added to fumaric and maleic derivatives carrying an enantiopure alcohol moiety as chiral auxiliary [128]. The best results were obtained when 8-phenylmenthol was used as chiral auxiliary **209** (Scheme 57) [129]. The primary adduct **V** was unstable and cyclized to the corresponding (S)-terebic ester **210**. The chiral induction was significantly improved when chiral furanones like **211** were used as substrate [130,131]. In this case, the facial differentiation was enhanced by the structural rigidity of the five-membered ring. The radical attack occurred specifically anti with respect to the menthyloxy substituent. The adduct **212** could also be transformed into (S)-terebic acid. Ketyl radicals, generated via electron transfer from tertiary amines to ketones, also added **211** with high stereoselectively [132].

Similarly, dioxolane was added successfully to the chiral dioxinone **213** (Scheme 58) [133]. The radical attack occurred preferentially syn with respect to the isopropyl substituent of the menthone moiety to produce **214** and **215**. This site differentiation was rationalized by the favored conformation of the substrate.

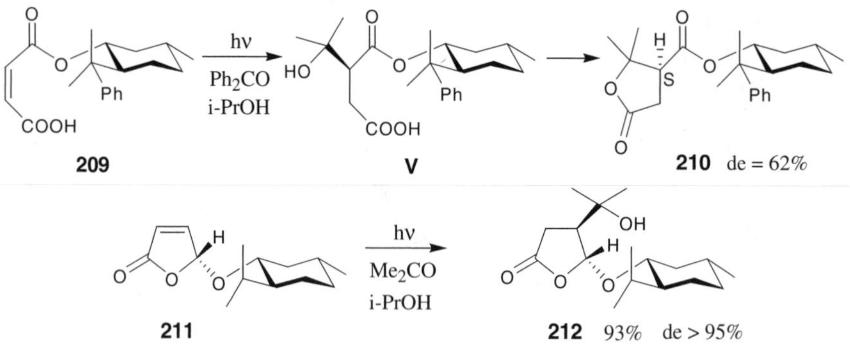

Scheme 57

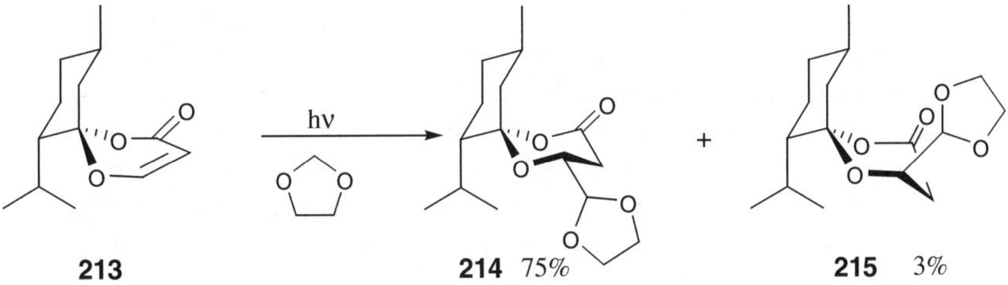
hν
O
213
214 75%
215 3%

Scheme 58

Recently, an efficient photocatalytic method has been reported for the radical addition of tertiary amines to electron-deficient double bonds (Scheme 59) [134]. *N*-methylpyrrolidine was added to the furanone **216**. The radical attack occurred specifically anti with respect to the menthyloxy substituent, and the adducts **217** and **218** were isolated in high yields in the presence of sensitizers having electron-rich substituents or in the presence of semiconductors like TiO_2. However, the configuration of the chiral center in the α-position of the nitrogen atom could not be controlled. Although the diastereoselectivity was almost the

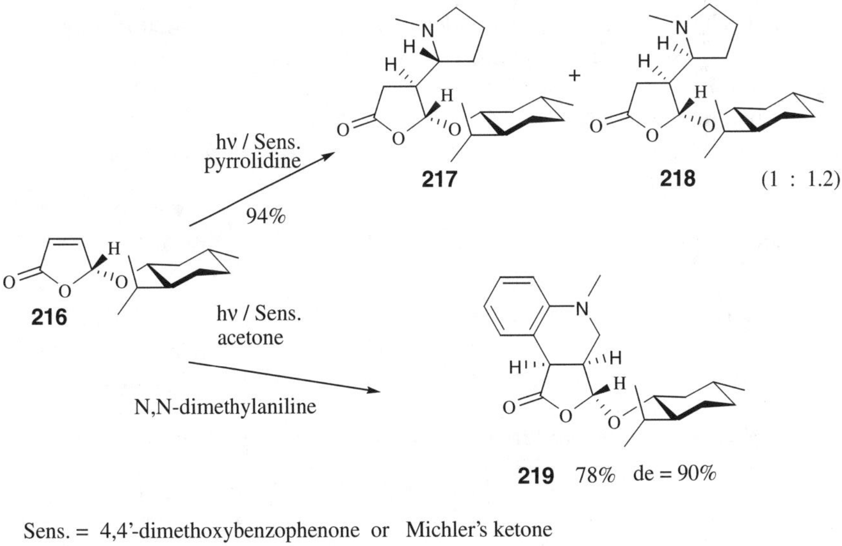
hν / Sens.
pyrrolidine
94%
217
218 (1 : 1.2)
216
hν / Sens.
acetone
N,N-dimethylaniline
219 78% de = 90%
Sens. = 4,4'-dimethoxybenzophenone or Michler's ketone

Scheme 59

same, yields were significantly lower, and considerable side reactions were observed with benzophenone or acetophenone as sensitizers [135]. The products could be transformed in a few steps into the pyrrolidine alkaloids such as laburnine and isoretronecanol. Under similar reaction conditions, a tandem addition–cyclization reaction could be carried out preferentially with aromatic tertiary amines [136]. Using the furanone **216** as chiral synthon, the tetrahydroquinoline derivatives **219** were formed in 90% *de* in the presence of *N,N*-dimethylaniline.

VI. CONCLUSIONS

A good control of the configuration of the newly created stereogenic centers is essential in synthesis. During the past 20 years, a better understanding of the factors gearing the diastereodifferentiation in photochemical reactions has led to more and more synthetic applications. In this review, we have presented various aspects of stereochemical control during photochemical reactions carried out in solution. When chiral starting materials are involved in [2 + 2], [4 + 2], or [4 + 4] photocycloaddition processes, high stereoselectivities can usually be obtained. Recently, the use of removable chiral auxiliaries in photocycloaddition reactions has provided the organic chemist involved in the synthesis of complex molecules with new and efficient tools. A good diastereodifferentiation can also be observed in solution for photooxygenations and various photorearrangement reactions.

Although an almost completely perfect enantioselection has recently been observed with photoreactions carried out in rigid media or in supramolecular assemblies, diastereoselective photoreactions in solution will remain a method of choice for synthetic applications. However, a great effort has still to be made to improve the diastereodifferentiation of photorearrangement processes and to control the selectivity of photoreactions involving polyfunctional substrates.

REFERENCES

1. Rau H. Chem Rev 1983; 83:535–547.
2. Inoue Y. Chem Rev 92:741–770, 1992. Everitt SRL, Inoue Y. In: Ramamurthy V, Schanze KS, eds. Organic Molecular Photochemistry, Marcel Dekker, 1999, Vol. 3, pp. 71–128. Pete JP. In: Adv Photochem 21:135–216, 1996.
3. For recent reviews see Bach T Synlett 1699-1707, 2000. Bach T. Synthesis 683–703, 1998. Mattay J, Conrads R, Hoffmann R. In: Helmchen G, Hoffmann RW, Mulzer J, Schaumann E, eds. Methods of Organic Chemistry (Houben-Weyl) E21, Vol 5. Stuttgart: Georg Thieme Verlag, 1996, pp. 3133–3178. Porco JA Jr, Schreiber SL. In: Trost BM, Fleming I, Paquette LA, eds. Comprehensive Organic Synthesis, 5 Oxford Pergamon Press 1991 pp. 151–192.

4. Freilich SC, Peters KS. J Am Chem Soc 107: 3819–3822,1985. Freilich SC, Peters KS. J Am Chem Soc 103: 6255–6527, 1981. Mattay J, Gersdorf J, Buchkremer K. Chem Ber 120: 307–318, 1987. Eckert G, Goez M. J Am Chem Soc 116: 11999–12009, 1994. Abe M, Shirodai Y, Nojima M. J Chem Soc, Perkin Trans I: 3253–3260, 1998. Gan CY, Lambert JN. J Chem Soc, Perkin Trans I: 2362–2372, 1998. de Lucas NC, Silva MT, Gege C, Netto-Ferreira JC. J Chem Soc, Perkin Trans 2: 2795–2801, 1999.

5. Jarosz S, Zamojski A. Tetrahedron 1982; 38:1453–1456.

6. Schreiber SL, Satake K. Tetrahedron Lett 1986; 27:2575–2578.

7. Zagar C, Scharf H-D. Chem Ber 1991; 124:967–969.

8. Turro NJ, Farrington GL. J Am Chem Soc 1980; 102:6056–6063.

9. Ong K-S, Whistler RL. J Org Chem 1972; 37:572–574.

10. Bach T, Jödicke K, Kather K, Fröhlich R. J Am Chem. Soc. 1997; 119:2437–2445.

11. D'Auria M, Racioppi R, Romaniello G. Eur J Org Chem 2000:3265–3272.

12. Bach T, Schröder J, Brandl T, Hecht J, Harms K. Tetrahedron 1998; 54:4507–4520.

13. Bach T, Brummerhop H, Harms K. Chem Eur J 2000; 6:3838–3848.

14. Gotthardt H, Lenz W. Angew Chem Int Ed Engl 1979; 18:868–868.

15. Koch H. Chirale Induction bei photochemischen Reaktionen; Untersuchungen zur chiralen Oxetan- und CyclobutanbildungRWTH Aachen, Aachen, 1984.

16. Nehrings A, Scharf H-D, Runsink J. Angew Chem Int Ed Engl 1985; 24:877–878.

17. Esser P, Buschmann H, Meyer-Stork M, Scharf H-D. Angew Chem Int Ed Engl 1992; 31:1190–1192.

18. Buschmann H. Von der asymmetrisch gesteuerten Paternò-Büchi-Reaktion zum Isoinversionsprinzip, einem allgemeinen Selektionsmodell in der Chemie. RWTH Aachen, Aachen, 1992.

19. Weuthen M, Scharf H-D, Runsink J. Chem Ber. 120: 1023–1026, 1987. Weuthen M, Scharf H-D, Runsink J, Vaen R. Chem Ber 121: 971–976, 1988.

20. Plath M. Mechanismus und Stereochemie der Photo-Aldol-Reaktion mit 2,3-Dihydro-1H-imidazolen. PhD diss. RWTH Aachen, Aachen, 1991.

21. Koch H, Scharf H-D, Runsink J, Leismann H. Chem Ber 1985; 118:1485–1503.

22. Koch H, Runsink J, Scharf H-D. Tetrahedron Lett 1983; 24:3217–3220.

23. Buschmann H, Scharf H-D, Hoffmann N, Plath MW, Runsink J. J Am Chem Soc 1989; 111:5367–5373.

24. Buschmann H, Hoffmann N, Scharf H-D. Tetrahedron Asymm 1991; 2:1429–1444.

25. Pelzer R, Jütten P, Scharf H-D. Chem Ber 1989; 122:487–491.

26. Jarosz S, Zamojski A. Tetrahedron 1982; 38:1447–1451.

27. Pelzer R, Scharf H-D, Buschmann H, Runsink J. Chem Ber 1989; 122:1187–1192.

28. Bach T, Bergmann H, Grosch B, Harms K. J Am Chem Soc 2002; 124:7982–7990.

29. Buschmann H, Scharf H-D, Hoffmann N, Esser P. Angew Chem Int Ed Engl 1991; 30:477–515.

30. Giesbeck AG, Fiege M, Ramamurthy IV, Schanze KS, eds. Organic, Physical, and Materials Photochemistry. New York: Marcel Dekker, 2000, pp. 33–100. Griesbeck AG, Mauder H, Stadtmüller S. Acc Chem Res 27: 70–75, 1994. Further examples: Dopp D, Fischer M-A. Recl Trav Chim Pays-Bas 114: 498–503, 1995. Abe M, Torri E, Nojima M. J Org Chem 65: 3426–3431, 2000. Abe M, Fujimoto K, Nojima

M. J Am Chem Soc 122: 4005–4010,2000. Griesbeck AG, Bondock S, Gudipati MS. Angew Chem Int Ed 40: 4684–4687, 2001. Zhang Y, Xue J, Gao Y, Fun H-K, Xu J-H. J Chem Soc, Perkin Trans 1: 345–352, 2002. Adam W, Stegmann VR. J Am Chem Soc 124: 3600–3607, 2002.

31. Griesbeck AG, Stadtmüller S. Chem Ber 1990; 123:357–362.
32. Salem L, Rowland C. Angew Chem Int Ed Engl 11 92–111, 1971. Kutateladze AG. J Am Chem Soc 123: 9279–9282, 2001.
33. Gotthardt H, Lenz W. Tetrahedron Lett 1979:2879–2880.
34. Brunne J. Beiträge zum Isoinversionsprinzip: Anwendungen des Isoinversionsprinzips auf die diasteroselektive thioanaloge Paternò-Büchi-Reaktion sowie auf die diastereoselektive Reduktion von cyclischen Ketonen mit Aluminiumhydridreagenzien, PhD diss. RWTH Aachen, Aachen, 1994.
35. Schuster DI, Lem G, Kaprinidis NA. Chem Rev 1993; 93:3–22.
36. Cf. Ref. 2c Crimmins MT. Chem Rev 88: 1453–1473 1988 Crimmins MT, Reinhold TL. In: Paquette LA, ed. Organic Reactions. New York: John Wiley, 1993, 44: 299–588.
37a. White JD, Gupta DN. J Am Chem Soc 1966; 88:5364–5365.
37b. Smith AB, Sulikowski GA, Fujimoto K. J Am Chem Soc 1989; 111:8039–8041.
37c. Baldwin SW, Fredericks JE. Tetrahedron Lett 1982; 23:1235–1238.
37d. Lange GL, Neidert EE, Orrom WJ, Wallace DJ. Can J Chem 1978; 56:1628–1633.
37e. Hansson T, Wickberg B. J Org Chem 1992; 57:5370–5376.
38a. Lange GL, Mc Carthy FC. Tetrahedron Lett 19: 4749–4750, 1978 Lange GL, Lee M, Synthesis 1986: 117–120.
38b. Baldwin SW, Mazzuckelli TJ. Tetrahedron Lett 1986; 27:5975–5978.
38c. Van Audenhove M, De Keukeleire D, Vandevalle M. Tetrahedron Lett 1980; 21: 1979–1982.
38d. Lange GL, Humber CC, Manthorpe JM. Tetrahedron: Asymmetry 2002; 13: 1355–1362.
39a. Nagaoka H, Miyaoka H, Yamada Y. Tetrahedron Lett 1990; 31:1573–1576.
39b. Neh H, Blechert S, Schnick W, Jansen M. Angew Chem Int Ed Engl 1984; 23: 905–906.
39c. Berkowitz WF, Perumattam J, Amarasekara A. Tetrahedron Lett 1985; 26: 3665–3668.
39d. Cargill RL, Morton GH, Bordner J. J Org Chem 1980; 45:3929–3930.
40a. Alibes R, Bourdelande JL, Font J. Tetrahedron Asymmetry 1991; 2:1391–1402.
40b. Alibes R, Bourdelande JL, Font J. Tetrahedron Lett 1993; 34:7455–7458.
40c. Sugihara Y, Morokoshi N, Murata I. Tetrahedron Lett 1977; 18:3887–3888.
40d. Alibes R, de March P, Figueredo M, Font J, Racamonde M, Rustullet A, Alvarez-Larena A, Piniella JF, Parella T. Tetrahedron Lett 2003; 44:69–71.
41a. Becker D, Haddad N. Tetrahedron Lett 1986; 27:6393–6396.
41b. Becker D, Haddad N, Sahali Y. Tetrahedron Lett 1989; 30:4429–4432.
41c. Becker D, Haddad N. Tetrahedron 1993; 49:947–964.
41d. Becker D, Nagler M, Harel Z, Gillon A. J Org Chem 1983; 48:2584–2590.
41e. Becker D, Denekamp C, Haddad N. Tetrahedron Lett 1992; 33:827–830.
42a. Srinivasan R, Carlough KH. J Am Chem Soc 1967; 89:4932–4936.

42b. Liu RSH, Hammond GS. J Am Chem Soc 1967; 89:4936–4944.

43a. Oppolzer W. Accounts of Chem Res 1982; 15:135–141.

43b. Pirrung MC. J Am Chem Soc 1981; 103:82–87.

44. Rao VB, George CF, Wolff S, Agosta WC. J Am Chem Soc 1985; 107:5732–5739.

45a. Connolly PJ, Heathcock CH. J Org Chem 1985; 50:4135–4144.

45b. Garibaldi P, Jommi G, Sisti M. Gazz Chim Ital 116: 291–301, 1986. Jommi G, Orsini F, Resmini M, Sisti M. Tetrahedron Lett 32: 6969–6972, 1991.

45c. Tanaka M, Tomioka L, Koga K. Tetrahedron Lett 1985; 26:3035–3038.

45d. Cruciani G, Margaretha P. Helv Chim Acta 73: 288–297, 1990. Altmeyer I, Margaretha P, Helv Chim Acta 60: 874–881, 1977.

45e. AG Schultz M Plummer AG Taveras RK Kullnig J Am Chem Soc 110 5547–5555 1988 AG Schultz AG Taveras Tetrahedron Lett 29 6881–6884 1988.

45f. see also Winkler JD, Hey JP, Williard PG. J Am Chem Soc 108:6425–6427,1986.

46a. Oppolzer W, Godel T. J Am Chem Soc 1978; 100:2583–2584.

46b. Oppolzer W, Godel T. Helv Chim Acta 1984; 67:1154–1167.

46c. Pearlman BA. J Am Chem Soc 1979; 101:6398–6404 and 6404–6408.

46d. Amougay A, Pete JP, Piva O. Tetrahedron Lett 1992; 33:7347–7350.

46e. Le Blanc S, Pete JP, Piva O. Tetrahedron Lett 1993; 34:635–638.

46f. Oppolzer W, Bird C. Helv Chim Acta 1979; 62:1199–1202.

46g. see also JD Winkler, MB Rouse, MF Greaney, SJ Harrison, YT Jeon. J Am Chem Soc 124: 9726–9728 2002.

47. Maradyn DJ, Weedon AC. J Am Chem Soc 1995; 117:5359–5360.

48a. Baldwin JE. J Chem Soc Chem Commun 1976:734–736.

48b. Baldwin JE, Cutting J, Dupont W, Kruse L, Silberman L, Thomas RL. J Chem Soc Chem Commun 1976:736–738.

49a. Crimmins MT, Jung DK, Gray JL. J Am Chem Soc 1993; 115:3146–3155.

49b. Winkler JD, Hershberger PM. J Am Chem Soc 1989; 111:4852–4856.

50a. Crimmins MT, Watson PS. Tetrahedron Lett 1993; 34:199–202.

50b. Crimmins MT, King BW, Watson PS, Guise LE. Tetrahedron 1997; 53:8963–8974.

51. Crimmins MT, Choy AL. J Am Chem Soc 1997; 119:10237–10238.

52. Wiesner K. Tetrahedron 1975; 31:1655–1658.

53a. Audley M, Geraghty WA. Tetrahedron Lett 1996; 37:1641–1644.

53b. Busqué F, De March P, Figueredo M, Font J, Margaretha P, Raya J. Synthesis 2001:1143–1148.

54. Carreira EM, Hastings CA, Shepard MS, Yerkey LA, Millward DB. J Am Chem Soc 1994; 116:6622–6630.

55. Bach T, Pelkmann C, Harms K. Tetrahedron Lett 1999; 40:2103–2104.

56a. Salomon RG. Tetrahedron 1983; 39:485–575.

56a. Salomon RG, Ghosh S, Raychaudhuri SR, Miranti TS. Tetrahedron Lett 1984; 25: 3167–3170.

57. Bach T, Spiegel A. Eur J Org Chem 2002:645–654.

58. Bach T, Kräger C, Harms K. Synthesis 2000:305–320.

59. Green BS, Hagler AT, Rabinsohn Y, Rejtö M. Isr J Chem 1976/77; 15:124–130.

60. Tolbert LM, Ali MB. J Am Chem Soc 1982; 104:1742–1744.

61. Haag D, Scharf HD. J Org Chem 1996; 61:6127–6135.

62.   Langer K, Mattay J. J Org Chem 1995; 60:7256–7266.
63.   Amougay A, Letsch O, Pete JP, Piva O. Tetrahedron 1996; 52:2405–2420.
64.   Lange GL, Decicco CP, Tan SL, Chamberlain G. Tetrahedron Lett 1985; 26:
      4707–4710.
64.   Lange GL, Lee M. Tetrahedron Lett 1985; 26:6163–6166.
65.   Lange GL, Decicco CP, Lee M. Tetrahedron Lett 1987; 28:2833–2836.
66a. Herzog H, Koch H, Scharf HD, Runsink J. Tetrahedron 1986; 42:3547–3558.
66b. Herzog H, Koch H, Scharf HD, Runsink J. Chem Ber 1987; 120:1737–1740.
67.   Kakiuchi K, Ikki T, Endo K. Pacifichem December 14–19, 2000(N°1493).
68.   Lange GL, Decicco CP. Tetrahedron Lett 1988; 29:2613–2614.
69.   Bonvalet C, Bouquant J, Feigenbaum A, Pete JP, Scholler D. Bull Soc Chim Fr
      1994; 131:687–692.
70a. Meyers AI, Fleming SA. J Am Chem Soc 1986; 108:306–307.
70b. Hoffmann N, Scharf HD, Runsink J. Tetrahedron Lett 1989; 30:2637–2638.
70c. Hoffmann N, Scharf HD. Liebigs Ann Chem 1991:1273–1277.
71.   Hoffmann N, Buschmann H, Raabe G, Scharf HD. Tetrahedron 1994; 50:
      11167–11186.
72.   Bertrand S, Hoffmann N, Pete JP. Tetrahedron 1998; 54:4873–4888.
73.   Demuth M, Palomer A, Sluma H-D, Dey AK, Kräger C, Tsay Y-H. Angew Chem
      Int Ed Eng 1986; 25:1117–1119.
73.   Sato M, Takayama K, Abe Y, Furuya T, Inukai N, Kaneko C. Chem PharmBull
      1990; 38:336–339.
74.   Sato M, Abe Y, Kaneko C. Heterocycles 1990; 30:217–221.
75a. Le Blanc S, Pete JP, Piva O. Tetrahedron Lett 1992; 33:1993–1996.
75b. Meyer C, Pete JP, Piva O. Rec Trav Chim Pays-Bas 1995; 114:492–497.
75c. Amougay A, Piva O, Pete JP. Tetrahedron Lett 1993; 34:5285–5286.
75d. Meyer C, Piva O, Pete JP. Tetrahedron 2000; 56:4479–4489.
76.   Piva-Le Blanc S, Pete JP, Piva O. Chem Commun 1998:235–236.
77.   Faure S, Piva O. Tetrahedron Lett 2001; 42:255–259.
78.   Faure S, Piva-Le Blanc S, Bertrand C, Pete JP, Faure R, Piva O. J Org Chem 2002;
      67:1061–1070.
79.   Faure S. Photocycloadditions et photoréarrangements asymétriques; applications à
      la synthèse de produits naturels Ph D diss., URCA, Reims, France, 1999.
80.   Chen C, Cheng V, Cai N, Duesler E, Mariano PS. J Am Chem Soc 2001; 123:
      6433–6434.
81a. Wender PA, Siggel L, Nuss JM. In Trost BM , Fleming I , Paquette LA, Eds.
      Comprehensive Organic Synthesis. Vol. 5. Oxford: Pergamon Press, 1991:645–673.
81b. DeKeukeleire D. Aldrichchimica Acta 1994; 27:59–69.
82.   Cornelisse J. Chem. Rev. 1993; 93:615–669.
83.   Cornelisse J, de Haan R. In Ramapurthy V., Schanze KS., eds. Molecular and
      Supramolecular Photochemistry,. Vol. 8. New York: Marcel Dekker, 2001:1–125.
84a. Bryce-Smith D, Gilbert A, Orger B, Tyrrell H. Chem Commun 1974:334–336.
84b. Leismann H, Mattay J, Scharf H-D. J Am Chem Soc 1984; 106:3985–3991.
84c. Mattay J. Tetrahedron 1985; 41:2393–2404.
84d. Mattay J. Tetrahedron 1985; 41:2405–2417.

85a. Bryce-Smith D, Gilbert A. Tetrahedron 1976; 32:1309–1326.
85b. Bryce-Smith D, Gilbert A. Tetrahedron 1977; 33:2459–2489.
85c. Houk KN. Pure Appl Chem 1982; 54:1633–1650.
85d. Hart JA van der, Mulder JJC, Cornelisse J. J Mol Struct (Theochem) 1987; 151: 1–10.
85e. van der Hart JA, Mulder JJC, Cornelisse J. J Photochem Photobiol A: Chem 1991; 61:3–13.
85f. Stehouwer AM, van der Hart JA, Mulder JJC, Cornelisse J. J Mol Struct (Theochem) 1992; 92:333–338.
85g. van der Hart JA, Mulder JJC, Cornelisse J. J Photochem Photobiol A: Chem 1995; 86:141–148.
86. Wender PA, Singh SK. Tetrahedron Lett 1990; 31:2517–2520.
87. Sugimura T, Nishiyama N, Tai A, Hakushi T. Tetrahedron Asymm 1994; 5: 1163–1166.
88. Wagner PJ. Acc Chem Res 2001; 34:1–8.
89. Hoffmann N, Pete J-P. Tetrahedron Lett 1995; 36:2623–2626.
90. Wagner PJ, McMahon K. J Am Chem. Soc 1994; 116:10827–10828.
91. Döpp D, Pies M. J Chem Soc Chem Commun:1734–1735, 187.
92. Döpp D. In Ramamurthy V , Schanze KS, Eds. Molecular and Supramolecular Photochemistry. Vol. Vol 6. New York: Marcel Dekker, 2000:101–148.
93. McN Sieburth S. In Harmata M, Ed. Advances in Cycloaddition. Vol. Vol 5. Stamford: JAI Press, 1999:85–118.
94a. Cochez Y, Jespers J, Libert V, Mislow K, Martin RH. Bull Soc Chim Belg 1975; 84:1033–1036.
94b. Cochez Y, Martin RH, Jespers J. Israel J Chem 1976/1977; 15:29–32.
95. Kodani T, Matsuda K, Yamada T, Kobatake S, Irie M. J Am Chem Soc 2000; 122: 9631–9637.
96. Yamaguchi T, Uchida K, Irie M. J Am Chem Soc 1997; 119:6066–6071.
97. Tanaka K, Osuga H, Suzuki H. Tetrahedron: Asymmetry 1993; 4:1843–1856.
98. Yokoyama Y, Uchida S, Yokoyama Y, Sugawara Y, Kurita Y. J Am Chem Soc 1996; 118:3100–3107.
99. Dugat D, Gramain JC, Dauphin G. J Chem Soc Perkin Trans 1990; 2:605–611.
100. Sato M, Katagiri N, Muto M, Haneda T, Kaneko C. Tetrahedron Lett 1986; 27: 6091–6094.
101. Wagner PG, Kelso AE, Kemppainen AE, McGrath JM, Schott HN, Zepp RG. J Am Chem Soc 1972; 94:7506–7512.
102a. Lindemann U, Wulff-Molder D, Wessig P. J Photochem Photobiol A Chemistry 1998; 119:73–83.
102b. Wessig P, Schwartz J. Helv Chim Acta 1998; 81:1803–1814.
102c. Griesbeck AG, Heckroth H. J Am Chem Soc 2002; 124:396–403.
103. Steiner A, Wessig P, Polborn K. Helv Chim Acta 1996; 79:1843–1862.
104. Wessig P, Wettstein P, Giese B, Neuburger M, Zehnder M. Helv Chim Acta 1994; 77:829–837.
105. Lindemann U, Reck G, Wulff-Molder D, Wessig P. Tetrahedron 1998; 54: 2529–2544.

106a. Lindemann U, Wulff-Molder D, Wessig P. Tetrahedron: Asymmetry 1998; 9: 4459–4473.

106b. Wessig P. Tetrahedron Lett 1999; 40:5987–5988.

107a. Weigel W, Schiller S, Reck G, Henning H-G. Tetrahedron Lett 1993; 34: 6737–6740.

107b. Weigel W, Schiller S, Henning H-G. Tetrahedron 1997; 53:7855–7866.

108. Griesbeck AG, Heckroth H. Synlett 2002:131–133.

109. Giese B, Wettstein P, Stähelin C, Barbosa F, Neuburger M, Zehnder M, Wessig P. Angew Chem Int Ed 1999; 38:2586–2587.

110. Griesbeck AG, Kramer W, Lex J. Angew Chem Int Ed 2001; 40:577–579.

111. Griesbeck AG, Kramer W, Bartoschek A, Schmickler H. Org Lett 2001; 3:537–539.

112. Pete JP. In Horspool WM , Song P-S, Eds. Handbook of Organic Photochemistry and Photobiology. Boca Raton: CRC Press, 1995:593–606.

113. Piva O, Pete JP. Tetrahedron Lett 1990; 31:5157–5160.

114a. Mortezaei R, Awandi D, Henin F, Muzart J, Pete JP. J Am Chem Soc 1988; 110: 4824–4826.

114b. Awandi D, Henin F, Muzart J, Pete JP. Tetrahedron; Asymmetry 1991; 2: 1101–1104.

115. Muzart J, Henin F, Pete JP, M'boungou-M'Passi A. Tetrahedron: Asymmetry 1993; 4:2531–2534.

116. Piva O, Pete JP. Tetrahedron: Asymmetry 1992; 3:759–768.

117. Bargiggia F, Dos Santos S, Piva O. Synthesis 2002:427–437.

118. Faure S, Piva O. Synlett 1998:1414–1416.

119a. Adam W, Prein M. J Am Chem Soc 1993; 115:3766–3767.

119b. Adam W, Peters EM, Peters K, Prein M, Georg von Schnering H. J Am Chem Soc 1995; 117:6686–6690.

120. Adam W, Güthlein M, Peters EM, Peters K, Wirth T. J Am Chem Soc 1998; 120: 4091–4093.

121. Adam W, Saha-Möller CR, Schambony SB. J Am Chem Soc 1999; 121:1834–1838.

122. Adam W, Bosio SG, Turro NJ. J Am Chem Soc 2002; 124:8814–8815.

123a. Adam W, Nestler B. J Am Chem Soc 1992; 114:6549–6550.

123b. Adam W, Bränker HG. J Am Chem Soc 1993; 115:3008–3009.

123c. Adam W, Nestler B. J Am Chem Soc 1993; 115:5041–5049.

123d. Bränker HG, Adam W. J Am Chem Soc 1995; 117:3976–3982.

124a. Adam W, Peters K, Peters EM, Schambony SB. J Am Chem Soc 2000; 122: 7610–7611.

124b. Adam W, Peters K, Peters EM, Schambony SB. J Am Chem Soc 2001; 123: 7228–7232.

125. Curran DP, Porter NA, Giese B. Stereochemistry of Radical Reactions. Weinheim: VCH, 1996.

126. Fraser-Reid B, Holder NL, Hicks DR, Walker DL. Can J Chem 1977; 55: 3978–3985.

127. Sakakibara T, Nakagawa T. Carbohydr Res 1987; 163:239–246.

127. Sakakibara T, Takaide A, Seda A. Carbohydr Res 1992; 226:271–278.

128. Horner L, Klaus J. Liebigs Ann Chem 1979:1232–1257.

129. Vassen R, Runsink J, Scharf H-D. Chem Ber 1986; 119:3495–3497.
130. Hoffmann N. Tetrahedron Asymm 1994; 5:879–886.
131. Mann J, Weymouth-Wilson A. Synlett 1992:67–69.
132. Brulé C, Hoffmann N. Tetrahedron Lett 2002; 43:69–72.
133. Graalfs H, Fröhlich R, Wolff C, Mattay J. Eur J Org Chem 1999:1057–1073.
134a. Bertrand S, Hoffmann N, Pete J-P. Eur J Org Chem 2000:2227–2238.
134b. Marinkocic S, Hoffmann N. Chem Commun 2001:1576–1577.
135. see also E Santiago de Alvarenga, CJ Cardin, J Mann. Tetrahedron 53: 1457–1466, 1997 .
136. Bertrand S, Hoffmann N, Humbel S, Pete J-P. J Org Chem 2000; 65:8690–8703.

6
Chirality in Photochromism

Yasushi Yokoyama and Masako Saito
Yokohama National University
Yokohama, Japan

I. INTRODUCTION

Photochromism has been defined as "a reversible transformation of a single chemical species being induced in one or both directions by electromagnetic radiation between two states having different distinguishable absorption spectra" by Dürr [1]. Photochromic compounds change their structures by photoirradiation. Change in color (or absorption spectrum) is one of the accompanying properties with the photochromic structural change (Fig. 1). Usually (but not always) one isomer is colorless and thermally more stable than the colored counterpart. The large difference in color (or absorption spectrum) of the isomers is required in order to achieve the highly biased ratio of the isomers at the photostationary states.

Studies on photochromic compounds in connection with chirality have been continuously done from the several viewpoints such as photochemical control of chirality-related liquid crystalline (LC) properties, photochemical control of biological activities by attaching a photochromic compound to biological macro-molecules, and use for the optical memory media, as well as the purely scientific interests.

In general, if a photochromic compound has an intrinsic chirality that is preserved during the photochromic reaction, it is just an extension of photochromism of achiral molecules unless the intrinsic chirality does not induce the generation of the second chirality in the biased ratio. If the intrinsic chirality induces another chirality selectively, it is called "diastereoselective photochromism" [2].

There are several categories of photochromic reactions: electrocyclization of conjugated polyenes [3], *E-Z* isomerization of a double bond [4], radical forma-

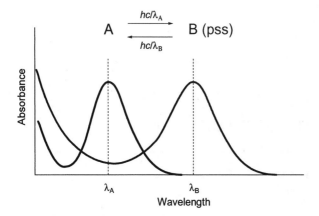

Figure 1 Photochromism.

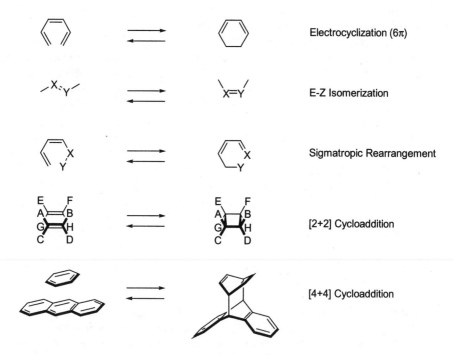

Scheme 1 Representative reaction modes of photochromism.

tion by bond scission [5], sigmatropic (usually prototropic) rearrangements [6], and [2 + 2] or [4 + 4] cycloadditions [7] (Scheme 1). As for the photochemical *E-Z* isomerizations, they seem to affect little the change in chirality. However, for azobenzenes as a representative of the *E-Z* isomerizing photochromic compounds, it has been reported that some can control the bulk or microenvironmental chirality by photoirradiation [8]. Another important class of *E-Z* isomerization is the highly crowded benzoanalogs of stilbene derivatives (overcrowded alkenes) [9,10]. On the other hand, electrocyclization, sigmatropic rearrangements, and cycloadditions may generate stereogenic centers or transfer chirality (e.g., from helicity to stereogenic center) as a result of photochromic reactions. To date, however, little is known about the chiral photochromic sigmatropic rearrangements because the transferring group is usually hydrogen (e.g., salicylideneaniline) [6], which induces neither generation nor transformation of chirality. As for cycloadditions and cycloreversion reactions, they can hardly constitute a photochromic system, probably because the cycloaddition products usually absorb much shorter wavelengths.

Consequently, the main class of photochromic compounds that shows change in chirality by photochromic transformation is that based on electrocyclization. Fulgides [11–13], diarylethenes [14–16], and spiropyrans [17,18] are included in this class. It may be somewhat strange to put spiropyrans into this category because the colored merocyanine forms are known to take zwitterionic structures, which do not cyclize through the electrocyclization mechanism. However, in order to simplify the classification, we consider spiropyrans to cyclize from the neutral dienone structures as the merocyanine forms.

II. CHIRAL PHOTOCHROMISM BASED ON ELECTROCYCLIZATION

A. Fulgide Derivatives

Fulgides are the representatives of 6π-electrocyclization-based photochromic compounds (Scheme 2). The most important feature is that it is thermally irreversi-

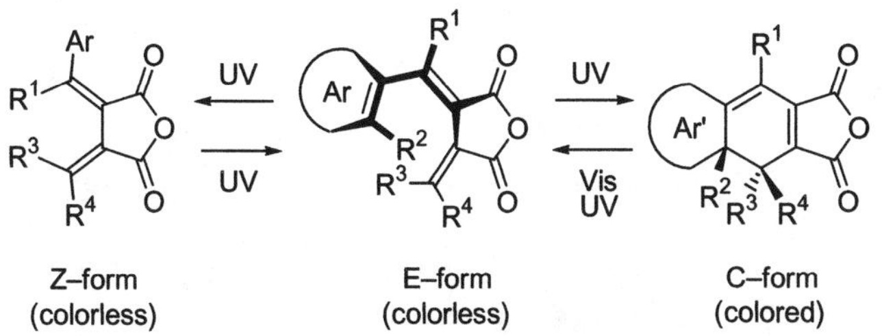

Z–form (colorless)	E–form (colorless)	C–form (colored)

Scheme 2 Photochromism of fulgides.

ble [19,20]. As the hexatriene moiety that undergoes 6π-electrocyclization is overcrowded if the double bonds are fully studded with bulky substituents, the triene moiety cannot take the planar, plane-symmetric conformation but is forced to take a helical conformation [21]. If a prochiral carbon atom bearing two identical substituents is introduced on the hexatriene skeleton, the substituents suffer different electronic and magnetic influences from the helical hexatriene, so that they behave as different substituents when observed by NMR spectroscopy. If the helical structure is labile and changes its conformation between right-handed (P) and left-handed (M) screw senses thermally, one can observe the coalescence of the signals of the two prochiral substituents as the evidence of the enantioisomerization process by variable-temperature NMR experiments.

It was indeed observed by Yokoyama et al. for a furylfulgide **F-1** [22,23]. The two methyl groups of the isopropyl group in **F-1E** appeared as two doublets at δ 0.88 and 1.33. Upon heating, those signals coalesced. This means that the racemization of the helical chirality is occurring with the time constant of NMR

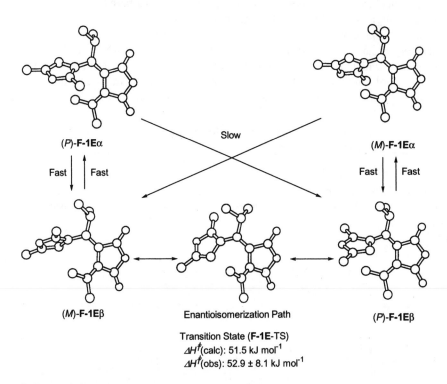

(P)-**F-1Eα**   Slow   (M)-**F-1Eα**

Fast | | Fast   Fast | | Fast

(M)-**F-1Eβ**   Enantioisomerization Path   (P)-**F-1Eβ**

Transition State (**F-1E-TS**)
$\Delta H^{\ddagger}$(calc): 51.5 kJ mol^{-1}
$\Delta H^{\ddagger}$(obs): 52.9 ± 8.1 kJ mol^{-1}

Scheme 3   Enantioisomerization process of **F-1E** calculated by AM-1 semiempirical MO calculation method.

measurements. The activation enthalpy and activation entropy, obtained from two independent coalescent experiments with two NMR machines possessing different magnetic fields, were 52.9 ± 8.1 kJ mol^{-1} and -36.5 ± 24.2 J mol^{-1} K^{-1}, respectively. As the helicity of **F-1** is the only chirality

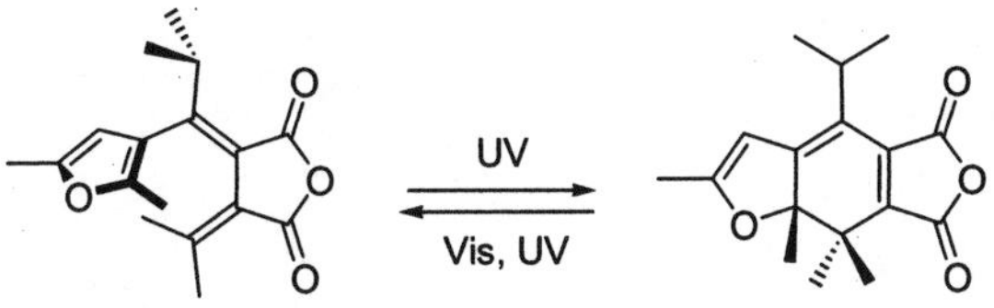

F-1E   F-1C

involved, this process induces the thermal racemization. The process was proved to occur by a "belly-roll" type motion of the furan ring over the isopropylidne group followed by passing of the furan by the isopropylidene methyl group. The activation enthalpy of this process (51.5 kJ mol^{-1}), calculated by the AM1 semiempirical MO method, was in good agreement with the experimental value (Scheme 3).

In order to obtain an optically resolvable fulgide, the introduction of sterically demanding substituents was required. Thermal stability against the racemization was examined with several fulgides by Yokoyama et al., and **F-2** was found to be optically resolvable [24]. Figure 2 shows the change in absorption spectra of **F-2** upon 405 nm light irradiation (Fig. 2). Although the racemization was negligible at room temperature, the racemization was observed at elevated temperature. The activation energy of thermal racemization in aromatic solvents was 107 kJ mol^{-1}. Racemization also occurred by continuous irradiation of UV light.

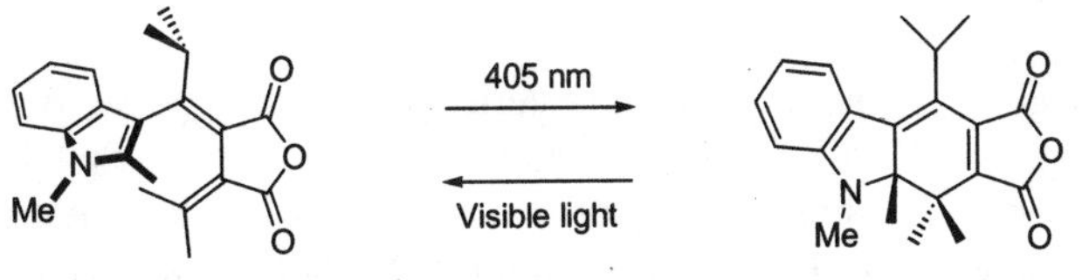

F-2E   F-2C

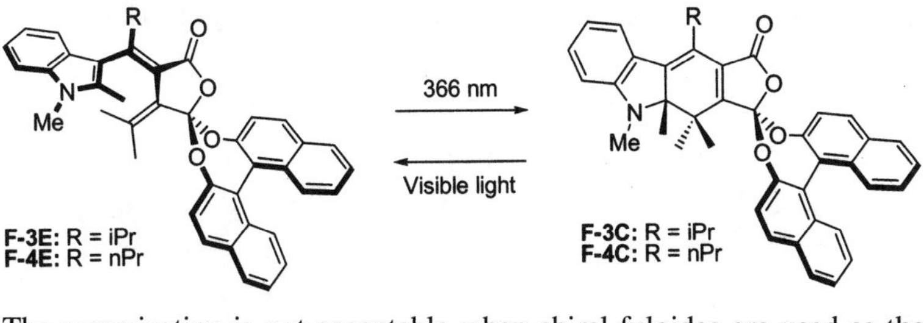

F-3E: R = iPr   F-3C: R = iPr
F-4E: R = nPr   F-4C: R = nPr

The racemization is not acceptable when chiral fulgides are used as the optical switches. If a photochemically as well as thermally stable intrinsic chirality is

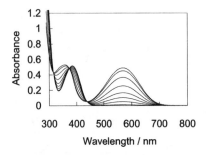

Figure 2   Absorption spectral change of **F-2** (E to C) upon irradiation with 405 nm light on toluene solution. Concentration: 8.42×10^{-5} mol dm^{-3}. Irradiation time (min): 0, 0.5, 2, 4.2, 7.4, 13, 23, 42, 71.

introduced in a photochromic molecule, and the chirality induces another chirality diastereoselectively during photochromic reaction, it is much more reliable. When an (*R*)-binaphthol was introduced to fulgide molecules in a form of acetal (**F-3** and **F-4**) by Yokoyama et al., one of the naphthalene rings pinned down the outer methyl group of the isopropylidene group so that the hexatriene moiety took (*P*)-helicity [2]. Therefore the photocyclization of **F-3E** and **F-4E** by UV irradiation generated **F-3C** and **F-4C**, respectively, in which the absolute configuration of the newly formed stereogenic carbon atom of the major diastereomer was supposed to be mostly *S*. Indeed, the diastereomeric excess (*de*) was 92% for **F-3** and 90% for **F-4** (Scheme 4). Upon iterative UV and visible light irradiation to a PMMA film containing **F-3** or **F-4**, reversible change of optical rotation was observed (Fig. 3) [2].

As **F-4** is chiral, it induced the cholesteric LC state by adding to nematic LC as a chiral dopant [25,26]. Upon light irradiation, the cholesteric pitch length was changed reversibly. A similar change was observed for **F-2**, though it showed much smaller changes than **F-3** and **F-4**. Similar cholesteric pitch change was observed by doping a racemic indolylfulgide to a cholesteric LC [27].

Similar diastereoselective photochromism was observed for **F-5**, in which a chiral xylenediol was condensed onto the acid anhydride moiety of an indolylfulgide [28].

F-5E   F-6E

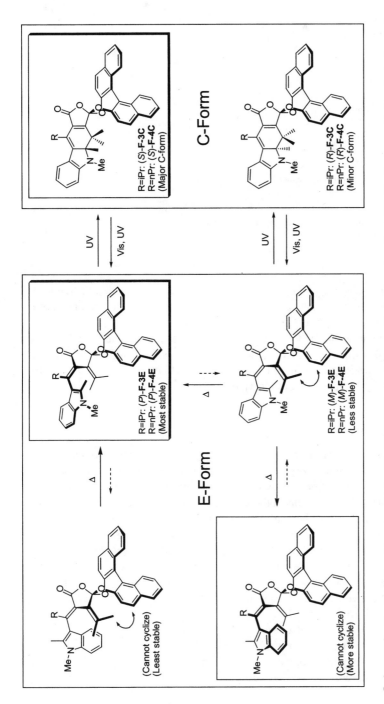

Scheme 4 Conformational isomers obtained by PM-3 MO calculation method and explanation of highly diastereoselective photochemical cyclization of **F-3** and **F-4**.

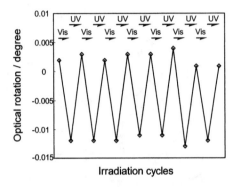

Figure 3 Optical rotation change of **F-3** in a PMMA film upon iterative irradiation of visible light (> 470 nm) and UV light (366 nm).

The diastereoselective photochromism was observed for bis(thienylethylidene)fulgide **F-6** by Yokoyama et al., although the resolution of the enantiomers of the colorless forms was impossible at room temperature [29].

The relationship between the absolute stereostructure and the CD spectral characteristics was elucidated by calculation for **F-2** and **F-4** [30]. Although the calculated wavelengths of the maximum Cotton effects were not very accurate, the shapes of the spectra were nicely reproduced.

B. Diarylethenes

Photochromic reactions of *cis*-1,2-diarylethenes are the extension of photochemical electrocyclization of *cis*-stilbene, which yields dihydrophenanthrene. When the aromatic stabilization energy (aromaticity) of at least one aryl group is low (such as furan, thiophene, benzothiophene) and the nonhydrogen substituents are located on the ring-forming carbon atoms, the thermally irreversible photochromism is observed. When the aromaticity of both aryl groups is high (such as phenyl, indolyl, or pyrryl), the diarylethene is thermally reversible [31].

As with fulgides, the photochromism of diarylethenes is also based on 6π-electrocyclization. Therefore the helical conformation of the open form is transferred to two newly formed stereogenic carbon atoms of the cyclohexadiene moiety by UV irradiation (Scheme 5).

The first example of diastereoselective photochromism of a diarylethene was reported by Irie's group for the bisbenzothienylethene **D-1** possessing a-(*l*)- or (*d*)-menthyloxy substitutent on C-2 of one of the two benzothiophene rings [32]. The diastereoselectivity was dependent on solvent polarity and temperature

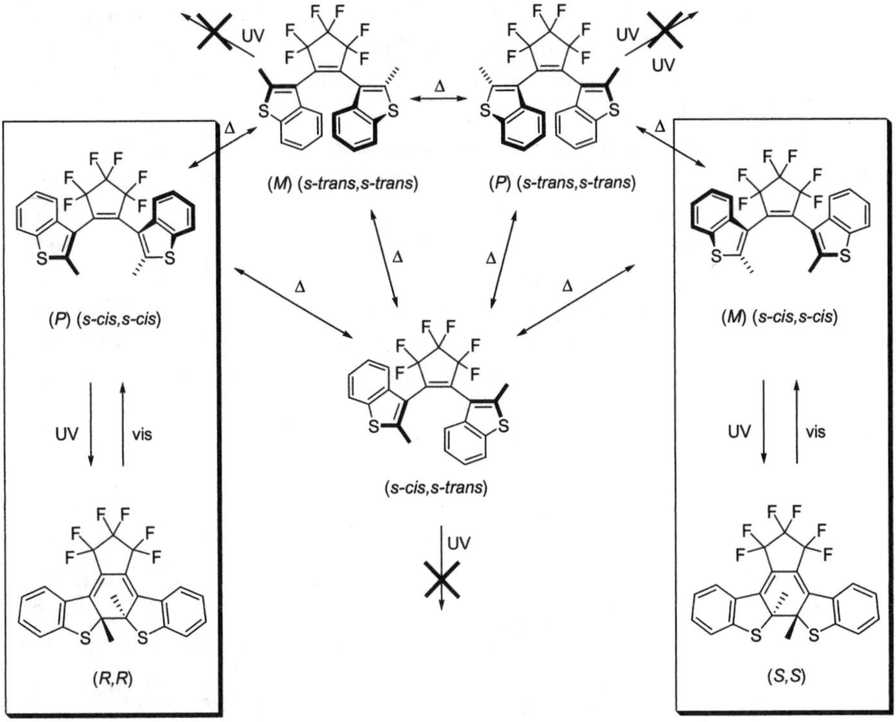

Scheme 5 Phtochochromism of diarylethenes on the basis of stereochemistry.

of photoirradiation. At room temperature, the highest diastereoselectivity reported was 62.8% in a hexane–THF mixture. In toluene at −40°C, the selectivity was as high as 86.6% *de*.

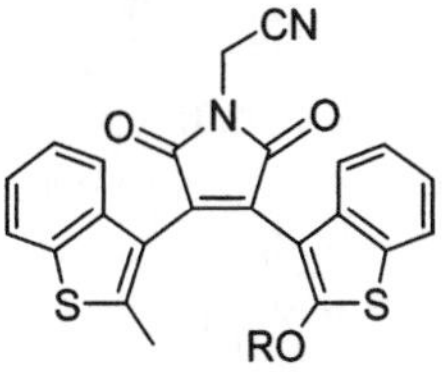

R = *d*- or *l*-menthyl

D-1

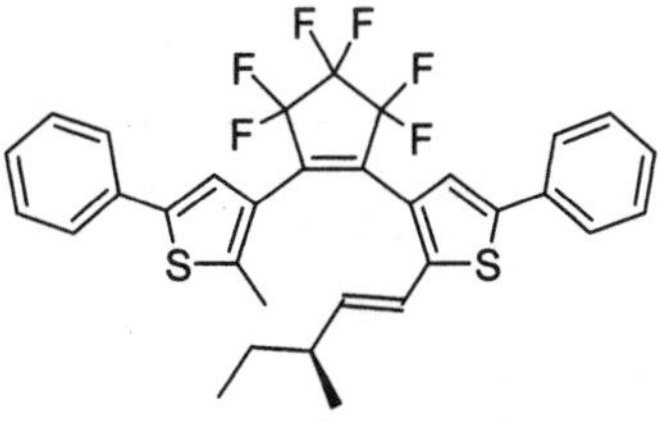

(*S*)-**D-2**

Irie et al. reported that a bisthienylethene possessing (*R*)- or (*S*)-3-methyl-1-penten-1-yl group on C-2 of one of the two thiophene rings ((*S*)-**D-2** and (*R*)-**D-**

2) showed a remarkable diastereoselective photochromic cyclization reaction in single crystals (almost 100% *de* when the conversion was not more than 6%). However, when the conversion ratio was greater than 10%, the diastereomer excess began to decrease [33].

When two diarylethene molecules are attached to (*1R,2R*)-bisdiaminocyclohexane through aldimine double bonds, the resulting **D-3** worked as the chiral dopant to generate the cholesteric LC state when doped in 4-cyano-4'-pentylbiphenyl (5CB), and the cholesteric pitch changed upon photoirradiation. However, the stereoselectivity of the ring closure was not described [34,35]. Similarly, two binaphthol-attached diarylethenes **D-4** and **D-5** were reported. Both of them have two diarylethene units on both naphthol rings. However, no description on diastereoselectivity was given for **D-4**, and no diastereoselectivity of ring closure for **D-5** was reported [36].

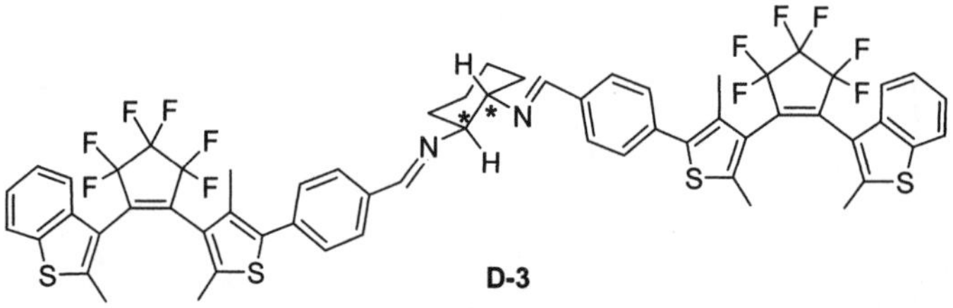

D-3

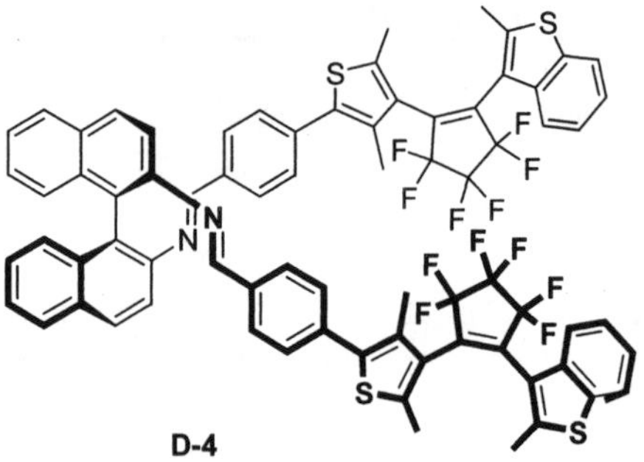

D-4

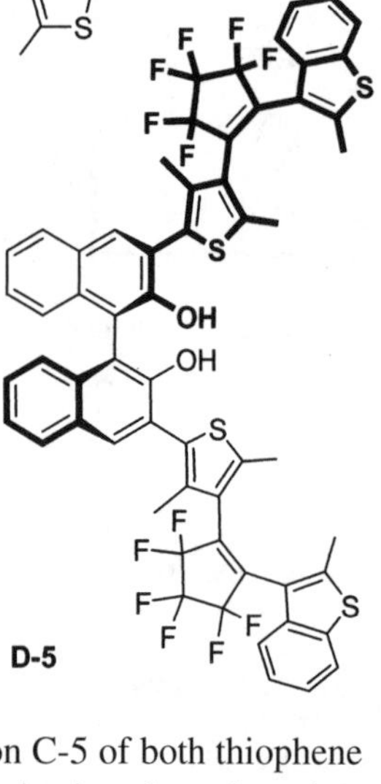

D-5

A bisthienylethene possessing a chiral oxazoline group on C-5 of both thiophene rings (**D-6**), reported by Branda et al., showed diastereoselective photochromism when Cu(I) salt was present [37]. It recorded 98% *de* at 10^{-3} mol dm^{-3} concentration, though decomposition was reported to occur. At 10^{-4} mol dm^{-3} level, the *de* was 86–89%.

A bisthienylethene **D-7** with imines of (*S*)-1-phenylethylamine on C-5 of both thiophene rings showed a slight diastereoselectivity. It was proved by the induction of cholesteric LC phase [38] or ^{1}H NMR spectra with a shift reagent [39].

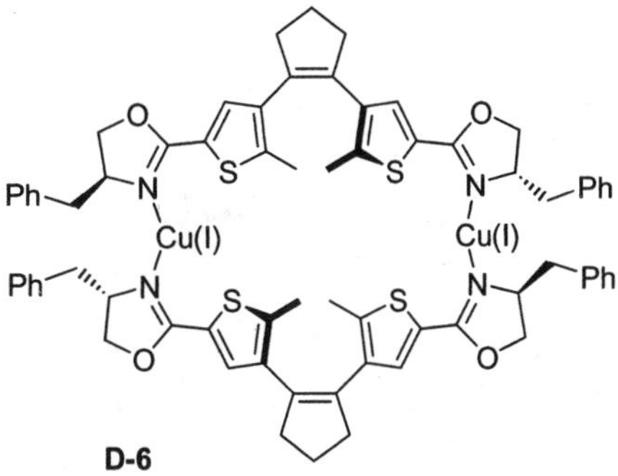

D-6

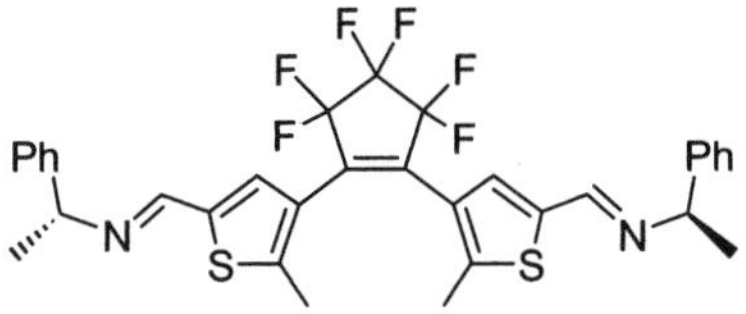

D-7

A unique metacyclophane-type 1,2-diphenylmaleimide **D-8** was prepared by Takeshita and Yamato. Because of steric congestion, **D-8** can be resolved into enantiomers possessing axial chirality. After optical resolution, the enantiomers were thermally stable and showed photochromism while preserving their chirality [40].

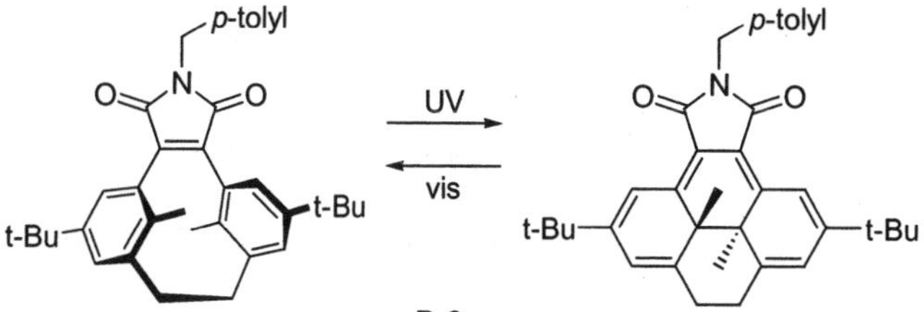

D-8

A diastereoselective photochromism of a bisbenzothienylethene was reported by Yokoyama et al. By employing $A^{1,3}$-strain around the double bond of one of the benzothiophene of **D-9**, the conformation of the stereogenic carbon atom with regard to the hexafluorocyclopentene ring was fixed. As a result, the second benzothiophene lay closer to the medium-sized substituent so that it would generate one diastereomer predominantly by photoirradiation. Indeed, the diastereoselectivity of photocyclization of **D-9** in toluene at room temperature was 88% *de* at 85% conversion [41].

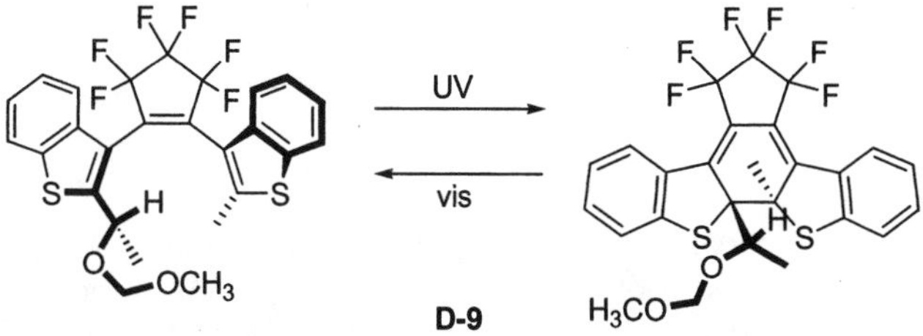

D-9

Two reports dealing with optical resolution and CD spectra of the colored forms

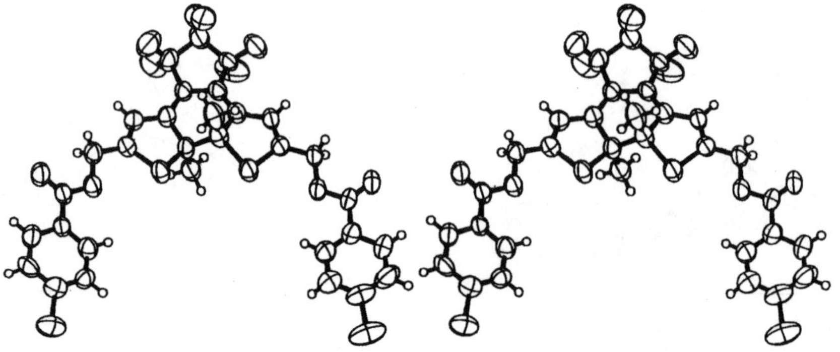

Figure 4   Stereoview of ORTEP drawing of bis(4-chlorobenzoate) of **D-10**.

of diarylethenes (**D-10** and **D-11**) have appeared (Chart 9) [42,43]. Figure 4 shows
the stereoview of bis(4-chlorobenzoate) of (*S,S*)-**D-10**.

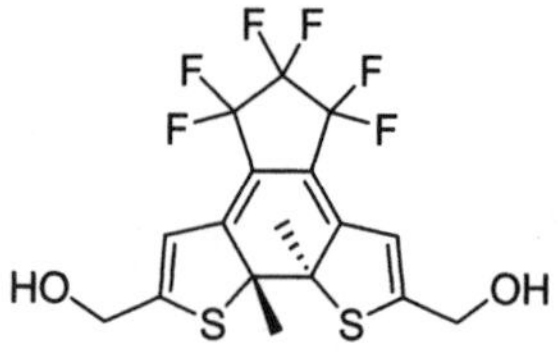

(*R,R*)-**D-10**

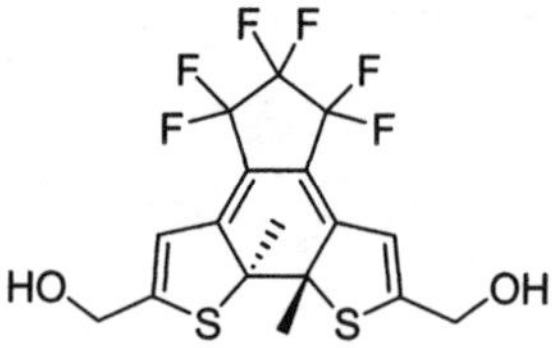

(*S,S*)-**D-10**

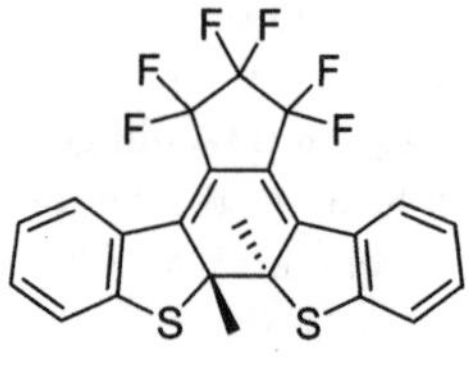

(*R,R*)-**D-11**

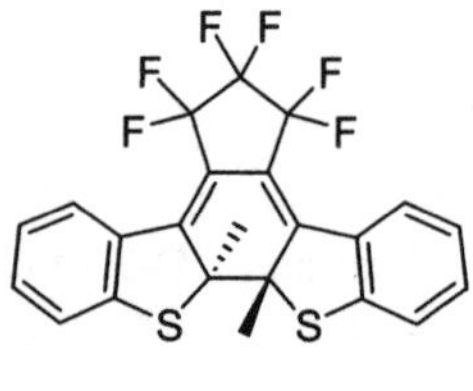

(*S,S*)-**D-11**

C.   Spiropyrans

The first example of optical resolution of a photochromic compound was done
with spiropyrans **SP-1**. During the photochemical ring opening and subsequent
thermal ring closing reactions, **SP-1** preserved chirality, although it had a chance
to racemize when it took merocyanine structures [44–46].

	R¹	R²	R³
SP-1a	CH$_3$	OCH$_3$	H
SP-1b	H	OCH$_3$	NO$_2$
SP-1c	H	CH$_3$	H
SP-1d	H	Cl	H

(S)-SP-1

Spiropyrans **SP-2**, reported by Miyashita et al., forming chromium complexes on their indoline phenyl ring, can possibly take two diastereomeric spiro forms [47]. Because of the favorable interaction of the pyran oxygen atom and the chromium atom, the thermal ring closure occurred diastereomerically to give only one diastereomer from the planar merocyanine form generated by photoirradiation.

a: R= H
b: R= CH$_2$=C(CH$_3$)COOCH$_2$

A fairly diastereoselective ring closure (1.6 : 1) was observed for an enantiomerically enriched spiropyran **SP-3** (optical purity not reported) [48]. The oxygen atom of the benzopyran ring was shown, by NOE experiments, to prefer the methyl side rather than the propyl side.

When the most common spiropyran **SP-4** was mixed in poly(γ-benzylglutamate) to prepare a film, a reversible change in optical rotation was observed upon 365 nm light irradiation and visible light irradiation [49]. The largest $\Delta\alpha$ before and after UV irradiation at sodium D-line (589 nm) was 1.225° for 4 mol% concentration. It was supposed that the change might have arisen from the chiral interaction between the chiral polypeptide environment and the colored merocyanine. Indeed, an induced CD was observed.

Like azobenzenes (III.A), polypeptides carrying spiropyrans have been studied to control the conformations of polypeptides. In polar solvents such as hexafluoropropanol, the colored merocyanine form is thermally more stable than the spiro form. Therefore it shows reverse (or negative) photochromism; irradiation by visible light on the colored species bleaches it, while the spiro form formed returns to the colored merocyanine thermally [8]. For example, poly(glutamate) attaching spiropyrans on the carboxyl side chain through the nitrogen of the indoline moiety (**SP-5**) showed reversible change of the conformations between the random coil (merocyanine form before visible light irradiation) and α-helix (spiro form after visible light irradiation), evidenced by CD spectral change [50–52]. The change in conformation was also affected by the solvent system employed [53,54].

Unlike poly(glutamate), spiropyran-modified poly(lysine) **SP-6** was always a random coil in hexafluoropropanol. However, when triethylamine was added, the change from random coil to α-helix was observed upon visible light irradiation [55,56]. This is because the ammonium group generated by rather acidic hexafluoropropanol, which prefers the extended random coil, owing to the cationic repulsion, was changed to the amino group, so that the repulsive interaction was removed.

III. CHIRAL PHOTOCHROMISM BASED ON *E-Z* ISOMERIZATION

A. Azobenzenes

Azobenzenes are the most extensively examined photochromic compounds (Scheme 6). As for the chiral photochromism of azobenzenes, however, studies

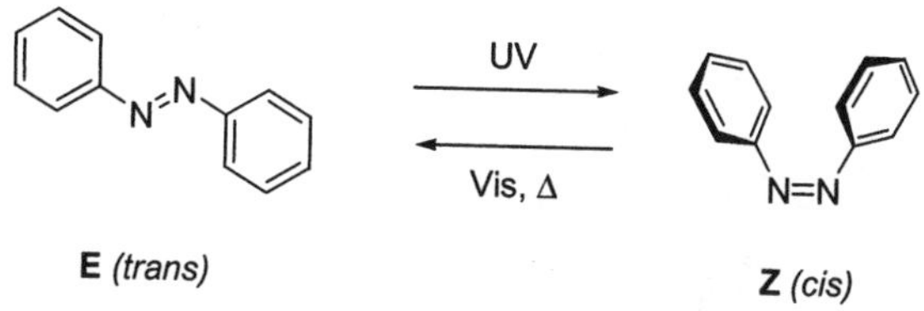

Scheme 6 Photochromism of azobenzene.

have been done mostly in two categories. One is for the control of LC states, the other for the control of helicity of macromolecules.

In 1971, Sackmann reported that the reflection light wavelength, and therefore the pitch, of cholesteric LC phase was changed by photochemical E-Z isomerization of the dopant azobenzene **A-1** [57]. Recently, the similar trials to induce the change in selective light reflection have been reported. Addition of a chiral dopant **A-2** and a chiral azobenzene **A-3** to a nematic cyanobiphenyl-type LC mixture brought about a cholesteric LC phase. Irradiation of 366 nm light caused E-Z isomerization of **A-3**, which induced the change in selective light reflection [58]. It returned to the original reflection within 1 h in the dark at ambient temperature. Irradiation of 420 nm light to cause Z-E isomerization induced quick recovery of the initial state.

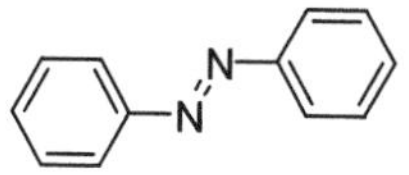

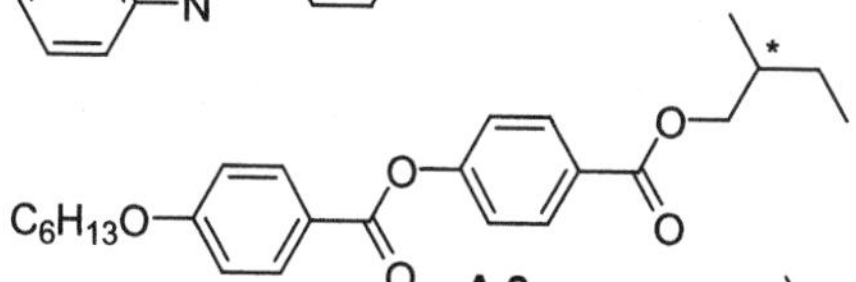

Similar results were reported for the combination of chiral azobenzenes doped in nematic liquid crystals [59,60].

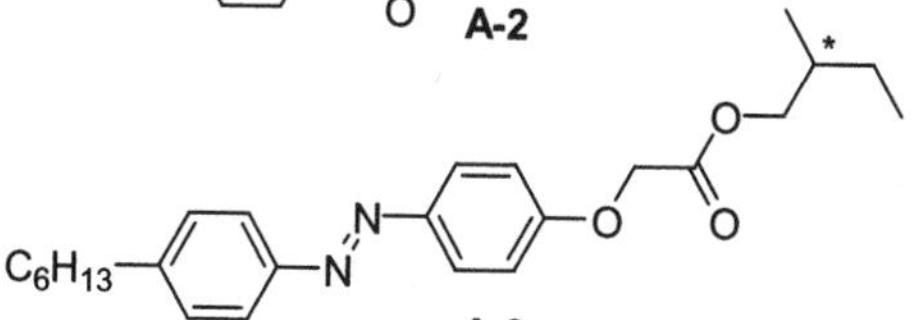

Photochemical control of properties of SmC* LC phase was achieved by doping azobenzene **A-4** possessing a chiral carbon atom to a ferroelectric LC **A-5** [61]. When the SmC* LC is in the surface stabilized state, the bulk dipole moment can be flipped by an external electric field. As the hysteresis curve for the Z form is narrower than that of E form, irradiation of UV light to cause E-

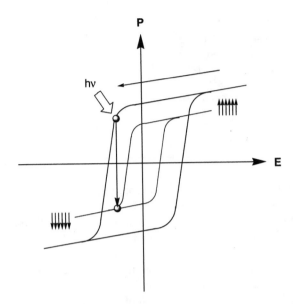

Figure 5 Polarization flip of ferroelectric LC (**A-5**) containing **A-4** by photochemical *E*/*Z* isomerization of **A-4**.

to-*Z* isomerization may cause photochemical flipping of the polarization when the LC system is under the strength of an applied electric field not sufficient to flip the polarization of the *E*-form-containing LC but sufficient to cause flipping of *Z*-form-containing LC (Fig. 5) [62].

A-4

A-5

The photochemical control of phase change of antiferroelectric LCs was also achieved either by using azobenzenes possessing chiral side chains as dopants (**A-6**) in an antiferroelectric LC or by using azobenzene-containing chiral antiferroelectric LCs (**A-7**, **A-8**) [63,64].

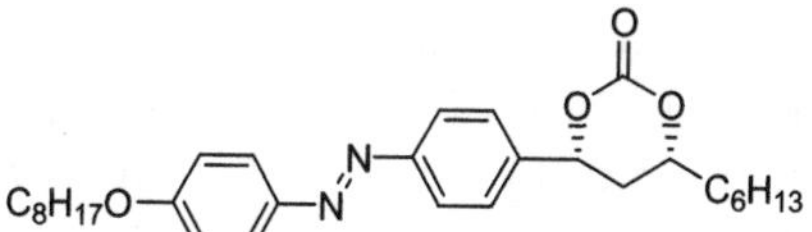

A-6

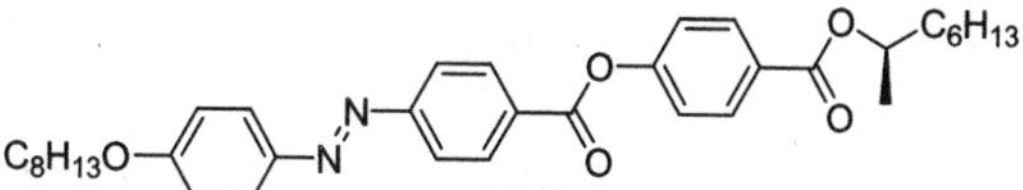

A-7

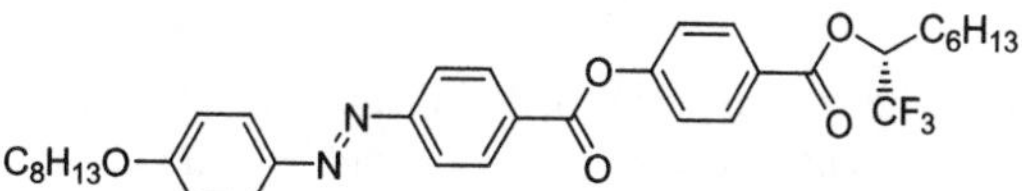

A-8

When azobenzenes are attached to polypeptides, photochromic reactions of azobenzenes can induce the change in helical properties of the polypeptides, which may be detected by CD spectrum as well as optical rotation. For 4-phenylazophenylamine-condensed poly(γ-glutamic acid) **A-9** containing up to 80 mol% of 4-phenylazophenylamide side chain, UV irradiation in organic solvents, such as

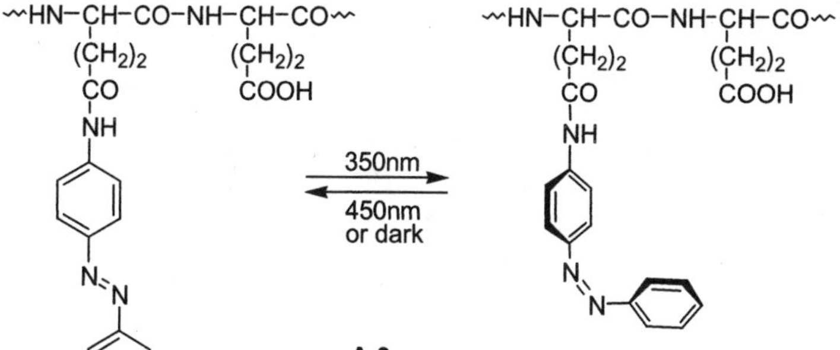

A-9

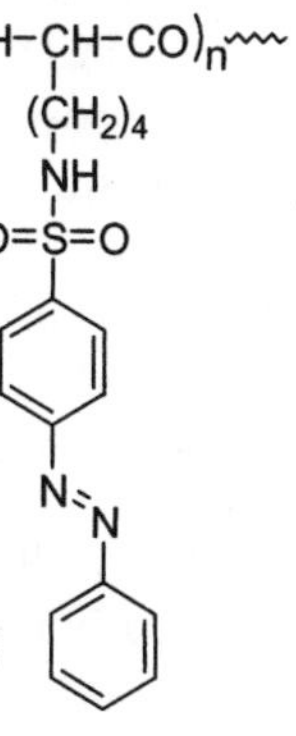

A-10

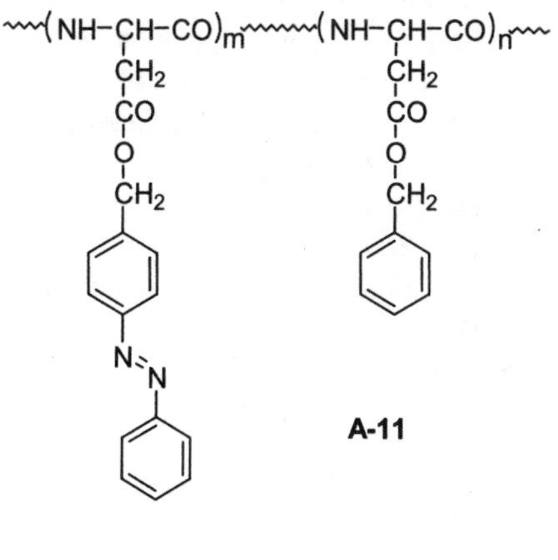

A-11

trimethylphosphate or trifluoroethanol, caused spectral changes in the region above 250 nm, which were attributed to the induced CD of the side chain, while no change was observed below 250 nm, implying that α-helix was not affected by the E-Z isomerization of the azobenzene moiety [65]. Similar results were obtained by Sisido et al. [66,67]. In aqueous solution at pH 5–7, however, UV-irradiation of **A-9** caused the E-Z isomerization, accompanying reversible helical structure changes [68,69].

As for the sulfonylazobenzene-modified poly(lysine) **A-10** at the free amino group, similar phenomena were observed in hexafluoro-2-propanol containing 2–15% methanol [70].

Azobenzene-modified poly(β-benzyl aspartate) **A-11** showed reversal of helical structure upon UV irradiation, evidenced by the sign of a CD band around 220 nm in dichloroethane, when the content of the azo-modified peptide unit was 59 or 81 mol% [71–74]. Extensive studies have been done on similar azobenzene-modified poly(aspartate) esters by Ueno and Osa [75–82].

A cyclic oligopeptide-containing azobenzene unit **A-12** took a β-turn-like structure after UV irradiation, while it was rather an extended structure before irradiation, as elucidated by NMR and molecular dynamics calculations [83].

A-12

Recently, Natansohn et al. reported that a poly(methacrylate) film **A-13**, carrying a push–pull type 4-amino-4′-nitorazobenzene derivative as the ester moiety, in the smectic A phase LC at room temperature showed mirror-imaged CD spectra upon iterative irradiation of circularly polarized light (488 nm) of opposite handedness [84,85]. When right-handed circularly polarized light was irradiated, a positive CD appeared at 488 nm. When the polarization of the light was switched to the left, it showed a negative CD within 50 seconds. After this change in CD sign was repeated several times, the amplitude of the change became smaller, probably because the orientation of some of the chromophores became perpendicular to the film surface. This phenomenon is attributed to the change in supramolecular structure in the orientation of the smectic LC domain. The optically active state can be erased either by the irradiation of the opposite circularly polarized light or by heating.

A-13

B. Overcrowded Stilbenes

It is well known that stilbene photochemically isomerizes between the E and the Z form and cyclizes to produce dihydrophenanthrene. The latter has been utilized in the photochromic diarylethene systems described above, while the former successfully yielded the overcrowded stilbenes. The first resolved overcrowded stilbene **St-1** was reported in 1991 by Feringa et al. [86]. The photochemical change in E/Z (that is P/M) ratio, though small, between (M,Z)-**St-1** and (P,E)-**St-1** and between (P,Z)-**St-1** and (M,E)-**St-1** upon irradiations at two different wavelengths, and the thermal helicity change (racemization process) between (M,Z)-**St-1** and (P,Z)-**St-1** and between (P,E)-**St-1** and (M,E)-**St-1**, were observed. The activation energy of racemization between (M,Z)-**St-1** and (P,Z)-**St-1** was 110 kJ mol^{-1}. Upon the photochemical change in E/Z ratio, the molar ellipticity of the CD spectra changed.

This work has been extended to compound **St-2**, which possesses electron-donating and electron-withdrawing substituents [87]. When (*M,Z*)-**St-2** was incorporated into a nematic LC, 4′-cyano-4-pentyloxybiphenyl, a cholesteric phase was induced. Irradiation by 365 nm light yielded another cholesteric phase, with the opposite handedness. Irradiation at 435 nm generated the cholesteric phase of the initial handedness. Upon irradiation at 313 nm, the nematic phase was induced, because of the generation of the same amount of (*M,Z*)-**St-2** and (*P,E*)-**St-2** [88].

When a nematic LC phase doped with racemic **St-3** was irradiated with left-handed circularly polarized 313 nm light, (*M*)-**St-3** was produced in slight excess as a result of *E-Z* isomerization. Upon irradiation of right-handed circularly polarized 313-nm light, (*P*)-**St-3** became slightly abundant. As a result, the nematic phase changed to the cholesteric phases with different handedness when irradiated with the circularly polarized light of opposite handedness. Although the enantiomeric excess was merely 0.07%, the induced cholesteric phases were visually recognized by the characteristic texture [89].

The overcrowded stilbenes were beautifully elaborated to build a unidirectional molecular rotor [90]. When the resolved (*P,P,E*)-**St-4** was irradiated with > 280 nm light, an *E-Z* isomerization occurred to give (*M,M,Z*)-**St-4**. However, as it was thermally rather unstable because of the steric conflict caused by the two methyl groups, it isomerized to thermally more stable (*P,P,Z*)-**St-4** at 20°C. Subsequent irradiation with > 280 nm light induced another *E-Z* isomerization to yield (*M,M,E*)-**St-4**, which is again thermally unstable. At 60°C, it changed its conformation to regenerate the starting material (*P,P,E*)-**St-4**. Although *E-Z* isomerization occurs reversibly by > 280 nm light and > 380 nm light irradiation, the thermal processes are irreversible. Therefore, the tetrahydrophenanthrene group rotates in the unidirectional manner with regard to the other tetrahydrophenanthrene moiety.

C. Thioindigo

A chiral photochromic thioindigo **T-1** was reported by Lemieux et al. When it was doped in racemic LC **T-2**, it successfully altered the spontaneous polarization of the induced ferroelectric LC (Sc*) by photochemical *E-Z* isomerization [91,92].

IV. CONCLUDING REMARKS

We have briefly surveyed the chirality in photochromism. As there are a variety of strategies to incorporate chirality in photochromic systems, it is difficult to describe them in a systematic manner. In this chapter, we categorized the chiral photochromic systems in terms of the reaction modes and compounds in order to show the latest trends of the relevant photochromic compounds. Those who wish to study this field should refer to general reviews and books on photochromism [1,93,94] and become familiar with the compounds first. The book by Feringa [95] should be consulted also.

To date, organic photochromic compounds have been used for the autoregulation of ophthalmic plastic lenses and for little else. But chiral photochromic compounds hold more information than nonchiral or racemic photochromic compounds, so they have a great advantage over them. When a photochromic compound is used as a switch, its chiral derivative will be even more useful.

REFERENCES

1.   Dürr H, Bouas-Laurent H. Photochromism: Molecules and Systems. Amsterdam: Elsevier, 1990.
2.   Yokoyama Y, Uchida S, Yokoyama Y, Sugawara Y, Kurita Y. J Am Chem Soc 1996; 118:3100–3107.
3.   Dürr H, Bouas-Laurent H. Photochromism: Molecules and Systems. Amsterdam: Elsevier, 1990, pp. 193–513.
4.   Dürr H, Bouas-Laurent H. Photochromism: Molecules and Systems. Amsterdam: Elsevier, 1990, pp. 64–192.
5.   Dürr H, Bouas-Laurent H. Photochromism: Molecules and Systems. Amsterdam: Elsevier, 1990, pp. 713–737.
6.   Dürr H, Bouas-Laurent H. Photochromism: Molecules and Systems. Amsterdam: Elsevier, 1990, pp. 654–712.
7.   Dürr H, Bouas-Laurent H. Photochromism: Molecules and Systems. Amsterdam: Elsevier, 1990, pp. 514–560.
8.   Ciardelli F, Pieroni O. Photoswitchable polypeptides In Feringa BL, Ed. Molecular Switches. Weinheim: Wiley-VCH, 2001:399–441.
9.   Feringa BL, van Delden RA, Koumura N, Geertsema EM. Chem Rev 2000; 100: 1789–1816.
10. Feringa BL, van Delden RA, ter Wiel MKJ. Chiroptical molecular switch In Feringa BL, Ed. Molecular Switches. Weinheim: Wiley-VCH, 2001:123–163.
11. Yokoyama Y. Chem Rev 2000; 100:1717–1739.
12. Yokoyama Y. Molecular switch with photochromic fulgides In Feringa BL, Ed. Molecular Switches. Weinheim: Wiley-VCH, 2001:107–121.
13. Fan MG, Yu L, Zhao W. Fulgide family compounds synthesis photochromism and applications In Crano JC , Guglielmetti RJ, Eds. Organic Photochromic and Ther-

mochromic Compounds. Main Photochromic Families. Vol. Vol. 1. New York: Plenum Press, 1999:141–206.

14. Irie M. Chem Rev 2000; 100:1685–1716.

15. Irie M. Photoswitchable molecular systems based on diarylethenes In Feringa BL, Ed. Molecular Switches. Weinheim: Wiley-VCH, 2001:37–62.

16. Irie M. Diarylethenes with Heterocyclic Aryl Groups In Crano JC, Guglielmetti RJ, Eds. Organic Photochromic and Thermochromic Compounds. Main Photochromic Families. Vol. Vol. 1. New York: Plenum Press, 1999:207–222.

17. Berkovic G, Krongauz V, Weiss V. Chem Rev 2000; 100:1741–1753.

18. Bertelson RC. Spiropyrans In Crano JC, Guglielmetti RJ, Eds. Organic Photochromic and Thermochromic Compounds. Main Photochromic Families. Vol. Vol. 1. New York: Plenum Press, 1999:11–83.

19. Heller HG, Megit RM. J Chem Soc Perkin Trans 1974; 1:923–927.

20. Darcy PJ, Heller HG, Strydom PJ, Whittall J. J Chem Soc Perkin Trans 1981; 1: 202–205.

21. Kaftory M. Acta Crystalogr 1984; 40:1015–1019.

22. Yokoyama Y, Iwai T, Yokoyama Y, Kurita Y. Chem Lett 1994:225–226.

23. Yokoyama Y, Ogawa K, Iwai T, Shimazaki K, Kajihara Y, Goto T, Yokoyama Y, Kurita Y. Bull Chem Soc Jpn 1996; 69:1605–1612.

24. Yokoyama Y, Shimizu Y, Uchida S, Yokoyama Y. J Chem Soc Chem Commun 1995:785–786.

25. Yokoyama Y, Sagisaka T. Chem Lett 1997:687–688.

26. Sagisaka T, Yokoyama Y. Bull Chem Soc Jpn 2000; 73:191–196.

27. Janicki SZ, Schuster GB. J Am Chem Soc 1995; 117:8524–8527.

28. Yokoyama Y, Okuyama T, Yokoyama Y, Asami M. Chem Lett 2001:1112–1113.

29. Yokoyama Y, Sagisaka T, Yamaguchi Y, Yokoyama Y, Kiji J, Okano T, Takemoto A, Mio S. Chem Lett 2000:220–221.

30. Ankai E, Sakakibara K, Uchida S, Uchida Y, Yokoyama Y, Yokoyama Y. Bull Chem Soc Jpn 2001; 74:1101–1108.

31. Nakamura S, Irie M. J Org Chem 1988; 53:6136–6138.

32. Yamaguchi T, Uchida K, Irie M. J Am Chem Soc 1997; 119:6066–6071.

33. Kodani T, Matsuda K, Yamada T, Kobatake S, Irie M. J Am Chem Soc 2000; 122: 9631–9637.

34. Yamaguchi T, Nakazumi H, Uchida K, Irie M. Chem Lett 1999:653–654.

35. Yamaguchi T, Inagawa T, Nakazumi H, Irie S, Irie M. Mol Cryst Liq Cryst 2000; 345:287–292.

36. Yamaguchi T, Inagawa T, Nakazumi H, Irie S, Irie M. J Mater Chem 2001; 11: 2453–2458.

37. Murguly E, Norsten TB, Branda NR. Angew Chem Int Ed 2001; 40:1752–1755.

38. Denekamp C, Feringa BL. Adv Mater 1998; 10:1080–1082.

39. Fernandez-Acebes A. Chirality 2000; 12:149–152.

40. Takeshita M, Yamato T. Angew Chem Int Ed 2002; 41:2156–2157.

41. Yokoyama Y, Shiraishi H, Tani Y, Yokoyama Y, Yamaguchi Y. J Am Chem Soc 2003; 125:7194–7195.

42. Yokoyama Y, Hosoda N, Osano YT, Sasaki C. Chem Lett 1998:1093–1094.

43. Yamaguchi T, Tanaka Y, Nakazumi H. Enantiomer 2001; 6:309–311.
44. Stephan B, Mannschreck A, Voloshin NA, Volbushko NV, Minkin VI. Tetrahedron Lett 1990; 31:6335–6338.
45. Stephan B, Zinner H, Kastner F, Mannschreck A. Chimia 1990; 44:336–338.
46. Leiminer A, Stephan B, Mannschreck A. Mol Cryst Liq Cryst 1994; 246:215–221.
47. Miyashita A, Iwamoto A, Kuwayama T, Shitara H, Aoki Y, Hirano M, Nohira H. Chem Lett 1997:965–966.
48. Eggers L, Buss V. Angew Chem Int Ed Engl 1997; 36:881–883.
49. Suzuki Y, Ozawa K, Hosoki A, Ichimura K. Polym. Bull 1987; 17:285–291.
50. Cooper TM, Obermeier KA, Ntarajan LV, Crane RL. Photochem Photobiol 1992; 55:1–7.
51. Angelini N, Corrias B, Fissi A, Pieroni O, Lenci F. Biophys J 1998; 74:2601–2610.
52. Pachter R, Cooper TM, Natarajan LV, Obermeier KA, Crane RL. Biopolymers 1992; 32:1129–1140.
53. Fissi A, Pieroni O, Ciardelli F, Fabbri D, Ruggeri G, Umezawa K. Biopolymers 1993; 33:1505–1517.
54. Sato M, Fujii Y, Kato F, Komiyama J. Biopolymers 1991; 31:1–10.
55. Pieroni O, Fissi A, Viegi A, Fabri D, Ciardelli F. J Am Chem Soc 1992; 114: 2734–2736.
56. Fissi A, Pieroni O, Ruggeri G, Ciardelli F. Macromolecules 1995; 28:302–309.
57. Sackmann E. J Am Chem Soc 1971; 93:7088–7090.
58. Lee HK, Doi K, Harada H, Tsutsumi O, Kanazawa A, Shiono T, Ikeda T. J Phys Chem B 2000; 104:7023–7028.
59. Ruslim C, Ichimura K. J Chem Phys B 2000; 104:6529–6535.
60. Kurihara S, Nomiyama S, Nonaka T. Chem Mater 2001; 13:1992–1997.
61. Ikeda T, Sasaki T, Ichimura K. Nature 1993; 361:428–430.
62. Sasaki T, Ikeda T, Ichimura K. J Am Chem Soc 1994; 116:625–628.
63. Negishi M, Tsutsumi O, Ikeda T, Hiyama T, Kawamura J, Aizawa M, Takeshita S. Chem Lett 1996:319–320.
64. Negishi M, Kanie K, Ikeda T, Hiyama T. Chem Lett 1996:583–584.
65. Houben JL, Fissi A, Bacciola D, Rosato N, Pieroni O, Ciardelli F. Int J Biol Macromol 1983; 5:94–100.
66. Sisido M, Ishikawa Y, Itoh K, Tazuke S. Macromolecules 1991; 24:3993–3998.
67. Sisido M, Ishikawa Y, Harada M, Itoh K. Macromolecules 1991; 24:3999–4003.
68. Pieroni O, Houben JL, Fissi A, Costantino P, Ciardelli F. J Am Chem Soc 1980; 102:5913–5915.
69. Ciardelli F, Pieroni O, Fissi A, Houben JL. Biopolymers 1984; 23:1423–1437.
70. Fissi A, Pieroni O, Balestreri E, Amato C. Macromolecules 1996; 29:4680–4685.
71. Ueno A, Anzai J, Osa T, Kadoma Y. J Polym Sci Polym Letters 1977; 15:407–410.
72. Ueno A, Anzai J, Osa T, Kadoma Y. Bull Chem Soc Jpn 1977; 50:2995–2999.
73. Ueno A, Anzai J, Osa T, Kadoma Y. Bull Chem Soc Jpn 1979; 52:549–554.
74. Ueno A, Anzai J, Osa T. J Polym Sci Polym Letters 1979; 17:149–154.
75. Ueno A, Takahashi K, Anzai J, Osa T. Macromolecules 1980; 13:459–460.
76. Ueno A, Takahashi K, Anzai J, Osa T. Bull Chem Soc Jpn 1980; 53:1988–1992.
77. Ueno A, Takahashi K, Anzai J, Osa T. Chem Lett 1981:113–116.

78. Ueno A, Takahashi K, Anzai J, Osa T. Macromol Chem 1981; 182:693–695.
79. Ueno A, Takahashi K, Anzai J, Osa T. J Am Chem Soc 1981; 103:6410–6415.
80. Ueno A, Nakamura J, Adachi K, Osa T. Makromol Chem Rapid Commun 1989; 10:683–686.
81. Ueno A, Adachi K, Nakamura J, Osa T. J Polymer Sci: Polymer Chem 1990; 28: 1161–1170.
82. Ueno A, Takahashi K, Anzai J, Osa T. Makromol Chem Rapid Commun 1984; 5: 639–642.
83. Ulysse L, Cubillos J, Chmielewski J. J Am Chem Soc 1995; 117:8466–8467.
84. Iftime G, Labarthet FL, Natansohn A, Rochon P. J Am Chem Soc 2000; 122: 12646–12650.
85. Natansohn A, Wu Y, Rochon P. Polym Prep Am Chem Soc Div Polym Chem 2002; 43:55–56.
86. Feringa BL, Jager WF, de Lange B, Meijer EW. J Am Chem Soc 1991; 113: 5468–5470.
87. Jager WF, de Jong JC, de Lange B, Huck NPM, Meetsma A, Feringa BL. Angew Chem Int Ed Engl 1995; 34:348–350.
88. Feringa BL, Huck NPM, van Doren HA. J Am Chem Soc 1995; 117:9929–9930.
89. Huck NPM, Jager WF, de Lange B, Feringa BL. Science 1996; 273:1686–1688.
90. Koumura N, Zijlstra RWJ, van Delden RA, Harada N, Feringa BL. Nature 1999; 401:152–155.
91. Dinescu L, Lemieux RP. J Am Chem Soc 1997; 119:8111–8112.
92. Saad B, Galstyan TV, Dinescu L, Lemieux RP. Chem Phys 1999; 245:395–405.
93. Chem. Special Thematic Issue, "Photochromism: Memories and Switches.". Vol. 100, 2000:1685–1890.
94. Crano JC, Guglielmetti RJ, eds. Organic Photochromic and Thermochromic Compounds. Vol. Vols. 1 and 2. New York: Plenum Press, 1999.
95. Feringa BL ed. Molecular Switches. Weinheim: Wiley-VCH, 2001.

7

Chiral Photochemistry with Transition Metal Complexes

Shigeyoshi Sakaki
Kyoto University
Kyoto, Japan

Taisuke Hamada
Okinawa National College of Technology
Okinawa, Japan

I. INTRODUCTION

One of the challenging research subjects in chemistry is to perform highly stereo-selective reactions. In particular, it is not easy to achieve high stereoselectivity in photoinduced electron transfer reactions, since the contact between substrates is in general very weak in such an outer-sphere electron transfer reaction as the photoinduced electron transfer reaction.

It was not very long ago that stereoselectivity in the thermal outer-sphere electron transfer reaction was reliably observed by Geselowitz and Taube (1980) [1]. Before their study, stereoselectivity had not been clearly observed even in the thermal outer-sphere electron transfer reaction. For instance, the stereoselectivity was reported in a thermal outer-sphere electron transfer reaction between $[Co(phen)_3]^{3+}$ and $[Cr(phen)_3]^{2+}$ [phen = 1,10-phenanthroline; see Eq.(1)] [2], but the stereoselectivity was not observed by the different group [3], where Δ- and Λ-forms of $[M(phen)_3]^{3+}$ are shown in Scheme 1.

$$[Co(phen)_3]^{3+} + [Cr(phen)_3]^{2+} \rightarrow [Co(phen)_3]^{2+} + [Cr(phen)_3]^{3+} \quad (1)$$

The absence of stereoselectivity was explained in terms of the racemization of $[Cr(phen)_3]^{2+}$ [Eq. (2a)] and the self-exchange reaction between $[Cr(phen)_3]^{2+}$

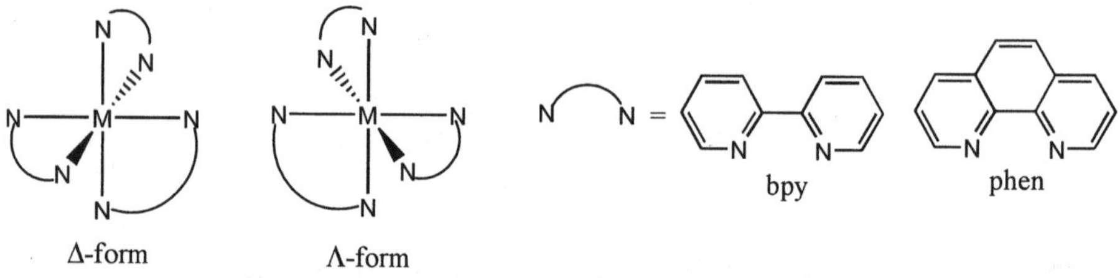

Δ-form   Λ-form   bpy   phen

Scheme 1

and $[Cr(phen)_3]^{3+}$ [Eq. (2b)], as follows:

$$\Delta\text{-}[Cr(phen)_3]^{2+} \rightarrow \Delta \text{ and } \Lambda\text{-}[Cr(phen)_3]^{2+} \tag{2a}$$

$$\Delta\text{-}[Cr(phen)_3]^{3+} + [Cr(phen)_3]^{2+} \rightarrow \Delta\text{-}[Cr(phen)_3]^{2+} + [Cr(phen)_3]^{3+} \tag{2b}$$

Even if $\Delta\text{-}[Cr(phen)_3]^{3+}$ was formed more rapidly than $\Lambda\text{-}[Cr(phen)_3]^{3+}$ in Eq. (1), $\Delta\text{-}[Cr(phen)_3]^{3+}$ underwent the self-exchange reaction with $[Cr(phen)_3]^{2+}$ [Eq. (2b)] to afford $\Delta\text{-}[Cr(phen)_3]^{2+}$, which racemized to Δ- and Λ-forms, as shown in Eq. (2a). As a result, $\Delta\text{-}[Cr(phen)_3]^{3+}$ formed in excess converts to Δ- and $\Lambda\text{-}[Cr(phen)_3]^{3+}$. Thus the stereoselectivity disappears.

In 1980, Geselowitz and Taube carried out an elegant experiment in which they oxidized $\Delta\text{-}[Os(bpy)_3]^{2+}$ (bpy = 2,2′-bipyridine) with $S_2O_8^{2-}$ and synthesized $\Delta\text{-}[Co(edta)]^-$ from Co(II) and $Na_2(edta)\cdot2HClO_4$ (edta^{4-} = ethylenediamine tetraacetate) with $\Delta\text{-}[Os(bpy)_3]^{3+}$, as shown in Eq. (3) [1].

$$(1/2)S_2O_8^{2-} + \Delta\text{-}[Os(bpy)_3]^{2+} \rightarrow SO_4^{2-} + [Os(bpy)_3]^{3+} \tag{3a}$$

$$\Delta\text{-}[Os(bpy)_3]^{3+} + Co^{2+} + edta^{4-} \rightarrow \Delta\text{-}[Os(bpy)_3]^{2+} + [Co(edta)]^- \tag{3b}$$

The enantiomer excess (*ee*) was not large (2.9%), where the optical purity of Δ-$[Os(bpy)_3]^{2+}$ was 59%. From this result, the *ee* value was estimated to be 5.0% if optically pure $\Delta\text{-}[Os(bpy)_3]^{2+}$ was employed. The stereoselectivity was observed in similar synthesis of $[Co(edta)]^-$ from Co(II) and $Na_2(edta)\cdot2HClO_4$ with $\Delta\text{-}[Ru(bpy)_3]^{2+}$, where the *ee* value (0.4 %) was much smaller than that of the reaction with $\Delta\text{-}[Os(bpy)_3]^{2+}$. Though these syntheses have not been determined to occur through the outer-sphere electron transfer reaction, the stereoselectivity was certainly experimentally observed. In particular, the use of

$[Ru(bpy)_3]^{2+}$ in these syntheses is worthy of note because this complex is a typical photosensitizer. Thus we can expect to construct the stereoselective photo-induced electron transfer reaction with optically pure $[Ru(bpy)_3]^{2+}$.

Though it is not clear that the above-mentioned reactions are the outer-sphere electron transfer reaction, the stereoselectivity of the outer-sphere electron transfer reaction was clearly reported in the reaction between di-μ-oxo-μ-(propyl-ene diamine tetra acetato)bis[oxo molybdate(V)] and μ-amido-μ-hyper oxo-bis-[bis(ethylene diamine)cobalt(III)] [see Eq.(4)] [4]:

$$[Mo_2^V O_4(R,S\text{-pdta})]^{2-} + 2\ \Delta,\ \Delta\ \text{-}[(en)_2Co^{III}(\mu\text{-NH}_2,O_2^{(-)})Co^{III}(en)_2]^{4+}$$
$$\rightarrow 2\ \text{``Mo}^{VI}\text{pdta''} + 2\ \Delta,\ \Delta\ \text{-}[(en)_2Co^{III}(\mu\text{-NH}_2,O_2^{(2-)})Co^{III}(en)_2]^{4+}$$

$$(4)$$

Other examples have been reviewed recently [5].

These results show clearly that the stereoselective electron transfer reaction of transition metal complexes should be carefully investigated; the stereoselectivity is small even in the thermal electron transfer reaction.

To perform a highly stereoselective photoinduced electron transfer reaction, we need an optically pure photosensitizer and a substrate that do not cause photo-induced racemization. In addition, we need the substrate of which the self-exchange reaction occurs slowly or does not occur. If not, the stereoselectivity disappears, as was mentioned above.

There are several excellent photosensitizers; one of them is $[Ru(bpy)_3]^{2+}$ [6]. There are two optical isomers in this complex: one is $\Delta\text{-}[Ru(bpy)_3]^{2+}$ and the other is $\Lambda\text{-}[Ru(bpy)_3]^{2+}$, as shown in Scheme 1. Thus one can expect to perform the stereoselective electron transfer reaction with Δ- and Λ-$[Ru(bpy)_3]^{2+}$. Unfortunately, however, the racemization of $[Ru(bpy)_3]^{2+}$ is induced photochemically [7]. The reasonable way to suppress the photoracemization of this complex is to introduce the optically active organic functional group into the transition metal complexes, as will be discussed in Sec. II.B. The other photosensitizer that is useful for the photoinduced electron transfer reaction is the copper(I) complexes with 1,10-phenanthroline and their derivatives [8,9]. Zinc(II) porphyrin is also an excellent photosensitizer for photoinduced electron transfer reaction [10]. In these complexes, molecular chirality does not exist, unlike in $[Ru(bpy)_3]^{2+}$. Thus one must introduce some chiral functional group into these compounds, to use these complexes as chiral photosensitizers.

In this chapter, we will discuss what type of chiral photosensitizer is applied to stereoselective photoinduced electron transfer reaction, what type of substrate is used, and what type of chiral functional group is introduced to these compounds to construct a chiral photosensitizer. We will describe also the stereoselective photocatalytic reactions.

II. CHIRAL RUTHENIUM(II) COMPLEXES AND THEIR APPLICATION TO STEREOSELECTIVE PHOTOINDUCED ELECTRON TRANSFER REACTION

A. Characteristic Features of Ruthenium(II) Complexes in the Excited State

First, we will mention the characteristic features of $[Ru(bpy)_3]^{2+}$. The electronic structure of the excited state is schematically shown in Scheme 2, and its absorption and emission properties are summarized in Table 1. In Table 1 and Scheme 2, several important features are found: (1) the metal-to-ligand charge transfer (MLCT) absorption is observed in the visible region, while the d-d absorption is not observed in the visible region; (2) in the MLCT excited state, one-electron excitation occurs from Ru $4d_\pi$ orbital to the π^* orbital of bpy; (3) the triplet MLCT (^{3}MLCT) excited state is at a higher energy than the triplet d-d excited state, (4) the photoreactive state is the ^{3}MLCT excited state, because the intersystem crossing easily takes place owing to the large spin-orbit coupling constant of the ruthenium center; (5) the ^{3}MLCT excited state is long-lived; and (6) $[Ru(bpy)_3]^{2+}$ possesses a highly negative redox potential in the ^{3}MLCT excited state, enough to reduce protons to hydrogen gas.

These features deeply relate to the excellent photoreactivity of $[Ru(bpy)_3]^{2+}$. The first point is that the MLCT absorption is observed in the visible region, being at a lower energy than the d-d absorption. In the usual transition metal complexes, d-d absorption is observed in the visible region, but the CT absorption is observed in the near-UV to UV region. This means that the energy of the visible light is well utilized by $[Ru(bpy)_3]^{2+}$ but not by the usual transition metal complexes, because the d-d absorption exhibits a much smaller molar extinction coefficient than that of the CT absorption. The second point is that the ^{3}MLCT excited state is at a lower energy than the ^{3}d-d excited state in $[Ru(bpy)_3]^{2+}$. In the ^{3}MLCT excited state, the excited electron exists in the outer

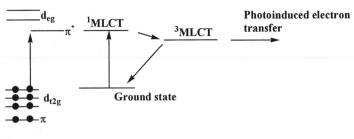

Scheme 2

Table 1 Photochemical Properties of $[Ru(bpy)_3]^{2+}$ and Its Derivatives

	Absorption λ_{max} $(\varepsilon)^a$	Emission λ_{max} $(\tau)^b$	Redox potential^c		
			$E^{2+/+}$	$E^{3+/2+}$	$E^{3+/2+*}$
$[Ru(bpy)_3]^{2+d}$	$452\ (14600)^e$	$607^e\ (600)^f$	-1.33^e	$+1.29^e$	-0.81^e
$[Ru(dmp)_3]^{2+g,h}$	459^i	$618^i\ (931)^i$	-1.46	$+1.10$	-0.58^i
$[Ru(dmp)_2(dcbpy)]^{2+g,j}$	492^i	$694^i\ (853)^i$	-1.03	$+1.30$	
$[Ru(dmp)(dcbpy)_2]^{2+g,j}$	483^i	$658^i\ (1415)^i$	-0.96	$+1.44$	-0.44^k
$[Ru(decby)_3]^{2+g,j}$	467^i	$629^i\ (2230)^i$	-0.91	$+1.55$	-0.42^k

^{a} The absorption maximum in nm. In parentheses is the molar extinction coefficient ($dm^3\ mol^{-1}$ cm^{-1}).

^{b} The emission maximum in nm. In parentheses is the lifetime in ns.

^{c} $E^{2+/+}$, $E^{3+/2+}$, and $E^{3+/2+*}$ represent the redox potentials (in V vs. SCE) for $[Ru(bpy)_3]^{2+/+}$, $[Ru(bpy)_3]^{3+/2+}$, and $[Ru(bpy)_3]^{3+/2+*}$, respectively.

^{d} In water.

^{e} Ref. 11.

^{f} Ref. 12.

^{g} In dichloromethane.

^{h} dmp = 4,7-dimethyl-2,2'-bipyridine.

^{i} Ref. 13.

^{j} dcbpy = 2,2'-bipyridyl-4,4'-dicarboxylic acid.

^{k} Ref. 14.

part of the excited $*[Ru(bpy)_3]^{2+}$ because it is in the π^* orbital of the ligand. Such an excited electron easily transfers to the substrate, because of the good overlap between the π^* orbital of bpy and the acceptor orbital of the substrate. In the ^{3}d-d excited state, the excited electron exists in the metal d orbital, which cannot overlap well with the acceptor orbital of the substrate because the ligands block the approach of the substrate to the metal center. Thus the electron transfer to the substrate occurs with a difficulty in the d-d excited state. In the usual transition metal complex, the ^{3}MLCT excited state easily converts to the ^{3}d-d excited state through internal conversion, even if the ^{3}MLCT excited state emerges by the photoexcitation, and therefore the ^{3}MLCT excited state is not long-lived and contributes little to the photoreaction. The third point is that the redox potential of the ^{3}MLCT excited state is very negative, because of the presence of the excited electron in the π^* orbital of bpy. At the same time, the MLCT excited state has a strong oxidizing ability because a hole exists in the Ru $4d_\pi$ orbital. Thus both the oxidative quenching and the reductive one can occur in the photochemistry.

B. Stereoselective Photoinduced Electron Transfer Reaction of Ruthenium(II) *tris*-Bipyridine Complexes

As shown in Scheme 1, there are two enantiomers in $[Ru(bpy)_3]^{2+}$. Since racemic $[Ru(bpy)_3]^{2+}$ can be separated into two enantiomers [15], one can investigate the steeroselectivity in a photoinduced electron transfer reaction with such enantiomers. Porter and Sparks successfully carried out stereoselective electron transfer reaction between Δ-$[Ru(bpy)_3]^{2+}$ and $Co(acac)_3$ (acac = acetylacetonato) in water [16]. In this reaction, $Co(acac)_3$ is photochemically reduced to Co(II) and $acac^-$ by Δ-$[Ru(bpy)_3]^{2+}$, as shown in Eq. (5);

$$\Delta-[Ru(bpy)_3]^{2+} + Co(acac)_3 \xrightarrow{hv} \Delta-[Ru(bpy)_3]^{3+} + Co(acac)_2 + acac^-$$

(5)

This reaction easily occurs because Δ-$[Ru(bpy)_3]^{2+}$ is excited by visible light (about 450 nm) to afford the ^{3}MLCT excited state, and the ^{3}MLCT-excited Δ-*$[Ru(bpy)_3]^{2+}$ has a great ability to reduce the substrate, as discussed above. If the ruthenium(III) complex is not reduced by some reagent, back electron transfer from $Co(acac)_2$ to Δ-$[Ru(bpy)_3]^{3+}$ would take place to afford Δ-$[Ru(bpy)_3]^{2+}$ and $Co(acac)_3$. In the experiment, no reducing reagent was added, but the authors discussed that the acetylacetonato anion played a role of reducing reagent to afford Δ-$[Ru(bpy)_3]^{2+}$. The reducing ability of acetylacetonato anion will be shown below in more detail. When the mixture of Δ-$[Ru(bpy)_3]^{2+}$ and $Co(acac)_3$ was irradiated with visible light (455 nm), the CD spectrum showed a positive peak at around 650 nm, and a negative peak at around 580 nm. The CD spectrum indicates that Δ-$Co(acac)_3$ exists in excess in the solution after the photoirradiation. This means that Λ-$Co(acac)_3$ is more easily reduced than Δ-$Co(acac)_3$. The enantiomer excess is only 4%, which shows that the electron transfer to the Δ and Λ isomers occurs in the ratio 1.3 to 1.2. Though this value is not large, the stereoselective photoreduction of $Co(acac)_3$ certainly takes place. One of the reasons for the very low *ee* value is that not only the photoreduction of $Co(acac)_3$ but also the photoracemization of Δ-$[Ru(bpy)_3]^{2+}$ occurs in the reaction. Also, it is noted that stereoselective energy transfer was involved in this reaction [17]. Because of this energy transfer, the quantum yield of the photoreduction is very small [18].

The similar but more systematic study of stereoselective quenching reaction of Δ-$[Ru(bpy)_3]^{2+}$ with $[Co(edta)]^-$ was reported by Kaizu and collaborators [19]. Remember that $[Co(edta)]^-$ exhibits molecular asymmetry, as shown in Scheme 3.

$$\Delta-[Ru(bpy)_3]^{2+} + Co(edta)^- \xrightarrow{hv} \Delta-[Ru(bpy)_3]^{3+} + Co(II) + edta^-$$

$$\Lambda-[Co(edta)^- > \Delta-[Co(edta)]^-$$

(6)

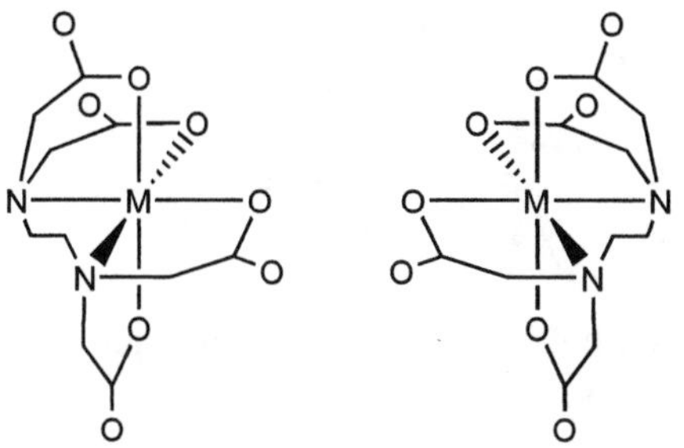

Δ-form   Λ-form

Scheme 3

This reaction occurs easily because the redox potential ($E^{3+/2+} = -0.84$ V vs. NHE in water) of $[Ru(bpy)_3]^{2+}$ in the photoexcited state is significantly negative to the reduction potential ($E_{1/2} = 0.37$ V vs. NHE in water) of $[Co^{III}(edta)]^-$/ $[Co^{II}(edta)]^{2-}$. In this experiment too, the reducing reagent was not used. However, the edta anion plays the role of reducing reagent, as reported [20]. The rate constant of the quenching reaction is 9.7×10^9 M^{-1} s^{-1} for Δ-[Co(edta)]$^-$ and 10.3×10^9 M^{-1} s^{-1} for Λ-[Co(edta)]$^-$ at 15°C in 90% methanol-water. Thus the selectivity is 1.06 at 15°C but increases to 1.23 at 45°C, surprisingly. This is against our expectation that the stereoselectivity increases with decrease in the reaction temperature. Significant solvent dependence of the selectivity was not observed; the selectivity was 1.19 in 50% methanol-water, 1.11 in 81% methanol-water, and 1.19 in 90% methanol-water at 25°C. Later, several reports indicate that the quenching reaction of photoexcited ruthenium(II) complex by cobalt(III) complexes takes place not only through photoinduced electron transfer but also through energy transfer, as will be discussed below in greater detail. If so, the stereoselectivity reported in this work is an averaged value of the photoinduced electron transfer and energy transfer reactions.

Since the above-described reaction would involve not only the electron transfer process but also the energy transfer process that would provide the smaller stereoselectivity, the ratio of the quenching reaction rate constants gives the smaller stereoselectivity than the true selectivity of the photoinduced electron transfer reaction. One way to avoid the energy transfer quenching is to use a substrate that does not have any absorption in the range of the emission spectrum of the photosensitizer. Since methylviologen (see Scheme 4A) does not have any absorption in the visible region, only the electron transfer process participates in the quenching reaction between methylviologen and the photoexcited ruthenium(II) complex of bipyridine and its derivatives.

(A)

Methylviologen

(B)

**1-methyl-1'-[(3S)-(-)-3-pinanylmethyl]-
4,4'-bipyridinium**

Scheme 4

If some chiral group is introduced into the viologen compound, one can perform a stereoselective photoinduced electron transfer reaction of chiral ruthenium(II) tris-bipyridine complex with the chiral viologen. An interesting chiral viologen, 1-methyl-1'-[(3S)-(−)-3-pinanylmethyl]-4,4'-bipyridinium, S-PMV^{2+}, was synthesized by Rau and Ratz [21], as shown in Scheme 4B. They applied this chiral viologen to the quenching reactions of Δ- and Λ-[Ru(bpy)$_3$]$^{2+}$ [Eq. (7)]. The observed stereoselectivity is surprisingly large, $k_q^\Lambda/k_q^\Delta = 1.66$, where k_q is a quenching reaction rate constant obtained from Stern–Volmer relation of the dynamic quenching mechanism shown in Scheme 5.

$$\Delta\text{- or }\Lambda\text{-}[Ru(bpy)_3]^{2+} + \text{S-PMV}^{2+} \longrightarrow h\nu\,\Delta\text{- or }\Lambda\text{-}[Ru(bpy)_3]^{3+} + \text{S-PMV}^+ \tag{7}$$

In their experiments, optical purities of Δ- and Λ-[Ru(bpy)$_3$]$^{2+}$ were 92% and 54%, respectively. If their optical purities were 100%, the stereoselectivity increased to 1.95. In our understanding, this value is very large. However, the photoreaction of S-PMV^{2+} with Δ- and Λ-[Ru(bpy)$_3$]$^{2+}$ involves many elementary steps, as shown in Scheme 5. Not only the quenching reaction but also the formation of S-PMV$^{\cdot+}$ was monitored to estimate the stereoselectivity of the

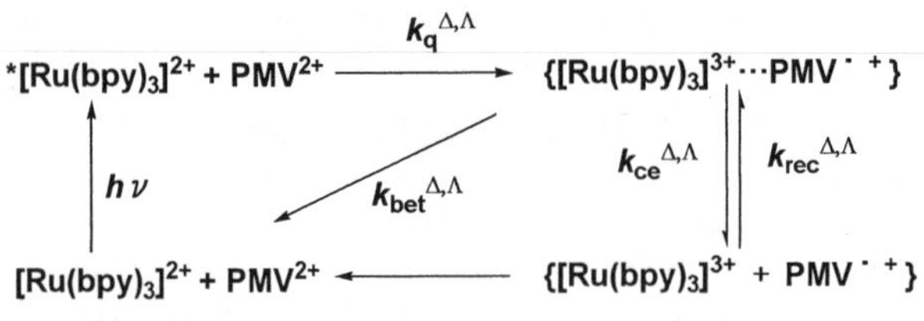

Scheme 5

overall photoreduction. Its selectivity was 1.32 if optically pure Δ- or Λ-$[Ru(bpy)_3]^{2+}$ was used. From these values, the authors concluded that the considerably large stereoselectivity in the quenching reaction is lost in the competing back electron transfer and ion pair dissociation.

The stereoselective luminescence quenching of the ^{3}MLCT excited Δ-*$[Ru(bpy)_3]^{2+}$ was also carried out in water (I = 0.01 mol dm^{-3}) with the different chiral viologen derivatives by Tsukahara and collaborators [22]. In chiral viologens, 1,1′-bis(1-phenylethylcarbamoylmethyl)-4,4′-bipyridinium (OAV^{2+}), 1,1′-bis(1-naphthylethylcarbamoylmethyl)-4,4′-bipyridinium ($POAV^{2+}$), and 1-(1-phenylethyl-carbamoylmethyl)-1′-(1-naphthylethylcarbamoylmethyl)-4,4′-bipyridinium ($NPOAV^{2+}$), the optically active groups, are introduced to the 1,1′-positions through the amido group, as shown in Scheme 6.

The quenching behavior is different between OAV^{2+} and the others ($NOAV^{2+}$ and $NPOAV^{2+}$), as follows: In the presence of $NOAV^{2+}$ and $NPOAV^{2+}$, the emission spectrum of the excited *$[Ru(bpy)_3]^{2+}$ exhibits red shift by 4 nm, and the decay curve of the excited state involves two components, the fast component and the slow one. In the quenching by OAV^{2+}, on the other hand, such a red shift of the luminescence spectrum is not observed, and no fast component is involved in the decay curve. From these results, the authors suggest that the fast component of the decay corresponds to the static quenching of Δ-$[Ru(bpy)_3]^{2+}$ by $NOAV^{2+}$ and $NPOAV^{2+}$, and that the quenching by OAV^{2+} occurs through only the dynamic quenching mechanism, as shown in Scheme 7. The difference is interpreted as that $NOAV^{2+}$ and $NPOAV^{2+}$ possess hydropho-

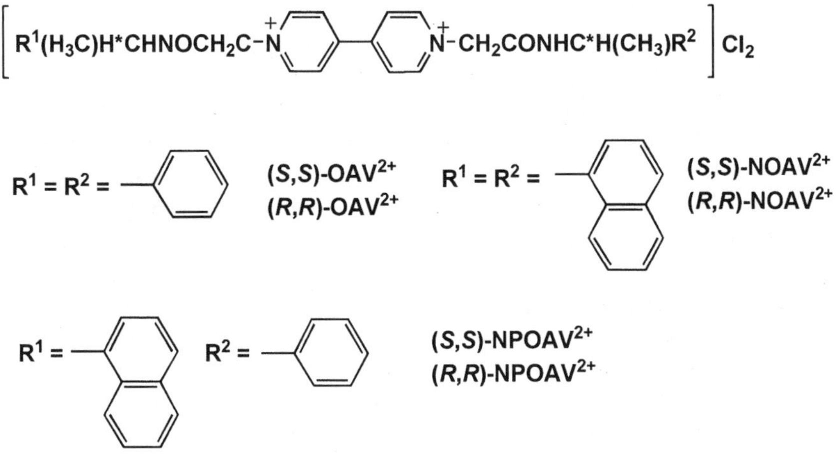

Scheme 6

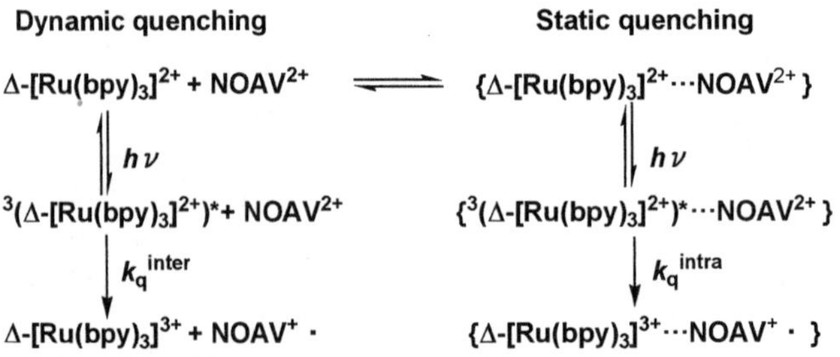

Dynamic quenching   Static quenching

Δ-[Ru(bpy)$_3$]$^{2+}$ + NOAV^{2+} $\rightleftharpoons$ {Δ-[Ru(bpy)$_3$]$^{2+}\cdots$NOAV^{2+}}

$\quad\Big\Vert\ h\nu$ $\qquad\qquad\qquad\qquad\qquad\Big\Vert\ h\nu$

$^3(\Delta$-[Ru(bpy)$_3$]$^{2+})^*$+ NOAV^{2+} {$^3(\Delta$-[Ru(bpy)$_3$]$^{2+})^*\cdots$NOAV^{2+}}

$\quad\Big\downarrow\ k_q^{\,inter}$ $\qquad\qquad\qquad\qquad\quad\Big\downarrow\ k_q^{\,intra}$

Δ-[Ru(bpy)$_3$]$^{3+}$ + NOAV$^+\cdot$ {Δ-[Ru(bpy)$_3$]$^{3+}\cdots$NOAV$^+\cdot$}

Scheme 7

bic naphthyl and phenyl groups, respectively, which are favorable for the hydrophobic interaction between these viologens and the bpy ligand of Δ-[Ru(bpy)$_3$]$^{2+}$. In OAV^{2+}, the adduct formation is difficult because of the absence of hydrophobic substituent, and therefore only the dynamic quenching takes place. In both cases, homochiral selectivity is observed. The stereoselectivity (k_q(S,S)/k_q(R,R)) is 1.33 for NOAV^{2+} and 1.13 for NPOAV^{2+} in the fast component and 1.27 for OAV^{2+}, 1.13 for NOAV^{2+}, and 1.09 for NPOAV^{2+} in the slow component, where the k_q is the rate constant of the quenching reaction (see Scheme 7). Apparently, the static quenching reaction leads to the larger stereoselectivity.

One of the interesting features of these chiral viologens is that this kind of viologen can be systematically synthesized with a variety of bulky substituents. Their redox potentials moderately depend on the aryl substituent, and they are observed in the range of -0.17 to -0.24 V ($E^{3+/2+}$ vs. SCE in water) [23].

Since tris(2,2'-bipyridine)ruthenium(II) causes racemization under irradiation [7] this complex is not a good photosensitizer for the stereoselective photoreaction. To avoid this weak point, Ohkubo et al. synthesized a new chiral ruthenium(II) complex by introducing a chiral functional group into 2,2'-bipyridine derivative [24], and Sakaki et al. synthesized a chiral copper(I) complex by using a chiral phosphine ligand [25]. As shown in Scheme 8A, the chiral ruthenium(II) complex, [Ru(($-$)-menbpy)$_2$(bpy-C$_{12}$)]$^{2+}$, has a bpy derivative (bpy-C$_{12}$) with long alkyl chains besides chiral menthyl groups. These long alkyl chains are useful in the adsorption of the metal complex to the micelle surface. [Ru(($-$)-menbpy)$_2$(bpy-C$_{12}$)]$^{2+}$ was applied to the photoreduction of [Co(edta)]$^-$ in the presence of such micelles as cationic cetyltrimethyl ammonium bromide (CTAB), anionic sodium dodecyl sulfate (SDS), and neutral octylphenol (Triton X). Because of the presence of the hydrophobic long alkyl chain on the bpy-C$_{12}$ ligand, the chiral ruthenium(II) complex is well bound to the micelle independently of the charge of the micelle surface. However, [Co(edta)]$^-$ cannot approach the SDS micelle, but it is well adsorbed with the CTAB micelle, since this complex

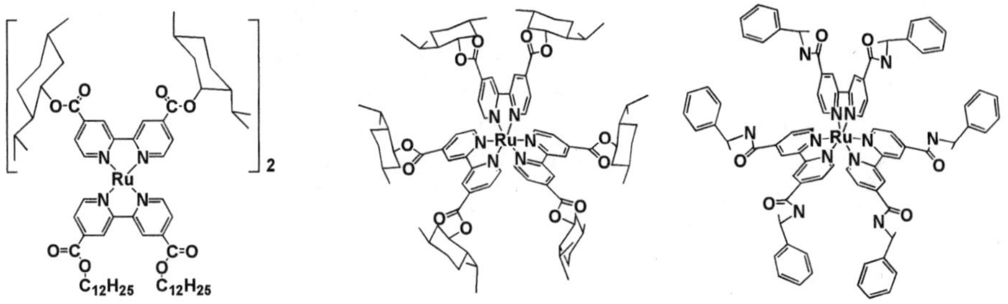

A $[Ru((-)\text{-menbpy})_2(bpy\text{-}C_{12})]^{2+}$ B $[Ru((-)\text{-menbpy})_3]^{2+}$ C $[Ru((S)\text{-PhEtbpy})_3]^{2+}$

Scheme 8

is hydrophilic and is negatively charged. The photoreduction of [Co(edta)]⁻ efficiently occurs in the CTAB micelle, moderately in the Triton X micelle, but little
in the SDS micelle. The stereoselectivity, defined as the ratio of the rate constant
of photoreduction of each isomer, also depends on the micelle charge; the stereoselectivity is the greatest (2.14) in the CTAB micelle, moderate (1.16) in the
Triton X micelle, and not at all in the SDS micelle.

Slightly later, Ohkubo and collaborators synthesized the tris(4,4′-bis-
((1R,2S,5R)-(−)menthylcarboxy)-2,2′-bipyridine)ruthenium(II) complex, [Ru-
((−)-menbpy)₃]²⁺ (see Scheme 8B), and applied this complex to the photoreduction of Co(acac)₃ [26]. After the photoreaction, the CD spectrum corresponding
to Λ-Co(acac)₃ appears, which clearly shows that Δ-Co(acac)₃ more rapidly
undergoes photoreduction with [Ru((−)-menbpy)₃]²⁺. The enantioselectivity
(k_Δ/k_Λ) is 1.33 for both the photoreduction and the quenching reaction.

$$\left[Ru((-)-menbpy)_3\right]^{2+} + Co(acac)_3 \xrightarrow{h\nu}$$

$$\left[Ru((-)-menbpy)_3\right]^{3+} + Co(acac)_2 + acac^-$$

$$\frac{k_\Delta}{k_\Lambda} = 1.33 \text{ in } 90\% \text{ ethanol} - \text{water}$$

$$\frac{k_\Delta}{k_\Lambda} = 1.05 \text{ in } 70\% \text{ to } 50\% \text{ ethanol} - \text{water} \tag{8}$$

The similar ruthenium(II) complex, [Ru(S(−)-PhEtbpy)₃]²⁺ ((S(−)-PhEtbpy =
4,4′-bis[(S)-(−)-1-phenylethylaminocarbonyl]-2,2′-bipyridine), was synthesized
by the same authors [27]. In this complex, chiral phenethylamine is introduced
into 2,2′-bipyridine through the amido bridge (see Scheme 8C). This complex
was also applied to stereoselective photoreduction of Co(acac)₃. In this reaction,
the stereoselectivity is reverse to that observed in the reaction by [Ru((−)-men-

bpy)$_3$]$^{2+}$. However, the reasons are still ambiguous. To clarify the reasons, we should know with what orientation and with what interaction Co(acac)$_3$ approaches the ruthenium(II) complex. To consider the reasons, the review of stereoselective thermal electron transfer reaction is useful [5]. The stereoselectivity significantly depends on the solvent composition of ethanol-water; when ethanol

$$\left[Ru((-)-PhEtbpy)_3\right]^{2+} + [Co(edta)]^- \xrightarrow{h\nu}$$
$$\left[Ru((-)-menbpy)_3\right]^{3+} + Co(acac)_2 + acac^-$$
$$k_\Lambda$$
$$k_\Lambda = 1.54 \text{ in } 90\% \text{ ethanol} - \text{water} \tag{9}$$

is 90%, the selectivity is 1.54 but decreases to 1.05 when ethanol is 70 to 50% [28].

Several properties of these ruthenium(II) complexes are shown in Table 2. Apparently, the absorption and emission maxima of [Ru((−)-menbpy)$_3$]$^{2+}$ and [Ru(S(−)-PhEtbpy)$_3$]$^{2+}$ exhibit considerably large red shifts, compared to those of [Ru(bpy)$_3$]$^{2+}$. A similar red shift was observed in [Ru(dmp)$_n$(dcbpy)$_{3-n}$]$^{2+}$ (dcbpy = 2,2'-bipyridyl-4,4'-dicarboxylic acid), as was shown in Table 1. These red shifts are easily understood in terms of the introduction of electron-withdrawing substituents at the 4 and 4' positions of 2,2'-bipyridine. The other important feature is that the lifetime of the ^{3}MLCT excited state becomes much longer than that of [Ru(bpy)$_3$]$^{2+}$. One of the important reasons is the increase in the energy difference between the triplet d-d (^{3}d-d) and ^{3}MLCT excited states, as follows [13,29]: Since the electron-withdrawing substituent of 2,2'-bipyridine stabilizes the π^* orbital of 2,2'-bipyridine, the ^{3}MLCT excited state becomes lower in energy, but the ^{3}d-d excited state is little influenced in energy by the substituent.

Table 2 Photochemical Properties of [Ru(menbpy)$_3$]$^{2+}$ and [Ru(PhEtbpy)$_3$]$^{2+}$

	Absorption λ_{max} (ε)[a]	Emission λ_{max} (τ)[b]	Redox potential[c]		
			$E^{2+/+}$	$E^{3+/2+}$	$E^{3+/2+}*$
[Ru((−)-menbpy)$_3$]$^{2+d}$	466 (27200)	623 (1550)	−0.90	+1.55	−0.45
[Ru(S(−)-PhEtbpy)$_3$]$^{2+d}$	464 (21200)	620 (1800)	—	+1.40	−0.60

[a] The absorption maximum in nm. In parentheses is the molar extinction coefficient (dm^3 mol^{-1} cm^{-1}).
[b] The emission maximum in nm. In parentheses is the lifetime in ns.
[c] $E^{2+/+}$, $E^{3+/2+}$, and $E^{3+/2+}*$ represent the redox potentials (in V vs. SCE) for [Ru(menbpy)$_3$]$^{2+/+}$, [Ru(menbpy)$_3$]$^{3+/2+}$, and [Ru(menbpy)$_3$]$^{3+/2+}$, respectively.
[d] In 90% ethanol-water. Ref. 28.

As a result, the introduction of the electron-withdrawing substituent increases the energy difference between ^{3}d-d and ^{3}MLCT excited states, to suppress the deactivation of the ^{3}MLCT excited state via the ^{3}d-d excited state. Thus the ^{3}MLCT excited states of $[Ru((-)-menbpy)_3]^{2+}$ and $[Ru(S(-)-PhEtbpy)_3]^{2+}$ are longer-lived than that of $[Ru(bpy)_3]^{2+}$. The electron-withdrawing substituent is also responsible for the more positive redox potential $E^{3+/2+}$ and the less negative redox potential $E^{3+/2+*}$ than those of $[Ru(bpy)_3]^{2+}$.

The photoreduction of cobalt(III) complexes by these complexes takes place catalytically, in which one-electron oxidized $[Ru((-)-menbpy)_3]^{3+}$ and $[Ru(S(-)-PhEtbpy)_3]^{3+}$ are reduced by ethanol in the solvent [26]. This is because these complexes have more positive reduction potential than that of $[Ru(bpy)_3]^{2+}$.

Since $[Ru((-)-menbpy)_3]^{2+}$ and $[Ru(S(-)-PhEtbpy)_3]^{2+}$ possess chiral groups in the 2,2'-bipyridine ligand and molecular asymmetry in the metal frame, as shown in Scheme 1, these complexes are diastereomer. Thus Δ- and Λ-isomers can be separated with silica gel column chromatography [28]. Interestingly, these Δ- and Λ-isomers have larger helical structure than does $[Ru(bpy)_3]^{2+}$, as shown in Scheme 9.

The other interesting feature of these complexes is that the Δ- and Λ-isomers do not cause racemization under visible light irradiation, probably because the chiral groups introduced into the ligand are bulky enough to suppress the photoinduced racemization; the quantum yield for the racemization of these complexes is 4.1×10^{-6} for $[Ru((-)-menbpy)_3]^{2+}$, 7.6×10^{-7} for $[Ru(S(-)-PhEtbpy)_3]^{2+}$ in ethanol [28] and 2.88×10^{-5} for $[Ru(bpy)_3]^{2+}$ in water [7]. These large helical complexes were applied to photoreduction of Co(acac)$_3$ [28,30–32]. When Δ-$[Ru((-)-menbpy)_3]^{2+}$ was used as a photosensitizer, the stereoselectivity, defined by the ratio of reduction rate constants, k^Δ/k^Λ, was very large (14.7) in 90% ethanol-water. On the other hand, the stereoselectivity was only 1.60, when $\Delta + \Lambda$-$[Ru((-)-menbpy)_3]^{2+}$ was used as a photosensitizer. These results indicate that the chirality of the metal center considerably participates in the stereoselection. Though a reducing reagent was not used, the ethanol of the solvent plays the role of reducing reagent, as discussed above. The catalytic cycle is proposed, as shown in Scheme 10.

In spite of the very large stereoselectivity of the photoreduction of Co(acac)$_3$, the quenching reaction of Δ-$[Ru((-)-menbpy)_3]^{2+}$ by Δ- and Λ-Co(acac)$_3$ takes place with the smaller stereoselectivity (k_q^Δ/k_q^Λ) of 1.28. Of course, the quenching reaction of $\Delta + \Lambda$-$[Ru((-)-menbpy)_3]^{2+}$ by Δ- and Λ-Co(acac)$_3$ occurs with the further smaller stereoselectivity (1.14) than that of Δ-$[Ru((-)-menbpy)_3]^{2+}$. Ohkubo and collaborators suggested that the very large stereoselectivity of the photoreduction arose from the reverse reaction, as follows: In the photoreduction of rac-Co(acac)$_3$ with Δ-$[Ru((-)-menbpy)_3]^{2+}$, Δ-Co(acac)$_3$ more rapidly reacts with Λ-$[Ru((-)-menbpy)_3]^{2+}$ than does Λ-Co(acac)$_3$, lead-

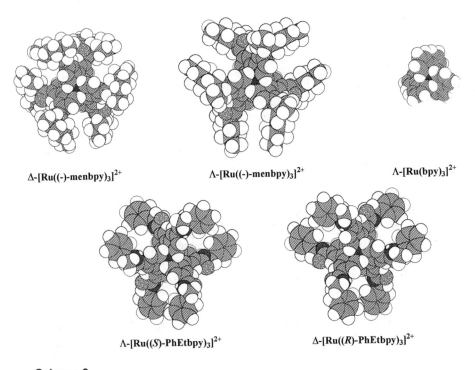

Δ-[Ru((-)-menbpy)₃]²⁺   Λ-[Ru((-)-menbpy)₃]²⁺   Λ-[Ru(bpy)₃]²⁺

Λ-[Ru((S)-PhEtbpy)₃]²⁺   Δ-[Ru((R)-PhEtbpy)₃]²⁺

Scheme 9

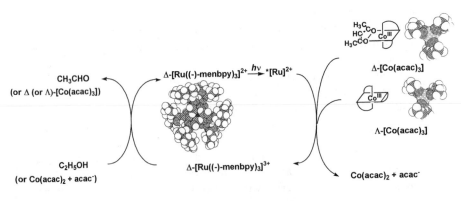

Scheme 10

ing to the larger concentration of Δ-Co(acac)$_3$ than that of Λ-Co(acac)$_3$. Δ-[Ru((−)-menbpy)$_3$]$^{3+}$ can be reduced by either ethanol in the solvent or Co(acac)$_2$, because this ruthenium(III) complex is easily reduced owing to the presence of electron-withdrawing substituents on bpy (see Table 1). If this ruthenium(III) complex is reduced by only ethanol, the stereoselectivity is determined by the quenching and the charge-separation steps. However, this ruthenium(III) complex is also reduced by Co(acac)$_2$, to afford Co(acac)$_3$, where Λ-Co(acac)$_3$ is more preferentially formed than the Δ-enantiomer, as shown in Scheme 10 (see parentheses of this scheme). Thus the stereoselectivity increases very much by this reverse reaction, where the reverse reaction represents the reaction between the ruthenium(III) complex and the cobalt(II) complex and the back electron transfer is the electron transfer to the one-electron oxidized photosensitizer from the one-electron reduced reactant in the encounter complex, hereafter. This explanation suggests that the stereoselectivity significantly depends on the reaction conditions by which the reverse reaction is influenced. One of them is the concentration of acac$^-$, because acac$^-$ is necessary for the thermal reverse reaction. Actually, the k^{Δ}/k^{Λ} value increases very much with an increase in the concentration of Hacac; its value is 14.7 when [Hacac]/[Ru] = 0, but 40.1 when [Hacac]/[Ru] = 2.0, and finally it increases to 91.9 when [Hacac]/[Ru] = 10.0, where [Ru] means the concentration of Δ-[Ru((−)-menbpy)$_3$]$^{3+}$ [32]. These results are consistent with the mechanism of Scheme 10.

The different group also reported that the overall stereoselectivity is influenced by the reverse reaction in the photoreduction of cobalt(III) complexes with Δ-[Ru(bpy)$_3$]$^{2+}$ [33]. Kato and collaborators carried out the stereoselective photoreduction of tris(oxalate)cobaltate(III), [Co(ox)$_3$]$^{3-}$, and tris(acetyl acetonato) cobalt(III) with Δ-[Ru(bpy)$_3$]$^{2+}$.

$$\Delta - [Ru(bpy)_3]^{2+} + [Co(ox)_3]^{3-} \xrightarrow{h\nu} \Delta - [Ru(bpy)_3]^{3+} + Co(II) + 3\ ox^-$$

$$\Delta - [Co(ox)_3]^{2+} > \Lambda - [Co(ox)_3]^{2+} \text{ in water}$$

$$\Delta - [Co(ox)_3]^{2+} < \Lambda - [Co(ox)_3]^{2+} \text{ in 30\% to 50\% methanol - water}$$

$$(10)$$

The reduction of [Co(ox)$_3$]$^{3-}$ exhibits the homochiral (Δ-Δ or Λ-Λ) preference in water, while the stereoselectivity is not large: 5% *ee* after a 180 min reaction (70% conversion of [Co(ox)$_3$]$^{3-}$). Interestingly, the stereoselectivity disappears at 20% methanol-water, and becomes reverse in 30 and 50% methanol-water, i.e., the heterochiral (Δ-Λ) preference was observed in these solvent systems. On the other hand, the reduction of Co(acac)$_3$ always exhibits only the heterochiral preference independent of the methanol content.

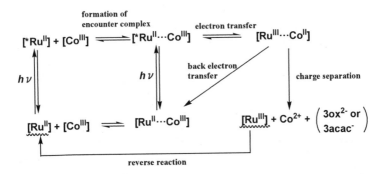

Scheme 11

$$\Delta-[Ru(bpy)_3]^{2+} + Co(acac)_3 \xrightarrow{h\nu} \Delta-[Ru(bpy)_3]^{3+} + Co(acac)_2 + acac^-$$

$$\Lambda-Co(acac)_3 > \Delta-Co(acac)_3 \text{ in water} \qquad (11)$$

In the quenching reaction of Δ-[Ru(bpy)$_3$]$^{2+}$ by [Co(ox)$_3$]$^{3-}$ and Co(acac)$_3$, only the homochiral preference was observed in water, whereas the stereoselectivity of the quenching by [Co(ox)$_3$]$^{3-}$ becomes reverse in 80% methanol-water. These results suggest that the stereoselectivity is determined not only by the photoinduced electron transfer but also by the different elementary step such as the reverse reaction. The photoreduction of the cobalt(III) complex by the ruthenium(II) complex involves various elementary steps, as shown in Scheme 11. Considering this scheme, one can easily understand that the overall photoreduction of the cobalt(III) complexes is determined by not only the quenching process but also the reverse reaction between the reduced Co(II) complexes and the oxidized ruthenium(III) complex. This conclusion is essentially the same as that reported by Ohkubo and his collaborators.

Recently, the pulse radiolysis of Δ-[Ru((−)-menbpy)$_3$]$^{2+}$ with Δ- and Λ-Co(acac)$_3$ was investigated in detail [34]. In this reaction, Δ-[Ru((−)-menbpy)$_3$]$^{2+}$ is reduced by the ethanol radical CH$_3$C·HOH, to afford Δ-[Ru((−)-menbpy)$_3$]$^+$, where the ethanol radical is produced by the reactions of H· and OH· with ethanol in N$_2$O-saturated solution. In Δ-[Ru((−)-menbpy)$_3$]$^+$, an extra odd electron occupies the π^* orbital of the menbpy ligand. This situation is similar to that of the ^{3}MLCT excited state, because the odd electron exists in the π^* orbital of menbpy in both the ^{3}MLCT excited Δ-*[Ru((−)-menbpy)$_3$]$^{2+}$ and the one-electron reduced Δ-[Ru((−)-menbpy)$_3$]$^+$. Δ-[Ru((−)-menbpy)$_3$]$^+$ reacts with Co(acac)$_3$ to afford [Co(acac)$_3$]$^-$, which further decomposes to Co(acac)$_2$ + acac$^-$. In this reaction, Δ-[Ru((−)-menbpy)$_3$]$^+$ preferentially reduces Δ-Co(acac)$_3$ with the considerably large stereoselectivity of 2.7.

$$\Delta - [Ru((-) - menbpy)_3]^+ + Co(acac)_3$$

$$\rightarrow \Delta - [Ru((-)-menbpy)_3]^{2+} + Co(acac)_2 + acac^-$$

$$\Delta - Co(acac)_3 > \Lambda - Co(acac)_3 \text{ in } 85\% \text{ ethanol - water} \tag{12}$$

Δ-[Ru((−)-menbpy)$_3$]$^+$ possesses two kinds of chiral center; one is in the menbpy ligand and the other is in the metal center (Δ or Λ). It is interesting which chirality plays an important role in stereoselection. If only the chiral centers of the menbpy ligand participate in the stereoselection, Δ-[Ru((−)-menbpy)$_3$]$^+$ and the Λ-isomer provide the same stereoselectivity. If only the chiral center of the metal–ligand flame participates in the stereoselection, the stereoselectivity becomes reverse between Δ- and Λ-[Ru((−)-menbpy)$_3$]$^+$. Δ-[Ru((−)-menbpy)$_3$]$^+$ gives the stereoselectivity of 1.7 (Δ/Λ) at [phosphate] = 0.2 mol dm^{-3} in 85% ethanol-water, while Λ-[Ru((−)-menbpy)$_3$]$^+$ yields the reverse but smaller selectivity of 1.25 (Λ/Δ) under the same reaction conditions. This means that the chirality of the metal center plays a more important role than the chirality in the menbpy ligand. Of course, the chirality in the menbpy ligand plays a considerable role in enhancing the stereoselection. These results present valuable knowledge of the stereoselectivity in the photoinduced electron transfer reaction.

The stereoselective quenching reaction of the excited Δ-*[Ru((−)-menbpy)$_3$]$^{2+}$ by Δ- and Λ-Co(acac)$_3$ was also investigated by the same authors. The selectivity (k^Δ/k^Λ) was 1.2 for Co(acac)$_3$ and 1.1 for [Co(edta)]$^-$ in 50% ethanol-water. One of the important results is that no long-lived species, such as [Ru((−)-menbpy)$_3$]$^{3+}$, [Co(acac)$_3$]$^-$, or Co(acac)$_2$, was observed by the transient absorption spectrum measurement, where a laser with a pulse width of 6 ns was used. This means that the oxidative quenching process (Scheme 12A) does not mainly participate in the overall quenching reaction. Moreover, the driving force (ΔG^0) is −0.11 V for the quenching reaction, while it is −0.56 V for the thermal electron transfer reaction between Δ-[Ru((−)-menbpy)$_3$]$^+$ and Co(acac)$_3$. According to Marcus theory [35], the electron transfer reaction rate constant k_{et} is represented by the equation

$$k_{et} = \left(\frac{2\pi}{h}\right) H_{AB}^2 (4\pi\lambda RT)^{-\frac{1}{2}} \exp\left[\frac{-(\Delta G^0 + \lambda)^2}{(4\lambda RT)}\right] \tag{13}$$

where ΔG^0 is the driving force of the reaction, H_{AB} is the electron-coupling matrix element, and λ is the reorganization energy. In the quenching reaction of Δ-*[Ru((−)-menbpy)$_3$]$^{2+}$ and the thermal electron transfer reaction of Δ-[Ru((−)-menbpy)$_3$]$^+$, one electron transfers from the same π^* orbital of the menbpy ligand to the Co(III) complex. This suggests that the H_{AB} value is similar in these two reactions. The relative rate constant of the quenching reaction is surprisingly large, compared to the thermal electron transfer reactions, whereas the driving force for the thermal electron transfer reaction between Δ-[Ru((−)-menbpy)$_3$]$^+$

(A) Oxidative quenching

$$[Ru(menbpy)_3]^{2+} \xrightarrow{h\nu} {}^*[Ru(menbpy)_3]^{2+}$$

$${}^*[Ru(menbpy)_3]^{2+} + [Co(acac)_3] \underset{k_{-d}}{\overset{k_d}{\rightleftharpoons}} [{}^*Ru(menbpy)_3{}^{2+} \cdots Co(acac)_3]$$

$$[{}^*Ru(menbpy)_3{}^{2+} \cdots Co(acac)_3] \xrightarrow{k_{et}} [Ru(menbpy)_3{}^{3+} \cdots Co(acac)_3{}^-]$$

$$[Ru(menbpy)_3{}^{3+} \cdots Co(acac)_3{}^-] \underset{k'_d}{\overset{k'_{-d}}{\rightleftharpoons}} [Ru(menbpy)_3]^{3+} + [Co(acac)_3]^-$$

$$[Co(acac)_3]^- \rightleftharpoons Co(acac)_2 + acac^-$$

(B) Quenching through energy transfer

$$[Ru(menbpy)_3]^{2+} \xrightarrow{h\nu} {}^*[Ru(menbpy)_3]^{2+}$$

$${}^*[Ru(menbpy)_3]^{2+} + [Co(acac)_3] \underset{k_{-d}}{\overset{k_d}{\rightleftharpoons}} [{}^*Ru(menbpy)_3{}^{2+} \cdots Co(acac)_3]$$

$$[{}^*Ru(menbpy)_3{}^{2+} \cdots Co(acac)_3] \xrightarrow{k_{et}} [Ru(menbpy)_3{}^{2+} \cdots {}^*Co(acac)_3]$$

$$[Ru(menbpy)_3{}^{2+} \cdots {}^*Co(acac)_3] \underset{k_d}{\overset{k_{-d}}{\rightleftharpoons}} [Ru(menbpy)_3]^{2+} + [Co(acac)_3]$$

Scheme 12

and $Co(acac)_3$ is larger than that of the quenching reaction. Though the effects of the λ value are not mentioned in the discussion, these results strongly suggest that the quenching reaction mainly occurs through the energy transfer (see Scheme 12B). As a result, the stereoselectivity of the overall photoreduction differs from that of the quenching reaction.

In summary, the stereoselectivity was certainly observed in the photoinduced electron transfer reactions of chiral ruthenium(II) complexes with chiral viologen and Co(III) complexes. However, not only the photoinduced electron transfer reaction but also the charge separation in the encounter complex and the reverse reaction between the ruthenium(III) complex and $Co(acac)_2 + acac^-$ participate in the stereoselection. In the reactions between the ruthenium(II) and Co(III) complexes, the energy transfer also contributes to the quenching reaction, which makes difficult the observation of stereoselectivity in the quenching reaction.

C. Application of Chiral Ruthenium(II) Complexes to Stereoselective Photocatalytic Reactions

One-electron oxidized photosensitizer is in general a highly oxidizing reagent, and it is easily reduced by the reducing reagent. If the reducing reagent is absent in the reaction solution, the one-electron oxidized photosensitizer is reduced by the one-electron reduced product after the photoinduced electron transfer reaction, as discussed in Sec. II.B. If the purpose is to carry out photoreduction of some substrate, the reducing reagent should be added to the solution enough to suppress the reverse reaction. However, if the reducing reagent is not added sufficiently, one can construct several interesting photoreactions by utilizing the strong oxidation function of the one-electron reduced photosensitizer. In this section, we wish to report such photoreaction systems.

An interesting photocatalytic system was reported by Hamada and collaborators [36]. They applied the helical chiral ruthenium(II) complex, $[Ru((-)-men-bpy)_3]^{2+}$, to the asymmetric photosynthesis of bi-2-naphthol from 3-substituted-2-naphthol.

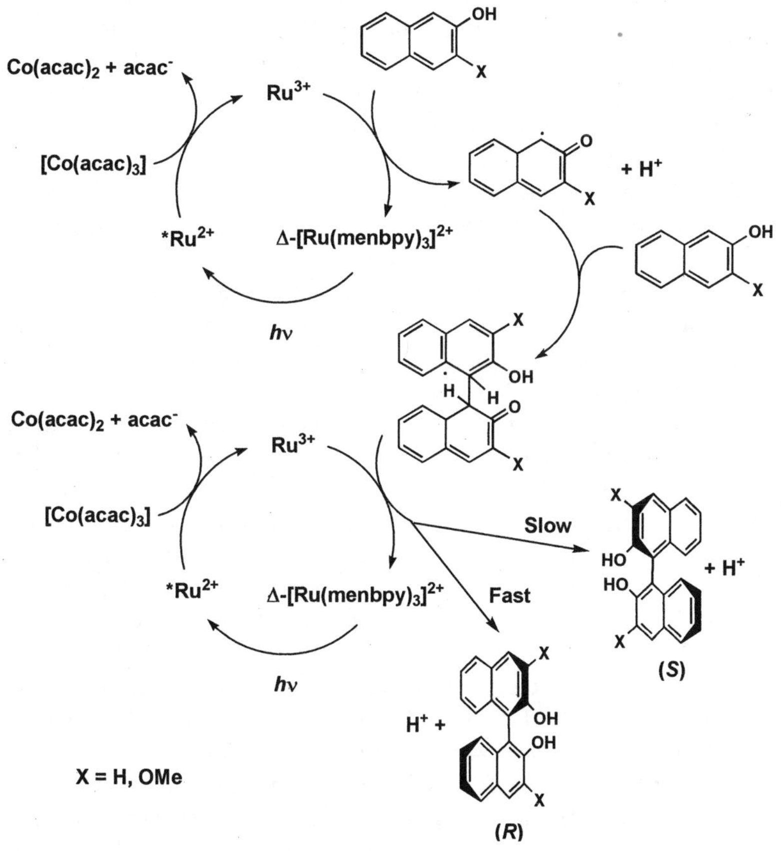

Scheme 13

$$2 \quad \text{(naphthol, X = H, OMe)} \quad \xrightarrow{\Delta\text{-[Ru(menbpy)}_3]^{2+},\ h\nu} \quad 2H^+ + \text{(binaphthol)} \quad (14)$$

X = H, OMe

(R) > (S)

In this reaction, the photoexcited *[Ru((−)-menbpy)$_3$]$^{2+}$ rapidly undergoes the oxidative quenching by Co(acac)$_3$ to afford [Ru((−)-menbpy)$_3$]$^{3+}$, as shown in Scheme 13. Then [Ru((−)-menbpy)$_3$]$^{3+}$ oxidizes 2-naphthol to afford the 2-naphthol cation radical with reproduction of [Ru((−)-menbpy)$_3$]$^{2+}$. The 2-Naphthol cation radical undergoes the coupling reaction with the other 2-naphthol to yield the adduct radical. Since this adduct radical still has reducing ability, it reduces [Ru((−)-menbpy)$_3$]$^{3+}$, remaining in the solution to regenerate [Ru((−)-menbpy)$_3$]$^{2+}$ with formation of bi-2-naphthol. The oxidation potential of 2-nathrol derivatives is +1.34 V and +1.32 V vs. SCE for X = H and OMe, respectively, in MeCN, where X represents the substituent at the 3-position of naphthol. The redox potential of Δ-[Ru((−)-menbpy)$_3$]$^{2+}$ ($E^{3+/2+}$ = +1.55 V vs. SCE in MeCN) is much more positive than that (+1.29 V vs SCE) of [Ru(bpy)$_3$]$^{2+}$. Because of the considerably large positive redox potential, Δ-[Ru((−)-menbpy)$_3$]$^{3+}$ can easily oxidize 2-naphthol derivatives. Note that usual [Ru(bpy)$_3$]$^{3+}$ can not oxidize 2-naphtol because of its less positive redox potential. This means that Hamada and collaborators succeeded in constructing this interesting catalytic cycle by employing [Ru((−)-menbpy)$_3$]$^{2+}$, which possesses the electron-withdrawing substituents in the menbpy ligand. The turnover numbers (66) are not large after a 16 h reaction, and the enantiomer excess is 16.2% for X = H and 3.95% for X = OMe. Though the reaction efficiency and the enantiomer excess are not very good, we think that there are interesting points to be noted in this reaction: the first is the application of the inorganic photosensitizer to organic synthesis, and the second is the utilization of the high oxidizing ability of the one-electron oxidized photosensitizer. The third is the achievement of asymmetric photosynthesis, and the fourth is the utilization of both the reduction by [Ru((−)-menbpy)$_3$]$^{2+}$ and the oxidation by [Ru(menbpy)$_3$]$^{3+}$. Also, it should be noted that radical coupling occurs under the influence of the chiral ruthenium(II or III) complexes. This means that the 2-naphthol cation radical exists near to the ruthenium(II) complex probably because of the hydrophobic interaction. Thus the solvent used would play an important role in stereoselection.

If bi-2-naphthol and Co(acac)$_3$ exist in excess in the reaction solution, the photoexcited ruthenium(II) complex undergoes oxidative quenching by Co(acac)$_3$ to afford the ruthenium(III) complex, as shown by the catalytic cycle of Scheme

14. The resultant ruthenium(III) complex oxidizes bi-2-naphthol to some unknown oxidation product, as shown in Scheme 14 [37].

If bi-2-naphtol was stereoselectively oxidized, the kinetic resolution of bi-2-naphtol was achieved. Actually, the S enantiomer of bi-2-naphtol is more rapidly consumed than the R enantiomer with an enantiomer excess of 15.2% after 1 h and 2 h reactions and 11.0% after a 12 h reaction.

When the ruthenium(II) complex is irradiated under an oxygen atmosphere, the oxidative quenching of the photoexcited ruthenium(II) complex by an oxygen molecule rapidly occurs, to afford the ruthenium(III) complex. Because the ruthenium(III) complex is a powerful oxidizing reagent, it can oxidize some substrate. Such a catalytic cycle was applied to the asymmetric photosynthesis of $Co(acac)_3$ from $Co(acac)_2$ + $acac^-$. Δ- and Λ-[Ru((−)-menbpy)$_3$]$^{2+}$ and Δ- and Λ-[Ru(R(or S)-PhEtbpy)$_3$]$^{2+}$ were used as photosensitizers for such asymmetric photosynthesis [38,39].

$$Co(acac)_2 + acac^- \xrightarrow[\Lambda-Co(acac)_3 > \Delta-Co(acac)_3]{[Ru((-)-menbpy)_3]^{2+}, O_2, h\nu} Co(acac)_3 \qquad (15a)$$

As shown in Scheme 15, [Ru((−)-menbpy)$_3$]$^{3+}$, which is formed through the oxidative quenching by the O_2 molecule, easily oxidizes $Co(acac)_2$ to $Co(acac)_3$. The photosensitizer is a mixture of Δ- and Λ-forms of the metal center, but it still possesses six chiral centers on each menbpy ligand. Because of the presence

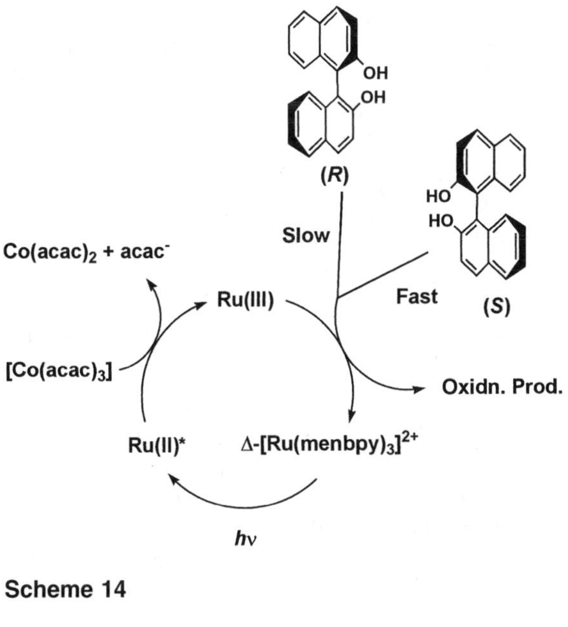

Scheme 14

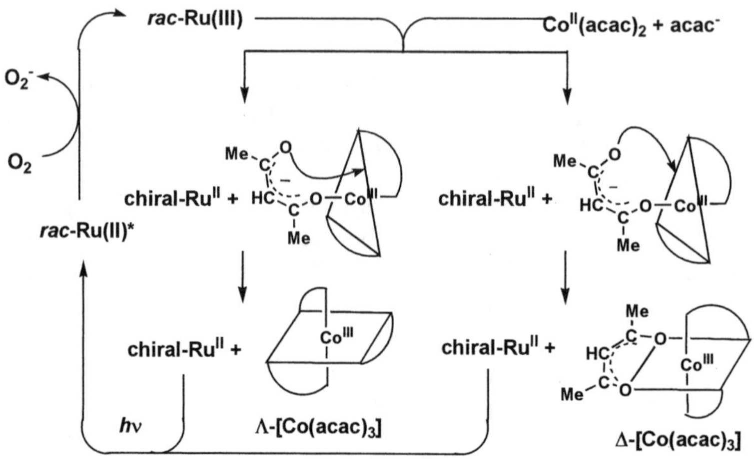

Scheme 15

of these chiral centers, the oxidation of Co(acac)$_2$ occurs stereoselectively. When [Ru(menbpy)$_3$]$^{2+}$ is used as a photosensitizer, Λ-Co(acac)$_3$ is formed to a greater extent than the Δ-enantiomer; the enantiomer excess is 10% in 70% ethanol-water at the 40% yield of Co(acac)$_3$ when the ratio of Hacac/Co(acac)$_2$ is 100 : 1. However, the enantiomer excess decreases to 4.4% when the ratio of Hacac/Co(acac)$_2$ is 1.0, while the yield increases to 55%. In addition, the yield and the enantiomer excess depend on the solvent composition; for instance, the enantiomer excess is very small (1.9%) in ethanol but increases to 10% in 75% ethanol-water, and little changes in 70% to 50% ethanol-water solvents. From these results, the authors suggested that the hydrophobic interaction enhanced the contact between Co(acac)$_2$ and the ruthenium(III) complex, which led to an increase of the enantiomer excess with the decrease in the ethanol content.

When [Ru(R(or S)-PhEtbpy)$_3$]$^{2+}$ was used as a photosensitizer, however, the enantiomer excess was negligibly small, while the yield was similar to that of the reaction with [Ru((−)-menbpy)$_3$]$^{2+}$ [39]. This complex provides the reverse stereoselectivity to that of the photoreduction of Co(acac)$_3$ by [Ru((−)-menbpy)$_3$]$^{2+}$ (see above). These reasons are ambiguous.

$$Co(acac)_2 + acac^- \xrightarrow[\Lambda-Co(acac)_3 \approx \Delta-Co(acac)_3]{[Ru((-)-PhEtbpy)_3]^{2+}, O_2, h\nu} Co(acac)_3 \qquad (15b)$$

It is necessary to make further investigation to clarify how the chiral ruthenium(II) complex approaches Co(acac)$_3$. Though this type of investigation is not easy, the works reported by Lappin et al. [5] presents meaningful information about the chirality selection. The other attempt to be carried out is to use one enantiomer of Δ- and Λ-[Ru((−)-menbpy)$_3$]$^{2+}$ and that of Δ- and Λ-[Ru(R(or S)-PhEt-bpy)$_3$]$^{2+}$. In these studies, only the chiral center on the bpy derivatives was utilized, while the molecular asymmetry of the metal center was not used because a mixture of Δ and Λ isomers was used for reaction. If either isomer is used for the reaction, the stereoselecitivty should become larger.

Besides asymmetric photosynthesis and optical resolution, photoinduced deracemization has been reported so far. The term deracemization means that one enantiomer of a racemic mixture converts to the other enantiomer without change in the total concentration; i.e., without any loss of total compounds. This deracemization is very interesting, because the entropy decreases in the reaction; in other words, the increase in enthalpy of the reaction system induces the decrease in entropy. The first example is the deracemization of tris(oxalate)chromate(III), [Cr(ox)$_3$]$^{3-}$, with circularly polarized light [40]. When [Cr(ox)$_3$]$^{3-}$ is irradiated with right-handed circularly polarized light at 546 nm in water, optical activity appears; the best value is about 10% at 1.4°C, while the enantiomer excess decreases upon increase in the temperature. Tris(acetylacetonato)chromium(III), Cr(acac)$_3$, also undergoes deracemization by irradiation of circularly polarized light at 546 nm, at which Cr(acac)$_3$ exhibits d–d abosorption [41]. The observed optical purity was small, too. Thus the irradiation of circularly polarized light does not induce efficiently the photoinduced deracemization of transition-metal complexes, at this moment.

However, very efficient deracemization was reported in tris(4,4′-dimethyl-2,2′-bipyridine)iron(II), [Fe(4,4′-Me$_2$bpy)$_3$]$^{2+}$, with tris(trichlorobenzenediolato)phosphate (trisphat), where trisphat is shown in Scheme 16 [42]. Though this is not a photoreaction but a thermal reaction, we briefly discuss this result because this deracemization is very interesting and provides an interesting idea of photoinduced deracemization.

The essence is simple; one enantiomer isomerizes to the other one under the influence of a chiral environment, if the racemic molecule is configurationally

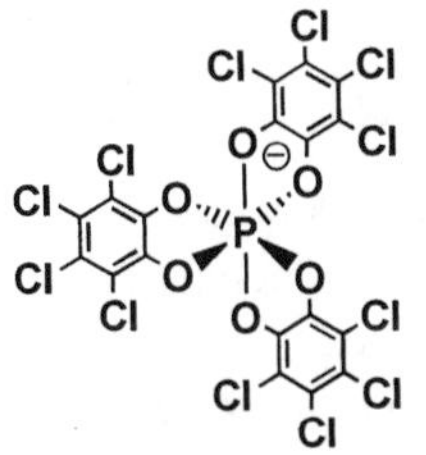

trisphat

Scheme 16

labile. This idea was proposed previously, but the induced asymmetry was considerably small [43]. In the deracemization of $[Fe(4,4'-Me_2bpy)_3]^{2+}$, the very large enantiomer excess over 96% was observed in $CDCl_3$, while the enantiomer excess lowers upon increasing the solvent polarity. This is because trisphat has a chiral center like $[Ru(bpy)_3]^{2+}$ and forms an ion-pair adduct with $[Fe(4,4'-Me_2bpy)_3]^{2+}$ in a nonpolar solvent but does not do so well in a polar solvent. In the ion pair, $[Fe(4,4'-Me_2bpy)_3]^{2+}$ exists under the influence of the chiral trisphat. Since $[Fe(4,4'-Me_2bpy)_3]^{2+}$ is labile, this complex converts to one enantiomer, which forms a stable diastereomeric ion-pair adduct. As a result, the deracemization easily takes place. Though this idea is not useful when the compound is not labile, many coordination compounds become labile in a photo-excited state or in a one-electron reduced state. If so, we can expect that the photoinduced deracemization can take place in the presence of an appropriate chiral species that forms an ion-pair adduct with the compound.

This idea was realized in recent photoinduced deracemization of $Co(acac)_3$ with $\Delta\text{-}[Ru((-)\text{-menbpy})_3]^{2+}$ in the absence of oxygen [44]. When the solution of rac-$Co(acac)_3$ and $\Delta\text{-}[Ru((-)\text{-menbpy})_3]^{2+}$ was irradiated with visible light corresponding to the MLCT absorption of this ruthenium(II) complex, $Co(acac)_3$ was little consumed, but the CD spectrum exhibited a positive peak at 580 nm and a negative one at 660 nm. This CD spectrum is the same as that of Λ-$Co(acac)_3$. This result clearly shows that rac-$Co(acac)_3$ converts to Λ-$Co(acac)_3$ with $\Delta\text{-}[Ru(menbpy)_3]^{2+}$.

$$rac-Co(acac)_3 \xrightarrow{h\nu,\ \Delta-[Ru(menbpy_3)_2]^+} \Lambda-Co(acac)_3 \qquad (16)$$

The concentration of each enantiomer of $Co(acac)_3$ changes, as shown in Fig. 1, and the maximum enantiomer excess is near to 50%.

This value is much larger than that observed in the previously reported deracemizations of $[Cr(ox)_3]^{3-}$ and $Cr(acac)_3$. In this reaction, the basic condition is necessary, and the addition of Hacac increases the enantiomer excess, for which the reason will be discussed below. The reaction mechanism shown in Scheme 17 was proposed. In the mechanism, the ^{3}MLCT excited $\Delta\text{-*}[Ru((-)\text{-menbpy})_3]^{2+}$ is oxidatively quenched by $Co(acac)_3$ to form an exciplex with $Co(acac)_3$ followed by electron transfer to $Co(acac)_3$ from $\Delta\text{-*}[Ru((-)\text{-menbpy})_3]^{2+}$, which leads to the formation of a successor complex, $[\Delta\text{-}Ru^{III}((-)\text{-menbpy})_3^{3+} \cdots Co^{II}(acac)_3^-]$. This successor complex dissociates to $\Delta\text{-}[Ru^{III}((-)\text{-menbpy})_3]^{3+}$, $Co(acac)_2$, and $acac^-$. If the reducing reagent is absent or the reducing reagent does not effectively reduce the ruthenium(III) complex, $Co(acac)_2$ reduces $\Delta\text{-}[Ru^{III}((-)\text{-menbpy})_3]^{3+}$ to $\Delta\text{-}[Ru^{II}((-)\text{-menbpy})_3]^{2+}$ concomitantly with the formation of $Co(acac)_3$. As discussed in Sec. II.A., the photoreduction of $Co(acac)_3$ occurs stereoselectively. In addition, the oxidation of $Co(acac)_2$ to $Co(acac)_3$ occurs stereoselectively, because $Co(acac)_2$ reacts with the chiral ruthen-

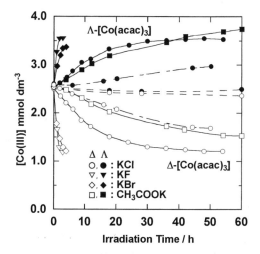

Figure 1 The concentration changes of Δ- and Λ-[Co(acac)$_3$)] in acetonitrile-water (7: 3 v/v) at 25°C. Δ-[Ru(menbpy)$_3$]$^{2+}$ = 32 μmol dm^{-3}, [Co(acac)$_3$] = 5.0 mmol dm^{-3}, [Hacac] = 50 mmol dm^{-3}, [TEA] = 0.5 mmol dm^{-3}. Broken lines (---) represent the reaction in the absence of both Hecac and TEA, dot-dashed lines ([—·—]) that in the absence of Hecac.

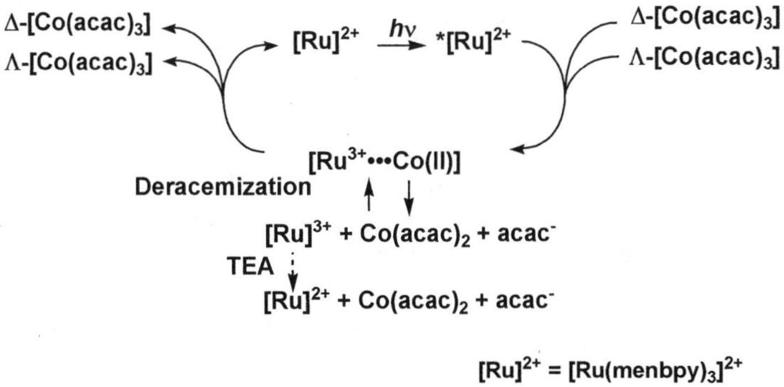

Δ-[Co(acac)$_3$]   Λ-[Co(acac)$_3$]

Λ-[Co(acac)$_3$]   [Ru]$^{2+}$ $\xrightarrow{h\nu}$ *[Ru]$^{2+}$ Λ-[Co(acac)$_3$]

[Ru^{3+}···Co(II)]

Deracemization

[Ru]$^{3+}$ + Co(acac)$_2$ + acac$^-$

TEA

[Ru]$^{2+}$ + Co(acac)$_2$ + acac$^-$

[Ru]$^{2+}$ = [Ru(menbpy)$_3$]$^{2+}$

Scheme 17

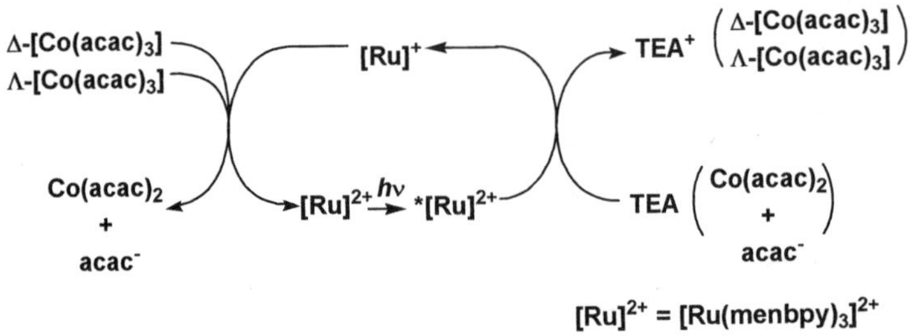

Scheme 18

ium(III) complex, Δ-[Ru^{III}(($-$)-menbpy)$_3$]$^{3+}$. If Δ-Co(acac)$_3$ is more rapidly reduced to Co(acac)$_2$ than Λ-Co(acac)$_3$ in the quenching step, and if Λ-Co(acac)$_3$ is more rapidly formed from Co(acac)$_2$ than Δ-Co(acac)$_3$ in the reverse reaction, Λ-Co(acac)$_3$ accumulates in the reaction. The other reaction mechanism (Scheme 18) is considered possible, in which the photoexcited Δ-*[Ru(($-$)-menbpy)$_3$]$^{2+}$ is reductively quenched by either triethylamine (TEA) or the reduction products, Co(acac)$_2$ and acac$^-$, to afford Δ-[Ru^{I}(($-$)-menbpy)$_3$]$^{+}$. Since Δ-[Ru^{I}(($-$)-menbpy)$_3$]$^{+}$ has a high reducing ability, the reduction of Co(acac)$_3$ easily takes place stereoselectively. This reaction was certainly observed in the pulse radiolysis experiment, as was discussed in Sec. II.B [34]. However, this mechanism was excluded, as follows: If the photoinduced deracemization takes place through the mechanism of Scheme 18, the ^{3}MLCT-excited Δ-*[Ru(($-$)-menbpy)$_3$]$^{2+}$ is reduced by acac$^-$ to afford Δ-*[Ru(($-$)-menbpy)$_3$]$^{+}$. The quenching reactions by Co(acac)$_2$ and TEA can be neglected under the experimental conditions, because the concentrations of Co(acac)$_2$ and TEA are much lower than that of acac$^-$. Then [Ru(($-$)-menbpy)$_3$]$^{+}$ reduces Co(acac)$_3$ to Co(acac)$_2$. This means that Co(acac)$_2$ accumulates in the solution. However, the concentration of Co-(acac)$_3$ little decreases in the reaction. It should be concluded that the deracemization takes place through the mechanism of Scheme 17.

The stereoselectivity in the oxidative quenching should be the reverse of that in the reoxidation process to accumulate one of the enantiomers in the reaction, as shown in Scheme 17. Actually, the oxidative quenching occurs more rapidly for the homochiral pair, Δ-*[Ru(($-$)-menbpy)$_3$]$^{2+}$ $\cdots$ Δ-Co(acac)$_3$, than that of the heterochiral pair, Δ-*[Ru(($-$)-menbpy)$_3$]$^{2+}$ $\cdots$ Λ-Co(acac)$_3$. In the reoxidation step, Λ-Co(acac)$_3$ is more easily produced than Δ-Co(acac)$_3$ [38]. Thus the deracemization takes place smoothly. The next issue to be discussed is the reason that the stereoselectivity of the oxidative quenching is the reverse of that of the reoxidation reaction. The authors explained the reason as follows.

Since the oxidative quenching of a homochiral pair occurs more rapidly than that of a heterochiral pair, the encounter complex of homochiral pair would be more stable than that of the heterochiral pair, as shown in Scheme 19. In the reoxidation step, the situation is reversed; the energy difference between a homochiral pair and a heterochiral one is smaller in the transition state and the product than that in the reactant, because the contact is stronger in the encounter complex, $[Ru((-)-menbpy)_3{}^{3+} \cdots Co(acac)_3{}^-]$, owing to the stronger electrostatic attraction than that in the product complex, $[Ru((-)-menbpy)_3{}^{2+} \cdots Co(acac)_3]$. This means that the activation energy of the reoxidation is smaller in the heterochiral pair than in the homochiral pair. As a result, the stereoselection in the quenching step is the reverse of that of the reoxidation step.

At the end of this section, we wish to mention the reasons that this photoinduced deracemization is interesting. The deracemization is interesting itself, of course. Besides, this reaction is of considerable interest from the point of view of photochemistry, because both the highly reducing ability of the photosensitizer and the highly oxidizing ability of the one-electron oxidized photosensitizer are utilized in the reaction. This type of photoreaction provides a wider application of photochemistry.

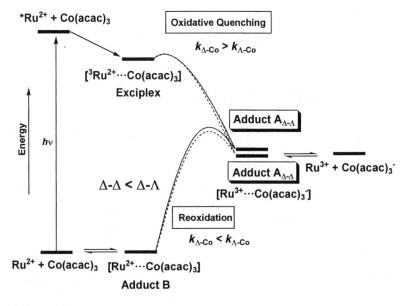

Scheme 19

(A) Usual metal complex

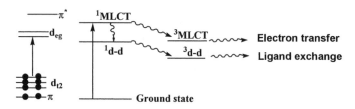

(B) Copper(I) complex

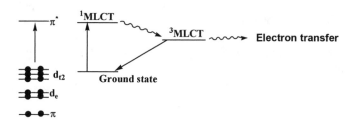

Scheme 20

III. CHIRAL COPPER(I) COMPLEXES AND THEIR APPLICATION TO STEREOSELECTIVE PHOTOINDUCED ELECTRON TRANSFER REACTION

A. Electronic Structure of the Copper(I) Complex in the Excited State

First, we will briefly discuss the electronic structure of the copper(I) complex in the excited state, because the electronic structure is less well known than that of the ruthenium(II) complex. In the usual transition-metal complexes, the d–d excited state exists in a lower energy than the MLCT excited state, as shown in Scheme 20A. This feature is not favorable for the electron transfer reaction, as was discussed in Sec. II.B. However, the d–d excited state does not exist in the copper(I) complexes, because the copper(I) atom takes a d^{10} electron configuration.

If the ligand has a π^* orbital like 2,2′-bipyridine, 1,10-phenanthroline, and their derivatives, only the MLCT excitation occurs. As a result, the triplet ^{3}MLCT excited state is formed as the lowest energy excited state in such copper(I) complexes, as shown in Scheme 20B. This electron configuration in the excited state is essentially the same as that of $[Ru(bpy)_3]^{2+}$ shown in Scheme 2. Thus we can expect that similar photochemical reactions are performed with the copper(I) photosensitizer. Photochemical properties of copper(I) complexes have been investigated well by McMillin and his collaborators [8], as shown in Table 3. They reported one important difference between copper(I) and ruthenium(II) complexes: in the copper(I) complex, the conjugated ligand such as 2,2′-bipyridine and 1,10-phenanthroline must have substituents at the neighbor positions to the coordinating site, as shown in Scheme 21.

If these substituents are not introduced to the ligand, the lifetime of the excited state becomes too short to perform a photoreaction. This is understood in terms of the solvation in the excited state, as follows: In the ground state, the copper(I) complex takes a tetrahedral (Td) or pseudo-Td structure because of its d^{10} electron configuration [45]. In the MLCT excited state, however, the copper center becomes similar to copper(II), which takes a d^9 electron configuration.

Table 3 Absorption and Emission Maxima, Lifetime of the ^{3}MLCT Excited State, and Redox Potential of Copper(I) Complexes

	Absorption λ_{max} $(\varepsilon)^a$	MLCT emission λ_{max} $(\tau)^b$	Redox potential^c	
			$E^{2+/+}$	$E^{2+/+}*$
$[Cu(dmp)_2]^{+d}$	454 (7950)	730		-1.4
$[Cu(bpy)(PPh_3)_2]^{+e}$		620 (4 μs at 90K)		
$[Cu(bpy)(PPh_3)_2]^{+f}$	355(2820)	650 at 295K		
$[Cu(phen)(PPh_3)_2]^{+e}$		608 (115 μs at 90K)		
$[Cu(dmp)(PPh_3)_2]^{+e}$	365 (1.40)	545 (225 μs at 90K)	ca.0.7	-0.9
$[Cu(dmp)(PPh_3)_2]^{+g}$		560 (330 ns at 298K)		
$[Cu(phen)(PPh_3)_2]^{+h}$		680 (220 ns at 298K)		
$[Cu(dmp)_2]^{+i}$	460(16500)	750 (90 ns at 298K)		
$[Cu(tmdcbpy)(PPh_3)_2]^{+j}$	352(1910)	550 (256 ns at 298K)		

[a] The absorption maximum in nm. In parentheses is the molar extinction coefficient (dm^3 mol^{-1} cm^{-1}).

[b] The emission maximum in nm. In parentheses is the lifetime in ns.

[c] The $E^{2+/+}*$ value represents the redox potential (in V vs. SCE) for the copper(I) complex.

[d] Dmp = 2,9-dimethyl-1, 10-phenanthroline. In dichloromethane. Ref. 8c.

[e] Ref. 8e. In methanol-ethanol (1 : 4) glass.

[f] Ref. 8g. In methanol.

[g] Ref. 8q. In methanol.

[h] Ref. 8q. In dichloromethane.

[i] Ref. 8s. In ethanol-methanol (4 : 1) for absorption and in dichloromethane for emission.

[j] Tmdcbpy = 4,4′-6,6′-tetramethyl-5,5′-bis[(S)-(−)-1-phenylethylcarbamoyl]-2,2′-bipyridine. Ref. 53.

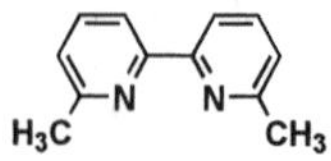

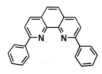

6,6'-dimethyl-2,2'-bipyridine

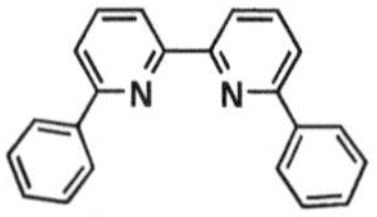

6,6'-diphenyl-2,2'-bipyridine

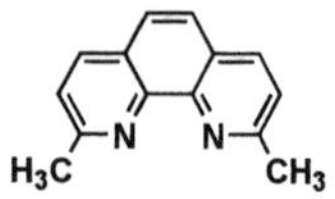

2,9-dimethyl-1,10-phenanthroline

2,9-diphenyl-1,10-phenanthroline

Scheme 21

This was experimentally ascertained with resonance Raman spectroscopy [46] and theoretically with the ab initio MO method [47]. Because of this electron configuration, the copper(I) complex tends to take a square planar structure in the ^{3}MLCT excited state [47], in which the solvent molecule can easily approach the copper(II) center to induce deactivation through the energy transfer to vibration modes of solvent molecule [8h,8m,8s]. However, the substituents that are introduced at the neighbor positions to the coordination sites suppress the geometry change to a square planar structure, because such substituents induce steric repulsion with each other in the planar structure. As a result, the energy transfer to the solvent molecules from the excited copper(I) complex becomes difficult, and the lifetime of the excited state becomes longer in the copper(I) complex with the ligands shown in Scheme 21.

Because the copper(I) complexes with 2,2'-bipyridine, 1,10-phenanthroline, and their derivatives have a triplet MLCT excited state like that of [Ru(bpy)$_3$]$^{2+}$, the copper(I) complexes exhibit similar photocatalyses to those of [Ru(bpy)$_3$]$^{2+}$; for instance, the trans–cis isomerization of stylbene through energy transfer [48] and the photoreduction of viologen compounds [9b,c,e,49,50] were successfully carried out with the copper(I) complexes. Also, a Grätzel-type solar cell was constructed with the copper(I) complexes, recently [51,52].

Thus one can expect that the copper(I) complexes with 2,2'-bipyridine, 1,10-phenanthroline, and their derivatives are successfully applied to asymmetric photoreactions, as with chiral ruthenium(II) complexes, if the optically active moiety is introduced to the ligand, as discussed above (see introduction).

B. Stereoselective Photoinduced Electron Transfer Reaction of Copper(I) Complexes

Since the copper(I) complexes, $[Cu(NN)_2]^+$ and $[Cu(NN)(PR_3)_2]^+$ (NN = 1,10-phenanthroline, 2,2'-bipyridine, and their derivatives) were applied to stoichiometric and catalytic photoreduction of cobalt(III) complexes [8a,b,e,9a,d], one can expect to perform the asymmetric photoreduction system with the similar copper(I) complexes if the optically active center is introduced into the copper(I) complex. To construct such an asymmetric photoreaction system, we need chiral copper(I) complex. Copper(I) complex, however, takes a four-coordinate structure. This means that the molecular asymmetry around the metal center cannot exist in the copper(I) complex, unlike in six-coordinate octahedral ruthenium(II) complexes. Thus we need to synthesize some chiral ligand in the copper(I) complexes.

The first example of a chiral copper(I) photosensitizer is [Cu(dmp)((R,R-diop))]$^+$ [R,R-diop = (R,R)-2,3-O-isopropylidene-2,3-dihydroxy-1,4-bis(diphenylphosphino)-butane; dmp = 2,9-dimethyl-1,10-phenanthroline], in which two chiral centers are introduced in the (R,R)-diop ligand. This complex was applied to the stereoselective photoreduction of [Co(edta)]$^-$ [25]. After the reaction, the CD spectrum exhibits a positive peak at 590 nm and a negative one at 515 nm, which indicates the presence of excess Δ-[Co(edta)]$^-$. This means that Λ-[Co(edta)]$^-$ more rapidly reacts with the photoexcited copper(I) complex than does the Δ-enantiomer, where the stereoselectivity, defined as the ratio of the conversion rate, is 1.17. However, the photoreduction of Co(acac)$_3$ and [Co(bpy)$_3$]$^{3+}$ occurs without stereoselectivity. This is probably because the electrostatic attraction between [Cu(dmp)((R,R-diop))]$^+$ and [Co(edta)]$^-$ is favorable for the stereoselection, but such interaction does not exist between [Cu(dmp)((R,R-diop))]$^+$ and the other cobalt(III) complexes.

$$\Delta + \Lambda - [Co(edta)]^- \xrightarrow[\Lambda-[Co(edta)]^- > \Delta-[Co(edta)]^-, \text{in ethanol-water}]{[Cu(dmp)((R,R-diop))]^+,\, h\nu} Co(II) + edta^{4-} \qquad (17a)$$

$$\Delta + \Lambda - [Co(acac)_3] \xrightarrow[\Lambda-[Co(acac)_3] \approx \Delta-[Co(acac)_3] \text{ in ethanol-water}]{[Cu(dmp)((R,R-diop))]^+,\, h\nu} Co(acac)_2 + acac \qquad (17b)$$

$$\Delta + \Lambda - [Co(bpy)_3]^{3+} \xrightarrow[\Lambda-[Co(bpy)_3]^{3+} \approx \Delta-[Co(bpy)_3]^{3+} \text{ in ethanol-water}]{} Co(II) + 3 \text{ bpy} \qquad (17c)$$

In this copper(I) complex, there is one weak point: the chiral centers that exist in (R,R)-diop are distant from the dmp ligand on which an excited electron mainly exists. This means that the reaction site is far from the chiral center. It is desirable to introduce the chiral center on the conjugated ligand, like menbpy. However, we need to remember that the methyl or phenyl substituents should be introduced to the neighbor positions to the coordinating sites. Thus menbpy and PhEtbpy, which were used to construct chiral ruthenium(II) complexes, are not useful for the copper(I) complex.

Considering these conditions, a new conjugated ligand, 4,4'-6,6'-tetramethyl-5,5'-bis[(S)-(−)-1-phenyl-ethyl-carbamoyl]-2,2'-bi-pyridine (tmdcbpy), was synthesized, as shown in Scheme 22 [53]. In this ligand, two chiral groups are introduced to 2,2'-bipyridine flame through the amido bridge, and two methyl groups are introduced to 4 and 4' positions.

As expected, the copper(I) complex of tmdcbpy, $[Cu(tmdcbpy)(PPh_3)_2]^+$, exhibits a large MLCT absorption in the near UV or visible region because this ligand has π and π^* orbitals. Also this copper(I) complex has a long-lived ^{3}MLCT excited state, as shown in Table 3. This complex shows the CD spectrum in which a positive peak is observed at 300 nm and a negative one at 360 nm. The chiral center is involved in the organic group, which does not have any absorption around this region. Since the CD spectrum is observed in the range of the MLCT absorption, the chiral center induces the circular dichroism in the π^* orbital of 2,2'-bipyridine frame probably through the conjugation between the π^* orbital and the σ orbital of the –CH(Me)Ph moiety.

The photoreduction of $[Co(edta)]^-$ by $[Cu(tmdcbpy)(PPh_3)_2]^+$ occurs under irradiation of near UV light (350 to 400 nm). After the reaction, the new CD spectrum is observed around 450 to 600 nm in which a positive peak is found at 500 nm. This result clearly shows that Λ-$[Co(edta)]^-$ is more rapidly consumed than the Δ enantiomer, since Δ-$[Co(edta)]^-$ exhibits a positive peak at 500 nm in its CD spectrum.

$$\Delta + \Lambda - [Co(edta)]^- \xrightarrow[\Lambda-[Co(edta)]^- > \Delta-[Co(edta)]^- \text{ in ethanol-water}]{[Cu(tmdcbpy)(PPh_3)_2]^+, h\nu} Co(II) + edta^{4-} \quad (18)$$

The enantiomer excess depends on the solvent system; isopropanol-water $<$ ethanol-water $\sim$ methanol-water. The best values are 42% in 75% ethanol-water at 25°C and 40% in 75% methanol-water. The reaction mechanism was investigated kinetically. In general, photoreaction takes place via either a dynamic quenching mechanism or a static quenching mechanism, as shown in Scheme 23.

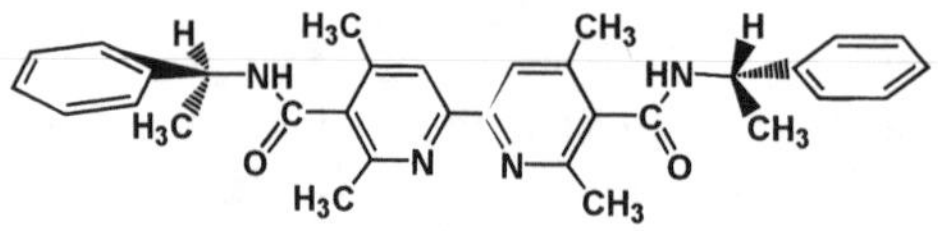

4,4',6,6'-tetramethyl-5,5'-bis[(S)-(-)-1-
phenylethylcarbamoyl]-2,2'-bipyridine

Scheme 22

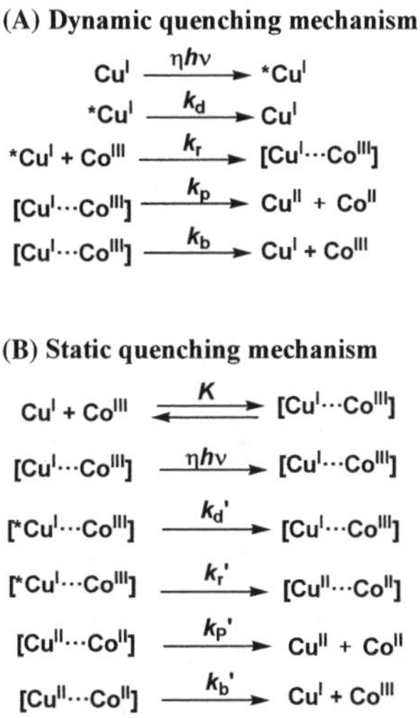

(A) Dynamic quenching mechanism

$$Cu^I \xrightarrow{\eta h\nu} {}^*Cu^I$$

$${}^*Cu^I \xrightarrow{k_d} Cu^I$$

$${}^*Cu^I + Co^{III} \xrightarrow{k_r} [Cu^I{\cdots}Co^{III}]$$

$$[Cu^I{\cdots}Co^{III}] \xrightarrow{k_p} Cu^{II} + Co^{II}$$

$$[Cu^I{\cdots}Co^{III}] \xrightarrow{k_b} Cu^I + Co^{III}$$

(B) Static quenching mechanism

$$Cu^I + Co^{III} \xrightleftharpoons{K} [Cu^I{\cdots}Co^{III}]$$

$$[Cu^I{\cdots}Co^{III}] \xrightarrow{\eta h\nu} [{}^*Cu^I{\cdots}Co^{III}]$$

$$[{}^*Cu^I{\cdots}Co^{III}] \xrightarrow{k_d'} [Cu^I{\cdots}Co^{III}]$$

$$[{}^*Cu^I{\cdots}Co^{III}] \xrightarrow{k_r'} [Cu^{II}{\cdots}Co^{II}]$$

$$[Cu^{II}{\cdots}Co^{II}] \xrightarrow{k_p'} Cu^{II} + Co^{II}$$

$$[Cu^{II}{\cdots}Co^{II}] \xrightarrow{k_b'} Cu^I + Co^{III}$$

Scheme 23

In the dynamic quenching mechanism (Scheme 23A), the Stern–Volmer relation is represented by Eq. (19), where τ and τ_0 mean the lifetime in the presence and the absence of quencher, respectively, and $[Co^{III}]$ represents the concentration of the cobalt(III) complex.

$$\frac{\tau_0}{\tau} = 1 + k_r\, \tau_0 [Co^{III}] \tag{19}$$

In the static quenching mechanism (Scheme 23B), the Stern–Volmer equation based on emission intensity is given by Eq. (20), where I and I_0 mean the emission intensities in the presence and the absence of quencher, respectively.

$$\frac{I_0}{I} = 1 + K[Co^{III}] \tag{20}$$

Even when the quenching reaction takes place through the static quenching mechanism, some of the photosensitizer remains free from adduct formation with the quencher, since the adduct between the photosensitizer and the quencher is not in general stable very much. This means that the dynamic quenching takes place concomitantly with the static quenching. For the dynamic quenching reaction, Eq. (19) is also kept. Thus the τ_0/τ value is not the same as the I_0/I value in the static quenching mechanism. In the quenching reaction of [Cu(tmdc-bpy)(PPh$_3$)$_2$]$^+$ with [Co(edta)]$^-$, the slope of the Stern–Volmer relation based on the I_0/I value is much different from that of the Stern–Volmer relation based on the τ_0/τ value. It should be clearly concluded that the photoreduction of $[Co(edta)]^-$ proceeds through the static quenching mechanism. The quenching reaction was investigated with Δ- and Λ-$[Co(edta)]^-$ to clarify the stereoselectivity in the quenching step. However, little difference between Δ- and Λ-$[Co(edta)]^-$ is observed in both Stern–Volmer relations based on τ_0/τ and I_0/I values. From this result, the authors concluded that the stereoselectivity does not arise from the quenching process, i.e., the photoinduced electron transfer process, but from the charge-separation and/or back-electron transfer process. Also, we need to remember the possibility that the energy transfer is involved in the quenching reaction and the energy-transfer would occur with much smaller stereoselectivity.

Later, this copper(I) complex was applied to photoasymmetric synthesis of $[Co(edta)]^-$ from cobalt(II) salt under near UV light irradiation in the presence of oxygen gas [54]. The catalytic cycle is shown in Scheme 24; the first step is photoexcitation of $[Cu(tmdcbpy)(PPh_3)_2]^+$, the second is the oxidative deactivation of *$[Cu(tmdcbpy)(PPh_3)_2]^+$ by molecular dioxygen to afford the copper(II) complex, $[Cu(tmdcbpy)(PPh_3)_2]^{2+}$, and the last is oxidation of Co(II) salt by $[Cu(tmdcbpy)(PPh_3)_2]^{2+}$. The last step occurs stereoselectively because of the presence of chiral ligand in the copper(II) complex.

$$CoX_2 + L_4edta \xrightarrow{[Cu(dmp)((R,R\text{-}diop))]^+, \, O_2, \, h\nu} [Co(edta)]^-$$

$$Na_4edta \quad \Delta-[Co(edta)]^- > \Lambda-[Co(edta)]^- \text{ in ethanol-water}$$
$$H_4edta \quad \Lambda-[Co(edta)]^- > \Delta-[Co(edta)]^- \text{ in ethanol-water}$$
$$(21)$$

The stereoselectivity depends on the kind of edta salt; when H_4edta was used with $Co(OAc)_2$, Λ-$[Co(edta)]^-$ was formed with the enantiomer excess of 7%. However, when Na_4edta was used with $Co(OAc)_2$, the Δ-enantiomer was formed with the same enantiomer excess. The stereoselectivity also depends on the kind of cobalt(II) salt. When $Co(NO_3)_2$ was used, the selectivity did not depend very much on the kind of edta salt; the Λ-enantiomer was formed with the enantiomer excess of 9% and 5% in H_4edta and Na_4edta, respectively. However, when $CoCl_2$ was used, the reaction does not proceed at all in the presence of H_4edta but

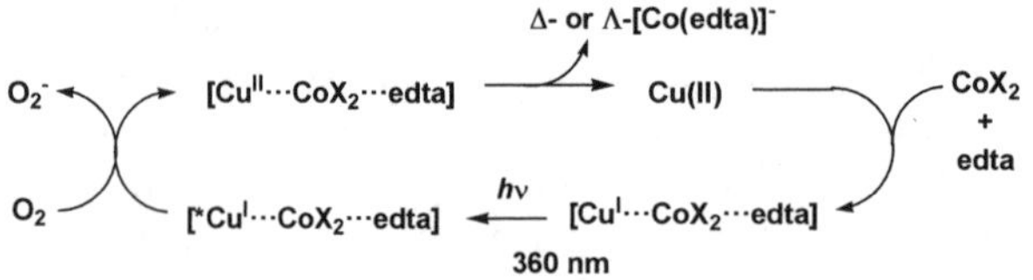

Scheme 24

proceeds a little with no stereoselectivity in the presence of Na_4edta. These results suggest that the interaction between Co(II) and the $edta^-$ anion is important and that the reaction does not take place when the anion of cobalt(II) salt inhibits this interaction. The details are still ambiguous.

IV. STEREOSELECTIVE PHOTOINDUCED ELECTRON TRANSFER REACTION OF ZINC(II)-SUBSTITUTED MYOGLOBIN

A. Electron Transfer Reactions of Metalloproteins and the Stereoselectivity

As is well known, the electron transfer reaction deeply relates to the function of metalloenzymes. In this regard, many experimental studies have been carried out concerning the electron transfer reactions of metalloenzymes [55]. Since metalloenzymes involve many amino acid residues, we expect that the stereoselectivity should be observed in electron transfer reactions. In spite of this expectation, reports of the stereoselective electron transfer reactions of metalloenzymes have been limited, so far. Here, we wish to mention several works in which thermal stereoselective electron transfer reaction was investigated.

The first observation was reported in the electron transfer reaction between spinach plastocyanin with optically active iron(II) complexes, [Fe(S,S)-alamp] (Λ-form) and its (R,R)-isomer (Δ-form), where alamp is 2,6-bis[3-(S)- or 3-(R)-carboxy-2-azabutylpyridine (see Scheme 25) [56].

$$\text{Spinach plastocyanin} + [\text{Fe(alamp)}]$$
$$\longrightarrow [\text{Spinach plastocyanin}]^+ + [\text{Fe(alamp)}]^-$$
$$k_{obs}(R,R) > k_{obs}(S,S) \tag{22}$$

In this complex, there are two optically active sites. Spinach plastocyanin is a type I copper protein, in which two reactive sites have been identified on its surface, at least. The electron transfer reaction occurs with significantly large stereoselectivity; the ratio of the observed reaction rate constant (k_Δ/k_Λ) is 1.6 to 2.0. The difference in the activation enthalpy, $\Delta\Delta H\ddagger_{\Delta-\Lambda}$, is 3.0 kJ mol^{-1}, and the difference in the activation entropy, $\Delta S\ddagger_{(\Delta-\Lambda)}$, is 15 J mol^{-1} K^{-1}. This means that the stereoselectivity arises from the entropy term.

One year later, the stereoselective electron transfer reaction between ferrocytochrome c and tris(acetylacetonato)cobalt(III) was reported independently [57].

$$\text{Ferrocytchrome c(II)} + \text{Co(acac)}_3$$
$$\longrightarrow \text{Ferricytchrome c(III)} + \text{Co(II)} \quad k(\Delta) > k(\Lambda) \tag{23}$$

$$\text{Ferrocytchrome c(II)} + [\text{Co(bpy)}_3]^{3+}$$
$$\longrightarrow \text{Ferricytchrome c(III)} + \text{Co(II)} \quad k(\Lambda) > k(\Delta) \tag{24}$$

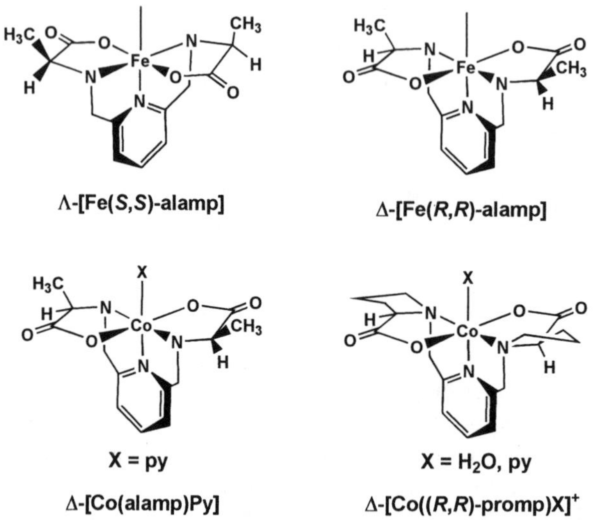

Λ-[Fe(S,S)-alamp]   Δ-[Fe(R,R)-alamp]

X = py   X = H₂O, py

Δ-[Co(alamp)Py]   Δ-[Co((R,R)-promp)X]⁺

Scheme 25

In this reaction, rac-Co(acac)₃ was used as an oxidant, and Λ(−)-Co(acac)₃ remained in excess after the reaction. The enatiomer excess is not large; 2 to 6%. This enantiomer excess depends on pH and ethanol content of the solvent; the selectivity becomes larger when pH decreases and the ethanol content increases. Acetate buffer yields larger enantiomer excess than does phosphate buffer. The same protein was applied to the stereoselective electron transfer reaction with [Co(bpy)₃]³⁺ [58]. The stereoselectivity is reverse to and larger than those observed in the reaction between ferrocytochrome c(II) and Co(acac)₃, and the best enantiomer excess is 12%. The stereoselectivity moderately increases upon increase of the ethanol content but little depends on pH. The larger enantiomer excess is interpreted as that the bpy ligand can penetrate into the hydrophobic crevice of ferrocytochrome c(II), which is an active site for the electron transfer reaction. The reaction rate was analyzed via the Michaelis–Menten scheme. The K_M value is smaller in the reaction with [Co(bpy)₃]³⁺ than in the reaction with Co(acac)₃. This means that [Co(bpy)₃]³⁺ forms the activate adduct with ferrocytochrome c(II) to a greater extent than does Co(acac)₃. This result is consistent with the explanation that the bpy ligand can penetrate into the hydrophobic crevice.

Spinach ferredoxine was applied to the stereoselective electron transfer reaction with optically active cobalt(III) complexes that are similar to the iron complexes that were applied to the stereoselective electron transfer reaction of plastocyanine [59]. Ferredoxins are [2Fe-2S] proteins involved in the photosynthetic electron transport of photosynthetic systems. Thus the electron transfer reaction of ferredoxins is an important subject of research. Δ-[Co((R,R)-promp)X]$^+$, where promp is NN'-[(pyridine-2,6-diyl)bis(methylene)]bis[(S)- or (R)-proline] (see Scheme 25), reacts with spinach ferredoxin more rapidly than the Λ-enantiomer; the stereoselectivity is 1.32 for [Co(promp)(H$_2$O)]$^+$, 1.17 for [Co(promp)(py)]$^+$, and 1.27 for [Co(alamp)(py)]$^+$.

$$[Co((R,R)\text{-}promp)X]^+ + ferredoxin$$
$$\longrightarrow \quad Co(II) + promp + X + [ferredoxin]^+$$
$$\Delta\text{-}[Co((R,R)\text{-}promp)X]^+ > \Lambda - [Co((R,R)\text{-}promp)X]^+ \quad (25)$$

The $\Delta\Delta H\ddagger_{(\Delta-\Lambda)}$ value is -11.8 kJ mol^{-1} for [Co(alamp)(py)]$^+$ and -5.6 kJ mol^{-1} for [Co((R,R)-promp)(H$_2$O)]$^+$. These values are largely compensated by the $\Delta\Delta S\ddagger_{(\Delta-\Lambda)}$ value. Interestingly, the $\Delta H\ddagger$ value linearly depends on the $\Delta S\ddagger$ value. This result shows that the enthalpy–entropy compensation certainly occurs in the reaction. The compensation is discussed, as follows: Since this reaction involves outer-sphere-type electron transfer, the activation enthalpy is necessary to bring the reactants close enough together to allow electron transfer and to break up the hydration shell of the reactants. The entropy term contains the contribution due to rearrangement of the solvent molecules. As a result, the enthalpy–entropy compensation occurs in the reaction.

The stereoselective reduction of spinach plastocyanin with several cobalt cage complexes (Scheme 26) has been reported, too [60]. These cage complexes are very useful for investigation of outer-sphere electron transfer reactions because of their inertness to hydrolysis and to loss of ligands in the redox reaction.

$$Plastocyanin(II) + [Co^{II}((NR_2)_2 - sar)]^{4+}$$
$$\longrightarrow \quad Plastocyanin(I) + [Co^{III}((NR_2)_2 - sar)]^{5+} \quad k(\Lambda) > k(\Delta) \quad (26)$$

The stereoselectivity (k_Λ/k_Δ) is 1.7 for [Co((N(CH$_3$)$_3$)$_2$-sar)]$^{4+}$ [(NMe$_2$)-sar = 1,8-bis(trimethylammonium)-3,6,10,13,16,19-hexaazabicyclo[6,6,6]icosane] [60]. This stereoselectivity depends on pH, but it is independent of the ionic strength. The dependence is discussed in terms of the protonation of the active site. On the other hand, the reduction of plastocyanin by [Co((NH$_2$)$_2$-sar)]$^{2+}$ (NH$_2$-sar = 1,8-diamino-3,6,10,13,16,19-hexaazabicyclo[6,6,6]icosane) occurs without stereoselectivity. This is probably because the negatively charged plastocyanin ($-9e$ at pH = 7.0) easily forms an ion-pair adduct with [Co((N(CH$_3$)$_3$)$_2$-sar)]$^{4+}$ but does not easily form the adduct with [Co((NH$_2$)$_2$-sar)]$^{2+}$. The reduction of plastocyanine(II) by [Co((NH$_2$)$_2$-sar)]$^{2+}$ and [Co((NO$_2$-capten))$^{2+}$] [NO$_2$-capten = 1-methyl-8-nitro-3,13,16-trithia-6,10,19-triazabicyclo[6,6,6]icosane] occurs without stereoselectivity. Also, the reduction of ferricytochrome C takes place without stereoselectivity. This result is interpreted in terms of the highly positive charge ($+7$ at pH = 7.0) of the protein, too, as follows: Because of the

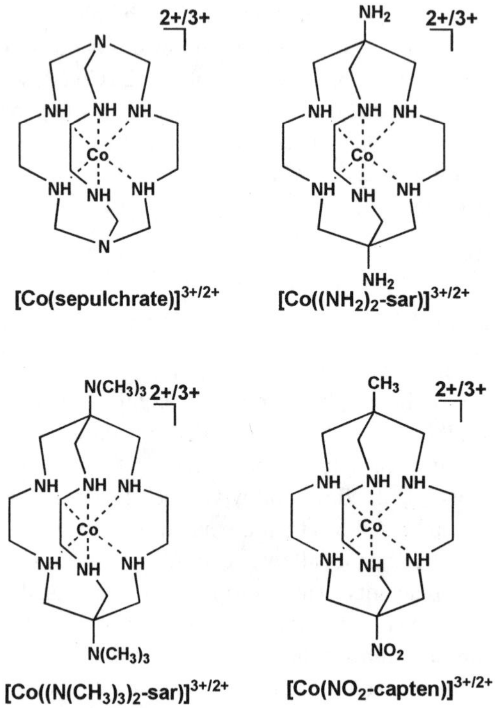

[Co(sepulchrate)]$^{3+/2+}$ [Co((NH$_2$)$_2$-sar)]$^{3+/2+}$

[Co((N(CH$_3$)$_3$)$_2$-sar)]$^{3+/2+}$ [Co(NO$_2$-capten)]$^{3+/2+}$

Scheme 26

electrostatic repulsion between the protein and the cobalt(II) complex, the ion-pairing association constant should be small, which leads to the weak contact between the protein and the Co(II) complexes, when the ion-pair adduct is easily formed.

From these thermal electron transfer reactions, one can expect that the photoinduced electron transfer reaction of the proteins would occur stereoselectively, although such a reaction has not been investigated yet.

B. Photoexcited States of Zinc(II) Porphyrin

As is well known, metalloporphyrins exhibit intense Soret bands around 420 nm and weaker Q bands between 500 and 600 nm. These absorption bands are assigned to the π–π* excitation in origin [61]. Because these absorptions are observed in the visible region and their molar extinction coefficients are very large, metalloporphyrins are potentially useful as photosensitizers. In particular, zinc(II) porphyrin has been used as an efficient photosensitizer, because of its very long lifetime and considerably negative redox potential of the excited state.

We wish to discuss briefly the photoexcited state of zinc(II) porphyrin. Zinc(II) porphyrin has several singlet excited states that are formed by visible light absorption [62]. Of those singlet excited states, the first and the second, S_1 and S_2, have been investigated well [63]. Energy levels of these excited states are schematically shown in Scheme 27.

The lifetime of the S_2 state is 2.35 ps. In a different paper, the slightly longer lifetime (3.6 ps) was reported [64]. The lifetime of the S_2 state correlates with the rise-time of the S_1 state. The lifetime of the S_1 state is 2.7 ns. It is noted that the $S_1 \rightarrow S_0$ conversion is not important, and the intersystem crossing from the S_1 state to the triplet T_1 state occurs with an efficiency near to unity [65]. The triplet state is very long-lived (see Table 4), and its redox potential is sufficiently negative, like that of *$[Ru(bpy)_3]^{2+}$ in the ^3MLCT excited state [66,67].

Apparently, one can expect that zinc(II) porphyrin is an excellent photosensitizer because of the very large absortion band in the visible region, the long lifetime, and the negative redox potential in the triplet excited state. The other interesting feature is that there are three types of zinc(II) porphyrins; Zn(TPP) (TPP = tetraphenylporphyrin) is neutral, Zn(TPPS) (TPPS = tetra(p-sulfoxyphenyl)porphirin) is negatively charged, and Zn(TMPyP) (TmyPyP = tetra(p-methylpiridino)prophyrin) is positively charged. One can expect to investigate the charge effects in the photochemical reactions with these zinc(II) porphyrins. The other important feature is that zinc(II) porphyrin is useful to substitute iron porphyrin in heme proteins. The proteins modified with zinc(II) porphyrin are photoreactive and can be used in photoinduced electron transfer reactions.

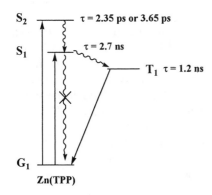

S_2 $\tau = 2.35$ ps or 3.65 ps

S_1 $\tau = 2.7$ ns

T_1 $\tau = 1.2$ ns

G_1

Zn(TPP)

Scheme 27

Table 4   The Lifetimes of the Lowest Energy Triplet Excited State and the Redox
Potentials of Metalloporphyrins in the Singlet and Triplet States

	Triplet excited state τ(ms)	Singlet excited state		Triplet excited state	
		$E^*(P)^a$	$E^*(P)^b$	$E(P^+/P^*)^c$	$E(P^*/P^-)^c$
Zn(TPP)	1.20	2.06	1.59	−0.55	+0.48
Zn(TPPS)	1.40	2.05	1.61	−0.75	+0.45
Zn(TMPyP)	1.30	1.98	1.63	−0.45	+0.78
Zn(TPPC)	1.30	2.03	1.60	−0.80	+0.35
H₂TPP	1.5	1.91	1.44	−0.20	+0.58
H₂(TPPS)	0.42	1.92	1.44	−0.34	
H₂(TMPyP)	0.165	1.83	1.44		
H₂(TPPC)	1.30	1.91	1.44		
Pd(TPP)	0.430	2.21	1.80	−0.56	+0.89
Pd(TPPS)	0.350	2.21	1.78	−0.60	+0.74
Pd(TMPyP)	0.144	2.18	1.81	−0.41	+1.18
Pd(TPPC)	0.285	2.21	1.77		

[a] Energy level of the S_2 state.
[b] Energy level of the S_1 state.
[c] Redox potential. Data for TPP species are measured in CH_2Cl_2 and converted to values vs. NHE by adding 0.24 V. The other data are measured in water.
Sources: Refs. 66, 67.

C.   Photoinduced Electron Transfer Reaction of Zinc(II) Myoglobin with Chiral Viologen

Myoglobin is a well-known heme enzyme. Since the heme moiety is not photoreactive, myoglobin itself cannot be applied to a photoinduced electron transfer reaction. However, the iron ion can be substituted for the zinc(II) ion [56], and such zinc(II)-substituted myoglobin (hereafter, this is represented as ZnMb) can be applied to a photoinduced electron transfer reaction, since zinc(II) porphyrin is one of the good photosensitizers, as discussed in Sec. IV.B. The stereoselective photoinduced electron transfer reaction between ZnMb and chiral viologen derivatives was successfully carried out by Tsukahara and collaboratos [68]. They synthesized chiral viologen derivatives, 1,1′-bis(1-phenylethylcarbamoylmethyl)-4,4′-bipyridinium (OAV^{2+}) and its 1-naphthyl derivative ($NOAV^{2+}$), by introducing chiral amine groups into two —CH_2COOH groups that are bound with the N atom at the 1 and 1′ positions, as shown in Scheme 6. Because these viologen derivatives possess two optically active sites, they are named (R,R) or (S,S) isomers when both optically active sites have R or S configuration, respectively. They investigated the quenching reaction of triplet excited zinc(II) myoglobin ^{3}(ZnMb)* with these viologen derivatives.

$$^3(ZnMb)* + ROAV^{2+} \rightarrow ZnMb^+ + ROAV^+ \qquad (27)$$

The (S,S) isomer more rapidly quenches ^{3}(ZnMb)* than does the (R,R) isomer, and the stereoselectivity, which is defined as the ratio of quenching reaction rate constants, $k_q(S,S)/k_q(R,R)$, depends on the R group of the chiral amine moiety; it is 1.5 in quenching by NOAV^{2+}, 1.3 in quenching by OAV^{2+}, and 1.2 in quenching by CHOAV^{2+}, where the ionic strength is 0.02 mol dm^{-3} (NaCl). When the ionic strength increases to 0.32 mol dm^{-3}, the reaction rate increases but the stereoselectivity decreases to 1.4 in the quenching by NOAV^{2+} and disappears in the quenching by OAV^{2+}. These results were interpreted in terms of the electrostatic interaction, as follows: The electrostatic attraction becomes weak in high ionic strength, because of the insertion of chloride anion into the region between ZnMb and chiral viologen in higher ionic strength. As a result, the discrimination of the chiral environment around the reaction site would become unclear, leading to a decrease of stereoselectivity. The k_q value of the quenching reaction by the (R,S) isomer is intermediate between those of the (R,R) and (S,S) isomers, and the stereoselectivity disappears, of course. The authors suggested that the stereoselectivity is induced by the interaction between the polypeptide chain of myoglobin and the optically active sites of viologen. Also, the authors deduced that the S configuration of the polypeptide chain fitted well to the S configuration of phenylethylcarbamoyl and naphthylethylcarbamoyl moieties and that the interaction was stronger in the naphthylethylcarbamoyl moiety than that in the phenylethylcarbamoyl moiety because of the bulky size. It is noted that the k_q value in the quenching by NOAV^{2+} is larger than that by OAV^{2+}. This result does not arise from the difference in the redox potential, since the redox potentials of both OAV^{2+} and NOAV^{2+} are similar (-0.2 V vs. NHE). The authors said that because of the steric bulk, the naphthyl group of NOAV^{2+} induces the larger conformational change of the myoglobin flame to accelerate the quenching reaction and to lead to the larger k_q value than the phenyl group.

Viologens tend to form a dimeric adduct in the high-concentration condition. From the investigation of adduct formation, Tsukahara and collaborators presented experimental support that the S configuration fitted to the S configuration, as follows [69]: They observed that the adduct formation occurred more favorably in the (R,R)-(R,R) [or (S,S)-(S,S)] combination of two optically active viologens than in the (S,R)-(S,R) combination. This result comes from the steric repulsion of the methyl group of the phenyl- and naphthylcarbamoylmethyl moieties, RCH(CH$_3$)- (R = phenyl or naphthyl); in the (S,R)-(S,R) combination, there is the possibility that the S configuration moiety approaches the R configuration moiety, which induces the steric repulsion of the methyl group with the phenyl or naphthyl group to suppress the adduct formation due to the steric repulsion.

Other evidence was reported by Tsukahara and collaborators from the disproportionation reaction of the radical cation of the optically active bis-viologens, (S)-1-(1-naphthyl ethyl)carbamoyln methyl-1'-(4-(4-(1-naphthyl ethyl)carbamoyl methyl-pyridinio)pyridinio-4,4'-bi pyridinium tetra chloride, $NBVPR^{2+}$, 1-phenyl analog, $PBVPR^{2+}$, and 1-cyclohexylethyl analog $CHBVPR^{2+}$ (see Scheme 28) [70].

$$2RBVPR^{\cdot\,3+} \;\overset{K}{\rightleftharpoons}\; RBVPR^{\cdot\,2+} + RBVPR^{4+}$$
$$R=1\text{-naphthyl, 1-phenyl, or 1-cyclohexyl} \tag{28}$$

The disproportionation constant K of the (S,S)-(S,S) combination [or (R,R)-(R,R) combination] is much larger than that of the (S,R)-(S,R) combination; the ratio of the constant $K[(S,S)-(S,S)]/K[(R,S)-(R,S)] = 4.0$ for $R = 1$-naphthyl, 1.6 for $R = 1$-phenyl, and 1.8 for $R = 1$-cyclohexyl. Since the disproportionation is mainly controlled by the comproportionation rate constant, which is small for the strong interaction in a closed conformation [71], the reported results suggest that the homochiral pair adduct involves a stronger intermolecular interaction than the heterochiral pair adduct.

These bis-viologenes were also applied to the stereoselective photoinduced electron transfer reactions with ZnMb [68]. Though the quenching reaction of

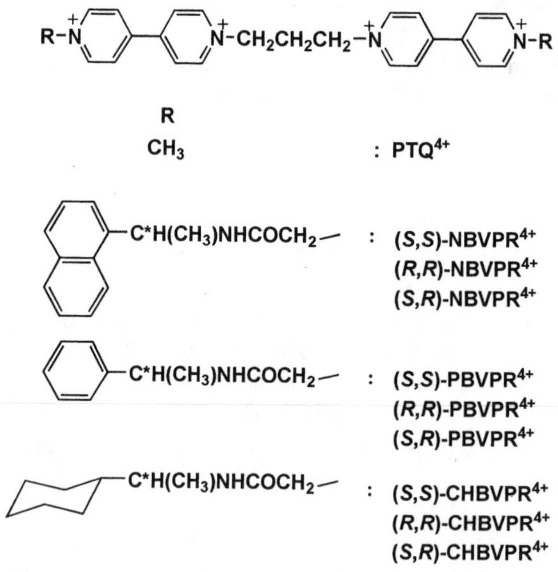

Scheme 28

ZnMb is carried out more efficiently by the (S,S) isomer than that by the (R,R) isomer, the stereoselectivity becomes smaller than that in the reaction by monoviologens; the selectivity is 1.2 for R = 1-naphthyl, 1.1 for R = 1-phenyl, and 1.1 for R = 1-cyclohexyl. The smaller stereoselectivity arises because bis-viologen has only one optically active substituent on one viologen moiety and an achiral methylene group on the other moiety.

The thermal back electron transfer reaction between one-electron reduced viologen and one-electron oxidized ZnMb was also investigated [68]. The stereoselectivity is similar to that observed in the quenching reaction; $k_r(S,S)/k_r(R,R) = 1.1$ to 1.4.

There are two possibilities for the origin of the stereoselectivity; in the first one, the reactant loosely associates or collides with the proteins in a stereoselective way. In the other one, the reactant stereoselectively diffuses through the protein. Though both possibilities provide reasonable explanation of the stereoselectivity, the authors suggested that the former was plausible because the cationic quencher attacked the positively charged Lys and/or Arg residue(s) in the heme pocket, and the bulky quenchers might not diffuse into the heme pocket; in other words, the reaction would occur at the surface of the protein [68].

These chiral viologens were applied to the reduction of metmyoglobin, too [72]. In this reaction, the one-electron reduced viologen radical cations were photochemically produced by tris(2,2′-bipyridine)ruthenium(II) in the presence of disodium salt of ethylene diamine tetra acetic acid (Na₂H₂edta), and the cation radical reacts with metmyoglobin to reduce it. The (S,S) isomer more rapidly reacts with metmyoglobin than the (R,R) isomer. The reaction rate is analyzed with Michaelis–Menten mechanism, as shown in Scheme 29.

Apparently, both k_{intra} and K values of the (S,S) isomer are larger than those of the (R,R) isomer. In this reaction, the reactivities of OAV^{2+} and $CHOAV^{2+}$ are similar to each other. This means that the aromatic group does not play an important role in the interaction, but the carbamoyl group would interact with the carboxylate and/or ammonium groups of metmyoglobin.

V. STEREOSELECTIVE ENERGY TRANSFER REACTIONS

Stereoselectivity has been observed not only in the photoinduced electron transfer reaction but also in the energy transfer reaction, whereas few reports have been presented for the stereoselective energy transfer reaction, except for several pioneering works, as follows: The stereoselective energy transfer reactions of lanthanide(III) complexes, [Tb(dpa)₃]³⁻ and [Eu(dpa)₃]³⁻ (dpa = 2,6-piridinedicarboxylate anion), with ruthenium(II) and cobalt(III) complexes, were investigated by Metcalf and collaborators [73]. Slightly later, the similar stereoselective energy transfer reactions of lanthanide complexes, [Eu(cda)₃]⁶⁻ and [Tb(cda)₃]⁶⁻ (cda

$$\text{metMb} + \text{OAV}^{+\cdot} \xrightleftharpoons{K} \{\text{metMb}\cdots\text{OAV}^{+}\}^{\cdot}$$

$$\xrightarrow{k_{\text{intra}}} \{\text{deoxyMb}\cdots\text{OAV}^{2+}\} \xrightarrow{\text{fast}} \text{deoxyMb} + \text{OAV}^{2+}$$

$$k_{\text{obsd}} = k_{\text{intra}}K[\text{metMb}]_0/(1+K[\text{metMb}]_0)$$

Table Intracomples electron-transfer rate constants and association constants for the reaction of metMb with viologen radical cation at 25°C and pH 7.0 (0.020 M phosphate buffer)

Viologen radical	I/M	$k_{\text{intra}}/10^2$ s^{-1}		$K/10^5$ M^{-1}	
		(S,S)-	(R,R)-	(S,S)-	(R,R)-
OAV$^{+\cdot}$	0.040	4.0±0.2	3.3±0.2	1.1±0.1	0.78±0.05
	0.10	4.8±0.3	3.9±0.3	2.0±0.1	1.7±0.1
	0.50	6.1±0.3	6.0±0.3	3.6±0.2	2.4±0.1
CHOAV$^{+\cdot}$	0.10	3.0±0.2	2.2±0.1	1.2±0.1	0.94±0.05
	0.50	4.1±0.2	3.9±0.2	1.5±0.1	1.0±0.1

Scheme 29

= 4-oxo-2,6-pyridinedicarboxylate anion), with cobalt(III) complexes, [Co(en)$_3$]$^{3+}$, [Co(en)$_2$(R,R-chxn)]$^{3+}$, [Co(en)(R,R-chxn)$_2$]$^{3+}$, and [Co(R,R-chxn)$_3$]$^{3+}$ (en = ethylene diamine; (R,R)-chxn = trans-(1R,2R)-1,2-diamino-cyclohexane), were investigated by the same group [74]. These complexes are schematically shown in Scheme 30. In these complexes, there are significantly large diffenrences in electronic structure and charge between Eu^{3+} (4f^6) and Tb^{3+} (4f^8), while the size, shape, and stereochemical properties are very similar to each other (see Scheme 30). The emitting state is ^{5}D$_0$ and the most intense luminescence is observed in the ^{5}D$_0 \rightarrow {}^7$F$_2$ deactivation and ^{5}D$_0 \rightarrow {}^7$F$_4$ deactivation, while ^{5}D$_0 \rightarrow {}^7$F$_1$ deactivation exhibits the strongest chiroptical properties. The emission spectrum is observed at 540 to 550 nm for Tb complexes and at 585 to 620 nm for Eu complexes [73b].

These cobalt(III) complexes have the same charge (+3) and essentially the same electronic state to each other, but they differ with respective to size, shape, and stereochemical properties: for instance, [Co(en)$_3$]$^{3+}$ has only one molecular asymmetry about the metal center, but [Co(en)$_2$(R,R-chxn)]$^{3+}$, [Co(en)(R,R-chxn)$_2$]$^{3+}$, and [Co(R,R-chxn)$_3$]$^{3+}$ have chiral centers on the ligand besides the chirality about the metal center. In the excited state of [Eu(dpa)$_3$]$^{3-}$ and [Eu(cda)$_3$]$^{6-}$, Λ form $\leftrightarrow \Delta$ form interconversion occurs with the first order rate constants of 15.8 and 29.6 s^{-1}, respectively. These values are much smaller

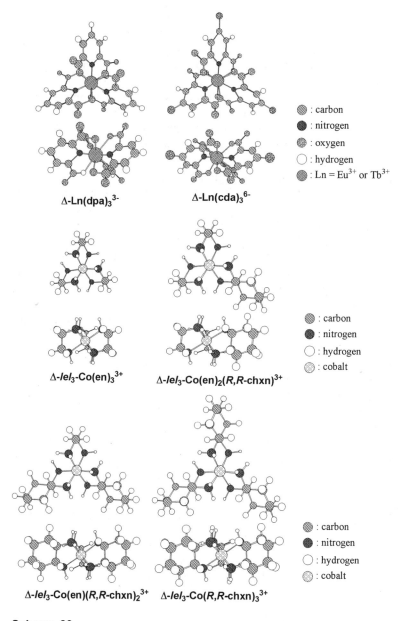

Δ-Ln(dpa)$_3$$^{3-}$ Δ-Ln(cda)$_3$$^{6-}$

: carbon
: nitrogen
: oxygen
: hydrogen
: Ln = Eu^{3+} or Tb^{3+}

Δ-*lel*$_3$-Co(en)$_3$$^{3+}$ Δ-*lel*$_3$-Co(en)$_2$(*R,R*-chxn)$^{3+}$

: carbon
: nitrogen
: hydrogen
: cobalt

Δ-*lel*$_3$-Co(en)(*R,R*-chxn)$_2$$^{3+}$ Δ-*lel*$_3$-Co(*R,R*-chxn)$_3$$^{3+}$

: carbon
: nitrogen
: hydrogen
: cobalt

Scheme 30

than the luminescence decay of the $^5D_0$ state. Thus these complexes are considered not to be stereochemically labile in the $^5D_0$ state. The same situation would exist in the Tb analogs. In solution, these complexes exist in racemic mixture. However, nonracemic populations of Eu^{3+} and Tb^{3+} complexes emerge in the excited state by irradiation with circularly polarized light (465 nm for the Eu^{3+} complexes and 488 nm for Tb^{3+} complexes). Time-resolved chiroptical luminescence measurement was made to investigate the stereoselective quenching reaction by the cobalt(III) complex, where the quenching reaction occurs through the energy transfer. Larger stereoselectivity was observed in the quenching reactions of $[Tb(dpa)_3]^{3+}$ with $[Ru(phen)_3]^{2+}$ and $[Co(phen)_3]^{3+}$ than in the reactions with $[Ru(bpy)_3]^{2+}$ and $[Co(bpy)_3]^{3+}$. Interestingly, a large enantiomer excess is observed for Δ-$[Co(R,R\text{-}chxn)]^{3+}$, and the selectivity increases upon going from $[Co(en)_3]^{3+}$ to Δ-$[Co(R,R\text{-}chxn)]^{3+}$, as shown in Table 5. This result is interpreted in terms of the increase in the surface irregularity, where the term "surface irregularity" is used to represent the bulkiness of the substituent of chxn, which relates to the optical asymmetry of $[Co(R,R\text{-}chxn)]^{3+}$, in other words, the irregularity of the molecular surface. The stereoselectivity becomes larger upon going from the conformationally flexible en ligand to the larger R,R-chxn ligand, as the surface irregularity increases. One of the interesting features is that the stereoselectivity of the quenching reaction by $[Co(en)_3]^{3+}$ is the reverse of that of the

Table 5   Stereoselectivity in the Energy Transfer Between Lanthanide Complexes and Cobalt(III) Complexes

(a) Enantiomeric preference ratio (it was not determined which isomer exists in excess)

	$[Tb(dpa)_3]^{3-}$	$[Eu(dpa)_3]^{3-}$
$[Ru(phen)_3]^{3+}$	1.41	—
$[Ru(bpy)_3]^{3+}$	1.04	—
$[Co(phen)_3]^{3+}$	1.53	—
$[Co(en)_3]^{3+}$	1.25	1.99

Source: Ref. 73b.

(b) The k_q^-/k_q^+ value

	$[Eu(dpa)_3]^{3-}$	$[Eu(cda)_3]^{6-}$	$[Tb(dpa)_3]^{3-}$	$[Tb(dpa)_3]^{6-}$
$[Co(en)_3]^{3+}$	0.78	0.75	0.87	0.71
$[Co(en)_2(R,R\text{-}chxn)]^{3+}$	1.67	1.33	1.58	1.42
$[Co(en)(R,R\text{-}chxn)_2]^{3+}$	3.02	2.01	2.57	2.00
$[Co(R,R\text{-}chxn)_3]^{3+}$	4.01	2.77	3.40	3.21

Source: Ref. 73c.

quenching reaction by the other Co(III) complexes. There are two possible factors to determine the stereoselectivity: one is the stereoselective diffusion, which leads to the formation of an encounter complex (or an exciplex), and the other is the stereoselective energy transfer in the encounter complex (or the exciplex). The authors proposed that the stereoselective energy transfer in the encounter complexes was an origin of these stereoselective quenching reactions, since ionic strength, viscosity, and temperature, which induce changes in the diffusion rate, little influence the stereoselectivity.

Lanthanide complexes, $[Ln(dpa)_3]^{3+}$ (Ln = Eu^{3+} or Tb^{3+}), were also applied to the stereoselective quenching reaction by cobalt(III) nucleotide complexes, shown in Scheme 31 [75].

Though there is a conformational chirality in the metal center, the authors did not separate these enantiomers and used the mixture of Δ- and Λ-[Co(N-$H_3)_4$(nucleotide)]. Thus the stereoselectivity arises from the configurational chirality on the nucleotide. The selectivity significantly depends on the kind of nucleotides; the largest values (13.7 and 13.0) were observed for Co(III)-atp complex (atp = adenosine triphosphate) in quenching reactions of $[Tb(dpa)_3]^{3+}$ and $[Eu(dpa)_3]^{3+}$, respectively. The next values (6.8 and 7.1) were observed for Co(III)-gtp complex (gtp = guanosine triphosphate), while the stereoselectivity is the reverse of that of the reaction with Co(III)-atp complex. These results suggest that there are significantly large differences in the interaction between the nucleotide residue and $[Ln(dpa)_3]^{3+}$. The details, however, are not clear.

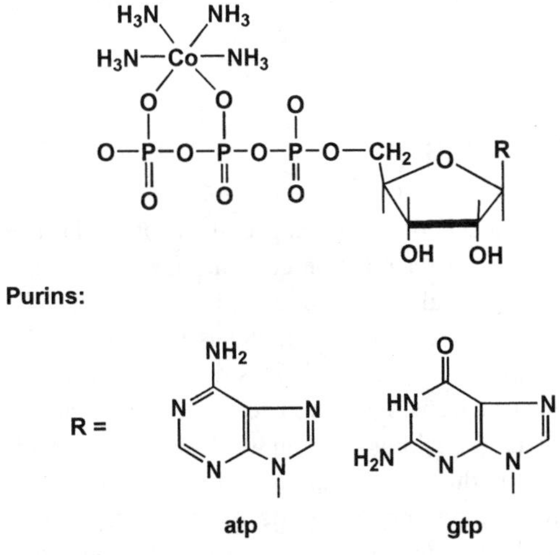

Purins:

R =

atp   gtp

Scheme 31

We expect that this kind of quenching reaction deeply relates to bioinorganic chemistry.

Chirality effects are also reported in the energy transfer from the ruthenium(II) polypyridyl complex to the osmium(II) complex in Langumuir–Blodgett (LB) films [76]. In this experiment, the LB film was prepared with [Ru(dp-phen)$_3$]$^{2+}$ (dp-phen = 4,7-diphenyl-1,10-phenanthroline) and stearic acid, where the molar ratio was 1:1 to 1:4. The quenching reaction of the ruthenium(II) complex was carried out with optically pure osmium(II) complex, [Os(dp-phen)$_3$]$^{2+}$. This reaction consists of photoexcitation of the ruthenium(II) and osmium(II) complexes [Eqs. (29) and (30)], spontaneous decays of the excited ruthenium(II) and osmium(II) complexes [Eqs. (32) and (33)], and the energy transfer between the exited ruthenium(II) complex and the osmium(II) complex [Eq. (31)].

$$\text{Ru(II)} + h\nu \rightarrow {}^*\text{Ru(II)} \quad \text{excitation of photosensitizer} \tag{29}$$

$$\text{Os(II)} + h\nu \rightarrow {}^*\text{Os(II)} \quad \text{direct excitation} \tag{30}$$

$${}^*\text{Ru(II)} + \text{Os(II)} \rightarrow \text{Ru(II)} + {}^*\text{Os(II)} \quad \text{energy transfer } (k_{tr}) \tag{31}$$

$${}^*\text{Os(II)} \rightarrow \text{Os(II)} + h\nu \quad \text{spontaneous decay } (k_{os}) \tag{32}$$

$${}^*\text{Ru(II)} \rightarrow \text{Ru(II)} + h\nu \quad \text{spontaneous decay } (k_{ru}) \tag{33}$$

The experimentally estimated k_2 value ($= k_{os}k_{tr}[\text{Os(II)}]/k_{ru}$) ($36 \times 10^7$ s^{-1}) for the Λ-*Ru(II)/Λ-Os(II) pair is much larger than that (8.5×10^7 s^{-1}) for the Λ-*Ru(II)/Δ-Os(II) pair. This result is consistent with the chirality effect on the surface pressure versus molecular area (π-A) isotherms, in which the Λ-ruthenium(II) complex forms a more compact monolayer than the racemic mixture of the ruthenium(II) complex; in other words, the racemic mixture induces some steric repulsion between Δ- and Λ-ruthenium(II) complexes.

VI. SUMMARY AND PERSPECTIVES

Summarizing the above results, we reach the following conclusions: (1) The stereoselective photoinduced electron transfer reaction certainly takes place between ruthenium(II) complexes and chiral viologen compounds. (2) In the quenching reactions of ruthenium(II) and copper(I) complexes with cobalt(III) complexes, however, the stereoselectivity is still small. This does not mean that the stereoselective photoinduced electron transfer reaction does not occur, because the less stereoselective energy transfer quenching reaction would occur concomitantly with the stereoselective photoinduced electron transfer reaction in the quenching process. (3) Several interesting photocatalytic reactions have been presented, in which not only the high reducing ability of the excited photosensitizer but also the high oxidizing ability of the one-electron oxidized photosensitizer is utilized for the reactions. We believe that this type of reaction should be applied to many photoreactions and photoasymmetric syntheses. (4) Also, photoinduced deracemization is very interesting, because both the high reducing ability of the excited state of the photosensitizer and the high oxidizing ability of the one-electron oxidized photosensitizer are utilized for the reaction.

Apparently, the stereoselective energy transfer reaction has not been investigated well except for several pioneering works. The relation between the stereoselectivity and the energy transfer mechanism (Dexter or Förster mechanism) has not been investigated yet, despite its significant interest. Since it is not easy to determine through which mechanism the quenching reaction takes place, and what factor determines the stereoselectivity, there remain many issues to be investigated in stereoselective energy transfer reactions.

The combined systems of the photosensitizer with micelle, LB film, protein, and so on are interesting, because we can expect that new functions will be found in such systems. In some pioneering works, such ideas were applied to stereoselective photoinduced electron transfer reactions. We believe that one can construct the new photoreaction with such combined systems and also expect the enhancement of stereoselectivity in such systems.

One of the most important issues to be investigated is to clarify the stereoselective electron transfer reaction and the energy transfer reaction from the viewpoint of physical chemistry. For instance, the parameters of Marcus theory have not been reported yet for the stereoselective electron transfer reaction.

We believe that a new frontier will open in the chiral photochemistry of transition metal complexes.

REFERENCES

1. Geselowitz DA, Taube H. J Am Chem Soc 1980; 102:4525.
2. Sutter JH, Hunt JB. J Am Chem Soc 1967; 91:3107.
3. Kane-Maguire NAP, Tollinson RM, Richardson DE. Inorg Chem 1976; 15:499.
4. Kondo S, Sasaki Y, Saito K. Inorg Chem 1981; 20:429.
5. Lappin AG, Marusak RA. Coord Chem Rev 1991; 109:125.
6. Juris A, Balzani V, Barigelletti F, Campagna S, Belser P, Von Zelewsky A. Coord. Chem. Rev 1988; 84:85.
7. Portor G B, Sparks RH. J Photochem 1980; 13:123.
8a. McMillin DR, Buckner MT, Ahn BT. Inorg Chem 1977; 16:943.
8b. Ahn BT, McMillin DR. Inorg Chem 1978; 17:2253.
8c. Buckner MT, Matthews TG, Lyttle FE, McMillin DR. J Am Chem Soc 1979; 101: 5846.
8d. Blaskie MW, McMillan DR. Inorg Chem 1980; 19:3519.
8e. Ahn BT, McMillin DR. Inorg Chem 1981; 20:1427.

8f.   Rader RA, McMillin DR, Buckner MT, Matthews TG, Casadonte DJ, Lengel RK, Whittaker SB, Darmon LM, Lytle FE. J Am Chem Soc 1981; 103:5906.
8g.   Breddels PA, Berdowski PAM, Blasse G, McMillin DR. J Chem Soc Faraday Trans 1982; 78:595.
8h.   Del AA, Paggio D, McMillin DR. Inorg Chem 1983; 22:691.
8i.   Dietrich-Buchecker CO, Marnot PA, Sauvage JP, Kirchhoff JR, McMillin DR. J Chem Soc Chem Commun 1983; 513.
8j.   Kirchhoff JR, Gamache RE, Blaskie MW, Paggio AAD, Lengel RK, McMillin DR. Inorg Chem 1983; 22:2380.
8k.   Gamache RE, Rader RA, McMillin DR. J Am Chem Soc 1985; 107:1141.
8l.   Casadonte RJ, McMillin DR. J Am Chem Soc 1987; 109:331.
8m.   Goodwin KV, McMillin DR. Inorg Chem 1987; 26:875.
8n.   Casadonte J, McMillin DR. J Am Chem Soc 1987; 109:331.
8o.   Casadonte J, McMillin DR. Inorg Chem 1987; 26:3950.
8p.   Palmer CEA, McMillin DR. Inorg Chem 1987; 26:3837.
8q.   Crane DR, DiBenedetto J, Palmer CEA, McMillin DR, Ford PC. Inorg Chem 1988; 27:3698.
8r.   Palmer CEA, McMillin DR. Inorg Chem 1987; 26:3837.
8s.   Palmer CEA, McMillin DR, Kirmaier C, Holten D. Inorg Chem 1987; 26:3167.
8t.   Ichinaga AK, Kirchhoff JR, McMillin DR, Dietrich-Buchecker CO, Marnot PA, Sauvage JP. Inorg Chem 1987; 26:4290.
8u.   Gushurt AKI, McMillin DR, Dietrich-Buchecker CO, Sauvage JP. Inorg Chem 1989; 28:4070.
8v.   Stacy EM, McMillin DR. Inorg Chem 1990; 29:393.
8w.   Everly RM, Ziessel R, Suffert J, McMillin DR. Inorg Chem 1991; 30:559.
8x.   Eggleston MK, Fanwick PE, Pallenberg AJ, McMillin DR. Inorg Chem 1997; 36:4007.
8y.   Cunningham KL, McMillin DR. Inorg Chem 1998; 37:4114.
8z.   Crane DR, Ford PC. J Am Chem Soc 1991; 113:8510.
9a.   Sakaki S, Koga G, Sato F, Ohkubo K. J Chem Soc Dalton Trans 1985:1959.
9b.   Sakaki S, Koga G, Ohkubo K. Inorg Chem 1986; 25:2330.
9c.   Sakaki S, Koga G, Hinokuma S, Hashimoto S, Ohkubo K. Inorg Chem 1987; 26:1817.
9d.   Sakaki S, Hashimoto S, Koga G, Ohkubo K. J Chem Soc Dalton Trans 1988:1641.
9e.   Sakaki S, Mizutani H, Kase Y, Arai T, Hamada T. Inorg Chim Acta 1994; 225:261.
10.   For instance, Lammi RK, Ambroise A, Balasubramanian T, Wagner RW, Bocian DF, Holten D, Lindsey JS. J Am Chem Soc 122; 7579:2000 and references therein. Imahori H, Tamaki K, Guldi DM, Luo C, Fujitsuka M, Ito O, Sakata Y, Fukuzumi S. J Am Chem Soc 123: 2607, 2001 and references therein.
11.   Lin CT, Boettcher W, Chou M, Creutz C, Sutin N. J Am Chem Soc 1976; 98:6536.
12.   Bock CR, Connor JA, Gutierrez AR, Meyer TJ, Whitten DG, Sullian BP, Nagle JN. J Am Chem Soc 1979; 101:4815.
13.   Wacholtz WF, Auerbach RA, Schmell RH. Inorg Chem 1986; 25:227.
14.   Elliott M, Freitag RA, Blaney DD. J Chem Soc Chem Commun 1993:1423.
15.   Dwyer FP, Gyarfas EC. Proc R Soc N S W 1949; 83:170.

16. Porter GB, Sparks RH. J Chem Soc Chem Commun 1979:1094.
17. Irie M, Yorozu T, Hayashi K. J Am Chem Soc 1978; 100:2236.
18. Creaser II, Gahan LR, Geue RJ, Launikonis A, Lay PA, Lydon JD, McCarthy MG, Mau AWH, Sargeson AM, Sasse WHF. Inorg Chem 1985; 24:2671.
19. Kaizu Y, Mori T, Kobayashi H. J Phys Chem 1985; 89:332.
20. Krishnan CV, Brunschwig BS, Creutz C, Sutin N. J Am Chem Soc 1985; 107:2005.
21. Rau H, Ratz R. Angew Chem Int Ed Engl 1983; 22:550.
22. Tsukahara K, Kaneko J, Miyaji T, Hara T, Kato M, Kimura M. Chem Lett 1997: 455.
23. Tsukahara K, Kaneko J, Miyaji T, Abe K, Matsuoka M, Hara T, Tanase T, Yano S. Bull Chem Soc Jpn 1999; 72:139.
24. Ohkubo K, Arikawa Y. J Mol Cat 1985; 33:65.
25. Sakaki S, Satoh T, Ohkubo K. New J Chem 1986; 10:145.
26. Ohkubo K, Hamada T, Inaoka T, Ishida H. Inorg Chem 1989; 28:2021.
27. Ohkubo K, Hamada T, Ishida H. J Chem Soc Chem Commun 1993:1423.
28. Ohkubo K, Ishida H, Hamada T, Inaoka T. Chem Lett 1989:1545.
29. Hamada T, Tanaka S, Koga H, Sakai Y, Sakaki S. Dalton 2003:692.
30. Ohkubo K, Hamada T, Ishida H, Fukushima M, Watanabe M. J Mol Cat 1994; 89: L5.
31. Ohkubo K, Hamada T, Ishida H, Fukushima M, Watanabe M, Kobayashi H. J Chem Soc Dalton Trans 1994:239.
32. Ohkubo K, Fukushima M, Ohta H, Usui S. J Photochem Photobiol A Chemistry 1996; 98:139.
33. Kato M, Sasagawa T, Ishihara Y, Yamada S, Fujitani S, Kimura M. J Chem Soc Dalton Trans 1994:583.
34a. Hamada T, Sakaki S, Brunschwig BS, Fujita E, Wishart JF. Chem Lett 1998:1259.
34b. Hamada T, Brunschwig BS, Eifuku K, Fujita E, Korner M, Sakaki S, van Eldik R, Wishart JF. J Phys Chem A 1999; 103:5645.
35. Marcus RA, Sution N. Biochim Biophsica Acta 1985; 811:265.
36. Hamada T, Ishida H, Usui S, Watanabe Y, Tsumura K, Ohkubo K. J Chem Soc Chem Commun 1993:909.
37. Hamada T, Ishida H, Usui S, Tsumura K, Ohkubo K. J Mol Cat 1994; 88:L1.
38a. Ohkubo K, Hamada T, Watanabe M. J Chem Soc Chem Commun 1993:1070.
38b. Ohkubo K, Hamada T, Watanabe M, Fukushima M. Chem Lett 1993:1651.
39. Ohkubo K, Watanabe M, Ohta H, Usui S. J Photochem Photobiol A Chemistry 1996; 95:231.
40. Stevenson KL, Verdieck JF. J Am Chem Soc 1968; 90:2974.
41. Stevenson KL. J Am Chem Soc 1972; 94:6652.
42. Lacour J, Jodry JJ, Ginglinger C, Torche-Haldimann S. Angew Chem Int Ed Engl 1998; 37:2379.
43a. Kirschner S, Ahmad N, Munir C, Pollock RJ. Pure Appl Chem 1979; 51:913.
43b. Owen DJ, Schuster GB. J Am Chem Soc 1996; 118:259.
43c. Owen DJ, VanDerveer D, Schuster GB. J Am Chem Soc 1998; 120:1705.
44a. Hamada T, Ohtsuka H, Sakaki S. Chem Lett 2000:364.
44b. Hamada T, Ohtsuka H, Sakaki S. J Chem Soc Dalton Trans 2001:928.

45a. Kirchhof JR, McMillin DR, Robinson WR, Powell DR, McKenzie AT, Chen S. Inorg Chem 1985; 24:3928.

45b. Klemens FK, Fanwick PE, Bibler JK, McMillin DR. Inorg Chem 1989; 28:3076.

46a. Bell SEJ, McGarvey JJ. Chem Phys Lett 1986; 124:336.

46b. McGarvey JJ, Bell SE J, Bechara JN. Inorg Chem 1986; 25:4327.

47. Sakaki S, Mizutani H, Kase Y. Inorg Chem 1992; 31:4575.

48. Sakaki S, Okitaka I, Ohkubo K. Inorg Chem 1984; 23:198.

49. Edel A, Marnot PA, Sauvage JP. Nouv J Chim 1984; 8:495.

50. Ruthkosky M, Kelly CA, Zaros MC, Meyer GJ. J Am Chem Soc 1997; 119:12004.

51. Vante NA, Nieregarten JF, Sauvage JP. J Chem Soc Dalton Trans 1994:1649.

52. Sakaki S, Kuroki T, Hamada T. J Chem Soc Dalton Trans 2002:840.

53. Sakaki S, Ishikura H, Kuraki K, Tanaka K, Satoh T, Arai T, Hamada T. J Chem Soc Dalton Trans 1997:1815.

54. Sakaki S, Horita R, Kuraki K, Hamada T. Chem Lett 1998:827.

55. K Bernauer. Electron-transfer reactions in metalloproteins. Metal Ion in Biological Systems 27. H. Sigel, A. Sigel, eds. Marcel Dekker, New York, 1991, p. 265. RA Marusak, TP Shields, AG Lappin. Inorganic compounds with unusual propoerties III. electron tranfser in biology in the solid state. Advances in Chemistry Series 226. MK Johnson, RB King, DM Kurts Jr, C Kutal, ML Norton, RA Scott. American Chemical Society. Washington DC 237, 1990.

56. Bernauer K, Sauvain JJ. J Chem Soc Chem Commun 1988:353.

57. Sakaki S, Nishijima Y, Koga H, Ohkubo K. Inorg Chem 1989; 28:4061.

58. Sakaki S, Nishijima Y, Ohkubo K. J Chem Soc Dalton Trans 1991:1143.

59. Bernauer K, Monzione M, Schurmann P, Viette V. Hel Chim Acta 1990; 73:346.

60. Pladziewicz JR, Accola MA, Osvath P, Sargeson AM. Inorg Chem 1993; 32:2525.

61. M Gouterman. In: The Porphydins, Vol. 3. D Dolphin, ed. Academic Press, New York, 1978, p. 1. M Gouterman, GH Wagniere. J Mol Spectrosc 11:108, 1963.

62a. Kurabayashi Y, Kikuchi K, Kokubun H, Kaizu Y, Kobayashi H. J Phys Chem 1984; 88:1308.

62b. Tobita S, Kaizu Y, Kobayashi H, Tanaka I. J Chem Phys 1984; 81:2962.

62c. Ohno O, Kaizu Y, Kobayashi H. J Chem Phys 1985; 82:1779.

62d. Kaizu Y, Asano M, Kobayashi H. J Phys Chem 1986; 90:3906.

62e. Kaizu Y, Maekawa H, Kobayashi H. J Phys Chem 1986; 90:4234.

63. Gurzadyan GG, Tran-Thi TH, Gustavsson T. J Chem Phys 1998; 108:385.

64. Tsvirko MP, Stelmakh GF, Pyatosin VE, Solovyov KN, Kachura TF. Chem Phys Lett 1980; 73:80.

65. Harriman A. J Chem Soc Faraday I 1981; 77:1281.

66. Harriman A, Porter G, Richoux MC. J Chem Soc Faraday Trans 1981; 77:833, 1939.

67. Kalyabnasundaram K, Spallart M N. J Phys Chem 1982; 86:5163.

68a. Tuskahara K, Kimura C, Kaneko J, Hara T. Chem Lett 1994:2377.

68b. Tsukahara K, Kimura C, Kaneko J, Abe K, Matsui M, Hara T. Inorg Chem 1997; 36:3520.

69. Tsukahara K, Kaneko J, Miyaji T, Abe K. Tetrahedron Lett 1996; 37:3149.

70. Tsukahara K, Abe K, Matsuoka M. Chem Lett 1997:941.

71. Altherton SJ, Tsukahara K, Wilkins RG. J Am Chem Soc 1985; 108:3380.

72. Tsukahara K, Goda M. Chem Lett 1998:929.
73a. Metcalf DH, Snyder SW, Wu S, Hilmes GL, Michl JP, Demas JN, Richardson FS. J Am Chem Soc 1989; 111:3082.
73b. Metcalf DH, Snyder SW, Demas JN, Richardson FS. J Am Chem Soc 1990; 112: 5681.
74a. Richardson FR, Metcalf DH, Glover DP. J Phys Chem 1991; 95:6249.
74b. Bolender JP, Metcalf DH, Richardson FS. Chem Phys Lett 1993; 213:131.
74c. Metcalf DH, Bolender JP, Driver MS, Richardson FS. J Phys Chem 1993; 97:553.
75. Metcalf DH, McD Stewart JM, Grisham CM, Richardson FS. Inorg Chem 1992; 31: 2445.
76. Okamoto K, Takahashi M, Taniguchi M, Yamagishi A. Chem Lett 2000:740.

8

Template-Induced Enantioselective Photochemical Reactions in Solution

Benjamin Grosch and Thorsten Bach
Technische Universität München
München, Germany

I. INTRODUCTION

Enantioselective reactions are defined as transformations in which a prochiral substrate is converted into a chiral product such that one of the two enantiomers is formed in significant excess. The degree of enantioselectivity is measured by the enantiomeric excess (*ee*), as defined in Scheme 1. In this schematic example the prochiral substrate S^1, represented by a triangle, is converted into the two enantiomeric, chiral tetrahedral products P^1 and *ent*-P^1 (enantioface-differentiating reaction). Alternatively, but less commonly used, enantioselectivity can be induced by the differentiation of enantiotopic substituents, as depicted for S^2 and P^2/*ent*-P^2.

In an enatioface-differentiating reaction the enantiotopic faces (*re/si*) of the prochiral substrate have to be distinguished. A variety of strategies have been developed for photochemical reactions to achieve this goal. General reviews related to this area have been published by Rau [1] and by Inoue et al. [2,3]. The conventional way to synthesize enantiomerically enriched photoproducts is covalently to attach a chiral auxiliary [4] to the prochiral substrate. The two faces of the substrate become diastereotopic, and the subsequent reaction is diastereoselective; the auxiliary is then removed. In recent years, there have been efforts to achieve enantioselective photochemical reactions going directly from **S** to either **P** or *ent*-**P** with high *ee*. Four main strategies can be categorized based on the way the chiral information is provided: chiral sensitizers, chiral templates, circularly polarized light (CPL), and magnetochiral anisotropy. Chiral sensitizers

315

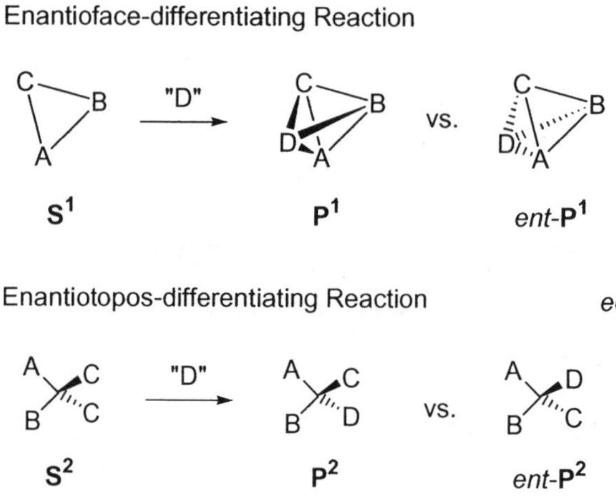

Enantioface-differentiating Reaction

Enantiotopos-differentiating Reaction

$$ee = \frac{\left|[P] - [ent\text{-}P]\right|}{[P] + [ent\text{-}P]}$$

Scheme 1

transfer the chiral information to the substrate by short-lived interactions in the excited state, i.e., during the lifetime of an exciplex between a reaction intermediate and the chiral sensitizer [5]. The chirality is multiplied, and only catalytic amounts of the optically active sensitizer are required. In contrast, chiral templates differentiate the enantiotopic faces of the photosubstrate in the ground state. An equilibrium is set up between the noncovalently bound substrate and the complexing agent. The substrate can be the starting material itself or a photochemically produced intermediate that undergoes further thermal conversion, e.g., protonation. Finally, the chiral information can be provided by a chiral electromagnetic environment. With CPL the preferential formation of one enantiomer in the photochemical conversion of a prochiral substrate has been observed in some instances [6]. Another external source of chirality is magnetochiral anisotropy. Unpolarized light in a parallel magnetic field can give rise to enantiomeric excess in photochemical reactions [7].

This review concentrates on chiral templates that transfer the chiral information in a ground state complex in solution, and it considers results published before the beginning of 2003. Reviews focusing on chiral auxiliaries, chiral sensitizers, CPL as well as on enantioselective photochemistry in the solid state, in confined media, or in molecular aggregates can be found elsewhere in this book. Moreover, it should be mentioned that this account deals with synthetic templates. Natural templates such as cyclodextrins, bovine serum albumin, and DNA will be accounted for separately.

The effectiveness of a chiral template is determined by two parameters: the binding of the substrate and the quality of the face differentiation in the complex. The chiral template desirably forms a stable and specific complex with the ground state substrate or the photochemically produced reactive intermediate employing noncovalent interactions. All kinds of electrostatic or hydrophobic interactions can be utilized, such as hydrogen bonding, dipolar, ionic, van der Waals, charge transfer, and electron donor–acceptor interactions. The complex equilibrium is governed by the association constant K_A, the first important parameter to characterize the effectiveness of a chiral template. The ideal chiral inductor provides perfect differentiation of the enantiotopic faces of the substrate. In this case the enantiomeric excess is limited only by the concentration of the complex in the equilibrium. The expected enantioselectivity of a perfect face differentiating host can be easily deduced according to Scheme 2. Unbound starting material is converted to racemic product, while bound substrate is specifically transformed to one of the enantiomers. The major product, enantiomer **P**, is therefore formed from the entire bound and half of the unbound substrate, while the minor enantiomer *ent-***P** results from the other half of the unbound substrate. This leads to the intuitively sensible result that the expected enantiomeric excess reflects the relative amount of substrate that is complexed by the chiral template. The value is readily calculated from K_A and the starting concentrations of the substrate and the chiral host compound. Strictly speaking, this relationship is only valid at the beginning of the reaction. If the products have binding properties that differ markedly from those of the substrate, the relative amount of bound substrate

$$\text{Substrate + Template} \underset{}{\overset{K_a}{\rightleftharpoons}} \{\text{Substrate/Template}\}$$

$$\Big\downarrow h\nu \qquad\qquad\qquad\qquad \Big\downarrow h\nu$$

no enantioselectivity   perfect enantioselectivity

$$ee = \frac{\big|[P] - [ent\text{-}P]\big|}{[P] + [ent\text{-}P]} = \frac{c(\{\text{Substrate/Template}\}) + c(\text{Substrate})/2 - c(\text{Substrate})/2}{c(\{\text{Substrate/Template}\}) + c(\text{Substrate})}$$

$$ee = \frac{c(\{\text{Substrate/Template}\})}{c_0(\text{Substrate})}$$

Scheme 2

shifts in the course of the reaction. If the binding strengths of the substrate and the products to the chiral host are comparable, the equation gives a reasonable estimate for the enantioselectivity that can be achieved in the case of perfect differentiation with a given association behavior.

The face differentiation is accomplished either through passive shielding of one of the faces reducing its accessibility to chemical attack or through active chemical conversion favoring one face, e.g., through selective protonation of one of the two enantiofaces. The ability of the chiral inductor to ensure face differentiation is assessed by evaluating the net face differentiation. This corresponds to the induced enantiomeric excess divided by the relative amount of bound substrate, i.e., the actual enantiomeric excess divided by the enantiomeric excess in the case of perfect face differentiation. Thus the chiral template can be characterized according to the effectiveness of its chiral induction by two parameters, the association constant K_A and the net face differentiation factor.

Ideally, the binding of the template to the photoproduct is significantly weaker than to the starting material. Following chiral induction and dissociation of the product, the template is available for the next chirality transfer. In this case of catalytic turnover only substoichiometric amounts of the chiral agent are necessary. This effect is dramatically enforced if the substrate in the complex can be selectively excited. Even if chiral amplification is not achieved through catalytic use, a photostable complexing agent used in stoichiometric or excessive amounts can be recovered and reused. Generally, the complexing agent should be readily separable from the reaction mixture without an additional and possibly damaging bond cleavage step, as in the case of a covalently bound auxiliary.

The chiral templates that will be dicussed within this review are classified into four categories: chiral solvents that provide a chiral environment without specific binding (Sec. II), chiral Brønsted acids that act as proton donor in the enantiodifferentiating step (Sec. III), chiral transition metal complexes that catalyze the enantiodifferentiating step (Sec. IV), and chiral complexing agents that accomplish face differentiation in a specific complex without chemically converting the substrate (Sec. V).

II. ASYMMETRIC INDUCTION BY CHIRAL SOLVENTS

In the 1970s several attempts were made to direct photoreactions by irradiating the prochiral starting material in a chiral solvent, sometimes mixed with an achiral cosolvent. In contrast to chiral Brønsted acids and chiral complexing agents, chiral solvents do not play an active role in the enantiodifferentiating step as proton or hydrogen bond donors or acceptors. They provide a rather undefined and passive chiral environment, being present in large excess of the photosubstrate. Some weak hydrogen bonding or other noncovalent interactions may be involved, but no defined face differentiating complex with the substrate of the photoreaction is thought to exist. For practical purposes, this strategy differs from rationally based approaches discussed in Sec. V in that the chiral template is used in large excess.

After early unsuccessful attempts to direct the photoreduction of ketones with chiral secondary alcohols [8–10]. Weiss et al. examined the sensitized cis–trans photoisomerization of 1,2-diphenylcyclopropane in chiral solvents but obtained the product without detectable optical rotation [11]. Seebach and co-workers were the first to achieve asymmetric induction for a photochemical reaction by a chiral solvent [12–15]. They examined the photopinacolization of aldehydes and ketones in the chiral solvent (S,S)-$(+)$-1,4-bis(dimethylamino)-2,3-dimethoxybutane (DDB, **4**). Irradiation of acetophenone in the presence of 7.5 equiv. of DDB yielded a mixture of chiral d,l-pinacols **3**/ent-**3** and achiral meso-pinacol **2**. At 25°C pinacol **3** was obtained with 8% ee, with the (R, R)-$(+)$-enantiomer prevailing. At lower temperatures the asymmetric induction was more effective, up to 23% ee at -78°C in a 1 : 5 mixture of DDB and pentane (Scheme 3).

The mode of asymmetric induction can be rationalized from the mechanism of the photopinacolization in the presence of aliphatic amines. The electron transfer from the amine to the excited triplet ketone furnishes charge transfer complex **5**, from which a radical pair is formed by proton transfer. The weakly coordinated chiral amine seems to favor the dimerization of radical **6** from the si face leading to the (R, R)-enantiomer **3**. The much lower selectivities observed with methanol as the cosolvent (3% ee at 27°C) indicate dipolar or hydrogen bonding interactions between the chiral diamine and the prochiral radical (Scheme 4).

The photopinacolization of other carbonyl compounds—benzaldehyde, propiophenone, pivalophenone, 1-tetralone, and 3-benzoylpyridine—in the pres-

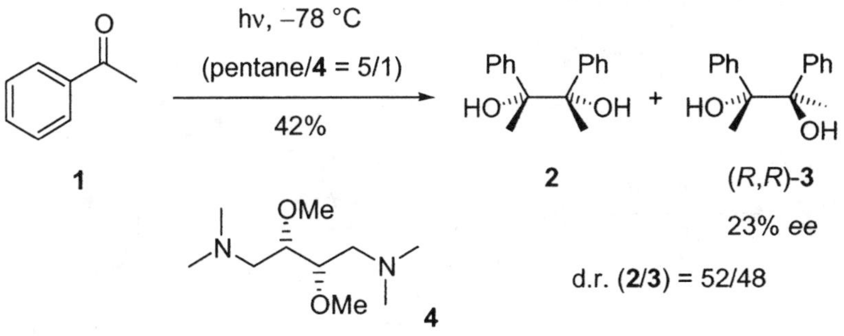

Scheme 3

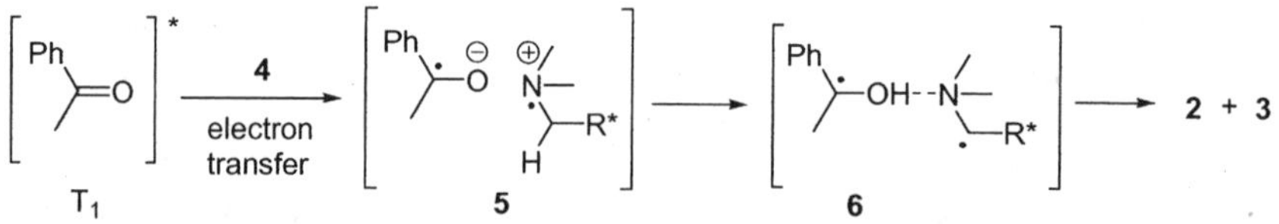

Scheme 4

ence of DDB gave the corresponding *d, l*-pinacols in optically active form, but no *ee* values were reported [15].

Horner and Klaus examined the same photopinacolization reaction in the presence of chiral lactates [16]. Radical **6** is formed by hydrogen transfer from the lactic acid hydroxy group to the triplet excited acetophenone. At room temperature, 4% *ee* were observed in the presence of an equimolar amount of *l*-menthyl *l*-lactate. With *l*-menthyl *d,l*-lactate no asymmetric induction was obtained. This shows that the stereocenter of the lactic acid influences the enantiodifferentiating dimerization step.

The photochemical synthesis of optically active oxaziridines from prochiral nitrones was driven by an interest in examining the configurational stability of the oxaziridin nitrogen atom. Nitrone **7** was irradiated at −78°C in a 1 : 1 mixture of (+)- or (−)-2,2,2-trifluorophenylethanol (**8**) and fluorotrichloromethane yielding nitrone **9**/*ent*-**9** in 30% *ee* (absolute stereochemistry unknown) [17,18]. At room temperature this selectivity decreased to 5% *ee*. Less bulky residues also reduced the selectivity. An exchange of *t*-butyl by *i*-propyl led to 20% *ee* at −78°C (Scheme 5).

The juxtapositioning of the nitrone and the chiral solvent in complex **10** was postulated to rationalize the observed enantioselectivities. The complex is stabilized by a hydrogen bond between the acidic hydroxy group and the basic oxygen atom of the nitrone and the interaction of the acidic carbinol proton with the π system of the aryl ring.

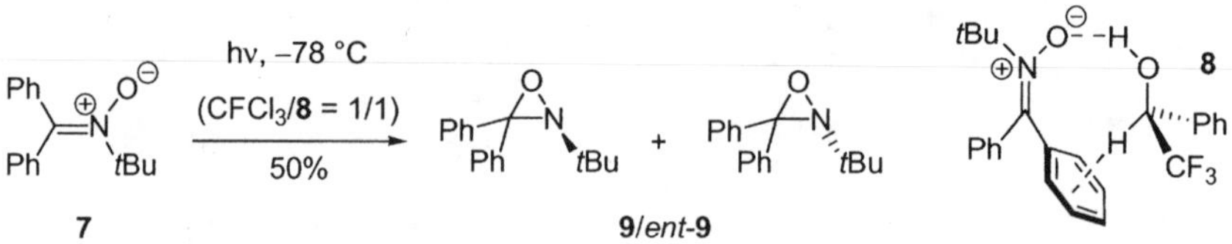

Scheme 5

Only low enantioselectivities were obtained for the oxidative photocyclization of *cis*-2-styrylbenzo[c]phenanthrene (**11**) to chiral hexahelicene (**12**) in chiral
solvents. The conformational equilibrium of alkene **11** shows some preference
for the chiral, prehelical cis–syn conformations **11a** and *ent*-**11a** over the cis–anti
conformation **11b** (Scheme 6). These equilibria are temperature and solvent dependent. In achiral solvents, both enantiomers of **11a** are present in equal amounts.
Chiral solvents favor one of the cis–syn enantiomers according to different equilibrium constants and consequently lead to enantiomerically enriched hexahelicene (**12**). In an early study, Laarhoven and Cuppen employed numerous lactates,
tartrates, and mandelates as chiral solvents [19,20]. The best asymmetric induction
of 2.1% *ee* was achieved by use of (*S*)-ethyl mandelate. Though of limited preparative use, this value is significantly higher than the 0.6% *ee* achieved by circularly
polarized light [21]. Recently, the same group examined this reaction again using
more accurate methods to determine the enantiomeric excess of the product [22].
They essentially confirmed their previous results. In the same study they conducted the reaction in chiral cholesteric or solid phases, a topic that will not be
discussed in detail here. The enantioselectivity, however, remained low (<7%
ee).

A similar selectivity was obtained by Nakazaki and coworkers for the
cis–trans isomerization of doubly bridged alkenes in diethyl-(+)-tartrate. Upon
irradiation at room temperature in this chiral solvent, *cis*-bicyclic α,β-unsaturated
ketone **13** was transformed into optically active (−)-*trans*-ketone **14** (Scheme

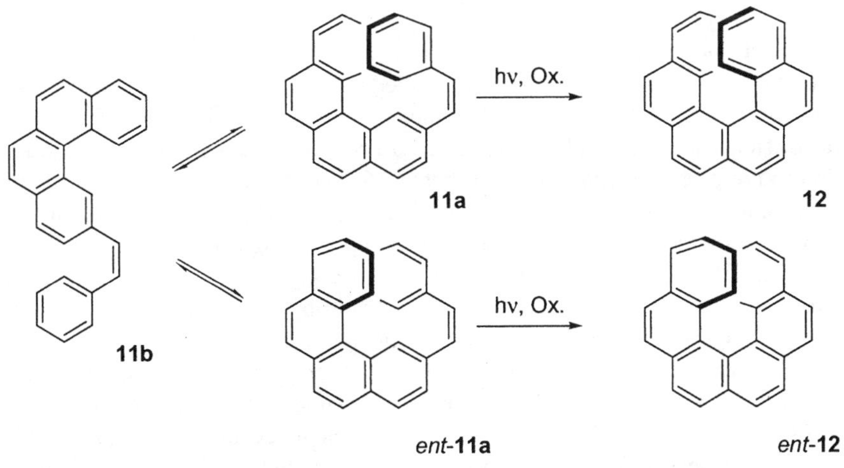

Scheme 6

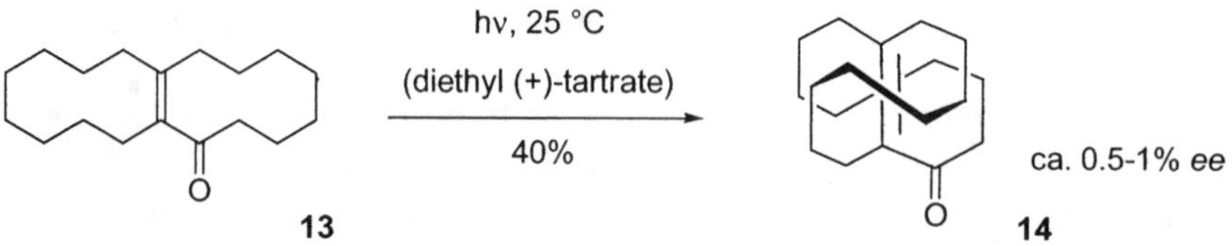

Scheme 7

7) [23]. The enantiomeric excess was estimated to be in the range of 0.5–1% *ee* [24].

In summary, chiral solvents have only induced limited enantioselectivity into different types of photochemical reactions as pinacolization, cyclization, and isomerization reactions. These studies are nevertheless very important, because they are among the early examples of chiral induction by an asymmetric environment. Based on our classification of chiral solvents as chiral inductors that only act as passive reaction matrices, effective asymmetric induction by this means seems to be intrinsically difficult. From the observed enantioselectivities it can be postulated that defined interactions with the prochiral substrate during the conversion to the product are a prerequisite for effective template induced enantioselectivity.

III. ASYMMETRIC INDUCTION BY CHIRAL BRØNSTED ACIDS

The most important work in this field has been conducted by Pete, Piva, Muzart and coworkers, who have extensively studied the enantioselective photodeconjugation reaction mediated by chiral aminoalcohols. Upon irradiation α,β-unsaturated esters and lactones of type **15** undergo a Norrish type II γ-hydrogen abstraction to the corresponding prochiral dienol **16**. Thermal proton transfer leads to the chiral deconjugated ester **17**. This tautomerization has been shown to be facilitated by the presence of a weak acid as protonating agent [25]. With chiral protonating agents the protonation of the two enantiofaces should not occur at the same rate, and the deconjugated ester should be formed enantioselectively. With chiral alcohols, chiral amines, or mixtures of both, no asymmetric induction was observed. With chiral amino alcohols, however, impressive enantioselectivities were obtained. In the presence of 0.1 equivalents of (+)-ephedrine (**18**) at −78°C benzyl ester **15** was converted to the photodeconjugated product with 37% *ee*. In several studies, a variety of chiral amino alcohols under various conditions were tested [26–38] Ephedrine analogs with bulkier *N*-alkyl substituents and camphor-derived amino alcohols, such as compounds **19** (79% *ee* at −40°C) and **20** (91% *ee* at −55°C) in CH$_2$Cl$_2$ as the solvent, proved to be most effective [39,40]. A selection of the best results is depicted in Scheme 8.

The synthetic applicability is somewhat limited in that the asymmetric induction is very substrate dependent. Esters other than benzyl esters showed lower enantiomeric excesses. The substitution pattern at the γ position has a drastic effect on the efficiency of the asymmetric induction. Monosubstitution led to enantioselectivities around 35% *ee* (with inductor **19**, −40°C). In the case of unsubstituted γ position, the induction went down to 10% *ee* [39]. For lactones, enantioselectivities up to 43% *ee* were reported (inductor **20**, −55°C) [40].

The asymmetric induction increased with the amount of chiral inductor up to 0.1 equivalents and then levelled off. Thus substoichiometric amounts of the chiral Brønsted acid achieved good enantiomeric excess with significant amplification of chirality. This was rationalized by the assumption that under the applied irradiation conditions at any moment there is a large excess of the chiral inductor with respect to the photodienol. The synergy between the amino and the hydroxyl group might be indicative of a cyclic transition state illustrated by complex **21**. In this transition state, the deprotonation of the enol by the amino group is concerted with the protonation of the enol by the hydroxyl group. Evidently, the aminoalcohol interacts strongly enough with the intermediate dienol to ensure

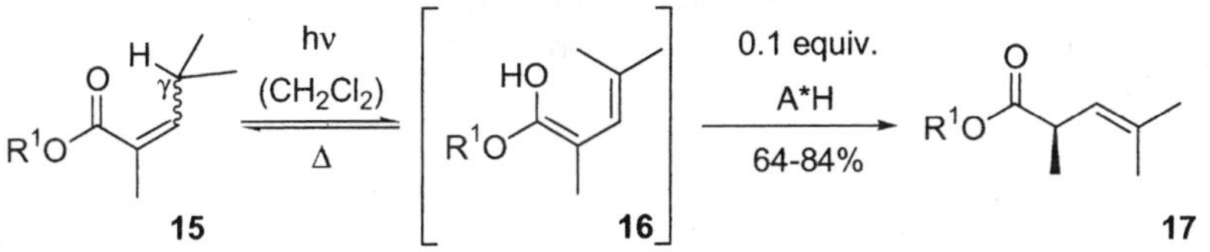

R^1	T [°C]	A*H	ee [%]
Bn	−40	18	37 (*ent*-17)
Bn	−40	19	70 (17)
Bn	−55	20	91 (17)
*i*Pr	−40	19	68 (17)
Et	−40	19	40 (17)

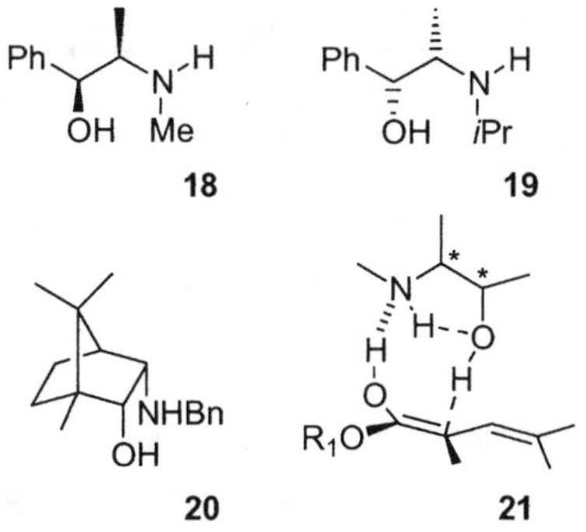

Scheme 8

efficient face differentiation but at the same time does not form stable complexes with the keto forms of the esters and turnover is observed. The absolute configuration of the product seems to be determined by the carbon atom bearing the amine function.

The groups of Pete and Rau also employed chiral amino alcohols for the enantioselective protonation of simple enols **23a–c** that were photochemically generated from 2-*i*-butyl indanones and tetralones **22a–c** by a Norrish type II photoelimination (Scheme 9) [41,42]. Best enantioselectivities were obtained at $-40°C$ in acetonitrile with 0.1 equivalent of the chiral amino alcohol. In the case of indanone **22a**, the selectivity reached 49% *ee* with (−)-ephedrine (*ent*-**18**) and could not be further enhanced by the camphor derived inductor **20**. With this amino alcohol, enantioselectivities over 80% *ee* were induced in the case of tetralone **22b**. A benzyl substituent in place of the methyl group led to substantial decrease of the selectivity to 47% *ee*. Linear ketones gave low yields and enantioselectivities around 9% *ee*.

The enantiodifferentiating protonation by aminoalcohols was further utilized in the partial kinetic resolution of *trans*-1-acetylcyclooctene (**26**), which is formed by irradiation of the cis cyclooctene **25** (Scheme 10) [43,44]. The trans alkene is stable in carefully dried acetonitrile. Protic species catalyze the isomerization back to the cis alkene. When the irradiation was carried out in the presence of (+)-ephedrine, a new CD signal appeared that was assigned to the trans alkene **26**. The addition of cyclopentadiene gave optically active Diels–Alder product **27**, which was isolated with at least 22% *ee*.

Ninomiya and Naito established an enantioselective variant of their enamide photocyclization based on chiral proton donors. The enamide photocyclization

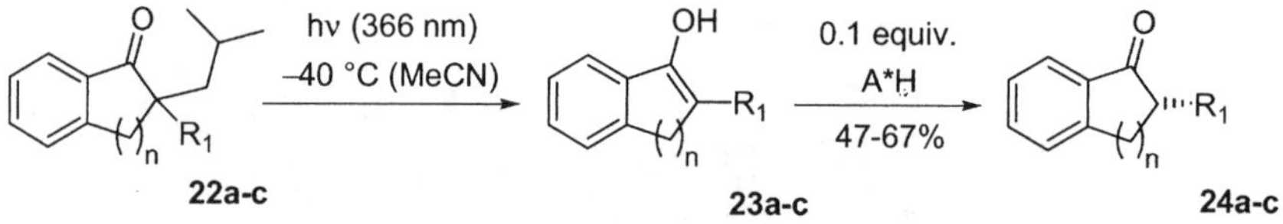

	n	R_1	A*H	ee [%]
22a	1	Me	*ent*-18	49 (24a)
22b	2	Me	*ent*-18	54 (24b)
22b	2	Me	20	83 (24b)
22c	2	Bn	20	47 (24c)

Scheme 9

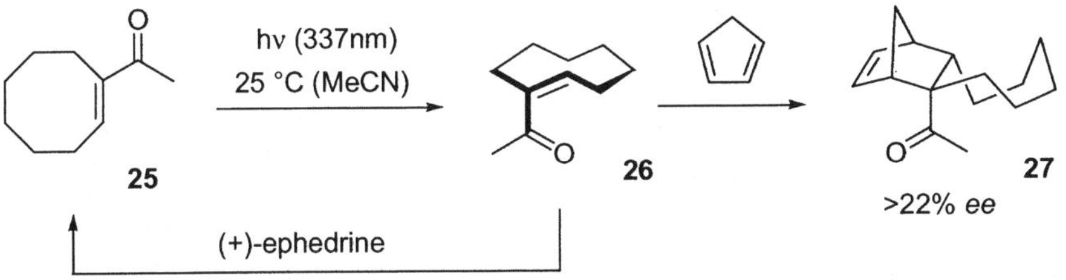

Scheme 10

comprises a photochemically allowed conrotatory [6π]-electrocyclization to an intermediate of type **29** followed by a rearomatizing hydrogen shift to cis and trans lactams **30** and **31** [45]. As deuteration experiments revealed, this hydrogen shift is of a concerted nature in aprotic solvents, but in protic media it takes place to a great extent as a deprotonation/protonation sequence. In the presence of a chiral proton source, the stereogenic center that is defined in the protonation step should be built up enantioselectively. To this end, enamides **28a/b** were cyclized in the presence of several chiral acids [46]. The best results were obtained with (−)-di-*p*-toluoyltartaric acid (**32**). The major diastereoisomers, cis-quinolones **30a/b**, were obtained in 0 and 11% *ee*, the minor diastereoisomers, trans quinolones **31a/b**, in 26 and 38% *ee* (Scheme 11). The absolute configuration of the products was determined by comparison with synthetic material obtained from enantiomerically pure precursors.

Hydrogen bonds were postulated between the two acid hydroxy groups and both the nitrogen atom and the protonated carbon atom of the intermediate to explain the observed predominating absolute stereochemistry. Finally, it should be mentioned that the replacement of the cyclohexene ring by a prop-1-enyl group led to only 14% *ee*.

In a more recent study, the enamide photocyclization with very similar photosubstrates was examined in the presence of chiral amino alcohols and chiral amines as asymmetric inductors [47]. The achieved enantioselectivities are in the same range as the ones reported by Ninomiya and Naito, but in this approach the asymmetric induction was more effective for the cis products. In cyclopentane at − 40°C, 0.1 equivalents of the most effective inductor, (−)-ephedrine (*ent*-**18**), gave the cis cyclization products with up to 37% *ee* and the trans products with only 2% *ee*. The role of the chiral inductor as a Brønsted acid was supported by flash photolysis experiments. The presence of the chiral amino alcohol led to an increase in the rate of disappearance of a transient that was assigned to the primary cyclization intermediate of type **29**, i.e., the chiral inductor accelerates the protonation/deprotonation sequence that reestablishes the aromatic ring.

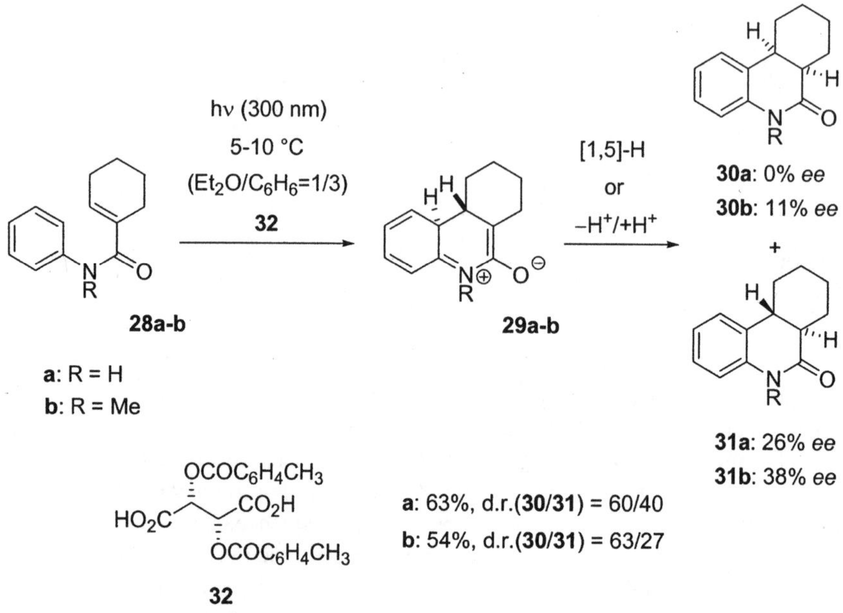

Scheme 11

In summary, chiral Brønsted acids have been used to protonate prochiral intermediates that were photochemically produced. Optimization of the chiral amino alcohols and the conditions led to excellent enantioselectivities of up to 91% *ee* employing only 0.1 equivalent of chiral inductor with respect to the substrate. These results demonstrate that the desirable goal of chiral amplification can be reached with chiral templates. The asymmetric induction varies strongly for different substrates, and therefore the general applicability as a synthetic tool seems to be limited.

IV. ASYMMETRIC INDUCTION BY CHIRAL TRANSITION METAL COMPLEXES

Chiral transition metal complexes have been employed in the enantioselective [2 + 2]-photocycloaddition reaction, in asymmetric electron transfer reactions and photooxidations/reductions. In the enantiodifferentiating step of the latter reaction type the chiral transition metal complex is involved in an electron transfer, i.e., the metal is converted from an excited oxidative state to a more stable one. This type of reaction is therefore a borderline case between chiral sensitizers that interact in the photochemically excited state and chiral templates that form a ground state complex. This review confines itself to enantioselective photooxidations and photoreductions of only organic substrates. Studies on the asymmetric synthesis of inorganic metal complexes like Δ/Λ-Co(acac)₃ by chiral Ru complexes [48–54] as well as asymmetric fluorescence quenching reactions with chiral Ru complexes [55] will not be discussed in detail.

Mattay et al. employed asymmetric copper(I)-catalyzed intramolecular [2 + 2]-photocycloaddition reactions in a synthetic approach to (+)- and (−)-grandisol [56]. Racemic dienol **33** was irradiated in the presence of CuOTf and a chiral ligand to yield mainly cyclobutanes **34** and *ent*-**34** as a mixture of enantiomers. Other 1,6-dienes were also employed. A number of chiral nitrogen-containing bidentate ligands were tested, the most effective of which, (4*S*,4′*S*)-4,4′-diisopropyl-2,2′-bisoxazoline (**35**) and (4*R*,4′*R*)-4,4′-diethyl-2,2′-bisoxazoline (**36**), ensured a minor enantiomeric excess of <5% *ee* (Scheme 12). The coordination of the diene to the chiral Cu(I) complex under formation of a complex of type **37** was proved by CD analysis. The authors suggest a lower reactivity of the chiral complex compared to the copper ion coordinated to solvent molecules as the reason for the low enantioselectivities observed.

Ohkubo and coworkers studied the photooxidative racemic resolution of *rac*-1,1′-bi-2-naphthol (*rac*-**38**) with the axially chiral ruthenium complex Δ-[Ru(menbpy)₃]²⁺ (**39**, menbpy = 4,4′-dimenthoxycarbonyl-2,2′-bipyridine)

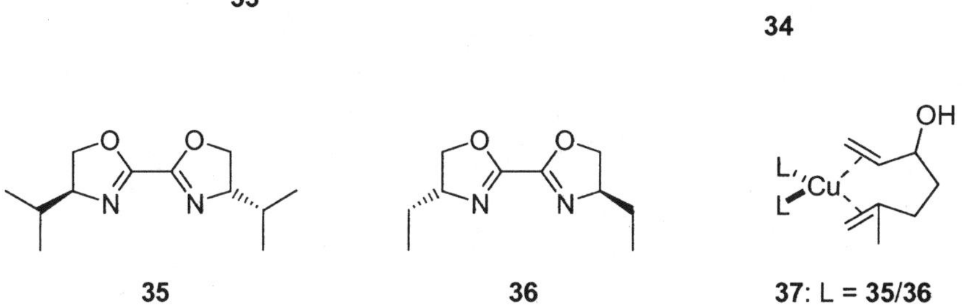

Scheme 12

[57]. They set up the catalytic reaction system depicted in Scheme 13. $[Ru(menbpy)_3]^{3+}$ is generated by oxidative quenching of $[Ru(menbpy)_3]^{2+*}$ by the oxidant $[Co(acac)_3]$ (**40**, Hacac = pentane-2,4-dione). By oxidation of binaphthol **38**, $[Ru(menbpy)_3]^{3+}$ is converted to $[Ru(menbpy)_3]^{2+}$. The oxidation products have not been established yet. With chiral Δ-$[Ru(menbpy)_3]^{3+}$ (*S*)-($-$)-1,1'-bi-2-naphtol (**38**) was converted faster than the (*R*)-($+$) enantiomer (*ent*-**38**), so that after 3 h binaphthol *ent*-**38** was obtained in 15% *ee*.

The same chiral Ru complex **39** was applied in the photooxidative dimerization of 2-naphthol to produce (*R*)-($+$)-1,1'-bi-2-naphthol in 16.2% *ee* [58].

Other approaches of enantiodiscriminating oxygenations in the presence of molecular aggregates [59] or with cyclodextrin-linked iron or manganese porphyrins [60] are beyond the scope of this review, as mentioned in the introductory remarks.

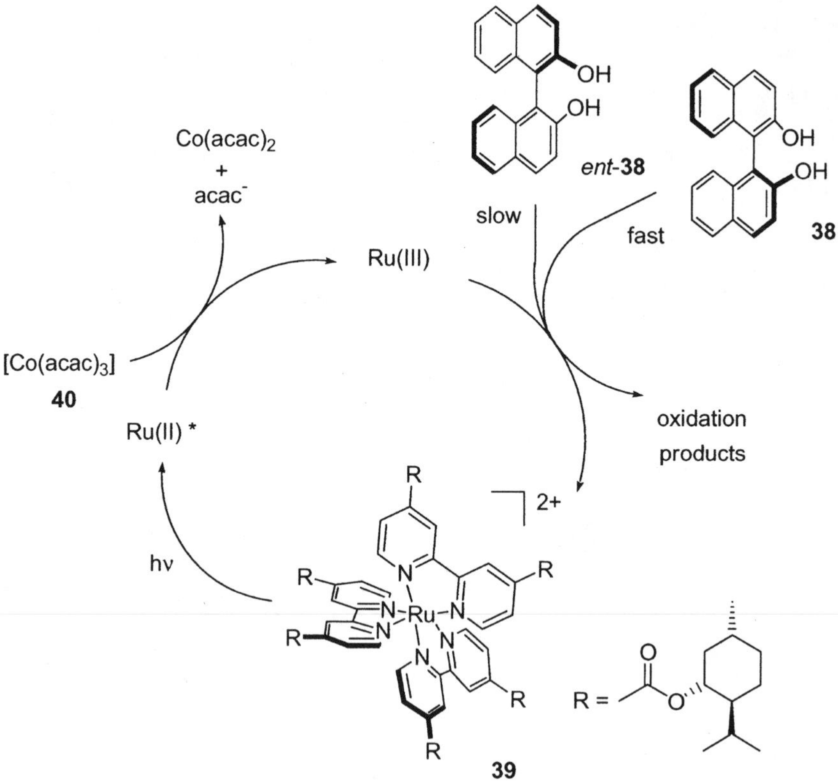

Scheme 13

V. ASYMMETRIC INDUCTION BY CHIRAL COMPLEXING AGENTS

Chiral complexing agents bind the substrate in a defined ground state complex employing various noncovalent interactions. As mentioned above (cf. Sec. I), the ability of the complexing agent to induce enantioselectivity depends on two parameters, the association constant K_A and the net face differentiation factor. If the association constant of the chiral complexing agent with the photoproduct is significantly lower than that with the starting material, even catalytic turnover and chiral amplification is possible, because after chiral induction and dissociation of the product, the host is available for the next chirality transfer. In this case only a substoichiometric amount of the chiral complexing agent is necessary. For high enantioselectivity at a substoichiometric level, another prerequisite must be met. The photochemical reaction must be significantly accelerated in the complex with the chiral host, because unbound starting material is converted unspecifically and dilutes the enantiomeric excess of the product. The acceleration can be accomplished through electronic or steric effects.

The first complexing agents employed in enantioselective photoreactions were natural templates such as cyclodextrins or bovine serum albumin. These intriguing studies are described in detail in another chapter of this book. In recent years, rationally designed chiral complexing agents were introduced. These chiral hosts serve as stoichiometric complexing agents, which differentiate the enantiotopic faces of the starting material by binding to it in a defined orientation via hydrogen bonds. The conformational flexibility is constrained by a rigid backbone that offers an easily accessible binding site and at the same time exhibits a sterically demanding substituent that effectively shields one face of the bound substrate. In early studies, chiral lactams **42** and **43** with a menthoxycarbonyl residue as the shielding group were examined. Better results were obtained by the more elaborate chiral lactams **44** and *ent*-**44** in which the hydrogen acceptor and hydrogen donor site define a plane parallel to the flat tetrahydronaphthyl residue. This shield is photochemically stable in contrast to further extended aromatics. The required spatial alignment is provided by Kemp's triacid (**41**), from which these hosts were synthesized (Scheme 14). Racemic host *rac*-**44** is accessible in 49% overall yield. The optical resolution to chiral benzoxazoles **44** and *ent*-**44** is readily achieved via the *N*-(−)-menthyloxycarbonyl derivative by column chromatographic separation and subsequent cleavage of the menthyloxycarbonyl group [61].

The efficiency of these chiral host compounds has been shown in highly enantioselective photocyclization and photocycloaddition reactions of prochiral lactams. These substrates, for example 2-quinolone derivatives, are expected to coordinate to lactam **44** with its NH-group as the hydrogen donor and the carbonyl group as the hydrogen acceptor, as depicted in Scheme 15. In this complex, any

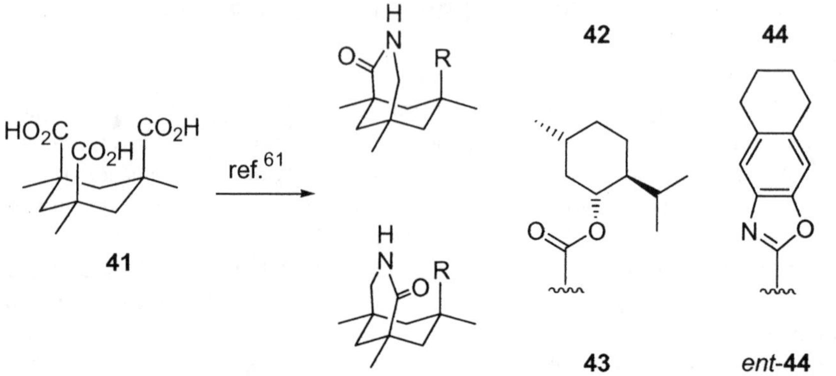

Scheme 14

intra- or intermolecular attack at the quinolone double bond can occur exclusively from the *re* face relative to carbon atom C-3. The other face is shielded by the bulky tetrahydronaphthalene unit. Aprotic nonpolar solvents like pentane, benzene, or toluene are essential to establish stable complexes with only two hydrogen bonds. At the same time, sufficient solubility has to be ensured. In most cases toluene proved to be the best choice in this trade-off between strong binding and sufficient solubility. The association constant of 2-quinolone (**45**) and host **44** in toluene was determined by ^{1}H-NMR titration experiments as $K_a \cong$ 500 M^{-1} at 30°C, and a 1 : 1 complex stoichiometry was proved [62,63–64]. Since the complex formation was found to be exothermic, an enhanced association

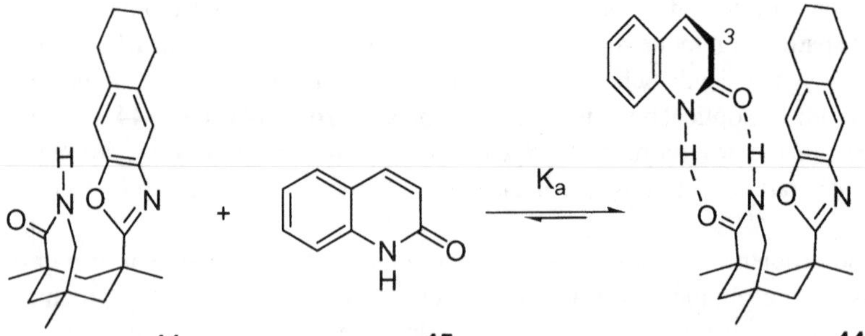

Scheme 15

was expected at lower temperatures, and indeed much higher enantioselectivities were observed, as the following examples will demonstrate.

Density functional theory calculations (Becke3LYP/6–31G*) were employed to elucidate the geometry of the complex between chiral host *ent*-**44** and 2-quinolone (**45**) as a representative and simple guest molecule [65]. The optimized geometry depicted in Fig. 1 confirms the assumed almost parallel alignment of the tetrahydronaphthalene shield and the guest molecule. The bond lengths for the two NH···O bonds were calculated to 1.87 and 1.80 Å.

The capability of hosts **42–44** to induce asymmetry was examined with various substrates in a number of photocyclization and photocycloaddition reactions. Imidazolidinone **46** cyclizes in a Norrish–Yang type cyclization via intramolecular hydrogen abstraction and subsequent radical combining ring closure to form bicyclic diastereoisomer **47** (exo) with a decent diastereomeric ratio (d.r.) of 80/20 and in good yields (Scheme 16). When the irradiation was carried out at −45°C in toluene in the presence of 2.5 equivalents of hosts **42** or **44**, the exo product **47** was obtained in 26% and 60% *ee*, respectively, i.e., host **44** proved to be superior to **42** with regard to enantioface differentiation [66,67]. Neither host changed the diastereoselectivity of the reaction. Host **43** under the same conditions favored the other enantiomer of the exo product (*ent*-**47**) with 25% *ee*, i.e., host **42** and **43** behaved as if they were enantiomers. Because of decreased host–substrate association, the *ee* dropped significantly at elevated temperatures

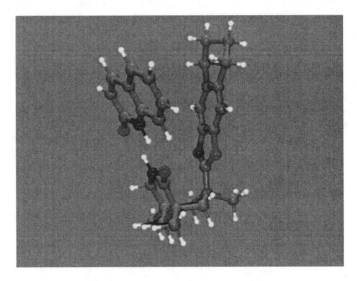

Figure 1   DFT calculation (Becke3LYP/6–31G*) of complex [*ent*-**44** · **45**].

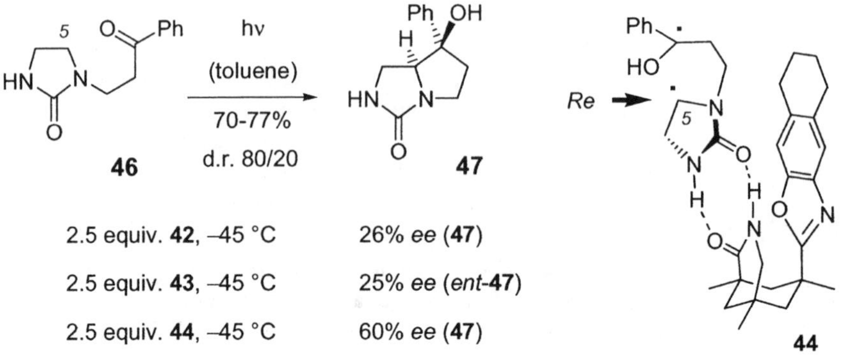

Scheme 16

or upon reducing the molar ratio of host to substrate. The absolute configuration
of the products was determined by x-ray crystallography. The stereochemistry
of the predominating enantiomer proved that in the presence of host **44** the cyclization of the intermediate biradical indeed occurred from the unshielded *re* face of
the imidazolidinone, as also depicted in Scheme 16.

The photochemical electrocyclic [4π]-ring closure of 2-pyridones in the
presence of chiral lactam *ent*-**44** was found to be very slow at low temperatures.
As a consequence, the potential of the host could not be fully exploited. At − 20°C
after 96 h of irradiation in toluene, bicyclic β-lactam **49** was obtained in 51%
yield. The presence of 1.5 equivalents of host *ent*-**44** led to 23% *ee* (Scheme 17)
[68].

The third type of cyclization examined in the presence of chiral lactam
hosts is the [6π]-cyclization of enamide **28a**, the protonation step of which has
been already enantioselectively directed with up to 38% *ee* by chiral proton donors, as described in Sec. III [46]. When the cyclization of enamide **28a** was

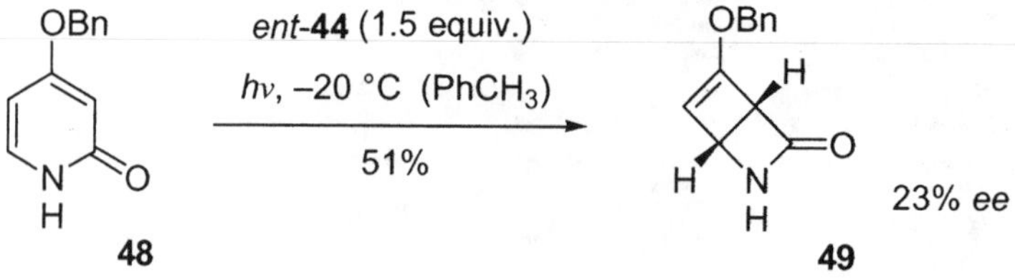

Scheme 17

carried out in the presence of host *ent*-**44**, deuteration experiments revealed a complicated interplay between enantioselective ring closure and enantioselective protonation [69]. At −55°C the conrotatory cyclization was controlled with 34% *ee*, favoring the depicted enantiomer of intermediate *ent*-**29a** (Scheme 18). The subsequent H rearrangement was also influenced by the chiral lactam, which this time acted as a chiral Brønsted acid favoring the protonation from the shielded face. This led to the trans product *ent*-**31a** in 57% *ee*, because the two modes of action of the chiral template are complementing each other. Cis product *ent*-**30a**, however, was obtained in only 30% *ee*, because cyclization and protonation favor different enantiomers. This led to the astonishing effect of a decrease in enantio-selectivity upon lowering the reaction temperature.

The selectivity of the cyclization step can be rationalized by the complex [*ent*-**44** · **28a**]. In this arrangement, the cyclohexene moiety of the amide is located directly opposite to the tetrahydronaphthalene moiety of the host. The phenyl group is not in the immediate vicinity of the sterically bulky environment. Upon conrotatory ring closure, the cyclohexene ring can either turn in the direction of the tetrahydronaphthalene or it can move away and open its *si* face to an attack from the benzene ring. According to simple molecular models, the latter move-ment, as schematically drawn in Scheme 19, appears to be favored for steric reasons. As a consequence, intermediate *ent*-**29a** is formed as the major product. Dissociation of the guest, which is a necessary requirement for protonation by the host, opens the *re* face for a potential proton attack. A potential transition state that would be accessible by rotation around the O⋯HN hydrogen bond is also depicted in Scheme 19.

The applicability of the developed chiral complexing agents was further demonstrated by asymmetrically directing [2 + 2]- and [4 + 4]-photocycloaddition

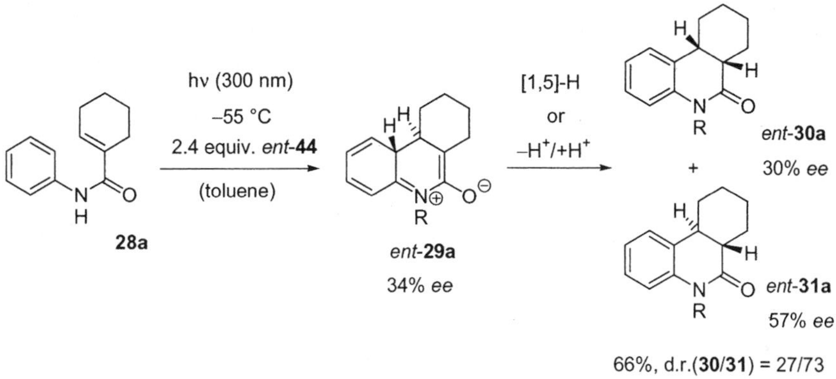

Scheme 18

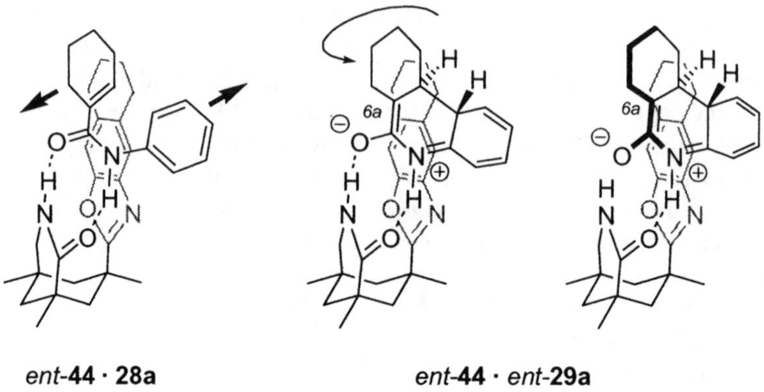

ent-**44** · **28a**   **ent-44** · **ent-29a**

Scheme 19

reactions. In this context, the [4 + 4]-photocycloaddition of 2-pyridone (pyridine-2(1*H*)-one, **50**) with cyclopentadiene was examined [68]. 2-Pyridone was irradiated at −50°C in the presence of host *ent*-**44** and cyclopentadiene furnishing a 3 : 2 mixture of endo and exo products **51** and **52**. Both diastereoisomers were obtained in high enantiomeric excess, the endo cycloadduct **51** with 87% *ee* and the exo cycloadduct **52** with 84% *ee* (Scheme 20). These enantioselectivities clearly surpass the asymmetric induction achieved in the cyclization reactions. The absolute configuration of the predominating enantiomer was elucidated by single x-ray crystallography. As expected, the photoexcited 2-pyridone was attacked by cyclopentadiene from the *si* face relative to the prostereogenic carbon atom C-3, because the *re* face is shielded following complexation with host *ent*-**44**.

Even higher enantioselectivities were accomplished in the case of intra- and intermolecular [2 + 2]-photocycloadditions of 2-quinolone derivatives. Upon

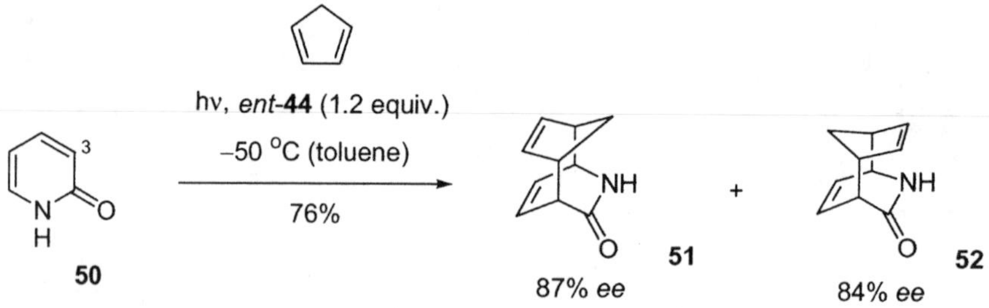

hν, *ent*-**44** (1.2 equiv.)

−50 °C (toluene)

76%

87% *ee*   **51**

84% *ee*   **52**

50

Scheme 20

irradiation with long wavelength UV light, the 4-alkenyloxy-2-quinolones **53** and **55** cyclized in an intramolecular [2 + 2]-photocycloaddition diastereoselectively to the crossed and straight cycloadducts **54** and **56**, respectively. In the presence of 2.6 equivalents of host compound **44** or *ent*-**44**, high enantioselectivities were achieved in a nonpolar solvent at low temperatures [62,70]. At − 60°C in toluene, **54** and **56** were obtained with > 90% *ee* (Scheme 21). Higher temperatures (84% *ee* at − 15°C, 39% *ee* at 30°C) and more polar solvents (4% *ee* at 30°C in acetonitrile) significantly reduced the enantioselectivity under otherwise identical conditions. The host is photostable under the irradiation conditions ($\lambda \geq 300$ nm) and was recovered after each irradiation by column chromatography in > 95% yield.

Again, the absolute configuration of photoproducts **54** and **56** was proved by single-crystal x-ray crystallography and is in accordance with the *si* face attack with respect to C-3, as the *re* face is shielded in the complex with chiral host *ent*-**44**. The enantioselectivities observed at low temperatures can be satisfactorily rationalized by the simple model that was decribed in the introduction. Under the assumption of perfect face differentiation and based on the association constants, the theoretical enantioselectivities were calculated to 97% and 90% *ee* at − 60 and − 15°C, respectively. These values are in good accordance but slightly higher than the selectivities recorded for quinolone **53** (93% and 84% *ee*) [62]. This indicates a net face differentiation slightly below unity.

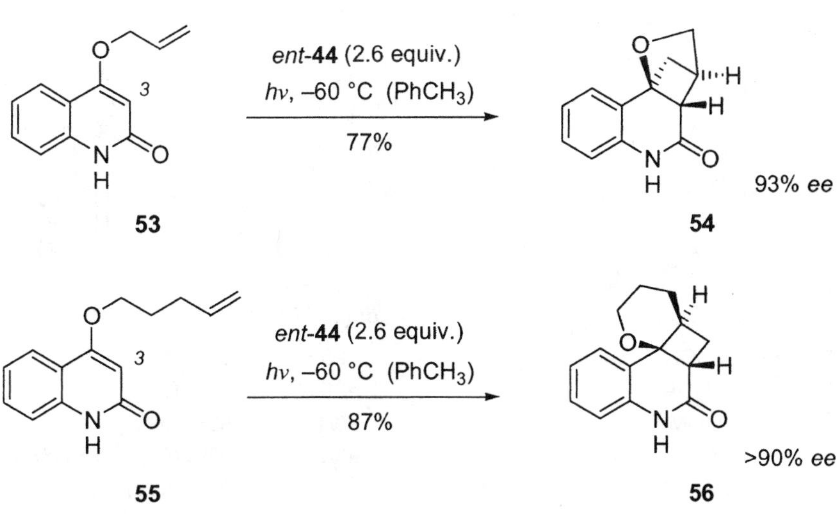

Scheme 21

The scope of this approach was widened by the observation of excellent enantioselectivities in intermolecular [2 + 2]-photocycloaddition reactions with various alkenes [62,71]. In the presence of an excess amount of alkene, 4-me-thoxy-2-quinolone (57) was converted with high chemo- and regioselectivity to the exo and endo cyclobutanes 59 and 60. With 4-penten-1-ol (58a), allyl acetate (58b), methyl acrylate (58c), and vinyl acetate (58d), the exo diastereomers 59a–d were formed with high simple diastereoselectivity and in high yields (80–89%). Under optimized irradiation conditions (2.4 eq. of host 44 or ent-44, −60°C), high enantiomeric excesses were achieved in all instances, as depicted in Scheme 22. These enantiomeric excesses are unprecedented for an intermolecular photo-chemical reaction.

As in the intramolecular case, host 44 induced re attack at carbon C-3 and host ent-44 induced a si attack.

The examined reactions suggest that the chiral lactams 44 and ent-44 can be considered versatile and effective complexing agents. The enantioselectivities observed in the case of photocyclization reactions were good but not in the range of being immediately applicable to organic synthesis. When employed in photo-cycloaddition reactions of prochiral flat lactams, however, excellent enantioselec-tivities were obtained in all cases if the reaction center was in the vicinity of the binding site and the photoreaction was conducted at low temperatures in nonpolar

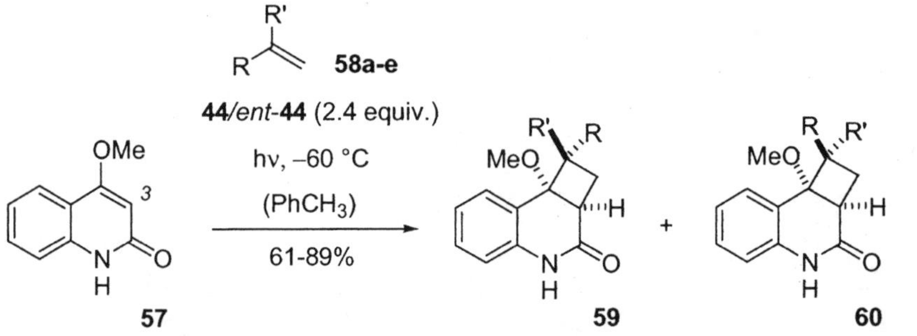

			dr (59/60)	ee
58a	R = CH$_2$CH$_2$CH$_2$OH	R' = H	>95/5	81% (59a)
58b	R = CH$_2$OAc	R' = H	>95/5	92% (59b)
58c	R = COOMe	R' = H	90/10	82% (59c)
58d	R = OAc	R' = H	63/27	93% (59d), 98% (60d)
58e	R = CH$_2$CH$_3$	R' = CH$_2$CH$_3$	-	92% (59e)

Scheme 22

solvents. This renders lactam **44** a potentially generally applicable synthetic tool to asymmetrically direct photocycloadditions of prochiral lactams.

VI. CONCLUSION

The ideas generated to address the challenge of an enantioselective photochemical transformation in solution have matured in recent years and have led to many promising results. Each method shows distinct advantages and disadvantages. It is consequently important to evaluate critically the scope and limitations before a method is selected for an application to a synthetic target. Chiral solvents generally provide comparably low enantioselectivities. Their advantage is the fairly broad scope that they offer; their disadvantage is the limited predictability of both extent and direction of the face differentiation. With chiral Brønsted acids, excellent chiral induction (up to 91% *ee*) has been achieved, even catalytically. The lack of specific interactions, however, narrows the number of substrates and makes it hard to define the structural prerequisites that a substrate must fulfill to be amenable to good asymmetric induction. Transition-metal-based templates promise great potential as they can be readily tuned by an appropriate choice of the chiral ligand. Contrary to their widespread use in conventional synthetic organic chemistry, they have not yet lived up to their expectations in photochemistry. Significant enantioselectivities have not been achieved so far. Rationally designed chiral templates that bind the substrate in a defined and specific way represent a new promising approach to accomplishing highly enantioselective photoreactions. Specific binding requires the substrate to exhibit a binding site complementary to the binding site of the chiral template and inherently imposes restrictions on the range of employable substrates. Currently, the method is limited to lactams and amides. In several cases, high chemical yields and enantioselectivities (up to 98% *ee*) have been reported in photochemical transformations of substrates that exhibit an NHCO-linkage. Chiral templates for other substrate classes of interest remain to be developed.

Given the potential of the available methods, and the many conceivable extensions of the methodology, the area of enantioselective photochemical transformation in solution will undoubtedly continue to grow in importance. As previously happened in conventional synthetic chemistry, catalytic photochemical methods will be developed from stoichiometric template-based photochemical procedures. Several of the compounds that can already be produced with high enantioselectivity are potential intermediates in the synthesis of more complex biologically active compounds. We are therefore convinced that the topic will remain for many years to come one of the most challenging and most intriguing subjects in organic photochemistry.

REFERENCES

1. Rau H. Chem Rev 1983; 83:535–547.
2. Everitt SRL, Inoue Y. In: Ramamurthy V , Schanze KS, Eds. Molecular and Supramolecular Photochemistry: Organic Molecular Photochemistry. Vol. 3. New York: Marcel Dekker, 1999:71–130.
3. Inoue Y. Chem Rev 1992; 92:741–770.
4. Roos G. Compendium of chiral auxiliary applications. San Diego: Academic Press, 2002.
5. Asaoka S, Horiguchi H, Wada T, Inoue Y. J Chem Soc, Perkin Trans 2000, and references cited therein; 2:737–747.
6. Nishino H, Kosaka A, Hembury GA, Aoki F, Miyauchi K, Suitomi H, Onuki H, Inoue Y. J Am Chem Soc 2002, and references cited therein; 124:11618–11627.
7. Rikken GLJA, Raupach E. Nature 2000, and references cited therein; 405:932–935.
8. Weizmann C, Bergmann E, Hirshberg Y. J Am Chem Soc 1938; 60:1530–1533.
9. Pitts JN, Letsinger RL, Taylor RP, Patterson JM, Recktenwald G, Martin RB. J Am Chem Soc 1959; 81:1068–1077.
10. Cohen SG, Laufer DA, Sherman WV. J Am Chem Soc 1964; 86:3060–3068.
11. Faljoni A, Zinner K, Weiss RG. Tetrahedron Lett 1974; 15:1127–1130.
12. Seebach D, Daum H. J Am Chem Soc 1971; 93:2795–2796.
13. Seebach D, Dörr H, Bastani B, Ehrig V. Angew Chem, Int Ed 1969; 8:982–983.
14. Seebach D, Oei H-A. Angew Chem, Int Ed 1975; 14:634–636.
15. Seebach D, Oei H-A, Daum H. Chem Ber 1977; 110:2316–2333.
16. Horner L, Klaus J. Liebigs Ann Chem 1979:1232–1257.
17. Boyd DR, Neill DC. Chem Commun 1977:51–52.
18. Boyd DR, Campbell RM, Coulter PB, Grimshaw J, Neill DC, Jennings WB. J Chem Soc, Perkin Trans 1985; 1:849–855.
19. Laarhoven WH, Cuppen TJHM. Chem Commun 1977:47.
20. Laarhoven WH, Cuppen TJHM. J Chem Soc, Perkin Trans 1978; 1:315–318.
21. Bernstein WJ, Calvin M, Buchardt O. J Am Chem Soc 1973; 95:527–532.
22. Prinsen WJC, Laarhoven WH. Recl Trav Chim Pays-Bas 1995; 114:470–475.
23. Nakazaki M, Yamamoto K, Maeda M. Chem Commun 1980:294–295.
24. Nakazaki M, Yamamoto K, Maeda M. J Org Chem 1980; 45:3229–3232.
25. Henin F, Mortezaei R, Pete JP. Synthesis 1983:1019–1021.
26. Henin F, Mortezaei R, Muzart J, Pete J-P. Tetrahedron Lett 1985; 26:4945–4948.
27. Mortezaei R, Henin F, Muzart J, Pete J-P. Tetrahedron Lett 1985; 26:6079–6080.
28. Pete J-P, Henin F, Mortezaei R, Muzart J, Piva O. Pure Appl Chem 1986; 58:1257–1262.
29. Mortezaei R, Piva O, Henin F, Muzart J, Pete J-P. Tetrahedron Lett 1986; 27:2997–3000.
30. Piva O, Henin F, Muzart J, Pete J-P. Tetrahedron Lett 1986; 27:3001–3004.
31. Piva O, Henin F, Muzart J, Pete J-P. Tetrahedron Lett 1987; 28:4825–4828.
32. Henin F, Mortezaei R, Muzart J, Pete J-P, Piva O. Tetrahedron 1989; 45:6171–6196.

33. Awandi D, Henin F, Muzart J, Pete J-P. Tetrahedron: Asymmetry 1991; 2: 1101–1104.
34. Henin F, Muzart J, Pete J-P, Piva O. New J Chem 1991; 15:611–613.
35. Muzart J, Henin F, Pete J-P, M'Boungou-M'Passi A. Tetrahedron: Asymmetry 1993; 4:2531–2534.
36. Piva O. Synlett 1994:729–731.
37. Piva O. J Org Chem 1995; 60:7879–7883.
38. Henin F, Letinois S, Muzart J. Tetrahedron: Asymmetry 2000; 11:2037–2044.
39. Piva O, Mortezaei R, Henin F, Muzart J, Pete J-P. J Am Chem Soc 1990; 112: 9263–9272.
40. Piva O, Pete J-P. Tetrahedron Lett 1990; 31:5157–5160.
41. Henin F, Muzart J, Pete J-P, M'Boungou-M'Passi A, Rau H. Angew Chem, Int Ed 1991; 30:416–418.
42. Henin F, M'Boungou-M'Passi A, Muzart J, Pete J-P. Tetrahedron 1994; 50: 2849–2864.
43. Henin F, Muzart J, Pete J-P, Rau H. Tetrahedron Lett 1990; 31:1015–1016.
44. Henin H, Muzart J, Pete J-P. New J Chem 1992; 16:979–985.
45. Ninomiya I, Naito T. Heterocycles, 1981.
46. Naito T, Tada Y, Ninomiya I. Heterocycles 1984; 22:237–240.
47. Formentin P, Sabater MJ, Chrétien MN, García H, Scaiano JC. J Chem Soc, Perkin Trans 2002; 2:164–167.
48. Porter GB, Sparks RH. Chem Commun 1979:1094–1095.
49. Ohkubo K, Ishida H, Hamada T, Inaoka T. Chem Lett 1989:1545–1548.
50. Ohkubo K, Hamada T, Inaoka T, Ishida H. Inorg Chem 1989; 28:2021–2022.
51. Hamada T, Ishida H, Kuwada M, Ohkubo K. Chem Lett 1992:1283–1286.
52. Ohkubo K, Hamada T, Watanabe M, Fukushima M. Chem Lett 1993:1651–1654.
53. Ohkubo M, Hamada T, Ishida H. Chem Commun 1993:1423–1425.
54. Ohkubo K, Hamada T, Ishida H, Fukushima M, Watanabe M. J Mol Catal 1994; 89:L5–L10.
55. Wörner M, Greiner G, Rau H. J Phys Chem 1995; 99:14161–14166.
56. Langer K, Mattay J. J Org Chem 1995; 60:7256–7266.
57. Hamada T, Ishida H, Usui S, Tsumura K, Ohkubo K. J Mol Catal 1994; 88:L1–L5.
58. Hamada T, Ishida H, Usui S, Watanabe Y, Tsumura K, Ohkubo K. Chem Commun 1993:909–911.
59. Ohkubo K, Yamashita K, Ishida H, Haramaki H, Sakamoto Y. J Chem Soc, Perkin Trans 1992; 2:1833–1838.
60. Weber L, Imiolczyk I, Haufe G, Rehorek D, Hennig H. Chem Commun 1992: 301–303.
61. Bach T, Bergmann H, Grosch B, Harms K, Herdtweck E. Synthesis 2001:1395–1405.
62. Bach T, Bergmann H, Grosch B, Harms K. J Am Chem Soc 2002; 124:7982–7990.
63. Bach T, Bergmann H, Harms K. J Am Chem Soc 1999; 121:10650–10651.
64. Bach T, Bergmann H, Brummerhop H, Lewis W, Harms K. Chem Eur J 2001; 7: 4512–4521.
65. Grosch B, Strassner T, Bach T. unpublished results, 2002.

66. Bach T, Aechtner T, Neumüller B. Chem Commun 2001:607–608.
67. Bach T, Aechtner T, Neumüller B. Chem Eur J 2002; 8:2464–2475.
68. Bach T, Bergmann H, Harms K. Org Lett 2001; 3:601–603.
69. Bach T, Grosch B, Strassner T, Herdtweck E. J Org Chem 2003; 68:1107–1116.
70. Bach T, Bergmann H, Harms K. Angew Chem, Int Ed 2000; 39:2302–2304.
71. Bach T, Bergmann H. J Am Chem Soc 2000; 122:11525–11526.

9

Supramolecular Asymmetric Photoreactions

Takehiko Wada and Yoshihisa Inoue
Osaka University, Suita, and Japan Science and Technology Agency, Kawaguchi, Japan

I. INTRODUCTION

Asymmetric synthesis is currently one of the most vital areas of chemistry, in which a variety of catalytic and enzymatic methodologies have been developed in the last three decades [1–10], although until recently the photochemical branch has not been very extensively or intensively explored [11–16]. However, the photochemical approach to asymmetric synthesis possesses advantages over the thermal ones: as photoprocesses proceed through the electronically excited state, they often yield strained and/or thermally difficult-to-access products of unique structures in a single step [17]. Furthermore, photochemical reactions, being free from the fetter of activation energy, are run over a wide range of temperatures without an accompanying retardation of the reaction rate at lower temperatures or undesirable side reactions at elevated temperatures.

Photochemistry also has drawbacks: the excited-state interactions are weak and short-lived and are therefore difficult to control; also the detection/observation of transient species and the subsequent elucidation of the reaction mechanism are in general more difficult [17]. Consequently, it has long been believed that the critical and precise control of asymmetric photoreactions is a hard task, and that the optical yields obtained therefrom are low. To overcome this, two strategies have been developed in the evolution of asymmetric photochemistry, or photochirogenesis: (1) the introduction of a chiral handle to the photosubstrate (diastere-odifferentiating photoreaction) and (2) entropic control by environmental factors (enantiodifferentiating photoreaction). Both strategies have been successful in enhancing the optical yields of photoproducts through close and long-lived weak intra/intermolecular interactions in the excited state, both of which are reviewed in Chaps. 5 and 4, respectively. More recently, a new strategy has been employed in photochirogenesis; supramolecular interactions in the ground and/or excited state are employed in order to modify the original photoreactivity, so to trigger a hidden or forbidden photoreaction or enhance the intra/intermolecular contacts in confined media.

Supramolecular chemistry, using conventional hosts such as cyclodextrin, calixarene, and micelles, has been extensively investigated and applied to various areas of science and technology [18]. These host–guest systems are expected to have novel physical and chemical properties through the formation of unique supramolecular, or self-assembled, structures. One of the distinctive and crucial features of a supramolecular system is the creation of ''confined space'' for critically controlling, and often enhancing, the noncovalent interactions between the host and the guest. In connection with this, a solvent cage may also be regarded as a passive confined medium, which is more dynamic in time and less defined in space.

Most of the earlier studies on supramolecular chemistry were concentrated on the molecular recognition of target guests in the ground state, while little effort was devoted to the active control of supramolecular interactions and reactions in the ground and electronically excited states [19–22]. In the last decade, the role of structural freezing, or confinement, has attracted much attention not only in thermal asymmetric synthesis but also in the photochemical counterpart [23]. As a result, increasing attention has been directed toward the new methodology of photochirogenesis using various supramolecular systems, such as chirally modi-fied zeolites, cyclodextrins, DNA, and proteins. In the confined space of a supra-molecular system, noncovalent weak interactions, including hydrogen bonding, dipole–dipole, ion–dipole, π–π, cation–π, CH–π, van der Waals, and hydro-phobic interactions, can be enhanced or modified to influence significantly the chemo-, regio-, and/or stereoselectivity of the (photo)reactions occurring in the confined media [24,25].

In this chapter we will review the recent advances of supramolecular photo-chirogenesis in various confined media, excluding micelles, chiral solvents, liquid crystals, metal complexes, polymer matrices, clays, and crystals. Micelles are a typical supramolecular assembly with an internal hydrophobic core which shows a unique boundary effect, e.g., enhanced radical recombination of geminate radi-cal pairs produced by ketone photolysis [26], but essentially no asymmetric photo-reaction has hitherto been reported in micelles. Photochemical asymmetric induc-tion in chiral solvents [27,28] and chiral liquid crystals [29,30] have been known for many years, but the obtained optical yields are generally low, and the chirogen efficiency is extremely low in the absence of specific interactions between the chiral media and the substrate. These results render this strategy less attractive from the asymmetric synthetic viewpoint and will not be discussed in this chapter. The solid-state asymmetric photoreactions of chiral crystals of achiral molecules [31–34] and of achiral molecules cocrystallized with photochemically inert chiral inductors [35,36] are reviewed in Chap. 12. Photochemical asymmetric induction using (supramolecular) chiral metal complexes [37] has been actively investigated in recent years and is reviewed in Chap. 10. Clay interlayers have also been used in supramolecular photochemical and photophysical studies [38,39], but no efficient asymmetric photoreactions have been reported. Asymmetric induction with imprinted polymers provides a new and interesting strategy [40], which is however beyond the scope of this chapter. In this review, we will concentrate on supramolecular photochirogenesis in confined media, which employ chirally modified zeolites, cyclodextrins, chiral molecular hosts, proteins, and DNA, as classified in Table 1.

II. SUPRAMOLECULAR PHOTOCHIROGENESIS WITH CHIRALLY MODIFIED ZEOLITES

Zeolites are crystalline aluminosilicates ($M_{x/n}^{x+}[Al_xSi_yO_{2(x+y)}]^{x-} \cdot zH_2O$) with uniform channels and cavities of molecular-size dimensions (7–13 Å). The aperture of the zeolite channel and cavity can be classified as small, medium, or large, depending on the number of O atoms [41–45]. The negative charge of the zeolite framework, which is generated by a trivalent Al substituting tetravalent Si in the silicate framework, is neutralized by a countercation such as alkali, alkaline earth, or (alkyl)ammonium cations. The size and properties of the zeolite cavity, often called "supercage," strongly depend on the nature of countercation. Hence the cavity dimension is a critical function of the countercation, giving a variable free volume available for the adsorbed molecule from 873 Å^3 for NaX to 732 Å^3 for CsX. A variety of organic molecules are included and immobilized in zeolite supercages accessing through a channel/window between cavities, although there is a limiting size for guest molecules determined by the channel diameter. Thus the zeolite supercages can recognize the size, shape, and volume of reactant, transition state, and/or product upon inclusion through the portal and cavity sizes [46,47]. Exploiting these features as tools for controlling the ground- and excited-state interactions, much work on supramolecular photoreactions in zeolite supercages has been done [46,47]. In particular, supramolecular photochirogenesis in chirally modified zeolite media has been a field of considerable recent interest and is currently investigated actively with the aim of achieving a precise control of product chirality and enantiomeric/diastereomeric excess [19–22].

Table 1 Supramolecular Photochirogeneses

Chiral Host	Type of Reaction	Excitation Mode	Differentiation Mechanism	Reference
Chirally modified zeolite	Norrish/Yang Type II Reaction	Direct excitation	Enantiodifferentiation	49–53
			Diastereodifferentiation	54
	Radical Recombination	Direct excitation	Enantiodifferentiation	55
	Photorearrangement	Direct excitation	Enantiodifferentiation	56
			Diastereodifferentiation	63
	Photoisomerization	Direct excitation	Enantiodifferentiation	68
			Diastereodifferentiation	68–70
		Sensitization	Enantiodifferentiation	69,90
			Diastereodifferentiation	69
	Photocyclization	Direct excitation	Enantiodifferentiation	71–75
			Diastereodifferentiation	71–75
	Photoreduction	Direct excitation	Enantiodifferentiation	79
Cyclodextrin	Radical Recombination	Direct excitation	Enantiodifferentiation	101
	Photocyclization	Direct excitation	Enantiodifferentiation	102, 103
	Photodimerization	Direct excitation	Enantiodifferentiation	104–106,
				109–111, 120
	Photocycloaddition	Direct excitation	Enantiodifferentiation	121
	Photoisomerization	Direct excitation	Enantiodifferentiation	107, 122
		Sensitization	Enantiodifferentiation	123, 124
	Photooxidation	Sensitization	Enantiodifferentiation	125
Synthetic chiral template	Photocycloaddition	Direct excitation	Enantiodifferentiation	134–138
		Sensitization	Enantiodifferentiation	139, 148
Protein	Photodestruction	Direct excitation	Enantiodifferentiation	141–144
	Photocycloaddition	Direct excitation	Enantiodifferentiation	146
Nucleoside/dsDNA	Photoisomerization	Sensitization	Enantiodifferentiation	151, 152

Pioneering work on the photochemical diastereocontrol in zeolite supercages was reported by Turro and coworkers in 1991 [48]. They investigated the diastereoselective photodecarbonylation of 2,4-diphenyl-3-pentanone (DPP) adsorbed in various cation-exchanged X and Y zeolites to find that the diastereoselectivity of *d,l*- over *meso*-2,3-diphenylbutane increases in the order: LiX ~ NaX < LiY ~ NaY < KY. In 1996, Ramamurthy and coworkers reported the first example of photochemical asymmetric induction in chirally modified zeolites [49], where they employed the Norrish/Yang type II reaction of *cis*-4-*tert*-butyl-cyclohexyl aryl ketones to the corresponding cyclobutanols. Since then, a variety of asymmetric photoreactions in zeolite supercages have been reported as reviewed below.

A. Enantiodifferentiating Norrish/Yang Type II Reactions in Chirally Modified Zeolites

Ramamurthy, Scheffer and coworkers and other investigators reported the enantiodifferentiating Norrish/Yang photocyclization [49–53] of aryl *cis*-4-*tert*-butyl-cyclohexyl ketones **1** to cyclobutanols **2** (Scheme 1) in chirally modified zeolite supercages [49].

They used the zeolites as "microreactors," in which a non-light-absorbing optically active molecule is immobilized as a chiral inductor and then a photoreactive substrate is introduced into the same or a nearby supercage. Although photolyses of **1a–c**, dissolved in ephedrine-containing hexane or immobilized in unmodified zeolite supercages, led to racemic products **2a–c**, irradiation of **1a–c** immobilized in (−)- or (+)-ephedrine-modified NaY zeolite gave antipodal enantiomerically enriched **2a–c** in up to 30% enantiomeric excess (*ee*) for **2c**. In this system, no enantioselective adsorption of photoproduct **2** by chirally modified zeolites was observed, and therefore enantiodifferentiating photoisomerization in chirally modified zeolites is thought to be responsible for the supramolecular photochirogenesis. Interestingly, only a slight temperature effect was observed on the product *ee*, demonstrating a less-important role of the entropy factor in the

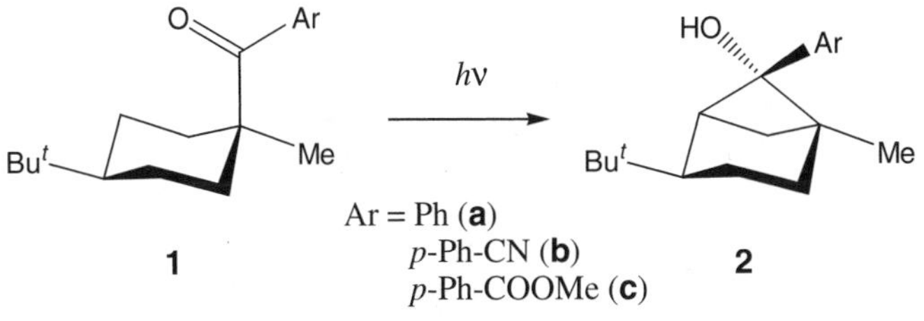

Ar = Ph (**a**)
   p-Ph-CN (**b**)
   p-Ph-COOMe (**c**)

Scheme 1

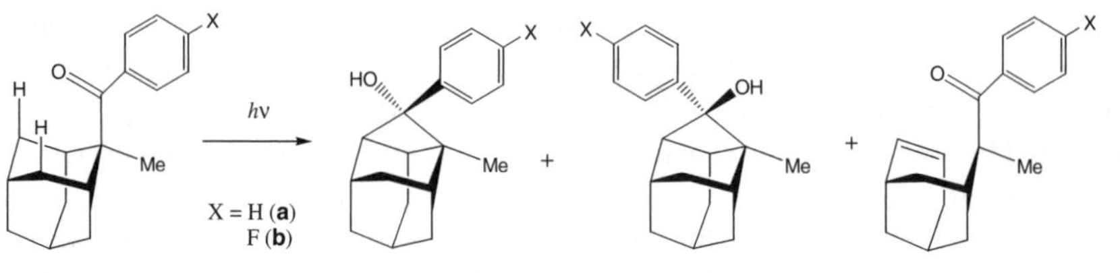

Scheme 2

supramolecular photoreaction of the substrate coincluded with a chiral inductor in a supercage. This is one of the important general features of supramolecular photoreactions.

They further investigated the enantio- and diastereodifferentiating Yang photocyclizations of achiral 2-benzoyladamantanes **3** and optically active menthyl and isomenthyl 2-benzoyl-2-adamantanecarboxylates **7** in chirally modified zeolites (Scheme 2 and Scheme 3) [54].

Irradiation of **3** in zeolites yielded *endo*-cyclobutanols **4** as the only photoproduct via γ-hydrogen abstraction. Moderate *ees* of up to 32% (**4a**) and 30% (**4b**) were obtained upon irradiation in (−)-pseudoephedrine-immobilized zeolites, while irradiations of **3** in a solution containing (−)-pseudoephedrine or in unmodified zeolites gave racemic **4**. On the other hand, photolyses of menthyl and isomenthyl esters **7** gave the cyclobutanol derivatives **8** and **9**, which however suffered the retro-aldol reaction to give diastereomeric δ-benzoylcarboxylates (**10**, **11**). Although the diastereomeric excesses (*des*) of **10** and **11** obtained in the solution-phase photolyses were low (< 15%), the *de* value was significantly enhanced up to 79% upon irradiations of **7a** in LiY zeolite and **7b** in NaY. *Ab initio* calculations revealed that coordination of a reactant ketone to a cation in

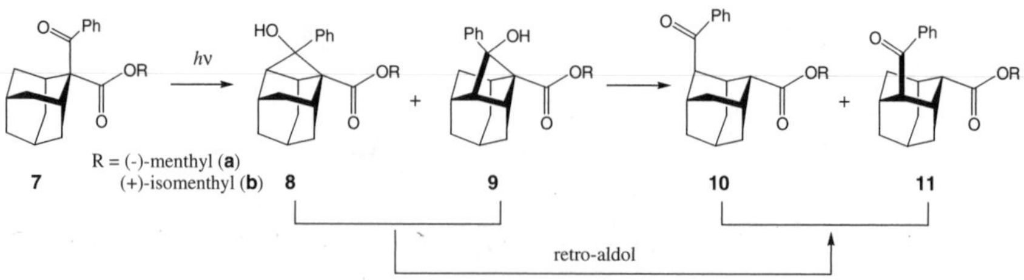

Scheme 3

the zeolite could be the origin of the enhanced *de*, and further that the alkali ion–organic guest interaction can be a powerful tool for controlling (asymmetric) photochemical and thermal reactions in zeolites and serve as a complementary method to the solid-state approach.

B. Enantiodifferentiating Radical Recombination in Chirally Modified Zeolites

Turro and coworkers reported the photolysis of racemic benzoin methyl ether **12** within chiral inductor-immobilized zeolite supercages [55].

Photolysis of **12** produces a geminate radical pair *via* Norrish type I α-cleavage and a biradical *via* type II γ-hydrogen abstraction not only in homogeneous solutions but also in zeolite supercages (Scheme 4). The geminate radical pair recombines at the original position to regenerate **12** or at the *para* position of methoxybenzyl radical to give **15**, or alternatively cage-escapes to give homocoupling products **13** and **14** in 23% and 70% yield, respectively, as major products obtained in solution (Scheme 4). In zeolite, the homocoupling was strongly suppressed and the geminate recombination was much enhanced to give **15** in up to 70% yield. Upon irradiation of racemic **12** in NaY zeolite supercages modified by D-diethyl tartrate or (+)-ephedrine, the recovered **12** was 9.2% *S*-rich or 4.9% *R*-rich, respectively.

C. Enantiodifferentiating Photorearrangement: Benzonorbornadiene to Benzonorbornene in Chirally Modified TIY Zeolites

Ramamurthy and coworkers reported the enantiodifferentiating photorearrangement (Scheme 5) of achiral benzonorbornadiene (**16**) to chiral benzonortricyclene (**17**) in chirally modified cation-exchanged zeolites [56].

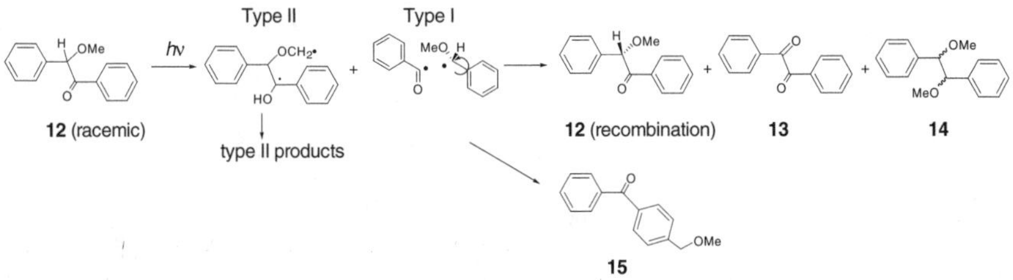

Scheme 4

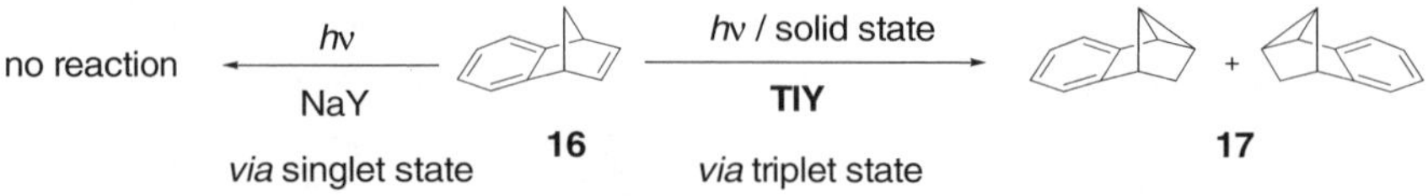

Scheme 5

This photorearrangement is known to proceed through the triplet manifold [57,58], and indeed no photoproduct was obtained upon direct irradiation in homogeneous solution. Interestingly, direct irradiation of **16** immobilized in NaY zeolite gave no photoproduct, whereas the use of TlY zeolite greatly enhanced the photoreactivity of **16** to give **17** in 40% yield, most probably through the accelerated intersystem crossing to triplet **16** in the presence of a heavy Tl cation in the supercage. Then the asymmetric photorearrangement of **16** was examined in (+)-ephedrine-immobilized TlY zeolite, which gave optically active **17** as the single product in modest *ees* of up to 14%. This is the first example utilizing the synergetic effects of a heavy cation and a chiral inductor within zeolite supercages for supramolecular photochirogenesis.

D.   Photorearrangement of α-Oxoamides into β-Lactams in Chirally Modified Zeolites

Ramamurthy and coworkers reported the enantio- and diastereodifferentiating Yang photocyclizations [59–62] of α-oxoamides **18** to β-lactams **19** (Scheme 6) in chirally modified and unmodified zeolites, respectively [63].

Enantiodifferentiating photocyclizations of prochiral **18a–e** in zeolites modified with ephedrine, norephedrine, pseudoephedrine, menthol, and 1-phenylethylamine gave **19a–e** in moderate *ees* of up to 44% (obtained for **18a** in (−)-norephedrine-immobilized NaY). Diastereodifferentiating photocyclizations

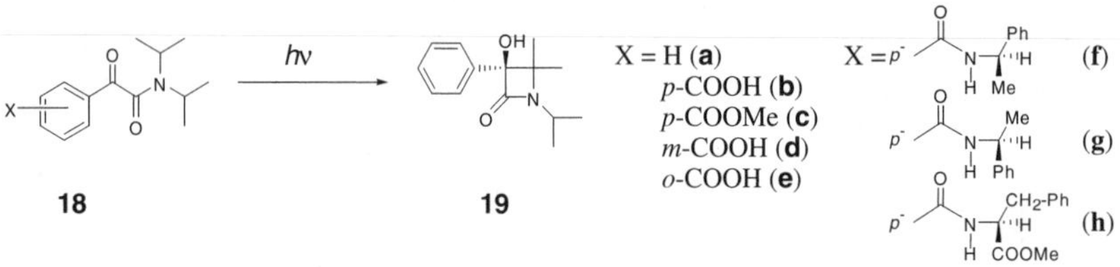

Scheme 6

of optically active **18f–h** gave modest *de*s of < 28% in homogeneous solution, but much higher *de*s of up to 82% were obtained upon irradiation in unmodified zeolites. Thus, in both enantio- and diastereodifferentiating photoreactions, the confinement in zeolite supercages is useful in enhancing the level of photochemical asymmetric induction, although the best *ee* and *de* reported are not as high as those obtained for photochirogenesis in crystals, which afford 99% *ee* for **18b** and >99% *de* for **18f**. These results undoubtedly reflect the greater flexibilities and vibrational motions of substrates in the zeolite interior compared to that in crystals. It is also revealed that chiral auxiliaries covalently bonded to NaY zeolite are more effective in asymmetric induction than external chiral inductors immobilized in zeolites, for which unmodified supercages left after chiral modification would be responsible. They also demonstrated that the nature of the cation in the zeolite strongly affects the product *ee* and *de*, and in some cases even leads to a complete reversal of enantio- or diastereoselectivity. For example, upon irradiation of **18h** in NaY zeolite, one of the two diastereomers of **19h** was predominantly produced in 82% *de*, whereas the other diastereomer was favored in KY in 54% *de*. It is believed that this diastereoselectivity switching is attributable to the possible interaction of the zeolite's cation with the chiral amino inductor or α-oxoamides in the supercage, which may be utilized as a factor for controlling photochirogenesis in confined media.

E. Enantio- and Diastereodifferentiating cis–trans Photoisomerization of 1,2-Diphenylcyclopropane in Chirally Modified Zeolites

Achiral *cis*-1,2-diphenylcyclopropane photoisomerizes to the chiral *trans* isomer upon singlet- or triplet-photosensitized irradiation [64–67]. It is expected that the reactant and chiral inductor immobilized in a zeolite supercage interact intimately with each other to afford more efficient photochirogenesis. Ramamurthy and coworkers reported that the enantio- and diastereodifferentiating photoisomerizations of *cis*-2β,3β-diphenyl-1α-cyclopropanecarboxylates **20** (Scheme 7) in chirally modified zeolite supercages lead to the corresponding chiral *trans* isomer **21** [68].

The enantiodifferentiating photoisomerization of achiral ethyl ester **20a** was performed in NaY zeolite supercages, which were chirally modified with ephedrine, pseudoephedrine, norephedrine, diethyl tartrate, alaninol, phenylalaninol, valinol, menthol, and bornylamine, to give trans product **21a** in relatively low *ee*s; modification with diethyl tartrate led to the highest *ee* of 12%, while the other inductors afforded *ee*s less than 5%. They also investigated the diastereodifferentiating photoisomerization of chiral esters **20b–d** in unmodified zeolites. Photoirradiation of menthyl, neomenthyl, isomenthyl, fenchyl, isopinocamphyl, and 2-methylbutyl 2β,3β-diphenyl-1α-cyclopropanecarboxylates **20b–d** in LiY,

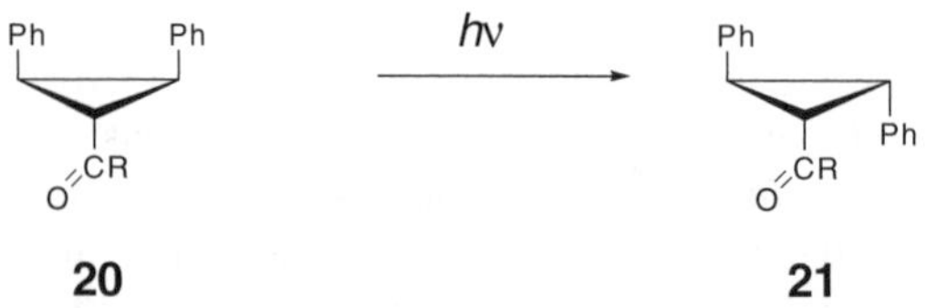

Ph   Ph   *hv*   Ph

                                                                R = -*O*-ethyl (**a**)
                                              Ph   -*O*-menthyl (**b**)
   CR   CR   -*O*-2-methyl-1-butyl (**c**)
O   O   -NH-1-phenylethyl (**d**)

20   **21**

Scheme 7

NaY, KY, RbY, and CsY zeolites efficiently produced the corresponding *trans* isomers **21b–d**. In sharp contrast to the very low *de*s (0–5%) obtained upon irradiation of **20b–d** in isotropic dichloromethane/hexane solutions, photolyses of the same substrates in zeolites gave much improved *de*s of up to 55% for **21b** in NaY. The *de* of **21b** critically depends on the countercation of Y zeolite, affording 50% for LiY, 55% for NaY, 30% for KY, 22% for RbY, and 5% for CsY. In LiY 1-phenylethyl amide **20d** afforded **21b** in *de* as high as 80%. Ramamurthy emphasized the importance of the cation–substrate interaction within the supercage for controlling the diastereodifferentiating photoreaction [69,70].

F. Photocyclization in Chirally Modified Zeolites

The most successful supramolecular photochirogenesis with chirally modified zeolites is the enantiodifferentiating photocyclization of tropolone derivatives in zeolites modified with (−)-ephedrine or (−)-norephedrine, giving bicyclo[3.2.0] heptadienes in high *ee*s. It is interesting to compare this result with the highly efficient diastereodifferentiating photocyclization of chirally modified tropolone derivatives in unmodified zeolites [71–75]. A wide variety of substrate-inductor-zeolite combinations were examined by Ramamurthy's group, and these works have recently been reviewed [74,75]; hence we will mention only the essential parts of their work in this section.

  They proposed four general methodologies for enantio- and diastereodifferentiating photocyclizations in zeolites, which employ (1) a chiral inductor immobilized in a zeolite supercage, (2) a chiral auxiliary bonded to the substrate, (3) a combination of methods 1 and 2, and (4) a chiral inductor as a reagent; most of which worked well and gave reasonable *ee*s and *de*s as shown in Scheme 8.

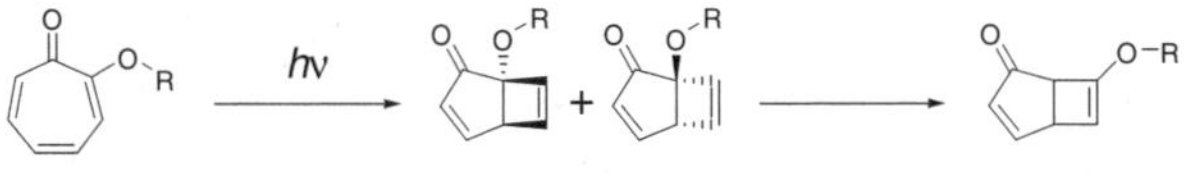

R = -CH_3 (**a**)
-CH_2CH_2Ph (**b**)
-(*S*)-2-methyl-butyl (**c**)

22 **23** **24**

Scheme 8

Diastereodifferentiating photocyclization behavior of chiral tropolone esters **22** was compared in homogeneous solutions and in zeolites [73]. Photolysis of **22** gives chiral bicyclo[3.2.0]heptadienone **23** via a 4π-electron disrotatory ring closure [76–78], which further rearranges to **24** upon prolonged irradiation (Scheme 9). Irradiation of a hexane slurry of (*S*)-tropolone-2-methylbutyl ether **22c** in dry unmodified NaY zeolite gave **23c** in 51% *de* in favor of diastereomer A at room temperature, with the *de* increasing to 68% at −20°C. In contrast, photolysis of **22c** in an isotropic hexane solution afforded racemic **23c** at the same temperature, demonstrating the positive effects of confined media for asymmetric induction. Quite interestingly, the diastereoselectivity was a critical function of the cation size in Y zeolite, affording the isomer A of **29c** in 51% *de* in NaY, 21% *de* in KY, 12% *de* in RbY, and finally the antipodal isomer B in 17% *de* in CsY. This is a clear demonstration of cation control of the diastereoselectivity, which is enhanced from practically zero in solution at room temperature up to 68% in unmodified zeolite at −20°C.

NaY zeolites modified with (−)-ephedrine or (−)-norephedrine showed even better performance in the photolysis of chiral tropolone ether **22c** immobilized in the supercage, affording **23c** in 90% *de*. In contrast, irradiation of **22c** in solution or on a silica surface in the presence of enantiopure ephedrine or norephedrine gave 1 : 1 diastereomeric mixtures.

Enantiodifferentiating photocyclizations of tropolone methyl and phenethyl ethers **22a** and **22b** were also examined in isotropic and anisotropic media (Scheme 8). Although the photolyses of **22a** and **22b** gave racemic **23a** and **23b**

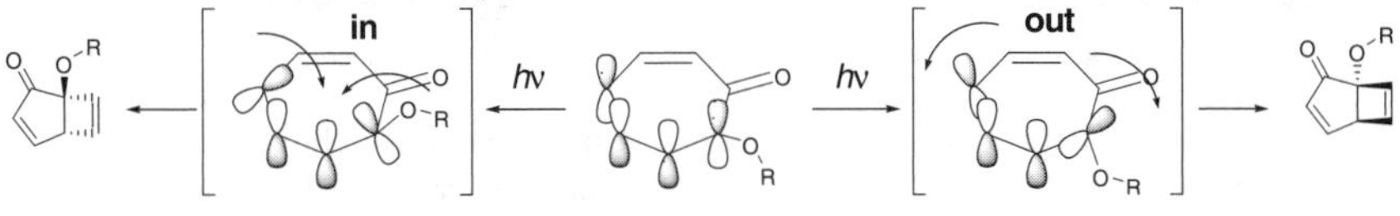

23 **22** **23**

Scheme 9

not only in homogeneous solutions containing chiral inductors but also on chirally modified silica surfaces, irradiation of **22b** in a hexane slurry of (−)-ephedrine-immobilized NaY zeolites gave **23b** in 78% *ee*. A model was proposed for the chiral interaction on zeolite surfaces, as shown in Fig. 1, where the three-point interaction is formed between the prochiral reactant and the chiral inductor, incorporating the two hydrogen bonds between the carbonyl and ether oxygens of **22** and the hydroxyl and amino hydrogens of ephedrine and the cation–π interaction of NaY zeolite with **22** and/or ephedrine. This model is justified in that chiral inductors with a single functional group such as borneol, menthol, bornylamine, and methylbenzylamine give racemic products, while those with three functionalities such as ephedrine, norephedrine, pseudoephedrine, and diethyl tartrate consistently afford *ee*s higher than 30%.

It was emphasized that, in comparison to the short-lived and weak interactions in isotropic solution, more intimate and long-lived interactions are expected to occur between a chiral inductor and an achiral, as well as chiral, substrate, leading to an enhanced enantio- or diastereoselectivity in confined zeolite supercages.

G. Enantioselective Photoreduction of Aryl Alkyl Ketones in Chirally Modified Zeolites

Achieving efficient photochemical chiral induction within zeolite supercages requires a geometrical distribution of reactant and chiral inductor molecules in the same supercage (type V in Fig. 2). By restricting a photoreaction to a supercage that contains both a reactant and a chiral inductor, and by avoiding the reaction in the absence of the chiral inductor, efficient chiral induction should be achieved

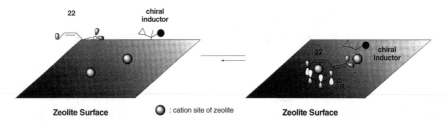

Figure 1   An adsorption (top)–desorption (bottom) model for chiral induction on a zeolite surface, incorporating a reactant (tropolone alkyl ether, shown at the left), a chiral inductor (with four different substituents, at the right), and a cation (small ball on the surface). Tropolone's carbonyl and ether oxygens hydrogen-bond to chiral inductor, while its π system interacts with zeolite's cation ion.

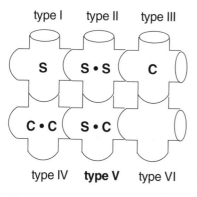

type I   type II   type III

S   S•S   C

C•C   S•C

type IV   **type V**   type VI

Figure 2

S: photoreactive substrate
C: chiral inductor

Maximum two molecules are able to immobilized
into a zeolite supercage due to space limitation.

[74,75]. Such a desirable condition was fulfilled in the photoreduction of an aryl alkyl ketone in zeolites modified with chiral amine [79].

Ramamurthy and coworkers reported the photochemical electron-transfer-initiated intermolecular hydrogen transfer reaction [80,81] of phenyl cyclohexyl ketone **25** in chiral amine-immobilized zeolite cavities.

In this enantiodifferentiating photoreduction, the chiral amine plays two roles, as a chiral inductor and as an electron donor. Irradiation of **25** (Scheme 10) in a hexane slurry of unmodified NaY zeolite gave only the intramolecular hydrogen abstraction product **26**. However, photolysis of **25** coimmobilized with ephedrine, pseudoephedrine, or norephedrine in NaY supercages afforded the reduction product **27** along with **26**. It is clear that the immobilized amine plays the decisive role in the photoinduced electron-transfer reduction of **25**, since **27** was not formed in unmodified or (−)-diethyl tartrate-modified zeolites. Consequently, the *ee* of obtained **27** was independent of the loading level of the chiral inductor.

This is a clever invention to limit the photoreduction to the when both the substrate and chiral inductor reside in the same cavity. The *ee* obtained by this

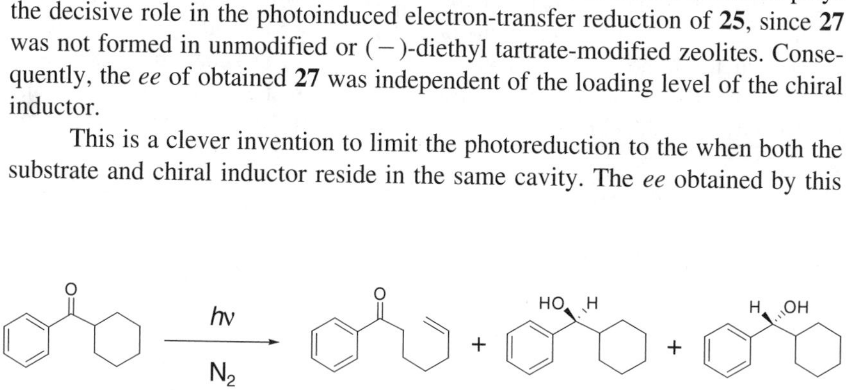

25 →(*hv*, N₂)→ **26** + **27A** + **27B**

Scheme 10

strategy is noteworthy. Earlier attempts at achieving chiral induction in the photo-reduction of aryl alkyl ketones by chiral amines gave *ee*s of < 8% at room temperature, whereas the use of (+)-norephedrine-immobilized zeolite led to the formation of **27A** in 61% *ee*. As indicated above, the product *ee* was almost independent of the loading level of the chiral inductor but was dramatically re-duced to 2% just by adding a small amount of water to the chirally modified zeolite. This result clearly shows the crucial role of weak interactions, which can be broken in the presence of trace water, in asymmetric induction in chirally modified zeolite supercages. In this supramolecular photochirogenesis, the entire reaction takes place within the chirally modified supercages, but the product *ee*s are not very high (< 61%). This was ascribed to the multistep nature of the reaction, which involves at least two distinct enantiodifferentiation processes (Scheme 11). Both the reactant ketone and the intermediate radical **29** (Scheme 11) possess prochiral enantiofaces. In solution, the addition of hydrogen to both enantiofaces of **29** are equally likely, thus affording a racemic product. Chirally modified zeolite supercages may allow the enantioface-selective adsorption of reactant ketone, with appreciable asymmetric induction likely to occur if this preference is retained until the completion of the reaction. It is believed that the cations present in the zeolite framework work as an anchor for the ketone/radical throughout the reaction. Hence the magnitude of the chiral induction depends on the ability of the chiral inductor to enantioface-selectively adsorb the ketone and also on the ability of the cation to keep it anchored from the same face.

H. Photosensitization in Zeolite Supercages

All of the foregoing photochirogenesis strategies using zeolite supercages are not "catalytic" but rather "stoichiometric," necessitating at least an equimolar, or even larger, amount of chiral inductor per substrate molecule. However, if the chiral inductor immobilized in zeolite can photosensitize an enantiodifferentiating process, this can be a versatile chirogen-efficient method for transferring and multiplying molecular chirality through supramolecular interactions in the excited state [11–13].

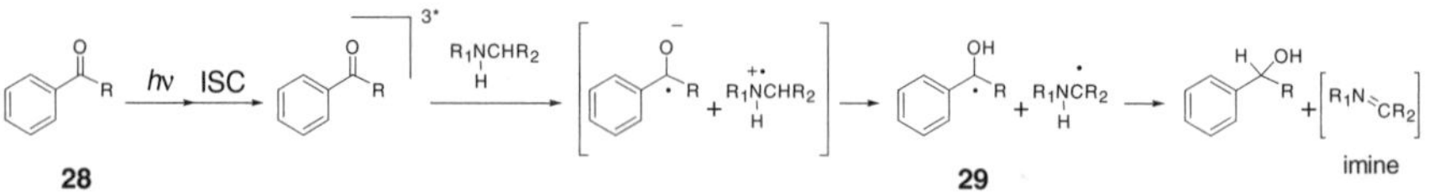

Scheme 11

The enantiodifferentiating photosensitization in homogeneous phase has a long history since the first report by Hammond and Cole in 1965 [82], and the *ee* obtained has been improved greatly in recent years [11–13,83–89]. In particular, the enantiodifferentiating geometrical photoisomerization of (Z)-cyclooctene sensitized by optically active (poly)alkyl benzene(poly)carboxylates gives (E) isomer in good *ee*, and the enantioselectivity is dramatically affected by environmental factors, such as temperature, pressure, and solvent, accompanied by a switching of product chirality in several cases [83–89].

An intriguing logical extension of such work is the construction of a supramolecular host modified with a chiral sensitizer for executing enantiodifferentiating photosensitization. Inoue and coworkers reported a novel supramolecular photochirogenic system, in which an optically active sensitizer immobilized in zeolite supercages sensitizes the enantiodifferentiating photoisomerization of an excess amount of substrate dissolved in bulk solution [90].

Such a supramolecular "catalytic" enantiodifferentiating photosensitization system may be related to an enzymatic reaction, where the substrate molecule diffuses to the binding site of the enzyme and a substrate–enzyme complex is formed prior to the reaction. In analogy, the substrate molecule freely diffuses through the zeolite supercages without any interaction with the immobilized sensitizer in the ground state, but once the light is switched on the excited sensitizer forms an exciplex with the substrate, leading to the "catalytic" geometrical photoisomerization (Fig. 3)

The photoisomerization of (Z)-cyclooctene (**30**) (Scheme 12) to the (E)-isomer (**31**) was sensitized by enantiopure alkyl benzenecarboxylates immobilized in zeolite to give modest *ee*s. The use of an antipodal sensitizer pair of (R)- and (S)-1-methylheptyl benzoates, **32d** and **32e**, yielded enantiomeric **31** in −5% and +5% *ee*, respectively, while the same sensitizers gave practically racemic **31** upon irradiation in homogeneous solutions. This small, but apparent, enhancement of the product *ee* observed upon irradiation in modified zeolite supercages is likely to arise from the decreased conformational freedom of the adsorbed sensitizer, the hindered approach of **30** to the sensitizer, and/or the different exciplex structure in confined media. In this context, it is interesting to examine the effect of temperature on the supramolecular photochirogenesis in modified zeolites and to compare the results with those obtained in the homogeneous phase. Such an examination will reveal the distinctly different role of entropy in confined media, which should be clarified in a future study.

In this study, they demonstrated for the first time that chirally modified zeolites not only function as supramolecular photosensitizing media but also enhance the original enantiodifferentiating ability of the chiral photosensitizer. This newly developed methodology should not be restricted to this particular system but be readily expandable to a wide variety of supramolecular photochirogenesis reactions.

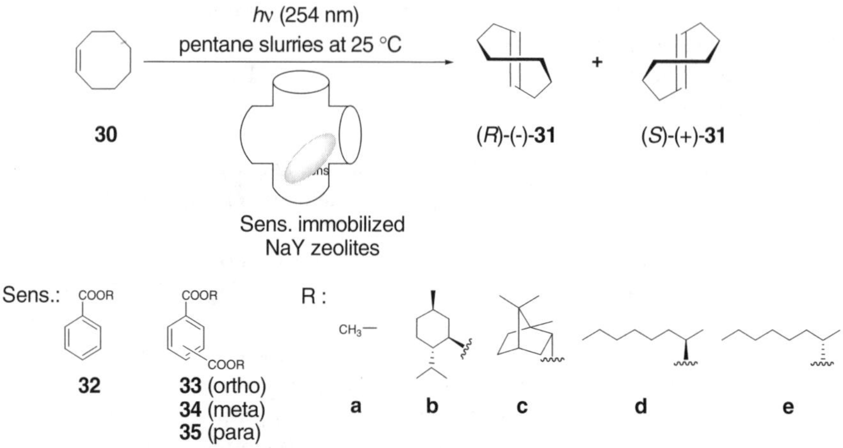

Scheme 12

III. SUPRAMOLECULAR PHOTOCHIROGENESIS WITH NATIVE AND MODIFIED CYCLODEXTRINS

Cyclodextrins (CDx) are a series of cyclic oligosaccharides composed typically of 6–8 glucopyranose units, which are called α-, β-, and γ-cyclodextrin, respectively. Possessing a truncated cone-shaped hydrophobic cavity, CDx forms inclusion complexes with a wide variety of neutral and charged organic guests and is indeed one of the most frequently used supramolecular hosts [91–94]. The cavity diameter varies from 4.5 Å for α-CDx to 7.0 Å for β-CDx and then to 8.5 Å for γ-CDx, while the cavity depth remains at a constant at 7.9 Å throughout the series. The secondary hydroxyl groups at the 2- and 3-positions of the glucopyranose unit are located at the rim of the wider opening of the cavity, and the primary hydroxyls at the 6-position are on the rim of the narrower opening. The 2-hydroxyl group successively hydrogen bonds to the 3-hydroxyl of the adjacent glucopyranose unit, forming a hydrogen-bond network surrounding the cavity. In aqueous solution, the hydrophobic cavity is occupied by "high-energy," or energetically disfavored, water molecules, which are readily replaced by appropriate hydrophobic guest molecule(s). The stoichiometry of CDx–guest complex is typically 1 : 1, but multiple host and/or guest molecules may associate to form more complicated 1 : 2, 2 : 1, 2 : 2, or even higher-order complexes. The inclusion complexes formed are stable in general and isolable as amorphous or (micro)crystalline solids. Once these CDx complexes are dissolved in aqueous solution, an equilibrium is established between the associated and dissociated species, which is quan-

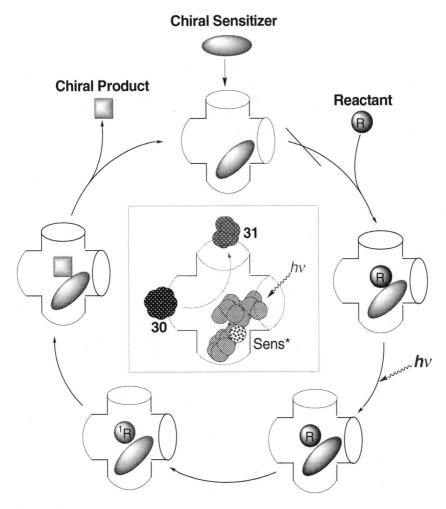

Chiral Sensitizer

Chiral Product

Reactant

Figure 3

titatively described by a complex stability constant (K_a) and the relevant thermodynamic parameters ($\Delta H°$ and $\Delta S°$). The major driving forces for inclusion complexation by CDx include the hydrophobic interaction of organic guests with the CDx cavity and the hydrogen bonding interactions between the CDx's hydroxyl groups and the guest's hydroxyl or heteroatoms. Hence, the property of solvents used is a crucial factor in controlling the supramolecular system. The most important feature of CDx, which is essential for photochirogenesis, is its inherently chiral cavity, which can recognize and discriminate enantiomeric guests upon inclusion complexation. Indeed, this feature of CDx has been applied to the chiral separation of various racemic compounds by gas and liquid chromatography incorporating a CDx derivatized stationary phase [95]. Supramolecular photochemistry in CDx cavities has also been recently investigated [96–120], and such works on the control and modification of chirality by the CDx have been reviewed by Monti et al., although the asymmetric photochemistry has not explicitly been described in detail [96–98]. Here, we will place more emphasis on the supramolecular photochirogenesis in CDx cavities.

In the early 1980s, pioneering work on photochirogenesis using CDx was carried out by two Japanese research groups [99,100]. Takeshita and coworkers investigated the valence photoisomerization of tropolone derivatives included in α- and β-CDx [99]. Irradiations of achiral tropolone and 2-methoxytropolone incorporated in 1 : 1 complex with α- and β-CDx gave chiral 1-hydroxy- and 1-methoxybicyclo[3.2.0]hepta-3,6-dien-2-one, both of which were demonstrated to be optically active by circular dichroism spectroscopy. However, no further examinations were made to determine the *ee* of the photoproducts. Inouye and coworkers reported the asymmetric photocyclization of aryl nitrone in the presence of β-CDx and amino acids added as chiral inductor in aqueous mixed solvents. The optically active oxaziridines were obtained, and the optical yields of < 1.9% were estimated from the specific rotation. The same enantiomers were obtained in the presence of L- and D-alanine [100]. Thus these works unambiguously demonstrate that the CDx cavities are usable for photochemical asymmetric induction, although the reported optical yields are low and the critical chiral discrimination step and mechanism are not very clear.

A. Radical Recombination in CDx Cavities

Rao and Turro reported the solid-state photolysis of the β-cyclodextrin complex of benzaldehyde **36**, which affords optically active benzoin **37** [101]. In view of the free radical nature of the photolysis revealed by photo-CIDNP studies, it is likely that excited triplet benzaldehyde abstracts the aldehyde hydrogen from the ground-state benzaldehyde to form a radical pair, which recombines to give benzoin **37**. In the absence of CDx, photoirradiation of **36** in ethanol gave 1,2-diphenyl-1,2-ethanediol as the major product, along with a trace amount of benzoin. Benzaldehyde **36** forms a 1 : 1 complex with α- and β-CDx, and a 2 : 1 complex with γ-CDx, both of which are isolable and stable under air or oxygen atmospheres. Irradiation of the 1 : 1 complex of **36** with β-CDx in the solid state gave benzoin **37** and 4-benzoylbenzaldehyde **38** in a 7 : 3 ratio in 80% combined yield (Scheme 13). A moderate *ee* of 15% was reported for **37**. Interestingly, a similar irradiation of the α-CDx complex produced no photoproducts, while the γ-CDx complex gave racemic **37** and **38** in a 55 : 45 ratio in a 78% combined

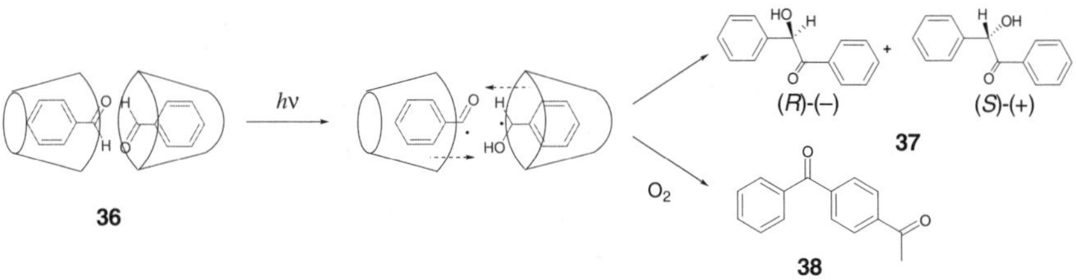

Scheme 13

yield. In aqueous solution, photolyses of α- and β-CDx complexes of **36** gave only trace amounts of racemic **37**.

The observed photobehavior of the benzaldehyde-CDx complexes in the solid state is unique and completely different from that of these complexes in aqueous solution and also from that of benzaldehyde **36** in organic solvents. The substantial formation of 4-benzoylbenzaldehyde **38** upon irradiation in β- and γ-CDx cavities indicates that these medium-sized CDx's provide the radical pair within a fairly spacious "supercage" environment, thus allowing the *para*-rearrangement (Scheme 13). The formation of practically racemic **37** upon irradiation of the γ-CDx complex may also be attributed to the looser orientation of benzaldehyde **36** in the γ-CDx cavity than in the β-CDx cavity. It was thus demonstrated that the chiral hydrophobic cavity of native cyclodextrins not only modifies the photoreactivity of the included guest but also functions as a chiral supramolecular environment for photochirogenesis, albeit resulting in only modest *ee*s.

B. Photocyclization of Tropolone Derivatives in the CDx Cavities

As briefly mentioned above, Takeshita et al. reported the first enantiodifferentiating photoisomerization of tropolone derivatives to the optically active bicyclo[3.-2.0]heptadienones in 1980 [99], but the product *ee* and detailed chiral discrimination mechanism have not been determined until recently. Ramamurthy and coworkers reinvestigated the photobehavior of tropolone alkyl ethers in α-, β-, and γ-CDx cavities [102].

Irradiation of tropolone alkyl ether **22** (Scheme 14) led to a 4π-disrotatory ring closure to yield bicyclo[3.2.0]heptadienone **23** with two chiral centers, while prolonged irradiations led to the formation of a secondary product **24** [76–78]. As the same photocyclization was performed in chirally modified zeolites, it is interesting to compare the asymmetric photochemical behavior of **22** in the distinctly different chiral confined media of zeolites and cyclodextrins. Even in the

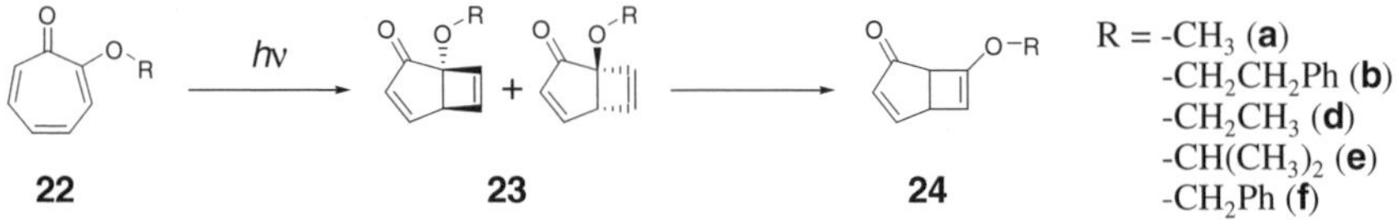

Scheme 14

presence of CDx, irradiation of **22** in aqueous solution led to the production of racemic **23**, whereas the solid-state photolysis of the CDx complex of **22** gave **23** in moderate *ee*s of up to 33%. This result is consistent with the long-standing belief that a rigid medium is a crucial factor for efficient chiral induction in thermal, as well as photochemical, asymmetric synthesis. The *ee*s obtained in the solid-state photolysis of CDx complexes critically depend on the combination of CDx and tropolone derivative **22**. For example, in the case of tropolone methyl ether **22a**, the use of α-CDx led to moderate *ee*s of up to 28%, while β- and γ-CDx gave much lower *ee*s of 5% and 0%, respectively, demonstrating the importance of cavity size in chiral induction. Although tropolone methyl ether **22a** and ethyl ether **22d** included in α-CDx gave moderate *ee*s, the α-CDx complex of isopropyl ether **22e** gave a low *ee* (5%). Consequently, the ability of the guest molecule to manage to fit into the host cavity appears to be a crucial factor controlling the photochemical chiral induction within the CDx cavity. Then the close matching in size and shape between the tropolone guest and the CDx cavity is essential for obtaining high enantioselectivity. It is also reasonable that the alkyl bulkiness of tropolone derivatives plays a major role, since the chiral centers are located mostly on the rims of CDx. Upon solid-state photolyses of 2 : 1 complexes of benzyl and 2-phenylethyl ethers (**22f, 22b**) with β-CDx, and of 2-phenylethyl ether **22b** with γ-CDx, similar *ee*s of 20–33% were obtained, suggesting that the conformations of the tropolone derivatives resemble each other in the CDx cavities.

C. Enantiodifferentiating Photocyclization of *N*-Alkyl Pyridones in CDx Cavities

Very recently, Ramamurthy and coworkers reported contrasting photobehavior of β-cyclodextrin complexes of *N*-alkylpyridones (**39**) in aqueous solution and in the solid state [103]. Upon irradiation, the achiral *N*-alkylpyridone was *trans*-formed to chiral 2-azabicyclo[2.2.0]hex-5-en-3-one (**40**) (Scheme 15) [104–106]. Thus the solid-state irradiation of **39** afforded bicyclic β-lactam **40** in 59% *ee*, whereas the solution-phase irradiation of **39** in the presence of an excess amount

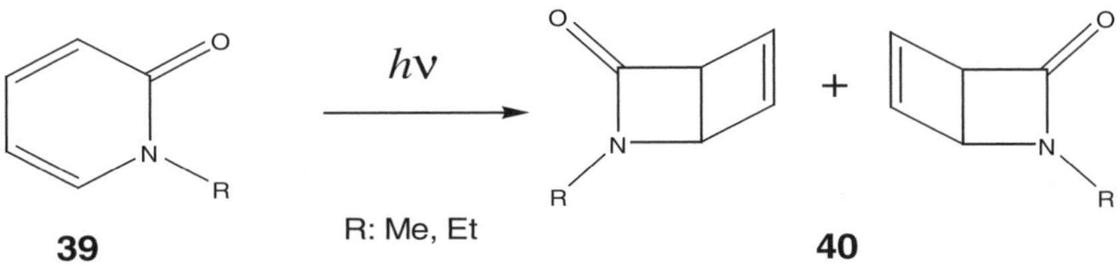

Scheme 15

of β-CDx gave almost racemic product (< 5% *ee*). Interestingly, the complex was prepared by a mechanical mixing of substrate **40** with β-CDx. It is also noted that, despite the high solubilities in water, solid-state complexes of *N*-methyl- and *N*-ethylpyridone with β-CDx can be prepared by mechanically mixing the host and the guest without the help of solvents. The solid-state complex can also be prepared by recrystallization from a hexane-methanol mixture (95 : 5) as solvent. Both methods gave essentially the same complex, exhibiting the identical x-ray powder diffraction patterns and solid-state NMR spectra. It was further demonstrated that the water contents of the crystals plays a crucial role in the asymmetric induction process. Thus the original "wet" sample of **39**-CDx complex, which was not specifically dried and contained 9% water, gave **40** in 59% ee as mentioned above, while the same sample dried in vacuo (containing 2% water) afforded the *antipodal* product in a substantially decreased *ee* of 26%.

D. Enantio- and Diastereoselective Geometric Photoisomerization of Diphenylcyclopropans in CDx Cavities

Direct and triplet-sensitized enantiodifferentiating geometrical photoisomerizations of *cis*-diphenylcyclopropane **41a** and *cis*-diphenylcyclopropane-1-carboxylic ester **41b** and amides **41c,d** (Scheme 16) were performed in both solution and solid state in the presence of CDx as a chiral environment for supramolecular photochirogenesis [107]. In solution, the enantio- or diastereoselectivities of obtained *trans* isomers were quite poor, giving almost 1 : 1 enantiomeric or diastereomeric mixtures. Although all the substrates formed solid-state complexes with β-CDx, the complex stoichiometry depended on the substrate employed, thus affording a conventional 1 : 1 complex with **41a** and **41b** but a 1 : 3 complex with **41c** and **41d**. Direct and triplet-sensitized photoisomerization of the solid-state complexes with β-CDx gave the *trans* isomers in modest enantiomeric excesses of up to 13% and diastereomeric excesses of up to 30%.

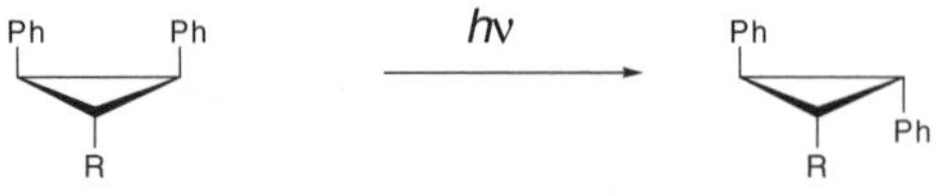

R = -H(**a**)
-COO-Et (**b**)
-CONH-1-phenylethyl (**c**)
-CONH-1-cyclohexylethyl (**d**)

41 **42**

Scheme 16

As illustrated in Scheme 17, the C2–C3 bond cleavage, followed by the rotation around the C1–C2 or C1–C3 bond and the subsequent recombination, completes the geometrical isomerization from *cis* to *trans*. In achiral environment, the two modes of bond rotation occur at equal probability to yield a racemic *trans*, whereas one of the two modes is favored in the chiral β-CDx cavity, as a consequence of different steric hindrances experienced by the substituent during the rotation. Intriguingly, the favored stereoisomeric photoproducts **42a,b** are in agreement with the preferentially included stereoisomers by β-CDx in aqueous solution. This observation may indicate that the chiral environment of the CDx cavity affects the photoisomerization of **41a** and **41b** even at a very early stage, biasing the isomerization toward the stereoisomer preferred by the cavity. This is not unlikely in this particular case, since the stereochemistry of final product is already determined at the stage of the precursor 1,3-biradical. If such a relationship between the thermodynamic and kinetic preference is widely established with a variety of host–guest combinations, we can predict the stereochemistry of favored photoproduct prior to the photoreaction by examining the ground-state complexation of racemic photoproduct with the specific host.

E. Photocyclodimerization of Anthracene Derivatives in CDx Cavities

β-CDx and in particular γ-CDx are known to accommodate two aromatic moieties under certain circumstances. Hence if two appropriate prochiral guest molecules are included in the same CDx cavity, regio- and enantioselective bimolecular photoreactions are expected to occur [108]. Indeed, it is known that the presence of CDx not only accelerates the rate but also modifies the product distribution of photocyclodimerizations of anthracene derivatives [109–111], coumarin derivatives [113–115], stilbene derivatives [116,117], stilbazole [118], and tranilast [119]. For instance, Tamaki and coworkers reported significantly enhanced quantum yields of photodimerization of anthracenesulfonates and anthracenecarboxylates in the presence of β- and γ-CDx [109–111]. These anthracene derivatives form 2 : 2 and 2 : 1 guest–host complexes with β- and γ-CDx, respectively, and

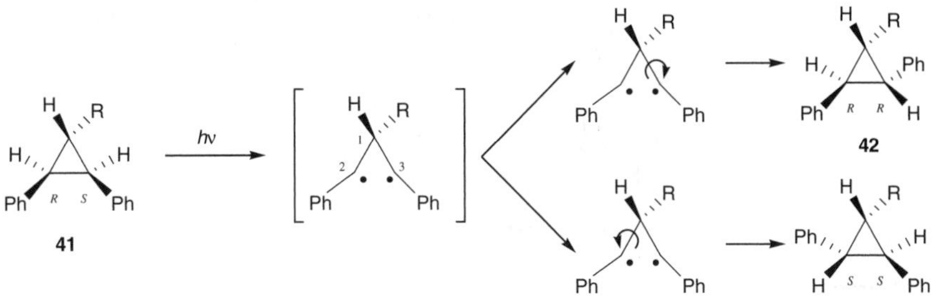

Scheme 17

the photodimerization in aqueous solution is greatly accelerated by the complex formation. Although the product distribution among four stereoisomeric cyclodimers was not appreciably altered in the presence of CDx, the obtained cyclodimers exhibited appreciable circular dichroism, indicating the formation of optically active photodimers. However, no further examination to evaluate the product ee was done at that time.

Recently, Nakamura and Inoue have succeeded in determining the *ee* of the chiral photocyclodimers of 2-anthracenecarboxylic acid (**43**) (Scheme 18) by using chiral HPLC [120]. They quantitatively analyzed the complexation behavior of **43** with γ-CDx and performed the photoreaction in aqueous buffer solutions in the presence and absence of γ-CDx to obtain two chiral (**45** and **46**) and two achiral (**44** and **47**) cyclodimers. 2-Anthracenecarboxylate **43** forms very stable stereoisomeric 2 : 1 complexes with γ-cyclodextrin, which instantaneously photocyclodimerize to **44–47** in the CDx cavity upon irradiation [120]. The product distribution is governed by the initial population of the precursor complexes in the ground state before photoreaction, because there is not enough time and space for included **43** molecules to change the orientation in the cavity. Moderate to good *ee*s of up to 32% at 25°C and 41% at 0°C were obtained. This emphasized that the use of CDx as a chiral template is an effective and potentially versatile strategy for supramolecular photochirogenesis.

F. Intramolecular *Meta*-Photocycloaddition

Eycken and coworkers [121] employed β-CDx as a chiral template for asymmetric induction in the intramolecular *meta*-photocycloaddition [126,127] of 4-phenoxy-1-butenes **48** and **50** in the solid state. (Scheme 19.)

The complexation behavior of **48** and **50** critically depended upon the shape of the substrate (guest); consequently, linear *para*-substituted **48** and bent *meta*-

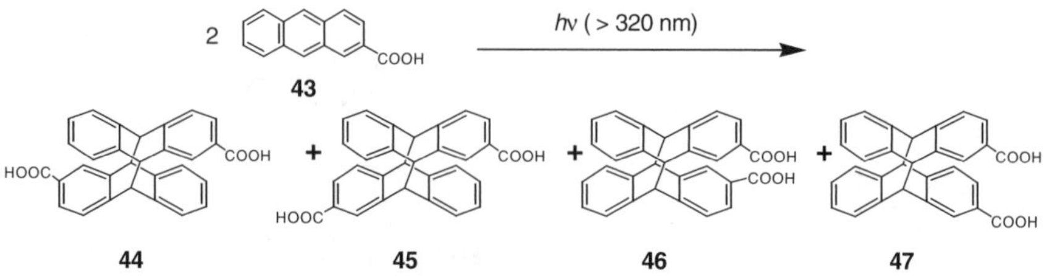

Scheme 18

substituted **50** formed 2 : 1 and 1 : 1 host guest complex with β-CDx, respectively. Solid-state photolysis of the 2 : 1 complex of **48** gave only a small amount of racemic *meta*-photocycloaddition product **49**. In contrast, irradiation of the 1 : 1 complex of **50** yielded two regioisomeric *meta*-adducts **51** and **52** in a 1 : 3 ratio; moderate *ee*s of up to 17% were obtained for **51**, while the major adduct **52** was almost racemic (*ee* ≤ 2%). No chiral discrimination was observed upon inclusion of racemic photoproducts by β-CDx, and hence the observed *ee*s are thought to originate solely from the photocycloaddition in CDx. The poor photoreactivity and enantioselectivity of *para*-substituted **48** were attributed to the substrate conformation stretching through the center of the CDx cavity. Under such circumstances, the *meta*-photoaddition is substantially hindered and, if it occurs at all, is less enantioface-selective, since the alkenyl side chain is only loosely bound

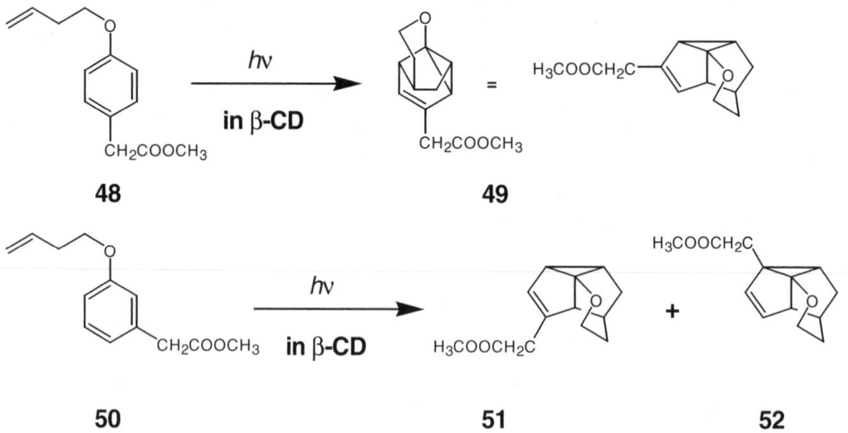

Scheme 19

in the cavity. NMR spectral examination of the 1 : 1 complex of *meta*-substituted **50** with β-CDx revealed that both the aromatic H-2 and the olefinic H-4 of **50** were located near the β-CDx's H-3 or the rim. They proposed a plausible complex structure depicted in Fig. 4. In this complex structure, two approaches are allowed for the alkenyl side-chain, i.e., one from the less-hindered side, leading to the major product **52**, and another from the more-hindered side, giving the minor product **51**. In the less-hindered approach, the alkenyl attack on the aromatic moiety does not induce any steric interactions with the chiral CDx rim, leading to the predominant formation of **52** in low *ee*, as actually observed. On the contrary, the hindered approach causes significant conflicts with the chiral CDx rim, giving a lower yield and a higher *ee* for **51**. Since the moderate *ee* of 17% is rather acceptable for the first trial on the *para*-photocycloaddition and the above interpretations sound reasonable in general and are compatible with the experimental results, further efforts along this line, using homologous substrates and modified CDx may enhance the product *ee* in the near future.

G. Photosensitized Geometrical Photoisomerization of Cycloalkenes in CDx Cavities

In most supramolecular photochirogenesis, a prochiral substrate is first introduced into the chiral environment of a natural or synthetic host to form a host–guest complex, which is then photoirradiated to give an optically active product(s). In such a strategy, the turnover number of the chiral host, or the chirogen efficiency, is inherently low, unless the substrate absorption shows an appreciable bathochromic shift upon complexation and the selective excitation of such a band is possible. In this context, asymmetric photosensitization is an attractive and unique methodology for transferring and multiplying chirality through the sensi-

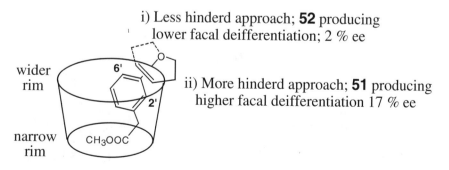

Figure 4

tizer–substrate interactions in the excited state, which may be regarded as a photochemical version of asymmetric catalysis (see Chap. 4). A similar, even more sophisticated, strategy can be employed in supramolecular photochirogenesis by introducing a sensitizer moiety into a supramolecular host. Inoue and coworkers synthesized a series of arenecarboxylate-appended α-, β-, and γ-cyclodextrins and used them as chiral sensitizing hosts for the enantiodifferentiating geometrical photoisomerization of cycloalkenes [122–124].

The fundamental idea and mechanism of supramolecular asymmetric photosensitization using modified cyclodextrins as chiral sensitizing hosts is illustrated in Chart 1. In this strategy, substrate molecules located in the bulk solution are not accessible to the excited sensitizer moiety included in the CDx cavity, so no "exterior" photosensitization, which in principle gives low *ee*, takes place. However, once the substrate is introduced into the chiral cavity, the sensitizer moiety, being close to the included substrate, can transfer its energy to the substrate in the chiral environment, thus realizing efficient supramolecular photochirogenesis.

Prior to such a sophisticated attempt, a more straightforward strategy was examined in order to evaluate the chiral discrimination ability of the CDx cavity. Thus the direct photoisomerization at 185 nm of a 1 : 1 complex of (*Z*)-cyclooctene **30** with β-CDx was carried out in the solid state to give an *E–Z* mixture of *E/Z* = 0.47 [122]. The *ee* of the obtained (*E*)-isomer **31** was low (0.24 %) [123], but this work paved the way for the supramolecular photosensitization of **30** with chromophore-modified β-CDx derivatives **53–55** in solution [123,124].

There are several advantages of using CDx as a host: (1) the cavity is inherently chiral, (2) the tethered sensitizer promotes tight packing and conformational fixing of the substrate/sensitizer pair in the cavity, and most importantly, (3) undesirable sensitization outside the chiral cavity is automatically prohibited

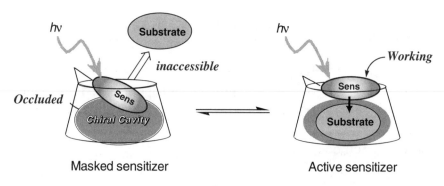

Chart 1

by the occlusion of the sensitizing group in the absence of included substrate, and therefore the efficient photosensitization is designed to occur only in the chiral cavity. Circular dichroism spectral examination revealed that the aroyl group in **54a–d**, tethered to β-CDx through a short ester linkage, is residing on the narrow rim of the cavity.

It was shown that, although both **30** and **31** are better accommodated in the β-CDx cavities of **54a–e** rather than of the α-CDx derivative **53a**, these CDx derivatives possess only poor abilities of chiral recognition for **31** ($< 1.8\%$ ee) at least in the ground state [123] (Scheme 20).

The supramolecular enantiodifferentiating photoisomerization of **30** to chiral **31** through inclusion and sensitization by α-, β-, and γ-cyclodextrins with benzoate (**53a, 54a, 55a**) and isomeric phthalate (**54b–d**) chromophores was investigated in aqueous methanol solutions at varying temperatures. The sensitization by α- and γ-CDx benzoates (**53a, 55a**), which possess a too small or too large cavity, gave **31** in poor ees of less than 3% and 5%, respectively, while the β-CDx derivatives (**54a–e**) afforded much higher ees of up to 24% (for **54b**), depending on the sensitizer moiety and solvent. It was thus demonstrated that the modification of cyclodextrin with a sensitizer group, which does not appear to contribute to the chiral recognition ability in the ground state, greatly enhances the product ee through the *excited-state* supramolecular interaction within the cavity.

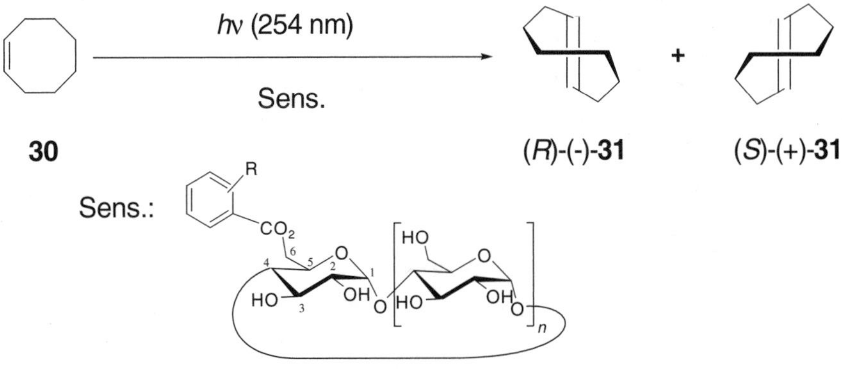

$$ hv\ (254\ nm) $$
Sens.

30 (R)-$(-)$-**31** (S)-$(+)$-**31**

Sens.:

53a $n = 5$, R = H
54a $n = 6$, R = H
54b $n = 6$, R = o-CO$_2$Me
54c $n = 6$, R = m-CO$_2$Me
54d $n = 6$, R = p-CO$_2$Me
55a $n = 7$, R = H

Scheme 20

In Fig. 5a, the product *ee*s obtained with **54a** at 25°C are plotted as a function of the host occupancy (the percentage of occupied host), which is determined by the solvent composition and temperature employed, to give an excellent straight line. The extrapolation of the regression line to zero occupancy gives the *ee* value expected for exterior sensitization, while the extrapolated *ee* at 100% occupancy is the ultimate *ee* value for interior sensitization. Thus the sensitization outside the cavity of **54a** leads to an extremely low *ee* of − 1%, whereas the sensitization exclusively within the **54a** cavity gives − 10% *ee*. Unexpectedly, the product *ee* is not sensitive to the temperature but is governed predominantly by the water content or more strictly by the host occupancy. Thus all the *ee* values obtained upon sensitization with **54b** at different temperatures ranging from − 40 to + 55°C fall on the *same* straight line (Fig. 5b). It is interesting that the product *ee*s obtained with benzoate **54a** and methyl phthalate **54b** are not a simple function of either temperature or solvent but are nicely correlated with the host occupancy.

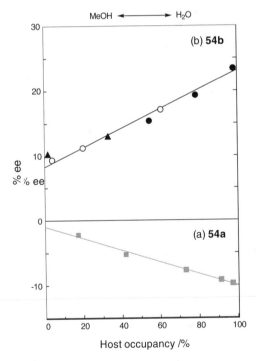

Figure 5   % *ee* as a function of last occupancy for the enantiodifferentiating photosensitization of 30 with (a) 54a at 28°C (■) and (b) 54b at .58 (▲) 25°, and − 40°C (●) in methanol-water mixture of various compositions.

This result indicates that the entropy factor plays a minor role in this supramolecular photochirogenesis, as is exactly the case with the solid-state photochemistry, but is in sharp contrast to the decisive role of entropy in the conventional (nonsupramolecular) counterpart performed in isotropic media, where an inversion of product chirality by temperature variation occurs [83–89].

It may be concluded therefore that supramolecular photochirogenesis, employing modified cyclodextrins as sensitizing hosts, is a promising strategy not only for efficiently transferring the environmental (supramolecular) chirality to the molecular chirality of the photoproduct through excited-state interactions but also for enhancing the original photoenantiodifferentiating ability of native cyclodextrins. Indeed, the extremely poor *ees* ($< 1\%$) obtained upon direct photoisomerization of **30** in the native β-CDx cavity is increased up to 24% *ee* upon sensitization with methyl phthalate–tethered β-CDx. This supramolecular strategy is not restricted to the cyclodextrin host but is widely applicable to a variety of chiral as well as achiral supramolecular systems by introducing a chiral/achiral chromophore.

H. Photooxidation of Linoleic Acid by Cyclodextrin-Sandwiched Porphyrin Sensitization

Ogoshi and coworkers reported regio- and stereospecific oxidation of linoleic acid by singlet oxygen, which was generated by CDx-sandwiched porphyrin sensitization [125] (Scheme 21).

CDx-sandwiched porphyrin **56** (an equimolar mixture of diagonal type isomers) [128,129] was synthesized and used as a chiral $^1O_2$-generating host, while

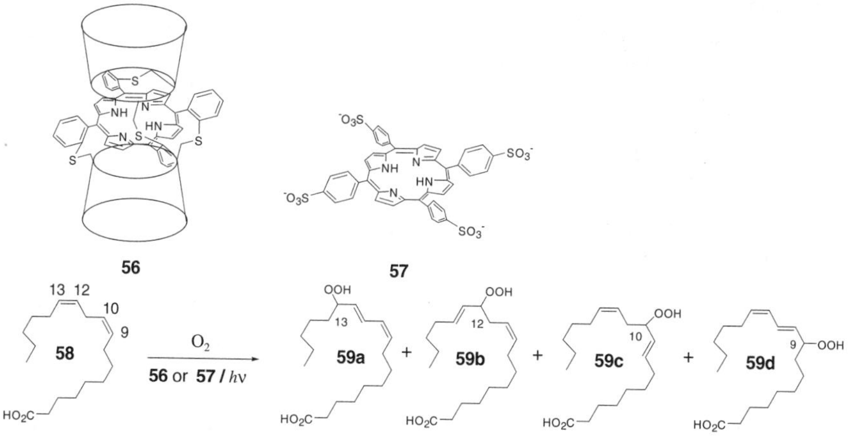

Scheme 21

tetrakis(p-sulfonatophenyl)porphyrin 57 was employed as a reference sensitizer without a substrate binding site (Scheme 21). Although the reference sensitizer 57 resulted in nonselective hydroperoxidation of linoleic acid 58 at the 9,10- and 12,13-double bonds in varying 59a/59b and 59c/59d ratios [130–133], CDx-sandwiched porphyrin 56 led to the selective hydroperoxidation at the 12,13-double bond to give 59a and 59b in 82% yield. The product distribution of 59a/59b/59c/59d (formed in a 51/31/11/7 ratio) indicates that the attack of singlet oxygen occurs preferentially at the positions distant from the carboxylic group. These results suggest that the selective hydroperoxidation occurs deep in the CDx cavity where the hydrophobic alkandienyl chain and the $^1O_2$-generating porphyrin moiety are located. This idea may be supported by the fact that the main products 59a and 59b obtained upon sensitization by 56 are optically active (up to 20% *ee*). Thus the chiral hydrophobic cavity of 56 regulates not only the regioselectivity but also the enantioselectivity of the singlet oxygen attack.

IV. SUPRAMOLECULAR PHOTOCHIROGENESIS WITH SYNTHETIC CHIRAL TEMPLATES

Photochemical asymmetric synthesis with native and modified CDx's is certainly one of the most potentially successful examples of supramolecular photochirogenesis, which is applicable in principle to most photoreactive substrates of appropriate size and shape. Despite this wide applicability, the enantiodifferentiating ability of CDx and hence the product *ee* obtained are not sufficiently high in many cases. Furthermore, the elucidation of the chiral discrimination mechanism and the rationalization of the product chirality and *ee* are generally more difficult in photochemical asymmetric induction. For more simple and well-defined host–guest interactions (in the ground as well as excited states), several approaches to photochemical asymmetric induction with newly designed synthetic chiral hosts have been reported.

A. Enantiodifferentiating [2 + 2] Photocycloaddition with a Chiral Template

Bach and coworkers reported a new strategy to achieve enantioselective photoreactions in solution based on the use of a chiral host derived from Kemp's acid [134–138]. They designed and synthesized a hydrogen bonding chiral template with a lactam functionality 60, which can bind prochiral amide substrates through two hydrogen bonds in an enantioface-selective fashion with the aid of a bulky group fixed in the 1,3-axial position, as illustrated in Scheme 22.

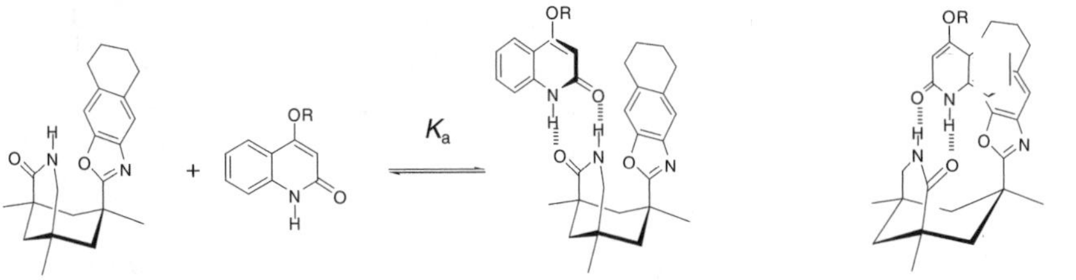

60 61 60 • 61 complex *ent*-60 • 61 complex

Scheme 22

In the host–guest complex of template **60** with substrate **61**, the two enantiofaces of quinolone **61** are distinctly discriminated. As one of the enantiofaces of **61** is blocked by the benzotriazole moiety of **60**, photochemical attack to the substrate is expected to occur from the open face. Indeed, in the presence of **60** and its enantiomer (*ent*-**60**), highly enantioselective intramolecular [2 + 2] photocycloaddition of allyl quinolonyl ether **62** [134] and intermolecular [2 + 2] photocycloaddition of quinolone **61** to alkenes **63** [135] were reported to occur in solution (Scheme 23). The intermolecular photocycloaddition of **63** to **61**, as well as the intramolecular photocycloaddition of **62** proceeded with excellent enantioselectivities (81–98% *ee*) and in high yields (61–89%).

B. Enantiodifferentiating [2 + 2] Photocycloaddition with Sensitizing Receptor

A similar but advanced approach using a sensitizing host, or receptor, has been reported very recently. Krische and coworkers synthesized a supramolecular host

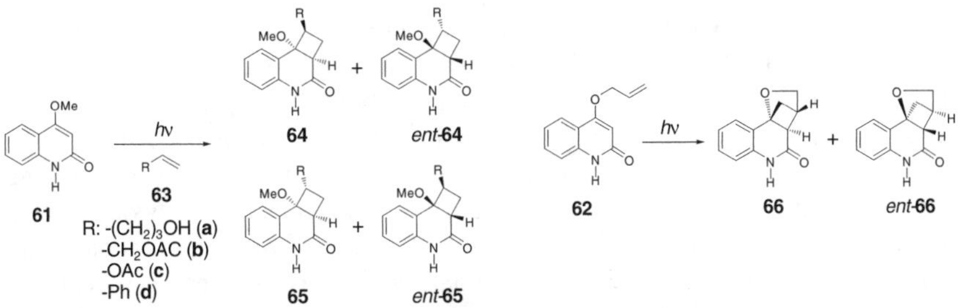

Scheme 23

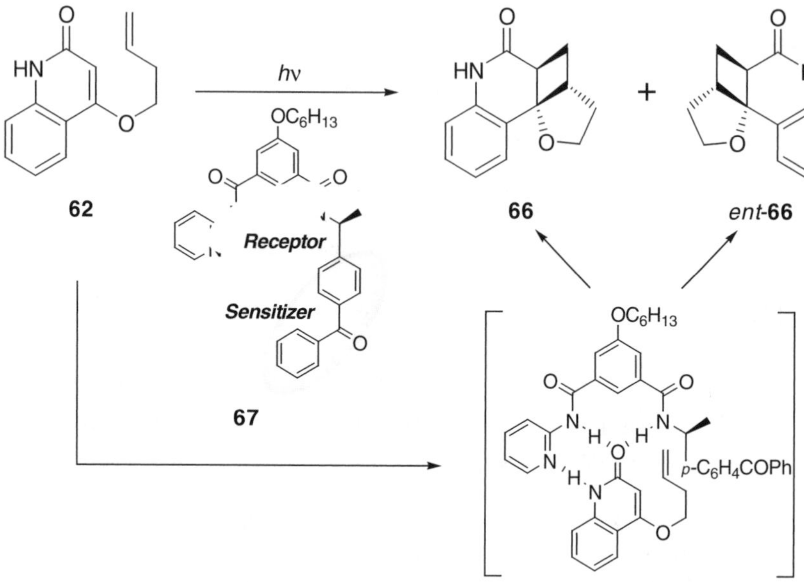

Scheme 24

67, which possesses an amide receptor and a triplet sensitizer in the same molecule. This sensitizing receptor was designed to bind 3-butenyl quinolonyl ether **62** through three hydrogen bonds in the ground state and then to use the closely located benzophenone moiety to photosensitize the enantiodifferentiating [2 + 2] photocycloaddition of bound **62** to enantiomeric **66** (Scheme 24) [139]. The association constant is relatively low (log K_a = 2.5 in CDCl$_3$), at least at 23°C. Nevertheless, substoichiometric amounts (as low as 25% of substrate **62**) of the sensitizing receptor **67** can bind the complementary lactam **62** to form a supramolecular complex **68**, which upon irradiation undergoes the intracomplex sensitized cycloaddition of **62** to **66**. The product **66** obtained at 30°C was racemic (even in the presence of two equivalents of **67**), probably because of the low binding constant and the involvement of the intermolecular sensitization by free **67**. However, the product ee was enhanced at lower temperatures, ultimately giving 19–22% ee at −70°C [139]. In this system, the product ee is mainly determined by the enantioface selectivity of the butenyl moiety of **68** upon complexation with **67**, which is not sufficient to discriminate effectively between the two diastereomeric forms of complex **68**. The introduction of a more bulky substituent at the tether group may improve the enantioface selectivity.

V. SUPRAMOLECULAR PHOTOCHIROGENESIS WITH PROTEINS

Possessing inherently chiral binding sites for organic molecules, biopolymers can be used as versatile chiral hosts for supramolecular photochirogenesis, although such an opportunity has not extensively been explored. In contrast, the control of thermal reactivity through complexation with proteins has been investigated more frequently, and many effective cases have been reported. For such purposes, serum albumin, among the vast variety of proteins, has frequently been employed, as this is the most abundant component of blood plasma, less expensive, and one of the most well-characterized water-soluble proteins. More importantly in the context of supramolecular photochirogenesis, serum albumins are known to bind and transport various endo/exogenous hydrophobic compounds, such as fatty acids, bilirubins, hormones and hydrophobic drugs in vivo [140]. Bovine serum albumin (BSA) possesses the ability to bind not only relatively water-insoluble endogenous substrates but also fused aromatic carboxylates and alcohols *in vitro* in its chiral hydrophobic binding pockets [141].

A. Enantiodifferentiating Photodestruction with BSA

Zandomeneghi and coworkers demonstrated for the first time the usefulness of BSA in supramolecular photochirogenesis. They reported the highly enantioselective photodecomposition of racemic binaphthol **69** and ketoprofen **70** in the presence of BSA. [141–143,145].

They revealed that BSA binds (*S*)-(−)-**69** (Scheme 25) preferentially over the (*R*)-(+)-isomer by a factor of five and that the bound **69** shows an induced circular dichroism with an accompanying bathochromic shift of the absorption band at 350–400 nm, which is ascribable to the deprotonation of naphthol. Irradiation of this new band selectively excites (*S*)-(−)-**69** bound to BSA, resulting in racemization. In contrast, photoracemization of BSA-bound (+)-**69** to (−)-**69** was not observed. Consequently, the photolysis of racemic **69** in the presence of BSA led to the preferential photodestruction, as well as the photoracemization, of (−)-**69**, both of which contribute to the enrichment of (+)-**69** in the remaining substrate up to 89% *ee* upon 57% decomposition and ultimately 99% *ee* upon 77% decomposition. The facts that (1) the bathochromic shift is observed upon complexation with BSA and that (2) the photoracemization occurs upon irradiation of (−)-**69** in basic CH$_3$CN jointly indicate that naphthol **69** is bound to a basic binding site of BSA in the photoreactive naphtholate anion form. From comparative photolysis experiments using racemic and nearly enantiopure **69** in the presence of BSA, they concluded that the preferential photodestruction of (−)-**69** is due to both the stronger light absorption by (−)-**69** in BSA (5 times

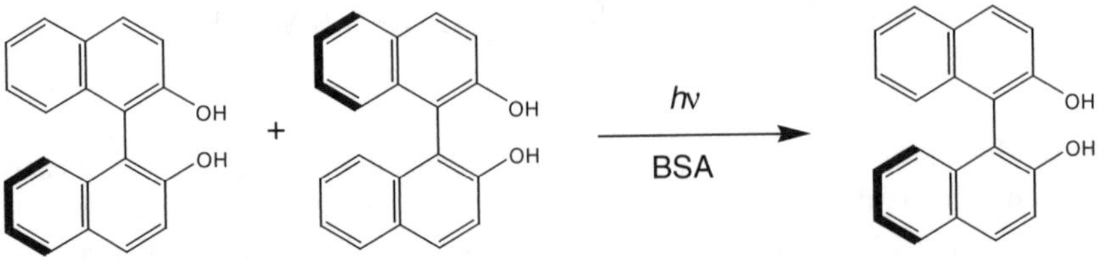

(*R*)-(+)   (*S*)-(−)   99.5% ee at 77% conversion
         69   **69**

Scheme 25

larger than that of the antipode) and the higher quantum yield of the subsequent photodestruction and/or photoracemization ($\Phi^-/\Phi^+ \sim 1.7$).

They also investigated the photodestruction of racemic 2-[(3-benzoyl) phenyl]propionic acid (ketoprophene) **70** (Scheme 26) in the presence of BSA [143,144]. Upon complexation with BSA, **70** exhibited a new absorption at longer wavelengths, irradiation of which at 365 nm led to the enantioselective decomposition of **70**; but a high *ee* was obtained only after nearly complete consumption of the substrate; i.e., 80% *ee* after 99.8% photodestruction.

B. Enantiodifferentiating Photocyclodimerization of Anthracene Derivatives with BSA

Although Zandomeneghi *et al.* elegantly demonstrated the effective use of BSA to effect enantioselective binding and photodecomposition/racemization of chiral substrates bound to BSA, [140–145], no new chiral center is created in the photoreactions. Very recently, Inoue, Wada, and coworkers proposed a more dynamic supramolecular photochirogenesis using BSA, in which stereogenic centers are *generated* in the photoproduct through supramolecular interactions in the excited

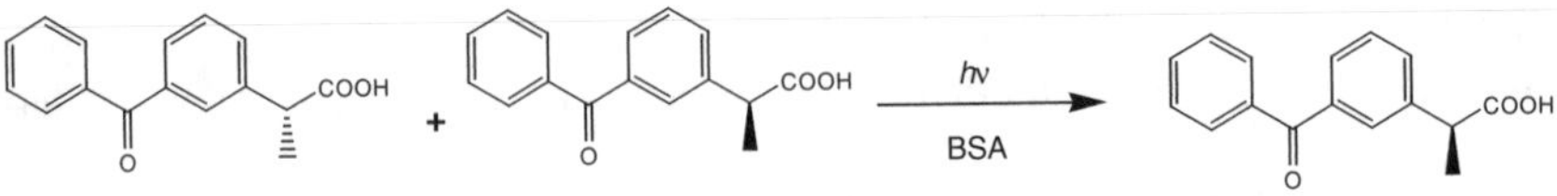

(*R*)-(+)   (*S*)-(−)   80% ee at 99.8% conversion
         70   **70**

Scheme 26

state [146]. They intended to expand the range of supramolecular photochirogenesis with BSA from the unimolecular to the bimolecular system and from asymmetric photodestruction to photogeneration, and also to elucidate the factors and mechanisms governing the product ratio and *ee* in the BSA-mediated bimolecular photochirogenesis. For these ends, the photocyclodimerization of 2-anthracenecarboxylic acid **43** was employed, since it is well documented that photoirradiation of **43** produces four stereoisomeric [4 + 4] cyclodimers, *i.e.* *anti-* and *syn-head-to-tail* (HT) dimers (**44** and **45**) and *anti-* and *syn-head-to-head* (HH) dimers (**46** and **47**), of which only **45** and **46** are chiral (Scheme 27).

The ground-state interactions and complexation behavior of **43** with BSA in an aqueous buffer solution were fully elucidated by UV-Vis, circular dichroism, and fluorescence spectral titration experiments. It was revealed that there are four independent binding sites for **43** in BSA, which bind 1, 3, 2, and 3 guest molecules with binding constants of 4.0×10^7, 1.3×10^5, 1.4×10^4, and $3.0 \times 10^3 \, M^{-1}$, respectively. Remarkably, the simultaneous binding of multiple **43** molecules at the latter three binding sites causes no appreciable CD exciton coupling, indicating that these **43** molecules, though bound to the same binding site, are not very closely located to each other. Then the enantiodifferentiating photodimerization of **43** was performed in the presence of varying amounts of BSA. With increasing BSA/**43** ratios from zero to unity (keeping [**43**] constant), the HH/HT ratio, i.e., ([**46**] + [**47**])/([**44**] + [**45**]), was dramatically enhanced from 0.28 to 4.3, clearly indicating that the HH dimers are highly preferred upon photodimerization in BSA binding sites, probably owing to the attractive electrostatic interaction of the carboxylate anion of **43** with a cationic amino acid residue of BSA, such as lysine. The *ee*s of **45** and **46** were also enhanced up to 29% and 41% *ee*, respectively, at the highest BSA/**43** ratio of 0.8. It was deduced that the optically active HT and HH cyclodimers **45** and **46** are produced in the chiral BSA binding pockets through two independent paths: the external attack of a free **43** to a BSA-bound **43** gives HT dimer **45** (and **44**), while the intrasite reaction of two BSA-bound **43** molecules affords the HH dimer **46** (and **47**).

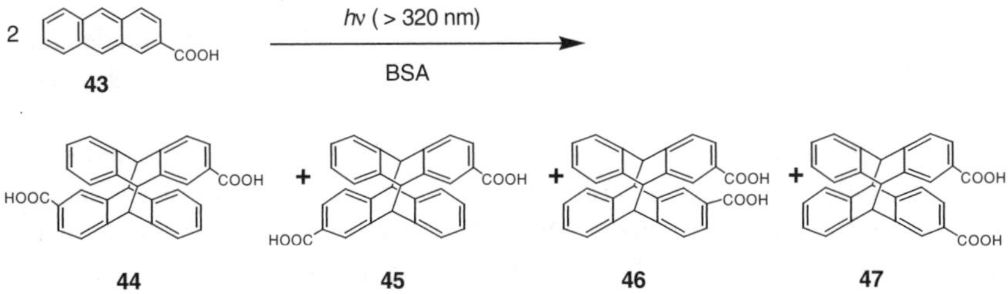

Scheme 27

This first supramolecular photochirogenesis using an achiral substrate demonstrates (1) that BSA possesses four independent binding sites for **43** of different affinity, stoichiometry, and chiral environment for photoreactions, (2) that BSA-mediated photodimerization of **43** switches the original regioselectivity from HT to HH, and (3) that it affords optically active products **45** and **46** in up to 29% and 41% *ee*, respectively. It was emphasized that the selective excitation of bound substrate, utilizing the spectral shift upon complexation with BSA, is not a prerequisite for efficient photochirogenesis using biomolecules. This is highly encouraging in expanding the range of substrates and the scope of supramolecular photochirogenesis with biomolecules.

VI. SUPRAMOLECULAR PHOTOCHIROGENESIS WITH DNA

With rapid development of molecular biology and genetic engineering, the chemistry of nucleic acids has recently received much attention. Photochemistry of nucleic acids has also been intensively investigated, particularly in connection with the causes of skin cancer [147]. In the last decade, photoinduced electron transfer through double stranded DNA has been one of the most exciting topics not only for photochemists but also for bio- and physicochemists, and a number of papers on the direct and sensitized photolyses of DNA have been reported [148]. In contrast to the well-documented photosensitized reactions of DNA/RNA [149,150], the (chiral) photosensitizing ability of these biomolecules has not been well understood until recently. Possessing both a chromophoric nucleobase and an optically active furanose, these finely structured double-helical biomolecules with hydrophobic grooves are expected to provide a chiral environment for supramolecular asymmetric photosensitization.

VII. ENANTIODIFFERENTIATING PHOTOSENSITIZATION BY NUCLEOSIDES AND DOUBLE STRANDED DNA

Inoue, Wada, and coworkers employed ds-DNA and its component nucleosides as chiral supramolecular photosensitizers to effect the enantiodifferentiating geometrical isomerization of (Z)-cyclooctene **30** (Scheme 28), producing the chiral (E)-isomer **31** [151,152]. Although the photosensitized reactions [148] and modifications [149,150] of DNA and RNA have extensively been investigated, nucleosides and DNA have rarely been employed as photosensitizers, probably owing to their photolabile nature [153]. In this first attempt to use nucleosides and ds-DNA as chiral sensitizing hosts, they employed the photoisomerization of **30** as a benchmark test system for examining their ability to transfer supramolecular

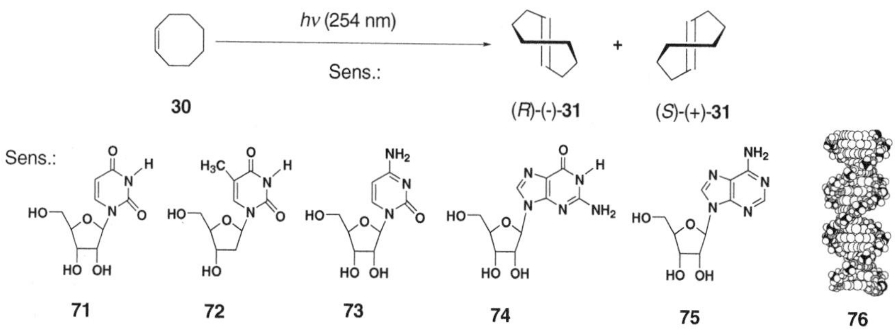

Scheme 28

chirality through excited state interactions, since common nucleosides possess similar singlet energies ($E_S \sim 100$ kcal/mol) [154] as frequently used benzenecarboxylates [83] and are therefore expected to act as chiral photosensitizers.

UV spectral examinations of the photostability of nucleosides revealed that thymidine (**72**), adenosine (**75**), and guanosine (**74**) are highly stable under the irradiation conditions employed, and the initial absorptions at 260 nm are retained almost completely (showing only a 2% decrease) after 1 h of photoirradiation, at least in the presence of the energy acceptor **30**. The photosensitizations with uridine (**71**) and thymidine (**72**) gave remarkably high photostationary-state E/Z ratios, $(E/Z)_{pss}$, of 0.33 and 0.66, respectively, which are much greater than those (~ 0.25) obtained with the conventional singlet sensitizers such as menthyl and DAG benzoates [84,89], indicating an efficient singlet-energy transfer from **71**/**72** to **30**. In contrast, photosensitizations with adenosine (**75**) and guanosine (**74**) afforded very low $(E/Z)_{pss}$ of ca. 0.005. This contrasting behavior cannot simply be attributed to the sensitizer energy but rather to the structural and electronic differences between the pyrimidine and purine nucleobases. Calf thymus DNA (**76**) gave a slightly better $(E/Z)_{pss}$ of 0.014. As all of the nucleobases are paired and stacked tightly in ct-DNA, the usual sensitization mechanism through the exciplex formation [83–89] occurs with difficulty. Although the sensitization with the purine nucleosides, **74** and **75**, resulted in the formation of the racemic product with very low $(E/Z)_{pss}$, the use of the pyrimidine nucleosides, **72** and **71**, as chiral sensitizers gave (S)-(−)-**31** in 5.2 and 3.1% ee, respectively. These ees are not particularly high, but they are appreciable improvements compared to the previous values (0–2.7% ee) obtained in the photosensitization with a variety of chiral alkyl benzoates after a decade-long effort [83–90]. Probably the shorter tether connecting the chromophoric and chirogenic moieties, and the reduced conformational freedom of the pyrimidine nucleosides compared to those of the chiral benzoates, are jointly responsible for the enhanced ees.

Interestingly, the photosensitization with ct-DNA **76** gave antipodal (R)-
(−)-**31** with a higher 9.2% ee at 25°C and 15.2% ee at 5°C. Since the monomeric
nucleosides used as chiral sensitizers consistently afford (S)-(+)-**31**, the same
enantiodifferentiating photosensitization mechanism cannot rationalize the formation of the antipodal product upon sensitization with **76**. In a separate spectroscopic examination under comparable conditions, the addition of **30** to an aqueous
solution of ct-DNA gave rise to appreciable changes in the circular dichroism
spectrum. Since the resulting spectrum still keeps the original shape of B-form
DNA, the small changes observed were concluded to originate from the weak
interaction of **30** with ds-DNA **76** probably in the hydrophobic grooves, rather
than a more hydrophilic site such as phosphate backbone. The small CD spectral
changes upon addition of **30** to a **76** solution, the lower ees at higher temperatures,
and the formation of the antipodal product indicate that supramolecular complex
formation between **30** and **76** in the ground state, followed by the photoisomerization to chiral **31**, is the essential process for achieving the highly enantiodifferentiating photosensitization. They also performed the photosensitization with ct-DNA
76 in 50% aqueous methanol solution, in which **30** is highly soluble and therefore
no supramolecular interactions between **76** and **30** are expected to occur, with
76 maintaining its original B-form. The ee obtained in aqueous methanol solution
greatly decreased to 0.9%, clearly indicating the crucial role of supramolecular
interaction with ct-DNA **76** in this enantiodifferentiating photosensitization.

From these results, they claimed that this newly developed function of
nucleosides and ds-DNA should not be restricted to this particular system but
could be expanded to a variety of supramolecular photochirogenesis.

VIII. DISCUSSION AND CONCLUSIONS

In this chapter, we have reviewed representative recent literature relating to supramolecular asymmetric photoreactions (SMAP). This is certainly a rapidly growing
and attractive area of photochemistry that utilizes a variety of noncovalent weak
interactions working in a wide variety of chiral supramolecular systems, ranging
from chirally modified porous inorganic materials (zeolites) to highly sophisticated biomolecules (BSA and DNA). Presently, most of the SMAP processes
employ a stoichiometric amount of chiral inductor or host, with the exception of
a few cases, to give modest-to-excellent optical yields in varying chemical yields.
To realize higher chirogen efficiency for synthetic purposes and also to expand
the applicability of the SMAP strategy, it is a must to increase considerably
the turnover number by introducing a smooth complexation-photochirogenesis-
decomplexation cycle in the supramolecular system. The use of sensitizing chiral
hosts automatically fulfills this condition, as the complexation-photochirogenesis-
decomplexation process occurs in a consecutive "quasi-single" step in the excited

state, while the SMAP employing the initial ground-state complexation may require some special device to excite selectively the bound species and to accelerate the decomplexation of photoproduct from the chiral host. Catalytic turnover in SMAP enables us to reduce the amount of chirogen and to use an excess amount of substrate dissolved in the bulk solution.

The less significant role of the entropy factor found in supramolecular photochirogenesis using CDx derivatives is an interesting finding. This is rather unexpected and intriguing in light of the vital role of entropy widely demonstrated in asymmetric photosensitization in isotropic homogeneous solutions [83–89]. This may be rationalized by the greatly reduced translational, rotational, conformational, and vibrational freedoms of the guest molecule in a supramolecular complex, where the entropy of the guest is inherently reduced and thus cannot easily be done in the subsequent photochirogenesis process. Thus such a low-entropy environment with reduced motional freedoms appears to be one of the important features of supramolecular photochirogenesis via structurally defined host–guest complexes or in confined media, leading to the enthalpic control of photochirogenesis through enhancement of noncovalent weak interactions.

As stated in the introduction and demonstrated in many examples, confinement is a key concept in supramolecular chemistry, but its nature is not clear. The confinement of a molecule in a host of molecular dimensions causes the following phenomena: (1) positional, orientational, and conformational freezing, (2) enhancement of the critically distance-sensitive weak interactions, such as van der Waals interactions (where energy is expressed as a function of distance (r) by the equation $U = -ar^{-6} + br^{-9 \text{ to } -12}$ and as hydrogen bonding and dipole-induced dipole interactions $U = -\mu^2\alpha/(4\pi\varepsilon)^2 r^6$, where μ is the dipole moment, α the polarizability, and ε the dielectric constant), (3) enhancement of the local concentration, and (4) increase of the effective microenvironment polarity felt by an immobilized guest, for example, in a zeolite supercage, which may favor polarity-related interactions. All of these effects in confined reaction spaces cooperatively and synergetically strengthen the supramolecular interactions and reduce the freedom of a guest.

In conclusion, all of the supramolecular photochirogenic approaches described in this chapter are conceptually new and mechanistically interesting, but synthetically not very successful in many cases. However, supramolecular photochirogenesis using natural and synthetic chiral hosts provides us with a method for transferring chirality to a prochiral substrate through ground- and/or excited-state host–guest interactions in a low-entropy environment. Hence even though the factors and mechanisms that govern the product chirality and ee are not sufficiently clear, and the optical yields obtained are not satisfactorily high at present, it should be emphasized that supramolecular photochirogenesis, though in its embryonic period, has already shown success in some cases and exhibits several phenomena that are completely different from conventional photochirogenesis performed in isotropic media, all of which are certainly worthy of future research. Further expansion of supramolecular systems employed for photochirogenesis and the global understanding of their features, as well as the elucidation of the detailed controlling factors and mechanisms, are crucial targets for the continued study of supramolecular photochirogenesis.

ACKNOWLEDGMENT

We would like to thank Dr. Guy A. Hembury for assistance in the preparation of this manuscript.

REFERENCES

1. Hayashi T, Tomioka K, Yonemitsu O. Asymmetric Synthesis: Graphical Abstracts and Experimental Methods. Tokyo/Amsterdam: Kodansha/Gordon & Breach, 1998.
2. Gawley R, Aub J. In Baldwin JE , Magnus FRS , Magnus PD, Eds. Principles of Asymmetric Synthesis. Oxford: Pergamon Press, 1996.
3. Ager DJ, East MB. Asymmetric Synthetic Methodology. Boca Raton: CRC Press, 1996.
4. Seyden-Penne J. Chiral Auxiliaries and Ligands in Asymmetric Synthesis. New York: John Wiley, 1995.
5. Nogradi M. Stereoselective Synthesis, 2nd ed.. Weinheim: VCH, 1995.
6. Noyori R. Asymmetric catalysis in organic synthesis. New York: John Wiley, 1994.
7. Wong CH, Whitesides GM. Enzymes in Synthetic Organic Chemistry. Oxford: Pergamon Press, 1994.
8. Collins AN, Sheldrake GN, Crosby J. Chirality in Industry. Vol. 1. Chichester: John Wiley, 1992.
9. Collins AN, Sheldrake GN, Crosby J. Chirality in Industry. Vol. 2. Chichester: John Wiley, 1997.
10. Sheldon RA. Chirotechnology—Industrial Synthesis of Optically Active Compounds. New York: Marcel Dekker, 1993.
11. Rau H. Chem Rev 1983; 83:535.
12. Inoue Y. Chem Rev 1992; 92:741.
13. Everitt SRL, Inoue Y. In Ramamurthy V , Schanze K., Eds. Organic Molecular Photochemistry. New York: Marcel Dekker, 1999.
14. Pete JP. Adv Photochem 1996; 21:135.
15. Buschmann H, Scharf HD, Hoffmann N, Esser P. Angew. Chem 1991; 103:480.
16. Buschmann H, Scharf HD, Hoffmann N, Esser P. Angew Chem Int Ed Engl 1991; 30:477.
17. Turro NJ. Modern Molecular Photochemistry: Menlo Park: University Science Press, 1990.
18. Lehn JM. Supramolecular Chemistry. Weinheim: VCH, 1995.

19. Ramamurthy V ed. Photochemistry in Organized and Constrained Media. New York: VCH, 1991.
20. Schneider HJ, Durr EH, eds. Frontiers in Supramolecular Organic Chemistry and Photochemistry, VCH. 1991.
21. Turro NJ. Acc Chem Res 2000; 33:637.
22. Turro NJ. Proc Nat Acad Sci USA 2002; 99:4805.
23. Burnet E. Chirality 2002; 14:135.
24. Joy A, Ramamurthy V. Chem Eur J 2000; 6:1287.
25. Shailaja J, Sivaguru J, Uppili S, Joy A, Ramamurthy V. Microporous and Mesoporous Materials 2001; 48:319.
26. Baretz BH, Turro NJ. J Am Chem Soc 1983; 105:1309.
27. For example, Faljoni A, Zinner K, Weiss RG. Tetrahedron Lett 1974; 15:1127.
28. For example, Laarhovewn WH, Cuppen TJHM. Chem Soc, Perkin Trans 2: 1978; 315, 1978 .
29. For example; Eskenazi C, Nicoud JF, Kagan HB. J Org Chem 1979; 44:995.
30. For example; Janiki SZ, Schuster GB. J Am Chem Soc 1995; 117:8524.
31. Gamlin JN, Jones R, Leibovitch M, Patrick B, Scheffer JR, Trotter J. Acc Chem Res 1996; 29:203.
32. Leibovitch M, Olovsson G, Scheffer JR, Trotter J. Pure Appl Chem 1997; 69:815.
33. Toda F. Acc Chem Res 1995; 28:480.
34. Desiraju G. Crystal Engineering, the Design of Organic Solids. Amsterdam: Elsevier, 1989.
35. Ramamurthy V, Venkatesan K. Chem Rev 1987; 87:433.
36. Vaida M, Popovitz-Biro R, Leserowitz L, Lahav L. In Ramamurthy V., Ed. Photochemistry in Organized and Constrained Media. New York: VCH, 1991.
37. For example, T Hamada, H Ishida, S Usui, K Tsumura, K Ohkubo, J Mol Catl 1994; 88: L1.
38. For example, Turro NJ, Grauer Z, Kumar CV, Barton JK. Langmuir 1987; 3:1056.
39. For example, Takagi K, Shichi T. Mol Supramol Photochem 2000; 5:31.
40. Sellergren B, Karmalkar RN, Shea KJ. J Org Chem 2000; 65:4009.
41. Scaiano JC, Garcia H. Acc Chem Res 1999; 32:783.
42. Bekkum HV, Flanigen EM, Jansen eds. JC. Introduction to Zeolite Science and Practice. Amsterdam: Elsevier, 1991.
43. Breck DW. Zeolite Molecular Sieves: Structure, Chemistry and Use. New York: John Wiley, 1974.
44. Dyer A. An Introduction to Zeolite Molecular Sieves. New York: John Wiley, 1988.
45. Bekkum HV, Flanigen EM, Jansen JC. Introduction to Zeolite Science and Practice. Amsterdam: Elsevier, 1991.
46. Joy A, Ramamurthy V. Chem Eur J 2000; 6:1287.
47. Shailaja J, Sivaguru J, Uppili S, Joy A, Ramamurthy V. Microporous and Mesoporous Materials 2001; 48:319.
48. Rao VP, Zimmtt MB, Turro NJ. J Photochem Photobiol A: Chem 1991; 57:7.
49. Leibovitch M, Olovsson G, Sundarababu G, Ramamurthy V, Scheffer JR, Trotter J. J Am Chem Soc 1996; 118:1219.
50. Wagner PJ, Park BS. Org Photochem 1991; 11:227.

51. Yang NC, Yang DH. J Am Chem Soc 1958; 80:2913.
52. Wagner PJ. Acc Chem Res 1989; 22:83.
53. Scaiano JC. Acc Chem Res 1982; 15:252.
54. Natarajan A, Joy A, Kaanumalle LS, Scheffer JR, Ramamurthy V. J Org Chem 2002; 67:8339.
55. Kaprinidis N, Landis MS, Turro NJ. Tetrahedron Lett 1997; 38:2609.
56. Joy A, Robbins RJ, Pitchumani K, Ramamurthy V. Tetrahedron Lett 1997; 38: 8825.
57. Edman JR. J Am Chem Soc 1969; 91:7103.
58. Hahn RC, Johnson RP. J Am Chem Soc 1977; 99:1508.
59. Akermark B, Johanson NG, Sjoberg B. Tetrahedron Lett 1969; 10:371.
60. Aoyama H, Hasegawa T, Omote Y. J Am Chem Soc 1979; 101:5343.
61. Aoyama H, Sakamoto M, Kuwabara K, Yoshida K, Omote Y. J Am Chem Soc 1983; 105:1958.
62. Chesta CA, Whitten DG. J Am Chem Soc 1992; 114:2188.
63. Natarajan A, Wang K, Ramamurthy V, Scheffer JR, Brian P. Org Lett 2002; 4: 1443.
64. Hammond GS, Cole RS. J Am Chem Soc 1965; 87:3256.
65. Aratani T, Nakanisi Y, Nozaki H. Tetrahedron 1970; 26:1675.
66. Ouannès C, Beugelmans R, Roussi G. J Am Chem Soc 1973; 95:8472.
67. Inoue Y, Yamasaki N, Shimoyama H, Tai A. J Org Chem 1993; 58:1785.
68. Cheung E, Chong KCW, Jayaraman S, Ramamurthy V, Scheffer JR, Trotter J. Org Lett 2000; 2:2801.
69. Kaanumalle LS, Sivaguru J, Sunoj RB, Lakshminarasimhan PH, Chandrasekhar J, Ramamurthy V. J Org Chem 2002; 67:8339.
70. Chong KCW, Sivaguru J, Shichi T, Yoshimi Y, Ramamurthy V, Scheffer JR. J Am Chem Soc 2002; 124:2858.
71. Joy A, Scheffer JR, Corbin DR, Ramamurthy V. Chem Comm 1998:1379.
72. Joy A, Scheffer JR, Ramamurthy V. Org Lett 2000; 2:119.
73. Joy A, Uppili S, Netherton MR, Scheffer JR, Ramamurthy V. J Am Chem Soc 2000; 122:728.
74. Joy A, Ramamurthy V. Chem Eur J 2000; 6:1287.
75. Koodanjeri S, Joy A, Ramamurthy V. Tetrahedron 2000; 56:7003.
76. Dauben WG, Koch K, Chapman OL. J Am Chem Soc 1963; 85:2616.
77. Takashita H, Kumamoto M, Koino I. Bull Chem Soc Jpn 1980; 53:1006.
78. Toda F, Tanaka K. J Chem Soc, Chem Commun 1986:1429.
79. Shailaja J, Ponchot KJ, Ramamurthy V. Org Lett 2000; 2:937.
80. Lewis FD, Johnson RW, Johnson DE. J Am Chem Soc 1974; 96:6090.
81. Stocker JH, Kern DH. J Chem Soc, Chem Commun 1969:204.
82. Hammond GS, Cole RS. J Am Chem Soc 1965; 87:3256.
83. Inoue Y, Yokoyama T, Yamasaki N, Tai A. J Am Chem Soc 1989; 111:6480.
84. Inoue Y, Yamasaki N, Yokoyama T, Tai A. J Org Chem 1992; 57:1332.
85. Inoue Y, Yamasaki N, Yokoyama T, Tai A. J Org Chem 1993; 58:1011.
86. Tsuneishi H, Hakushi T, Inoue Y. J Chem Soc, Perkin Trans 1996; 2:1601.
87. Inoue Y, Matsushima E, Wada T. J Am Chem Soc 1998; 120:10687.

88. Hoffmann R, Inoue Y. J Am Chem Soc 1999; 121:10702.
89. Inoue Y, Ikeda H, Kaneda M, Sumimura T, Everitt SRL, Wada T. J Am Chem Soc 2000; 122:406.
90. Wada T, Shikimi M, Inoue Y, Lem G, Turro NJ. Chem Comm 2001:1864.
91. Bender ML, Komiyama M. Cyclodextrin Chemistry. New York: Springer-Verlag, 1978.
92. Szejtli J. Chem Rev 1998; 5:1743.
93. Takahashi K. Chem Rev 1998; 98:2013.
94. Rekharsky MV, Inoue Y. Chem Rev 1998; 98:1875.
95. Easton CJ, Lincoln SF. Chem Soc Rev 1996:163.
96. Bortolus P, Monti S. Adv Photochem 1996; 21:1.
97. Bortolus P, Monti S. In Neckers DC, Volman DH, Bunau GV, Eds. Photochemistry in Cyclodextrin Cavities: Advances in Photochemistry. John Wiley. New York, 1996.
98. Bortolus P, Grabner G, Kohler G, Monti S. Coord Chem Rev 1993; 125:261.
99. Takeshita H, Kumamoto M, Kouno I. Bull Chem Soc Jpn 1980; 53:1006.
100. Nakamura I, Sugimoto T, Oda J, Inouye Y. Agric Biol Chem 1981; 45:309.
101. Rao VP, Turro NJ. Tetrahedron Lett 1989; 30:4641.
102. Koodanjeri S, Joy A, Ramamurthy V. Tetrahedron 2000; 56:7003.
103. Shailaja J, Karthikeyan S, Ramamurthy V. Tetrahedron Lett 2002; 43:9335.
104. Corey EJ, Streith J. J Am Chem Soc 1964; 86:950.
105. De Selms RC, Schleigh WR. Tetrahedron Lett 1972; 34:3563.
106. Szejtli ed. J. Cyclodextrins (Comprehensive Supramolecular Chemistry, Vol. 7). New York: Pergamon Press, 1996.
107. Koodanjeri S, Ramamurthy V. Tetrahedron Lett 2002; 43:9229.
108. Rideout DC, Breslow R. J Am Chem Soc 1980; 102:7816.
109. Tamaki T. Chem Lett 1984:53.
110. Tamaki T, Kokubu. J Inclusion Phenom 1984; 2:815.
111. Tamaki T, Kokubu T, Ichimura K. Tetrahedron 1987; 43:1485.
112. Tamagaki S, Fukuda K, Maeda H, Mimura N, Tagaki W. J Chem Soc, Perkin Trans 2, 1995; 389.
113. Moorthy JN, Venkatesan K, Weiss RG. J Org Chem 1992; 57:3292.
114. Brett TJ, Alexander JM, Stezowski JJ. J Chem Soc, Perkin Trans 2, 2000; 1095.
115. Brett TJ, Alexander JM, Stezowski JJ. J Chem Soc, Perkin Trans 2, 2000; 1105.
116. Herrmann W, Wehrle S, Wenz G. J Chem Soc, Chem Comm 1997:1709.
117. Rao KSSP, Hubig SM, Moorthy JN, Kochi JK. J Org Chem 1999; 64:8098.
118. Banu HS, Lalitha A, Pitchumani K, Srinivasan C. J Chem Soc, Chem Comm 1999; 607.
119. Utsuki T, Hirayama F, Uekema K. J Chem Soc, Perkin Trans 2; 1993; 109.
120. Nakamura A, Inoue Y. J Am Chem Soc 2003; 125:966.
121. Vizvardi K, Desmet K, Luyten I, Sandra P, Hoornaert G, Eycken VE. Org Lett 2001; 3:1173.
122. Inoue Y, Kosaka S, Matsumoto K, Tsuneishi H, Hakushi T, Tai A, Nakagawa K, Tong L. J Photochem Photobio A: Chem 1993; 71:61.
123. Inoue Y, Dong F, Yamamoto K, Tong L, Tsuneishi H, Hakushi T, Tai A. J Am Chem Soc 1995; 117:11033.

124. Inoue Y, Wada T, Sugahara N, Yamamoto K, Kimura K, Tong LH, Gao XM, Hou ZJ, Liu Y. J Org Chem 2000; 65:8041.
125. Kuroda Y, Sera T, Ogoshi H. J Am Chem Soc 1991; 113:2793.
126. Wender PA, Siggel L, Nuss JM. In Padwa A, Ed. Organic Photochemistry. New York: Marcel Dekker, 1989.
127. Wender PA, Dore TM. Tetrahedron Lett 1998; 39:8589.
128. Kuroda Y, Hiroshige T, Sera T, Shiroiwa Y, Tanaka H, Ogoshi H. J Am Chem Soc 1989; 111:1912.
129. Kuroda Y, Hiroshige T, Sera T, Ogoshi H. Carbohydr Res 1989; 192:347.
130. Thomas MJ, Pryor WA. Lipids 1980; 15:544.
131. Terao J, Matsushita S. J Food Process Preserv 1980; 3:329.
132. Gunstone FD. J Am Oil Chem Soc 1984; 61:441.
133. Chacon JN, McLearie J, Sinclair RS. Photochem Photobiol 1988; 47:647.
134. Bach T, Bergmann H, Harms K. Angew Chem, Int Ed 2000; 39:2302.
135. Bach T, Bergmann H. J Am Chem Soc 2000; 122:11525.
136. Bach T, Bergmann H, Grosch B, Harms K. J Am Chem Soc 2002; 124:7982.
137. Bach T, Bergmann H, Harms K. Org Lett 2001; 3:601.
138. Bach T, Aechtner T, Neumuller B. Chem Comm 2001:607.
139. Cauble DF, Lynch V, Krische MJ. J Org Chem 2003; 68:15.
140. Peters T. All About Albumin: Biochemistry, Genetics, and Medical Applications, Academic Press. 1996.
141. Zandomeneghi M. J Am Chem Soc 1991; 113:7774.
142. Levi-Minzi N, Zandomeneghi M. J Am Chem Soc 1992; 114:9300.
143. Cavazza M, Festa C, Lenzi A, Levi-Minzi N, Veracini CA, Zandomeneghi M. Gazz Chim Ital 1994; 124:525.
144. Festa C, Levi-Minzi N, Zandomeneghi M. Gazz Chem Ital 1996; 126:599.
145. Ouchi A, Zandomeneghi G, Zandomeneghi M. Chirality 2002; 14:1.
146. Wada T, Fujisawa T, Sugahara N, Nishijima M, Mori T, Nakamura A, Inoue Y. J Am Chem Soc 2003; 125:7492.
147. Kochevar I. J Invest Dermatol 1981; 77:59.
148. Turro C, Hall DB, Chen W, Zuilhof H, Barton JK, Turro NJ. J Phys Chem A 1998; 102:5708.
149. Norris C, Meisenheimer P, Koch TH. J Am Chem Soc 1996; 118:5796.
150. Sutherland BM, Randesi M, Wang K, Conlon K, Epling GA. Biochem 1993; 32: 1799.
151. Wada T, Sugahara N, Kawano M, Inoue Y. Nucleic Acids Res 2000; 44:115.
152. Wada T, Sugahara N, Kawano M, Inoue Y. Chem Lett 2000:1174.
153. For a recent review, see MW Grinstaff. Angew Chem, Int Ed Eng 38: 3629, 1999.
154. Vigny PCR. Seances Acad Sci, Ser D 1971; 272:3206.

10
Circular Dichroism in the Solid State

Reiko Kuroda
The University of Tokyo, Tokyo, and Japan Science and Technology Agency, Kawaguchi, Japan

I. INTRODUCTION

The phenomenon of optical activity was first discovered in quartz crystals. When Arago inserted a quartz plate (cut perpendicular to the optic axis) between a polarizer and a calcite analyzer and rotated the polarizer or the analyzer, a spectrum of colored images was observed [1]. His colleague Biot showed [2] that this effect arises from the rotation of the plane of polarization, and that there are two forms of quartz, one dextrorotatory and the other levorotatory. Investigations of optical activity, thus started in the solid state, have extended to the liquid state, first to natural products such as turpentine oil, lemon oil, and solutions of sucrose, and then to solutions of many man-made compounds including metal coordination complexes and organic compounds. The phenomenon of optical activity gave a great impetus to the establishment of modern stereochemistry, but it is based on the liquid state, rather than on solid-state experiments.

Most chemical reactions and spectroscopic work are carried out in solution. However, recent investigations have shown great potential for solid-state chemistry. First, supramolecular systems consisting of more than one kind of molecule exhibit characteristics that are different from those of the individual molecules. We have recently reported inverted supramolecular chirality of bis(zinc octaethylporphyrin), an enantiopure monoamine system in solution and in the solid phase [3]. Secondly, solid-state reactions occur more frequently than previously envisaged [4–7], and they may give products in high yield that are unobtainable by reaction in solution. They are kind to the environment because of their solvent-free nature and hence are currently attracting attention from the industrial sector as well. Compared with thermal or chemical reactions, photochemistry is particularly suitable for solid-state reactions. Photoreactions in the solid state may produce diastereoisomers that are different from those obtained in solution reactions [8], and such reactions are often highly enantiospecific [4]. This may be relevant to the "homochirality of the biological world," as discrimination of molecular handedness is orders of magnitude larger in the solid state than in solution.

When chirality is involved, information on solid-state structures and supramolecular properties must be obtained by solid-state circular dichroism (CD) spectroscopy, as certain characteristics may be lost upon dissolution. However, extreme care is required to obtain artifact-free solid-state CD spectra. This is because CD spectra in the solid state (except for special homogeneous cases [9,10]) are inevitably accompanied by parasitic signals that originate from the macroscopic anisotropies of a sample such as LD (linear dichroism) and LB (linear birefringence) [11–16]. We have been working in the field of solid-state chirality for the last 30 years and recently developed a novel universal chiroptical spectrophotometer, UCS: J-800KCM, for the measurement of true CD and circular birefringence (CB) spectra in the solid state [17].

This article describes the unique features and the theory of solid-state optical activity, and the instrumentation of the UCS: J-800KCM. Several examples of chiral photochemistry and solid-state optical activity of inorganic and organic compounds are presented.

II. FEATURES OF SOLID-STATE CD

Generally, optical activity is observed when electrons are displaced along chiral paths by an applied electric field. This does not necessarily require the involvement of dissymmetric molecules. Quartz crystals, in which the first observation of optical activity was achieved, have no chiral molecules or ions, and their optical activity arises simply from the helical placement of atoms in the crystal. Likewise, a crystal composed of achiral untwistable molecules may become optically active if there are strong helical interactions between neighboring molecules.

Solid-state CD can provide information on solute–solvent interactions when compared with the solution spectra in various solvents. The effects of solvents on the rotatory power are often the results of the formation of some kind of coordination compound between the solvent and the optically active molecules concerned in solution [10,18]. This may affect the optical activity of the molecule by way of conformation alteration in the case of flexible compounds, or through vicinal effects. In contrast, in the solid state, molecules are densely packed and are under a much stronger influence of neighboring molecules. In one sense, this situation can be regarded as an extreme of the "solvent effect" [11]. Thus an unusual conformation of a chiral molecule that is unstable in solution may be realized and studied in the solid state. This kind of optical activitiy is specific to the crystalline state and will be lost immediately in liquid states, thus it has to be measured in the solid state.

Samples can be single crystals, microcrystallines, or films. One big advantage for studying the CD of crystals is that the location of relevant atoms and molecules can be revealed through x-ray structure analyses. Crystals can be classified as isotropic or anisotropic, and the latter further into uniaxial and biaxial classes. Cubic crystals belong to the isotropic category. Uniaxial crystals of tetragonal, trigonal, or hexagonal crystal systems have one optical axis, which is identical to the principal dielectric and crystal axes. Biaxial crystals of triclinic, monoclinic, and orthorhombic systems have two optic axes, which may have no simple relation with their crystal axes. The measurement of optical activity of crystals along directions other than the optical axes suffers from serious birefringence, which is generally three orders of magnitude larger than the optical activity. This is why single-crystal measurements have been regarded as almost impossible.

In microcrystallines where crystals are randomly oriented, the macroscopic anisotropies can be cancelled, and thus in principle their CD can provide reasonable solid-state spectra. However, artifacts of different types may exist in such cases (see V.B). Films usually exhibit large macroscopic anisotropies.

Generally, substantial artifacts accompany solid state CD, and hence the measurement of solid-state chiroptical properties using a commercially available CD spectrophotometer has to be carried out with extreme care. Because of this difficulty, there have been few publications regarding solid-state CD. Recently, the situation has changed, and the technique has suddenly become popular especially among organic chemists. Unfortunately, some of this work appears to have been carried out without due regard to the above-mentioned difficulties. Thus a special instrument that can deal with the fundamental difficulties specific to the solid state has been much sought after. We have designed and constructed a universal chiroptical spectrophotometer (UCS: J-800KCM), which can measure artifact signals and extract true CD signals from the recorded spectrum. Another CD instrument which is designed for biological specimens and for diffuse reflectance spectra is currently being developed in our laboratory.

The HAUP (high accuracy universal polarimeter), a life work of Jinzo Kobayashi [19], was designed to minimize the intrinsic strain of analyzer and polarizer and measures simultaneously the optical birefringence and the optical rotatory dispersion (ORD) of single crystals of any crystal system with macroscopic anisotropy. The HAUP is excellent in this regard, but it is limited to ORD measurement at nonabsorbing wavelengths. More informative LD and CD, which are localized at particular transitions, cannot be obtained. The experimental procedure is quite difficult, and thus the HAUP's use is limited to specialists.

III. THEORETICAL BACKGROUND

We employed the Mueller matrix method [20] and the Storks vector, $S = [S_0, S_1, S_2, S_3]$, where S_0, S_1, S_2, and S_3 denote total intensity, plus 45-degree preference, right circular preference, and horizontal preference of light, respectively. The Storks vector is useful in expressing the polarization state of light [21]. The Mueller matrix is the 4×4 matrix that was devised to express the optical characteristics of optical elements and samples [14,15,17–24]. Using the matrix calculation of $\hat{D} \cdot \hat{S} \cdot \hat{M} \cdot \hat{P} \cdot \hat{I}_0$, signals observed of a sample having CD, CB, LD, and LB can be expressed [17] as

$$
\begin{aligned}
I_d &= \hat{D} \cdot \hat{S} \cdot \hat{M} \cdot \hat{P} \cdot \hat{I}_0 \\
&= (1/2)I_0(P_x{}^2 + P_y{}^2)\{1 + (1/2)(\mathrm{LD'}^2 + \mathrm{LD}^2) \\
&\quad + [\mathrm{CD} + 1/2(\mathrm{LD'LB} - \mathrm{LB'LD})]\sin(\delta + \alpha) \\
&\quad + (\mathrm{LD'}\sin 2\theta - \mathrm{LD}\cos 2\theta)\cos(\delta + \alpha)\} + (1/2)I_0(P_x{}^2 - P_y{}^2)\sin 2a\{-(\mathrm{LD'}\cos 2\theta \\
&\quad + \mathrm{LD}\sin 2\theta) + (\mathrm{LB'}\sin 2\theta - \mathrm{LB}\cos 2\theta)\sin(\delta + \alpha) + [\mathrm{CB} + (1/4)(\mathrm{LD}^2 \\
&\quad + \mathrm{LB}^2 - \mathrm{LD'}^2 - \mathrm{LB'}^2)\sin 4\theta + (\mathrm{LD'LD} + \mathrm{LB'LB})\cos 4\theta]\cos(\delta + \alpha)\} \\
&\quad + (1/2)I_0(P_x{}^2 - P_y{}^2)\cos 2a\{(\mathrm{LD}\sin 2\theta \quad \mathrm{LD}\cos 2\theta) \\
&\quad + (\mathrm{LB'}\cos 2\theta - \mathrm{LB}\sin 2\theta)\sin(\delta + \alpha) + [1 + (1/4)(\mathrm{LD}^2 - \mathrm{LB'}^2 - \mathrm{LD'}^2 \\
&\quad + \mathrm{LB}^2)\cos 4\theta + (1/2)(\mathrm{LD'LD} + \mathrm{LB'LB})\sin 4\theta]\cos(\delta + \alpha)\}.
\end{aligned}
\tag{1}
$$

Here, I_0 and I_d are the intensity of incident light and light at the detector, respectively. $\hat{D}$, $\hat{S}$, $\hat{M}$, and $\hat{P}$ are Mueller matrices for the photomultiplier (PM), sample, photoelastic modulator (PEM), and polarizer, respectively. CB, LD', and LB' are circular birefringence, 45-degree linear dichroism, and 45-degree birefringence, respectively. $P_x{}^2$ and $P_y{}^2$ are the transmittances of the photomultiplier along the x and y directions, and a is the azimuth angle of its optical axis with respect to the x axis. θ is the rotation angle of the sample, and α is the residual static birefringence of the PEM. δ is the periodic phase difference between the x and y axes of the PEM operating at frequency $\omega_m/2\pi$ and is adjusted so as to act as a quarter-wave plate.

$$
\delta = \delta_m^0 \sin \omega_m t
\tag{2}
$$

Here, δ_m^0 is the peak modulator retardation, and $\omega_m/2\pi$ is usually set to 50 kHz. We can expand $\cos \delta$ and $\sin \delta$ in a Fourier series,

$$
\sin(\delta_m^0 \sin \omega_m t) = 2J_1(\delta_m^0)\sin \omega_m t + 2J_3(\delta_m^0)\sin 3\,\omega_m t + \cdots
\tag{3}
$$

$$
\cos(\delta_m^0 \sin \omega_m t) = J_0(\delta_m^0) + 2J_2(\delta_m^0)\cos 2\,\omega_m t + \cdots
\tag{4}
$$

$J_0(\delta_m^0)$, $J_1(\delta_m^0)$, and $J_2(\delta_m^0)$ are the Bessel function of the zeroth, first, and second order, respectively. Hence, the 50 kHz signal detected by the lock-in amplifier can be expressed as

$$\text{Signal}_{50\text{kHz}} = G_1\{(P_x^2 + P_y^2)\;[\text{CD} + (1/2)(\text{LD'LB} - \text{LB'LD}) + (\text{LD'}\sin 2\theta$$
$$- \text{LD}\cos 2\theta)\sin\alpha]\} + G_1(P_x^2 - P_y^2)\sin 2a\{\text{LB'}\sin 2\theta - \text{LB}\cos 2\theta$$
$$+ [-\text{CB} + (1/4)(\text{LD}^2 + \text{LB}^2 - \text{LD'}^2 - \text{LB'}^2)\sin 4\theta$$
$$+ (\text{LD'LD} + \text{LB'LB})\cos 4\theta]\sin\alpha\}$$
$$+ G_1(P_x^2 - P_y^2)\cos 2a\{\text{LB'}\cos 2\theta - \text{LB}\sin 2\theta$$
$$+ [1 + (1/2)(\text{LD'}^2 - \text{LB'}^2)\sin^2 2\theta + (1/2)(\text{LD}^2 - \text{LB'}^2)\cos^2 2\theta$$
$$- 2(\text{LD'LD} + \text{LB'LB})\sin 4\theta]\sin\alpha\} \tag{5}$$

G_1 is the apparatus constant related to the sensitivity of the spectrometer at 50 kHz. Terms multiplied by $\sin\alpha$ are negligibly small as a PEM having a smaller residual static birefringence ($\alpha = 0.2$) was used in our CD spectrophotometer (see IV.A.5). We can also neglect the contribution of the term containing $\cos 2a$ as the PM's azimuth angle was set so as to make $\cos 2a = 0$ in the baseline calibration (IV.A.5). Thus the 50 kHz signal is written as

$$\text{Signal}_{50\text{kHz}} = G_1(Px^2 + Py^2)[\text{CD} + (1/2)(\text{LD'LB} - \text{LDLB'})]$$
$$+ G_1(Px^2 - Py^2)\sin 2a(-\text{LB}\cos 2\theta + \text{LB'}\sin 2\theta). \tag{6}$$

Similarly, the 100 kHz component of the photocurrent detected by the lock-in amplifier is given as

$$\text{Signal}_{100\text{kHz}} = G_2(Px^2 + Py^2)(\text{LD'}\sin 2\theta - \text{LD}\cos 2\theta)$$
$$+ G_2(P_x^2 - P_y^2)\sin 2a\{-\text{CB} + (1/2)(\text{LD}^2 + \text{LB}^2 - \text{LD'}^2$$
$$- \text{LB'}^2)\sin 4\theta + (\text{LD'LD} + \text{LB'LB})\cos 4\theta\} \tag{7}$$

Here G_2 is the apparatus constant related to the sensitivity of the spectrometer at 100 kHz.

When an analyzer is inserted into the optical path, from a matrix calculation of $\hat{A} \cdot \hat{D} \cdot \hat{S} \cdot \hat{M} \cdot \hat{P} \cdot \hat{I}_0$, the 50 kHz signal detected by the lock-in amplifier can be expressed as

$$J_d = \hat{D} \cdot \hat{A} \cdot \hat{S}(\theta) \cdot \hat{M} \cdot \hat{P} \cdot \hat{I}_0$$
$$= 1/4\; e^{-Ae} I_0 \{(P_x^2 + P_y^2) + (P_x^2 - P_y^2)\sin 2a\}\{1 + (1/2)(\text{LD}^2$$
$$+ \text{LD'}^2) - (\text{LD'}\cos 2\theta + \text{LD}\sin 2\theta) + [\text{CD} + (1/2)(\text{LD'LB}$$
$$- \text{LDLB'}) - (\text{LB}\cos 2\theta + \text{LB'}\sin 2\theta)]\sin(\delta + \alpha)$$
$$- [(\text{LD'}\sin 2\theta + \text{LD}\cos 2\theta) - \text{CB} + (1/2)(\text{LD}^2 + \text{LB}^2 - \text{LD'}^2$$
$$- \text{LB'}^2)\sin 4\theta + (\text{LDLD'} + \text{LBLB'})\cos 4\theta]\cos(\delta + \alpha)\}. \tag{8}$$

G_3 is the apparatus constant related to the sensitivity of the spectrometer with the analyzer inserted, and $\hat{A}$ is the Mueller matrix for the analyzer. The above equation can be approximated as

$$\text{Signal}_{50\text{kHz}} = G_3\{(Px^2 + Py^2) + (Px^2 - Py^2)\sin 2a\}$$
$$\times \{\text{CD} + (1/2)(\text{LD'LB} - \text{LDLB'}) - \text{LB}\cos 2\theta + \text{LB'}\sin 2\theta\}. \tag{9}$$

Similarly, the 100 kHz signal is given as follows:

$$\text{Signal}_{100\text{kHz}} = G_4\{-LD'\sin 2\theta + LD\cos 2\theta + CB - (1/2)(LD^2 + LB^2 - LD'^2$$
$$- LB\sin 4\theta - (LDLD' + LBLB')\cos 4\theta\}$$
$$= G_4\{CB + (LD^2 + LD'^2)^{1/2} \cos(2\theta + \eta) - (1/2)[(LB^2$$
$$+ LB'^2) \sin(4\theta + \gamma) + (LD^2 + LD'^2) \sin(4\theta + \zeta)]\} \tag{10}$$

where

$$\eta = \tan^{-1} [LD/LD'] \qquad \gamma = \tan^{-1} [LBLB'/ (1/2)(LB^2 + LB'^2)]$$
$$\zeta = \tan^{-1} [LDLD'/ (1/2)(LD^2 - LD'^2)]$$

Here G_4 is the apparatus constant related to the sensitivity of the spectrophotometer at 100 kHz.

Equation (5) or (6) represents what we normally detect as a CD signal. It is obvious from these equations that in addition to the desired CD term, there are many other terms that arise from the macroscopic anisotropy of the sample. In solution, these terms vanish. Signals given by Eqs. (7), (9), and (10) are what we normally use for LD, LB, and CB (= ORD) measurements, respectively, all of which also contain artifact terms. As is easily seen in Eqs. (5)–10, there exist artifact terms that are independent of rotation of the sample about the incident light beam. Thus even if CD spectra are unaltered on rotating the sample about the light beam, this is not sufficient to prove that there are no artifacts in the spectra. In the past, it has been proposed that an average be taken of CD spectra recorded at different rotation angles θ to cancel LD and LB to give the true CD [25], but this assumption is incorrect [26]. It is recommended to measure LD and LB of a nonhomogeneous sample by an instrument such as the UCS:J-800KCM in order to evaluate the extent of macroscopic anisotropies and to remove these extra artifact terms.

IV.   INSTRUMENTATION OF UCS: A NEW INSTRUMENT WHICH IS CAPABLE OF MEASURING CD, LD, LB AND CB.

A.   Outline and Specific Points

Figure 1 shows a block diagram of UCS: J-800KCM together with the orientation axis of the optical and electric components. Specific points of improvement are as follows.

1. It houses two lock-in amplifiers to record two spectra simultaneously.
2. It has a specially designed sample rotation holder and a stage controller, which are required for the process of eliminating artifacts from the apparent CD spectra. The stage, which is driven by a pulse motor, can rotate in a plane perpendicular to the transmitted light.

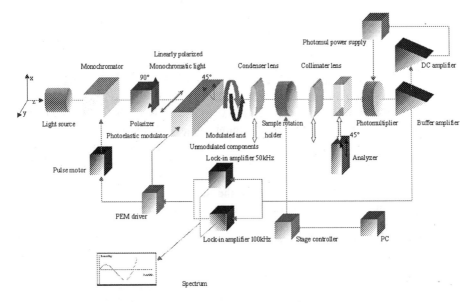

Figure 1 Block diagram of the solid-state dedicated CD spectrophotometer (J-800KCM). P: polarizer; M: photoelastic modulator (PEM); S: sample and rotation stage; L1, L2: lenses; A: analyzer (Glan–Tayler) D: photomultiplier (R376); MINI 12P: sample stage controller; lock-in amp1 JASCO; lock-in amp2: Stanford SRS 830; Recorder: SEKONIC SS-250F; PC: personal computer.

3. A sample holder was designed and built to enable measurement from both face and back sides of a solid sample. To cope with small solid samples, we designed and manufactured a two-lens unit that condenses the light to increase the light intensity that passes through the samples.

4. An analyzer and an analyzer stage are added for the simultaneous measurements of LB, LD, and CD. The analyzer can be rotated and easily removed from the light path.

5. A good quality PM and PEM are selected. For air blank, i.e., when CD = CB = LD = LB = 0, the CD baseline can be expressed as

$$CD_{base} = G_1(Px^2 - Py^2)\cos 2a \cdot \sin \alpha \qquad (11)$$

and the LD baseline is given as

$$LD_{base} = G_2(Px^2 - Py^2)\cos 2a \cdot \cos \alpha. \qquad (12)$$

In order to minimize the CD and LD baseline shifts that influence the performance of the instrument, it is necessary to select a PM having the least polarization characteristics and a PEM having the least residual static birefringence. We have selected the best PEM out of many, whose residual static birefringence was 0.2 degree. After selecting the best optical elements, the optical axis of the PM was set at 45 degrees with respect to the x and y axes, i.e., cos $2a = 0$. By this method, it is possible to minimize the CD and LD baselines shifts. The LD baseline shift is larger than that of CD as cos α nearly equals to 1 [Eqs. (11), (12)]. Thus we recommend minimizing the 100 kHz signal by carefully rotating the PM.

6. Addition of a phased-lock-loop (PLL) circuit to the PEM driver: the PLL circuit made especially for the UCS: J-800KCM is able to reduce the phase difference between the load current and the PEM-driving voltage to zero. The PEM is thus efficiently driven at lower voltage, and the lag of resonance frequency related to thermal change is decreased. Consequently, the stability of the PEM driver was raised [27].

B. Methods of Measurement Using UCS: J-800KCM

As seen above, signals for CD [Eq. (6)], LD [Eq. (7)], LB [Eq. (9)], and CB [Eq. (10)] contain artifacts that are either independent of or dependent of rotation of the sample about the incident light. If the CD value changes on rotating the sample 360 degrees around the z axis, it is clear that the macroscopic anisotropies contribute to the CD spectrum. From the measurements, we can determine the degree of contribution of LD and LB signals to the observed CD signals. As is obvious from Eq. (6), if LB and LD signals are negligibly small compared with CD, or if there are no macroscopic anisotropies as in solution samples, the signal observed is a true CD. If the 50 kHz spectra obtained with and without the analyzer are different, the 50 kHz signal detected without the analyzer contains an apparent CD signal and that detected with the analyzer is an LB signal. If the signal changes with cos 2θ periodicity at the wavelength where there is no absorption of the sample, it is obvious that the contribution of LB is substantial. If the wavelength-scan spectra with and without the analyzer are virtually the same, Signal$_{50kHz}$ gives a CD without artifacts. If they are different, Signal$_{50kHz}$ with an analyzer gives an LB signal and that without an analyzer gives an artifacts-contained CD.

1. CD Measurement

As seen in Eq. (6), the apparent CD component of $(1/2)(LD'LB-LDLB')$ is independent of the rotation of the sample, whereas terms including the coupling of LB with the polarization characteristic of the photomultiplier change with rotation of the sample. We have developed a set of procedures for the measurement of true CD, as follows (17).

Step 1: With an analyzer, LB measurement was carried out while rotating the sample through 360 degrees in the (X–Y) plane at the wavelength of an absorption maximum, λ_{max}. Then the LB spectrum was obtained by the wavelength scanning at the LB_{max} position.

Step 2: Without an analyzer, LD was measured with rotation of the sample at λ_{max}. The LD spectrum was obtained by wavelength scanning at the LD_{max} position.

Step 3: From the data obtained in step 1, we can locate the LB_{max} and LB_{min} positions. The sample was rotated 45 degrees from the LB_{max} position, whereby the LB value becomes 0 and the LB′ value a maximum. The wavelength scan was then carried out without an analyzer. From Eq. (6), the apparent CD signal of the face side is given as

$$[appCD]_{face} = G_1\{(P_x^2 + P_y^2)[CD - (1/2)LDLB'] + (P_x^2 - P_y^2)\sin 2aLB'\} \quad (13)$$

Step 4: The sample was then rotated by 180 degrees about the Y axis and a wavelength scan was carried out. This corresponds to the backside measurement. By this rotation, the CD and LD do not change their signs, but LB′ becomes $-$LB′. Hence the apparent CD signal of the backside becomes

$$[appCD]_{back} = G_1\{(P_x^2 + P_y^2)[CD + (1/2)LDLB'] - (P_x^2 - P_y^2)\sin 2aLB' \quad (14)$$

It is obvious that the average of Eqs. (13) and (14) gives Eq. (15), i.e.,

$$G_1(P_x^2 + P_y^2)\ CD. \quad (15)$$

Thus measurement of $[appCD]_{face}$ and $[appCD]_{back}$ spectra using our special sample holder, and taking an average of the two spectra, should give the true CD spectrum.

2. CB Measurement

The 2ω (100 kHz) signal with an analyzer contains not only CB but also LD and LB terms and changes with θ during the sample rotation at an arbitrary wavelength. If the contribution of LD is much larger than LB, the signal changes with $\cos 2\theta$ periodicity, whereas if the contribution of LB is much larger than LD, the change follows $\sin 4\theta$ periodicity [Eq. (10)]. Generally, LD is ten times smaller than LB, and thus Eq. (10) can be approximated as

$$\text{Signal } 2\omega\ (100\text{ kHz}) = G_0\{CB - (1/2)(LB^2 + LB'^2)\sin(4\theta + \gamma)\} \quad (16)$$

As the CB signal is usually 10^2 to 10^3 times smaller than LB and LB′ signals, and thus is buried under large macroscopic anisotropy signals, it is difficult to detect. However, we have found that the following set of measurements makes this possible. At its (+) extremum position during sample rotation, a signal 2ω becomes

$$(+)\text{ext } [\text{Signal } 2\omega] = G_0'\{CB + (1/2)(LB^2 + LB'^2)\} \quad (17)$$

At the negative extremum position, or at the position rotated 45° from the positive extremum position if the signal changes with sin 4θ periodicity, the signal detected is given as

$$(-)\text{ext [Signal } 2\omega] = G_0'\{CB - (1/2)(LB^2 + LB'^2)\} \tag{18}$$

Thus if we average the $(+)$ext [Signal 2ω] and $(-)$ext [Signal 2ω] spectra, we can obtain the CB.

V. APPLICATION

A. Single Crystals

As optic and crystal axes are common in the case of uniaxial crystals, single-crystal CD experiments are generally limited to these kinds of crystals. For cases where moderate macroscopic anisotropies exist, these can be studied on the UCS: J-800-KCM as is described in V.A.3.

1. Uniaxial Single Crystals: Comparison with Solution CD

Optical activity of single crystals is expected to provide valuable information such as band assignments and the magnitude of rotatory strengths, R. Transition metal complexes of D_3 symmetry often crystallize in a uniaxial system [9,10]. When all the complex ions in the crystal are oriented with the molecular trigonal axis parallel to the optic axis, circularly polarized light propagating along the optic axis excites only the E component, providing the rotational strength of the doubly degenerate $^1A_1 \rightarrow ^1E$, $R(E)$. Figure 2 compares the single-crystal and the corresponding solution CD for the tris-bidentate metal complex [Co(R-pn)$_3$]Br$_3$·3H$_2$O (pn = propylenediamine) and bis-bidentate complex [Co((2R-Me)-tacn)$_2$]I$_3$·5H$_2$O ((2R-Me)-tacn = (2R)-2-methyl-1,4,7-triazacyclononane). The intensities of the solid- state spectra are an order of magnitude larger than those of the solution spectra. In the case of solution CD of [Co(R-pn)$_3$]Br$_3$·3H$_2$O, there is one negative and one positive band in the first d-d transition region, with the negative band at lower energy. Single-crystal CD exhibits one negative peak [9,10]. Thus the longer wavelength band was conclusively assigned as having E symmetry, which is positive for the absolute configuration Λ, and negative for Δ. The values of $\Delta\varepsilon_{max}$ and $R(E)$ for crystals are an order of magnitude larger than those for the solution spectrum, clearly revealing that the solution spectrum is a residual wing absorption resulting from extensive cancellation of large positive and large negative peaks. In solution, $| R(A_2) |$ is generally smaller than $| R(E) |$.

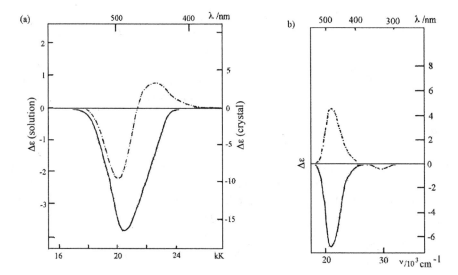

Figure 2 Comparison of a single-crystal (—) and an aqueous solution (······) CD of (a) δ − [Co(R-pn)$_3$]Br$_3$·H$_2$O and (b) [Co((2R-Me)-tacn)$_2$ I$_3$·5H$_2$O]. The CD of the single crystal recorded with radiation along the crystal c axis.

Single crystal CD of the bis-tridentate complex, [Co((2R-Me)-tacn)$_2$] I$_3$·5H$_2$O, exhibits one negative band near 21,000 cm^{-1}, in contrast to showing only one positive peak in the solution CD spectrum (Fig. 2b), as if the absolute configuration were opposite in the two cases [28]. This striking result was explained by the x-ray crystal structure of the complex [29]. The coordination octahedron Co^{III}N$_6$ is significantly distorted from a regular octahedron (i.e., a polar elongation and a clockwise azimuthal rotation of the coordination triangles), and the ligand groups are predominantly located in the polar region of the D_3 chromophore rather than the equatorial region as is found in many tris(diamine)cobalt(III) complexes. Thus the unusual relation of $| R(A_2) | > | R(E) |$ holds in the first absorption region, and the week E band is completely submerged under the intense positive A_2 band. In contrast, single-crystal CD along the c axis excites only the E bands.

2. Decomposition of CD Spectra into the E and A_2 Components

When the threefold axis of a complex ion is inclined at an angle α with respect to the principal axis of a uniaxial crystal, both the E and A_2 bands are excited with light propagated along the optic axis. We have proposed and successfully applied a method to obtain ν_{max} and R for both of the components by coupling the CD of oriented and randomly oriented samples, i.e., single-crystal and micro-crystalline CD spectra of uniaxial crystals [9,10]. In general, observed rotatory strengths of the decomposed E and A_2 components agree well with theoretical calculations, particularly in the case of five-membered chelate ring systems [9,10,30].

3. Single Crystals with Macroscopic Anisotropies

Single crystals with modest macroscopic anisotropies have been studied [17,31]. To test our instrumentation and measurement procedures described in Sec. IV.B, macroscopic anisotropies were introduced into a uniaxial crystal, α-Ni(-H$_2$O)$_6$·SO$_4$, which belongs to the enantiomorphous tetragonal space group of either $P4_12_12$ or $P4_32_12$ [32]. Under a polarizing microscope, single crystals of α-Ni(H$_2$O)$_6$.SO$_4$ were polished to make a thin plate perpendicular to the fourfold screw axis. The plate planes were then slightly tilted from the plane perpendicular to the unique axis of the crystals so as to introduce macroscopic anisotropies ($LB_{max}^{390} = 14.2°$, case II). A sample with the smallest possible degree of tilting was used as a reference (case I, $LB_{max}^{390} = 0.6°$). Crystals were mounted on the special solid sample holder and measured with the UCS: J-800KCM.

CD Measurement [17]. We followed the methods described in Sec. IV.B.1 to obtain the true CD. Figures 3a and 3b show the LB and LD spectra obtained at the LB$_{max}$ and LD$_{max}$ positions, respectively, for case I ($LB_{max}^{390} = 0.6°$) and II ($LB_{max}^{390} = 14.2°$). The signal$_{50kHz}$ without an analyzer was recorded by rotating the sample at the absorption maximum. If the CD signal changes on rotating the sample 360 degrees around the z axis, it is clear that the macroscopic anisotropies contribute to the CD spectrum. From the measurements, we can estimate the extent of contribution of LD and LB signals to the observed CD signals.

Figure 3c shows a CD spectrum at an arbitrary position. As is obvious from Eq. (6), if LB and LD signals are negligibly small compared with the CD, or if there are no macroscopic anisotropies as in solution, the signal observed is a true CD. In the case of a larger LB sample (case II), by comparing Fig. 3IIa and 3IIc, we could recognize that the LB signal was larger than the CD signal, and the 2nd and 3rd terms in Eq. (6) contributed to the CD spectrum. By averaging the two spectra, [appCD]$_{face}$ and [appCD]$_{back}$, we successfully eliminated extra terms and obtained a true CD spectrum. The true CD and apparent CD spectra are compared in Fig. 3IIc.

In the case of a smaller LB sample (case I), both LD and LB signals are negligibly small compared with the CD signal. The apparent and true CD spectra are virtually the same (Fig. 3Ic), and the LB spectrum (Fig. 3Ia) is also almost

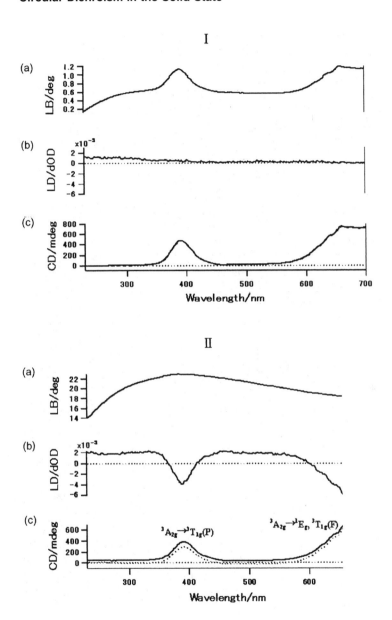

Figure 3 Spectra of a single crystal of α-Ni(H$_2$O)$_6$·SO$_4$ with small LB (case I) and substantial LB (case II). For the two cases, (a) LB spectrum at LB$_{max}$ position, (b) LD spectrum at LD$_{max}$ position, and (c) CD spectrum at arbitrary angle (dotted line) and true CD spectrum (solid line).

identical to the CD. The CD in case I is similar to the corrected CD of case II, assuring that our CD instrumentation and measurement procedures are correct. The method is best applied when the LB value is not too large, i.e., less than ~30° at 390 nm.

CB Measurement [31]. We have followed the methods described in Sec. IV.B.2 to obtain the true CB. In case **II** of $LB_{max}^{390} = 14.2°$, the 2ω signal with an analyzer (i.e., CB) during the sample rotation changed with cos 4θ periodicity (Fig. 4IIa), indicating that LB rather than LD is dominant in this case. In contrast, in the case where the crystal was mounted almost perpendicular to the light (case **I**), the 2ω signal hardly changed on rotation (note that the ordinate scale for Ia is more than 30 times that of Ib), with roughly cos 2θ periodicity (Fig. 4Ia).

With an analyzer, wavelength scans were carried out at the positive and negative maxima positions to give (+)ext (Signal 2ω) and (−)ext (Signal 2ω) spectra, respectively, and they are shown for the two cases in Fig. 4b. In case I, the (+)ext (Signal 2ω) and (−)ext (Signal 2ω) spectra are almost identical (Fig. 4Ib), whereas they are quite different in case II (Fig. 4IIb). The difference corresponds to the signal arising from the macroscopic anisotropy. Fig. 4Ic and 4IIc show the true CB obtained by averaging the (+)max and (−)max spectra.

Kramers–Kronig Relationship [31]. To test the reliability of our CD and CB spectra, we have checked the Kramers–Kronig relationship focusing on the $^3A_{2g} \rightarrow {}^3T_{1g}(P)$ Ni *d-d* transition at 390 nm. Drude's dispersion formula was employed to take into account all the contributions other than the *d-d* absorption concerned. Curve fitting of the CB spectra in the 250–600 nm region was carried out, using the formula CB $= A/(\lambda^2 - 0.1876^2) + B/(\lambda^2 - 0.6473^2) + C/(\lambda^2 - 0.7117^2) + D$, as shown in Fig. 4Ic and Fig. 4IIc. Here, 187.6, 647.3, and 711.7 nm are the bands associated with the oxygen transition [33], and the Ni *d-d* transitions $^3A_{2g} \rightarrow {}^3T_{1g}(F)$ and $^3A_{2g} \rightarrow {}^1E_g$ in the octahedral parentage [32], respectively. Thus the CB corresponding to the $^3A_{2g} \rightarrow {}^3T_{1g}(P)$ Ni *d-d* transition (at 390 nm) was calculated by eliminating the above: mentioned Drude's dispersion expression from the artifact-free CB spectrum. This is shown in Fig. 4Id and 4IId, together with true CD. The CB spectra in the *d-d* region obtained from the measurement agree well with CB spectra calculated by assuming the Kramers–Kronig relationship. The calculated CB spectra are also presented in the figures for comparison. Thus the Kramers–Kronig relation holds between the CB and the CD (Fig. 4Id and 4IId). The negative CB and positive CD around 550–600 nm are the tail of the transition $^3A_{2g} \rightarrow {}^3T_{1g}(F)$ at 647 nm. Similarly, the negative CB below 280 nm is the tail of the transition at 188 nm.

This is the first case for which the Kramers–Kronig relation has been demonstrated experimentally in the solid state. If the uncorrected CB spectrum was used, the Kramers–Kronig relation did not hold.

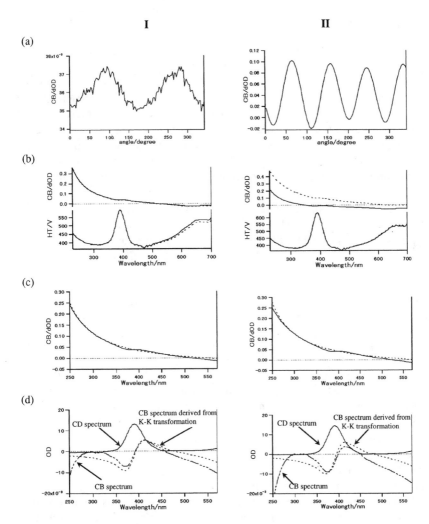

Figure 4 Spectra of a single crystal of α-Ni(H₂O)₆·SO₄ with small LB (case I) and substantial LB (case II). For the two cases, (a) change of 2ω signal at 300 nm during the rotation of the sample; (b) wavelength scan spectra at the (+)ext and (−)ext positions in (a); and (c) CB spectrum obtained by averaging the (+)ext and (−)ext spectra. The curve fitted to the Drude-type analytical expression, CB = $A/(\lambda^2 - 0.1876^2)$ + $B/$ $(\lambda^2 - 0.6473^2)$ + $C/(\lambda^2 - 0.7117^2)$ + D, in the 250–600 nm region is also shown. Contribution from the Ni d-d transition centred around 390 nm was omitted from the expression. A, B, C, and D are specific to a sample and its orientation with respect to light. They are 0.076981, 9.53 × 10⁻⁷, 0.000972, and −0.025034, respectively, for case I, and 0.0082421, 3.138 × 10⁻⁶, 0.0014136, and −0.020332, respectively, for case II. (d) Pure CD and CB spectra of α-Ni(H₂O)₆·SO₄ associated with the ³A₂g → ³T₁g(P) Ni d-d transition, satisfy the Kramers–Kronig relationship. The CB spectra calculated from the CD assuming the Kramers–Kronig relation are also shown for comparison.

B. Microcrystalline CD

Most organic compounds crystallize in the monoclinic or orthorhombic crystal systems, which contain substantial macroscopic anisotropies, and thus the single-crystal CD technique cannot be applied, although our method of measurement may be useful if the macroscopic anisotropies are not very large. An alternative way to obtain crystalline-specific information is to examine the microcrystalline state. Measurements can be usually made either in nujol mull or KBr disc form, where microcrystals are dispersed randomly either in nujol or in a KBr microcrystalline matrix. The method was developed and applied to inorganic complexes by Mason [34], Bosnich [35], and Kuroda [9,10], and since then its application to metal complexes has been carried out sporadically [36,37]. Recently, the technique has become popular in the field of organic chemistry as well, probably stimulated by our work [38]. An alternative technique recently developed by us is diffuse reflectance CD (DRCD) which will be briefly described in V.B.2.

1. Possible Artifacts

Although it is simple to record spectra, great care is needed to obtain spectra with as few artifacts as possible. Some possible artifacts are described here.

Dispersion Effects (9). Depolarization at grain boundaries of a sample may cause serious experimental problems. Even with an apparently translucent disk or homogenous looking nujol mull, there may be a large depolarization of the light beam due to reflection and refraction at the grain boundaries. This effect seems to increase with increasing inhomogeneity of grain size, grain size itself, and the turbidity. The smaller the particle size, the smaller is the depolarization of the circular polarized light during transmission through the disc [9,11]. In the case of the nujol mull method, generally, the particle size cannot be made smaller than the KBr matrix. Unfortunately, the effects of scattering are seen in many published microcrystalline CD spectra.

Figure 5a shows CD spectra of tartaric acid, which has an absorption in the short wavelength region and thus is prone to suffer from dispersion effects as compared with transition metal complexes. Two solution spectra in solvents of different polarity, water and dioxane, are similar to each other, but the CD of a nujol mull is quite different from that in solution. A KBr disc prepared to avoid dispersion effects gave a solid-state tartaric acid spectrum similar to that in solution (Fig. 5b). Thus the difference between the nujol mull CD and solution CD is not due to the different molecular conformation or intermolecular interaction in the two phases. Most likely, it is due to the dispersion effect in the case of the nujol mull form. Many nujol mull CD spectra of organic compounds have been reported recently, but most of them appear to suffer from substantial dispersion effects. It is to be noted that the dispersion terms for molecules of

a

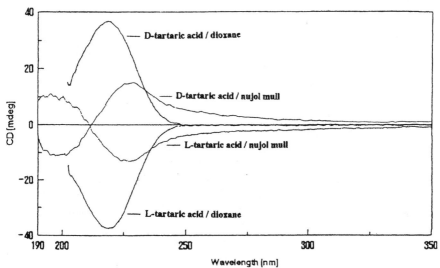

b

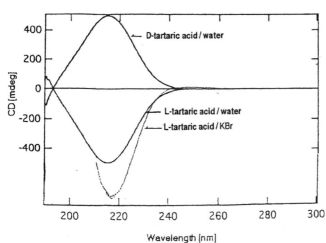

Figure 5 Comparison of solution CD and two microcrystalline CD spectra of d- and l-tartaric acid. (a) In dioxane solution and as a nujol mull, and (b) in aqueous solution and as a KBr matrix.

opposite handedness are mirror images of each other. Thus there is no guarantee that the recorded CD spectrum is a true one even when the enantiomer exhibits a mirror image spectrum.

Interaction with Matrix and Pressure Effect. Hydrophobic compounds may dissolve in nujol, although the solubility is only slight and may not be noticeable. When high-quality nujol mull CD spectra with a low noise level are obtained, it is worth checking whether the compound dissolves in nujol or not.

It has also to be borne in mind that the KBr matrix method may suffer from interaction of the compound concerned with the K^+ or Br^- ions in the matrix, or from the effects of pressure during the disc preparation. The effect of pressure applied during the KBr disc preparation was tested for an inclusion compound but this was not observed in this case [39]. The interaction with KBr has been regarded as rather rare, but an interesting paper was published recently by Braga et al. [40]. They report an unexpected solid–solid reaction upon preparation of KBr pellets: solid $[Co^{III}(\eta^5\text{-}C_5H_4COOH)(\eta^5\text{-}C_5H_4COO)]$, **1**, which carries a protonated and a deprotonated COO group, formed the supramolecular complex $\mathbf{1}_2\cdot K^+Br^-$. The K^+ cations were encapsulated within a cage formed by four zwitterionic molecules, dimerised via O–H $\cdots$ O hydrogen bonds.

LD and LB. As is seen in Eq. (6), even when the spectrum is independent of sample rotation, it does not necessarily mean that the recorded spectrum is a true CD. If the samples are well made, the order of LD was 3×10^{-3}, and LB less than 3 degrees, which can be regarded as negligibly small [41].

2. Diffuse Reflectance CD (DRCD)

Recently, Castiglioni, et al. have devised a new method, diffuse reflectance CD (DRCD), to measure the solid-state CD of a pure microcrystallite sample or one mixed with KBr [42,43] and checked their performance. An integrating sphere used for UV-Vis and IR spectroscopy was introduced (Fig. 6). The advantage of this technique is to remove possible artifact elements such as reactions with the KBr, the effect of pressure required for the pellet preparation, or sample dissolution in nujol. Further, samples can be recovered after the spectral measurement, in contrast to both KBr pellet and nujol mull methods.

Figure 7 shows DRCD spectra taken by our newly developed instrument which has about ten times more light efficiency than the instrument previously made [44]. D- and L-ammonium camphorsulfonate exhibited mirror-image spectra, and these are similar to their respective solution spectra.

3. Examples of Different Conformation in Solid and Solution

$[Co(R,R\text{-ptn})_3]^{3+}$ (ptn = 2,4-diaminopentane) shows quite different spectra in the solution and in the microcrystalline states (Fig. 8). Further, in solution, the

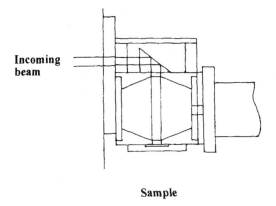

Incoming beam

Sample

Figure 6 Diffuse reflectance sphere attachment of a DRCD instrument.

spectra are dependent on the counterions: the perchlorate salt exhibits only one negative band, whereas the chloride salt exhibits an extra small positive peak at the high-energy side [45]. This is in sharp contrast to $[Co(S,S\text{-}chxn)_3]^{3+}$ (chxn = cyclohexanediamine), which exhibits similar spectra for microscrystalline CD and solution CD with the chloride and perchlorate counterions. The CoN_6 chromophore of $[Co(R,R\text{-}ptn)_3]^{3+}$ is not greatly distorted from the regular octahedron, and thus its optical activity arises from the flexible six-membered chelate ring atoms [46]. As a result, the CD of $[Co(R,R\text{-}ptn)_3]^{3+}$ is much weaker and is

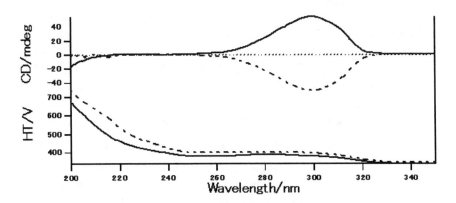

Figure 7 DRCD spectra of (+)- and (−)-ammonium camphorsulfonate.

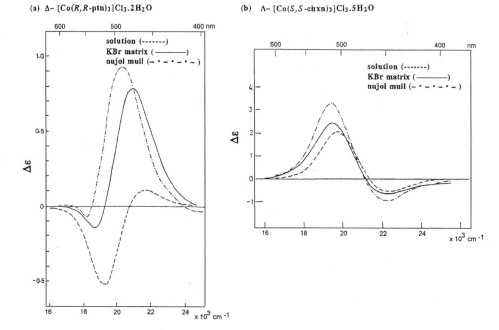

Figure 8 CD spectra of (a) $[Co(R,R\text{-ptn})_3]^{3+}$ and (b) $[Co(S,S\text{-chxn})_3]^{3+}$ in solution (---), in a KBr matrix (—), and as a nujol mull (- · - · -).

more easily influenced by the environment than the five-membered chelate ring complexes.

The conformation of this flexible ligand in solution must be different from that in the crystalline state. In general, to correlate CD spectra with structures that are revealed by single-crystal x-ray diffractometry, it is desirable to measure not only the solution but also the microcrystalline CD, although great care is required for the latter experiment. In this regard, our newly developed diffuse reflectance CD spectrophotometer is ideal.

C.   Chiral Photochemistry in the Solid State

1.   CD of Prochiral Compounds in the Solid State

2-Thioaryloxy-3-methylcyclohexen-**1**-ones, **2** (Scheme 1), are photocyclized to produce chiral dihydrobenzothiophene derivatives, **3**, on UV irradiation. When the reaction is carried out in solution, a racemic mixture of ($-$) and ($+$)-**3** is obtained. However, nonchiral molecules may become optically active in the solid

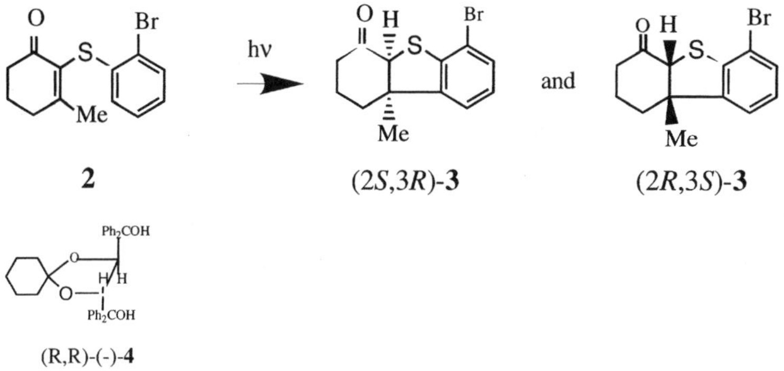

Scheme 1

state when the degrees of freedom of molecules, e.g., rotation and orientation about chemical bonds, are severely restricted by the formation of a crystal lattice. Several enantioselective photoreactions of prochiral compounds in the solid state [47] either in their own crystal lattices [48] or in inclusion compounds with optically active host molecules have been reported [49].

8-Bromo-2-thioaryloxy-3-methylcyclohexen-1-one, **2a**, is an achiral compound. Its conformation can be fixed in a chiral twisted form in the solid state, but it forms racemic crystals, and hence the solid-state reaction produces a racemic mixture of (−) and (+)**3a**. A method for achieving an enantioselective reaction is to form an inclusion complex with a chiral host compound, (−)-(R,R)-(−)-trans-2,3-bis(hydroxy diphenyl methyl)-1,4-dioxa spiro-[4,4]nonane, (−)-**4**. In a typical experiment, a benzene (5 mL) and hexane (5 mL) solution of (−)-**4** and **2a** left at room temperature for 24 h gave a 1 : 1 inclusion compound of (−)-**2a** and **4** as colorless crystals [38].

2a, as expected, did not show any CD peak in the 450–250 nm region in solution. The 1 : 1 inclusion crystal **2a**·(−)-**4** exhibited substantial CD peaks in a nujol mull or a KBr matrix (Fig. 9a). The two solid-state spectra resemble each other. When dissolved, the inclusion compound did not exhibit CD peaks in the 400–280 nm region even at higher concentration, suggesting that there is no interaction between the host ((−)-**4**) and the guest (**2a**) molecules in solution, and that the solid-state CD observed in the absorption region mainly originates from **2a** under the influence of molecules (−)-**4** in the solid state [50]. The KBr disc CD spectrum in the host region is similar to that in acetonitrile solution, assuring that the quality of our KBr disc is sufficient for the CD measurement. The small wavelength shift in the very short wavelength region (Fig. 9b) is most likely due to the dispersion effect of the disc. Its LD spectra were measured to

(a)

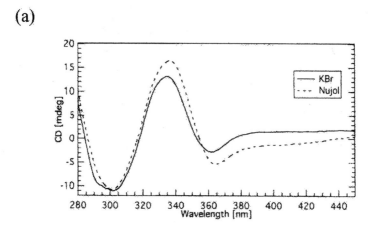

(b)

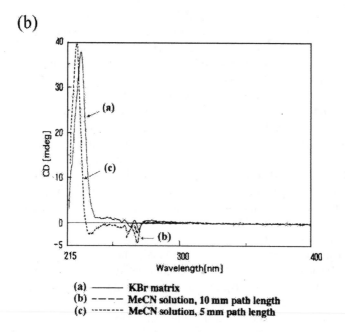

(a) ———— KBr matrix
(b) – – – – MeCN solution, 10 mm path length
(c) -------- MeCN solution, 5 mm path length

Figure 9 (a) Solid-state CD of **2a·(−)4** in a KBr matrix and as a nujol mull in the 280–450 nm region. (b) Comparison of solid-state CD of **2a·(−)4** in a KBr matrix and in an acetonitrile solution in the 215–400 nm region. For the shorter wavelength region, a 5 mm path length cell was used.

show that the intensities were three orders of magnitude smaller than those of the CD [50]. When cocrystallized with the host of opposite enantiomer, an almost mirror image solid-state CD spectrum was obtained.

The measured crystal optical activity, in general, can be either of molecular origin or due to the chiral helical arrangement of chiral or achiral molecules in the crystal, or both. The two factors are difficult to separate. Kobayashi defined a chirality factor $r = (\rho c - \rho s)/\rho c = 1 - \rho s/\rho c$, where ρc is the rotatory power per molecule of a randomly oriented crystal aggregate derived from the gyration tensors determined by HAUP, and ρs that in solution [51]. It is a measure of the "crystal lattice structural contribution" to the optical activity and represents the severity of the "crystal lattice structural contribution" to the optical activity, and represents the severity of the restriction of the freedom of molecular orientation by forming a crystal lattice. Quartz is a typical example of $r = 1$, as it does not contain chiral molecules or ions and its optical activity vanishes in random orientation ($\rho s = 0$).

2. X-Ray Structures

X-ray crystal structural analysis of the inclusion complex **2a**· $(-)$-**4** has been carried out [38]. Two independent host or guest molecules adopt similar molecular conformations. The crystal is made up of stacks of layers in a direction perpendicular to the crystal b axis, i.e., a layer of H1, a layer of G1 and G2, and a layer of H2 (H = host, G = guest). There are strong contacts between H1 and G1 and between H2 and G2. The two hydroxyl groups of the host seem to play a key role in the stereoselectivity of the photoreaction; the molecular conformation of the host is dictated by intramolecular hydrogen bonding between the two hydroxyl groups (O3–O4 = 2.62 Å for both H1 and H2). One of the hydroxyl groups additionally acts as a hydrogen bond donor to the carbonyl oxygen of a guest molecule (O4–O1G = 2.72 Å for H1-G1, 2.70 Å for H2-G2). As a consequence, the guest molecule is highly twisted in one direction with the central C3–C2–S1–C8 torsion angle of $-112.7°$ or $-116.6°$ (Fig. 10a). The photoreactive carbon C12 is either 3.71 or 3.74 Å away from the target C3 on one side of the cyclohexenone ring plane, favoring the R configuration at C3. A crystal structure of the reaction product, $(+)$-**3a**, unambiguously established the absolute configuration of the cis isomer as 2S,3R. Thus the conformation of the guest molecule imposed by the crystal structure explains the high enantioselectivity of the reaction. The average ring plane of the guest containing the carbonyl group is almost parallel to the neighboring phenyl group of another juxtaposed host molecule.

Compound **2a** usually crystallizes to produce racemic crystals. However, we have recently succeeded in obtaining chiral crystals of **2a** in the presence of certain organic compounds and called this "chirality scaffolding crystallization."

(a)

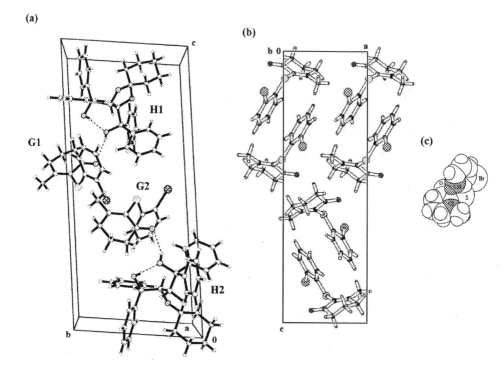

Figure 10 Molecular structure of **2a** and crystal packing of (a) the chiral inclusion crystal **2a**·(−)**4** and (b) the chiral crystal of **2a** on its own. (c) Space-filling representation of molecular structure of **2a** as found in the two crystal structures. Conformations are similar in the two cases.

Generally speaking, the mechanism of crystal formation is not well understood, and the scaffolding is no exception. Whether the added compounds affect the nucleation and/or growth steps is not known [53]. The molecular conformation in the chiral crystal is similar to that in the inclusion crystal and is twisted in one direction with a torsion angle of 107.1° (Fig. 10b). The photoreactive carbon C12 is 3.727 Å from the target C3 on one side of the cyclohexenone ring plane, favoring the *S* configuration at C3 in this crystal.

3. Solid-State Photoreaction

Irradiation of the inclusion crystals of **2a**· (−)-**4** suspended in H$_2$O (containing hexadecyl trimethylammonium bromide) with stirring for 30 h gave a crude reaction product, photocyclization products (+)-**3a**, as crystalline materials. This was filtered, air dried, and distilled at 250°C/2 mm Hg to give (+)-**3a** (82% enantiomeric excess (*ee*), 83% yield, mp 129–131°C) [38]. Although optical purities of **3a** obtained by the photocyclization reaction were not very high, the crystalline compounds can be easily purified by recrystallization to give optically pure samples. Thus recrystallization of (+)-**3a** of 83% *ee* (0.094 g) from *n*-hexane gave enantiopure (+)-**3a**. The corresponding solution reaction yielded racemic **3a**. The solid-state photoreaction of other derivatives of **2** gave *ee*s of 32 ~ 77% with an average of 59%.

Reactions using chiral crystals of **2a** were carried out using one single crystal in order to avoid the mixture of crystals with opposite chirality. The *ee* of several independent reactions was 56% on average [53]. The smaller value compared with that seen for the inclusion crystals must be due to weaker confinement of molecules in the crystal lattice compared with the case of the inclusion crystal, which has hydrogen bonds between host and guest molecules.

The solid-state photoreaction using chiral crystals is an absolute asymmetrical synthesis: crystallization of a nonchiral compound in a chiral crystal followed by a topochemical photoreaction. A similar interesting case of absolute asymmetrical synthesis has been reported by others including Addadi et al. [54].

4. Monitoring a Solid-State Reaction by Solid-State CD

Solid-state reactions can be monitored by measuring solid-state CD spectra at time intervals. An example is given for the photoreaction of the inclusion crystals (Fig. 11). A series of spectra obtained during the photoreaction with 365 nm light for different time periods in a KBr disc showed isosbestic points [50]. This monitoring method for a reaction in a KBr matrix is convenient and requires only a small amount of sample. From these good-quality spectra, information on the reaction mechanism may be acquired with the help of x-ray structure analysis. We have found that the solid-sate reaction proceeds in a single-crystal to single-crystal manner, the details of which will be reported elsewhere.

D. Films

Films, cast films, and LB membranes can provide useful information on aggregates, but these generally possess substantial anisotropies. For example, achiral PVA film dyed with Congo red exhibits substantial CD, as shown in Fig. 12. We have succeeded in eliminating the artifact CD signals [17]. A PVA film was stretched in order to make it optically homogeneous, and it was assumed that there is no face side-to-back side difference. In the homogeneous case, and when the detector is ideal, the 50 kHz signal detected by the photomultiplier is expressed as

$$\text{Signal}_{50\text{kHz}}^{\text{face}} = G_1 \, [\text{LD'LB} - \text{LB'LD}] \tag{19}$$

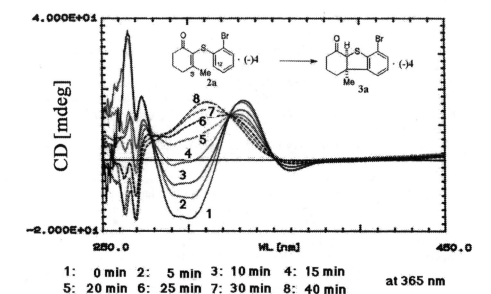

1: 0 min 2: 5 min 3: 10 min 4: 15 min
5: 20 min 6: 25 min 7: 30 min 8: 40 min   at 365 nm

Figure 11 Photochemical reaction of **2a** to **3a** in an inclusion complex **2a**·(−)**4** followed by CD in a KBr matrix.

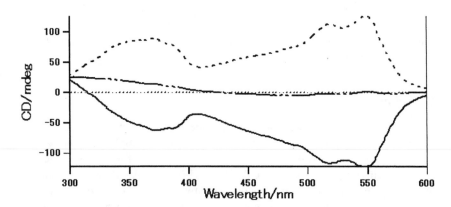

Figure 12 CD spectra of a stretched PVA film dyed with Congo red. The top and bottom spectra are the apparent CD of face side and back side, respectively. The middle trace is the true CD spectrum.

When we rotate the sample 180 degrees with respect to the vertical (y) axis, LD and LB do not change their signs, but LD' and LB' become $-$LD' and $-$LB', respectively. Therefore the 50 kHz signal of the back side detected by the photo-multiplier is expressed as

$$\text{Signal}^{\text{back}}_{\text{50kHz}} = G_1 \; [-\text{LD'LB} + \text{LB'LD}] \tag{20}$$

If we add Eqs. (19) and (20), the sum is 0. Thus the apparent CD spectra due to macroscopic anisotropies should cancel out by this procedure.

This is what we have actually observed. As shown in Fig. 12, the rather strong CD spectra observed for the face (top) and back (bottom) of the film are almost mirror images of each other, and they cancel out when added (middle). This proves that the observed CD for the PVA film is not a true CD but simply an artifact.

VI. CONCLUSION

This chapter has described the principles and application of solid-state CD. Solid-state CD can offer valuable information about the supramolecular nature of compounds and solid-state reactions, but it must be kept in mind that these suffer from substantial macroscopic anisotropies. We have shown that CD measurements in the solid state must be accompanied by measurement of corresponding LB and LD to evaluate the magnitudes of macroscopic anisotropies. We have developed an instrument and methods for obtaining the true CDs and CBs even for samples with modest macroscopic anisotropies. The technique is applied to solid-state photoreactions as well.

This new approach is valuable because strong molecular interactions often achieve chirally twisted molecular conformations and enantioselective reactions in the solid state, processes that may be relevant to the introduction of the homo-chiral biological world. Solid-state photochemistry and spectroscopy are promising research areas.

REFERENCES

1. Arago F. Mem Cl Sci Math Phys Inst 1811; 12:93–110.
2. Biot JB. Mem Inst France 1812; 1:1–32.
3. Borovkov VV, Harada T, Inoue Y, Kuroda R. Angew Chem Int Ed 2002; 41: 1378–1381.
4a. Tanaka K, Toda F. Chem, Rev. 2000 and references therein; 100:1025–1074.
4b. Toda F. Cryst Eng Commun 2002; 4:215–222.

5. Rothenberg G, Downie AP, Raston CL, Scott JL. J Am Chem Soc 2001; 123: 8701–8708.
6a. Nichols PJ, Raston CL, Steed JW. Chem Commun 2001:1062–1063.
6b. Pedireddi VR, Jones W, Chorlton AP, Docherty R. Chem Commun 1996:987–988.
6c. Patil AO, Curtin DY, Paul IC. J Am Chem Soc 1984 and references therein; 106: 348–349.
6d. Etter MC, Adsmond DA. Chem Commun 1990:589–590.
6e. Etter MC, Frankenbach GM. Chem Mat 1989; 1:10–12.
7a. Kuroda R, Imai Y, Sato T. Chirality 2001; 13:588–592.
7b. Imai Y, Tajima N, Sato T, Kuroda R. Chirality 2002; 14:604–609.
7c. Kuroda R, Imai Y, Tajima N. Chem Commun 2002:2848–2849.
8. Imai Y, Sato T, Tajima N, Kuroda R. To be submitted.
9. Kuroda R. Ph.D thesis, The University of Tokyo, 1975.
10. Kuroda R, Saito Y. Bull Chem Soc Jpn 1976; 49:433–436.
11. Kuroda R. In: Nakanishi K , Berova N , Woody JE, Eds. Circular Dichroism. New York: John Wiley, 2000:159–184.
12. Norden B. J Phys Chem 1977; 81:151–159.
13. Davidsson A, Norden N, Seth S. Chem Phys Lett 1980; 70:313–316.
14. Jensen HP, Schellman JA, Troxell T. Appl Spectrosc 1978; 32:192–200.
15. Shindo Y, Nakagawa M. Rev Sci Inst 1985; 56:32–39.
16. Ohmi Y. J Am Chem Soc 1985; 107:91–97.
17. Kuroda R, Harada T, Shindo Y. Rev Sci Inst 2001; 72:3802–3810.
18. Kuroda R, Saito Y. In: Nakanishi K , Berova N , Woody JE, Eds. Circular Dichroism. New York: John Wiley, 2000:563–599.
19. Kobayashi J, Uesu Y. J Appl Cryst 1983; 16:204–211.
20. Mueller H. J Opt Soc Am 1948; 38:661–673.
21. Born M, Wolf E. In: Principles of Optics, 2d ed.. Oxford: Pergamon Press, 1964: 1–70.
22. Schellman A. In: Samori B , Thulstrup EW, Eds. Polarized Spectroscopy of Ordered Systems. Dordrecht: Kluwer, 1988:231–274.
23. Schellman J, Jensen HP. Chem Rev 1987; 87:1359–1399.
24. Shindo Y, Nakagawa M, Ohmi Y. Appl Spectrosc 1985; 39:860–868.
25. Tunis-Schneider MJ, Maestre MF. J Mol Biol 1970; 52:521–541.
26. Shindo Y, Nishino M, Maeda S. Biopolymers 1990; 30:405–413.
27. Hayashi S. Jpn J App Phys 1989; 28:720–722.
28. Drake A, Kuroda R, Mason SF. J Chem Soc Dalton Trans 1979:1095–1100.
29. Mikami M, Kuroda R, Konno M, Saito Y. Acta Crystallogr 1977; B27:1485–1489.
30. Mason SF, Seal RH. Mol Phys 1976; 31:755–775.
31. Harada T, Shindo Y, Kuroda R. Chem Phys Lett 2002; 360:217–222.
32. Grinter R, Harding MJ, Mason SF. J Chem Soc A 1970:667–671.
33. Stadnica K, Glazer AM, Koralewski M. Acta Cryst B 1987; 43:319–325.
34. McCaffery AJ, Mason SF, Norman BJ. Chem Commun 1965:49–50.
35. Bosnich B, Harrowfield B. J Am Chem Soc 1972; 94:989–991.
36. Biscarini P, Kuroda R. Inorg Chim Acta 1988; 154:209–214.
37. Biscarini P, Franca R, Kuroda R. Inorg Chem 1995; 34:4618–4626.

38. Toda F, Miyamoto H, Kikuchi S, Kuroda R, Nagami F. J Am Chem Soc 1996; 118: 11315–11316.
39. Kuroda R, Honma T. Chirality 2000; 12:269–277.
40. Braga D, Maimi L, Polito M, Grepioni F. Chem Commun 2002:2302–2303.
41. Shindo Y, Kuroda R, Maeda T, Fukasawa T, 7th International Conference on Circular Dichroism, Mierki, 1999.
42. Castiglioni E, Albertini P. Chirality 2000; 12:291–294.
43. Bilotti I, Biscarini P, Castiglioni E, Feranti F, Kuroda R. Chirality 2002; 14:750–756.
44. Kuroda R. personal communication.
45. Kurod R, Fujita J, Saito Y. Chem Lett 1975:225–226.
46. Kobayashi A, Marumo F, Saito Y. Acta Crystallogr B 1973; 29:2443–2447.
47a. Toda F. Synlett 1993; 5:303–312.
47b. Toda F. Acc Chem Res 1995; 28:480–486.
48a. Toda F, Yagi M, Soda S. J Chem Soc Chem Commun 1987:1413–1414.
48b. Sekine A, Hori K, Ohashi Y, Yagi M, Toda F. J Am Chem Soc 1989; 111:697–670.
48c. Toda F, Tanaka K. Supramol Chem 1994; 3:87–88.
49a. Toda F, Tanaka K, Kakinoki O, Kawakami T. J Org Chem 1993; 58:3783–3784.
49b. Toda F, Miyamoto H, Kanemoto K. J Chem Soc Chem Commun 1995:1719–1720.
49c. Toda F, Miyamoto H, Kikuchi S. J Chem Soc Chem Commun 1995:621–622.
50. Kuroda R, Honma T. Chirality 2000; 12:269–277.
51. Asahi T, Utsumi H, Itagaki Y, Kagomiya I, Kobayashi J. Acta Cryst A 1996; 52: 766–769.
52. Kuroda R, First International Symposium on Asymmetric Photochemistry, Osaka, 2001.
53. Kuroda R, Imai Y, Honma T, Takeshita M, Sato T. To be submitted.
54. Addadi L, Lahav M. J Am Chem Soc 1982; 104:3422–3429.

11

Absolute Asymmetric Photochemistry Using Spontaneous Chiral Crystallization

Masami Sakamoto
Chiba University
Chiba, Japan

I. INTRODUCTION

Chirality is a concept well known to organic chemists and to all chemists concerned in any way with structure. The geometric property that is responsible for the nonidentity of an object with its mirror image is called chirality. A chiral object may exist in two enantiomorphic forms that are mirror images of one another. Such forms lack inverse symmetry elements, that is, a center, a plane, and an improper axis of symmetry. Objects that possess one or more of these inverse symmetry elements are superimposable on their mirror images; they are achiral. All objects belong to one of these categories.

The achievement of an asymmetric synthesis starting from an achiral reagent and in the absence of any external chiral agent has long been an intriguing challenge to chemists [1] and is also central to the problem of the origin of optical activity on Earth [2]. Stereospecific solid-state chemical reactions of chiral crystals formed by achiral materials are defined as ''absolute'' asymmetric synthesis [3]. This asymmetric synthesis involves two aspects: generating chiral crystals and performing topochemically controlled solid-state reactions which yield chiral products (Fig. 1) [4]. The crystallization of achiral molecules in chiral space groups, while rare and unpredictable, is well documented. Even rarer is to find materials exhibiting this behavior that also react chemically in the solid state and produce chiral products. As a result, very few successful absolute asymmetric syntheses have been reported, and those are mostly the result of serendipity rather

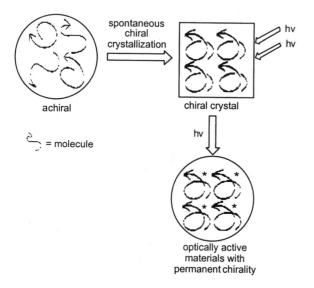

Figure 1   Absolute asymmetric synthesis in the chiral crystalline environment.

than planning. However, research on reactivity in the crystalline state has been extended in recent years to a variety of new systems such that it can now be regarded as an important branch of organic chemistry.

A.   Spontaneous Chiral Crystallization

Optically active molecules must crystallize into chiral space groups, but a racemic mixture in solution may either aggregate to form a nonchiral racemic crystal or undergo a spontaneous resolution in which the two enantiomers segregate into a conglomerate of enantiopure crystals. Nonchiral molecules may crystallize into either a nonchiral or a chiral space group. If they crystallize into the latter, the nonchiral molecules reside in a chiral environment imposed by the lattice. Most achiral molecules are known to adopt interconverting chiral conformations in fluid media, which could lead to a unique conformation upon crystallization. Crystals that have chiral space groups are characterized as enantiomorphrous. They exist in right-handed and left-handed forms that may or may not be visually distinguishable.

Chiral crystals, like any other asymmetric objects, exist in two enantiomorphous equienergetic forms, but careful crystallization of the material can induce the entire ensemble of molecules to aggregate into one crystal, of one handedness,

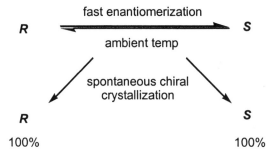

Scheme 1   Chiral crystallization of achiral materials.

presumably starting from a single nucleus. However, it is not uncommon to find both enantiomorphs present in a given batch of crystals from the same recrystallization (Scheme 1).

There are 230 unique space groups, which may be divided into two categories: (1) the chiral space groups, 65 in number, have only symmetry elements of the first kind, i.e., translations, rotations, and combinations of these; (2) the achiral space groups, such as a mirror plane or glide plane or center of inversion. Thus the unit cell of a compound belonging to an achiral space group will contain both the object and its mirror image.

For achieving asymmetric synthesis we should begin with a compound crystallizing in any one of the 65 chiral space groups. Of the 230 distinct space groups, the most commonly occurring are $P2_1/c$, $P\bar{1}$, $P2_12_12_1$, $P2_1$, $C2/c$, and Pbca, the chiral ones being $P2_12_12_1$, $P2_1$, and $P1$ (Table 1) [5].

The crystallization of achiral molecules in chiral space groups, while rare and unpredictable, is well documented. Molecules with a C2 symmetry axis tend to crystallize in chiral structures, according to Jacques and coworkers, but despite impressive work on crystal engineering, predictions of a correlation between crystal symmetry and molecular structures are still hard to make [6].

The asymmetric crystallization of achiral compounds is stimulated by autoseeding with the first crystal formed. Although the chiral sense of the spontaneously formed chiral crystals cannot be predicted, seed crystals of the preferred chirality can be added in a more practical procedure to obtain one enantiomorph of a crystal.

Kondepudi et al. [7] and McBride and Carter [7] indicate that stirred crystallization is effective in accelerating the enantiomeric excess (*ee*) of crystals in the recrystallization step. Kondepudi reported that the *ee* of the crystal greater than 95% can easily be obtained in stirred crystallization of achiral materials that crystallize in chiral form, such as sodium chlorate. Furthermore, stirred crystalli-

Table 1   The Most Common Space Groups of Organic Crystalline
Compounds Based upon a Survey of 29059 Crystal Structure Determinations

Order	Space group	Number	Percentage
1	$P2_1/c$	10450	36.0
2	$P\bar{1}$	3986	13.7
3	$P2_12_12_1$[a]	3359	11.6
4	$P2_1$[a]	1957	6.7
5	$C2/c$	1930	6.6
6	$Pbca$	1261	4.3
7	$Pnma$	548	1.9
8	$Pna2_1$	513	1.8
9	$Pbcn$	341	1.2
10	$P1$[a]	305	1.1

[a] Chiral space group.

zation of a 1,1′-binaphthyl melt also spontaneously generated a large *ee*, often greater than 90% [8]. That unstirred crystallization gives a Gaussian-like probability distribution in optical activity was demonstrated by Pincock et al., who also noted that an *ee* value of greater than 90% appears in roughly 1 in 150 crystallizations [9]. In stirred crystallization of binaphthyl, an *ee* value larger than 90% occurs almost always. The mechanism that results in such chiral symmetry breaking has been studied in some detail for the case of $NaClO_3$ [7]. A combination of both seeding and stirred crystallization may lead to the certainty for enantiomeric selectivity with high crystal enantiomeric purity.

II.   ABSOLUTE ASYMMETRIC TRANSFORMATION FROM NONCHIRAL MOLECULES IN CHIRAL CRYSTALS

A systematic solid state approach to asymmetric synthesis demands the design of chiral crystal structures having certain intermolecular or intramolecular features. Owing to the strict requirements, the discovery of new systems appropriate for such syntheses has been slow.

Penzien and Schmidt reported the first absolute asymmetric transformation in a chiral crystal [10]. They showed that enone 4,4′-dimethylchalcone **1**, although being achiral itself, crystallizes spontaneously in the chiral space group $P2_12_12_1$ with a highly twisted conformation (Scheme 2). When single crystals of this material are treated with bromine vapor in a gas–solid reaction, the chiral dibromide **2** is produced in 6–25% *ee*. In this elegant experiment, it is the reaction

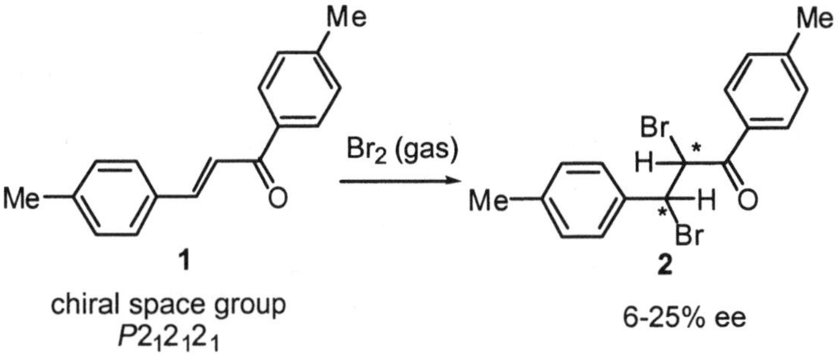

Scheme 2 First example of absolute asymmetric synthesis in the chiral crystal environment of 4,4'-dimethylchalcone **1**.

medium, the chiral crystal lattice, that provides the asymmetric influence that favors the formation of one product enantiomer over the other, and the chemist has done nothing but provide a nonchiral solvent (ethyl acetate) for the crystallization and a nonchiral reagent (bromine) for the reaction.

Since the concept of topochemically controlled reactions was established, various approaches to asymmetric synthesis using a solid-state reaction have been attempted, most actively by the research group at the Weismann Institute. Their studies have been concerned with the bimolecular reactions of chiral crystals in the solid state. In these studies, successful absolute asymmetric synthesis has been performed by using topochemically controlled four-centered photocyclodimerizations of a series of unsymmetrically substituted diolefin crystals. Research on reactivity in the crystalline state has been extended in recent years to a variety of new systems, and many absolute asymmetric syntheses have been provided. Successful examples of absolute asymmetric synthesis using chiral crystals are listed in Tables 2 to 4, which are divided into three categories: intermolecular photoreaction in the solid state (Table 2), intramolecular photoreaction in the solid state (Table 3, A–D), and asymmetric induction in the solid–gas and homogeneous reactions (Table 4).

A. Absolute Asymmetric Synthesis Involving Bimolecular Photochemistry

The strategy and the main results were reported by these investigators for the mixed crystal of 2,6-dichlorophenyl-4-phenyl-*trans,trans*-1,3-butadiene **3** with the 4-thienyl analog **4** (Scheme 3) [11]. These two materials crystallize in two isostructural arrangements in the chiral space group $P2_12_12_1$. Large mixed crystals

Table 2 Absolute Asymmetric Transformation Using Chiral Crystals via Intermolecular Reactions

Starting material (space group)	Reaction conditions	Product ee% (chemical yield %)		Reference
 3: Ar = 2,6-$C_6H_3Cl_2$ **4**: Th = 2-Thienyl ($P2_12_12_1$)	hv solid state conv. 29%	 **5**	~70% ee	11
 6a: R^1 = 3-Pen, R^2 = Me ($P2_1$) **6b**: R^1 = 3-Pen, R^2 = Et (P1) **6c**: R^1 = 3-Pen, R^2 = Pr (P1) **6d**: R^1 = (R,S)-s-Bu, R^2 = Et (P1) **6e**: R^1 = (R,S)-s-Bu, R^2 = Pr (P1)	hv solid state	**7a**: ~100% ee **7b**: 65% ee **7c**: 80% ee **7d**: 45% ee **7e**: 50% ee		12
 8 ($P2_12_12_1$)	hv solid state	 **9** 92–95% ee		13
 10 CT crystals **11** ($P2_1$)	hv solid state −70°C	 **12** 95% ee (91%)		14
 13 ($P2_12_12_1$)	hv solid state −30°C, conv. :71% −70°C, conv. :23%	 33% ee (28%) **14** 39% ee (34%)		15

Table 3A Absolute Asymmetric Transformation Using Chiral Crystals via Intermolecular Reactions

Starting material (space group)	Reaction conditions	Product *ee*% (chemical yield %)	Reference
15a: R = *i*-Pr ($P2_12_12_1$) **15b**: R = Et ($P2_12_12_1$)	h*v* solid state	**16a**: > 95% ee **16b**: variable	16
17 ($P2_12_12_1$)	h*v* solid state	**18** **19** 18°C, 9% conv. 34 %*ee* 96% ee -21°C, 51% conv. 22% ee 55% ee	17
20 Ar = 4-ClC$_6$H$_4$ ($P2_12_12_1$)	h*v* solid state	**21** : > 80% ee	16
22a: C$_6$H$_5$ ($P2_12_12_1$) **22b**: m-ClC$_6$H$_4$ (nd) **22c**: m-BrC$_6$H$_4$ (nd) **22d**: m-MeC$_6$H$_4$ (nd) **22e**: o-MeC$_6$H$_4$ (nd) **22f**: m,p-Me$_2$C$_6$H$_3$ (nd)	h*v* solid state	**23a**: 93% ee (74%) **23b**: 100% ee (75%) **23c**: 96% ee (97%) **23d**: 91% ee (63%) **23e**: 92% ee (54%) **23f**: 54% ee (62%)	19
24 ($P2_12_12_1$)	h*v* solid state	**25** 49–86% ee	20
26a: Ar = Ph ($P2_12_12_1$) **26b**: Ar = *p*-ClPh ($P2_12_12_1$)	(1) h*v* solid state −45°C (2) AcCl/Et$_3$N −78°C%	 **29a**: 84% ee (39%) **30a**: 50% ee (10%) **31a**: 20% ee (16%) **29b**: 70% ee (37%) **30b**: 40% ee (13%) **31b**: 13% ee (19%)	21

Table 3B Absolute Asymmetric Transformation Using Chiral Crystals via Intermolecular Reactions

Starting material (space group)	Conditions	Product ee% (chemical yield %)	Reference
32b : R^1 = H, R^2 = Me, R^3 = Me ($P2_12_12_1$) **32c**: R^1 = R^2 = -$(CH_2)_4$-, R^3 = H ($P2_1$)	hv solid state	**33b**: conv. : 81% 55% ee (55%) **33b**: conv. : 17 % 74% ee (95%) **33c**: conv. : 100% 81% ee (81%) **33c**: conv. : 20% 97% ee (97%)	22
34 ($P2_12_12_1$)	hv −40°C 70% conv.	**35**: 40% ee (70%)	23
36 ($P2_1$)	hv −78°C 15% conv.	**38**: 31% ee (90%)	24
39a: R = Pr^i ($P2_1$) **39b**: R = Bn (nd)	hv −78°C 100% conv.	**40a**: 99% ee (89%) **40b**: 91% ee (91%)	25
39g ($P2_1$)	hv solid state 0°C 66% conv.	**40g**: racemic (40%) **41g**: 88%ee (20%)	25

Table 3C Absolute Asymmetric Transformation Using Chiral Crystals via Intermolecular Reactions

Starting material (space group)	Conditions	Product $ee\%$ (chemical yield %)	Reference
42a: R = CH$_2$Ph $(P2_12_12_1)$ **42c**: R = o-(CH$_3$)$_3$C$_6$H$_4$CH$_2$ $(P2_1)$	hv solid state 0°C	**42a**: 82% ee (100%) **42c**: 100% ee (100%)	26
44 (nd)	hv solid state	**45**: 64% ee	27
46a: X = m-Cl $(P2_12_12_1)$ **46b**: X = m-Me $(P2_12_12_1)$ **46c**: X = m-OMe $(P2_12_12_1)$ **46d**: X = m-Br $(P2_12_12_1)$	hv solid state	**47a**: 69 - 78% ee **47b**: 58 - 71% ee **47c**: 90 - 100% ee **47d**: unstable	28
48 $(P2_1)$	hv solid state 0°C	**49** conv. 17% 46 %ee (17%) conv. 62% 21% ee (16%) / **50** 32 %ee (22%) 23% ee (18%)	29
51a: Ar^1 = o-tol $(P2_12_12_1)$ **51b**: Ar^1 = Ph $(P2_12_12_1)$ **51c**: Ar^1 = m-tol $(P2_12_12_1)$	hv solid state −78°C conv. 50% conv. 10% conv. 67%	**52a**: 65 %ee (92%) **52b**: 87% ee (95%) **52c**: 74% ee (95%)	30

Table 3D Absolute Asymmetric Transformation Using Chiral Crystals via Intermolecular Reactions

Starting material (space group)	Conditions	Product *ee*% (chemical yield %)	Reference

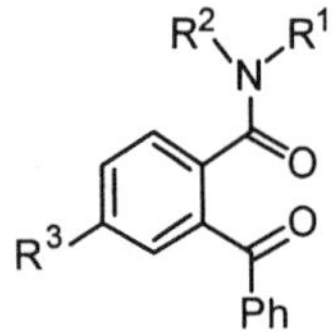

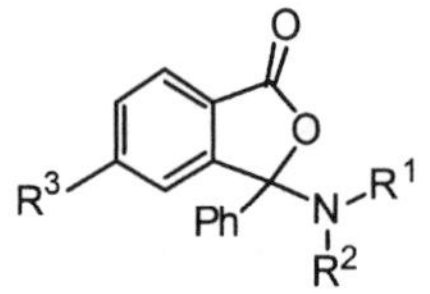

31

53a: R^1=R^2=Me, R^3=H (P2$_1$2$_1$2$_1$) **a**: powder 15°C conv. 0% **54a**: - (0%)
53b: R^1=Me, R^2=Ph, R^3=H (P2$_1$2$_1$2$_1$) **b**: crystal -50°C conv. 46% **54b**: 97% ee (100%)
53c: R^1=Me, R^2=Ph, R^3=Me (P2$_1$2$_1$2$_1$) **c**: powder -50°C conv. 25% **54c**: 87% ee (100%)

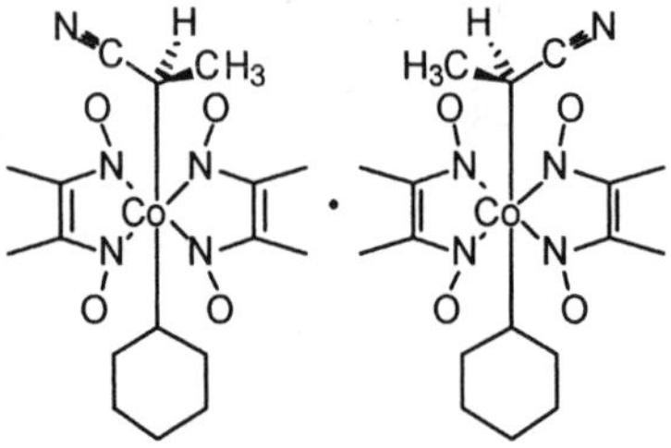

(R)-55 (S)-55

(R) : (S) = 50 : 50 (P2$_1$2$_1$2$_1$)

hv
solid state
70°C

(R)-**55** : (S)-**55** = 75 : 25

50% ee

32

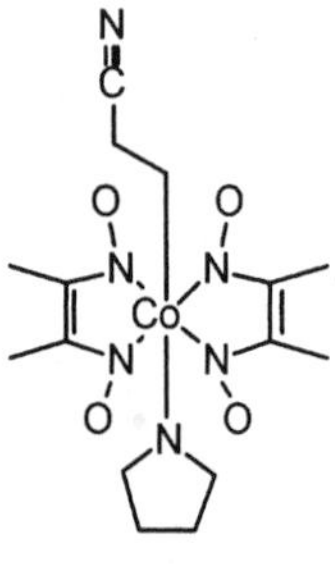

56 (P2$_1$2$_1$2$_1$)

hv
solid state
27°C

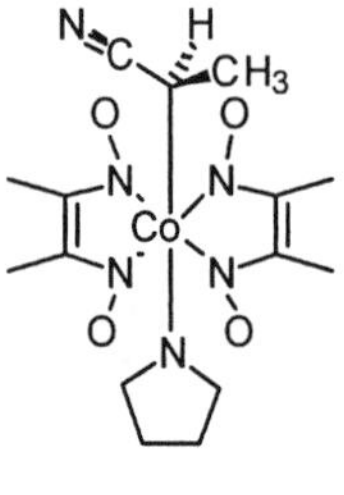

57

33

conv. 1.4% 21% ee
conv. 2.9% 19% ee
conv. 6.7% 20% ee
conv. 6.0% 20% ee
conv. 8.9% 17% ee

Table 4 Absolute Asymmetric Transformation in Solid–Gas or Homogeneous Media Using Chiral Crystals

Starting material (space group)	Conditions	Product ee% (chemical yield %)	Reference
1 ($P2_12_12_1$)	Br$_2$ solid - gas	**2**: 6-25% ee (>90%)	10
58 ($P2_12_12_1$)	Br$_2$ solid - gas	8% ee **59**	34
60 ($P3_121$)	$^1O_2$ RB / Resin	**61** $[\alpha]_{546}^{20} = 0.060°$ (30%)	35
62 ($P3_121$)	HCl solid - gas 80-120°C	**63**: 9% ee **64**: 22% ee	36
66 ($P2_1$)	PdCl$_2$(CH$_3$CN)$_2$ −78°C, in CH$_2$Cl$_2$	**67**: ~100% ee (93%)	38
68 ($P2_12_12_1$)	hv in THF −60°C	**69**: 87% ee (81%) **70**: 79% ee (19%)	39
68 ($P2_12_12_1$)	n-BuLi/TMEDA in THF, −80°C	**71**: 83% ee (46%) **72**: 81% ee (24%)	39

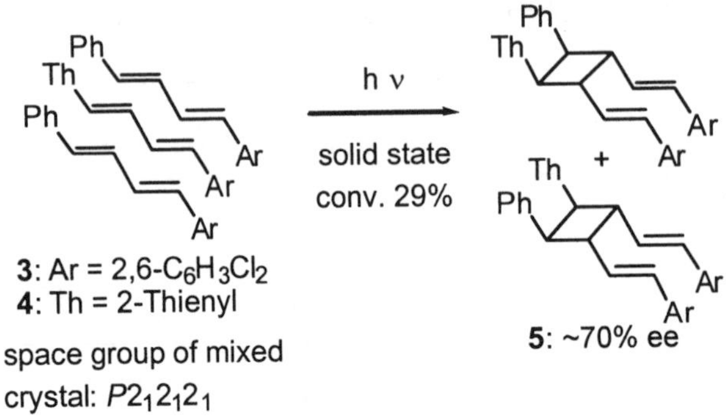

3: Ar = 2,6-C$_6$H$_3$Cl$_2$
4: Th = 2-Thienyl

space group of mixed
crystal: $P2_12_12_1$

Scheme 3   Asymmetric synthesis using mixed crystals.

of the phenyl material containing $\sim$ 15% of the thienyl derivative as guest were prepared. The latter absorbs light at a longer wavelength. As a result of this selective excitation, the thienyl reacts with a nearer phenyl neighbor to form a mixed cyclobutane dimer **5**. This dimer has been isolated to be optically active with an *ee* of $\sim$ 70%. As expected, some crystals gave left-handed and others right-handed cyclobutanes.

Several 1,4-disubstituted phenylenediacrylates **6** such as Me, Et, and Pr esters crystallized into chiral structures, and they photodimerized to either (*SSSS*)-cyclobutanes **7** or (*RRRR*)-cyclobutanes *ent*-**7** with medium to quantitative enantiomeric yields [12]. In addition to these dimers, the corresponding trimers and oligomers were also produced with high enantiomeric yields (Scheme 4).

Hasegawa et al. reported another example of a [2 + 2] asymmetric transformation in a chiral crystal (Scheme 5) [13]. Ethyl 4-[2-(pyridyl)ethenyl] cinnamate **8b** crystallizes in a chiral space group $P2_12_12_1$ and upon irradiation yields a chiral dimer with 92–95% *ee*; however, the methyl and *n*-propyl esters (**8a** and **8c**) crystallized in achiral space groups.

Suzuki et al. reported the photochemical reaction of CT crystals, in which cycloaddition reaction of bis(1,2,5-thiadiazolo)tetracyanoquinodimethane **10** (electron acceptor) and 2-divinylstylene **11** (electron donor) is efficiently induced (Scheme 6) [14]. The structural feature of the CT crystal is the asymmetric nature of the inclusion lattice because of the adoption of a chiral space group, $P2_1$. The [2 + 2] photoadduct was formed via the single crystal–to–single crystal transformation, and the optically active product with 95% *ee* was obtained.

Most of the bimolecular absolute asymmetric syntheses are limited to 2 + 2 cyclobutane formation or polymerization of olefins. Koshima et al. reported an example of a bimolecular reaction in which acridine **13** (Scheme 7) and diphenylacetic acid are assembled in a 1 : 1 molar ratio by hydrogen bonding and

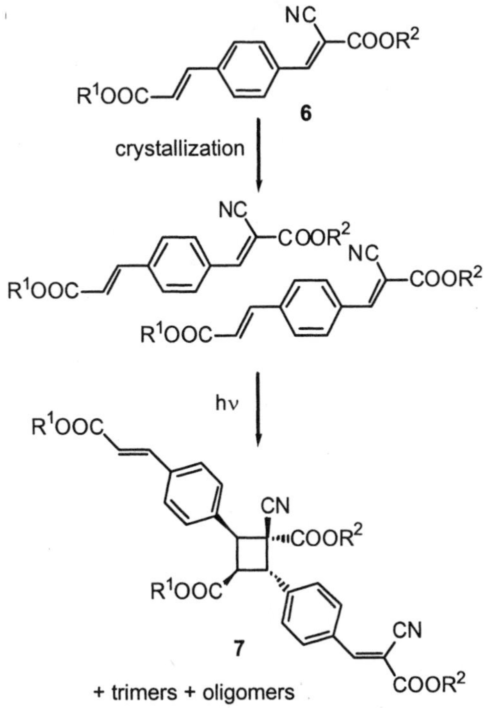

Scheme 4 Absolute asymmetric synthesis using diolefin crystals.

		space group	ee of photodimer **7**
6a: R^1 = 3-Pentyl, R^2 = Me		$P2_1$	~100%
6b: R^1 = 3-Pentyl, R^2 = Et		$P1$ or $P\bar{1}$	65 or 0%
6c: R^1 = 3-Pentyl, R^2 = Pr		$P1$	80%
6d: R^1 = (R,S)-s-Bu, R^2 = Et		$P1$	45%
6e: R^1 = (R,S)-s-Bu, R^2 = Pr		$P1$	50%

crystallized in a chiral space group, $P2_12_12_1$ [15]. Irradiation of the crystals caused stereospecific decarboxylating condensation to give chiral **14** in 33–39% *ee*.

B. Unimolecular Photochemistry in the Solid State

The concept of absolute asymmetric synthesis using a chiral crystal was applied to unimolecular photochemistry, and now many fine examples are reported of

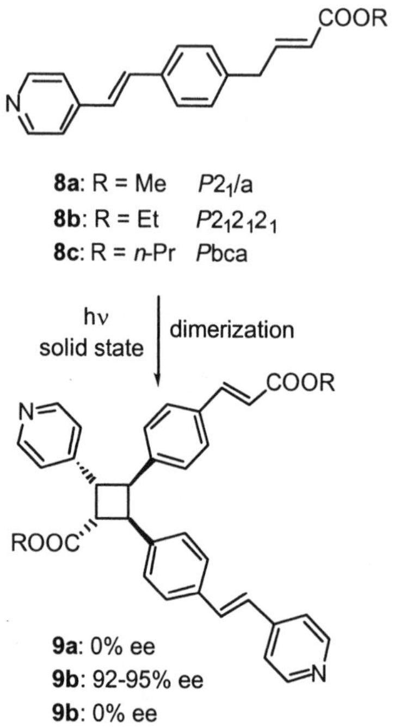

8a: R = Me $P2_1/a$
8b: R = Et $P2_12_12_1$
8c: R = n-Pr   Pbca

hv
solid state │ dimerization

9a: 0% ee
9b: 92–95% ee
9b: 0% ee

Scheme 5 Absolute asymmetric synthesis using diolefin crystals.

(a) di-π-methane rearrangement, (b) hydrogen abstraction by carbonyl, thiocarbonyl, and alkenyl carbon, (c) 2 + 2 thietane and oxetane formation, [4 + 4] cycloaddition, (d) electrocyclization, and (e) miscellaneous reactions involving a radical pair or zwitterionic intermediate.

1.   Photochemical Di-π-Methane Rearrangement

Scheffer et al. reported elegant unimolecular ''absolute'' asymmetric transformations (Scheme 8) [16]. This group demonstrated that the well-studied solution-phase di-π-methane photorearrangement can also occur in the solid state. The particular compounds investigated included dibenzobarrene-11,12-diester derivatives **15**. The corresponding diisopropylester **15a** is dimorphic, and one of the forms grown from the melt is chiral (space group $P2_12_12_1$), and another is nonchiral space group Pbca. Irradiation of the chiral crystals gave rearranged product **16a** in *ee*s greater than 95%. On the other hand, x-ray crystallography revealed that the Pbca crystals of the diester are racemic–composed of equal amounts of enantiomeric conformations related by crystal symmetry. Photolysis of these

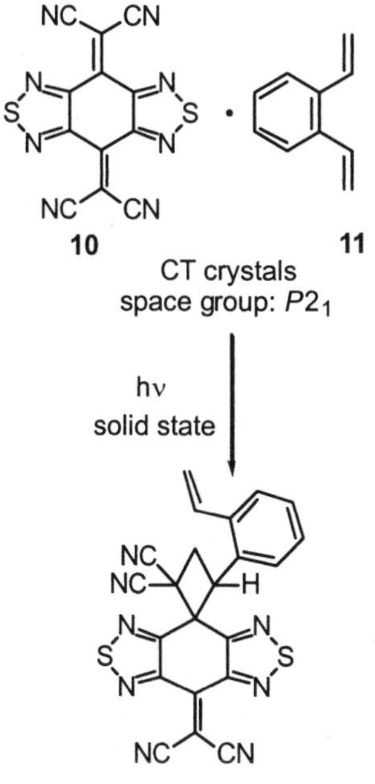

12: 95 %*ee*

Scheme 6 Asymmetric synthesis using CT crystals.

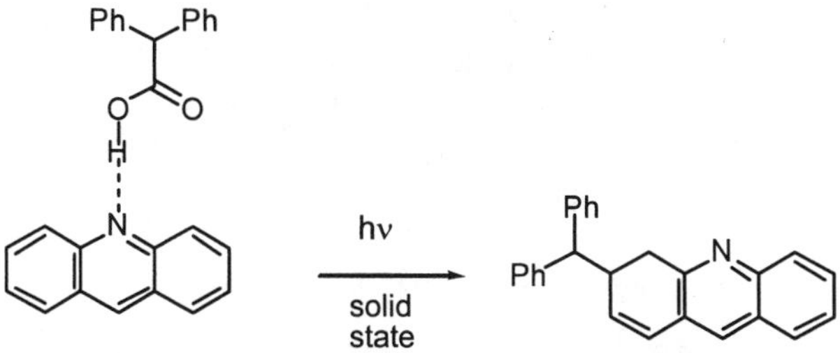

13

space group: $P2_12_12_1$

14: 33-39% ee

Scheme 7 Asymmetric synthesis involving decarboxylation.

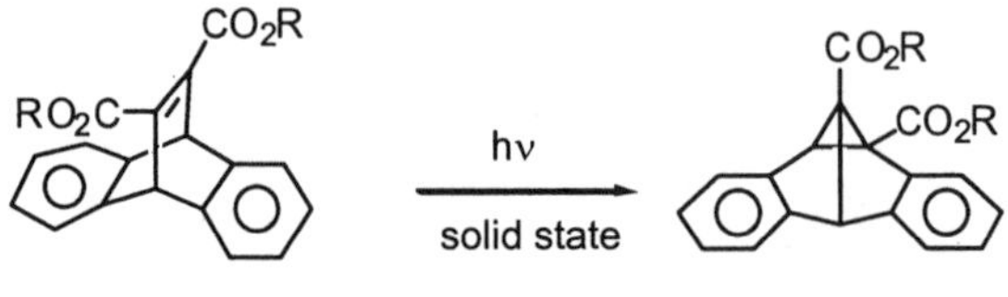

15 **16**

	space group	ee of **16**
15a: R = *i*-Pr	*P*$2_12_12_1$	> 95% ee
	*P*bca	0
15b: R = Et	*P*$2_12_12_1$	variable

Scheme 8 Asymmetric synthesis via di-π-methane rearrangement.

crystals gave racemic photoproducts. Of over 20 symmetrical and unsymmetric dialkyl 9,10-ethanoanthracene-11,12-decarboxylate **15**, only two compounds were found to undergo absolute asymmetric di-π-methane photorearrangement in the solid state. One is the diisopropyl diester **15a** and the other is the diethyl analog **15b**. Solid-state photolysis of **15b** also gave optically active **16b**; however, the *ee*s were variable under the reaction conditions. In general, lower photolysis temperature gave higher *ee*, and when the crystal melted or disintegrated during photolysis, a noticeably lower *ee* was observed.

Demuth et al. also have reported a solid-state di-π-methane type photorearrangement of seemingly homochirally crystallized starting material **17** to give preparative quantities of two rearranged products (Scheme 9) [17]. Considerable changes in product selectivity were observed from the rearrangement in homogeneous solution versus solid state. The differences are explained on the basis of the packing constraints in the crystal, because of the frozen geometry and the proximity of the reacting sites. The starting material adopts chiral packing, space group $P2_12_12_1$, and a helical molecular conformation. The rearrangements proceed so that the products, **18** and **19**, are obtained in respective *ee* of ≤44% and ≤96% and that the *ee*s decreased proportionately with conversion.

2. Absolute Asymmetric Transformation via Hydrogen Abstraction in the Solid State

Scheffer et al. provided another unimolecular asymmetric transformation involving the Norrish type 2 reaction, a well-known excited state process of ketones that is initiated by an intramolecular hydrogen atom transfer from carbon to

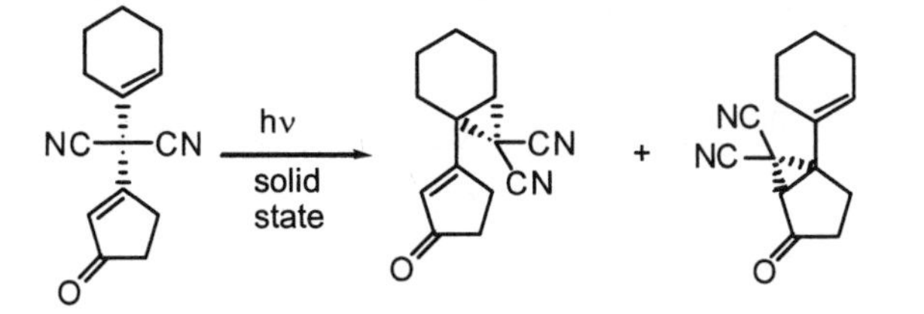

17
space group: $P2_12_12_1$

18: ≤ 44 % *ee*

19: ≤ 96 % *ee*

Scheme 9 Asymmetric synthesis via di-π-methane rearrangement.

oxygen through a six-membered transition state (Scheme 10) [16a]. An adamantyl ketone derivative **20** was found to crystallize from ethanol in very large prisms in the chiral space group $P2_12_12_1$. Upon irradiation of these crystals to approximately 10% conversion, the chiral cyclobutanol derivatives **21** were afforded as the major products in 80% *ee*.

The next example is a process that is similar to the photorearrangement of ketone **20** in that it is thought to involve initial hydrogen atom transfer from carbon to oxygen followed by coupling of the 1,4-diradical thus produced (Norrish type II process). This solid-state reaction was discovered by Aoyama et al. [18] and subsequently studied in more detail by Toda and coworkers [19]. This concerns the photochemical conversion of the achiral α-oxoamide derivative **22** (Scheme 11) into the chiral β-lactam **23**. When grown from benzene, crystals of **22** were found to occupy the chiral space group $P2_12_12_1$, and irradiation of one of the enantiomorphs was found to give (+)-**23** in 93% *ee*, while photolysis of the other enantiomorph gave (−)-**23**, also with the same *ee*. Samples of each enantiomorphous form of α-oxoamide were saved for use as seeds in later recrystallizations, and in this way, substantial quantities of either enantiomorph could be prepared as desired. The crystalline phase rearrangement of **22** to **23** is notewor-

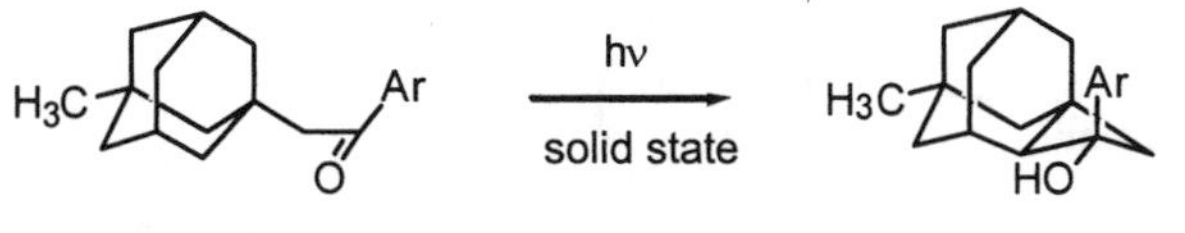

20: Ar = 4-ClC_6H_4
space group: $P2_12_12_1$

21: > 80 % *ee*

Scheme 10 Asymmetric synthesis via hydrogen abstraction.

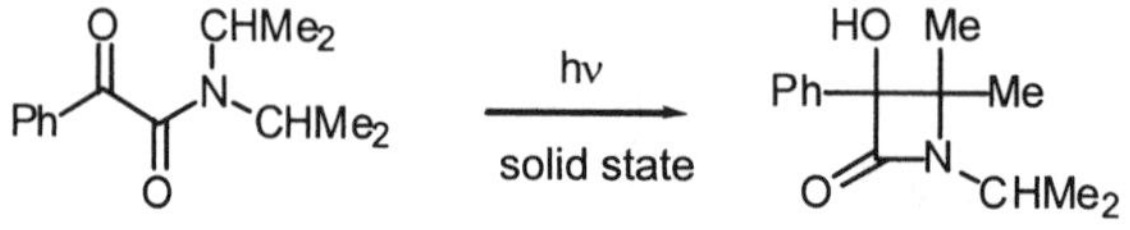

22
space group: $P2_12_12_1$

23: 93 % *ee*

Scheme 11   Asymmetric synthesis via Norrish type 2 reaction.

thy also for the high conversions (up to 75%) that can be tolerated without significant loss of enantioselectivity. As a result, the reaction has genuine preparative
utility. They also prepared seventeen α-oxoamide derivatives and investigated
the chiral spontaneous crystallization and the solid-state photoreaction, and found
six materials crystallized in a chiral space group [19c].

Irngartinger et al. reported another example of asymmetric synthesis involving δ-hydrogen transfer followed by cyclization [20]. Ketone **24a** (Scheme 12)
crystallized in the chiral space group $P2_12_12_1$, and the phenyl groups are fixed
in a syn conformation. Irradiation of single crystals of **24a** gave optically active
cis-1,2-diphenylacenaphthen-1-ol **25a** as the main product, with up to 97% *de*.
Only a small amount of trans acenaphthenol is also generated in the solid state,
presumably due to an increased thermal motion of the molecules in the crystal
during the irradiation (40–45°C). The photoreaction is frozen when the probe is
cooled to below 0°C. An excess of one enantiomer of **25a** is found in selected
single crystals after irradiation. In optimal cases, 86% *ee* of the first enantiomer
25a and 49% *ee* of the other enantiomer *ent*-**25a** were obtained. During the
irradiation, the elimination product 1,2-diphenylacenaphthylene is also generated.
In order to prevent the dehydration, they synthesized **24b**; however, it crystallized
in the achiral space group $P2_1/n$ and is therefore unsuitable for an asymmetric
synthesis.

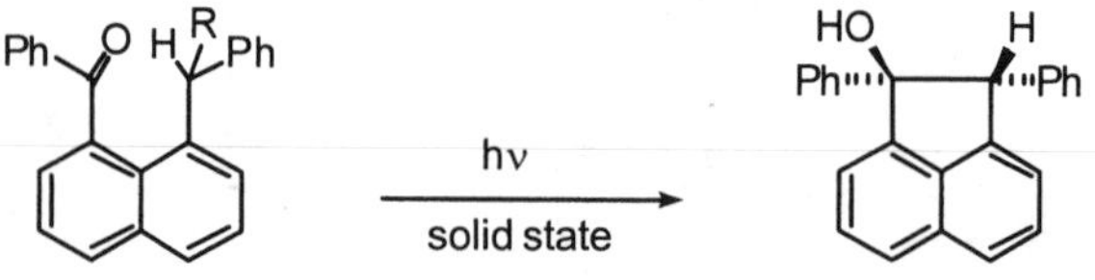

24a: R = H $(P2_12_12_1)$
24b: R = Ph $(P2_1/n)$

25a: 49–86% ee

Scheme 12   Asymmetric synthesis via δ-hydrogen abstraction by carbonyl group.

Sakamoto et al. provided an example of absolute asymmetric synthesis involving hydrogen abstraction by thiocarbonyl sulfur [21]. Achiral *N*-diphenylacetyl-*N*-isopropylthiobenzamide **26a** and *N*-diphenylacetyl-*N*-isopropyl(*p*-chloro)thiobenzamide **26b** crystallize in chiral space group $P2_12_12_1$ (Scheme 13). Photolysis of the chiral crystals in the solid state gave optically active azetidin-2-ones **27**, whereas achiral thioketones **28** were obtained as main products. When **26a** was irradiated in the solid state at $-45°C$ followed by acetylation (at $-78°C$), 2-acetylthio-3,3-dimethyl-1-diphenylacetyl-2-phenylaziridine (**30a**: 39% yield, 84% *ee*), 4-acetylthio-5,5-dimethyl-2-diphenylmetyl-4-phenyloxazoline (**31a**: 10% yield, 50% *ee*), 3,3-diphenyl-1-isopropyl-4-phenylazetidin-2-ones (**29a**: 16% yield, 20% *ee*), thioketone (**28a**: 15%) and *N*-isopropylthiobenzamide (15%)

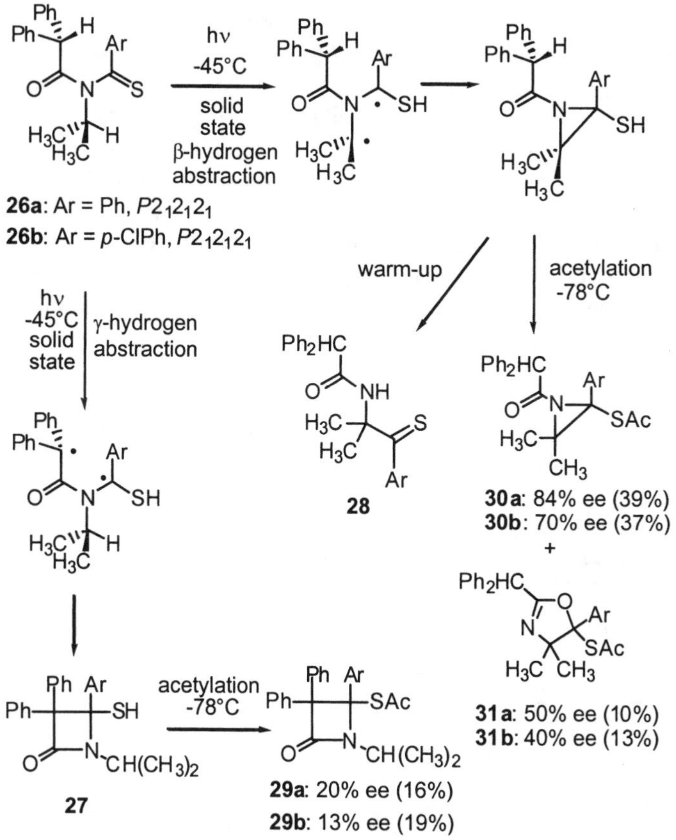

Scheme 13 Asymmetric synthesis using hydrogen abstraction by thiocarbonyl sulfur.

were obtained. In the case of the *p*-chlorothiobenzamide derivative **26b**, the corresponding aziridine (**30b**: 37% yield, 70% *ee*), oxazoline (**31b**: 13% yield, 40% *ee*), azetidin-2-ones (**29b**: 19% yield, 13% *ee*), thioketone (**28b**: 10%) and *N*-isopropyl-*p*-chlorothiobenzamide (18%) were obtained. Azetidine and oxazoline derivatives are formed via β-hydrogen abstraction by the thiocarbonyl sulfur atom; on the other hand, the mechanism to azetidinones involves γ-hydrogen abstraction followed by cyclization. In the monothioimides **26**, the $C=S\cdots H\beta$ contact is in the area from 2.37 to 2.57 Å, and the $C=S\cdots H\gamma$ distances are 3.79 and 3.84 Å, which is a little longer than the sum of the van der Waals radii of the sulfur atom and the hydrogen atom, 3.0 Å.

Next, Sakamoto et al. show an example which involves the hydrogen abstraction by the alkenyl carbon atom (Scheme 14) [22a–c]. Four α,β-unsaturated thioamides **32a–d** were subjected to x-ray crystallographic analysis; two of them were chiral, and the other two crystals were composed of a racemic space group. When the solid sample of **32a** was irradiated at 0°C, the crystals gradually changed to amorphous. At this point, only a mixture of (*E*)- and (*Z*)-isomers (*Z*/*E* = 1.5) was obtained. Photolysis of cycloheptenecarbothioamide **32d** in the solid state gave the corresponding β-thiolactam in the quantitative yield. As the x-ray crystallographic analysis revealed, the crystal is racemic (*P*2$_1$/n), and the isolated β-thiolactam **33d** was obtained as a racemate. 3-Methylcrotonoyl and 1-cyclohexenecarbonyl derivatives **32b** and **32c** gave a chiral crystal system. Figure 2 shows the UV and the mirror-imaged CD spectra of two enantiomorphic crystals of **32b** (Fig. 2, top) and of **32c** (Fig. 2, bottom) in KBr pellets, which were independently obtained by spontaneous resolution. These crystals gave specific curves in the region between 250 and 450 nm, which were mirror images designated as (+) and (−) at the wavelength of 370 nm.

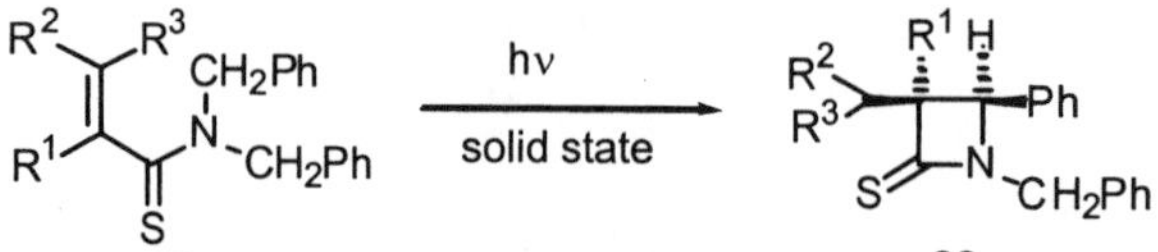

	R^1	R^2	R^3	space group
32a	Me	H	Me	*P*2$_1$/a
32b	H	Me	Me	*P*2$_1$2$_1$2$_1$
32c	-(CH$_2$)$_4$-		H	*P*2$_1$
32d	-(CH$_2$)$_5$-		H	*P*2$_1$/n

Scheme 14 Asymmetric synthesis involving hydrogen abstraction by alkenyl carbon.

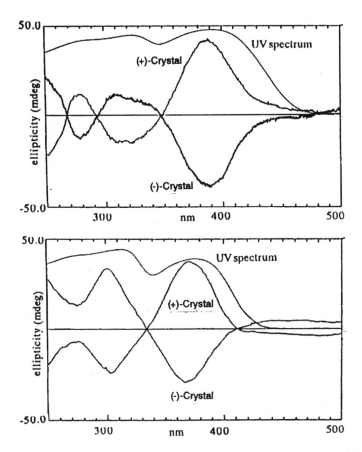

Figure 2 CD spectra of enantiomeric crystals of both antipodes of **32b** (up) and **32c** (down) in KBr, and reflected UV spectra.

    X-ray crystallographic analysis revealed that the crystals of thioamide **32b** are chiral and the space group is $P2_12_12_1$. Dextrorotatory crystals were irradiated at 0°C until the reaction conversion reached 81% yield. As expected, the asymmetric induction was observed, and optically active **33b** (55% *ee*) was isolated (Table 5). As a consequence of the suppression of the reaction conversion yield from 81% to 17%, the enantiomeric purity rose up to 74% *ee*. β-Thiolactam **33b** with only *Z*-configuration was isolated, and the absolute configuration of (+)-**33b** was determined as (*3S,4S*)-isomer by the x-ray anomalous scattering method.

    In the solid-state photoreaction of **32c**, a more chemoselective reaction occurred than in solution, and only β-thiolactam **33c** was obtained almost quanti-

Table 5   Photochemical Reaction of **32** in the Solid State

Entry	32	conv (%)	yield of 33(%)	*ee* (%) of 33
entry 1	**32a**	0[a]	0	0
entry 2		100	54	0
entry 3	**32b**	81	55	55
entry 4		50	60	59
entry 5		37	71	71
entry 6		17	95	74
entry 7	**32c**	100	81	81
entry 8		20	97	97
entry 9	**32d**	100	100	0

[a] *E-Z* isomerization occurred (*Z/E* = 1/1.5).

tatively. Furthermore, the x-ray crystallographic analysis and the solid-state CD spectra revealed that the crystals of **32c** are chiral, and the space group is $P2_1$. Dextrorotatory crystals were used for the solid state photoreaction.

Of particular importance is the finding that the solid-state photoreaction of **32c** involves a crystal-to-crystal nature where the optically active β-thiolactam **33c** is formed. Irradiation of (+)-crystals (defined by the CD spectrum) of **32c** at 0°C for 2 h exclusively gave optically active β-thiolactam **33c**, in 81% yield at 100% conversion (entry 7). As expected, the thiolactam (+)-**33c** showed optical activity (81% *ee*). This reaction exhibited good enantioselectivity throughout the whole reaction, where a small difference was observed in *ee* varying from 97 to 81% with increasing conversion from 20 to 100%. The absolute configuration was also determined as (*R*)-configuration by the x-ray anomalous scattering method. The solid-state photoreaction also proceeded without phase separation up to 100% reaction conversion. The crystal-to-crystal nature of the transformation was confirmed by x-ray diffraction spectroscopy.

Figure 3 shows the superimposed structure of each absolute structure of (+)-**32c** and (+)-**33c**, obtained by x-ray crystallography. This solid state photoreaction maintained a crystal-to-crystal nature throughout the reaction without phase separation. A really small movement of atoms was observed before and after the reaction. The maximum deviation was shown in the benzyl carbon atom, where the distance was 1.24 Å. The cyclohexane ring went into an original space where the ring had been placed. The shift of the benzyl group, which did not participate in the hydrogen abstraction and subsequent cyclization, also did not exceed 0.65 Å. Thus an extreme atom-transmigration before and after the reaction was not needed. Recently, Ohashi et al. reinvestigated the solid-state photoreaction of

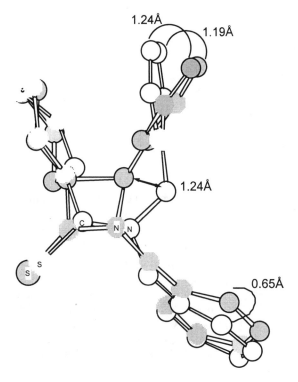

1.24Å

1.19Å

1.24Å

0.65Å

Figure 3 Superimposed structure of each absolute configuration of **32b** and **33b** obtained by x-ray crystallography.

32c. The mechanism of the transformation of **33c** and the chirality of the produced **33c** were explained from the reaction cavity [22d].

3. Photochemical [2 + 2] and [4 + 4] Cycloaddition

Sakamoto et al. reported intramolecular [2 + 2] thietane formation in the solid state (Scheme 15) [23]. Achiral N-(thiobenzoyl)methacrylamide **34a** formed (E,Z)-conformation of the imide moiety, crystallized in a chiral space group $P2_12_12_1$, and the photolysis of single homochiral crystals at room temperature resulted in the formation of an optically active thietane-fused β-lactam (**35a**,75%) with 10% ee. The solid state photoreaction proceeded even at $-45°C$ to give higher ee value, 40% ee (conv. 30%, yield 70%) (Table 6). By using seeding methods during recrystallization, both enantiomers could be obtained selectively in bulk. Three other thioamides **34b–c** crystallized in centrosymmetric space groups, and the irradiation of these crystals gave racemic β-lactams **35b–c**.

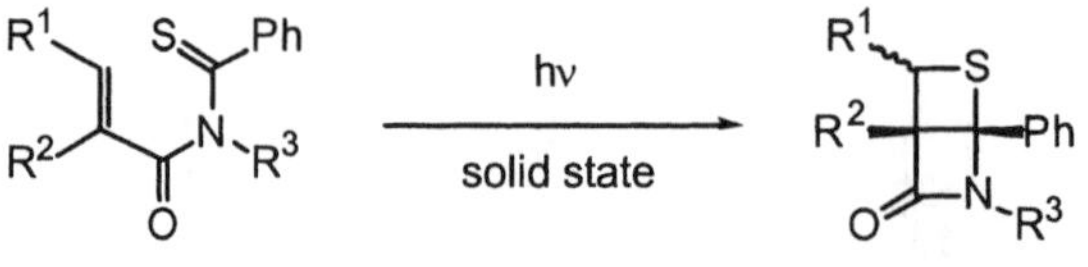

Scheme 15   Absolute asymmetric β-lactam synthesis.

Figure 4 shows the reflected UV and the CD spectra of the two enantiomorphic crystals of **34a** in KBr pellets, which were independently obtained by spontaneous crystallization. These crystals gave specific curves in the region between 300 and 600 nm, which were mirror images designated as (+) and (−) based on the CD sign at the wavelength of 470 nm. The monothioimide **34a** has no chiral center; but the absolute structure of the chiral conformation in the chiral crystal could be established by the x-ray anomalous scattering method, where the absolute structure of the chiral conformation of (−)-**34a**, which corresponds to the (−)-crystal in the solid-state CD spectra, was (M)-configuration.

By suppression of the reaction conversion to 30% and decrease of the reaction temperature to − 45°C, the enantiomeric purity rose up to 40% *ee*, as shown in Table 3. The absolute configuration of (+)-**35a** was determined as (1S,4R)-isomer by the x-ray anomalous scattering method, which was the major photoproduct obtained by the solid-state photoreaction of (M)-(−)-**34a**.

Scheme 16 shows the mechanism for the thietane formation, in which the six-membered 1,4-biradical **BR** is appropriate. There are two ways of cyclization to thietane, and each pathway gives an enantiomeric structure of thietanes, (1S,4R)- *or* (1R,4S)-**35a**, respectively [23b]. The answer to the question concerning the pathway was provided by a correlation study of the absolute structure before and after the reaction. Figure 5 shows the superimposed structure of both

Table 6   Solid State Photoreaction of Monothioimikdes **34**

	R¹	R²	R³	Space group	Yield of **35**	*ee* of **35**
34a	H	Me	Ph	$P2_12_12_1$	70%	40%
34b	H	Me	Pr^j	centrosymmetric	93	0
34c	Me	Me	Ph	$P2_1/n$	92	0
34d	Me	Me	Pr^j	$P2_1/c$	95	0

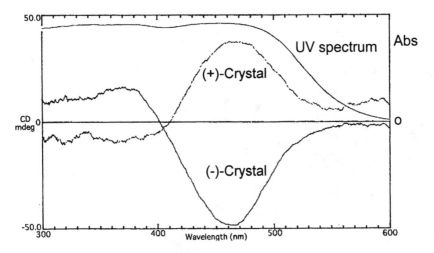

Figure 4 Reflected UV and solid state CD spectra of enantiomeric crystals of both antipodes of **34a** in KBr.

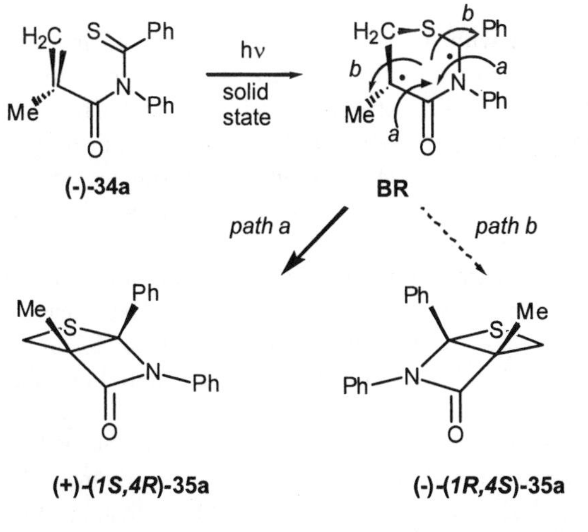

Scheme 16 The dynamic structural changes in the crystal.

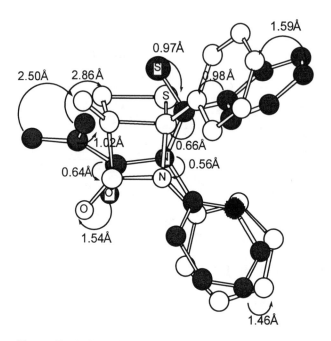

Figure 5 A front view of the superimposed structure of $(-)$-(M)-**34a** (black circle) and $(+)$-$(1S,4R)$-**35a** (empty circle) obtained by x-ray structural analysis using Overlay program in Chem3D.

absolute structures. It was drawn with the overlay program included in CSC Chem3D. The sulfur and the alkenyl carbon atoms are closely placed to make the C–S bond easily, and subsequent cyclization of biradical **BR** needs the rotation of the radical center like *path a* to yield $(1S,4R)$-**35a**. The molecular transformation from $(-)$-**34a** to $(+)$-**35a** needs considerable molecular rearrangement involving the rotation of the methacryl group. As the cyclization proceeded, the volume of the molecule was reduced, and this resulted in the formation of considerable spaces in the crystals. Furthermore, the molecular transmigration of the methacryl group resulted in making new defects in the crystal. The dynamic molecular rearrangement for cyclization was elucidated on the basis of direct comparison of the absolute configuration of both the starting material and the photoproduct.

Similar absolute asymmetric synthesis was demonstrated in the solid-state photoreaction of N-(β,γ-unsaturated carbonyl)-N-phenylthiocarbamate **36** [24]. Achiral O-methyl N-(2,2-dmethylbut-3-enoyl)-N-phenylthiocarbamate **36a** (Scheme 17) crystallized in chiral space group $P2_1$, and irradiation of these crys-

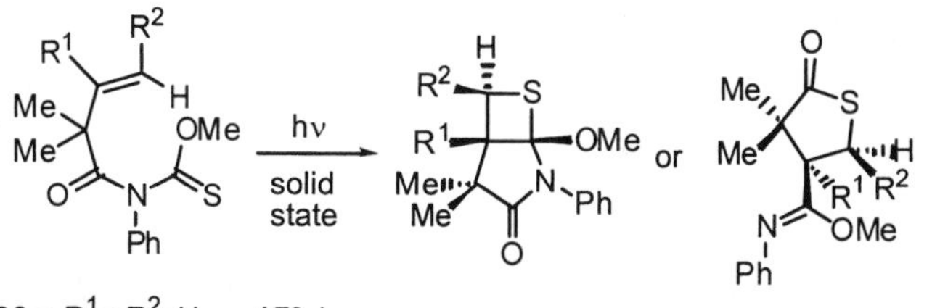

36a: $R^1 = R^2 = H$ (*P*2$_1$)
36b: $R^1 = R^2 = -(CH_2)_4-$ **37b**

38a: 10-31% ee

Scheme 17 Asymmetric synthesis via thietane formation.

tals gave optically active thiolactone in 10–31% *ee*. A plausible mechanism for the formation of **38** is rationalized on the basis that photolysis of **36** undergoes [2 + 2] cyclization to thietane **37** and is subsequently followed by rearrangement to thiolactone **38**. On the other hand, thiourethane **36b**, crystallized in the centro-symmetric crystal system, was led to the racemic thietane-fused γ-lactam **37** upon irradiation.

Sakamoto et al. also demonstrated the absolute oxetane synthesis in the solid-state photolysis of *N*-(α,β-unsaturated carbonyl)benzoylformamides **39** (Scheme 18) [25]. These compounds undergo intramolecular [2 + 2] cyclization to give bicyclic oxetanes **40** in homogeneous solution. The x-ray analysis of *N*-isopropyl substituted imide **39a** revealed that the crystal system was monoclinic and the space group *P*2$_1$ (Table 7). Crystals of **39a** were powdered and photolyzed at 0°C. The imide undergoes the [2 + 2] cycloaddition to afford the bicyclic oxetane **40a**, which is a mixture of diastereomers, namely, syn- and anti isomers at the C-7 position. In this reaction optically active *syn*-oxetane **40a** with 37% *ee* (84% chemical yield) and racemic *anti*-**40a** were obtained (Table 8). The solid-state photoreaction proceeded even at −78°C, and optically active *syn*-**40a**, which showed *ee* value as high as > 95% *ee* (conv 100%, chemical yield 89%), was

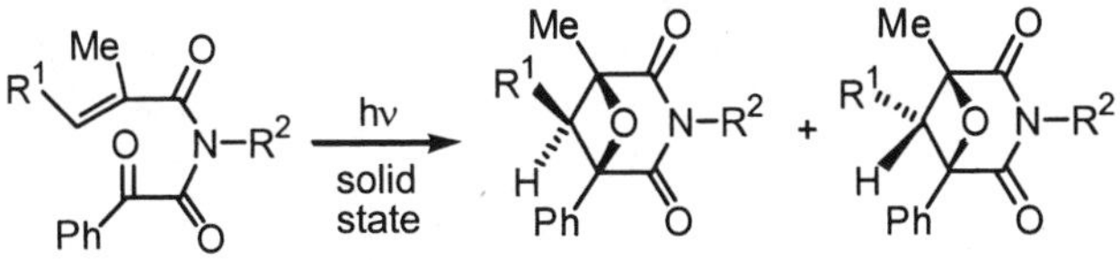

39   **syn-40**   **anti-40**

Scheme 18 Absolute asymmetric oxetane synthesis.

Table 7 Space Group and the Molecular Conformation of Imides **39a–j**

Imide	R^1	R^2	Space group	Conformation
39a	Me	Pr^i	$P2_1$	E,E
39b	Me	Bn	chiral	E,E
39c	Me	Ph	$Pca2_1$	E,Z
39d	Me	o-tol	$P2_1/a$	E,E
39e	Me	$2,6\text{-}Me_2C_6H_3$	$Pbca$	E,E
39f	H	Pr^i	$P2_1/a$	E,E
39g	H	Bn	$P2_1$	E,Z
39h	H	Ph	$P2_1/c$	E,Z
39i	H	$2,6\text{-}Me_2C_6H_3$	$P2_1/c$	E,Z
39j	H	$2,6\text{-}diCl_2C_6H_3$	$P2_1/c$	E,Z

formed in a higher diastereomeric ratio (syn/anti = 6.5). Under identical conditions N-benzyl substituted **39b** was irradiated in the solid state. The solid state photoreaction provided very high syn/anti stereoselectivity of **39b** (syn/anti ratio 2.1 in solution and 60 in the solid state, respectively). Furthermore, the solid state photoreaction was found to give the *syn*-**39b** as a chiral compound in 81% *ee*; however, the crystal system could not be determined.

Table 8 Photolysis of Imides **39a–39j** in the Solid State

Entry	Reaction temp (°C)	Yield of **40** (%) (syn/anti)	ee (%) of syn-**40** (**41g**)
39a	0	84 (3.7)	35
39a	−78	89 (6.7)	99
39b	15	100 (60)	91
39b	−78	100 (60)	91
39c	0	0[a]	—
39d	15	100 (27)	0
39e	15	100 (20)	0
39f	0	68 (31)[b]	0
39g	0	50 (20)[c]	0 (88)[d]
39h	15	0	—
39i	15	0	—
39j	15	0	—

[a] Only cis–trans isomerization took place, and the ratio of *cis*-**39c**/*trans*-**39c** was 1.3. [b] Chemical yield of α-cleavage reaction product. [c] Chemical yield of **41g**. [d] *ee* of **41g**.

In contrast to the photoreaction of **39a** and **39b**, the solid state photocycliza-tions of **39c** and **39h–j** to give the oxetanes did not occur. The x-ray analyses of these imides indicate that all molecules are apparently in unfavorable orienta-tion for the intramolecular cyclization, what was termed as (*E,Z*) conformation at the imide chromophores.

When crystals of **39d** and **39e** were irradiated at 20°C, mixtures of the corresponding oxetanes, **40d** and **40e**, were obtained in quantitative yields (con-version 100%), but in racemic form (Table 8). The solid-state photolysis of **39f** gave azetidine-2,4-dione via α-cleavage reaction leading to a radical pair interme-diate and was followed by cyclization. Molecules of **39g** (Scheme 19) adopt chiral packing (space group *P*2$_1$) and a helical molecular conformation and crystallize in (*E,Z*) conformation, which is unfavorable for oxetane formation. The solid-state irradiation of the substrate **39g** was found to give the oxetane **40g** and a β-lactam derivative **41g**. The β-lactam **41g** was revealed to be enantiomerically enriched to 88% *ee*, whereas the other photoproduct **40g** was racemic. The occurrence and mechanism of transformations of **39g** to **41g** involve hydrogen abstraction by the alkenyl carbon atom, which was described in Sec. II.B.2.

Kohmoto et al. reported a fine example of absolute asymmetric synthesis involving the intramolecular [4 + 4] photocycloaddition of 9-anthryl-*N*-(naph-thylcarbonyl)carboxamide derivatives **42a–g** in the solid state (Scheme 20) [26]. Out of seven carboxamides examined, **42a**, **42e**, and **42g** showed intramolecular photocycloaddition in the solid state to give the [4 + 4] cycloadducts **43a**, **43e**, and **43g** in almost quantitative yields after complete conversion and **42f** gave **43f** in 9% yield (Table 9). However, **42b**, **42c**, and **42d** were unreactive. The crystals

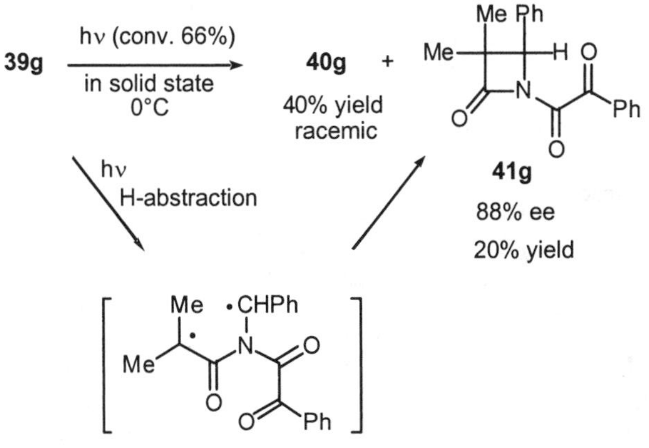

Scheme 19 Asymmetric synthesis via hydrogen abstraction by the alkenyl carbon atom.

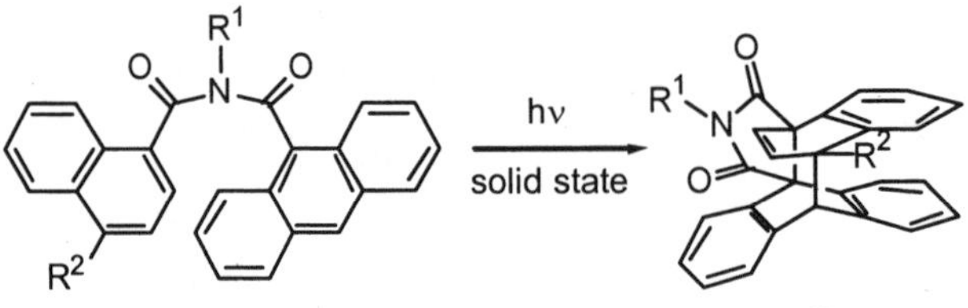

Scheme 20   Asymmetric synthesis via 4 + 4 cycloaddition.

of **42a**, **42b**, and **42e** have the chiral space groups $P2_12_12_1$, $P1$, and $P2_1$, respectively, and others belong to achiral space groups. Asymmetric induction in the photocycloaddition of **1a** and **1e** in single crystals was observed, and optically active products, **43a** (82% *ee*) and **43e** (100% *ee*), were obtained.

4.   Photochemical Electrocyclization

Toda et al. reported another example of absolute asymmetric photoreaction involving electrocyclization in the solid state; crystallization of 3,4-bis(phenylmethylene)-*N*-methylsuccinimide **44** gave chiral crystals as orange hexagonal plates and two racemic crystals as orange rectangular plates and yellow rectangular plates (Scheme 21) [27]. Irradiation of powdered enantiomeric crystals gave optically active photocyclization product **45** of 64% *ee*. Photolysis of eight other derivatives led to a racemic cyclization product in quantitative yield. Deeper understanding of the reaction pathway of this system must await a crystallographic study akin to that carried out for electrocyclization.

Table 9   Solid State Photoreaction of Imides **42** (%)

Imide	R^1	R^2	Space group	*ee* of **43** (yield) (%)
42a	CH_2Ph	H	$P2_12_12_1$	82 (100)
42b	$o\text{-}CH_3C_6H_4CH_2$	H	$P1$	(recovered)
42c	$p\text{-}CH_3C_6H_4CH_2$	H	$P\bar{1}$	(recovered)
42d	$p\text{-}FC_6H_4CH_2$	H	$P2_1/n$	(recovered)
42e	$o\text{-}(CH_3)_3C_6H_4CH_2$	H	$P2_1$	100 (100)
42f	$n\text{-}C_3H_7$	H	$P2_1/c$	0 (9)
42g	*iso*-Pr	CH_3	$P2_1/n$	0 (100)

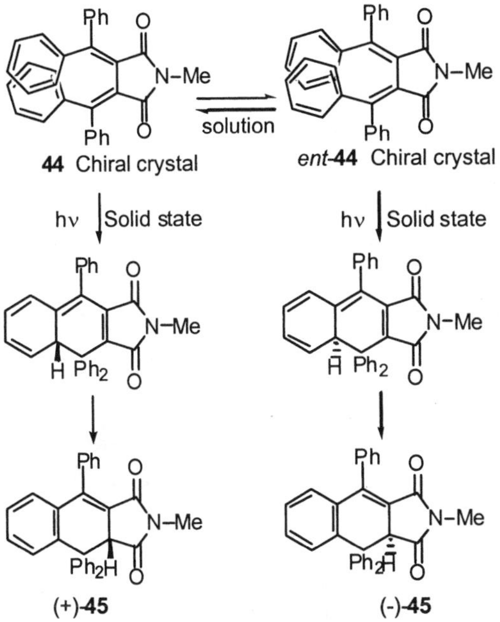

Scheme 21 Asymmetric synthesis using electrocyclization.

The major drawbacks of the crystal-field method, which prevent it from becoming a general method for absolute asymmetric synthesis, are the unpredictability of the crystallization of the achiral substrate, which as indicated can be circumvented by manual seeding, and the fact that achiral substrates seldom crystallize in chiral space groups. It was shown that it is possible to regulate this chiral crystallization by crystal engineering. In the case of the formation of the β-lactam, it was shown that the ability of a compound to crystallize in a chiral space group was increased by using compounds with *meta*-substituted aryl groups such as **46a–d** (Scheme 22) instead of their *ortho*- and *para*-substituted analogs **46e–h**; the absolute asymmetric photocyclization of the *m*-chloro-substituted benzyloxy pyridone resulted in the formation of **47a** with 78% *ee* [28] (Table 10).

In the case of α-oxoamides **22**, *meta*-substituted amides exhibit a relatively high tendency of crystallization with the chiral framework [19c]. If both observations that *meta*-substituted arenes prefer to crystallize in noncentrosymmetric space groups, in comparison to the *ortho*- and *para*-isomers, have some generality, then *meta*-substitution in an achiral compound could be an important structural

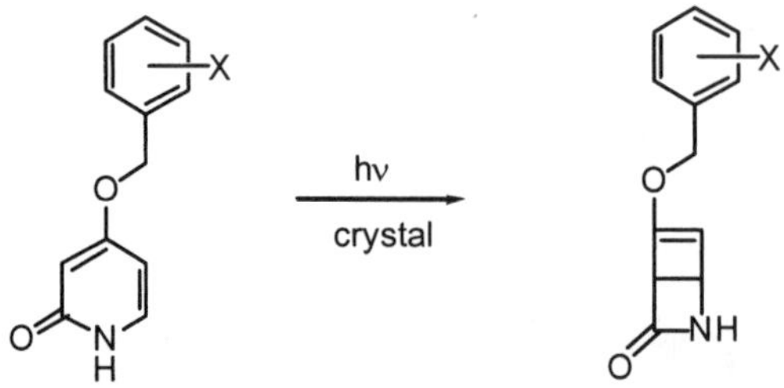

Scheme 22   Asymmetric synthesis via electrocyclization.

motif in the design of new absolute asymmetric synthesis, making it more predictable.

5.   Miscellaneous Photoreactions in the Solid State

Irradiation of a benzene solution of *S*-phenyl *N*-benzoylformyl-*N*-*p*-tolylthiocarbamate **48a** (Scheme 23) gave 5-phenyl-5-phenylthio-3-*p*-tolyloxazolidine-2,4-dione **49a** in 61% yield accompanied by oxazolidine-2,4-dione dimer (15%), *p*-tolyl isocyanate (22%), and diphenyl disulfide [29]. Photolysis of **48a** in the solid state gave oxazolidine-2,4-dione **49a** in 96% yield. For the *N*-methyl derivative **48b**, compared to the solution photochemistry in which only 8% of oxazolidinedione **49b** was obtained with a complex mixture, radical cyclization proceeds selectively to give oxazolidinedione in 75% yield in the solid state. Whereas *N*-*p*-tolyl and *N*-methyl derivatives, **48a** and **48b**, formed achiral crystals, the *N*-benzyl derivative **48c** crystallized in chiral space group *P*2₁. Photolysis of the chiral

Table 10   Solid State Photoreaction of Pyridones **46**

Pyridone **46**	Space group	*ee* of **47** (%)
46a X = *m*-Cl	$P2_12_12_1$	69–78
46b X = *m*-Me	$P2_12_12_1$	58–71
46c X = *m*-OMe	$P2_12_12_1$	90–100
46d X = *m*-Br	$P2_12_12_1$	unstable
46e X = H	$P2_1/n$	0
46f X = *o*-Cl	$P\bar{1}$	0
46g X = *p*-Cl	$P\bar{1}$	0
46h X = *p*-*t*-Bu	$P\bar{1}$	0

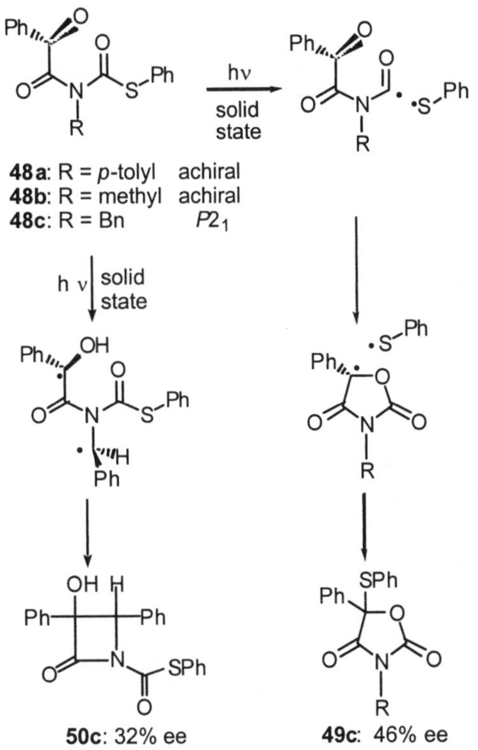

Scheme 23 Absolute asymmetric synthesis via Norrish type 2 reaction and radical pair intermediate.

crystals in the solid state gave optically active 1-benzyl-4-phenyl-4-phenylthioox-azolidine-2,4-dione (**49c**,16% chemical yield, 21% *ee*) and *cis*-3,4-diphenyl-3-hydroxy-1-(thiophenylcarbonyl)azetidin-2-one (**50c**, 18% chemical yield, 23% *ee*) in 62% conversion yield. Better optical purities were observed at low conversion (in 17% cov.), 46% *ee* for **49c**, and 32% *ee* for **50c**, respectively. A large quantity of chiral crystals of the thiocarbamate was prepared selectively and in bulk by seeding with a small amount of the chiral crystals during recrystallization.

The interatomic distance between the reacting carbonyl carbon and the other carbonyl oxygen for generating oxazolidine-2,4-dione is 3.37 Å, and the distance between the carbonyl carbon and the sulfur atoms is 4.72 Å. Furthermore, for the hydrogen abstraction of the *N*-benzyl derivative, the distance between the carbonyl oxygen and hydrogen atoms is 4.45 Å, which is much larger than the sum of the van der Waals radii (2.9 Å). This reaction provides an example of an

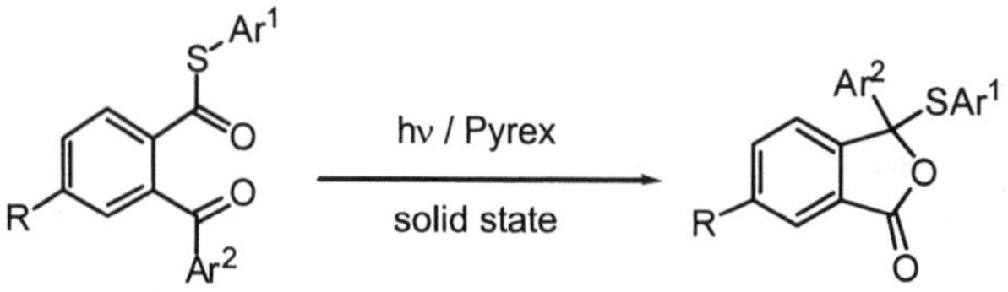

Space group of **51**

51a: R = H, Ar^1 = (*o*-tol), Ar^2 = Ph $P2_12_12_1$

51b: R = H, Ar^1 = Ph, Ar^2 = Ph $P2_12_12_1$

51c: R = H, Ar^1 = (*m*-tol), Ar^2 = Ph $P2_12_12_1$

51d: R = H, Ar^1 = (*p*-tol), Ar^2 = Ph $P2_1/c$

51e: R = Me, Ar^1 = Ph, Ar^2 = Ph $P\bar{1}$

51f: R = H, Ar^1 = Ph, Ar^2 = (*p*-tol) $P\bar{1}$

51g: R = H, Ar^1 = Ph, Ar^2 = (*p*-ClPh) $P\bar{1}$

Scheme 24   Absolute asymmetric phthalide synthesis via aryl migration.

absolute asymmetric synthesis via a radical pair intermediate in the crystalline state and also presents a rare example of two different types of reactions proceeding in the solid state holding chirality.

Next are the solid state photoreactions of *S*-aryl 2-aroylbenzothioates **51** (Scheme 24) which led to a remarkable conclusion that the major process involved was an unprecedented reaction sequence, starting with intramolecular cyclization, followed by aryl migration to the photoproducts **52** [30]. The absolute asymmetric generations were observed with good enantioselectivities, which seriously depended on the conversion and reaction temperature. The mechanistic studies based on stereochemical correlation before and after the reaction gave strong evidence for the putative reaction pathway as well as the geometrical analyses of the starting molecules in the crystal lattices. Thus the correlation study of both absolute structures before and after the reaction may offer a powerful tool for clarifying obscure reaction pathways during solid state reactions.

7-*S*-Aryl 2-aroylbenzothioates **51a–g** (Scheme 25) were subjected to the x-ray crystal analysis and it was revealed that three thioesters **51a–c** crystallized in a chiral space group, and the other crystals were racemic. Irradiation of powdered (+)-**51a** at 0°C led to the production of 3-phenyl-3-(*o*-tolylthio)phthalide **52a** as a major product (65% yield) with complete consumption of the starting material (Table 11). As expected, the asymmetric generation in **52a** was observed in 30% *ee*. Pronounced changes in the product profiles were recognized when

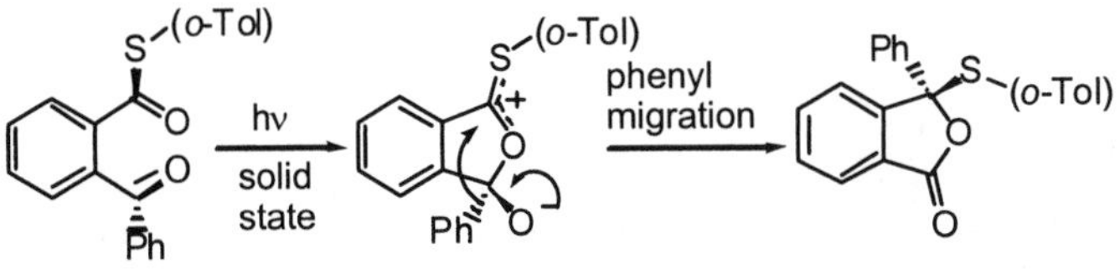

(P) -(+) - thioester **51a**
Space group: $P2_12_12_1$

(R) -(+) - phthalide **52a**
77% ee

Scheme 25 Mechanism for the phthalide formation via aryl migration.

the reaction was carried out at low (5%) conversion and at low temperature (−78°C). In the former case, a good *ee* value (77%) was obtained with exclusive product formation (> 95% conversion). Additionally, the reaction at −78°C gave similar results (50% yield, 65% *ee*) for the same reason: the molecules were strongly conformationally frozen in the crystal lattices. Obviously, the topochemical control was much more effective at lower temperature, since 65% *ee* was obtained even at 50% conversion, which value corresponded to about 39% *ee* at 0°C as an estimate from the changing profile. *S*-Phenyl and *S*-(*m*-tolyl) derivatives **51b** and **51c** formed chiral crystals (space groups $P2_12_12_1$), and their solid state photoreactions also led to optically active products with *ee* values varying with reaction conditions. Under well-optimized conditions, good *ees* were eventually

Table 11 Solid State Photoreactions of Thioesters **51**

Thioester **51**	Temp. (°C)	Conv. (%)	Yield of **52** (%)	*ee* of **52** (%)
51a	0	100	65	30
	0	5	> 95	77
	−78	50	92	65
51b	0	20	97	35
	0	10	> 95	71
	−78	10	> 95	87
51c	0	56	96	23
	0	10	95	71
	−78	67	> 95	74
51d	0	100	89	0
51e	0	67	78	0
51f	0	63	91	0
51g	0	71	85	0

obtained in the products as high as 87% *ee* for **52b** (10% conversion at −78°C) and 74% *ee* for **52c** (67% conversion at −78°C).

These intramolecular cyclization reactions in the solid state were also examined with other thioesters **51d–g** to give the corresponding phthalides **52d–g**, but as racemates. The failure of generation of optical activity in the products was well compatible with their centrosymmetric crystal systems such as $P2_1/c$ and $P\bar{1}$.

With regard to the mechanism of these solid state photoreactions, there are two possible pathways from the starting thioesters **53** to phthalides **54**. One mechanism involves arylthio group migration, initiated by the homolytic dissociation between C-S bonding to form a radical pair intermediate. Such a pathway is well recognized as the excited-state reaction of the thioester compounds. The other mechanism consists of a direct cyclization and subsequent phenyl migration sequence, in which zwitterionic intermediate is involved. To solve this mechanistic problem, stereochemical correlation studies before and after the solid state photoreactions were employed. Fortunately, both absolute configurations of the starting thioester in the chiral crystal and the optically active photoproduct were determined. Eventually, this reaction elucidated the stereochemical relationship from (*P*)-**1a** to (*R*)-**2a**. These facts clearly indicated that the solid state reaction did not involve arylthio group migration, but rolled in phenyl migration as shown in Scheme 25.

Absolute asymmetric synthesis was observed in the solid-state photoreaction of benzoylbenzamide **53** (Scheme 26) to phthalide **54**; the reaction mechanism, however, was completely different from that of thioester **51** [31]. Recrystallization of these amides **53a–c** from the chloroform-hexane solution afforded colorless prisms in all cases. X-ray crystallographic revealed that all prochiral amides **53a–c** adopted orthorhombic chiral space group $P2_12_12_1$ and were frozen in chiral and helical conformation in the crystal lattice.

The solid state photolysis of powdered **53b** gave a quantitative amount of 3-(*N*-methylanilino)-3-phenylphthalide **54b**, whereas **53a** was inert toward topochemical reaction upon irradiation. As expected, the asymmetric induction in **54b** was observed [Table 12]. As a result of the suppression of the reaction conversion yield and by decreasing the reaction temperature to −50°C, the enantiomeric purity rose up to 87% *ee* (at 25% conversion with quantitative yield). Furthermore, irradiation of single crystals of **53b** gave 97% *ee* of **54b** when the reaction conversion reached 46%.

With regard to the mechanism of the solid state photoreactions, there are two possible pathways from the starting amides **53** to phthalides **54**; either phenyl migration, which is the same as in the mechanism of thioester derivatives **51**, or the mechanism initiated by homolytic cleavage of the C(=O)–N bond to generate a radical pair intermediate. The radical mechanism has been confirmed in the photo-Fries rearrangement of aromatic amides. To answer the question of the

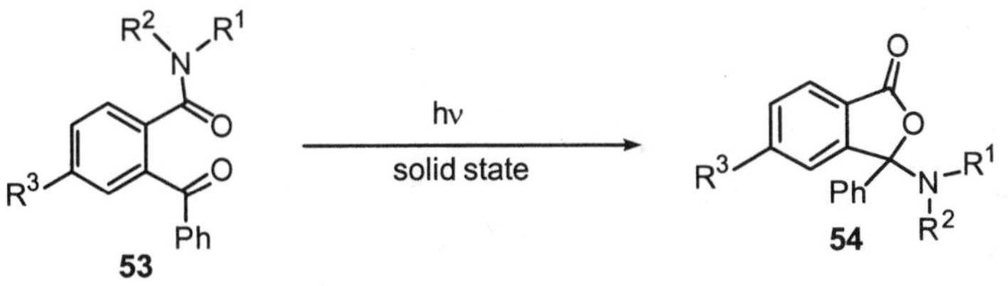

53a: $R^1 = R^2 = Me$, $R^3 = H$ $P2_12_12_1$
53b: $R^1 = Me$, $R^2 = Ph$, $R^3 = H$ $P2_12_12_1$
53c: $R^1 = Me$, $R^2 = Ph$, $R^3 = Me$ $P2_12_12_1$

Scheme 26 Absolute asymmetric phthalide formation via radical pair intermediate.

reaction mechanism involved, regiochemical correlation was examined using a methyl probe ($R^3 = Me$) on the central aryl ring in the starting material. The findings based on the methyl probe between the starting material and the final product provide straightforward proof for the determination of the pathway since the position of this substituent in the product can represent the reaction course.

When the crystal of **53c** (Scheme 27) was irradiated, a single regioisomer of **54c** was obtained quantitatively. As is apparent from the regiochemical correlation, the reaction pathway, which involves photo-Fries rearrangement, was indispensable for the rationalization of the product. X-ray structural analysis indicates that the crystal of **53c** is also chiral, the space group $P2_12_12_1$ (Scheme 26). The photoproduct **54c** showed an *ee* of 42% (Table 12). Furthermore, a higher *ee* value of **54** (87%) was obtained in the reaction at $-50°C$. The solid state photore-

Table 12 Solid State Photoreaction of **53** in Various Conditions

Amide	Condns. of solid	Temp. (°C)	Conv. (%)	Yield (%) of **54**	*ee* (%) of **54**
53a	powder	15	recovered	0	–
53b	powder	15	100	> 99	80
53b	powder	0	100	> 99	83
53b	powder	−50	47	> 99	87
53b	crystal	0	56	> 99	86
53b	crystal	−50	46	> 99	97
53c	powder	15	100	> 99	42
53c	powder	−50	25	> 99	87

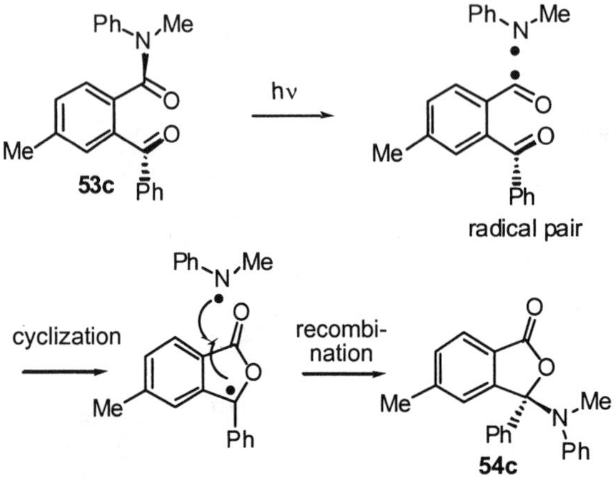

Scheme 27 Reaction mechanism for phthalide formation via radical pair intermediate.

action of *N,N*-disubstituted 2-benzoylbenzamides promoted intramolecular cyclization to phthalides via a radical pair intermediate, in which the "absolute" asymmetric generation into the prochiral starting materials in the chiral crystalline environment was performed with good enantioselectivity.

Ohashi et al. reported a fine example of racemic-to-chiral transformation in cobaloxime complex crystal by irradiation. A racemic mixture of (1-cyanoethyl)(piperidine)cobaloxime **55** (Scheme 28), which contains the chiral 1-cyanoe-

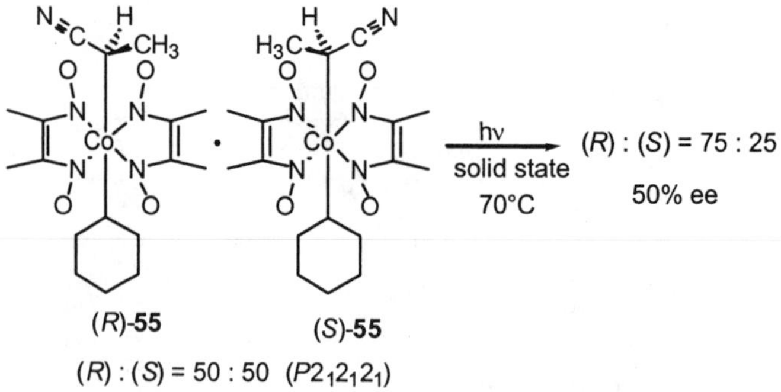

Scheme 28 Asymmetric transformation of racemic cobaloxime complex in the chiral crystalline environment.

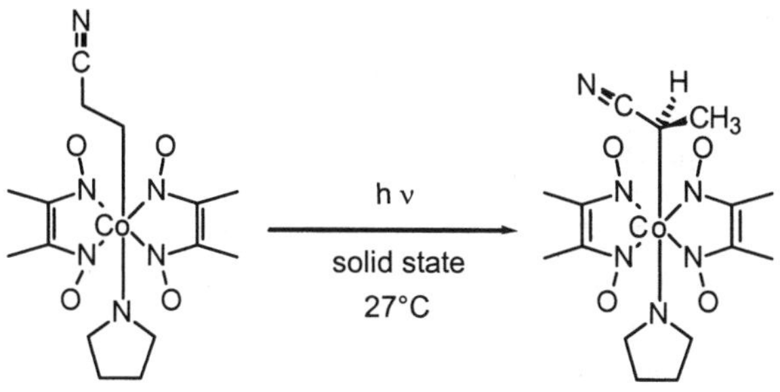

56 $P2_12_12_1$

57

conv.	ee
1.4%	21% ee
2.9%	19% ee
6.7%	20% ee
6.0%	20% ee
8.9%	17% ee

Scheme 29 Absolute asymmetric transformation of cobaloxime complex crystal.

thyl group, crystallized in a chiral space group ($P2_12_12_1$) in which *R* and *S* enantiomers occupy crystallographically distinct (diastereoisomeric) positions in the asymmetric unit cell. Single crystals were prepared of the Δ configuration. Irradiating the crystal with x-rays at 343K causes racemization of the *S* enantiomer while the *R* enantiomer remains unaltered. This racemization is thought to be due solely to the different volume constraints on the two enantiomers in their different crystal environments. The result is that the number of *R* molecules is increased at the expense of *S* molecules, and so the overall composition of the crystal changes from racemic to enriched in the *R* enantiomer.

Recently, Ohashi also reported the asymmetric induction by the irradiation of chiral crystals of achiral (2-cyanoethyl)(pyrrolidine)cobaloxime complex **56** (Scheme 29). In this case, the achiral cobaloxime complex crystallized in the $P2_12_12_1$ space group, and the solid state photolysis gave 1-cyanoethyl derivative **57** with optical activity. The *ee*s were variable in according with the conversion. The maximum *ee* (21%) was obtained at low conversion (1.4% yield).

III. ABSOLUTE ASYMMETRIC SYNTHESIS IN SOLID–GAS REACTIONS AND IN HOMOGENEOUS MEDIA USING CHIRAL CRYSTALS

A number of apparent achiral organic compounds have been obtained as chiral crystals, and the solid-state photoreaction gave many successful absolute asymmetric syntheses. Since most solid-state reactions proceed with the least atomic or molecular movement under topochemical control owing to well-defined atomic arrangements, selected and restricted reaction is allowed predominantly, and chemoselective or stereoselective reaction is observed in the solid-state reaction. In other words, there are many cases that we hoped would afford chiral crystals; regrettably they were unreactive for the solid state photoreaction.

Whereas some interesting examples of solid–gas reaction using chiral crystals were reported, the *ee*s were not satisfactory. Reaction of chiral crystals of compound **58** (Scheme 30) with bromine in connection with rearrangement gave optically active dibromide **59** in 8% *ee* [34].

Gerdil et al. reported two examples involving the solid–gas reactions of inclusion complexes of tri-*o*-thymonide with alkene or epoxycyclopentanone (Scheme 31). The complex **60** crystallized in $P3_121$ and the reaction with singlet oxygen gave endoperoxide **61**; however, the $[\alpha]_D$ value was low [35]. On the other hand, the reaction of chiral crystals of the complex **62** with hydrogen chloride gave two products **63** and **64**, whose *ee*s were 9% and 22%, respectively [36]. After all, intermolecular solid-state reactions are generally disadvantageous in achieving high enantioselectivity as compared with the solid-state photoreaction, because the reactions generally promote on the crystalline surface with breakdown of the lattice. If the chiral molecular information can be used in homogenous conditions, many chemical reactions including photochemical reactions will be applied to absolute asymmetric synthesis.

Azumaya et al. reported an interesting example of retention of the molecular chirality when the chiral crystal of 1,2-bis(*N*-benzoyl-*N*-methylamino)benzene **65** was dissolved in a cold solution (Fig. 6) [37]. Furthermore, Tissot et al. reported a fine example of the formation of optically active complex **67** ($\sim$ 100% *ee* in 93% yield) using axially chiral ligand **66** (Scheme 32) prepared by chiral crystallization [38]. Both achiral **65** and **66** exist as mixtures of many conformational isomers or diastereomers in solution at ambient temperature. If the molecular chirality is retained in homogeneous conditions like these examples, the frozen

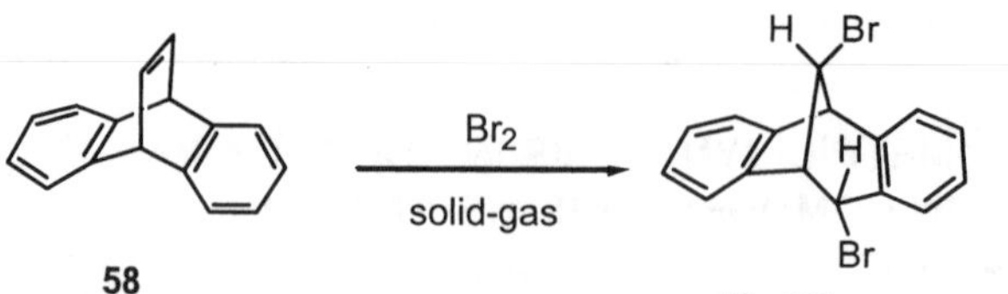

Scheme 30 Absolute asymmetric synthesis using solid–gas reaction.

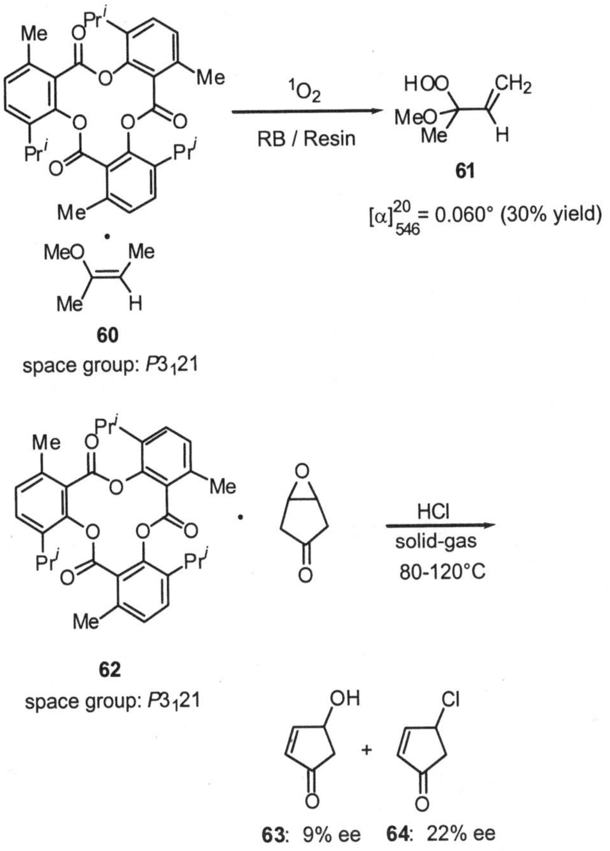

Scheme 31 Solid–gas reaction using inclusion complexes **60** and **62**.

chirality generated by spontaneous crystallization should be effectively transferred to the products by asymmetric reactions.

Recently, Sakamoto et al. reported an absolute asymmetric synthesis using the frozen chirality generated by chiral crystallization. Achiral asymmetricly substituted imide **68**, which bonds between the nitrogen atom and the tetrahydronaphtyl (TENAP) group, rotates freely at room temperature, crystallized in a chiral fashion, and the enantiomerization owing to the bond rotation was suppressed at low temperature (Scheme 33). Furthermore, the frozen molecular chirality could be transferred to optically active products in fluid solution [35]

The CD spectra of the THF solution of the imide **68** prepared by dissolving chiral crystals into cold THF using a cryostat apparatus were measured. When

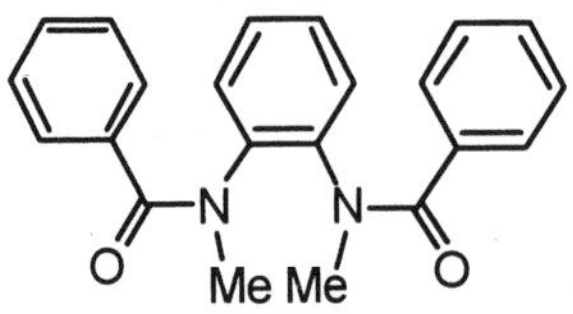

65
space group: $P2_12_12_1$

Figure 6 Retention of molecular chirality when the chiral crystal of 1,2-bis(N-benzoyl-N-methylamino)benzene **65** is dissolved in a cold solution.

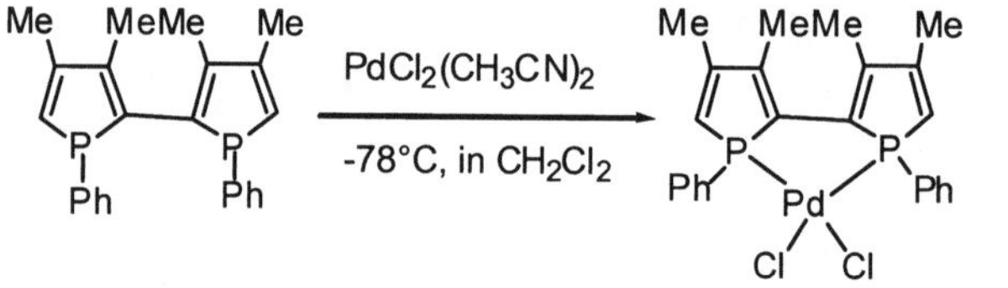

66
space group: $P2_1$

67
~100% ee (93% yield)

Scheme 32 Optically active complex formation in solution.

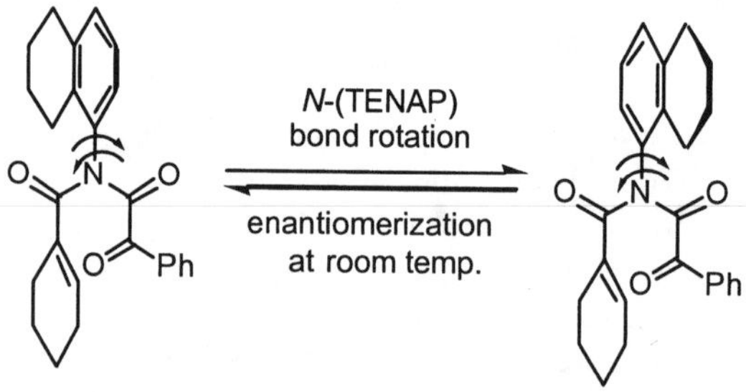

(S)-**68**   (R)-**68**

Scheme 33 Enantiomerization of acidic asymmetricly substituted imide **68**.

the crystals were dissolved at $-10°C$, the CD spectrum did not show any Cotton effect. However, optical activity was observed, when the crystals were dissolved at $-20°C$. The optical activity gradually decreased, and the half-life was 7.8 min. The half-life increased with lowering the temperature; $t_{1/2}$ was 31.6 min and 150.0 min at the temperatures of $-30°C$ and $-40°C$, respectively. The energy barrier ($\Delta G^{\neq}$) of enantiomerization was calculated from the temperature dependence of the kinetic constants to be 18.24–18.36 kcal mol^{-1} at 233–253K. These facts indicate that achiral imide **68** can retain the molecular chirality doped in the crystal lattice below $-20°C$ in homogeneous conditions, and the lifetime is enough for the next application to asymmetric synthesis.

When the bulk of the crystals of **68** (Table 13) in a test tube was irradiated at 15°C under an argon atmosphere, intramolecular [2 + 2] cyclization proceeded effectively without melting down, and two diastereomeric oxetanes, **69** and **70**, were obtained in 95 and 5% yield, respectively. The enantiomeric purity of the main product **69** was determined as > 99%. When **68** was irradiated after dissolving in THF at various temperatures, optically active oxetanes were isolated below $-20°C$, whereas the racemic oxetanes were naturally obtained from the photolysis above 0°C in THF. The photolysis in THF at $-60°C$ gave 87% *ee* of **69** and 62% *ee* of **70**, in 76 and 24% chemical yields, respectively. The memory of

Table 13 Absolute Asymmetric Synthesis in the Solid State and in Homogeneous Conditions Using Chiral Crystals

Temp. (°C)	Conditions	**69** : **70**	*ee* (%) of **69**	*ee* (%) of **70**
15	solid state	95 : 5	>99	–
0	THF	67 : 33	0	0
−20	THF	68 : 32	51	31
−40	THF	76 : 24	83	62
−60	THF	81 : 19	87	79

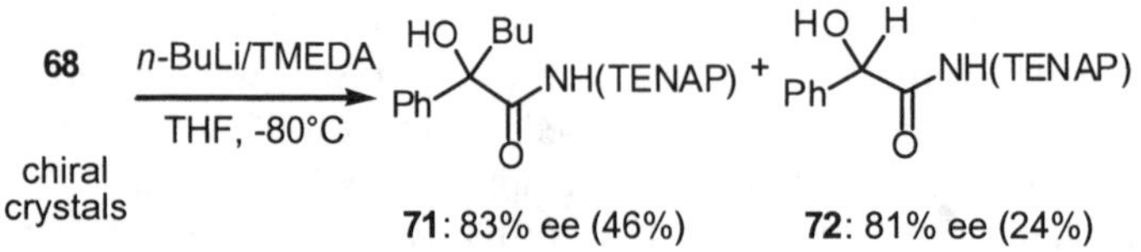

Scheme 34 Absolute asymmetric synthesis via nucleophilic reaction using chiral crystals.

chirality generated by chiral crystallization effectively transformed to permanent chirality in homogeneous conditions.

Apart from the photoreaction, absolute asymmetric synthesis in a nucleophilic reaction with *n*-butyllithium was also performed (Scheme 34). Commercially available hexane solution of *n*-butyllithium was added to the THF solution immediately after the powdered crystals of **68** were dissolved in the THF solution containing TMEDA at 20°C. After nucleophilic addition to the benzoyl carbonyl group occurred effectively, the adduct **71** and the hydrogenated product **72** were obtained in 88% yield, and the ratio **71** : **72** was 90 : 10. As a matter of course, molecules immediately lost chirality on dissolving the chiral crystals in a solvent, and racemic products were obtained. Surprisingly, when powdered crystals were added to the THF solution containing *n*-butyllithium and TMEDA at 20°C, optically active products **71** (41% *ee*) and **72** (5% *ee*) were isolated. At lowered temperatures, the enantioselectivity increased, and the maximum *ee* values, 83% for **71** and 81% for **72**, were obtained at -80°C. Controlling the activation parameters for enantiomerization resulted in the new absolute asymmetric synthesis using chiral crystals.

IV. CONCLUSIONS

The current status of organic transformations using solid state photoreactions using chiral crystals was summarized. The future of this field is intimately connected with progress in the general area of organic solid state chemistry as well as with deeper understanding of the molecular packing modes of crystals. If tailored crystals can be made, this solid state asymmetric synthesis will be extended to a variety of new systems, so that it can now be regarded as an important branch of organic chemistry. Furthermore, asymmetric synthesis, by the use of frozen molecular chirality memorized by chiral crystallization, can be applied to nontopochemical reactions; thus it is possible for many organic reactions to be available for absolute asymmetric synthesis.

REFERENCES

1a. Schmidt GMJ. Pure Appl Chem 1971; 27:647–678.
1b. Green BS, Lahav M, Rabinovich D. Acc Chem Res 1979; 12:191–197.
2a. Addadi L, Lahav M. In: Walker DC, Ed. Origin of Optical Activity in Nature, Elsevier. 1979.
2b. Mason SF. Nature (London) 1984; 311:19–23.
2c. Wlias WE. J Chem Educ 1972; 49:448–454.
 3. For reviews, see.
3a. Ramamurthy V. Tetrahedron 1986; 42:5753–5837.
3b. Scheffer JR, Garcia-Garibay M, Nalamasu O. In: Padwa A, Ed. Organic Photochemistry. Vol. 8. New York and Basel: Marcel Dekker, 1987:249–338.
3c. Ramamurthy V, Venkatesan K. Chem Rev 1987; 87:433–481.
3d. Vaida M, Popovitz-Bio R, Leiserowitz L, Lahav M. In Ramamurthy V, Ed. Photochemistry in Organized and Constrained Media. New York: VCH, 1991:249–302.
3e. Sakamoto M. Chem Eur J 1997; 3:384–389.
3f. Koshima H, Matsuura T. J Synth Org Chem Jpn 1998; 56:268–477.
3g. Feringa BL, Van Delden R. Angew Chem Int Ed 1999; 38:3419–3438.
 4. Asymmetric synthesis using nonchiral crystals was also performed.
4a. Chenchaiah PC, Holland HL, Richardson MF. J Chem Soc Chem Commun 1982: 436–437.
4b. Chenchaiah PC, Holland HL, Munoz B, Richardson MF. J Chem Soc Perkin Trans 2 1986:1775–1777 See also 16.
 5. Belsky VK, Zorkii PM. Acta Crystallogr Sect A 1977; 33A:1004–1006.
6a. Jacques J, Collet A, Wilen SH. In: Enantiomers, Racemates, and Resolutions, Kreiger. 1991.
6b. Collet A, Brienne M, Jacques J. Bull Soc Chim Fr 1972:127–142.
7a. Kondepudi DK, Kaufman R, Singh N. Science 1990; 250:975–976.
7b. McBride JM, Carter RL. Angew Chem Int Ed Engl 1991; 30:293–295.
7c. Kondepudi DK, Bullock KL, Digits JA, Hall JK, Miller JM. J Am Chem Soc 1993; 115:10211–10216.
8a. Kondepudi DK, Laudadio J, Asakura K. J Am Chem Soc 1999; 121:1448–1451.
8b. Asakura K, Soga T, Uchida T, Osanai S, Kondepudi DK. Chirality 2002; 14:85–89.
9a. Pincock RE, Perkins RR, Ma AS, Wilson KR. Science 1971; 174:1018–1020.
9b. Pincock RE, Wilson KR. J Am Chem Soc 1971; 93:1291–1292.
10a. Penzien K, Schmidt GMJ. Angew Chem 1969; 8:608–609.
10b. Green BS, Heller L. Sci 1974; 185:525–527.
11a. Elgavi A, Green BS, Schmidt GMJ. J Am Chem Soc 1973; 95:2058–2059.
11b. Warshel A, Shakked Z. J Am Chem Soc 1975; 97:5679–5684.
11c. Green BS, Lahav M, Rabinovich D. Acc Chem Res 19779; 12:191–197.
12a. Addadi L, Lahav M. J Am Chem Soc 1978; 100:2838–2844.
12b. Addadi L, Lahav M. J Am Chem Soc 1979; 101:2152–2156.
12c. Addadi L, Lahav M. Pure Appl Chem 1979; 51:1269–1284.
12d. Addadi L, Lahav M. J Am Chem Soc 1982; 104:3422–3429.
12e. van Mil J, Addadi L, Gati E, Lahav M. J Am Chem Soc 1982; 104:3429–3434.

12f. van Mil J, Addadi L, Lahav M, Leiserowitz L. J Chem Soc Chem Commun 1982: 584–587.

13. Hasegawa M, Chung CM, Murro N, Maekawa Y. J Am Chem Soc 1990; 112: 5676–5677.

13b. Chung CM, Hasegawa M. J Am Chem Soc 1991; 113:7311.

13c. Hasegawa M. Chem Rev: 1983; 83:507–518.

14. Suzuki T, Fukushima T, Yamashita Y, Miyashi T. J Am Chem Soc 1994; 116: 2793–2803.

14b. Suzuki T. Pure Appl Chem 1996; 68:281–284.

15. Koshima H, Ding K, Chisaka Y, Matsuura T. J Am Chem Soc 1996; 118: 12059–12065.

16a. Evans SV, Garcia-Garibay M, Omkaram N, Scheffer JR, Trotter J, Wireko F. J Am Chem Soc 1986; 108:5648–5649.

16b. Chen J, Garcia-Garibay M, Scheffer JR. Tetrahedron Lett 1989; 30:6125–6128.

16c. Chen J, Pokkuluri PR, Scheffer JR, Trotter J. Tetrahedron Lett 1990; 31:6803–6806.

16d. Gudmundsdottir AD, Scheffer JR. Tetrahedron Lett 1990; 31:6807–6810.

16e. Fu TY, Liu Z, Scheffer JR, Trotter J. J Am Chem Soc 1993; 115:12202–12203.

16f. Leibovitch M, Olovsson G, Scheffer JR, Trotter J. J Am Chem Soc 1997; 119: 1462–1463.

16g. Leibovitch M, Olovsson G, Scheffer JR, Trotter J. Pure Appl Chem 1997; 69: 815–823.

17. Roughton AL, Muneer M, Demuth M. J Am Chem Soc 1993; 115:2085–2087.

18. Aoyama H, Hasegawa T, Omote Y. J Am Chem Soc 1979; 101:5343–5347.

19a. Toda F, Soda S. J Chem Soc Chem Commun 1987:1413–1414.

19b. Sekine A, Hori K, Ohashi Y, Yagi M, Toda F. J Am Chem Soc 1989; 111:697–699.

19c. Toda F, Miyamoto H. J Chem Soc Perkin Trans 1 1993:1129–1132.

19d. Toda F, Miyamoto H, Koshima H, Urbanczyk-Lipkowska Z. J Org Chem 1997; 62: 9261–9266.

20. Irngartinger H, Fettel PW, Siemund V. Eur J Org Chem 1998:2079–2082.

21. Sakamoto M, Takahashi M, Shimizu M, Fujita T, Nishio T, Iida I, Yamaguchi K, Watanabe S. J Org Chem 1995; 60:7088–7089.

22a. Sakamoto M, Takahashi M, Moriizumi S, Yamaguchi K, Fujita T, Watanabe S. J Am Chem Soc 1996; 118:10664–10665.

22b. Sakamoto M, Takahashi M, Arai W, Mino T, Yamaguchi K, Watanabe S, Fujita T. Tetrahedron 2000; 56:6795–6804.

22c. Sakamoto M, Takahashi M, Hokari N, Fujita T, Watanabe S. J Org Chem 1994; 59: 3131–3134.

22d. Hosoya T, Ohhara T, Uekusa H, Ohashi Y. Bull Chem Soc Jpn 2002; 75:2147–2151.

23a. Sakamoto M, Hokari N, Takahashi M, Fujita T, Watanabe S, Iida I, Nishio T. J Am Chem Soc 1991; 115:7311.

23b. Sakamoto M, Takahashi M, Mino T, Fujita T. Tetrahedron 2001; 57:6713–6719.

24. Sakamoto M, Takahashi M, Arai T, Shimizu M, Yamaguchi K, Mino T, Watanabe S, Fujita T. J Chem Soc Chem Commun 1998:2315–2316.

25a. Sakamoto M, Takahashi M, Fujita T, Watanabe S, Iida I, Nishio T, Aoyama H. J Org Chem 1993; 58:3476–3477.

25b. Sakamoto M, Takahashi M, Fujita T, Watanabe S, Nishio T, Aoyama H. J Org Chem 1997; 62:6298–6308.

26. Kohmoto S, Ono Y, Masu H, Yamaguchi K, Kishikawa K, Yamamoto M. Org Lett 2001; 26:4153–4155.

27. Toda F, Tanaka K. Supramol Chem 1994; 3:87–88.

28. Wu LC, Cheer CJ, Olovsson G, Scheffer JR, Trotter J, Wang SL, Liao FL. Tetrahedron Lett 1997; 38:3135–3138.

29. Sakamoto M, Takahashi M, Fujita T, Nishio T, Iida I, Watanabe S. J Org Chem 1995; 60:4682–4683.

30a. Sakamoto M, Takahashi M, Yamaguchi K, Fujita T, Watanabe S. J Am Chem Soc 1996; 118:8138–8139.

30b. Takahashi M, Fujita T, Watanabe S, Sakamoto M. J Chem Soc Perkin Trans 2 1988: 487–491.

30c. Takahashi M, Sekine N, Fujita T, Watanabe S, Yamaguchi K, Sakamoto M. J Am Chem Soc 1998; 49:12770–12776.

31. Sakamoto M, Sekine N, Miyoshi H, Fujita T. J Am Chem Soc 2000; 122: 10210–10211.

32. Osano Y T, Uchida A, Ohashi Y. Nature 1991; 352:510–512.

33. Koura T, Ohashi Y. Tetrahedron 2000; 56:6769–6779.

34. Garcia-Garibay M, Scheffer J R, Trotter J, Wireko F. Tetrahedron Lett 1988; 29: 1485–1488.

35. Gerdil R, Barchietto G, Jefford C W. J Am Chem Soc 1984; 106:8004–8005.

36. Gerdil R, Huiyou L, Gerald B. Helv Chim Acta 1999; 82:418–434.

37. Azumaya I, Yamaguchi K, Okamoto I, Kagechika H, Shudo K. J Am Chem Soc 1995; 117:9083–9084.

38. Tissot K, Gouygou M, Dallemer F, Daran JC, Balavoine GGA. Angew Chem Int Ed 2002; 40:1076–1078.

39. Sakamoto M, Iwamoto T, Nono N, Ando M, Arai W, Mino T, Fujita T. J Org Chem 2003; 68:942–946.

12

The Solid-State Ionic Chiral Auxiliary Approach to Asymmetric Induction in Photochemical Reactions

John R. Scheffer
University of British Columbia
Vancouver, British Columbia, Canada.

I. INTRODUCTION

Strongly endothermic reactions (e.g., **A** → **B**, Fig. 1) are frequently unsuccessful in the ground state because the product, **B**, is kinetically unstable under the conditions required for its formation. By way of contrast, the formation of **B** from **A***, an electronically excited state of **A**, is often a downhill energy process, associated with a relatively low activation energy, and for this reason can be carried out under conditions where **B** is kinetically stable. This accounts for the common observation that, unlike their ground-state counterparts, photochemical reactions routinely afford highly strained, thermodynamically unstable products.

In seeking ways to capitalize on this particular advantage that photochemistry enjoys over ground-state chemistry, it would be desirable to be able to carry out the **A** → **B** transformation *enantioselectively*. Aside from the intellectual challenge posed by such a problem, the preparation of theoretically interesting, highly strained compounds in optically pure form could be of considerable interest in subsequent mechanistic studies of the chemical behavior of such species as well as in their use as synthons in total syntheses. The present volume, as well as a number of recent review articles and symposia [1], attest to the growing interest in the field of photochemical asymmetric synthesis.

There are two main factors that make asymmetric synthesis more difficult in the excited state than in the ground state. The first is that photochemistry is often incompatible with the type of external chiral auxiliaries used in the ground

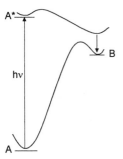

Figure 1   The advantage of photochemistry over ground-state chemistry in the formation of thermodynamically unstable products.

state, many of which are transition metal complexes with highly conjugated π-bond-containing ligands. The problem with such species is that they tend to absorb light in the same wavelength region as the reactants, and even when the reactants can be selectively excited, the chiral auxiliaries can act as quenchers, hydrogen atom donors, or electron transfer agents, thus effectively shutting down the desired photoreaction. A second impediment to photochemical asymmetric synthesis derives from the aforementioned high energy of electronically excited states and the corresponding low activation energy with which they react (Fig. 1). This translates into small activation energy differences between the diastereomeric transition states in the enantiodifferentiating step and correspondingly reduced levels of asymmetric induction.

There is no need to recount here the fascinating variety of methods that have been employed in attempting to achieve asymmetric induction in organic photochemistry; the history of the subject has been well reviewed [2]. Suffice it to say that the most successful approaches to date have come from the use of *organized and restricted media*—media that *preorganize* the constituent molecules in an arrangement favoring the formation of one enantiomer over the other, and which immobilize or *fix* the molecules in this arrangement during the time course of the photochemical event. Among such media, the most highly organized as well as the most restrictive is the *crystalline state*, and it is with this medium that we shall be concerned in the present chapter.

The first use of crystals to achieve asymmetric induction in a chemical reaction was reported by Penzien and Schmidt in 1969 [3]. In what the authors termed an "absolute asymmetric synthesis" because it occurs in the absence of any external source of optical activity, Penzien and Schmidt showed that the achiral compound 4,4′-dimethylchalcone **1** crystallizes spontaneously from ethyl acetate in the chiral space group $P2_12_12_1$, and when enantiomorphously pure single crystals of this substance are treated with bromine vapor in a gas–solid reaction, the chiral *trans*-dibromide **2** is formed in 6% enantiomeric excess (Scheme 1). In this beautiful experiment, it is the reaction medium—the homochiral crystal lattice—that provides the asymmetric influence favoring the formation of one product enantiomer over the other. Since this initial report, a large number of solid-state absolute asymmetric photochemical reactions have been reported, the great majority of which deal with unimolecular photorearrangements [4]. The first study of this type was carried out in our laboratory and was concerned with the dibenzobarrelene to dibenzosemibullvalene (di-π-methane) photorearrangement **3** to **4** shown in Scheme 1, a reaction that takes place with an *ee* of greater than 95% [5].

Unfortunately, nature is unreliable when it comes to providing chiral space groups for achiral molecules, and the great majority of achiral substances crystallize in centric or otherwise symmetric packing arrangements [6]. The question thus became not whether the chiral crystalline state would serve as a useful medium for asymmetric synthesis—this had already been demonstrated—but how a chiral space group could be *guaranteed* for the achiral compound whose photochemistry one wished to study.

The solution to this problem is straightforward: either chemically attach a homochiral auxiliary to the achiral reactant or cocrystallize the achiral reactant with an external homochiral auxiliary. In both cases, the crystals formed must be non-centrosymmetric, i.e., chiral, because it is not possible for homochiral objects to pack with a center of symmetry between them. In our research, we have pioneered the use of internal, built-in chiral auxiliaries to generate chiral

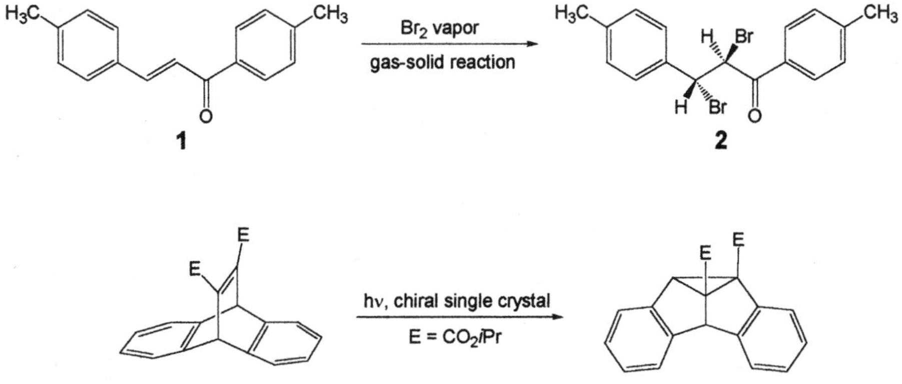

Scheme 1

space groups for photochemical asymmetric synthesis—work that forms the basis for the remainder of this article. The external (bimolecular) chiral auxiliary approach has been applied extensively by Toda, Tanaka and coworkers and will not be covered in the present chapter. The interested reader is referred to excellent review articles on this topic [7].

II. IONIC VS. COVALENT CHIRAL AUXILIARIES

There are two ways by which a chiral auxiliary can be attached to a molecule of interest in order to insure that it crystallizes in a chiral space group—covalently and ionically. The covalent approach requires little explanation, typical examples being ester or amide formation between the achiral carboxylic-acid-containing photoreactant and an optically pure alcohol or amine. The ionic attachment is similarly straightforward, consisting of salt formation between the carboxylic-acid-containing reactant and an optically pure amine. In this case, the resulting chiral ammonium ion is referred to as an *ionic chiral auxiliary*.

As we shall see, the only restriction as to where the carboxylic acid functional group can be located in the reactant is that it should not make it chiral. It is not necessary for the carboxylic acid group (and hence the chiral auxiliary) to be situated near the site of reaction; the role of the chiral auxiliary is simply to insure a chiral space group. A final point is that an amine functional group can serve equally well as the point of attachment of the chiral auxiliary to the reactant. In this case the ionic chiral auxiliary would be an optically pure carboxylate or sulfonate anion.

For a variety of conceptual and practical reasons, we have chosen to carry out most of our asymmetric synthesis work employing the ionic chiral auxiliary method. Conceptually, the idea was attractive because, as far as we were aware, it represents a new approach to asymmetric synthesis [8]. Furthermore, from the practical point of view, ionic solids have a number of advantages over purely molecular solids when it comes to studies in the crystalline state. Chief among these is that ionic solids have relatively high melting points, generally over 175°C. This is important when carrying out reactions in the solid state, because salts are less likely to melt as the reaction proceeds, and as a result, higher conversions are possible. It goes without saying that minimization of melting is critical in solid-state studies, since melting leads to loss of topochemical control and lower levels of asymmetric induction.

A second practical reason for choosing to work with ionic chiral auxiliaries is that they are easy to introduce via simple acid–base chemistry, and once the reaction is over, they are similarly easy to remove. The majority of our work has been carried out using ammonium ions as the ionic chiral auxiliaries, first because there are many optically pure amines of known absolute configuration available, and second because the reaction mixtures can be worked up with diazomethane to afford the photoproduct(s) as their methyl ester(s), which are easy to analyze by chiral GC and HPLC.

III. EXAMPLES OF THE SOLID-STATE IONIC CHIRAL AUXILIARY METHOD

In this section, emphasis will be given to reactions that have not been featured in our previous review articles on the subject. The interested reader may wish to consult these earlier review articles for additional examples and further discussion of the method [9].

A. The *trans,trans*-2,3-Diphenyl-1-Benzoylcyclopropane System

trans,trans-2,3-Diphenyl-1-benzoylcyclopropane and its derivatives (Scheme 2) provide ideal vehicles for testing the solid-state ionic chiral auxiliary approach to asymmetric synthesis. Early work by Zimmerman and Flechtner had shown that the parent unsubstituted compound, which is achiral, undergoes very efficient ($\Phi = 0.94$) photoisomerization in solution to afford the chiral *cis,trans* isomer [10]. Accordingly, we prepared the *p*-carboxylic acid derivative **5a** and treated this with a variety of optically pure amines to give salts **5b–5f**. Irradiation of crystals of these salts followed by diazomethane workup yielded methyl ester **6**, which was analyzed by chiral HPLC for enantiomeric excess. The results are summarized in Table 1 [11].

Three of the five salts investigated gave *ee*s in excess of 90%, a gratifying vindication of the solid-state ionic chiral auxiliary approach to asymmetric synthesis. There was a slight decrease in *ee* with increasing conversion, a feature that is characteristic of the method and is presumably due to the generation of lattice defects as reactants are converted to products. Also typical is the finding that not all ionic chiral auxiliaries gave excellent results. We shall have more to say about this later in the review. Finally, as an indication of the critical role played by the crystalline state, irradiation of the salts in solution was found to afford racemic **6**. As we shall see, this too is a characteristic result.

To what aspect of the solid-state environment can the high *ee*s be attributed? In general, one can imagine two scenarios: a *topochemical* effect, in which the course of the reaction is governed by steric effects between the reactant and its lattice neighbors, and a *conformational* effect, in which reactivity is controlled by the conformation of the individual reacting molecules in the crystal. Zimmerman and Flechtner showed that the photoisomerization of cyclopropyl ketone **5** most likely involves rupture of the C1–C2 (or C1–C3) bond followed by C2–C3

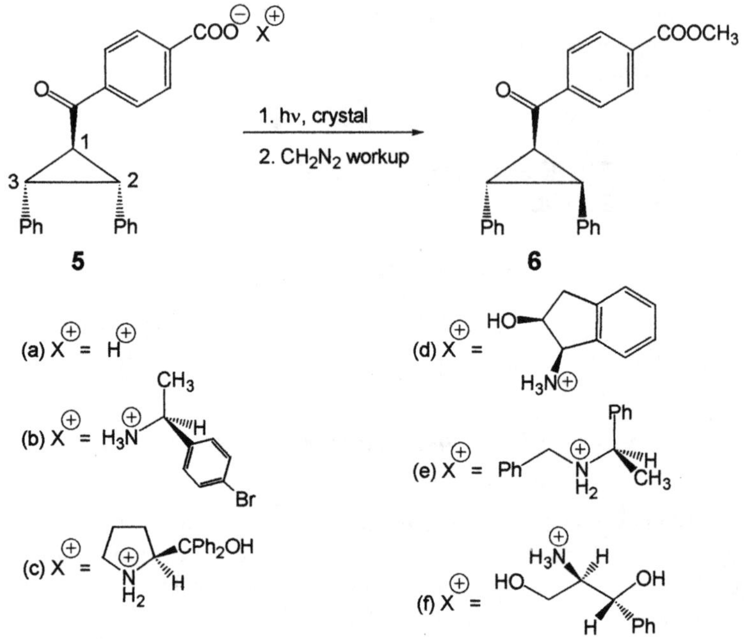

Scheme 2

Table 1 Enantiomeric Excesses Resulting from Irradiation of Crystalline salts **5b–f**

salt	conversion (%)	yield (%)[a]	ee (%)[b]
5b	25	99	99 (+)
	33	95	96(+)
	50	70	92 (+)
5c	25	99	99 (−)
	38	85	96 (−)
5d	33	99	91 (+)
5e	32	99	76 (−)
5f	18	99	54 (−)

[a] Proportion of **6** and/or *ent*-**6** other than starting material in the converted photoproduct mixture; conversions were kept ≤ 50% in order to minimize the formation of unidentified side products. [b] The designations (+) and (−) indicate the sign of rotation of the predominant enantiomer at the sodium D line; the absolute configuration was not determined.

bond rotation and reclosure [10], and it is easy to imagine that the rather large motions involved in this process could be topochemically less restricted in one direction and then the other by the homochiral crystal lattice. Alternatively, X-ray crystallography reveals that the methyl ester of keto-acid **5a** crystallizes in a conformation in which the p-orbital on the carbonyl carbon overlaps better with C1–C2 than with C1–C3, thus accounting for preferential cleavage of the former in the stereodifferentiating step. To date, salts **5b–f** have resisted all attempts to determine their crystal structures, so the answer to the question posed at the beginning of this paragraph remains unanswered. Theoretical calculations by Sevin and Chaquin, however, reveal that the preference for C1–C2 over C1–C3 cleavage in cyclopropyl ketones is extraordinarily sensitive to molecular conformation [12], and this represents our best current guess as to the origin of the asymmetric induction observed in the solid-state photochemistry of salts **5b–f**.

B. Linearly Conjugated Benzocyclohexadienones

Following the pioneering work of Schultz and others on the solution phase photochemistry of linearly conjugated cyclohexadienones [13], we prepared the achiral benzo derivative **7a** (Scheme 3) and studied the solid-state photochemistry of its salts with a series of optically pure amines [14]. Diazomethane workup afforded the expected chiral bicyclic photoproduct **8**, whose optical purity was determined by chiral GC. The results are summarized in Table 2.

One reason that salts of general structure **7** were chosen for investigation is that molecular mechanics calculations showed the benzocyclohexadienone framework to be planar. This planarity translates into an equal opportunity of forming either enantiomer of photoproduct **8**, or to put it another way, the photochemistry of these salts is conformationally unbiased with respect to enantioselectivity. As a result, any *ee* in the crystalline state would have to be due to an essentially pure topochemical effect. Only one of the salts investigated (**7b**) gave a respectable *ee* in the solid state, and unfortunately, it has not been possible to obtain an X-ray crystal structure of this material, thus precluding an analysis of the specific topochemical factors responsible for asymmetric induction in this case.

C. α-Mesitylacetophenone Derivatives

We turn next to a study in which it *has* been possible to pinpoint the factors responsible for asymmetric induction. The reaction in question is one that has been thoroughly investigated in solution by Wagner and coworkers—the photochemical conversion of achiral α-mesitylacetophenone derivatives into chiral cyclopentanols (Scheme 4) [15]. Following our usual protocol, we prepared the *para*-carboxylic acid derivative **9a** and treated this with a series of optically pure

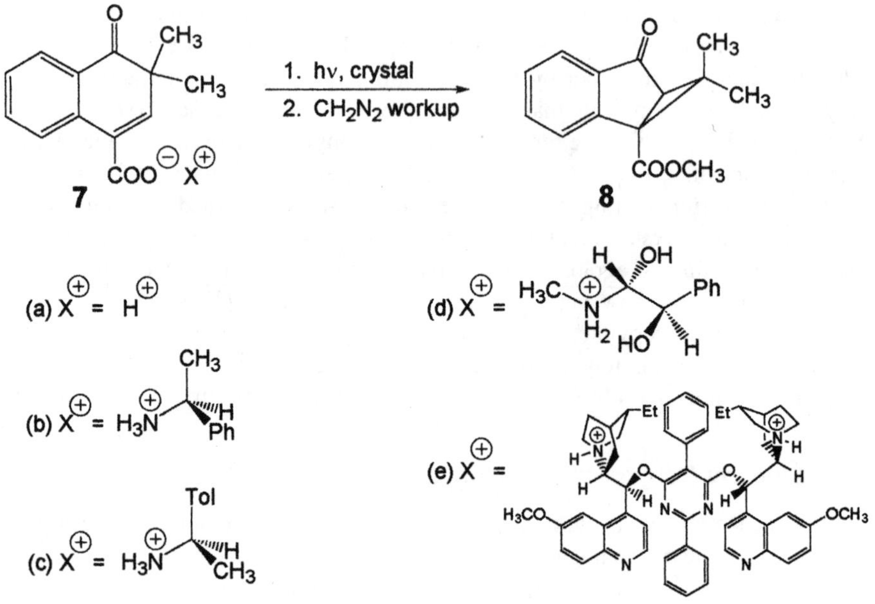

Scheme 3

Table 2 Solid-State Photolysis Results for Optically Pure Salts of Carboxylic Acid **7a**

Salt	Temp (°C)	Conversion (%)	ee (%)[a]	GC peak[b]
7b	room	25	81.0	2
	room	80	70.6	
	−78	30	86.5	
7c	room	09	50.1	1
	room	28	27.8	
	−78	15	47.6	
7d	room	11	38.1	1
	room	67	29.1	
7e[c]	room	36	35.6	2
	room	69	36.7	
	−78	53	52.5	

[a] Irradiation of salts **7b–e** in solution leads to racemic **8**. [b] Peak 1 indicates that the major enantiomer is the first peak eluted from chiral GC. [c] Stoichiometry: two equivalents acid, one equivalent base.

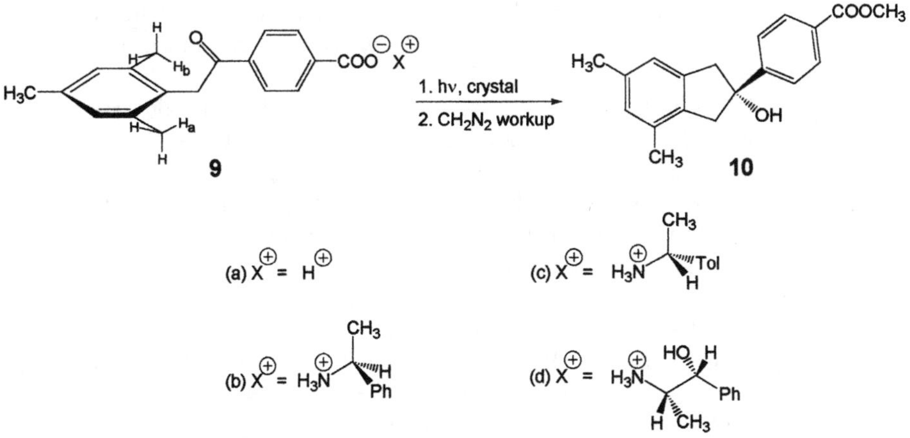

Scheme 4

amines to form the corresponding salts **9b–d** [16]. Irradiation of the salts in the crystalline state followed by diazomethane workup afforded cyclopentanol derivative **10**, whose *ee* was determined by chiral HPLC. The results are summarized in Table 3.

Once again the ionic chiral auxiliary method leads to respectable *ee*s at reasonably high conversions. As in every case studied to date, photolysis of the salts in solution leads to a racemic product, a result that emphasizes the critical role

Table 3 Solid-State Photolysis Results for Optically Pure Salts of Carboxylic Acid 9a

Salt	Temp (°C)	Conversion (%)	*ee* (%)[a]	HPLC peak[b]
9b	rt	16	94	1
	rt	69	69	
	−20	61	83	
9c	rt	89	83	1
9d	rt	12	90	2
	rt	80	80	
	−19	30	98	

[a] Irradiation of salts **9b–d** in solution leads to racemic **10**. [b] Peak 1 indicates that the major enantiomer is the first peak eluted from chiral HPLC. In the case of salts **9b** and **9d**, use of the optical antipode of the amine leads to the opposite enantiomer of photoproduct **10**.

played by the crystalline reaction medium. We note also that either enantiomer of photoproduct **10** can be prepared as desired by simply using the optical antipodes of the chiral auxiliary.

A variation of the Yang photocyclization reaction [17], the conversion of α-mesitylacetophenones into benzocyclopentenols, involves intramolecular δ-hydrogen atom abstraction from one of the *ortho*-methyl groups to form a 1,5-biradical, which closes to give the observed product [15]. The X-ray crystal structures of salts **9c** and **9d** reveal that asymmetric induction in the α-mesitylacetophenone system is conformationally rather than topochemically determined [16]. Consider, for example, the crystal structure of the norephedrine salt **9d** shown in Fig. 2. All of the reactive ions in the crystal are preorganized and frozen in a conformation that places the ketone oxygen atom much closer to δ-hydrogen atom Hb ($d_{C=O} \cdots _{Hb} = 2.69$ Å) than to Ha ($d_{C=O} \cdots _{Ha} = 3.68$ Å). We know from our extensive studies of the geometric requirements for γ-hydrogen atom abstraction in the Norrish type II reaction [18] that a difference in **d** of this magnitude translates into exclusive abstraction of Hb. Closure of the biradical formed by abstraction of Hb is thus predicted to lead to the (*R*)-enantiomer of photoproduct **10**. This prediction is based on the reasonable assumption that, with molecular motions severely restricted in the crystalline state, cyclopentanol formation occurs with "retention of configuration" at the carbonyl carbon. The analogous process of cyclobutanol formation with retention of configuration at the carbonyl carbon is a general feature of the Yang photocyclization reaction in the crystalline state [19].

The prediction that the (*1S,2R*)-(+)-norephedrine salt of acid **9a** should lead to the (*R*)-enantiomer of cyclopentanol **10** could not be verified directly

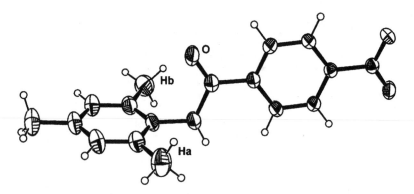

Figure 2 X-ray crystal structure showing the conformation of the carboxylate anion portion of salt **9d**.

owing to the difficulty of determining the absolute configuration of the latter. There is, however, convincing indirect evidence on this point, which comes from the X-ray crystal structure of the (R)-(+)-1-(p-tolyl)ethylamine salt **9c** [16]. In this case, abstraction of Ha ($d_{C=O} \cdots _{Ha} = 2.60$ Å) is favored over abstraction of Hb ($d_{C=O} \cdots _{Hb} = 3.66$ Å), which predicts that cyclopentanol **10** should be formed with the (S) absolute configuration in the crystalline state. The experimental finding that, as predicted, salts **9c** and **9d** lead to opposite enantiomers of photoproduct **10** (Table 3) provides strong support for the correctness of these crystal structure–solid state reactivity correlations.

D. Summary of Results for Other Systems

At this point, the basic strategy of the solid-state ionic chiral auxiliary approach to asymmetric synthesis should be clear. First, equip the reactant with a carboxylic acid group, taking pains to avoid placing it where it will make the molecule chiral. Second, react the acid with a half-dozen or so optically pure amines chosen more or less at random (1-phenylethylamine should be one of these, as it often gives good results) and then recrystallize the resulting salts from a suitable solvent. Finally, irradiate the salts in the crystalline state, work up the reaction mixtures with diazomethane to form the methyl esters of the reactant and product(s), and analyze the product(s) for enantiomeric excess by chiral GC or HPLC. It is also important that the irradiations be carried out for different lengths of time in order to determine how *ee* varies with conversion.

With these points in mind, we summarize in Table 4 all of the published examples of the use of the solid-state ionic chiral auxiliary method of which we are aware, including, for comparison purposes, the three reactions discussed above (entries 17–19). In order to conserve space, results are reported only for the chiral auxiliaries that gave the best results. The interested reader may wish to consult the original papers for a more detailed account of the work and to see what other chiral auxiliaries were tried.

E. Discussion of Results Presented in Table 4

Most of the ionic chiral auxiliary–mediated reactions presented in Table 4 occur with very respectable *ee*s, even at quite high conversions. The best of these in terms of overall optical and chemical yield is the reaction shown in entry 4, which gives an essentially quantitative GC yield of the β-lactam photoproduct with an *ee* of 99%. For a variety of practical reasons, and because modern methods of analysis do not require large sample sizes, all but one of the reactions in Table 4 were carried out on a microscale (< 10 mg). The lone exception was reaction 4, which could easily be scaled up to the 500 mg level to afford an *ee* of 99% with an *isolated* chemical yield of 91% [21]. This was done by irradiating the salt

Table 4 Photochemical Asymmetric Synthesis Using the Solid-State Ionic Chiral Auxiliary Method

Entry	Reaction	ee (%)	Conv (%)	Temp (°C)	Chiral auxiliary (CA)	Comments	Ref.
1		80	99	rt	(S)-(−)-1-phenyl-ethylamine	Single crystal-to-single crystal reaction	19b
2		92	85	rt	(R)-(+)-1-phenyl-ethylamine		19b
3		98	78	rt	(R)-(+)-bornyl amine		20
4		99	99	rt	L-prolinamide	Reaction run on 500 mg scale in hexane suspension	21

#	Reaction	ee (%)	Conv. (%)	Temp	Chiral auxiliary	Notes	Entry
5	$\xrightarrow[\text{CH}_2\text{N}_2\text{ workup}]{h\nu,\ \text{crystal}}$; COO⁻ CA⁺ → COOMe	92	100	15	(S)-(−)-1-phenyl-ethylamine	Single crystal-to-single crystal reaction	22
6	$\xrightarrow[\text{CH}_2\text{N}_2\text{ workup}]{h\nu,\ \text{crystal}}$; COO⁻ CA⁺ → COOMe	97	87	rt	(S)-(+)-pyrrollidine-methanol	Needle polymorph; plate polymorph gives 12% ee (see text)	23
7	$\xrightarrow[\text{CH}_2\text{N}_2\text{ workup}]{h\nu,\ \text{crystal}}$; COO⁻ CA⁺ → COOMe	99	100	rt	(R)-(+)-1-phenyl-ethylamine	60% yield; 2 other products formed	24
8	$\xrightarrow[\text{CH}_2\text{N}_2\text{ workup}]{h\nu,\ \text{crystal}}$; COO⁻ CA⁺ → COOMe	91	100	−20 rt	L-valine-4-benzoyl-phenyl ester	CA is both chiral auxiliary and triplet energy sensitizer	25
9	$\xrightarrow[\text{CH}_2\text{N}_2\text{ workup}]{h\nu,\ \text{crystal}}$; COOH·CA → COOMe	66	20	rt	L-proline		26

(Continued)

Table 4 Continued

Entry	Reaction		ee (%)	Conv (%)	Temp (°C)	Chiral auxiliary (CA)	Comments	Ref.
10			91	60	rt	(S)-(−)-malic acid	Substantial amounts of type 2 cleavage also observed	27
11			>95	20–40	−40	(S)-(−)-proline *tert*-butyl ester	Absolute config correlation established	28
12			>95	20–40	−40	(1S,2S)-(+)-pseudo-ephedrine	Absolute config correlation established	28b,c
13			68	20	−40	(R)-(−)-camphor sulfonic acid		29
14			63	95	rt	(1R,2S)-(+)-1-amino-2-indanol		30

Entry	Substrate / product (hv, crystal; CH$_2$N$_2$ workup)	No.	Chiral auxiliary	Temp (°C)	Yield	ee
15	(cyclopentyl phenyl ketone benzoate → bicyclic COOMe)	31	(S)-(−)-1-phenylethylamine	rt	67	42
16	(2,4,6-triisopropyl aryl ketone → COOMe cyclobutenol)	32	(S)-(+)-pyrrolidinemethanol	10	99	29
17	(diphenylcyclopropyl aryl ketone → COOMe)	11	(S)-(−)-1-p-bromophenylethylamine	rt	50	92
18	(naphthalenone → indanone COOMe)	14	(S)-(−)-1-phenylethylamine	−78	30	86.5
19	(mesityl ketone → indanol COOMe)	16	(R)-(+)-N-methyl-1-p-tolylethylamine	rt	89	83

crystals as a hexane suspension in a conventional immersion well apparatus—a technique that would allow this and any other ionic solid-state transformation to be conducted on the gram scale if desired.

Another feature of the results presented in Table 4 is that all but two of them feature salts in which the photoreactive component is a carboxylate anion and the chiral auxiliary is an ammonium ion. There is, of course, no reason why salts composed of reactive ammonium ions and optically pure carboxylate (or sulfonate) anions should not give equally good results, and entries 10 and 13 illustrate this point.

One of the reasons why we chose to work with ionic rather than with covalent chiral auxiliaries is that we hoped that the robust character of ionic solids might enhance the chances of observing single crystal–to–single crystal or topotactic transformations. Such reactions, in which perfect X-ray quality single crystals of the reactant are continuously and quantitatively converted into perfect single crystals of the product, are extremely rare and highly prized because they allow the transformations to be monitored by X-ray crystallography, not only at the beginning and end of the reaction, but at all intermediate stages. Two of the 19 reactions listed in Table 4 were found to be topotactic in nature (entries 1 and 5), a frequency that does not seem unusually high, and our preliminary conclusion is that topotacticity is probably no more likely in ionic crystals than in purely molecular crystals.

For four of the reactions shown in Table 4 (entries 1, 5, 11, and 12), absolute configuration correlations were established between the reactant and its solid-state photoproduct. Such correlations represent one of the most powerful methods available to the organic chemist for elucidating reaction mechanisms. In the Norrish/Yang type II reaction (entries 1 and 5), for example, it allows an unequivocal determination of which γ-hydrogen is abstracted, and for the di-π-methane rearrangement of dibenzobarrelene derivatives (entries 11 and 12), it tells us precisely which atoms are involved in the formation of the initial cyclopropyldicarbinyl diradical.

A final comment on Table 4 concerns the reaction shown in entry 8. Because the di-π-methane photorearrangement of benzonorbornadiene derivatives requires triplet energy sensitization, we could not use typical, passive amines such as (R)-$(+)$-1-phenylethylamine as chiral auxiliaries. We therefore prepared an optically pure amine to which a sensitizing benzophenone moiety was tethered, namely, the 4-benzoylphenyl ester of L-valine [25]. Photolysis of the salt of this amine at wavelengths where only the benzophenone chromophore absorbs led to the photoproduct in 91% *ee* at 100% conversion, a gratifying vindication of the concept. Optically active photosensitizers have been used in solution with limited success [33], but this represents the first example of simultaneous triplet–triplet energy transfer and asymmetric induction in the crystalline state.

F. Conformational Enantiomerism

As can be seen from Tables 1–3, not all ionic chiral auxiliaries lead to high enantiomeric excesses. While we have always managed to find at least one chiral auxiliary that leads to good (and in many cases excellent) *ee*, there are at least an equal number that do not, and the researcher trying the method for the first time should be prepared to try half a dozen or more chiral auxiliaries in order to have a reasonable chance of finding a good one. As mentioned earlier, 1-phenylethylamine should be one of these, but on occasion even this usually reliable amine fails.

What factors are responsible for low *ee*s in the solid-state ionic chiral auxiliary approach to asymmetric synthesis? We can rule out trivial explanations such as sample melting in low *ee* cases, since melting was not detectable upon examination of the crystals following photolysis, and the *ee*s were only slightly improved by using low temperatures and low conversions, where crystal melting should be minimal. In a number of cases, low *ee*s have been shown to be the result of a form of crystal packing termed "conformational enantiomerism" [34]. This refers to the situation in which the salt, apparently trying to achieve the packing efficiency of a racemic compound [35], crystallizes with equal amounts of two independent and mirror image related conformers of the reactant ion in the asymmetric unit. This was demonstrated by computationally inverting one of the conformers, superimposing it on the other, and calculating the root-mean-square error for the overlap of the heavy atoms of the two species, which proved to be very low.

An example of an ionic chiral auxiliary–mediated solid-state photoreaction in which conformational enantiomerism is responsible for low *ee*s is found in the α-oxoamide system shown in entry 4, Table 4. While L-prolinamide leads to 99% *ee* at 99% conversion, use of (*S*)-(−)-1-phenylethylamine as the chiral auxiliary affords the β-lactam photoproduct in only 3% *ee* [21]. X-ray crystallographic analysis of the 1-phenylethylamine salt revealed that it contains equal amounts of two independent mirror-image-related carboxylate anions in the asymmetric unit. Thus 50% of the ions have a conformation favorable for the formation of the (*R*)-β-lactam, whereas the other 50% are conformationally poised to give (*S*); it is therefore not surprising that low *ee*s are obtained. What is perhaps surprising at first sight is that a racemate is not formed. This can be understood when it is recalled that the two sets of conformers are not perfect mirror images of one another and that they are diastereomerically related when the presence of the (*S*)-ammonium ion is taken into account. As a result, the conformational enantiomers react at slightly different rates, and a small excess of one enantiomer of the photoproduct is obtained. Of the 19 separate systems studied by the solid-state ionic chiral auxiliary method (Table 4), conformational enantiomerism has been shown to occur in five, entries 3, 4, 6, 18, and 19.

G. Comparison of the Solid-State Ionic Chiral Auxiliary Method of Asymmetric Synthesis with the Pasteur Resolution Procedure

In principle, any of the photoproducts shown in Table 4 could have been prepared in enantiomerically pure form by irradiating their achiral precursors in solution to form a racemate and then separating the enantiomers by means of the classical Pasteur resolution procedure [36]. This sequence is shown in the lower half of Fig. 3. The top half of Fig. 3 depicts the steps involved in the solid-state ionic chiral auxiliary method of asymmetric synthesis. The difference between this approach and the Pasteur method is one of timing. In the ionic chiral auxiliary method, salt formation between the achiral reactant and an optically pure amine precedes the photochemical step, whereas in the Pasteur procedure, the photochemical step comes first and is followed by treatment of the racemate with an optically pure amine to form a pair of diastereomeric salts. The two methods are similar in that the crystalline state is crucial to their success. The Pasteur resolution procedure relies on fractional crystallization for the separation of the diastereomeric salts, and the ionic chiral auxiliary approach only gives good *ee*s when the photochemistry is carried out in the crystalline state.

Like other methods of asymmetric synthesis, the solid-state ionic chiral auxiliary procedure has an advantage over the Pasteur separation technique in product yield. The maximum yield of either enantiomer that can be obtained by resolution of a racemate is 50%, and in practice it is often considerably less [37]. On the other hand, the ionic chiral auxiliary approach affords a single enantiomer

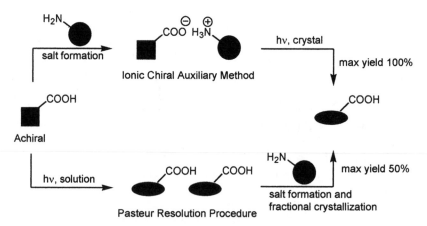

Figure 3 Comparison of the solid-state ionic chiral auxiliary method of asymmetric synthesis with the Pasteur resolution procedure.

of the photoproduct, often in chemical and optical yields of well over 90%. Furthermore, either enantiomer can be obtained as desired by simply using one optical antipode or the other of the ionic chiral auxiliary.

The success of the solid-state ionic chiral auxiliary approach to asymmetric synthesis can be analyzed in both thermodynamic and kinetic terms. We have seen that conformationally flexible reactants that have average planes of symmetry in solution crystallize in chiral conformations upon salt formation with optically pure amines. This thermodynamic aspect of the process biases the ensuing photochemistry toward a single diastereomer of the product salt by, for example, allowing abstraction of only one of two diastereotopic γ-hydrogen atoms (γ-hydrogens that were enantiotopic prior to salt formation). On the other hand, in the less common case of rigid reactants such as the linearly conjugated benzocyclohexadienone **7a** discussed earlier (entry 18, Table 4), or the benzonorbornadiene derivative of entry 8, conformational bias introduced at the crystallization stage does not play a significant role in determining which enantiomer of the photoproduct will be formed following removal of the chiral auxiliary. In such cases the preference for formation of one diastereomeric salt over the other is best viewed as a *kinetic* differentiation between diastereomeric transition states of unequal energy [27], the energy difference arising from steric interactions between the reactant and its chiral surroundings in the crystal lattice.

H. Summary

Having been demonstrated with 19 separate examples, there is little doubt that the solid-state ionic chiral auxiliary approach represents one of the most powerful and general methods of asymmetric synthesis in organic photochemistry. Chemical and optical yields are high, and we are convinced that the procedure has genuine synthetic utility. Although the method is not catalytic, requiring stoichiometric amounts of the chiral auxiliary, this is not a problem given the ready availability and low cost of a large number of optically pure amines and carboxylic acids from the chiral pool. The method has been applied to photochemical processes because of the paucity of techniques available for asymmetric synthesis in reactions of this type and because the temperature can be controlled so that the crystals are less prone to melting during reaction. In principle, however, there is no reason why the solid-state ionic chiral auxiliary approach could not be used for low-activation-energy ground-state reactions, and investigations along these lines are underway in our laboratory at present.

ACKNOWLEDGMENTS

JRS wishes to thank the Institute of Advanced Studies, University of Bologna, where this manuscript was written, for financial support in the form of a Senior

Guest Fellowship. We also thank the donors of the Petroleum Research Fund, administered by the American Chemical Society, for partial support of this research. Financial support by the Natural Sciences and Engineering Research Council of Canada is also gratefully acknowledged.

REFERENCES

1a. Everitt SRL, Inoue Y. In:. Molecular and Supramolecular Photochemistry Ramamurthy V , Schanze KS, Eds. Vol. 3. New York: Marcel Dekker, 1999:71.
1b. Pete JP. Adv Photochem 1996; 21:135.
1c. Inoue Y. Chem. Rev. 1992; 92:471.
1d. Buschmann H, Scharf HD, Hoffmann N, Esser P. Angew. Chem., Int. Ed. Engl. 1991; 30:477.
2.   Rau H. Chem. Rev. 1983; 83:535.
3.   Penzien K, Schmidt GMJ. Angew. Chem., Int. Ed. Engl. 1969; 8:608.
4a. Sakamoto M. Chem. Eur. J. 1997; 3:684.
4b. Koshima H. In:. Solid State Organic Reactions Toda F, Ed. Dordrecht: Kluwer, 2002, Chap. 5.
5.   Evans SV, Garcia-Garibay M, Omkaram N, Scheffer JR, Trotter J, Wireko F. J. Am. Chem. Soc. 1986; 108:5648.
6.   Jacques J, Collet A, Wilen SH. Enantiomers, Racemates and Resolutions. New York: John Wiley, 1981.
7a. Toda F. Acc. Chem. Res. 1995; 28:480.
7b. Toda F, Tanaka K, Miyamoto H. In:. Molecular and Supramolecular Photochemistry Ramamurthy V , Schanze K, Eds. Vol. 8. New York: Marcel Dekker, 2001:385.
8.   Corey and coworkers have recently shown that optically pure ammonium ions can be used to bring about asymmetric induction in a variety of ground-state reactions. See for example.
8a. Zhang F-Y, Corey EJ. Org. Lett. 2000; 2:1097.
8b. Zhang F-Y, Corey EJ. Org. Lett. 2000; 2:4257 and references cited therein.
9a. Gamlin JN, Jones R, Leibovitch M, Patrick B, Scheffer JR, Trotter J. Acc. Chem. Res. 1996; 29:203.
9b. Scheffer JR. Can. J. Chem. 2001; 79:349.
10. Zimmerman HE, Flechtner TW. J. Am. Chem. Soc. 1970; 92:6931.
11. Chong KCW, Sivaguru J, Shichi T, Yoshimi Y, Ramamurthy V, Scheffer JR. J. Am. Chem. Soc. 2002; 124:2858.
12. Sevin A, Paquin P. J. Org. Chem. 1982; 47:4145.
13. For reviews of the photochemistry of linearly conjugated cyclohexadienones, see.
13a. Schultz AG. In:. CRC Handbook of Organic Photochemistry and Photobiology Horspool WM , Song P-S, Eds. Boca Raton: CRC Press, 1991 Chap. 58.
13b. Quinkert G. Pure Appl. Chem. 1973; 33:285.
14. Cheung E, Netherton MR, Scheffer JR, Trotter J. Tetrahedron Lett. 1999; 40:8737.
15a. Wagner PJ, Meador MA, Zhou B, Park B-S. J. Am. Chem. Soc. 1991; 113:9630.
15b. Wagner PJ. Acc. Chem. Res. 1989; 22:83.

16. Cheung E, Rademacher K, Scheffer JR, Trotter J. Tetrahedron 2000; 56:6739.
17. Yang photocyclization refers to the photochemical formation of cyclobutanols following Norrish type II γ-hydrogen atom abstraction and was first reported by Yang NC, Yang DH. J. Am. Chem. Soc. 1958; 80:2913.
18. Ihmels H, Scheffer JR. Tetrahedron 1999; 55:885.
19a. Gudmundsdottir AD, Lewis TJ, Randall LH, Rettig SJ, Scheffer JR, Trotter J, Wu C-H. J. Am. Chem. Soc. 1996; 118:6167.
19b. Leibovitch M, Olovsson G, Scheffer JR, Trotter J. J. Am. Chem. Soc. 1998; 120: 12755.
20. Cheung E, Kang T, Raymond JR, Scheffer JR, Trotter J. Tetrahedron Lett. 1999; 40:8729.
21. Natarajan A, Wang K, Ramamurthy V, Scheffer JR, Patrick B. Org. Lett. 2002; 4: 1443.
22. Koshima H, Matsuhige D, Miyauchi M. Cryst. Eng. Comm. 2001; 33:1.
23. Jones R, Scheffer JR, Trotter J, Yang J. Tetrahedron Lett. 1992; 33:5481.
24. Cheung E, Netherton MR, Scheffer JR, Trotter J, Zenova A. Tetrahedron Lett. 2000; 41:9673.
25. Janz KM, Scheffer JR. Tetrahedron Lett. 1999; 40:8725.
26. Cheung E, Chong KCW, Jayaraman S, Ramamurthy V, Scheffer J, Trotter J. Org. Lett. 2000; 2:2801.
27. Cheung E, Netherton MR, Scheffer JR, Trotter J. J. Am. Chem. Soc. 1999; 121: 2919.
28a. Gudmundsdottir AD, Scheffer JR. Tetrahedron Lett. 1990; 31:6807.
28b. Gudmundsdottir AD, Scheffer JR, Trotter J. Tetrahedron Lett. 1994; 35:1397.
28c. Gudmundsdottir AD, Li W, Scheffer JR, Rettig S, Trotter J. Mol. Cryst. Liq. Cryst. 1994; 240:81.
29. Gudmundsdottir AD, Scheffer JR. Photochem. Photobiol. 1991; 54:535.
30. Scheffer JR, Wang L. J. Phys. Org. Chem. 2000; 13:531.
31. Kang T, Scheffer JR. Org. Lett. 2001; 3:3361.
32. Koshima H, Maeda A, Masuda N, Matsuura T, Hirotsu K, Okada K, Mizutani H, Ito Y, Fu TY, Scheffer JR, Trotter J. Tetrahedron Asymmetry 1994; 5:1415.
33a. Hammond GS, Cole RS. J. Am. Chem. Soc. 1965; 87:3256.
33b. Inoue Y, Matsushima E, Wada T. J. Am. Chem. Soc. 1998; 120:10687.
34. Cheung E, Kang T, Netherton MR, Scheffer JR, Trotter J. J. Am. Chem. Soc. 2000; 122:11753.
35. Brock CP, Schweizer WB, Dunitz JD. J. Am. Chem. Soc. 1991; 113:9811.
36. Pasteur L. C.R. Acad. Sci. 1853; 37:162.
37a. Eliel EL, Wilen SH. Stereochemistry of Organic Compounds. New York: John Wiley, 1994 Chapt. 7.
37b. Collet A. Angew. Chem. Int. Ed. 1999; 37:3239.

13

Chiral Solid-State Photochemistry Including Supramolecular Approaches

Hideko Koshima
Ehime University
Matsuyama, Japan

I. INTRODUCTION

In contrast to the great advances in asymmetric synthesis using thermal reactions during the last few decades [1–3], asymmetric photochemistry in solution and in the solid state still lies in the area of basic research. Despite the recent outstanding discoveries that reaction temperature and pressure can control enantioselectivity as well as switching of left and right handedness using the same photosensitizer by Inoue and coworkers, asymmetric photoreactions in the solution phase generally occur in low optical yields except for a few reactions [4–6]. On the other hand, recent intensive studies have revealed that crystalline state photoreactions lead to high enantio- and diastereodifferentiation because the motion of molecules is very restricted in the crystal lattice, and the chiral environnment is retained during reaction [7–10]. Crystals in which molecules are arranged at close positions in three-dimensional regularity therefore can be an ultimate reaction medium for asymmetric synthesis.

Solid-state chiral photochemistry has been developed based on spontaneous chiral crystallization and supramolecular approaches. Since the report of solid-state [2 + 2] photocycloaddition by Cohen and Schmidt in 1964 [11], the Weizmann Institute Group has been also a pioneer of solid-state chiral photochemistry. The research group first succeeded in absolute asymmetric [2 + 2] photocycloaddition by using chiral crystals spontaneously formed from achiral dienes in 1973 [12]. Thereafter, more than twenty successful examples of absolute asymmetric photoreactions have been reported by several research groups [9,10,13,14]. The spontaneous chiral crystallization approach is the best methodology for asymmetric synthesis without any external chiral source. The problem is that spontaneous chiral crystallization cannot be predicted at present. Nevertheless, it is a promising methodology for asymmetric synthesis.

Supramolecular approaches have led to recent advances in solid-state asymmetric photochemistry to a great extent. Toda and coworkers have applied host–guest cocrystals with chiral organic host compounds of the axle-wheel type or tartaric acid derivatives to achieve high enantiodifferentiation of a number of achiral compounds [7,15,16]. However, the applicable guest compounds are limited to relatively small molecules that can be included in the cavity of the host compounds. Ionic chiral auxiliaries utilizing salt crystals by Scheffer and coworkers have also resulted in good asymmetric induction. However, here only substrates having acid or base functional groups are applicable [8]. Most of the asymmetric photoreactions developed by using the host–guest crystals or the salt crystals are intramolecular reactions. We have targeted intermolecular (bimolecular) asymmetric photoreactions between two different molecules by using cocrystals [10]. The advantages of utilizing a cocrystal as a reactant are that even if a component molecule has no photoreactivity by itself, new photoreactivity and intermolecular reaction behavior can be induced by combining an electron donor and an electron acceptor in the crystal and exploiting photoinduced electron transfer. In fact, we have achieved enantio- and diastereoselective photodecarboxylative condensation, as well as absolute asymmetric photodecarboxylative condensation. Development of intermolecular reactions between different molecules should lead to extensions of solid-state chiral photochemistry. Recently, Ramamurthy and coworkers have applied zeolites that had been preloaded with optically pure chiral inductors [17,18]. However, the enantiodifferentiations were low except for a few reactions.

Solid-state asymmetric photoreactions have been already reviewed from various aspects [7–20]. However, the reviews do not seem to have given us a complete understanding of solid-state chiral photochemistry. In this chapter, solid-state asymmetric photoreactions are systematically reviewed by classification into supramolecular approaches and spontaneous chiral crystallization approaches since the beginning in the 1970s to the present.

II. CLASSIFICATION

Solid-state asymmetric photoreactions are classified in Table 1. A chiral crystal is first needed as a reactant medium in solid-state photoreaction. Chiral compounds necessarily crystallize in a chiral space group. Hence the most reliable method of bringing about an asymmetric reaction is to utilize a chiral crystal containing covalently chiral molecules. However, this kind of chiral crystal of a chiral mole-

Table 1 Classification of Solid-State Asymmetric Photoreactions

Reactant medium	Chiral source	Optical differentiation
Chiral crystal from chiral molecule	Chiral molecule	Moderate–high diastereodifferentiation
Supramolecular approach		
Host–guest crystal	Chiral host molecule	Moderate–high enantiodifferentiation
Salt crystal	Chiral acid or base molecule	Moderate–high enantiodifferentiation
Cocrystal	Chiral molecule	Moderate–high enantiodifferentiation
Modified zeolite	Chiral molecule	Low–moderate enantiodifferentiation
Spontaneous chiral crystallization approach		
Chiral crystal from achiral molecule	Chiral crystal lattice	Moderate–high enantiodifferentiation

cule may undergo diastereomeric photoreaction. In order to cause enantiomeric photoreaction from an achiral molecule to a chiral molecule, a chiral crystal involving an achiral photoreactant molecule must be submitted to reaction. Such a chiral crystal can be prepared by two methods; supramolecular crystallization between an achiral reactant molecule and a chiral molecule, and spontaneous chiral crystallization of an achiral molecule.

A. Supramolecular Approach

When an achiral photoreactant molecule is bound to a chiral compound through noncovalent bonds, a chiral supramolecular crystal is necessarily formed. It is the same situation as a covalently chiral molecule always forming a chiral crystal. The noncovalent bonds include hydrogen bonding, salt bridge, charge transfer, π–π interaction, C–H-π interaction, and halogen–halogen interaction. Host–guest crystals, salt crystals, and cocrystals are supramolecular crystals. Chirally modified zeolites are inorganic–organic hybrid host compounds, hence herein involved in the supramolecular approach.

Figure 1 shows typical chiral organic host compounds. Deoxycholic acid **1** [21] and cyclodextrin **2** [22] are well-known natural compounds. Tri-*o*-thymotide **3** is a conformationally chiral host compound [23]. The host compounds **1–3** were utilized in the early historical stages, but the enantioselectivities were not

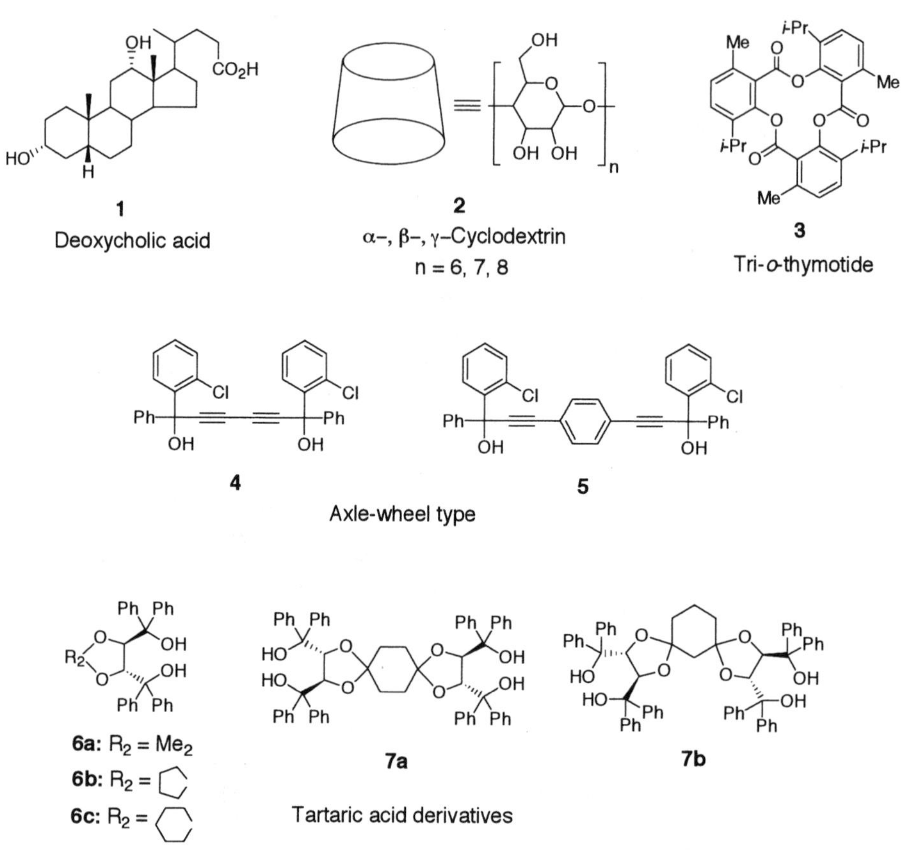

1

Deoxycholic acid

2

α–, β–, γ–Cyclodextrin
n = 6, 7, 8

3

Tri-o-thymotide

4

5

Axle-wheel type

6a: R$_2$ = Me$_2$

6b: R$_2$ = ☐

6c: R$_2$ = ⬡

7a

Tartaric acid derivatives

7b

Figure 1   Chiral organic host compounds.

excellent. Toda and coworkers have synthesized chiral axle-wheel-type host compounds **4**, **5** [24]. Compounds **6a–c** derived from tartaric acid were originally
prepared as catalysts by Zeebach and coworkers [25], but Toda and coworkers
utilized these as host compounds. They have achieved highly enantioselective
reactions of a number of achiral compounds by using the host–guest crystals
with **4–7** [7,15,16]. The host compounds include achiral reactant molecules in
their cavities and further bind through hydrogen bonding and other weak intermolecular interactions. The disadvantage of the host–guest approach is that substrates are limited to small guest molecules that can form inclusion complexes.
The reactivity and enantioselectivity are controlled by the molecular packing in

the host–guest crystal. In fact, this reaction type by Toda and coworkers is limited to almost only photocyclization of small molecules.

Since 1990, Scheffer and coworkers have applied ionic chiral auxiliary groups, which utilize a salt crystal of an achiral photoreactant with a carboxylic acid group and an optically pure amine or vice versa [8] to prepare chiral crystals. Figure 2 shows representative chiral amines. Irradiation of the chiral salt crystal can cause diastereomeric photoreaction. After reaction, removal of the amine or carboxylic acid gives the enantiomeric product, rather than diastereomers. They have achieved high enantiodifferentiation in the Norrish/Yang cyclization of adamantanes and the di-π-methane rearrangement of dibenzobarrelenes. Salts formed between carboxylic acids and amines tend to be high-melting crystalline materials that offer potential advantages for solid-state photoreactions. However, this approach is limited to substrates containing acidic or basic functional groups.

We have used chiral cocrystals composed of two different molecules [10]. When one of the two components is chiral, a cocrystal of a chiral space group is necessarily obtained. Combination of an electron donor molecule with an electron acceptor molecule, or a photoinert molecule with a photosensitizer molecule, in a cocrystal can induce new photoreactivity as well as intermolecular photoreaction.

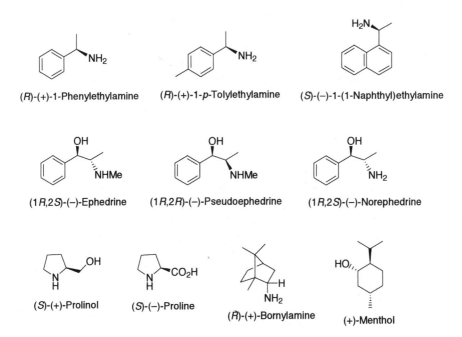

(*R*)-(+)-1-Phenylethylamine   (*R*)-(+)-1-*p*-Tolylethylamine   (*S*)-(–)-1-(1-Naphthyl)ethylamine

(1*R*,2*S*)-(–)-Ephedrine   (1*R*,2*R*)-(–)-Pseudoephedrine   (1*R*,2*S*)-(–)-Norephedrine

(*S*)-(+)-Prolinol   (*S*)-(–)-Proline   (*R*)-(+)-Bornylamine   (+)-Menthol

Figure 2   Chiral amines.

When two different achiral molecules form a chiral cocrystal by spontaneous chiral cocrystallization, the occurrence of absolute asymmetric intermolecular photoreaction can be expected. In fact, we have achieved enantio- and diastereo-selective photodecarboxylative condensation, as well as absolute asymmetric photodecarboxylative condensation. The development of intermolecular photoreactions leads to an extension of the scope of solid-state chiral photochemistry. Reactivity in a cocrystal is controlled by the crystal packing arrangement, so the key point is the preparation of photoreactive cocrystals.

Ramamurthy and coworkers have utilized zeolites modified with chiral organic compounds [17,18]. Zeolites are crystalline aluminosilicates with open framework structures. In this approach, the zeolite is first loaded with a chiral inductor and the compound to be photolyzed is then added in a second, separate adsorption step. Asymmetric induction ensues as a result of the close proximity enforced between reactant and chiral inductor in the confined space of the zeolite supercage. The zeolite method has the disadvantage that the size of the substrate is limited by the pore size of the zeolite being used. Most of the work using the chirally modified zeolite approach was compared with the ionic chiral auxiliary method by Scheffer and coworkers. The enantiodifferentiations by the zeolites are usually low to moderate.

B. Spontaneous Chiral Crystallization Approach

Even if a molecule is achiral, chiral crystals can form by spontaneous chiral crystallization [26]. The big advantage in utilizing a crystal as a reactant is that absolute asymmetric synthesis can be achieved by solid-state photoreaction of such a chiral crystal. The initial chiral environment in the crystal lattice is retained during the reaction process, owing to the low mobility of molecules in the crystalline state, and leads to an optically active product. The process represents transformation from crystal chirality to molecular chirality. This kind of absolute asymmetric synthesis does not need any external asymmetric source in the entire synthetic procedure [9–14].

A crystal lacking both a center of symmetry and a glide plane is defined as chiral. Such a chiral crystal must belong to a chiral space group. Of the 230 space groups theoretically obtained, there are 65 chiral space groups. The most frequent chiral space groups are $P2_1$ (#4) and $P2_12_12_1$ (#19). Although a such chiral crystallization had been considered a very rare phenomenon, our survey of the Cambridge Structural Database (190,000 structures) revealed that the statistical probability for the chiral crystallization of achiral compounds was around 8%, meaning that one can find one chiral crystal among a dozen achiral or optically inactive crystalline compounds [27]. In fact, several solid-state photochemists have found such chiral crystals by chance and have succeeded in using them in absolute asymmetric photoreactions. However, chiral cocrystals formed from two achiral compounds are scarcely known. At the present time, it is difficult to predict whether a given achiral organic compound can undergo chiral crystallization. Trial and error approaches are necessary to obtain new chiral crystals. We have prepared several series of helical-type and propeller-type chiral cocrystals by combining flexible achiral molecules (Fig. 3) [28–32].

There is no special crystallization technique required to prepare chiral crystals from achiral organic compounds. Ordinary crystallization techniques can be employed i.e., a hot saturated solution of a compound in a suitable solvent is slowly cooled, or a saturated solution is slowly evaporated to obtain crystals. Whether the molecule is chiral can be easily differentiated by the measurement of optical rotation in solution using a polarimeter. However, there are great difficulties in measuring the optical rotation of crystalline substances due to the large birefringence [33]. Solid-state circular dichroism (CD) spectral measurements of

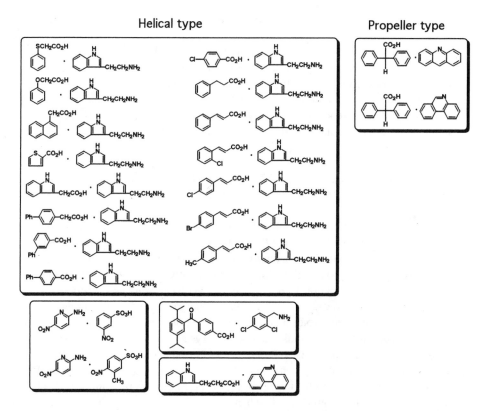

Figure 3 Chiral cocrystals formed from two different achiral molecules.

powdered crystals as Nujol mulls may be usable as a screening method [34]. However, it should be mentioned that, according to our experience, significant CD spectra were sometimes not observed for certain chiral crystals obtained from achiral compounds. The best way for designating the chirality of crystals is to know their space groups obtained by x-ray crystallographic analysis. When the space group of the single crystal of a compound is chiral, the crystal is designated as chiral. The absolute structure of the chiral crystal is determined by anomalous dispersion of slightly heavy elements such as chlorine and sulfur using Cu Kα radiation.

This kind of chiral crystallization of an achiral molecule always gives both right- and left-handed crystals. However, under ordinary crystallization conditions, enantiomorphous control is not possible. We have reported that a cocrystal is formed by the crystallization of a solution of a 1 : 1 mixture of 3-indolepropionic acid and phenanthridine in acetonitrile [28]. When the acetonitrile solution was divided into six parts and spontaneously evaporated in six containers, P-crystals appeared in four containers and M-crystals in two. If such chiral crystallization is carried out using a much larger number of flasks, the ratio of the formation of both enantiomeric crystals will approach 1 : 1.

When two enantiomorphous right- and left-crystals are separately obtained, one can conveniently use each crystal for the seeding of the selective chiral crystallization to either one of the two enantiomorphous crystals. More elegant pseudoseeding, based on utilizing different crystals with similar crystal structure as seed crystals, can enantiocontrol crystallization from solutions of tryptamine and achiral carboxylic acids [35].

III. DIASTEREOMERIC PHOTOREACTIONS OF CHIRAL MOLECULES IN THE CRYSTALLINE STATE

Chiral crystals formed from chiral molecules can undergo highly diastereoselective photoreactions, while diastereodifferentiation in solution is usually low. Here three types of diastereoselective photoreactions in the crystalline state are presented. Highly diastereoselective Norrish/Yang photocyclization of adamantane [36] and β-lactam formation from oxoamides [37] have been also reported.

The 4-(3′-butenyl)-2,5-cyclohexadien-1-one derivative **8** functionalized with a chiral substituent group produced two types of polymorphic crystals, the α- and β-forms. Solid-state irradiation of the α-crystal caused [2 + 2] photocyclization to give **10** in 80% diastereomeric excess at 60% conversion (Scheme 1) [38]. In contrast, photolysis of the β-crystals afforded almost complete reversal of the sense of diastereoselection resulting in the formation of **9** in 90% de at very low conversion. Photolysis of **8** in benzene solution resulted in low diastereodifferentiation.

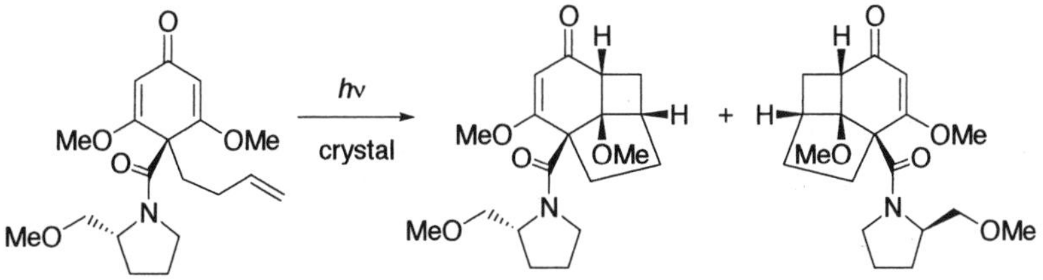

Scheme 1 Diastereoselective [2 + 2] photocyclization of **8**.

It is known that diarylethenes undergo photochromic cyclization/cyclore-version reactions in the crystalline state and in solution [39]. When a chiral substituent was introduced into the diarylethene **11**, crystals were formed in space group $P2_12_12_1$. Upon UV irradiation at 366 nm, the enantiopure crystal of (*S*)-**11** underwent highly diastereoselective [2 + 4] photocyclization to afford almost completely the one closed-ring diastereomer (*S*,*R*,*R*)-**12** (Scheme 2) [40]. Such diastereodifferentiation was not observed in solution, giving both (*S*,*R*,*R*)-**12** and (*S*,*S*,*S*)-**12**. Direct observation of the reaction process in the single crystal of the (*S*)-form by x-ray crystallography demonstrated it to be the (*S*,*R*,*R*)-cyclization product. Conversely, the (*R*)-crystal gave dominantly (*R*,*S*,*S*)-**12**. The closed-ring isomer was produced from the open-ring isomer by the minimal conrotatory rotation of the two thiophene rings in the crystal lattice.

A series of crystals of 2-cyanoethylcobaloxime complexes **13** coordinated with chiral axial ligands underwent diastereoselective photoisomerization, fol-

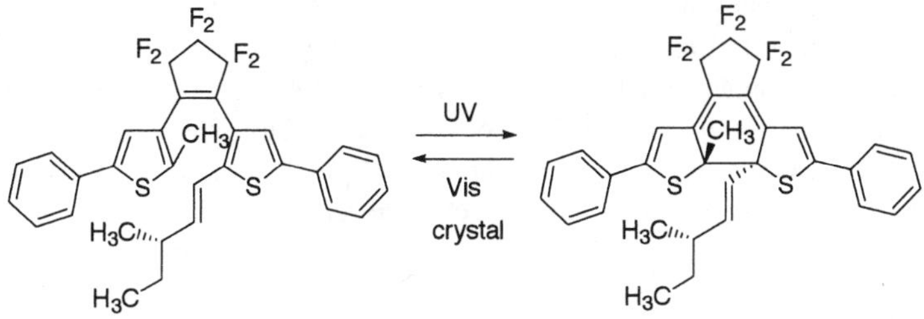

Scheme 2 Diastereoselective photochromic cyclization of the chiral diarylethene (*S*)-**11**.

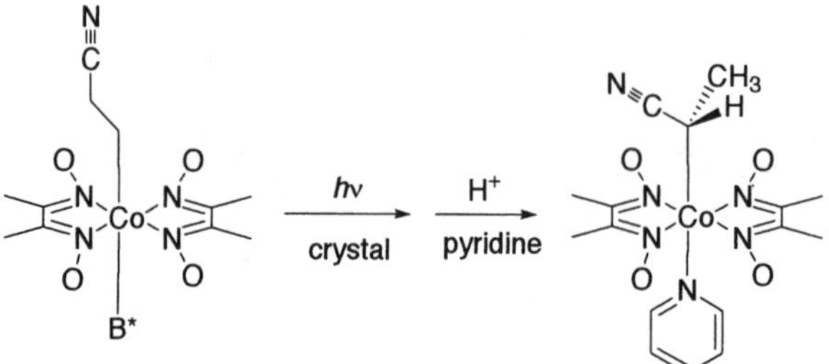

a: B* = (R)-1-Methylpropylamine
b: B* = (R)-1-(1-Naphthyl)ethylamine
c: B* = (R)-12-Phenylglycinol

Scheme 3 Diastereoselective photoisomerization of 2-cyanoethylcobaloxime complexes.

lowed by treatment with the achiral base pyridine, to afford optically active 1-cyanoethylcobaloxime complexes **14** in moderate to high enantioselectivities (Scheme 3) [41]. Precise x-ray crystallographic analysis of the complexes revealed that the handedness of the produced 1-cyanoethyl group depended on the asymmetric shape of the reaction cavity available for the 2-cyanoethyl group before irradiation.

IV. INTRAMOLECULAR ASYMMETRIC PHOTOREACTIONS IN SUPRAMOLECULAR SOLIDS

The host–guest and ionic chiral auxiliary approaches have been most intensively applied for solid-state asymmetric induction. A number of achiral organic compounds could be converted into chiral compounds in high enantioselectivities. However, all the photoreactions in themselves are well-known intramolecular photoreactions: photocyclization, [2 + 2] photocyclization, Norrish type II photocyclization, di-π-methane photorearrangement and photoisomerization. New types of asymmetric photoreactions have never been reported.

A. Photocyclization

It was known that 2-arylthio-3-methylcyclohexen-1-ones **15** underwent photocylization in solution via the thiocarbonyl ylide intermediates **16** to yield the dihydrobenzothiophene derivatives **17** [42]. However, asymmetric induction was difficult to carry out in solution. Toda and coworkers have achieved such enantiodifferentiation by solid-state photolysis of the 1 : 1 host–guest crystals formed with the optically active host compounds (*R*,*R*)-(−)-**6b,c** derived from tartaric acid (Scheme 4) [43]. The optical yields of (+)-**17a–h** were 30–80% *ee* in high chemical yields. Similarly, irradiation of the *N*-phenyl enaminone **18** in solution was known to produce the *N*-methylhexahydro-4-carbazolone **20** via the dipolar ionic intermediate **19** and involving a conrotatory ring closure [44]. Inclusion complexation of **18** with the optically pure host compound **6c** yielded dimorphic crystals. Irradiation of the prism-shaped crystals gave **20** in 87% *ee*, but the needlelike crystals were photochemically inert [45]. X-ray structure analysis revealed that two reaction sites, the phenyl and cyclohexenone groups of the reactant, are located in close and distant positions in the prism and the needle inclusion crystals, respectively.

Photocyclization of acrylanilide **21** to 3,4-dihydroquinolin-2(1*H*)-one **22** was applied in alkaloid synthesis, but enantiocontrol was unsuccessful in solution [46]. Solid-state photolysis of **21** in the 1 : 1 inclusion complexes with optically active host compounds **6a–c** gave **22** in almost perfect optical yields (Scheme 5) [34,47]. X-ray crystallographic analysis revealed that the configuration of the photocyclization products depended on slightly different lattice structures, which were controlled by the host molecules.

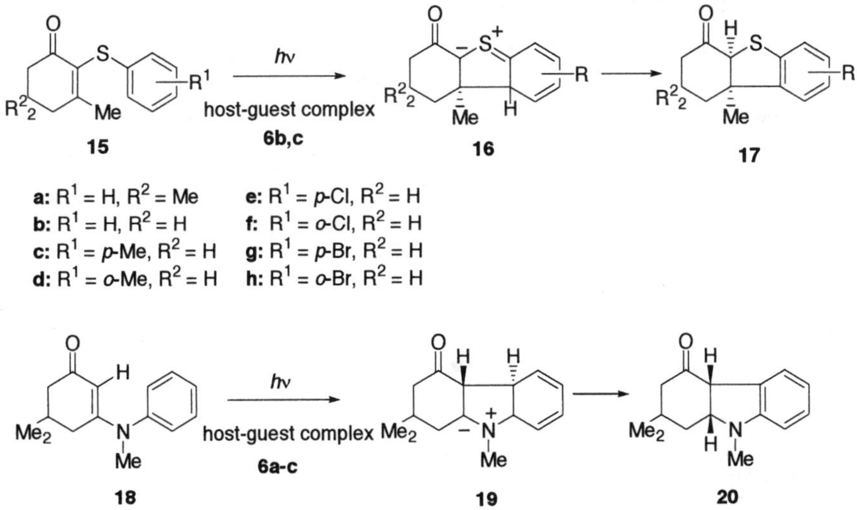
hν
host-guest complex
6b,c
15
16
17
a: R1 = H, R2 = Me
b: R1 = H, R2 = H
c: R1 = p-Me, R2 = H
d: R1 = o-Me, R2 = H
e: R1 = p-Cl, R2 = H
f: R1 = o-Cl, R2 = H
g: R1 = p-Br, R2 = H
h: R1 = o-Br, R2 = H
hν
host-guest complex
6a-c
18
19
20

Scheme 4 Enantioselective photocyclization via ylide intermediates in host–guest crystals.

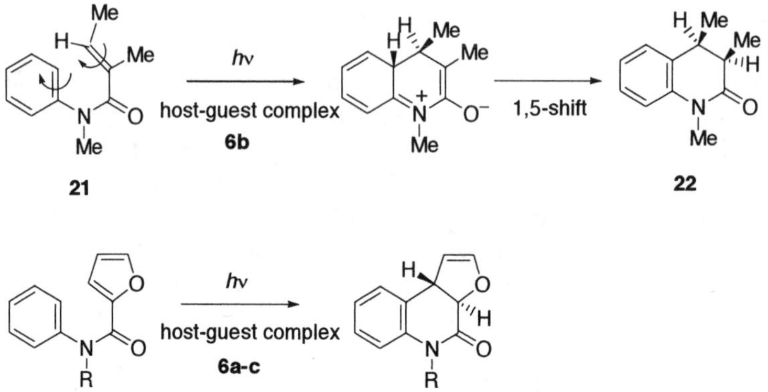

Scheme 5 Enantioselective photocyclization of acrylanilides in their host–guest crystals.

Irradiation of crystalline host–guest complexes of the nitrones **23** with the axle-wheel-type host compound (R,R)-$(-)$-**4** gave the optically active oxaziridines **24** (Scheme 6) [48]. In particular, the enantioselectivities for the formation of **24d** and **24e** were perfect.

B. [2 + 2] Photocyclization

Tropolone ethers have served as useful probes of asymmetric induction using several different supramolecular approaches [49]. Solid-state photolysis of 1 : 1 host–guest complexes of tropolone methyl ether **25** with (R,R)-$(-)$-**4** caused disrotatory [2 + 2] photocyclization to **26** followed by transformation to afford

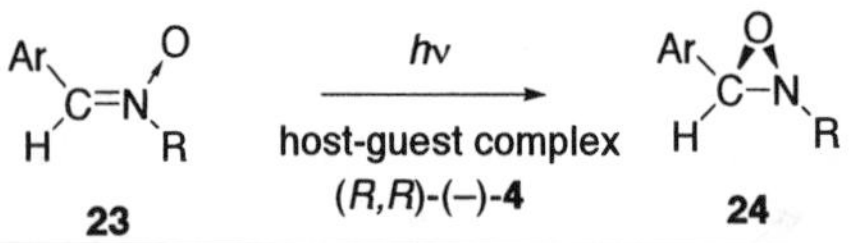

a: Ar = Ph, R = *i*-Pr
b: Ar = Ph, R = *t*-Bu
c: Ar = *p*-Cl-Ph, R = *t*-Bu
d: Ar = *o*-Cl-Ph, R = *t*-Bu
e: Ar = Ph, R = *i*-PrMeCH

Scheme 6 Enantioselective photocyclization of nitrones to oxaziridines in their host–guest crystals.

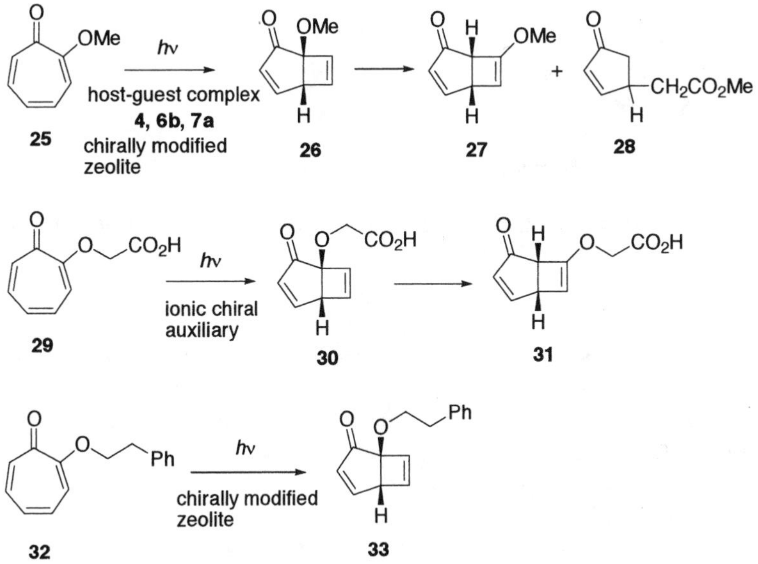

Scheme 7 Enantioselective [2 + 2] photocyclization of tropolones in supramolecular solids.

the chiral secondary product (1S,5R)-(−)-**27** of 100% *ee* in 11% yield (Scheme 7) [50]. Ring-opened product (S)-(+)-**28** of 91% *ee* in 26% yield was also obtained and was thought to be produced by the reaction of **26** with water. Other host compounds derived from tartaric acid were also utilized [51]. Irradiation of the 2 : 1 inclusion crystals of **25** with (−)-**6b** gave (−)-**27** in 91% *ee* and 10% yield, and (+)-**28** in 53% *ee* and 28% yield at 38% conversion. In the case of the 1 : 2 complex of **25** with (−)-**7a**, (+)-**28** was obtained as the sole product in 91% *ee* and 10% yield under prolonged irradiation for 70 h.

The ionic chiral auxiliary approach was also applied to the enantioselective photocylization of tropolone. Irradiation of salt crystals of tropolone ether carboxylic acid **29** with several chiral amines afforded the enantiomerically enriched secondary products **31** [52]. The best results were obtained with optically pure 1-phenylethylamine and 1-amino-2-indanol, which gave optical yields in the 60–80% *ee* range depending on the extent of conversion.

Furthermore, tropolone methyl ether **25** included in NaY and RbY zeolites that had been modified with optically pure norephedrine smoothly underwent enantioselective [2 + 2] photocyclization to afford the primary photoproduct **26** in around 40% *ee* (Scheme 7) [53]. The best results obtained are up to 78% *ee*

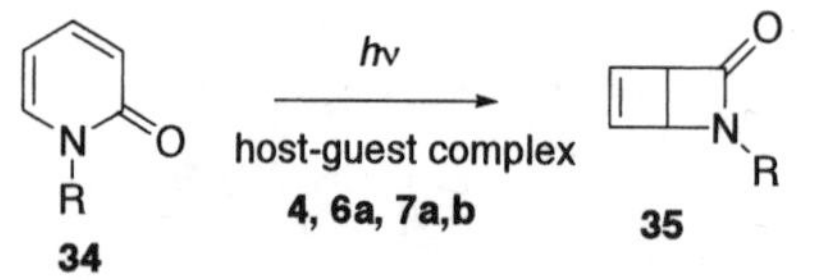

R = **a:** Me, **b:** Et, **c:** n-Pr, **d:** i-Pr, **e:** n-Bu, **f:** i-Bu

Scheme 8 Enantioselective [2 + 2] photocyclization of pyridones in their inclusion crystals.

of the primary photoproduct **33** in the photoreaction of tropolone ethyl phenyl ether **32** included within NaY zeolite preloaded with optically pure ephedrine.

Solid-state photolysis of host–guest complexes of 2-alkylpyridones **34** with the optically active host compounds **4, 6a, 7a,b** caused disrotatory [2 + 2] photocyclization to give chiral bicyclic β-lactams (+) or (−)-**35** with high enantioselectivities (Scheme 8) [54]. For instance, the 1 : 1 complex of 2-methylpyridone **34a** with (R,R)-(−)-**6a** produced the optically pure β-lactam (+)-**35a** in 93% chemical yield at 15% conversion.

Irradiation of host–guest crystals of the 2-[N-(2-propenyl)amino]cyclohex-2-enone derivatives **36** with chiral hosts **6a–c** caused intramolecular [2 + 2]

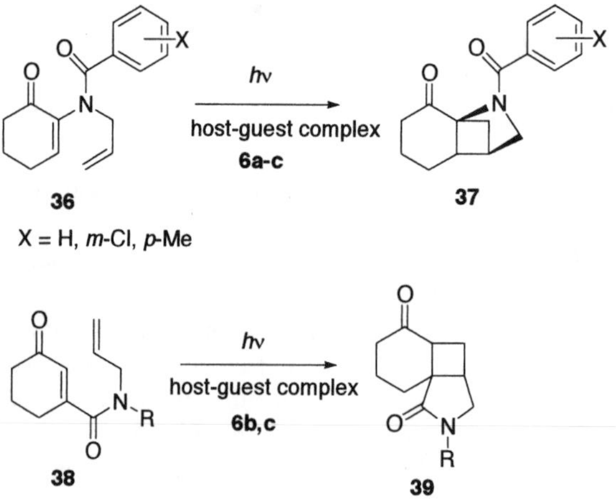

X = H, m-Cl, p-Me

R = n-Pr, PhCH₂, p-MePhCH₂,
     p-ClPhCH₂, o-ClPhCH₂

Scheme 9 Enantioselective [2 + 2] photocyclization of cyclohexenone derivatives in their host–guest crystals.

photocyclization to give the optically active products **37** (Scheme 9) [55]. Similarly, solid-state photolysis of host–guest crystals of *N*-allyl-3-oxo-1-cyclohexenecarboxamides **38** with **6b,c** afforded the [2 + 2] photocyclization products **39** in high optical yields [56].

C. Paternó-Büchi Reaction

When salt crystals of the aryl 1-phenylcyclopentyl ketone carboxylic acid **40** with chiral amines such as (+)-bornylamine or (−)-1-phenylethylamine were irradiated, the optically active exo- and endo-oxetanes **41** or **42** were formed in low to moderate enantiomeric excesses (Scheme 10) [57]. The formation of the oxetanes is believed to occur through Norrish type 1 cleavage and hydrogen abstraction, producing an alkene and an aldehyde, followed by a Paternó-Büchi reaction within the crystal lattice cage. In contrast, solution photolysis of **40** in acetonitrile afforded product **43** as the only isolable product via a typical Norrish type I α-cleavage followed by radical coupling.

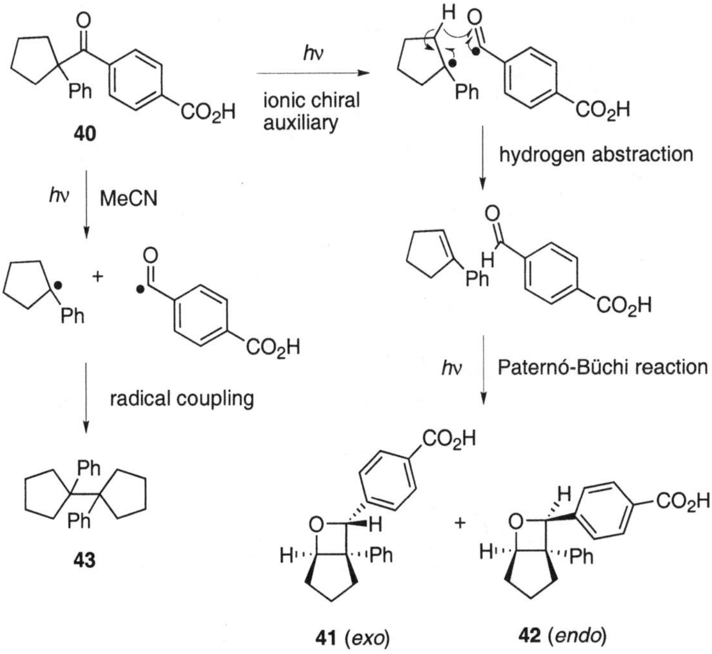

Scheme 10 Enantioselective Paternó-Büchi reaction by the ionic chiral auxiliary method.

D. Hydrogen Abstraction and Cyclization

Hydrogen abstraction by an excited carbonyl group is the most typical photoreaction in both solution and solid states. A number of intramolecular Norrish type II hydrogen abstraction reactions in the crystalline state are already known, and the geometric requirements have been precisely discussed by Scheffer [58]. Solid-state asymmetric induction in the Norrish type II photocyclization of carbonyl compounds using supramolecular approaches has been also intensively studied.

Solid-state photolysis of a salt crystal formed between a prochiral, photochemically reactive keto-acid of 2-benzoyladamantane **44** and a nonabsorbing optically active amine leads to enantioselective Norrish/Yang photocyclization giving the optically active cyclobutanol **45** (Scheme 11) [59–61]. Irradiation of a total of 17 salts gave moderate to near-quantitative enantiomeric excesses. One of the best results was obtained using the prolinol salts: **44** formed dimorphic salt crystals with with (S)-(+)-prolinol, of which the needle-shaped crystals gave

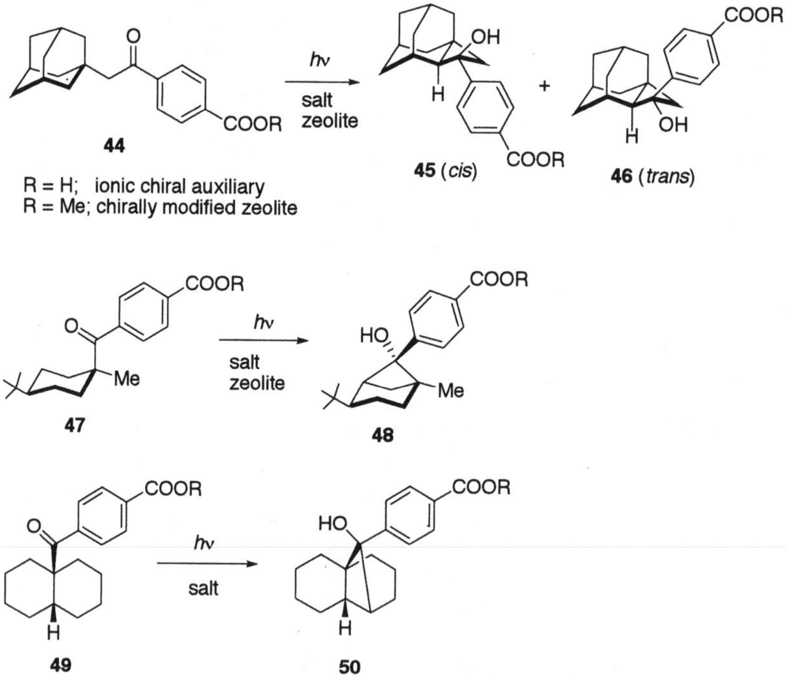

Scheme 11 Enantioselective Norrish/Yang photocyclization by using an ionic chiral auxiliary or a chirally modified zeolite.

cyclobutenol (+)-**45** in 97% *ee*. Photoreaction of the salt crystal with optically pure phenylethylamine proceeded via a single crystal–to–single crystal transformation [62].

Highly enantioselective Norrish/Yang photocyclization of *cis*-4-*tert*-butyl-1-benzoylcyclohexane **47** was also achieved in salt crystals using chiral amines like (*R*)-(+)-1-phenylethylamine or (−)-norephedrine [61,63]. Furthermore, *cis*-9-decalyl aryl ketones **49** in their salt crystals with chiral amines such as (*R*)-(+)-bornyl amine or (*S*)-(−)-1-phenylethylamine underwent highly regio-, diastereo- and enantioselective Norrish/Yang photocyclizations to give the cyclobutanol **50** alone in > 98% *ee* [64].

In contrast, the chirally modified zeolite approach led to poor results. For instance, irradiation of the methyl ester of 2-benzoyladamantane-2-carboxylic acid **44**, included in the NaY zeolite, which had been preloaded with optically pure ephedrine as a chiral inductor, gave the cyclobutanols **45** and **46** in 35 and 5% *ee*, respectively (Scheme 11) [65]. Similarly, the methyl ester of benzoylcyclohexane carboxylic acid **47** afforded the cyclobutenol **48** in 30% *ee* [63].

Irradiation of salt crystals of the 2,4,6-triisopropyl-4'-carboxybenzophenone **51** with various optically active amines caused regio- and enantioselective γ-hydrogen abstraction from the *o*-isopropyl group followed by cyclization to give the optically active cyclobutenol **52** in low to moderate enantiomeric excesses (Scheme 12) [66]. The solid-state photoreaction of the salt crystal with (*S*)-prolinol proceeded in a single crystal–to–single crystal manner without the decomposition of the crystal structure before and after irradiation. In the asymmetric units of both structures, there are two independent molecules, which are related by a pseudo glide plane. This pseudosymmetric arrangement of the two molecules is derived from the weak chirality of (*S*)-prolinol and is responsible for the enantiomeric excess (ca. 30%). The asymmetric induction was improved by using the salt crystal of 2,5-diisopropyl-4'-carboxybenzophenone **53** with (*S*)-(−)-1-phenylethylamine, which underwent enantiospecific single crystal–to–single crystal transformation to give (*R*)-(+)-cyclobutenol **54** in almost quantitative optical yield and 100% chemical yield [67].

Solid-state photolysis of salt crystals of α-mesitylacetophenone-*p*-carboxylic acid **55** and optically pure amines such as (*S*)-(−)-phenylethylamine or (1*R*,2*S*)-(−)-norephedrine caused δ-hydrogen abstraction from the methyl group, and then cyclization to produce the corresponding 2-indanol derivative **56** in excellent yield and high enantiomeric excess (Scheme 12) [68].

Optically active β-lactams, such as penicillin derivatives, are important antibiotics. Since the reports of Norrish type II photocyclization of α-oxoamides to racemic β-lactams in solution [69,70] and in the solid state [71], enantiocontrol of the photocyclization has been intensively studied by using supramolecular aproaches. First, host–guest complexes with deoxycholic acid **1** and cyclodextrin **2** as host compounds have been applied, but the transformation of α-oxoamides

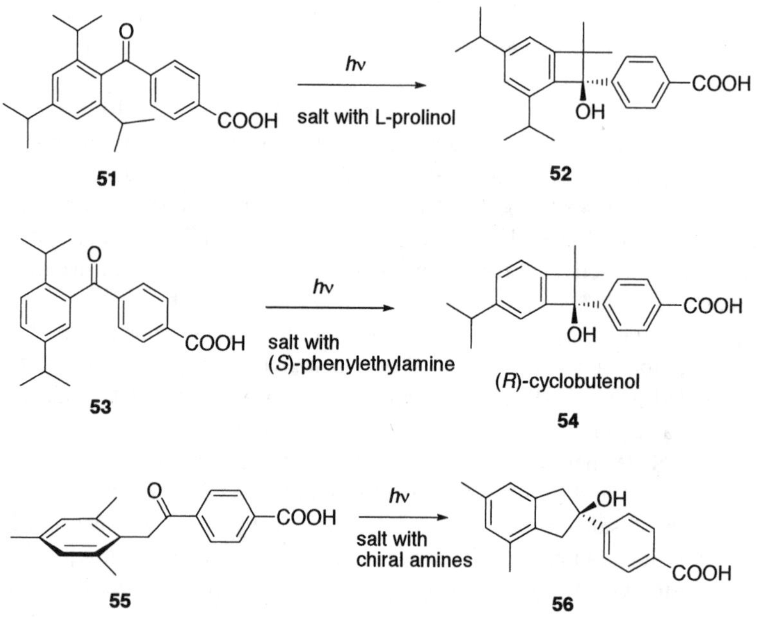

Scheme 12   Enantioselective γ- and δ-hydrogen abstraction and cyclization using the ionic chiral auxiliary method

57a,b to optically active β-lactams **58a,b** resulted in low enantiomeric excesses (Scheme 13) [72]. Subsequently, Toda and coworkers applied the host compounds **4, 6b,c**, to lead to high enantioselective transformations [73–75]. For instance, photolysis of the 1 : 1 complex of **57c** with (R,R)-$(-)$-**4c** gave enantiospecifically (S)-$(-)$-**58c** alone in 100% *ee* [73]. The x-ray crystal structure analysis revealed that the enantioselectivity of **57c** in this process was controlled by the conformation of the O=C–C=O single bond. The torsion angle is 101° and, in the absence of a mirror symmetry molecule, a single enantiomer is produced in a crystal. The left or right handedness of the chiral β-lactam is dictated by the absolute configuration of the host molecule. Similary, irradiation of the 1 : 2 complex of **57c** with (R,R)-$(-)$-**6c** afforded (S)-$(-)$-**58c** alone in 100% *ee* [74].

Salt crystals of the carboxylic-acid-containing α-oxoamide **59a** with optically pure amines were irradiated under nitrogen. (S)-Prolinamide was revealed to be the best chiral auxiliary, giving essentially optically pure photoproduct $(+)$-β-lactam **60a** in 99% *ee* and 94% chemical yield at almost complete conversion [76]. In contrast, the achiral oxoamide **59b**, included within chirally modified zeolites, which had been preloaded with optically pure chiral inductors like ephed-

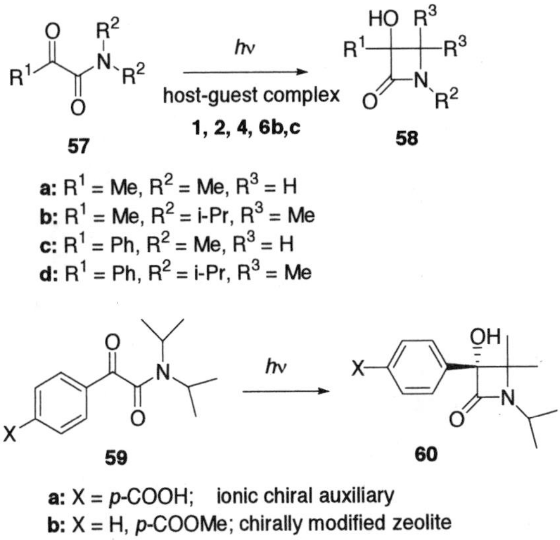
hν
host-guest complex
1, 2, 4, 6b,c
57
58
a: R1 = Me, R2 = Me, R3 = H
b: R1 = Me, R2 = i-Pr, R3 = Me
c: R1 = Ph, R2 = Me, R3 = H
d: R1 = Ph, R2 = i-Pr, R3 = Me
hν
59
60
a: X = p-COOH; ionic chiral auxiliary
b: X = H, p-COOMe; chirally modified zeolite

Scheme 13 Enantioselective conversion of oxoamides to β-lactams using supramolecular approaches.

rine, was converted into the corresponding β-lactam **60b** in only moderate or low optical yield.

Enantiodifferentiations of other lactams were also successful. Irradiation of the 1 : 1 host–guest crystals of *N*-(arylcarbonylmethyl)-2-piperidones **61** and the chiral host **6c** derived from tartaric acid caused enantioselective Norrish type II photocyclization to afford the chiral 7-aryl-7-hydroxyl-1-azabicyclo[4.2.0]octan-2-ones **62** in high optical yields (Scheme 14) [77]. Crystal structures and absolute configurations of the reactant and the product were determined by x-ray crystallographic analysis. The high enantioselectivity of the photocyclization was clearly explained by the two structures. Solid-state photolysis of inclusion crystals formed by 2-(*N*-acyl-*N*-alkylamino)cyclohex-2-enones **63** and the chiral hosts **7a–c** gave optically active spiro-β-lactams **64** in moderate to high enantiomeric excesses [78].

E. Di-π-Methane Photorearrangement

Scheffer and coworkers have intensively studied asymmetric induction for di-π-methane rearrangements using the ionic chiral auxiliary approach. Salt crystals of dibenzobarrelene carboxylic acid **65** with optically pure amines underwent enantioselective di-π-methane photorearrangement to afford chiral dibenzosemi-

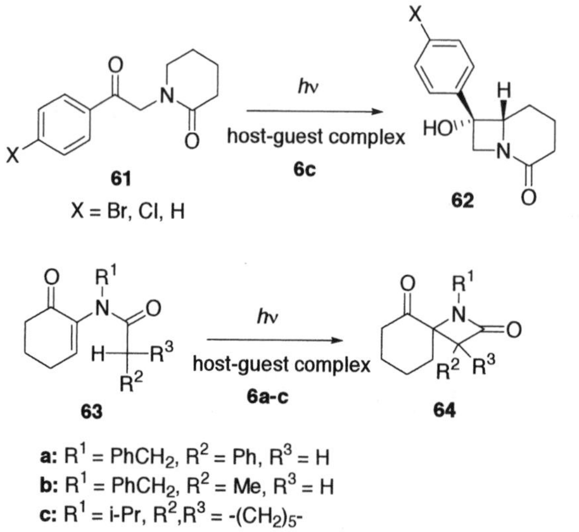

Scheme 14 Enantioselective Norrish type II photocyclization in host–guest crystals.

bullvalene derivatives (Scheme 15) [79,80]. In this reaction, two regioisomers **66** and **67** are possible. For example, however, the major photoproduct from the salt crystal with (R)-(+)-proline was (−)-**66** of 80% *ee* and 96% chemical yield. Initial benzovinyl bridging is favored at the carboxylate salt-bearing vinyl carbon atom. This result can be attributed to preferential radical formation at the ester-bearing vinyl carbon atom.

Salt crystals of the benzonorbornadiene carboxylic acid **68** with optically pure triplet sensitizer amines **70, 71** also underwent enantioselective di-π-methane photorearrangement to afford the optically active tetracyclo[5.4.0.0^{2,4}.0^{3,6}]undecane derivative **69** in high enantiomeric excess [81]. The 4-acetylbenzyl ester of L-phenylalanine **70** and the 4-benzoylphenyl ester of L-valine **71** perform the dual roles of asymmetric induction and triplet energy sensitization.

Irradiation of the salt crystals of benzocyclohexadienone carboxylic acid **72** with optically pure amines caused enantioselective oxa-di-π-methane photorearrangement to give the chiral product **73** [82]. The best results were obtained with (S)-(−)-1-phenylethylamine in around 80% *ee* (Scheme 16). The other amines gave optical purities of 20–50% *ee*.

Chirally modified zeolites were also applied to the asymmetric oxa-di-π-methane photorearrangement. Irradiation of a hexane slurry of compound **74** included in dry (−)-ephedrine-modified NaY zeolite gave product **75** enantiomerically enriched to around 30% [83]. However, photolysis of **74** in solution, in the presence of optically pure ephedrine, gave only a racemic product mixture.

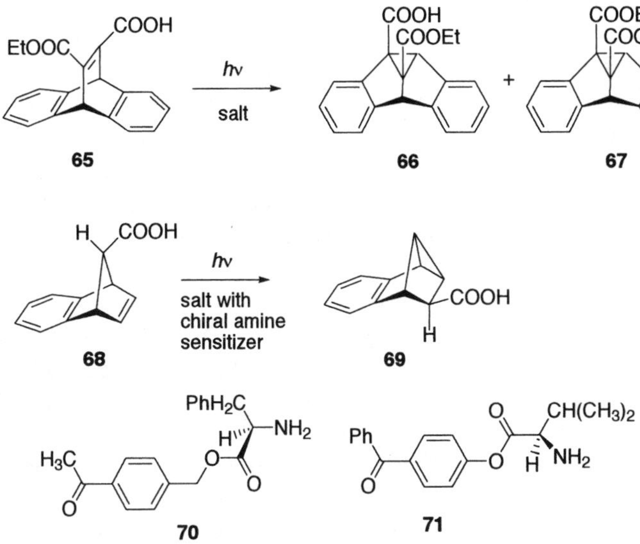

Scheme 15 Enantioselective di-π-methane rearrangement using the ionic chiral auxiliary method.

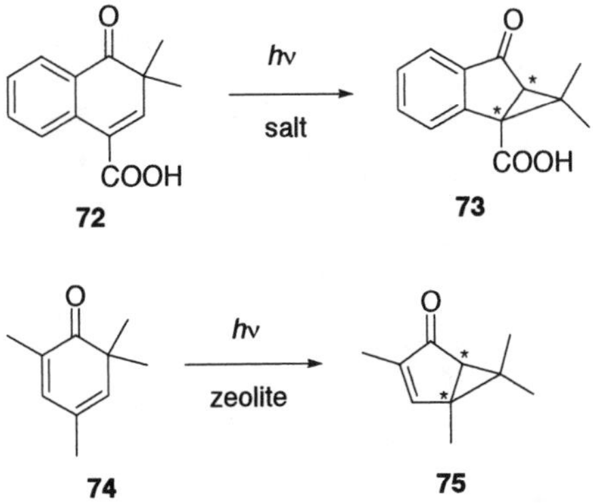

Scheme 16 Enantioselective oxa-di-π-methane rearrangement using the ionic chiral auxiliary and chirally modified zeolite method.

F.   Photoisomerization

Salt crystals were prepared between cyclopropane mesoacid **76** and ten different optically pure amines. Solid-state irradiation of these crystals caused photoisomerization affording the optically active *trans*-cyclopropane **77** in moderate enantiomeric excesses (Scheme 17) [84]. The best results were obtained using (*S*)-proline, which gave 79% *ee* at low conversion. When salt crystals of the diphenylbenzoyl acid **78** with several optically pure amines were irradiated, then the compounds **79** or **80** of near quantitative enantiomeric excesses were obtained, with a slight diminution in *ee* as the conversion was increased [85].

The methyl ester of acid **76** was adsorbed onto NaY zeolite, in which optically pure inductors like ephedrine and diethyl tartrate had been preloaded. In this instance, however, the results of photoreaction were disappointing [84]. Diethyl tartrate proved to be the best chiral inductor (12% *ee*); all the rest gave *ee* values of 5% or less. Photoisomerization of the methyl ester of the diphenylbenzoyl acid **78** using zeolites modified with chiral indicators also gave low *ee* values [85]. Norephedrine was found to be the best chiral indicator, resulting in 20% *ee*.

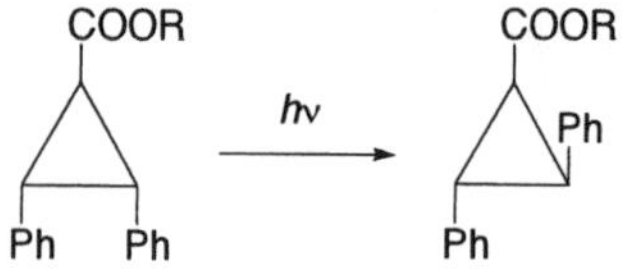

R = H;   ionic chiral auxiliary
R = Me; chirally modified zeolite

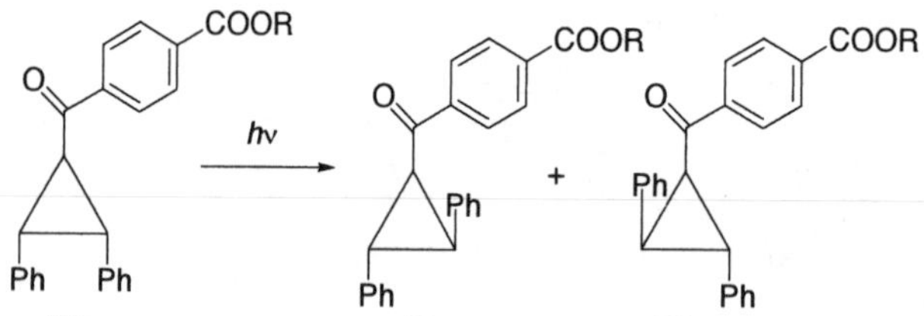

Scheme 17   Asymmetric photoisomerization of cyclopropanes using the ionic chiral auxiliary and chirally modified zeolite method.

V. INTERMOLECULAR ASYMMETRIC PHOTOREACTIONS IN CHIRAL SUPRAMOLECULAR CRYSTALS

In comparison to the variety of intramolecular asymmetric photoreactions, intermolecular asymmetric photoreactions have been scarcely reported. However, the reaction types do include [2 + 2] photodimerization, photoaddition, photodecarboxylative condensation, and photooxygenation. Herein the intermolecular reactions are more precisely described, including their reaction mechanisms and reaction paths in the crystals.

A. [2 + 2] Photodimerization

Irradiation of the 2 : 1 host–guest crystals of cyclohex-2-enone **81** with the axle-wheel-type host compound $(-)$-**5** as an aqueous suspension caused regio- and enantioselective $[2 + 2]$ photodimerization to afford the $(-)$-*anti-head-to-head* dimer **82** of 48% *ee* in 75% chemical yield (Scheme 18) [86]. Similarly, solid-state photolysis of the 3 : 2 complex of cycloocta-2,4-dien-1-one **83** with (R,R)-$(-)$-**4** gave the $(-)$-*anti-head-to-head* dimer **84** in moderate optical yield [87].

Irradiation of 1 : 1 host–guest crystals of coumarin **85a** with (R,R)-$(-)$-**6a** derived from tartaric acid gave the $(-)$-*anti-head-to-head* dimer **86a** of 96% *ee* [88]. Enantiospecific photodimerization of thiocoumarin **85b** gave optically pure $(+)$-*anti-head-to-head* dimer **86b** when the 1 : 1 complex with (R,R)-$(-)$-**6b** was used. X-ray structure analysis revealed that the distance between the two ethylenic double bonds was short enough (3.59 and 3.42 Å for **85a** and 3.73 and 3.41 Å for **85b**) for addition to occur and topochemically [89]. Further, both reactions were found to proceed *via* a single crystal–to–single crystal transformation.

B. Photoaddition

In contrast to the number of intramolecular asymmetric hydrogen abstraction reactions using supramolecular approaches (Sec. IV.D.), intermolecular asymmetric hydrogen abstraction reactions have been scarcely reported. Only two examples are presented; these occur in inclusion crystals of deoxycholic acid with ketones and crystals of cyclodextrin with acetophenone.

It is well known that deoxycholic acid **1** forms inclusion complexes with a number of guest molecules. The crystalline inclusion complexes **87a–c·1** were obtained from solutions of deoxycholic acid **1** with acetophenone **87a**, *m*-chloroacetophenone **87b**, or *p*-fluoroacetophenone **87c** in methanol. The host : guest molar ratios are 5 : 2, 3 : 1, and 8 : 3, respectively. The guest molecules are

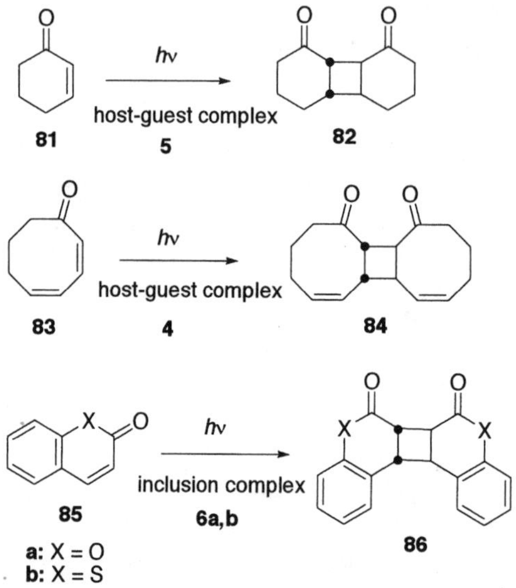

Scheme 18 Enantioselective [2 + 2] photodimerization of enones in their host–guest crystals.

included in cavities between the steroid molecules and occupy channel structures. UV irradiation (> 290 nm) of these crystals for about 30 days at room temperature gave the corresponding single diastereomeric photoadduct **88** formed from the host and guest molecules in moderate chemical yields (Scheme 19) [21]. A reasonable reaction process would be for the excited carbonyl oxygen of the acetophenone to abstract the hydrogen atom H–C5 of **1** and then for the steroid carbon radical C5 and the sp^3 carbon radical derived from the carbonyl carbon to couple. However, an important problem remained, since the newly generated chiral carbon had the (S)-configuration, which is opposite to that expected from the molecular arrangement. It indicates that photoaddition of the guest molecule to C5 takes place with a net rotation of 180° of the guest acetyl group.

X-ray crystallographic analysis of two complexes, **87a·1** and **87c·1**, before and after irradiation, were performed at low temperature (− 170°C). Fortunately, these reactions were revealed to proceed *via* a single crystal–to–single crystal transformation without decomposition of the initial crystal structures. Furthermore, the reaction pathway was successfully traced by x-ray analyses of crystals irradiated for different times up to 67 days. These revealed that the ketyl group rotated around the C(ketyl)–C5(phenyl) bond and finally coupled with the steroid C5 position to produce the adduct.

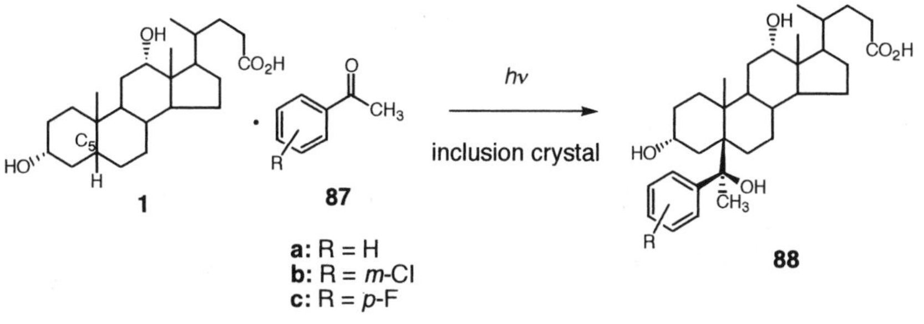

Scheme 19 Diastereomeric photoaddition between deoxycholic acid and acetophenones in their inclusion crystals.

Irradiation of solid 1 : 1 complexes of benzaldehyde **89** and β-cyclodextrin **2** resulted in enantioselective intermolecular hydrogen abstraction and radical coupling to afford (*R*)-(−)-benzoin **90** as a major product in up to 15% *ee* and 56% chemical yield. 4-Benzoylbenzaldehyde was obtained as a minor product (Scheme 20) [22]. This reactivity can be understood from the presence of guest-to-host (2 : 2) complexes generated by the head-to-head association of two 1 : 1 inclusion complexes.

C. Photodecarboxylative Condensation

Photodecarboxylations of organic carboxylic acids are well known to occur via photoinduced electron transfer [90]. Although a large number of photodecarboxylations in the solution phase have been studied by using various acceptors such as aza aromatic compounds [91] and polycyanoaromatics [92], the product selectivities are necessarily low due to the co-occurrence of subsequent radical coupling because of the free motion of radical species in solution. In contrast, solid

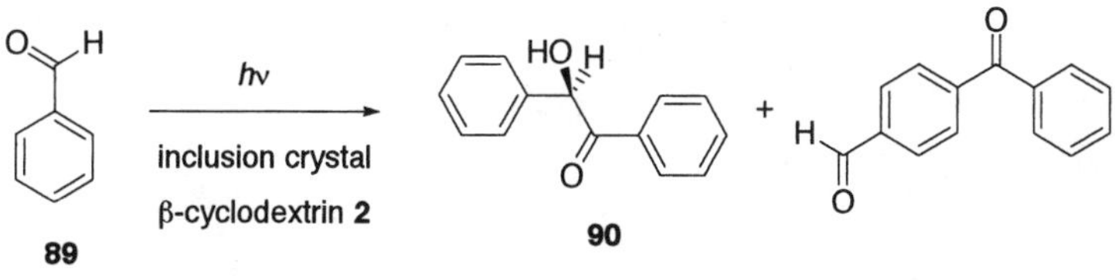

Scheme 20 Formation of optically active benzoin in the β-cyclodextrin complex.

state photoreactions in cocrystals combining chiral carboxylic acids and electron acceptors occur with regio- and enantioselectivity.

1,2,4,5-Tetracyanobenzene (TCNB) **91** and (S)-(+)-2-(6-methoxy-2-naph-thyl)propanoic acid (S)-(+)- **92** formed chiral 1 : 1 charge transfer (CT) crystals, for which a CT band was observed at 433 nm as a broad band. Irradiation of the crystals in the CT band afforded the optically active decarboxylative condensation product (+)-**93** of 21% *ee* in 95% chemical yield (Scheme 21) [93]. This implies that the chirality of **92** is transferred to product **93** in the decarboxylative coupling process. The radical species (Ar–MeHC·) formed in the solid-state reaction may retain its original hybridization state and configuration to some extent because of the restrictions of the crystal lattice when it reacts with the TCNB anion radical to form a new C–C bond. In contrast, solution photolysis between **91** and (S)-(+)-**92** in acetonitrile gave the racemic condensation product (±)-**93** in 72% yield. Both the optical rotation and enantiomeric excess were found to be zero. This suggests that the radical species (Ar–MeHC·) undergoes a change of hybridization from sp^3 to sp^2 and loses its original configuration in the free environment of the solution phase before it reacts with the TCNB anion radical to form a new C–C bond.

Acridine **94** and (R)-(−)-2-phenylpropionic acid (R)-**95** formed the chiral 3:2 cocrystal **94·**(R)-**95**. Irradiation of these cocrystals gave enantiomeric products **96** and **98** as well as the diastereomeric product **97** through photodecarboxylative condensation, with all positive $[\alpha]_D^{20}$ values (Scheme 22) [94]. Similarly, the opposite handed cocrystal afforded **96**–**98** with negative $[\alpha]_D^{20}$ values. In contrast, photolysis of a 1:1 solution of **94** and (R)-**95** in acetonitrile resulted in the formation of racemic **96** and biacridane **99**.

The crystal **94·**(R)-**95** belonged to space group $P2_1$, and its crystal structure and the absolute configuration were determined using the known handedness of (R)-**95**. In the crystal lattice, two independent hydrogen bond pairs of **94** and (R)-**95** are arranged in a columnar structure. A further additional aciridine molecule does not form a hydrogen bond. Upon irradiation of the cocrystals, decarboxyl-

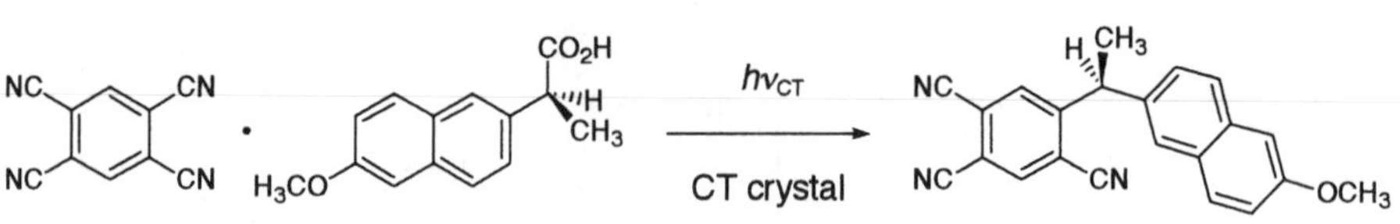

Scheme 21 Chirality memory for the photodecarboxylative condensation in the CT crystal.

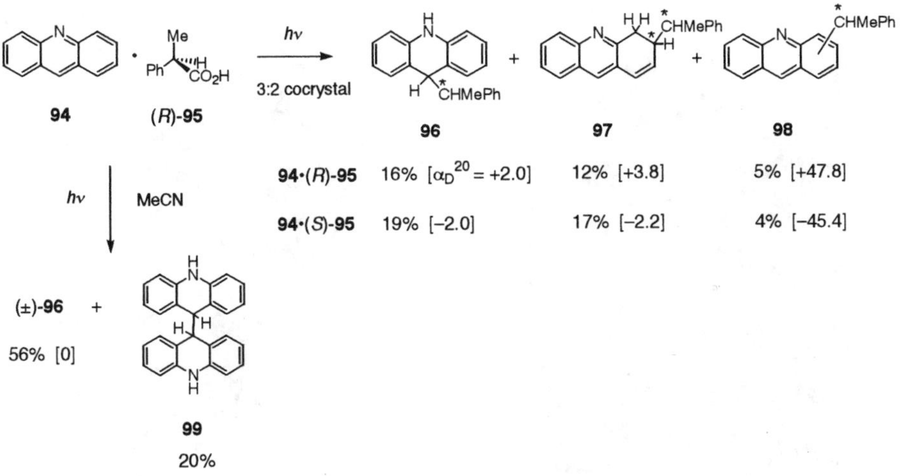

Scheme 22 Enantio- and diastereoselective photodecarboxylative condensation in the cocrystals.

ation occurs to give ·CHMePh and hydroacridine radical species. Next, C1–C2 radical coupling results as the highest priority event due to the short distance of 4.8 Å and the high charge density on the hydroacridine radical. The original configuration of ·CHMePh is retained to some extent owing to the restricted motion possible in the crystal lattice, this leading to the formation of optically active **96** (Fig. 4). The frontal C1–C3 attack by the reactive ·CHMePh radical across a distance of 4.6 Å should introduce another asymmetric carbon at the coupling position in **97**, which results in a diastereomeric compound having two chiral carbon centers. The C1–C4 coupling with a distance of 6.1 Å gives another diastereomer of **97** in much smaller yield.

D. Photooxygenation

(Z)-2-Methoxybut-2-ene **100** formed a 1:2 inclusion complex with tri-o-thymotide **3** [23]. Because the crystals are isotropic in space group $P3_121$, the two enantiomorphous single crystals were separated by sorting manually viewing through polarizers. The absolute configurations could be determined by x-ray anomalous dispersion to be the (P)-(+)- and (M)-(−)-forms. The (P)-(+)-crystals were mixed with ion exchange resin (IRA 401) to which rose bengal was fixed as a sensitizer. The mixture was tumbled in a stream of oxygen and irradiated at > 418 nm for 160 h to give chiral hydroperoxide **101** with optical rotation of +0.060°

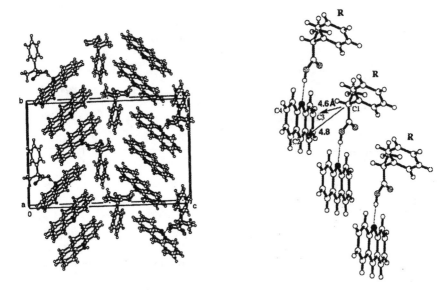

Figure 4   (a) Molecular packing and (b) radical coupling paths in the cocrystal **94·(R)-95**. Black and dotted atoms represent N and O atoms, respectively. (From Ref. 94 with permission. © 1997 Elsevier.)

in 30% chemical yield (Scheme 23). Photooxygenation of the (M)-(−)-crystals produced the opposite handed hydroperoxide with optical rotation of −0.051° in 27% chemical yield. However, the optical yields were not determined.

VI. ABSOLUTE ASYMMETRIC PHOTOREACTIONS BY THE SPONTANEOUS CHIRAL CRYSTALLIZATION APPROACH

Around 20 absolute asymmetric photoreactions have been reported by utilizing chiral crystals spontaneously formed from achiral molecules as reactants. Most

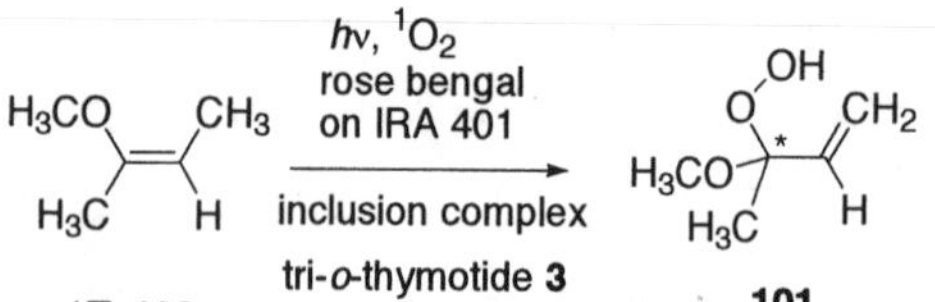

Scheme 23   Asymmetric photooxygenation in the inclusion crystal.

of the successful examples are intramolecular reactions in single: component crystals. They include photocyclization, [2 + 2] photocyclization, Norrish type II photocyclization, and di-π-methane photorearrangement. In contrast, intermolecular absolute asymmetric photoreactions have been scarcely reported. These are [2 + 2] photocycloadditions in single-component crystals and mixed crystals of diene compounds and the CT crystal of bis[1,2,5]thiadiazolotetracyanoquinodimethane and o-divinylbenzene, and the photodecarboxylative condensation in cocrystals of acridine and diphenylacetic acid.

A. Intramolecular Absolute Asymmetric Photoreactions

1. Photocyclization

Achiral 3,4-bis(diphenylmethylene)-N-methylsuccinimide **102** crystallized in three polymorphic forms, one of which was chiral in space group $P2_1$. Irradiation of the enantiomorphous crystal (+)-**102** gave the optically active photocyclization product (+)-**103** in 64% ee (Scheme 24) [95]. This enantioselective photoconversion consists of two steps, a conrotatory ring closure and a 1,5-hydrogen shift. The latter sigmatropic reaction occurs in the solid state in a suprafacial manner.

Sakamoto and coworkers have reported that the following two photocyclizations take place by different mechanisms. Achiral S-aryl 2-benzoylbenzothioates **104** spontaneously crystallized in chiral space group $P2_12_12_1$. Solid-state irradiation of the crystals (P)-**104** resulted in an intramolecular cyclization involving unprecedented phenyl migration to afford optically active corresponding 3-phenyl-3-(arylthio)phthalides (R)-**105** in good chemo- and enantioselectivities (Scheme 25) [96]. Stereochemical correlation based on the absolute configuration of the starting thioesters and the photoproducts confirmed the phenyl migration rather than the well-recognized radical mechanism.

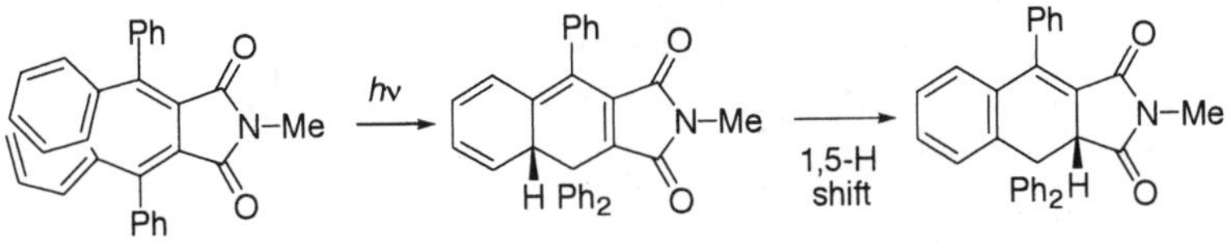

Scheme 24 Absolute asymmetric photocyclization of a succinimide derivative involving a 1,5-hydrogen shift.

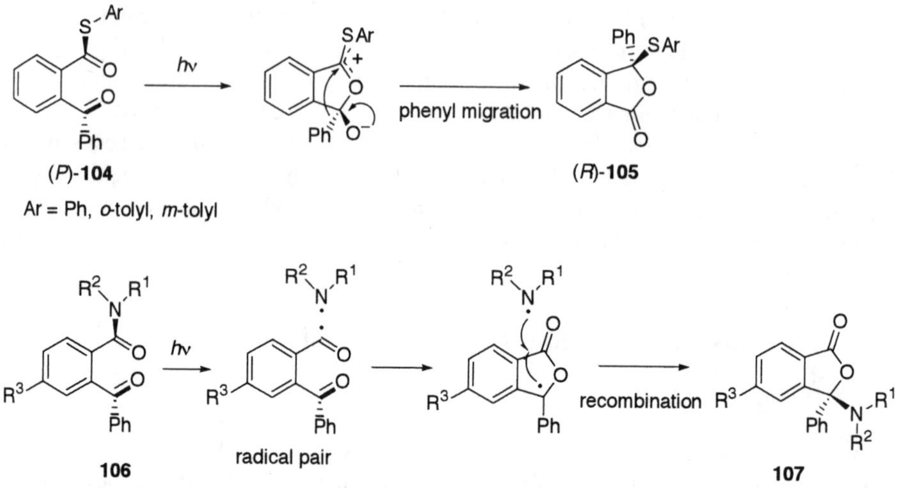

Scheme 25 Absolute asymmetric photocyclization involving phenyl migration and radical pair recombination.

N,N-Disubstituted 2-benzoylbenzamides **106** also formed chiral crystals in space group $P2_12_12_1$. Irradiation of these crystals gave optically active phthalides **107** in high optical and chemical yields [97]. It was concluded from the x-ray structural analysis that the photoreaction promoted intramolecular cyclization to phthalides *via* a radical pair intermediate, not phenyl migration.

2. [2 + 2] Photocyclization

Three examples of absolute asymmetric [2 + 2] photocyclizations are presented here and briefly summarized. 4-Benzoyl-2-pyridones **108**, whose phenyl groups are *meta*-substituted (*m*-Cl, *m*-Br, *m*-Me, *m*-OMe) revealed a significant tendency toward crystallization in chiral space groups. Irradiation of the single crystals led to optically active β-lactam derivatives **109** in high optical and chemical yields *via* an allowed disrotatory electrocyclization (Scheme 26) [98]. Solid-state photoreaction of *N*-(thiobenzoyl)methacrylanilide **110**, which formed chiral crystals of space group $P2_12_12_1$, gave the optically active thietane-fused β-lactam **111** in low enantiomeric excess [99]. *N*-Isopropyl-*N*-tiglylbenzoylformamide **112** formed chiral crystals in space group $P2_1$. Solid-state photolysis of these crystals resulted in intramolecular [2 + 2] cyclization to afford a chiral bicyclic oxetane **113** in high optical and chemical yield [100].

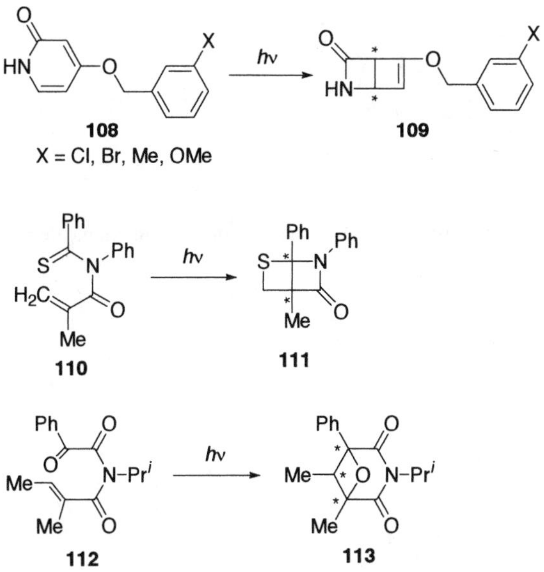

Scheme 26 Absolute asymmetric [2 + 2] photocyclization.

3. Hydrogen Abstraction and Cyclization

Several absolute asymmetric Norrish type II cyclizations have been reported in the solid state. Achiral α-(3-methyladamantyl)-p-chloroacetophenone **114** formed a chiral crystal in space group $P2_12_12_1$. Irradiation of these crystals caused Norrish/Yang cyclization to afford the cyclobutenol-type photoproduct **115** (six chiral centers) with respectable enantiomeric excess at low conversion (Scheme 27) [101]. Photolysis in solution phase led to mixtures of four of the six possible cyclobutanols, and no trace of optical activity could be detected in the mixtures.

Four absolute asymmetric Norrish type II cyclizations to produce β-lactams have been reported (Scheme 28). Achiral oxoamide **116a** spontaneously formed chiral crystals from solution. Irradiation of one type of the enantiomorphous crystals gave optically active (+)-β-lactam **117a** of 93% *ee* in 74% chemical yield [102]. The opposite handed crystals produced (−)-β-lactam **117a**. *m*-Chlorophenyl-*N,N*-diisopropylglyoxylamide **116b** was also found to crystallize spontaneously into the chiral space group $P2_12_12_1$. Irradiation of the enantiomorphous crystals gave (−)-β-lactam **117b** in 100% *ee* [103].

The *S*-phenyl *N*-(benzoylformyl)thiocarbamate **118** crystallized in chiral space group $P2_1$. Irradiation of the powdered crystals afforded optically active oxazolidinedione **119** and *cis*-β-lactam **120** in moderate enantiomeric excesses

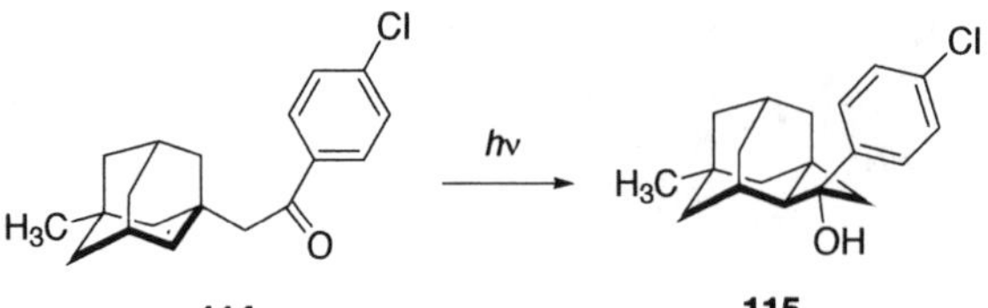

Scheme 27 Absolute asymmetric Norrish/Yang photocyclization of adamantane derivative **114**.

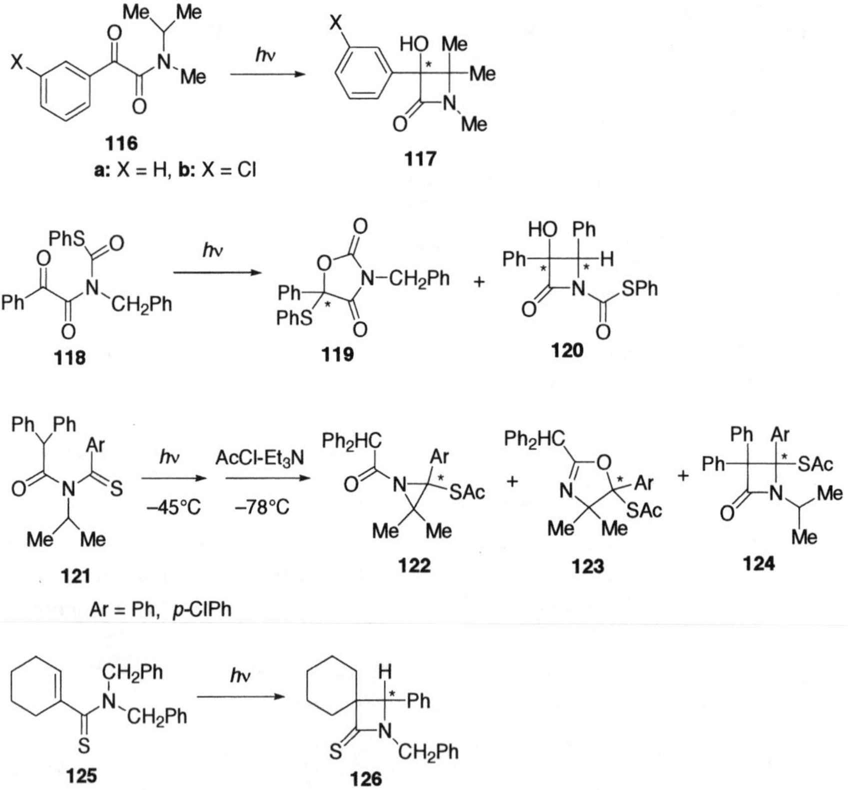

Scheme 28 Absolute asymmetric Norrish type II cyclization to β-lactams.

(Scheme 28) [104]. The achiral acyclic monothioimides **121** crystallized in chiral space group $P2_12_12_1$. Irradiation of the crystals at $-45°C$ followed by acetylation gave three optically active products [105]. β-Hydrogen transfer led to aziridine **122** and oxazoline **123** formation, in 84 and 50% *ee*, respectively. γ-Hydrogen transfer also occurred to give β-lactam **124** in 20% *ee* as a minor product. Solid-state photolysis of achiral α,β-unsaturated thioamide **125** as its chiral crystalline solid afforded optically active β-thiolactam **126** in high optical yield *via* a crystal-to-crystal process [106].

Recently, we have prepared the chiral salt crystal ($P2_12_12_1$) from achiral carboxylic acid 2,5-diisopropyl-4'-carboxybenzophenone **53** and achiral amine 2,4-dichlorobenzylamine. Irradiation of these chiral crystals afforded three optically active products: cyclopentenol **127**, cyclobutenol **54** and hydrol **128** in the ratio of 6 : 3 : 1 (Scheme 29) [107]. The enantiomeric excesses of the three products were excellent and constant over low to high conversions. The cyclopentenol **127** and cyclobutenol **54** are produced by δ- and γ-hydrogen abstraction from the *o*-isopropyl group followed by cyclization, respectively.

4. Di-π-Methane Photorearrangement

The dibenzobarrelene diisopropyl ester **129** formed dimorphic crystals in chiral space group $P2_12_12_1$ and achiral space group *Pbca*. Irradiation of the chiral crystals led to quantitative enantiomeric yields of the dibenzosemibullvalene **130** (Scheme 30) [108]. Another achiral ester-lactone derivative **131** crystallized spontaneously in the chiral space group $P2_12_12_1$. Solid-state photolysis afforded two regioisomeric di-π-methane products **132** and **133** in the ratio 87 : 13 [109]. One of the photoproducts **133** was formed in near-quantitative enantiomeric excess, whereas the other **132** was produced as a racemate. Compound **134** functionalized with diphenylphosphine oxide groups formed a 1:1 inclusion complex with the solvent molecule EtOH, for which the chiral space group was $P2_12_12_1$. The inclusion crystals underwent enantioselective di-π-methane photorearrangement to give the dibenzosemibullvalene **135** in high enantiomeric excess [110]. Because phosphorus is a sufficiently heavy atom, the absolute configuration of the reactant **134·EtOH**, as well as the photoproduct **135**, could be determined with a high degree of certainty by the Bijvoet method.

The achiral compound **136** crystallized in chiral space group $P2_12_12_1$. Enantioselective di-π-methane photorearrangement took place on irradiation to afford preparative quantities of **137** and **138** in respective enantiomeric excesses of $\leq 44\%$ and $\leq 96\%$, respectively (Scheme 31) [111]. The absolute configurations of the reactant and the products were not determined.

B. Intermolecular Absolute Asymmetric Photoreactions

1. [2 + 2] Photocycloaddition

The Weizmann Institute group first reported absolute asymmetric [2 + 2] photo-cycloaddition by using chiral crystals from achiral diene compounds in the 1970s

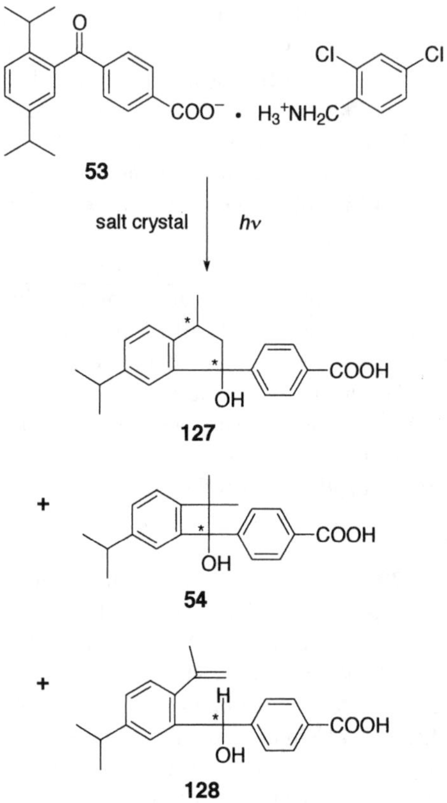

Scheme 29 Absolute asymmetric photocyclization in a salt crystal formed from two achiral components.

[12,112]. When unsymmetrically disubstituted dienes are arranged with the two different double bonds correctly juxtaposed, chiral crystallization sometimes occurs. For instance, the 1,4-disubstituted phenylenediacrylate **139** with two different substituents crystallized in chiral space group $P2_1$. The molecules of the transoid form are arranged in a head-to-head manner in the crystal. Irradiation of the crystal caused enantioselective intermolecular [2 + 2] photocycloaddition to afford almost optically pure dimer **140** (Scheme 32) [113]. Hasegawa and coworkers have also found that 4-[2-(pyridyl)ethyl]cinnamate **141** crystallizes in chiral space group $P2_12_12_1$. In the crystal, the molecules of cisoid form are arranged in head-to-tail packing. Irradiation of the crystal afforded the optically active dimer **142** in 92% *ee* [114].

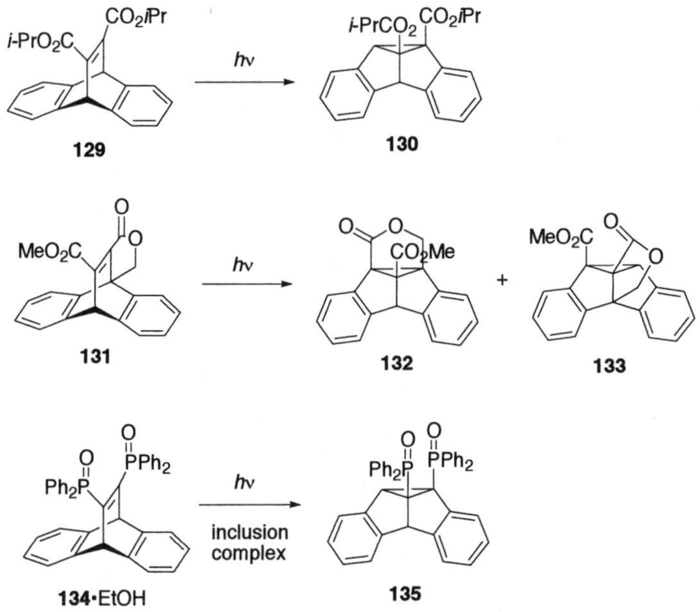

Scheme 30 Absolute asymmetric di-π-methane rearrangement of achiral dibenzobarrelenes.

Achiral butadienes **143** and **144** formed chiral mixed crystals (substitutional solid solutions) of space group $P2_12_12_1$ by cooling the melts of both components or on crystallization from ethanol solution. A single large-sized mixed crystal **143·144** was pulverized and irradiated resulting in [2 + 2] photocycloaddition, thereby giving the optically active heterodimers **145** as well as the achiral homodi-

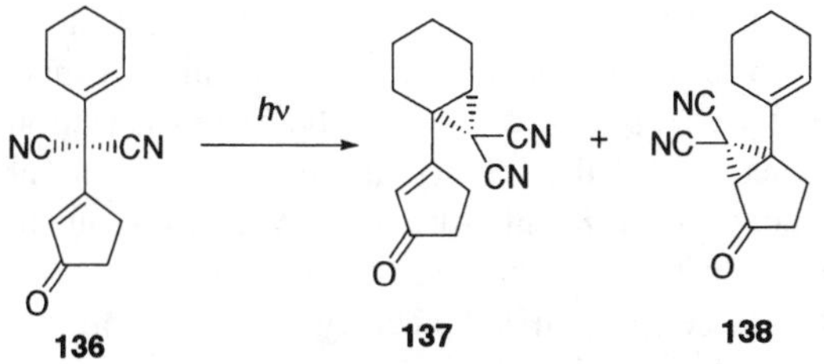

Scheme 31 Absolute asymmetric di-π-methane rearrangement.

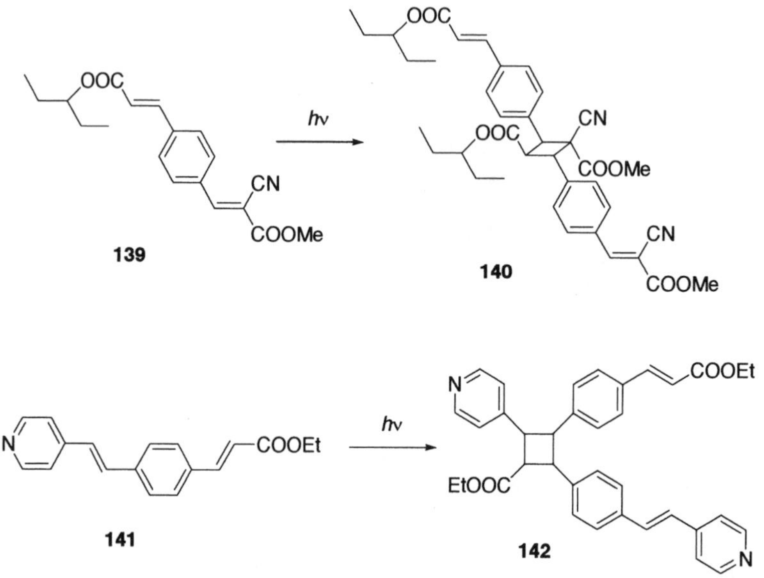

Scheme 32 Absolute asymmetric [2 + 2] photodimerization of dienes.

mers **146** and **147** (Scheme 33) [12]. As expected, some crystals gave predominantly left-handed **145** and others right-handed **145**. Selective excitation of the thiophene moieties, which absorb longer wavelength light than the phenyl groups, led to a decrease in the formation of the achiral homodimers **146** and **147**. In fact, irradiation at longer wavelength (> 400 nm) of the homochiral mixed crystals containing 15% of **144** and 85% of **143** gave the optically active heterodimer **145** with the relatively high optical yield of 70% *ee*. In contrast, irradiation at shorter wavelength (< 350 nm) resulted in lower optical yields, as expected owing to excitation of the phenyl moieties.

Absolute asymmetric [2 + 2] photocycloaddition in a crystalline CT complex was also found. A series of 1 : 1 CT crystals of bis[1,2,5]thiadiazolotetra cyanoquinodimethane (BTDA) **148** as an electron acceptor was obtained by mixing with neat arylolefins such as styrene, *o*-, *m*-, or *p*-divinylbenzene as electron donors. X-ray crystallographic analyses of the CT crystals revealed the chiral nature of the red crystals formed between **148** and *o*-divinylbenzene **oDV** due to their belonging to space group $P2_1$, while the other cases were achiral.

The single crystals of **148·oDV** were irradiated piece by piece to afford the optically active adduct **149** in an optical purity of 69–73% *ee* (optical rotation either positive or negative), thereby achieving success in absolute asymmetric

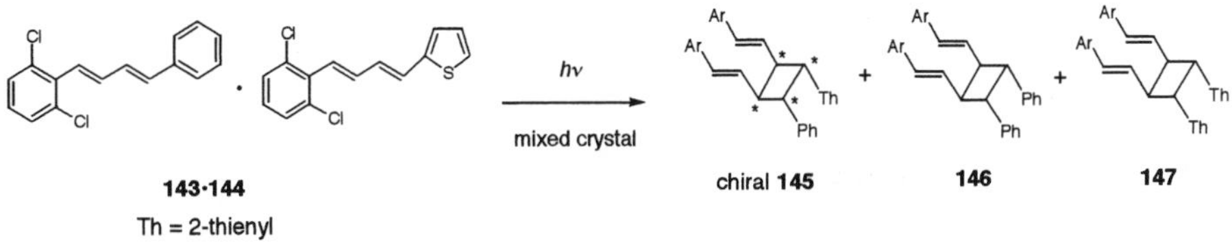

Scheme 33 Absolute asymmetric [2 + 2] photocycloaddition in the mixed crystals of butadienes.

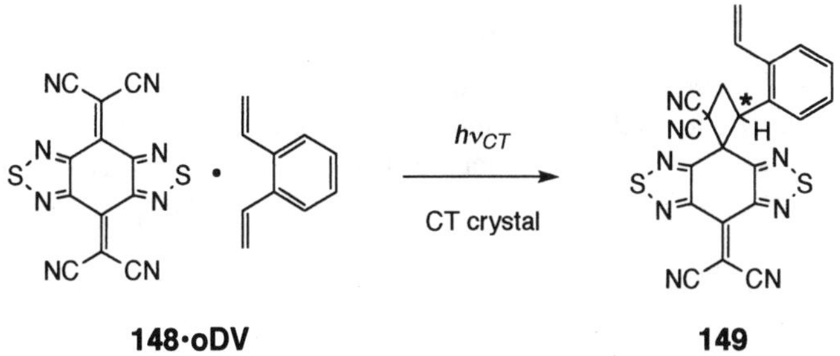

Scheme 34 Absolute asymmetric [2 + 2] photocycloaddition in the CT crystal **148·oDV**.

synthesis (Scheme 34) [115]. The *ee* values increased with lowering of the irradiation temperature; 62%, 71%, 83%, and 95% *ee* at 40°C, 15°C, −40°C, and −70°C, respectively. In contrast, solution photolysis of the CT complex of BTDA **148** and **oDV** in acetonitrile at > 450 nm (CT excitation) for 5 h gave the racemic mixture of adduct **149** in only 9% yield, showing that the apparent reactivity of the solid state is much higher than that in solution.

X-ray structure analysis before and after irradiation revealed that the reaction proceeded *via* single crystal–to–single crystal transformation without decomposition of the initial crystal structure. A one-dimensional columnar structure is formed, in which two types of molecular overlaps (type 1 and type 2) are repeated alternately (Fig. 5) [116]. Two olefinic groups are aligned nearly in parallel in both types of overlap. The interatomic distances are 3.4–3.9 Å, which are short enough to permit topochemical [2 + 2] cycloaddition in the solid state under the 4 Å rule. Direct evidence that the cycloaddition proceeded predominantly via the type 2 overlap in **148·oDV** could be obtained by comparison of the molecular arrangements.

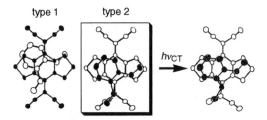

type 1   type 2

$h\nu_{CT}$

Figure 5 Adduct formation via the type 2 overlap in the CT crystal **148·oDV**. (From Ref. 115 with permission. © 1994 Am Chem Soc.)

2. Photodecarboxylative Condensation

Despite acridine and diphenylacetic acid being achiral molecules, it was found that these two compounds formed a chiral cocrystal on spontaneous crystallization, and further irradiation of this caused photodecarboxylation and then enantioselective radical coupling to give a chiral condensation product (Scheme 35) [31]. This kind of absolute asymmetric photodecarboxylation is not a topochemical reaction but is accompanied by gradual decomposition of the initial crystal structure as the reaction proceeds. This pathway is different from the four examples of [2 + 2] photocycloadditions described above.

The chiral cocrystal **150·DPA** was prepared by crystallization from a 1 : 1 solution of acridine **150** and diphenylacetic acid **DPA** in acetonitrile. The crystal

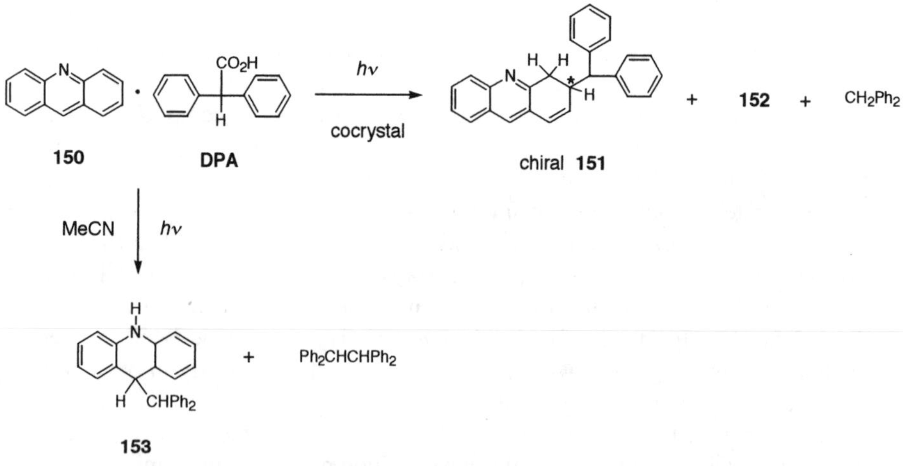

150   **DPA** $h\nu$ cocrystal   chiral **151**   +   **152**   + CH_2Ph_2

CO_2H

MeCN | $h\nu$

153   + $Ph_2CHCHPh_2$

Scheme 35 Absolute asymmetric photodecarboxylative condensation in the cocrystal of acridine and diphenylacetic acid.

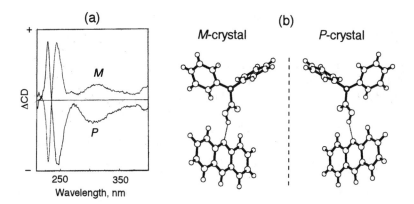

Figure 6 (a) Solid-state CD spectra and (b) molecular pairs in the *M*-crystal and *P*-crystal formed from acridine and diphenylacetic acid. (From Ref. 32 with permission. © 1997 Am Chem Soc.)

belongs to chiral space group $P2_12_12_1$. Both enantiomorphous crystals of the *M*- and *P*-forms were obtained by spontaneous crystallization from the solution. The crystal handedness was easily differentiated using the solid-state CD spectra recorded as Nujol mulls, the enantiomers showing an excellent mirror image relationship (Fig. 6) [32]. The absolute structure of reactant **150·DPA** was determined by x-ray anomalous dispersion.

A hydrogen bond pair of acridine and diphenylacetic acid molecules is present in the crystal lattice. The two phenyl planes and the carboxyl plane of the diphenylacetic acid molecule in the molecular pair form a propellerlike conformation of three blades (Fig. 6). The helicity around the C–H bond of the methine group is counterclockwise (minus), which is termed the *M*-crystal. The crystal chirality is generated from the existence of molecular pairs of counterclockwise propellers alone in the unit cell. Conversely, only the molecular pairs with the opposite conformation are observed in the *P*- crystal.

Irradiation of the *M*-crystals caused solid-state photodecarboxylation, and then enantioselective condensation occurred, to give the optically active condensation product (*S*)-(−)-**151** as the main product with $[\alpha]_D^{20} = -30$ in 35% *ee* and in 37% chemical yield (Scheme 35). Conversely, irradiation of the *P*-crystals resulted in formation of the opposite handed condensation product (*R*)-(+)-**151** with $[\alpha]_D^{20} = +30$ in 33% *ee* and in 38% chemical yield. For a comparison, solution phase photolysis of acridine **150** and diphenylacetic acid **DPA** in acetonitrile did not produce chiral product **151** but rather gave the achiral condensation product **153** in 74% as the major product at complete conversion of **DPA**.

The enantioselective reaction mechanism can be explained by considering the photochemical aspects and also the molecular arrangements (Scheme 36 and Fig. 7). Hydroacridine radical species **155** and **156**, which satisfy both such conditions, should be preferably produced with their higher stability. Next, decarboxylation of **154** gives the diphenylmethyl radical **150**. In the chiral crystal lattice of *M*-**150·DPA**, the molecular pairs of acridine and diphenyl acetic acid stack with an interplanar distance of 5.46 Å to form a columnar structure (Fig. 7). Radical coupling between ·C1 of **157** and ·C29 of **156** over the shortest distance of 5.11 Å can occur with highest priority to give condensation product (*S*)-(−)-**151** as the major enantiomer. Conversely, the (*R*)-(+)-product as a minor enantiomer is formed by the radical coupling between ·C1 of **157** and ·C21 of **156** over the longer distances of 6.79 Å. The radical coupling ratio giving S : R is calculated to be *ca.* 2 : 1 from the experimental *ee* value of around 35%. The radical coupling is necessarily accompanied by slightly larger movement of the radical species in the crystal lattice, in contrast to the well-known topochemical [2 + 2] photocycloaddition, for which a distance of less than 4 Å is indispensable [117]. Such decarboxylation and condensation reactions lead to the gradual decomposition of

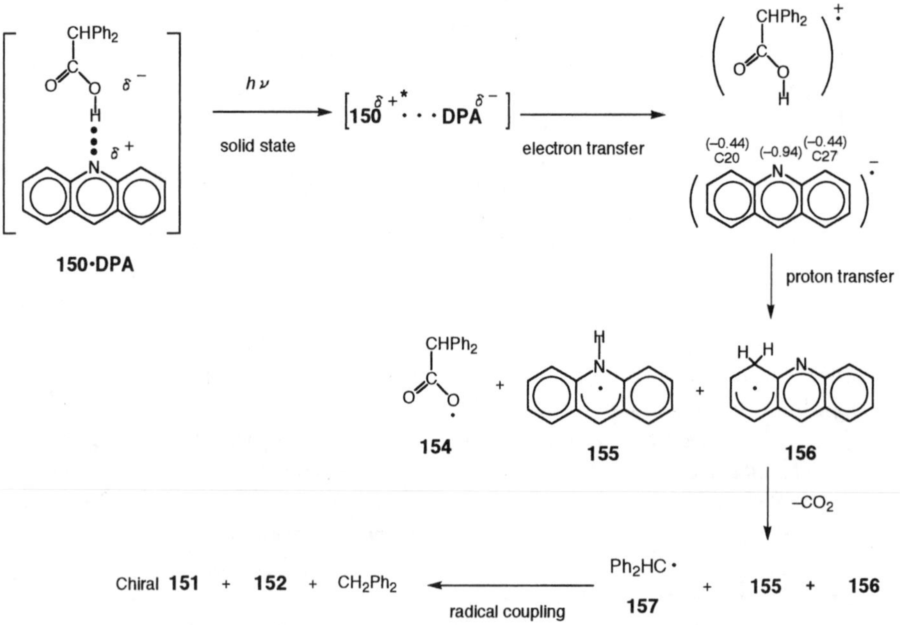

Scheme 36 Possible reaction mechanism on irradiation of the chiral cocrystal of acridine and diphenylacetic acid.

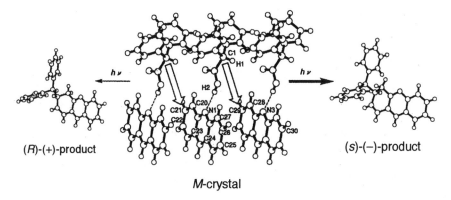

(R)-(+)-product   (S)-(−)-product

M-crystal

Figure 7   Enantioselective radical coupling path in the photoreaction of the cocrystal of acridine and diphenylactic acid. (From Ref. 31 with permission. © 1997 Am Chem Soc.)

the crystal structure of the reactant. Nevertheless, it is interesting that almost constant *ee* values of 35% were obtained over a wide range of conversions of acridine (6–52%) and diphenylacetic acid (13–71%).

VII.   CONCLUDING REMARKS

Solid-state chiral photochemistry has been reviewed since its start in the 1970s up to the present time. A number of asymmetric photoreactions have been developed in the solid state. One of the characteristic features of solid-state photoreactions is that the enantio- and diastereoselectivities are quite high. Absolute asymmetric photoreactions using chiral crystals spontaneously formed from achiral molecules can be accomplished in the crystalline state without using any external chiral source. Such absolute asymmetric synthesis is a promising methodology for asymmetric synthesis even if spontaneous chiral crystallization cannot be predicted at present. Supramolecular approaches using host–guest crystals, salt crystals and cocrystals are also excellent methodologies for asymmetric induction. Although only a small number of chiral host compounds are available at present, successful design of new chiral host compounds will lead to the extension of the reactant compounds that can be employed. In comparison to the number of intramolecular asymmetric photoreactions, intermolecular asymmetric phoreactions using cocrystals are relatively uncommon. Further development of intermolecular asymmetric photoreactions using cocrystals is desirable to extend the scope of solid-state chiral photochemistry.

REFERENCES

1. Nogradi M. In:. Enantioselective Reactions in Organic Chemistry. London: Ellis Horwood, 1995.
2. Noyori R. In:. Asymmetric Catalysis in Organic Synthesis. New York: John Wiley, 1994.
3. Ager DJ, East MB. In:. Asymmetric Synthetic Methodology. Boca Raton: CRC Press, 1996.
4. Inoue Y, Wada T, Asaoka S, Sato H, Pete JP. J Chem Soc, Chem Commun 2000: 251–259.
5. Everitt SRL, Inoue Y. In: Ramamurthy V , Schanze KS, Eds. Organic Molecular Photochemistry. New York: Marcel Dekker, 1999:71–130.
6. Inoue Y. Chem Rev 1992; 92:741–770.
7. Tanaka K. F Toda Chem Rev 2000; 100:1025–1074.
8. Gamlin JN, Jones R, Leibovitch M, Patrick B, Scheffer JR, Trotter J. Acc Chem Res 1996; 29:203–209.
9. Sakamoto M. Chem Eur J 1997; 3:684–689.
10. Koshima H. In: Toda F, Ed. Organic Solid-State Reactions, Dordrecht: Kluwer. 2002:187–268.
11. Cohen MD, Schmidt GM. J Chem Soc 1964:1996–2000.
12. Elgavi A, Green BS, Schmidt GMJ. J Am Chem Soc 1973; 95:2058–2059.
13. Green BS, Lahav M, Rabinovich D. Acc Chem Res 1979; 12:191–197.
14. Feringa BL, van Delden RA. Angew Chem Int Ed 1999; 39:3418–3438.
15. Toda F. Acc Chem Res 1995; 28:480–486.
16. Tanaka T, Toda F. In: Toda F, Ed. Organic Solid-State Reactions. Dordrecht: Kluwer, 2002:109–158.
17. Sivaguru J, Shailaja J, Uppili S, Ponchot K, Joy A, Arunkumar N, Ramamurthy V. In: Toda F, Ed. Organic Solid-State Reactions: Dordrecht: Kluwer, 2002:159–188.
18. Joy A, Ramamurthy V. Chem Eur J 2000; 6:1287–1293.
19. Ito Y. Synthesis, 1998:1–32.
20. Ito Y. In: Ramamurthy V , Schanze KS, Eds. Organic Molecular Photochemistry: New York: Marcel Dekker, 1999:1–70.
21. Lahav M, Leiserowitz L, Topovitz-Biro R, Tang CP. J Am Chem Soc 1978; 100: 2542–2544.
21a. Tang CP, Chang HC, Popovitz-Biro R, Frolow F, Lahav M, Leiserowitz L, Mcmullan RK. J Am Chem Soc 1985; 107:4058–4070.
21b. Chang HC, Popovitz-Biro R, Lahav M, Leiserowitz L. J Am Chem Soc 1987; 109: 3883–3893.
21c. Weisinger-Lewin Y, Vaida M, Popovitz-Biro R, Chang HC, Mannig F, Frolow F, Lahav M, Leiserowitz L. Tetrahedron 1987; 43:1449–.
22. Rao VP, Turro NJ. Tetrahedron Lett 1989; 30:4641–4644.
23. Gerdil R, Barchietto G, Jefford CW. J Am Chem Soc 1984; 106:8004–8005.
24. Toda F, Akagi K. Tetrahedron Lett 1968; 9:3695–3698.
24a. Toda F, Ward DL, Hart H. Tetrahedron Lett 1981; 22:3865–3868.
24b. Toda F, Tanaka K, Omata T, Nakamura K, Oshima T. J Am Chem Soc 1983; 105: 5151–5152.

25. Seebach D, Beck AK, Imwinkelried R, Roggo S, Wonnacott A. Helv Chim Acta 1987; 70:954–974.

26. Jacques J, Collet A, Wilen SH. In:. Enantiomers, Racemates, and Resolutions. New York: Wiley-Interscience, 1981.

27. Koshima H, Matsuura T. J Synth Org Chem 1998; 56:268–279.

27a. Koshima H, Matsuura T. J Synth Org Chem 1998; 56:466–477.

28. Koshima H, Hayashi E, Matsuura T, Tanaka K, Toda F. Tetrahedron Lett 1997; 38:5009–5012.

28a. Koshima H, Hayashi E, Matsuura T. Supramol Chem 1999; 11:57–66.

29. Koshima H, Khan SI, Garcia-Garibay MA. Tetrahedron: Asymmetry 1998; 9: 1851–1854.

29a. Koshima H, Honke S. J Org Chem 1999; 64:790–793.

29b. Koshima H, Honke S, Fujita J. J Org Chem 1999; 64:3916–3921.

30. Koshima H, Hamada M, Yagi I, Uosaki K. Cryst Growth Des 2001; 1:467–471.

31. Koshima H, Ding K, Chisaka Y, Matsuura T. J Am Chem Soc 1996; 118: 12059–12065.

32. Koshima H, Nakagawa T, Matsuura T, Miyamoto H, Toda F. J Org Chem 1997; 62:6322–6325.

33. Kobayashi J, Uesu Y. J Appl Crystallogr 1983; 16:204.

33a. Asahi T, Nakamura M, Kobayashi J, Toda F, Miyamoto H. J Am Chem Soc 1997; 119:3665–3669.

34. Toda F, Miyamoto H, Kanemoto H. J Org Chem 1996; 61:6490–6491.

35. Koshima H, Honke S, Miyauchi M. Enantiomer 2000; 5:125–128.

35a. Koshima H, Miyauchi M, Shiro M. Supramol Chem 2001; 13:137–142.

35b. Koshima H, Miyauchi M. Cryst Growth Des 2001; 1:355–357.

36. Cheung E, Netherton MR, Scheffer JR, Trotter J, Zenova A. Tetrahedron Lett 2000; 41:9673–9677.

37. Natarajan A, Wang K, Ramamurthy V, Scheffer JR, Patrick B. Org Lett 2002; 4: 1443–1446.

38. Schultz AG, Taveras AG, Taylor RE, Tham FS, Kullnig RK. J Am Chem Soc 1992; 114:8725–8727.

39. Irie M. Chem Rev 2000; 100:1685–1716.

40. Kodani T, Matsuda K, Yamada T, Kobatake S, Irie M. J Am Chem Soc 2000; 122: 9631–9637.

41. Ohgo Y, Arai Y, Hagiwara M, Takeuci S, Kogo H, Sekine A, Uekusa H, Ohashi Y. Chem Lett 1994:715–718.

41a. Ohashi Y, Sakai Y, Sekine A, Arai Y, Ohgo Y, Kamiya N, Iwasaki H. Bull Chem Soc Jpn 1995; 68:2517–2525.

41b. Koura T, Ohashi Y. Bull Chem Soc Jpn 1997; 70:2417–2423.

41c. Sekina A, Tatsuki H, Ohashi Y. J Organomet Chem 1997; 536–537:389–398.

41d. Hashizume D, Ohashi Y. J Phys Org Chem 2000; 13:415–412.

41e. Koura T, Ohashi Y. Tetrahedron 2000; 56:6769–6779.

42. Schultz AG. J Org Chem 1974; 39:3814–3815.

43. Toda F, Miyamoto H, Kikuchi S, Kuroda R, Nagami F. J Am Chem Soc 1996; 118:11315–11316.

44. Yamada K, Konakahara T, Ishihara S, Kanamori H. Tetrahedron Lett 1972; 25: 2513.
45. Toda F, Miyamoto H, Tamashima T, Kondo M, Ohashi Y. J Org Chem 1999; 64: 2690–2693.
46. Ogata Y, Tkaki K, Ishino I. J Org Chem 1971; 36:3975–3979.
46a. Ninomiya I, Naito T. In: Brossi A, Ed. The Alkaloids. Vol. XXII. San Diego: Academic Press, 1983:189–279.
46b. Ninomiya I, Naito T, Tada Y. Heterocycles 1984; 22:237.
47. Tanaka K, Kakinoki O, Toda F. J Chem Soc Chem Commun 1992:1053–1054.
47a. Toda F, Miyamoto H, Kanemoto K, Tanaka K, Takahashi Y, Takenaka Y. J Org Chem 1999; 64:2096–2102.
47b. Hosomi H, Ohba S, Tanaka K, Toda F. J Am Chem Soc 2000; 122:18181–1819.
48. Toda F, Tanaka K. Chem Lett 1987:2283–2284.
48a. Toda F, Tanaka K, Mak CW. Chem Lett 1989:1329–1330.
49. Dauben WG, Koch K, Smith SL, Chapman OL. J Am Chem Soc 1963; 85: 2616–2621.
50. Toda F, Tanaka K. J Chem Soc Chem Commun 1986:1429–1430.
51. Tanaka K, Nagahiro R, Urbanczyk-Lipkowska Z. Org Lett 2001; 3:1567–1569.
51a. Tanaka K, Nagahiro R, Urbanczyk-Lipkowska Z. Chirality 2002; 14:568–572.
52. Scheffer JR, Wang L. J Phys Org Chem 2000; 13:531–538.
53. Joy A, Scheffer JR, Corbin DR, Ramamurthy V. J Chem Soc Chem Commun 1998: 1379–1380.
53a. Joy A, Scheffer JR, Ramamurthy V. Org Lett 2000; 2:119–121.
54. Toda F, Tanaka K. Tetrahedron Lett 1988; 29:4299–4302.
54a. Fujiwara T, Tanaka N, Tanaka K, Toda F. J Chem Soc Perkin Trans 1989; 1: 663–664.
54b. Tanaka K, Fujiwara T, Urbanczyk-Lipkowska Z. Org Lett 2002; 4:3255–3257.
55. Toda F, Miyamoto H, Tanaka K, Matsugawa R, Maruyama N. J Org Chem 1993; 58:6208–6211.
56. Toda F, Miyamoto H, Kikuchi S. J Chem Soc Chem Commun 1995:621–622.
57. Kang T, Scheffer JR. Org Lett 2001; 3:3361–3364.
58. Scheffer JR. In: Desiraju GR, Ed. Organic Solid State Chemistry. Amsterdam: Elsevier, 1987:1–45.
59. Jones R, Scheffer JR, Trotter J, Yang J. Tetrahedron Lett 1992; 33:5481–5484.
59a. Jones R, Scheffer JR, Trotter J, Yang J. Acta Crysta 1994; B50:601–607.
60. Cheung E, Kang T, Netherton MR, Scheffer JR, Trotter J. J Am Chem Soc 2000; 122:11753–11754.
61. Leibovitch M, Olovsson G, Scheffer JR, Trotter J. J Am Chem Soc 1998; 120: 12755–12769.
62. Leibovitch M, Olovsson G, Scheffer JR, Trotter J. J Am Chem Soc 1997; 119: 1462–1463.
63. Leibovitch M, Olovsson G, Sundarababu G, Ramamurthy V, Scheffer JR, Trotter J. J Am Chem Soc 1996; 118:1219–1220.
64. Cheung E, Kang T, Raymond JR, Scheffer JR, Trotter J. Tetrahedron Lett 1999; 40:8729–8732.

64. Vishnumurthy K, Cheung E, Scheffer JR, Scott C. Org Lett 2002; 4:1071–1074.
65. Natarajan A, Joy A, Kaanumalle LS, Scheffer JR, Ramamurthy V. J Org Chem 2002; 67:8339–8350.
66. Koshima H, Maeda A, Masuda N, Matsuura T, Hirotsu K, Mizutani H, Ito Y, Fu TY, Scheffer JR, Trotter J. Tetrahedron: Asymmetry 1994; 5:1415–1418.
66a. Hirotsu K, Okada K, Mizutani H, Koshima H, Matsuura T. Mol Cryst Liq Cryst 1996; 277:99–106.
67. Koshima H, Matsushige D, Miyauchi M. Cryst Eng Comm 2001; 33:1–3.
68. Cheung E, Rademacher K, Scheffer JR, Trotter J. Tetrahedron Lett 1999; 40: 8733–8736.
68a. Cheung E, Rademacher K, Scheffer JR, Trotter J. Tetrahedron 2000; 56:6739–6751.
69. Akermak B, Johansson NG, Sjober B. Tetrahedron Lett 1969; 10:371–372.
70. Henry-Logan KR, Chen CG. Tetrahderon Lett 1973; 14:1103–1104.
71. Aoyama H, Hasegawa T, Omote Y. J Am Chem Soc 1979; 101:5343–5346.
72. Aoyama H, Miyazaki K, Sakamoto M, Omote Y. J Chem Soc Chem Commun 1983:333–334.
72a. Aoyama H, Miyazaki K, Sakamoto M, Omote Y. Tetrahededron 1987; 43: 1513–1518.
73. Toda F, Tanaka K, Yagi M. Tetrahedron 1987; 43:1495–1502.
73a. Kaftory M, Yagi M, Tanaka K, Toda F. J Org Chem 1988; 53:4391–4393.
74. Toda F, Miyamoto H, Matsukawa R. J Chem Soc Perkin Trans 1 1992:1461–1462.
75. Hashizume D, Uekusa H, Ohashi Y, Matsugawa R, Miyamoto H, Toda F. Bull Chem Soc Jpn 1994; 67:985–993.
75a. Toda F, Miyamoto H, Kanemoto K. J Chem Soc Chem Commun 1995:1719–1790.
75b. Toda F, Miyamoto H, Koshima H, Urbanczyk-Lipkowska Z. J Org Chem 1997; 62:9261–9266.
76. Natarajan A, Wang K, Ramamurthy V, Scheffer JR, Patrick B. Org Lett 2002; 4: 1443–1446.
77. Toda F, Tanaka K, Kakinoki O, Kawakami T. J Org Chem 1993; 58:3783–3784.
77a. Hashizume D, Ohashi Y, Tanaka K, Toda F. Bull Chem Soc Jpn 1994; 67: 2383–2387.
78. Toda F, Miyamoto H, Inoue M, Yasaka S, Matijasic I. J Org Chem 2000; 65: 2728–2732.
79. Gudmundsdottier AD, Scheffer JR. Tetrahedron Lett 1990; 31:6807–6810.
80. Chen J, Garcia-Garibay M, Scheffer JR. Tetrahedron Lett 1989; 30:6125–6128.
80a. Gudmundsdottier AD, Scheffer JR. Photochem Photobiol 1991; 54:535–538.
80b. Gudmundsdottier AD, Scheffer JR, Trotter J. Tetrahedron Lett 1994; 35: 1397–1400.
81. Janz KM, Scheffer JR. Tetrahedron Lett 1999; 40:8725–8728.
82. Cheung E, Netherton MR, Scheffer JR, Trotter J. Tetrahedron Lett 1999; 40: 8737–8740.
83. Uppili S, Ramamurthy V. Org Lett 2002; 4:87–90.
84. Cheung E, Chong KCW, Jayaraman S, Ramamurthy V, Scheffer JR, Trotter J. Org Lett 2000; 2:2801–2804.
85. Chong KCW, Sivaguru J, Shichi T, Yoshimi Y, Ramamurthy V, Scheffer JR. J Am Chem Soc 2002; 124:2858–2859.

86. Tanaka K, Toda F. J Chem Soc Perkin Trans 1 1992:307.
87. Toda F, Tanaka K, Oda M. Tetrahedron Lett 1988; 29:653–654.
87a. Fujiwara T, Nanba N, Hamada K, Toda F, Tanaka K. J Org Chem 1990; 55: 4532–4537.
88. Tanaka K, Kakinoki O, Toda F. J Chem Soc Perkin Trans 1 1992:943–944.
88a. Tanaka K, Toda F. Mol Cryst Liq Cryst 1998; 313:179–184.
89. Tanaka K, Toda F, Mochizuki E, Yasui N, Kai Y, Miyahara I, Hirotsu K. Angew Chem Int Ed 1999; 38:3523–3525.
89a. Tanaka K, Mochizuki E, Yasui N, Kai Y, Miyahara I, Hirotsu K, Toda F. Tetrahdedron 2000; 56:6853–6865.
90. Budac C, Wan P. J Phochem Photobiol A: Chem 1994; 67:135–166.
91. Noyori R, Kato M, Kawanishi M, Nozaki H. Tetrahedron 1969; 25:1125–1136.
91a. Brimage DRG, Davidson RS, Steiner PR. J Chem Soc, Perkin Trans 1 1973: 526–529.
91b. Okada K, Okubo K, Oda M. Tetrahedron Lett 1989; 30:6733–6736.
91c. Okada K, Okubo K, Oda M. J Phochem Photobiol A: Chem 1991; 57:265–277.
92. Tsujimoto K, Nakao N, Ohashi M. J Chem Soc, Chem Comun 1992:366–367.
93. Koshima H, Ding K, Chisaka Y, Matsuura T. Tetrahedron: Asymmetry 1995; 6: 101–104.
94. Koshima H, Nakagawa T, Matsuura T. Tetrahderon Lett 1997; 38:6063–6066.
95. Toda F, Tanaka K. Supramol Chem 1994; 3:87–88.
95a. Toda F, Tanaka K, Stein Z, Goldberg I. Acta Cryst 1995; C51:2722–2723.
95b. Toda F, Tanaka K, Stein Z, Goldberg I. Acta Cryst 1995; B51:856–863.
96. Sakamoto M, Takahashi M, Morizumi S, Yamaguchi K, Fujita T, Watanabe S. J Am Chem Soc 1996; 118:8138–8139.
96a. Takahashi M, Sekina N, Fujita T, Watanabe S, Yamaguchi K, Sakamoto M. J Am Chem Soc 1998; 120:12770–12776.
97. Sakamoto M, Sekine N, Miyoshi H, Mino T, Fujita T. J Am Chem Soc 2000; 122: 10210–10211.
98. Wu LC, Cheer CJ, Olovsson G, Scheffer JR, Trotter J, Wang SL, Liao FL. Tetrahedron Lett 1997; 38:3135–3138.
98a. Garcia-Garibay MA, Houk KN, Keating AE, Cheer CJ, Leibovitch M, Scheffer JR, Wu LC. Org Lett 1999; 1:1279–1281.
99. Sakamoto M, Hokari N, Takahashi M, Fujita T, Watanabe S, Iida I, Nshio T. J Am Chem Soc 1993; 115:818.
100. Sakamoto M, Takahashi M, Fujita T, Watanabe S, Iida I, Nishio T, Aoyama H. J Org Chem 1993; 58:3476–3477.
100a. Sakamoto M, Takahashi M, Fujita T, Watanabe S, Nishio T, Iida I, Aoyama H. J Org Chem 1997; 62:6298–6308.
101. Evans SV, Garcia-Garibay M, Omkaram N, Scheffer JR, Trotter J, Wirenko F. J Am Chem Soc 1986; 108:5648–5650.
102. Toda F, Yagi M, Soda S. J Chem Soc Chem Commun 1987:1413–1414.
102a. Sekine A, Hori K, Ohashi Y, Yagi M, Toda F. J Am Chem Soc 1989; 111:687–699.
103. Toda F, Miyamoto H. J Chem Soc Perkin Trans 1 1993; 1129–1132.
103a. Hashizume D, Kogo H, Sekine A, Ohashi Y, Miyamoto H, Toda F. J Chem Soc Perkin Trans 2 1995:61–66.

104. Sakamoto M, Takahashi M, Fujita T, Nishio T, Iida I, Watanabe S. J Org Chem 1995; 60:4682–4683.

105. Sakamoto M, Takahashi M, Shimizu M, Fujita T, Nishio T, Iida I, Yamaguchi K, Watanabe S. J Org Chem 1995; 60:7088–7089.

106. Sakamoto M, Takahashi M, Kamiya K, Yamaguchi K, Fujia T, Watanabe S. J Am Chem Soc 1996; 118:10664–10665.

107. Koshima H, Kawanishi H, Nagano M, Yu H, Shiro M. Submitted.

108. Garcia-Garibay M, Omkaram N, Scheffer JR, Trotter J, Wirenko F. J Am Chem Soc 1989; 111:4985–4986.

109. Chen J, Pokkuri PR, Scheffer JR, Trotter J. Tetrhedron Lett 1990; 31:6803–6806.

110. Fu TY, Liu Z, Scheffer JR, Trotter J. J Am Chem Soc 1993; 115:12202–12203.

111. Roughton AL, Muneer M, Demuth M. J Am Chem Soc 1993; 1115:2085–2087.

112. Addadi L, Lahav M. Pure Appl Chem 1979; 51:1269–1284.

113. Addadi L, van Mil J, Lahav M. J Am Chem Soc 1982; 104:3422–3429.

113a. van Mil J, Addadi L, Gati E, Lahav M. J Am Chem Soc 1982; 104:3429–3434.

114. Hasegawa M, Chung CM, Muro N, Maekawa Y. J Am Chem Soc 1990; 112: 5676–5677.

114a. Chung CM, Hasegawa M. J Am Chem Soc 1991; 113:7311–7316.

115. Suzuki T, Fukushima T, Yamashita Y, Miyashi T. J Am Chem Soc 1994; 116: 2793–2803.

116. Suzuki T. Pure Appl Chem 1996; 68:281–284.

117. Schmidt GM. Pure Appl Chem 1971; 27:647–678.

14

Racemic-to-Chiral Transformation and the Chirality Inversion Process in Cobaloxime Complex Crystals Only by Photoirradiation

Yuji Ohashi
Tokyo Institute of Technology
Tokyo, Japan

I. INTRODUCTION

There are two ways of obtaining chiral substances using a chiral crystal environment. One is to produce the chiral compounds from the prochiral ones, and the other is to obtain the chiral compounds from racemic ones. The former method is called absolute asymmetric synthesis, since the asymmetry is introduced from the physical conditions such as the chiral crystal environment. Several examples [1–7] have been reported since the first example of the chiral polymer produced in the photopolymerization of the chiral monomer crystal [8]. We also observed that chiral β-lactam compounds were produced from the prochiral oxoamide crystals [9,10].

On the other hand, there has been no report on the latter method until we found that the racemic crystal of a cobaloxime complex showed optical rotation after the crystal was exposed to visible light [11]. This indicates that only photoirradiation produced chiral substances in a racemic crystal.

Recently we found that the chiral alkyl group bonded to the cobalt atom in a cobaloxime complex crystal is almost (82%) inverted to the opposite configuration only by photoirradiation [12]. In this chapter we describe the processes of the racemic-to-chiral transformation and also the chirality inversion in the

533

cobaloxime complex crystals observed by x-ray crystal structure analyses and propose the mechanism on the basis of the reaction cavity for the reactive group.

II. CRYSTALLINE-STATE REACTION OF COBALOXIME COMPLEXES BY PHOTOIRRADIATION

Before the mechanism of the racemic-to-chiral transformation and inversion processes are examined, it may be better to explain the solid-state photoreaction with retention of the single crystal form, which is called a crystalline-state reaction. We found that the chiral 1-cyanoethyl group, bonded to the cobalt atom in a cobaloxime complex crystal, was racemized on exposure to x-rays or visible light [13]. Since the crystallinity was kept in the whole process of the racemization, the intensity data were collected at any stage of the reaction, and the process of the structural change was observed by x-ray crystal structure analysis. The change of the unit cell dimensions with time, which was well explained by first-order kinetics, corresponded to the rate of racemization [14] (Scheme 1).

When the axial base ligand was replaced with the other amines or phosphines, several types of racemization and different reaction rates were observed [15]. In order to explain the different types and reaction rates, we defined the reaction cavity for the reactive group as shown in Fig. 1 [14,16]. The reaction cavity is represented by the concave space limited by the envelope surface of the spheres, whose centers are positions of intra- and intermolecular atoms in the neighborhood of the reactive 1-cyanoethyl group, the radius of each sphere being

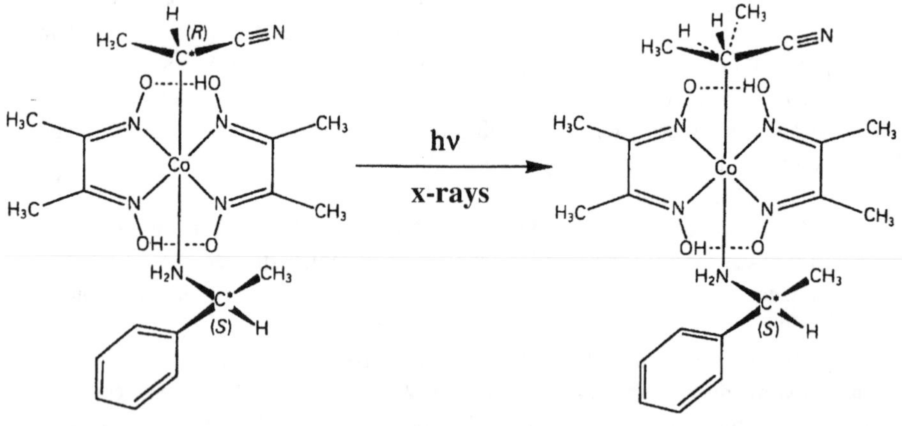

Scheme 1   The first example of crystalline-state racemization of (R)-1-cyanoethyl group.

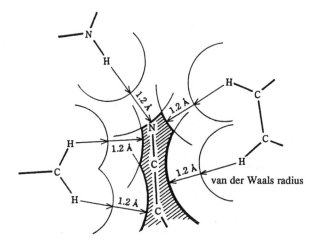

Figure 1 Definition of the reaction cavity.

greater by 1.2 Å than the van der Waals radius [17] of the corresponding atom. Any point in the cavity is considered to be accessed by the centers of atoms of the reactive group. The size of the reaction cavity explained the reason that the racemization occurred with different reaction rates when the axial ligand was replaced with different ones [15].

Such crystalline-state racemizations by photoirradiation were observed in the cobaloxime complexes not only with the chiral 1-cyanoethyl group but also with the chiral 1-(methoxycarbonyl)ethyl [18–20], 1-(ethoxycarbonyl)ethyl [12,21], 1,2-bis(methoxycarbonyl)ethyl [22], 1,2-bis(ethoxycarbonyl)ethyl [23], and 1,2-bis(allyloxycarbonyl)ethyl [24] groups as axial alkyl groups. The concept of a reaction cavity was also applicable to the racemization of the cobaloxime complexes with such bulky chiral alkyl groups.

Similar single crystal–to–single crystal reactions of organic compounds have been reported [25–30].

III. ABSOLUTE ASYMMETRIC SYNTHESES IN COBALOXIME COMPLEXES

As shown in Fig. 2, the chiral 1-cyanoethyl group bonded to the cobalt atom was produced from the prochiral 2-cyanoethyl group in some cobaloxime complex crystals with retention of the single-crystal form [31]. Only one enantiomer was observed in the crystal structure after the photoirradiation [32]. The chiral 1-

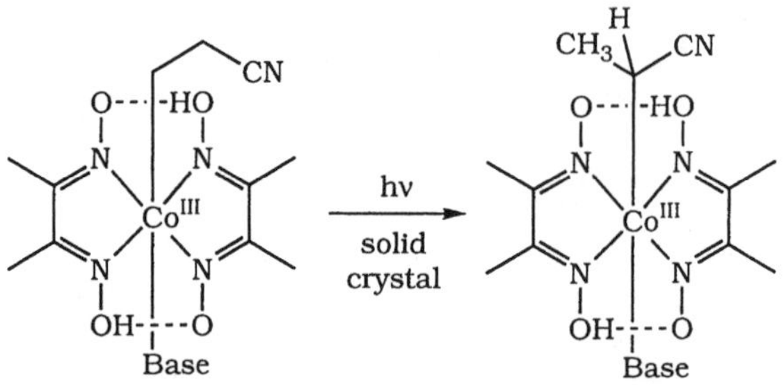

Figure 2 Solid state 2-1 photoisomerization of the 2-cyanoethyl group. The chiral 1-cyanoethyl group is produced after photoirradiation.

cyanopropyl group was produced from the prochiral 3-cyanopropyl group in some cobaloxime complex crystals with retention of the single-crystal form [33]. The isomerization process was clarified by neutron diffraction [34]. Moreover, the chiral 3-cyanobutyl, 2-cyanobutyl, and 1-cyanobutyl groups were successively produced from the prochiral 4-cyanobutyl group in the cobaloxime complex crystals [35]. In most of the above photoisomerizations, only one enantiomer was produced, that is, the absolute asymmetric reaction was observed. The chirality of the produced group was well explained by the shape of the reaction cavity for the prochiral group in the original crystal.

From the above observation, we proposed that the reaction rate should be explained by the size of the reaction cavity for the reactive group and that the chirality of the produced group should depend on the shape of the prochiral group. Using the concept of a reaction cavity, more complicated processes such as racemic-to-chiral transformation and chirality inversion will be made clear.

IV. RACEMIC-TO-CHIRAL TRANSFORMATION OF PIPERIDINE COMPLEX

A. Crystal Structures of Various *R:S* Compositions

The compound of (*racemic*-1-cyanoethyl)(piperidine)cobaloxime was prepared in a way similar to that reported previously [36,37]. The crystals were obtained from an aqueous methanol solution. Figure 3 shows the crystal structure viewed along the a axis. The space group belongs to $P2_12_12_1$, and there are two crystallographically independent molecules, A and B, in a unit cell. Figure 4 shows the molecular structure with the atomic numbering. The A molecule has a *R*-1-cyanoethyl group, whereas B has its *S* configuration. Therefore this crystal is racemic

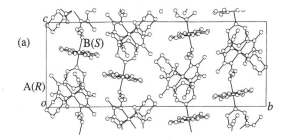

(a)

Figure 3 Crystal structure of pip-1 viewed along the a axis.

but has a chiral crystal environment. Such a racemic crystal with a chiral space group is very rare, since only five organic racemic crystals have been reported to have chiral space groups [38]. Figure 5 shows the crystal structure containing only one enantiomer, that is, the R-1-cyanoethyl group. The space group, $P2_12_12_1$, is the same as that of the racemic one. Moreover, the structure is isomorphous to the racemic one; that is, there are two crystallographically independent molecules, A and B. Both A and B have the R-1-cyanoethyl group. There is no pseudo symmetry between A and B. That the racemic crystal is isomorphous to the chiral one suggested that the mixed crystal with any $R:S$ ratio might be produced if the

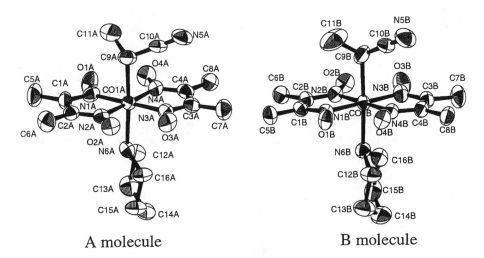

A molecule   B molecule

Figure 4 Molecular structures of A and B in pip-1. Thermal ellipsoids are drawn at the 50% probability level.

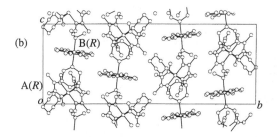

(b)

B(R)

A(R)

Figure 5 Crystal structure of pip-5 viewed along the a axis.

complexes with R-1-cyanoethyl and S-1-cyanoethyl groups were dissolved in the desired portions.

Five kinds of mixed crystals with different $R{:}S$ ratios were grown from aqueous methanol solutions in which the complexes with R-1-cyanoethyl and S-1-cyanoethyl groups were dissolved in the desired portions. Crystals with $R{:}S$ ratios 50:50, 57:43, 75:25, 87:13 and 100:0 were prepared; these are abbreviated hereafter pip-1, pip-2, pip-3, pip-4, and pip-5, respectively. Of course, the racemic and chiral ones described above are pip-1 and pip-5, respectively. These crystals are isostructural to each other. The real $R{:}S$ ratio in each crystal was determined by x-ray crystal structure analysis. The observed ratios in the crystals of pip-1, pip-2, pip-3, pip-4, and pip-5 were 50:50, 62.38, 74:26, 87:13, and 100:0, respectively, which are very similar to the corresponding ones in solutions.

Since the crystal of pip-1 is chiral, it should be either of the two enantiomeric crystals D and L. The absolute structures of 20 crystals obtained from a solution containing racemic compounds indicated that 12 crystals are D and 8 are L. When seed crystals with one of the enantiomeric structures, D or L, were added to the racemic solution, all the crystals showed the same enantiomeric structures as that of the seed crystals. The enantiomeric $D{:}L$ ratio of 20 crystals became 20:0.

The powdered sample with the same enantiomeric structures, D or L, was irradiated with a xenon lamp for 20 h and was dissolved in a chloroform solution. The specific rotation $[\alpha]_D$ of the chloroform solution was $+30°$. It is clear that the racemic-to-chiral transformation can be observed only by photoirradiation. Using the seed crystals, one of the enantiomeric crystals was selected in each experiment. This means that the A molecules have R configurations in the pip-1 to pip-5 crystals.

The crystal structure of pip-5 is shown in Fig. 5, which is isostructural to that of pip-1 in Fig. 3. Only the S-1-cyanoethyl group of the B molecule is replaced with the R-isomer. The A molecule occupies nearly the same position. In pip-2 to pip-4, the crystal structures are substantially the same as that of pip-

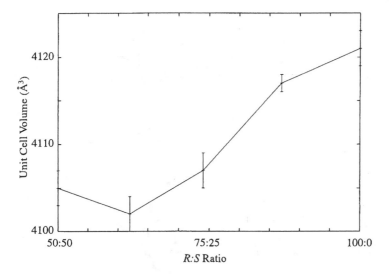

Figure 6 Relationship between unit cell volume and the R:S composition.

1, although the B molecules have the disordered structure composed of S-1-cyanoethyl group and its R-isomer. The disordered structures in pip-2 to pip-4 have the $R{:}S$ ratios of 12:38, 25:25, and 37:13, respectively. The unit cell volumes of pip-1 to pip-5 are plotted against the $R{:}S$ compositions in Fig. 6. The $R{:}S$ ratio in the crystals is nearly the same as those in the solutions except for pip-2, which has the $R{:}S$ ratio of 62:38 in spite of the ratio of 57:43 in solution. Figure 6 may indicate that the crystal with the $R{:}S$ ratio of 62:38 is the most stable.

B. Molecular Structures of pip-1 to pip-5

The molecular structures of A and B in the pip-1 crystal are shown in Fig. 4. The A molecules in the crystals of pip-1 to pip-5 are substantially the same, while the B molecules have different structures of the 1-cyanoethyl groups. Figure 7 shows the conformations of the 1-cyanoethyl groups viewed along the normal to the cobaloxime plane in pip-1, pip-3, and pip-5. The structures of the 1-cyanoethyl groups in pip-2 and pip-4 are very similar to that in pip-3 except for the $R{:}S$ ratio of the disordered structure.

C. Change of Cell Dimensions on Exposure to Visible Light

The changes of the c axis lengths of pip-1, pip-3, and pip-5 are shown in Fig. 8, showing the crystals when they were exposed to a xenon lamp. The cell dimen-

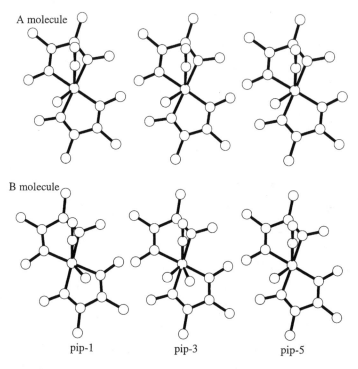

A molecule

B molecule

pip-1   pip-3   pip-5

Figure 7 The structure of the B cyanoethyl group of pip-1, pip-3, and pip-5, viewed along the normal to the cobaloxime plane.

sions appear to reach the same value after infinite exposure. The converged values are not the initial one of pip-3 but an intermediate one between pip-2 and pip-3.

D. Crystal and Molecular Structure After Irradiation

Figure 9 shows the crystal structure of pip-1 after 40 h exposure. The 1-cyanoethyl group of the B molecule changed to the disordered structure, while the A molecule remained unaltered. Almost the same structures were obtained for the crystals of pip-3 and pip-5. The $R{:}S$ ratios became 19:31, 19:31, and 21:29 for pip-1, pip-3, and pip-5, respectively. These values are identical to 20:30 within experimental error. It is noticeable that the converged value is not 25:25 but 20:30.

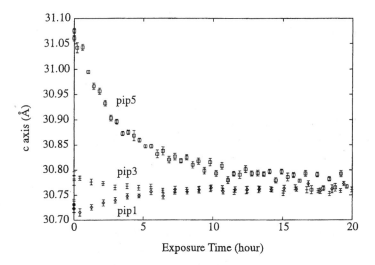

Figure 8 Changes of the c axis lengths of pip-1, pip-3, and pip-5 with exposure time.

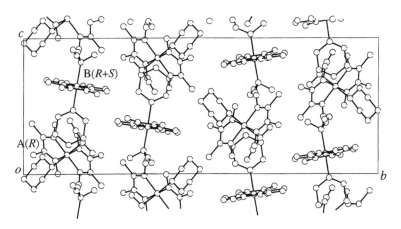

Figure 9 Crystal structure of pip-1 viewed along the a axis after 40 h exposure.

E. Racemic-to-Chiral Transformation Only by Photoirradiation

The *R:S* ratio in the racemic crystal of pip-1 before irradiation is 50:50, since the 1-cyanoethyl groups of the A and B molecules have *R* and *S* configurations. After the irradiation, the A molecule has the *R* configuration, while the B molecule became the disordered structure with *R:S* ratio of 20:30. This indicates that the *R:S* ratio in a crystal changed from 50:50 to 70:30. This is why the specific rotatory power of the chloroform solution containing the irradiated pip-1 crystal showed +30°.

F. Size and Shape of the Reaction Cavity

It must be explained why only the B molecule is partly inverted to create a disordered structure. The most important requirement for the movement of the reactive group is the void space around the reactive group. The void space is explained by the reaction cavity. Figure 10 shows the reaction cavities for the 1-cyanoethyl groups of the A and B molecules in pip-1. The volumes of the cavities for the A and B cyanoethyl groups were calculated to be 7.0 and 14.2 Å^3, respectively. These values are nearly the same in the crystals of pip-2 to pip-5. The volume of 7.0 Å^3 is too small for the inversion of the cyanoethyl group with retention of

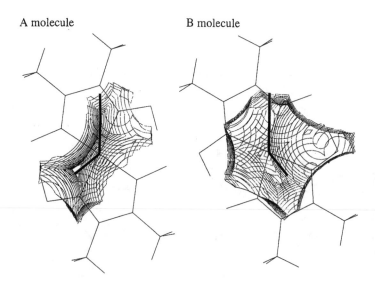

Figure 10 Reaction cavities for the A and B cyanoethyl groups viewed along the normal to the cobaloxime plane.

the single crystal form, whereas 14.2 Å^3 is large enough for the inversion, since the value of 11.5 Å^3 was proposed to be a threshold value for the crystalline state inversion [39]. Such different cavity volumes of the A and B cyanoethyl groups clearly explain the different behavior of the two groups: the A cyanoethyl group remained unaltered, whereas the B cyanoethyl group was partly inverted.

The final question why the converged *R:S* ratio of the B cyanoethyl group is not 25:25 but 20:30 must be explained. The reaction cavities for the B cyanoethyl groups in pip-1, pip-3, and pip-5 before and after the irradiation are shown in Fig. 11, in which each cavity is divided into two by the plane composed of the Co–C–C–N bonds. The volumes of the two parts were calculated. The left and right parts of the cavities before irradiation were 45:55, 56:44, and 63:37, respectively. These ratios became 48:52, 49:51, and 52:48, respectively after irradiation. These values are 50:50 within experimental error. This suggests that the inversion ratio of the B cyanoethyl group depends on the symmetry of the cavity. In other words, the inversion ratio is determined by the steric repulsion

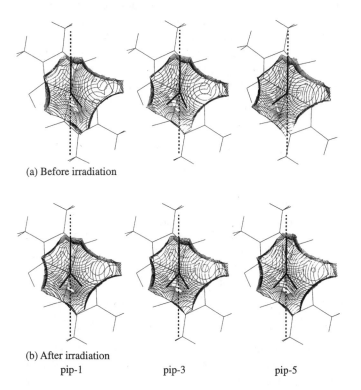

(a) Before irradiation

(b) After irradiation

pip-1 pip-3 pip-5

Figure 11 Reaction cavities of the B cyanoethyl groups in pip-1, pip-3, and pip-5 before and after irradiation. The dotted lines indicate the planes including the C–C–C–N bonds.

from the neighboring molecules around the B molecule after irradiation. Since the minimum unit cell volume is obtained at 62:38, which is close to 70:30, the enthalpy term of the crystal may be responsible to the ratio of 70:30. When the crystal after irradiation was warmed up to 343K, the $R:S$ ratio of the B molecule became 25:25. Although the reaction cavity was not calculated, because the precise positions of the hydrogen atoms were not obtained at 343K, it is adequate to assume that the reaction cavity has enough size to accommodate the cyanoethyl groups with both configurations, and the ratio of the left and right parts of the cavity would become 50:50.

V. PYRROLIDINE COMPLEX CRYSTAL

A. Preparation of Four Kinds of Crystals

For the cobaloxime complex with pyrrolidine as an axial base ligand, the crystalline state reaction was also observed when the crystal was exposed to a xenon lamp. Four kinds of crystals with different $R:S$ compositions were prepared. The crystals of pyrr-1 were obtained from the racemic solution. The pyrr-2 crystals were obtained from a solution that has the complex with an $R:S$ ratio of 75:25. The crystals of pyrr-3 and pyrr-4 were also obtained from solutions with the $R:S$ ratio of 80:20 and 90:10, respectively. The x-ray crystal analysis indicated that the four crystals are isostructural to each other. From solutions with the $R:S$ ratio greater than 9:1, pure enantiomeric crystals were obtained, which are not isostructural to the above crystals pyrr-1 to pyrr-4 [40].

B. Crystal Structure Before Irradiation

Figure 12 shows the crystal structures of pyrr-1 and pyrr-2 viewed along the c axis before irradiation. There are four crystallographically independent molecules, A, B, C, and D, in a P1 cell. In the crystal of pyrr-1, the A, B, C, and D molecules have R, R, S, and S configuration, respectively: therefore the crystal is racemic. In the crystal of pyrr-2 with the $R:S$ ratio of 75:25, the A, B, C, and D molecules have R, R, $R+S$, and $R+S$ configuration, respectively. The $R:S$ ratios of the C and D molecules are 10:15 and 16:9, respectively. As a whole, the pyrr-2 crystal has an $R:S$ ratio of 76:24, which is identical to the $R:S$ ratio of the solution, 75:25, within experimental error. The pyrr-3 and pyrr-4 crystals were also analyzed. The $R:S$ ratios of the A and B molecules are 25:0, whereas the $R:S$ ratios of the C and D molecules are 15:10 and 19:6 for pyrr-3 and 17:8 and 23:2 for pyrr-4, respectively. These values indicate that the $R:S$ ratios of the pyrr-3 and pyrr-4 crystals are 84:16 and 90:10, respectively, which are very similar to the $R:S$ ratios in the respective solutions. Since the crystals of pyrr-1 to pyrr-4 are all chiral, the chirality of the crystals was selected so that the A molecules have R-1-cyanoethyl groups.

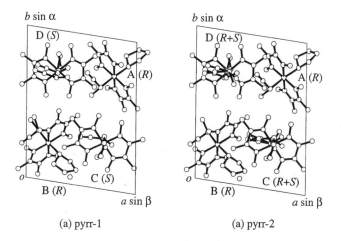

Figure 12 Crystal structures before irradiation viewed along the c axis: (a) pyrr-1 and (b) pyrr-2.

Figure 13 shows the molecular structures of A and C of pyrr-1. Figure 14 shows the structures of the 1-cyanoethyl groups viewed along the normal to the cobaloxime planes of the A to D molecules of pyrr-1. Although the cyanoethyl group of the D molecule is disordered, both of the disordered groups have the same chirality.

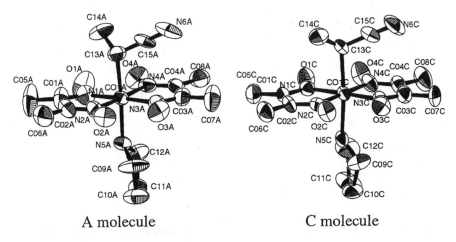

Figure 13 Molecular structures of A and C of pyrr-1 before irradiation. Thermal ellipsoids are drawn at 50% probability level.

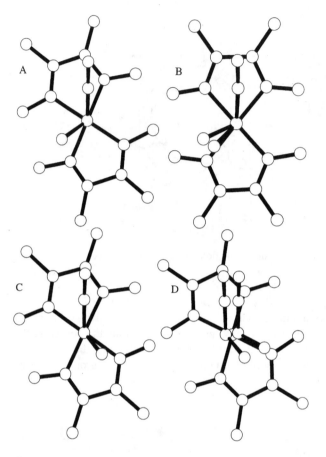

Figure 14 Structures of the cyanoethyl groups of A, B, C, and D of pyrr-1 viewed along the normal to the cobaloxime plane before irradiation.

C. Crystal and Molecular Structure After Irradiation

When pyrr-1 and pyrr-2 crystals were exposed to the xenon lamp, the cell dimensions were gradually changed. The changes appeared to converge to the same values after infinite exposure. Figure 15 shows the crystal structure of pyrr-1 after about 40 h exposure. The crystal structure is isostructural to that before irradiation except for $R:S$ ratios of the cyanoethyl groups of the C and D molecules. The structures of the cyanoethyl groups of C and D viewed along the

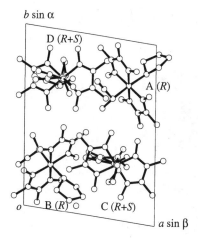

Figure 15   Crystal structure of pyrr-1 viewed along the c axis after 40 h exposure.

normal to the cobaloxime planes are shown in Fig. 16. The molecular structures of A and B after irradiation are almost the same as the corresponding ones before irradiation. The $R{:}S$ ratios became 8:17 and 14:11 for C and D. This indicates that the crystal, as a whole, changed from racemic to chiral, that is, the $R{:}S$ ratio changed from 50:50 to 72:28. The R:S ratios of pyrr-2, 3, and 4 crystals after the irradiation became 72:28, 72:28, and 71:29, respectively. The final ratio is 72:28 within experimental error. The pyrrolidine complex crystals are considered to converge to 72:28 after infinite irradiation.

D.   Reaction Cavities for A, B, C, and D Before Irradiation

In order to explain the reason that only the C and D molecules are partly inverted to the opposite configuration, the reaction cavities for the cyanoethyl groups of the A, B, C, and D molecules in pyrr-1 are drawn in Fig. 17. The volume of each cavity was calculated. The values of the A and B cavities, 9.8 and 9.5 Å^3, respectively, are smaller than the threshold value for the racemization of the 1-cyanoethyl group, 11.5 Å^3, proposed in the previous paper [39]. On the other hand, the volumes of the C and D cavities are 14.5 and 15.9 Å^3, which are significantly greater than the threshold value. The size of the reaction cavity for the cyanoethyl group well explains the above partial racemization.

E.   Change of Reaction Cavity After Irradiation

The next question is why the $R{:}S$ ratio after irradiation became 72:28. The volumes of the reaction cavities for the A, B, C, and D molecules after irradiation

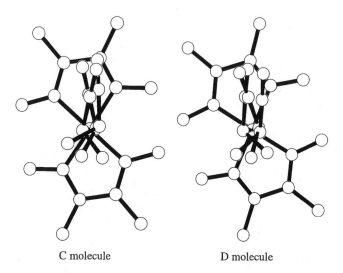

C molecule   D molecule

Figure 16   Structures of the cyanoethyl groups of C and D of pyrr-1 viewed along the normal to the cobaloxime plane after irradiation. The inverted groups are disordered.

were also calculated. They are 10.1, 8.4, 16.9, and 15.0, respectively, which are not so different from the corresponding ones before irradiation. However, it is impossible to divide the cavity of C and D into two by a plane composed of the Co–C–C–N bond, as was done in the piperidine complex, because the C and D groups after irradiation take very complicated disordered structures. Probably the ratios of the left and right halves for the C and D cavities would become 50:50 after irradiation, although the *R:S* ratio became 8:17 and 14:11 for C and D.

F.   Capability of Racemic-to-Chiral Transformation

Both of the racemic crystals of the piperidine and pyrrolidine complexes have chiral space groups before irradiation. The most important requirement for the racemic-to-chiral transformation is that the two molecules with *R* and *S* configurations crystallize in a chiral space group. Since the racemic compounds tend to make a pair around an inversion center in the process of crystallization, the racemic crystals, in general, have a center of symmetry. Otherwise, conglomerate crystals may be deposited from a racemic solution. Therefore, only several crystals with chiral space groups have been reported so far [38]. This may be a reason that such a racemic-to-chiral transformation has not been observed till now.

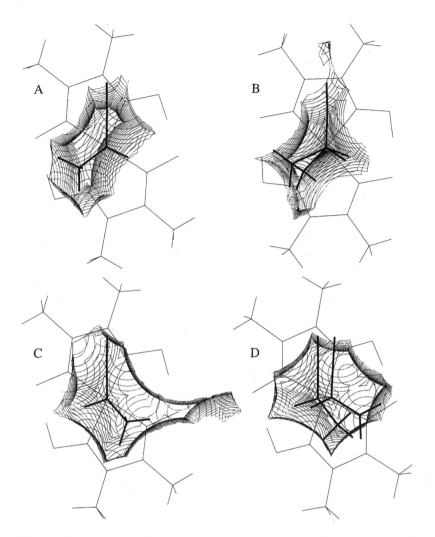

Figure 17 Reaction cavities for the A, B, C, and D cyanoethyl groups of pyrr-1 viewed along the normal to the cobaloxime plane before irradiation.

VI. RACEMIC-TO-CHIRAL TRANSFORMATION IN A DIASTEREOMERIC CRYSTAL

Recently a cobaloxime complex with a racemic 1-cyanoethyl group and methyl (S)-alaninate as axial ligands was prepared [41]. In order to resolve the diastereomeric pair, fractional crystallization was performed several times. However, the optical rotation of the complex did not increase after several times of crystallization. One of the crystals suitable for x-ray work was picked up and the structure was analyzed. To our surprise, there are two crystallographically independent molecules, which are a pair of diastereomers, in a chiral unit cell. This explains why the fractional resolution of the diastereomeric pair was impossible in the usual way.

Moreover, the cell dimensions gradually changed when the crystal was exposed to a xenon lamp. The optical rotation of the chloroform solution in which the irradiated crystals were dissolved increased significantly. This may indicate that the racemic-to-chiral transformation would also occur in a diastereomeric crystal.

A. Crystal and Molecular Structure Before Irradiation

Figure 18 shows the crystal structure before irradiation, viewed along the c axis. There are two crystallographically independent molecules, A and B, which are diastereomers to each other, in a unit cell of P1. The molecules are arranged head to tail along the a axis. Layers of the molecules of A and B are stacked alternately along the b axis. The cyano group of the 1-cyanoethyl group of B is hydrogen bonded to the amino group of the alaninate of A. The cyano group of A, on the other hand, has no hydrogen bond with the neighboring atoms.

Figures 19 show the molecular structures of A and B before irradiation. The configurations of the 1-cyanoethyl groups are S and R for A and B, respectively. The ester moieties of the two molecules are planar and take the syn conformation. The bond distance and angles are not significantly different from the corresponding ones in the related molecules.

B. Crystal and Molecular Structure After Irradiation

The crystal structure after irradiation is shown in Fig. 20. Only the 1-cyanoethyl group of B was changed, and the other moieties take nearly the same structures as those before irradiation. It must be emphasized that the nitrogen of the inverted 1-cyanoethyl group of B has no hydrogen bond with the neighboring atoms, although the original nitrogen atom forms a hydrogen bond through N$\cdots$H–N. This suggests that the entropy term is the driving force in this epimarization.

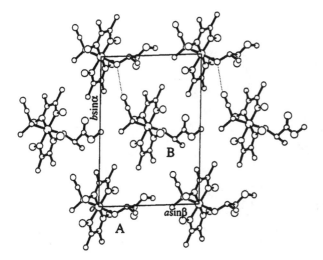

Figure 18 Crystal structure of ((R,S)-1-cyanoethyl)((S)-alaninate)cobaloxime before irradiation viewed along the c axis.

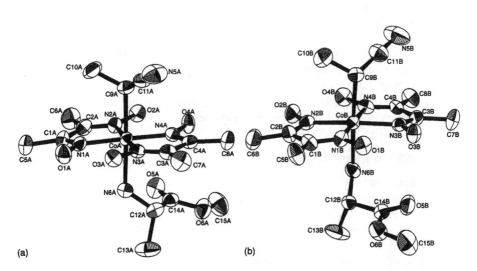

Figure 19 Molecular structures of (a) A (S isomer) and (b) B (R isomer) molecules before irradiation. The thermal ellipsoids are drawn at 50% probability level.

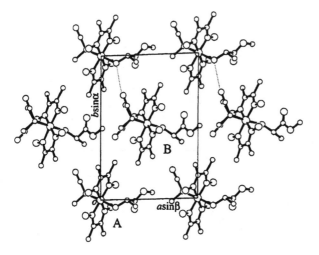

Figure 20   Crystal structure after irradiation viewed along the c axis.

The molecular structure of B after irradiation is shown in Fig. 21. Only the 1-cyanoethyl group was changed after irradiation. The inverted 1-cyanoethyl group takes a different conformation from those observed in the piperidine and pyrrolidine crystals; the methyl group is replaced with the cyano group in the alaninate crystal, whereas the methyl group is replaced with the hydrogen atom in the piperidine and pyrrolidine crystals. Such an inversion mode has been found only in a crystal of (1-cyanoethyl)(3-hydroxypyridine)cobaloxime [42].

The occupancy factors for the disordered R- and S-1-cyanoethyl groups were 0.68(1) and 0.32(1), respectively. This means that the ratio of the R and S enantiomers changed from 1:1 to 0.34:0.66 in a whole crystal. Thus the diastereomeric excess of the S-1-cyanoethyl complex became 32%.

C. Inversion Mechanism

In order to elucidate the reason why only B was epimerized, the cavities for the 1-cyanoethyl groups of A and B before irradiation were calculated, which are shown in Fig. 22. The volumes of A and B cavities are 11.5 and 13.1 Å^3, respectively. The A cavity is slightly small while the B is large enough for the inversion, considering from the threshold value. Since the inversion mode of the cyanoethyl group is different, the threshold value is probably greater than that observed for the ordinary inversion mode, 11.5 Å^3. The difference in cavity size well explains the different reactivities of the 1-cyanoethyl groups of A and B.

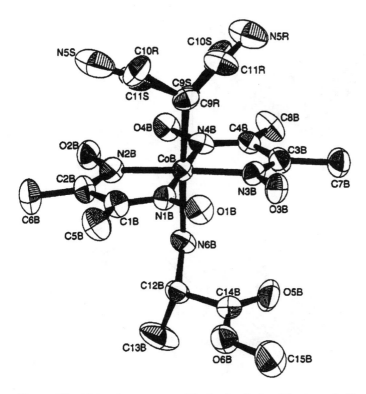

Figure 21 Molecular structure of B after irradiation. The thermal ellipsoids are drawn at 50% probability level.

The B cavity has a void space along the C–C(methyl) bond, which can accommodate the inverted cyano group. On the other hand, there is no void space around the methyne hydrogen atom of the cyanoethyl group. It seems impossible to accommodate the inverted methyl group in this area. Therefore it is impossible to replace the hydrogen atom with the methyl group in the process of inversion.

Figure 23 indicates that the inverted 1-cyanoethyl group is well accommodated in the cavity. After prolonged irradiation for 7000 min, the ratio of the *R*- and *S*-1-cyanoethyl groups of B became 68:32. To explain the reason, the B cavities before and after irradiation were divided into two by the plane composed of Co–C–H bonds, and the volumes of the two parts were calculated. The ratio of the two parts was 70:61 and 79:62 before and after irradiation, respectively. This indicates that the *R* isomer is favorable even after a part of the *R* isomer was inverted to *S*, since the ratio of the *R* isomer increased from 70 to 79. It remains uncertain why the final ratio became not 50:50 but 68:32.

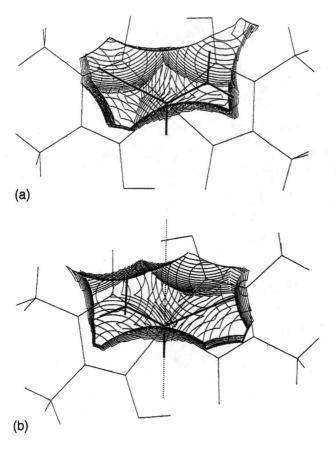

(a)

(b)

Figure 22 Cavities for the 1-cyanoethyl groups of (a) A and (b) B before irradiation. The broken line indicates the plane including the Co–C–H bond.

VII. CHIRALITY INVERSION PROCESS OF 1-ETHOXYCARBONYLETHYL GROUP ONLY BY PHOTOIRRADIATION

When the chiral alkyl group was replaced with the bulkier group than the 1-cyanoethyl group, a more interesting phenomenon, that is, chirality inversion only by photoirradiation, was observed [12]. The chiral alkyl group is S-1-(ethoxycarbonyl)ethyl, and the axial base ligand is S-1-cyclohexylethylamine, as shown in Fig. 24. On exposure to a halogen lamp, the S-1-(ethoxycarbonyl)ethyl group was inverted to the R isomer with retention of the single-crystal form. At the

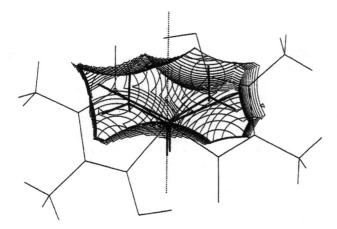

Figure 23 Cavity for the 1-cyanoethyl group of B after irradiation. The broken line indicates the plane including the Co–C–H bond.

final stage, the $R:S$ ratio of the 1-(ethyoxycarbonyl)ethyl group became about 4:1. To confirm this result, the other two crystals containing (R,S)- and R-1-(ethoxycarbonyl)ethyl groups were prepared. Both crystals are isostructural to that with the S-1-(ethoxycarbonyl)ethyl group. When the two crystals were exposed to the halogen lamp separately, their cell dimensions were gradually changed and converged to the values observed for the S isomer. The analyzed structures after the irradiation were the same as that of the S isomer after irradiation. In this section, the inversion process of the 1-(ethoxycarbonyl)ethyl group is described.

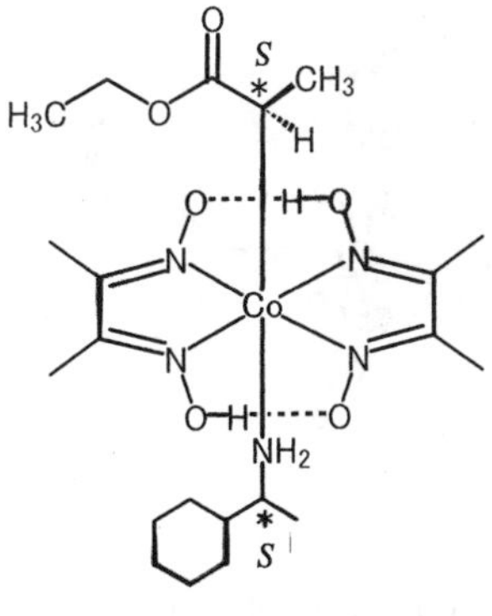

Figure 24 $((S)$-1-ethoxycarbonylethyl)$((S)$-1-hexylethylamine)cobaloxime.

A. Crystal and Molecular Structure of the *S* Isomer Before Irradiation

Figure 25 shows the crystal structure of the *S* isomer before irradiation. The space group belongs to the chiral $P2_12_12_1$. There are four molecules in a unit cell. This means that only one molecule is crystallographically independent. The chiral alkyl groups have contacts with each other as a ribbon along the 2_1 axis parallel to the a axis. Figure 26a shows the molecular structure with the atomic numbering. The methyl and carbonyl groups of the 1-(ethoxycarbonyl)ethyl group take a syn conformation to each other, and the ethyl group of the 1-(ethoxycarbonyl)ethyl group has an anti conformation to the carbonyl group, as shown in Fig. 26b.

B. Structural Change After Irradiation

When the crystal was exposed to a halogen lamp with a long-path filter (R64), the unit cell dimensions were gradually changed; the a and b axes slightly con-

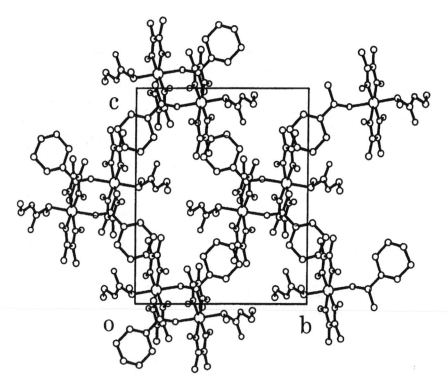

Figure 25 Crystal structure of ((*S*)-1-ethoxycarbonylethyl)((*S*)-1-hexylethylamine)cobaloxime before irradiation viewed along the a axis.

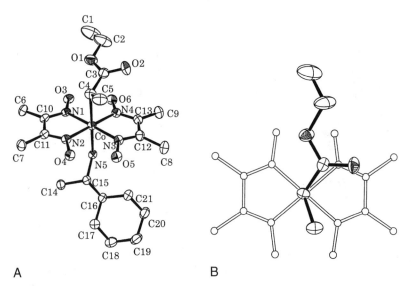

A   B

Figure 26 (a) Molecular structure with the atomic numbering. The thermal ellipsoids are drawn at 50% probability level. (b) The conformation of the (*S*)-1-ethoxycarbonylethyl group before irradiation viewed along the normal to the cobaloxime plane. The thermal ellipsoids of the atoms are drawn at 50% probability level.

tracted, but the c axis and the unit cell volume *V* significantly expanded. The crystal structure, however, is nearly the same as before except that the 1-(ethyoxy-carbonyl)ethyl group has a disordered structure owing to the partial inversion. Figures 27a and b show the molecular structure and the structure of the disordered 1-ethoxycarbonylethyl group viewed along the normal to the cobaloxime plane, respectively. Most of the 1-(ethoxycarbonyl)ethyl group was inverted from *S* to *R*. The *R:S* ratio became from 0:100 to 82:18 after 24 h exposure. The ratio of 82:18 was not changed after prolonged irradiation. It is clear that the chirality was inverted only by photoirradiation.

C. Crystal and Molecular Structure of *R* Isomer Before Irradiation

The crystal structure of the *R* isomer before irradiation is very similar to that of the *S* isomer (Fig. 25) except that the chiral 1-(ethoxycarbonyl)ethyl group has the *R* configuration. The molecular structure is shown in Fig. 28a. The conformation of the *R*-1-(ethoxycarnyl)ethyl group is shown in Fig. 28b, which is the

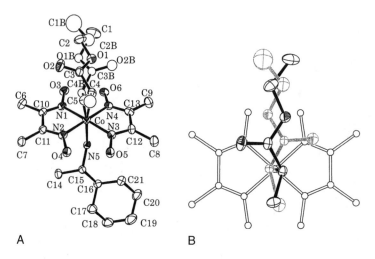

Figure 27 (a) Molecular structure after irradiation. The thermal ellipsoids are drawn at 50% probability level. (b) The conformation of the (S)-1-ethoxycarbonylethyl group after irradiation viewed along the normal to the cobaloxime plane. The thermal ellipsoids of the atoms are drawn at 50% probability level.

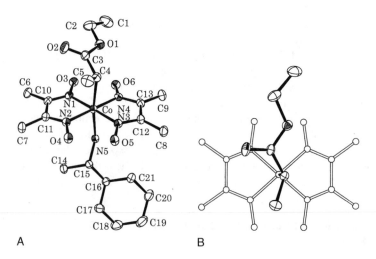

Figure 28 (a) Molecular structure of ((R)-1-ethoxycarbonylethyl)((S)-1-hexylethylamine)cobaloxime before irradiation with the atomic numbering. The thermal ellipsoids are drawn at 50% probability level. (b) The conformation of (R)-1-ethoxycarbonylethyl group viewed along the normal to the cobaloxime plane.

mirror image of the *S* isomer shown in Fig. 26b. Moreover, it is almost the same as that produced by photoirradiation shown in Fig. 27b.

D. Structural Change After Photoirradiation

When the crystal was irradiated with a halogen lamp, the cell dimensions were gradually changed. They converge to the values observed in the photoirradiated crystal with the *S* isomer. The structure analyzed by x-rays after 24 h exposure revealed that it is essentially the same as that of the photoirradiated crystal with the *S* isomer. The molecular structure became the disordered one with the *R*- and *S*-1-(ethoxycarbonyl)ethyl groups, which is the same as that shown in Fig. 27a. The *R:S* ratio converged to 82:18, which is also the same as that observed in the photoirradiated crystal with the *S* isomer.

In order to examine the final ratio more quantitatively, the mixed crystal with equal amounts of *S* and *R* isomers, that is, the (*R,S*) isomer, was prepared.

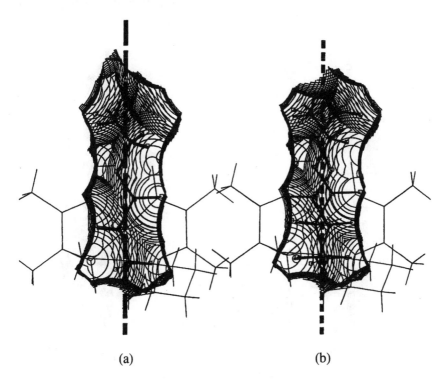

(a)   (b)

Figure 29 Cavities for the (*S*)-1-ethoxycarbonylethyl group (a) before and (b) after irradiation. The dotted line indicates the plane passing through the Co atom, perpendicular to the cobaloxime plane and parallel to the C–C bond of the cobaloxime in each figure.

When the crystal was irradiated with a halogen lamp, the cell dimensions were gradually changed, and the $R{:}S$ ratio of the disordered 1-(ethoxycarbonyl)ethyl group became from 50:50 to 82:18. It is clear that the $R{:}S$ ratio of 82:18 is the most stable when the crystal with any configuration of the chiral 1-(ethoxycarbonyl)ethyl group is exposed to the halogen lamp.

E. Why Is the Ratio of 82:18 the Most Stable?

The question why the ratio of 82:18 is the most stable should be answered. Figure 29 shows the reaction cavities for the 1-(ethoxycarbonyl)ethyl groups of the S isomer crystal before and after photoirradiation. Each cavity is divided into two by the plane, which passes through the Co atom, is perpendicular to the cobaloxime plane, and is parallel to the bond of C(10)–C(11). For the S isomer before irradiation, the ratio of the left and right halves of the cavity is 38:62. However, the ratio becomes 51:49 after irradiation, although the ratio of $R{:}S$ is 82:18. For the R isomer the cavities are shown in Fig. 30. The ratio of the left and right

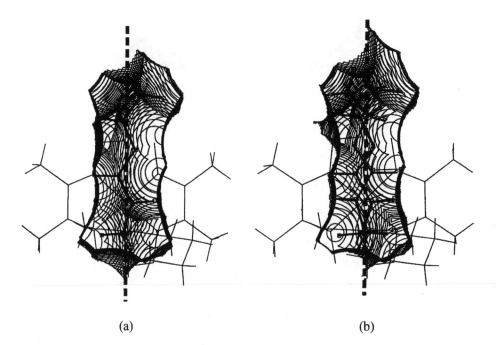

(a)   (b)

Figure 30 Cavities for the (R)-1-ethoxycarbonylethyl group (a) before and (b) after irradiation. The dotted line indicates the plane passing through the Co atom, perpendicular to the cobaloxime plane and parallel to the C–C bond of the cobaloxime in each figure.

halves of the cavity, 47:53, before irradiation, becomes 51:49 after irradiation. This suggests that the cavity becomes symmetric if the $R:S$ ratio is 82.18. For the (R,S) isomer, the same calculations were performed. The ratios of the left and right halves of the cavity are 46:54 and 51:49 before and after irradiation, respectively. The steric repulsion from the surrounding atoms would have equal effect on either of the groups with R and S configurations after irradiation. The enthalpy term of the 1-(ethoxycarbonyl)ethyl group plays an important role in the chirality inversion process.

VIII. SUMMARY

Molecules should be most closely packed in a crystal from the thermodynamical point of view. If, however, a molecule, has chiral groups or moieties, some void space will appear in the crystalline lattice. When any bond of the chiral group can be cleaved by photoirradiation and the crystallinity conserved, the chiral group will have two possibilities: to return to the original configuration, and to invert to the opposite configuration, if other reaction pathways are prohibited. All the above results indicate that some portion of the chiral group should be inverted so as to make the intermolecular interaction (or packing force) from the surrounding atoms equal to both of the configurations. If we can utilize such characteristics of the crystals, it may possible to obtain chiral substances only by photoirradiation, not only from the racemic substances but also from the chiral substances with the opposite configuration.

REFERENCES

1. Suzuki T, Fukushima T, Yamashita Y, Miyashi T. J Am Chem Soc 1994; 116: 2793–2803.
2. Saigo K, Sukegawa M, Maekawa Y, Hasegawa M. Bull Chem Soc Jpn 1995; 68: 2355–2362.
3. Gamlin JN, Jones R, Leibovitch M, Patrick B, Scheffer JR, Trotter J. Acc Chem Res 1996; 29:203–209.
4. Koshima H, Ding K, Chisaka Y, Matsuura T. J Am Chem Soc 1996; 118: 12059–12065.
5. Yu T, Liu Z, Olovsson G, Scheffer JR, Trotter J. Acta Cryst 1997; B53:293–299.
6. Ito. Y. Synthesis 1998; 1998:1–32.
7. Tanaka K, Toda F. Chem Rev 2000; 100:1025–1074.
8. Addadi L, van Mil J, Lahav M. J Am Chem Soc 1982; 104:3422–3429.
9. Sekine A, Hori K, Ohashi Y, Yagi M, Toda F. J Am Chem Soc 1989; 111:697–699.
10. Hashizume D, Kogo H, Sekine A, Ohashi Y, Miyamoto H, Toda F. J Chem Soc, Perkin Trans 1996; 2:61–66.

11. Osano YT, Uchida A, Ohashi Y. Nature 1991; 352:510–512.

12. Nitami T, Uekusa H, Ohashi Y. Bull Chem Soc Jpn. submitted.

13. Ohashi Y, Sasada Y. Nature 1977; 267:142–144.

14. Ohashi Y, Yanagi K, Kurihara T, Sasada Y, Ohgo Y. J Am Chem Soc 1981; 103: 5805–5812.

15. Ohashi Y. Acc Chem Res 1988; 21:268–274.

16. Ohashi Y, Uchida A, Sasada Y, Ohgo Y. Acta Cryst 1983; B39:54–61.

17. Bondi A. J Phys Chem 1964; 68:441–451.

18. Kurihara T, Uchida A, Ohashi Y, Sasada Y, Ohgo Y. J Am Chem Soc 1984; 106: 5718–5724.

19. Kurihara T, Uchida A, Ohashi Y, Sasada Y. Acta Cryst 1984; B40:478–483.

20. Sekine A, Saito M, Hashizume D, Uekusa H, Ohashi Y, Arai Y, Ohgo Y. Enantiomer 1998; 3:159–168.

21. Koike N. Master's thesis. Tokyo: Tokyo Institute of Technology, 1997.

22. Ohashi Y, Sakai Y, Sekine A, Arai Y, Ohgo Y, Kamiya N, Iwasaki H. Bull Chem Soc Jpn 1995; 68:2517–2525.

23. Sato H, Ohashi Y. Bull Chem Soc Jpn 1999; 72:367–375.

24. Sato H, Sakai Y, Ohashi Y, Arai Y, Ohgo Y. Acta Cryst 1996; C52:1086–1089.

25. Leibovitch M, Olovsson G, Scheffer JR, Trotter J. J Am Chem Soc 1997; 119: 1462–1463.

26. Foley JL, Li L, Sandman DJ, Vela MJ, Foxman BM, Albro R, Eckhardt CJ. J Am Chem Soc 1999; 121:7262–7263.

27. Honda K, Nakanishi F, Feeder N. J Am Chem Soc 1999; 121:8246–8250.

28. Tanaka K, Toda F, Mochizuki E, Yasui N, Kai Y, Miyahara I, Hirotsu K. Angew Chem Int Ed 1999; 38:3523–3525.

29. Hosomi H, Ohba S, Tanaka K, Toda F. J Am Chem Soc 2000; 122:1818–1819.

30. Hosomi H, Ito Y, Ohba S. Acta Cryst 1998; B54:907–911.

31. Uchida A, Ohashi Y, Sasada Y. Nature 1986; 320:51–52.

32. Sekine A, Tatsuki H, Ohashi Y. J Organomet Chem 1997; 536–537:389–398.

33. Sekine A, Yoshiike M, Ohashi Y, Ishida K, Arai Y, Ohgo Y. Mol Cryst Liq Cryst 1998; 313:321–326.

34. Ohhara T, Harada J, Ohashi Y, Tanaka I, Kumazawa S, Niimura N. Acta Cryst 2000; B56:245–253.

35. Vithana C, Uekusa H, Sekine A, Ohashi Y. Bull Chem Soc Jpn 2001; 74:287–292.

36. Schrauzer GN, Windgassen RJ. J Am Chem Soc 1967; 89:1999–2007.

37. Ohgo Y, Takeuchi S, Natori Y, Yoshimura J, Ohashi Y, Sasada Y. Bull Chem Soc Jpn 1981; 54:3095–3099.

38. Brock C, Schweizer W, Dunitz J. J Am Chem Soc 1991; 113:9811–9820.

39. Takenaka Y, Ohashi Y, Tamura T, Uchida A, Sasada Y. Acta Cryst 1993; B49: 272–277.

40. Takenaka Y, Ohashi Y, Tamura T, Uchida A, Sasada Y. Acta Cryst 1993; B49: 1015–1020.

41. Hashizume D, Ohashi Y. J Phys Org Chem 2000; 13:415–421.

42. Ohgo Y, Ohashi Y. Bull Chem Soc Jpn 1996; 69:2425–2433.

15
Chiral Photochemistry Within Zeolites

V. Ramamurthy, Arunkumar Natarajan, Lakshmi S. Kaanumalle, and S. Karthikeyan
Tulane University, New Orleans, Louisiana, U.S.A.

J. Sivaguru
Columbia University, New York, New York, U.S.A.

J. Shailaja
University of Colorado, Boulder, Colorado, U.S.A.

Abraham Joy
Georgia Institute of Technology, Atlanta, Georgia, U.S.A.

I. BACKGROUND: ASYMMETRIC PHOTOREACTIONS IN SOLUTION AND IN SOLID STATE

The last few decades have seen an abundance of methodologies to carry out ground state reactions stereoselectively. Meanwhile asymmetric photoreactions have not been attempted with the same kind of rigor as thermal asymmetric reactions and have seen a slower advance. It should also be stated that photoreactions involve highly energetic excited states, which generally have very short lifetimes. These short lifetimes are insufficient to develop an effective interaction between the substrate and the chiral source. In ground state reactions, a difference of a few kilocalories between the diastereomeric transition states is enough to give very high stereodifferentiation. The presence of small or negligible activation barriers in photochemical reactions makes them more difficult to control, and imaginative methods have to be developed to carry out these reactions. In spite of the apparent stumbling blocks in the way, advances have been made in the field of asymmetric photoreactions [1–7]. Several methods have been attempted to effect chiral induction in asymmetric photoreactions. Reactions have been carried out in vapor phase, solution phase, organized assemblies, and solid phase.

The chiral sources employed include circularly polarized light, chiral sensitizers, chiral solvents, chiral substituents, chiral host–guest assemblies, and chiral crystalline environments.

In the late 19th century, Le Bel and van't Hoff recognized the use of right- and left-circularly polarized light (CPL) to bring about an excess of one enantiomer in a photochemical reaction. Since then, there have been several attempts to use CPL as a chiral source [8–27]. In this photoprocess, an optically inactive racemic substrate is irradiated by a left or a right CPL. Differential interactions of this chiral light source with each enantiomer lead either to enantioselective destruction or to a shift in equilibrium between the two enantiomers or a fixation of enantiomeric ground state conformers. Both these processes lead to optically active products. A recent example of this process involving *trans*-cycloctene (**1**) is illustrated in Scheme 1 [28]. Typically enantioselective photodestruction or photoderacemization reactions using *r*- or *l*-CPL do not yield high stereoselectivity in the reaction, and the value obtained in this reaction was considered to be high for these types of reactions. More examples are to be found in Chap. 1.

Hammond and Cole reported the first asymmetric photosensitized geometrical isomerization with 1,2-diphenylcyclopropane (Scheme 2) [29]. The irradiation of racemic *trans*-1,2-diphenylcylcopropane **2** in the presence of the chiral sensitizer (*R*)-*N*-acetyl-1-naphthylethylamine **4** led to the induction of optical activity in the irradiated solution, along with the simultaneous formation of the cis isomer **3**. The enantiomeric excess of the *trans*-cyclopropane was about 7% in this reaction. Since then, several reports have appeared on this enantiodifferentiating photosensitization using several optically active aromatic ketones as shown in Scheme 2 [30–36]. The enantiomeric excesses obtained in all these reactions have been low. Another example of a photosensitized geometrical isomerization is the *Z*–*E* photoisomerization of cyclooctene **5**, sensitized by optically active (poly)alkylbenzene(poly)carboxylates (Scheme 3) [37–52]. Further examples and more detailed discussion are to be found in Chap. 4.

A more frequently employed source of chiral induction has been the use of a chiral substituent appended to a prochiral substrate as a chiral auxiliary [3,5]. A variety of photoreactions have been reported using this approach. These have

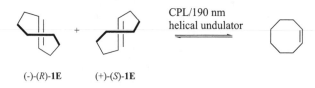

(-)-(*R*)-**1E**   (+)-(*S*)-**1E**

0.12% *ee* after prolonged irradiation

Scheme 1

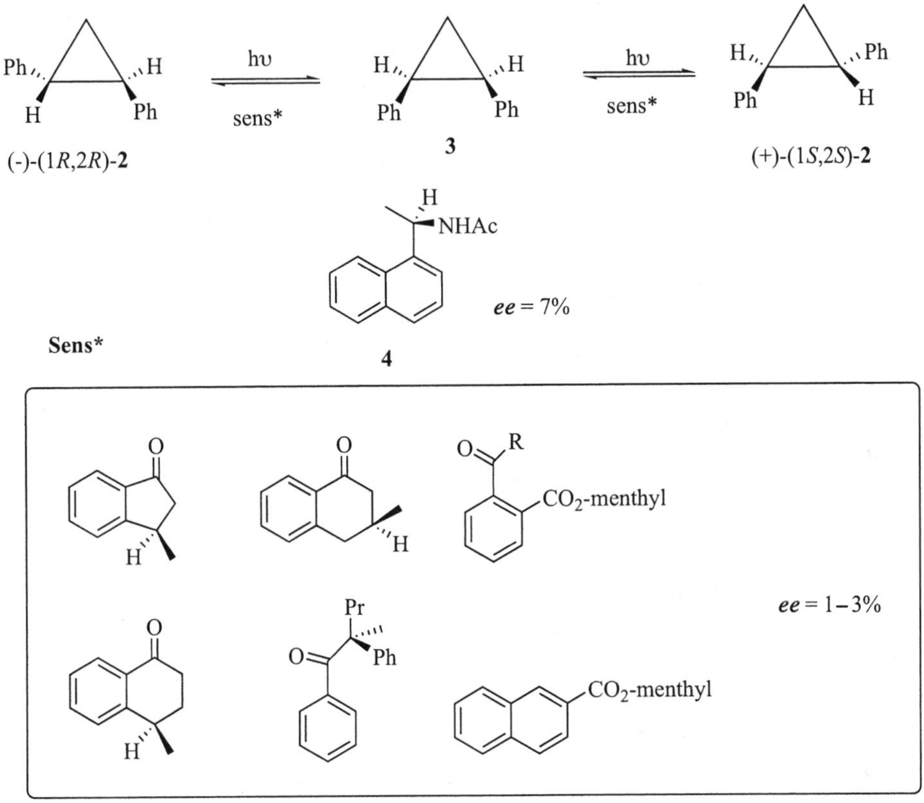

$(-)-(1R,2R)-2$ **3** $(+)-(1S,2S)-2$

Sens*

$ee = 7\%$

4

$ee = 1–3\%$

Scheme 2

included photocyclization, [2 + 2] photocycloaddition [53–87], Paterno-Büchi reaction [66,88–123], the Schenck reaction [124–127], [4 + 2] and hydrogen abstraction reaction [128–139]. Steiner et al. reported that the enantioselective cyclization of substituted *N*-(2-benzoylethyl) glycine esters **6** gave 3-hydroxyprolines
7 (Scheme 4) [139]. Intra- and intermolecular [2 + 2] photocycloadditions have
received a great deal of interest, possibly because they can be easily applied to
the single-step synthesis of complex ring systems that contain diverse functionalities. The intramolecular [2 + 2] photocyclodimerizations of cinnamates using *d*-
mannitol and *l*-erythritol as the chiral linkers were studied by Green et al. [54].
As exemplified for dicinnamate **9** (Scheme 5), the irradiation of the chiral polyols
and the subsequent ester exchange with acidic methanol gave two major head-
to-head cycloadducts—chiral δ-truxinate **10** and achiral β-truxinate **11**, along
with a small amount of neotruxinate **12**. High *des* up to 85% were reported for

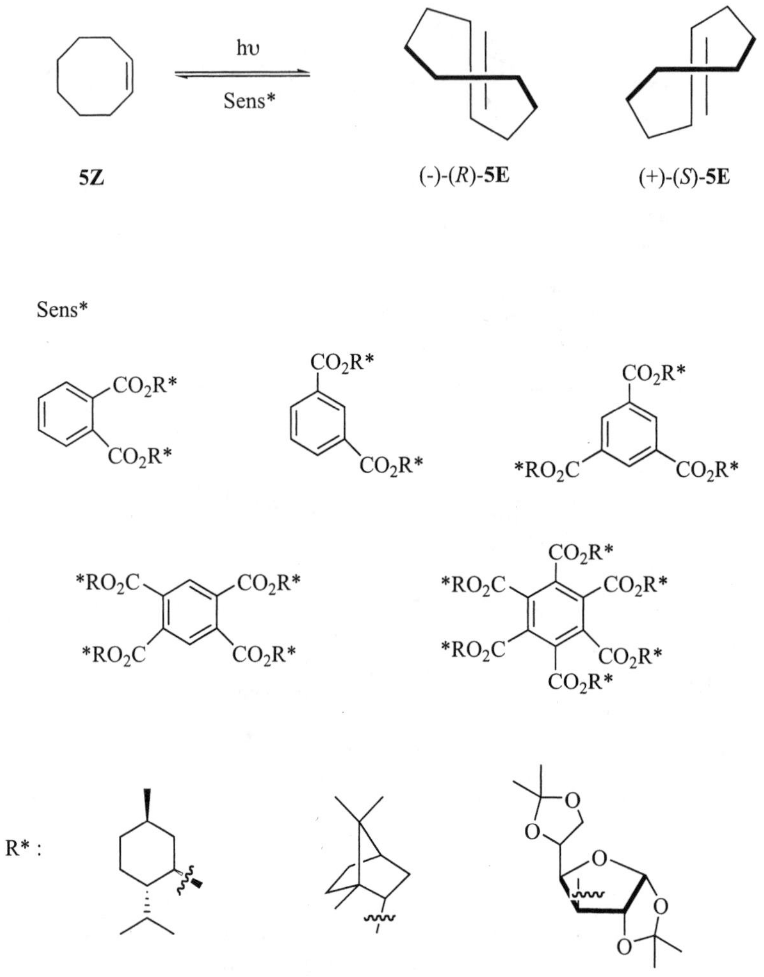

Scheme 3

δ-truxinate **10**, obtained in the photocyclizations of *d*-mannitol 1,6-dibenzoate 2,3,4,5-tetracinnamate and also of (−)-2,3,-di-*O*-methyl-L-erythritol 1,4-dicinnamate. Intermolecular [2+2] photocycloadditions between enones and alkenes have been reported by several groups. Moderate to high diastereoselectivities have been obtained in these systems. The Paterno-Büchi reaction consists of the reaction of a ketone or thioketone with an olefin, leading to the formation of the four membered oxetane or thiooxetane ring system. Chiral groups can be attached to either of the reacting species, and high stereoselectivities have been observed.

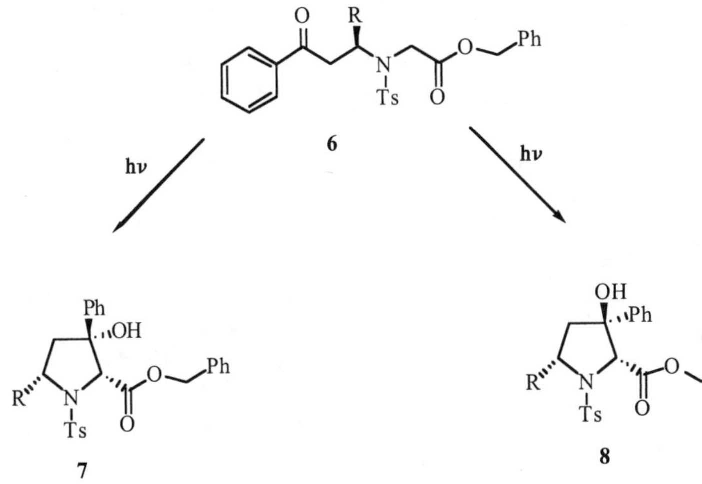

Scheme 4

7

R = CO₂Me (>98 % *de*)

R = CO₂Buᵗ (>98 % *de*)

8

R = t-Bu (>98 % *de*)

R* = D-mannitol or
L-erythritol

9

10 (δ)

11 (β)

12 (neo)

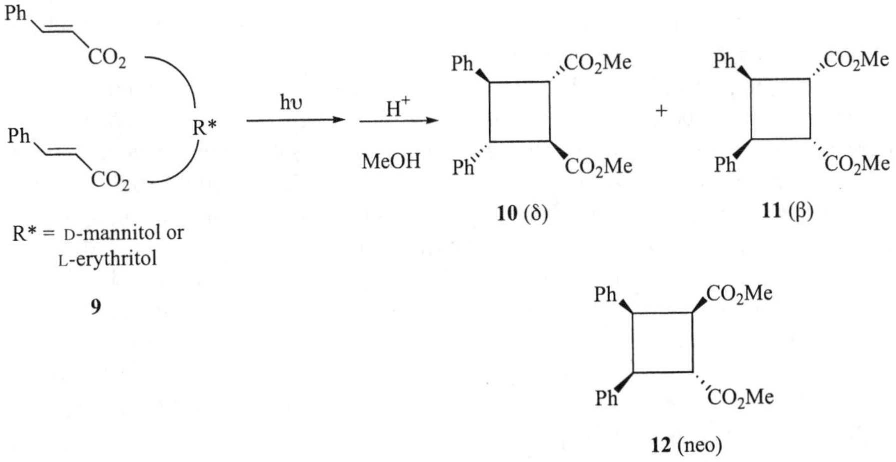

Scheme 5

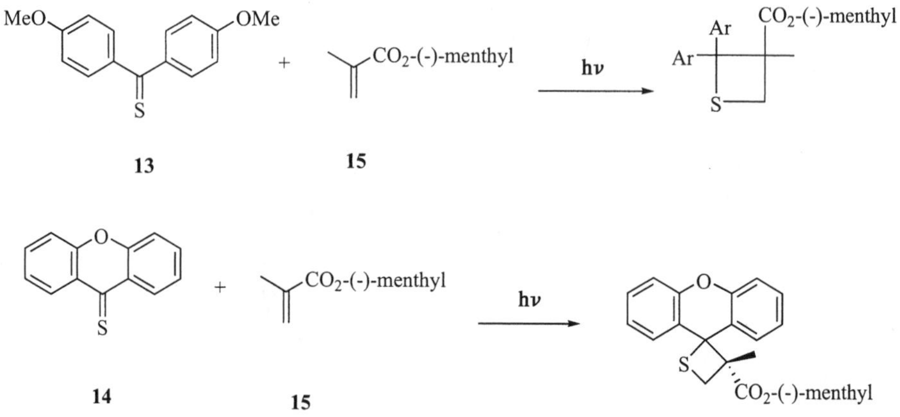

Scheme 6

Gotthardt and Lenz reported the diastereodifferentiating thietane formation in the photocycloaddition of thiobenzophenone **13** and xanthione **14** (Scheme 6) with (−)-menthyl methylacrylate **15** [88]. The photochemistry of thioketones is wavelength dependent, involving both excited triplet (T_1) and singlet (S_2) states that undergo independent processes. It was seen that the *des* obtained were greater (18%) for the product from the T_1 state than that from the S_2 state. In-depth discussion of this topic is to be found in Chap. 5.

The above are examples of the types of reactions carried out in isotropic media. The stereoselectivities attained in these types of reactions are usually low, except when the chiral source is covalently attached to the substrate. If the substrate and/or the chiral source are forced to be in the same neighborhood it might increase the probability of the substrate and the chiral source interacting within the lifetime of the excited state of the substrate. This idea is realized in the reactions carried out in host–guest assemblies and in solid-state reactions.

It has been known for over two decades that optically active products can be formed from achiral precursors without the intervention of preexisting optical activity. This was illustrated by Penzien and Schmidt when they showed that 4,4′-dimethylchalcone, although itself achiral, crystallizes spontaneously in a chiral space group [140]. When these crystals are treated with bromine vapor in a gas–solid reaction, a chiral dibromide is produced in 6% enantiomeric excess. Since then several research groups have carried out similar reactions wherein the achiral substrate crystallizes in a chiral space group [141–184]. Photolyses of these chiral crystals lead to optically active photoproducts. An example of these types of reactions is the di-π-methane rearrangement of substituted dibenzobarrelenes shown in Scheme 7 [151,178]. The dibenzobarrelene derivatives **16a** and

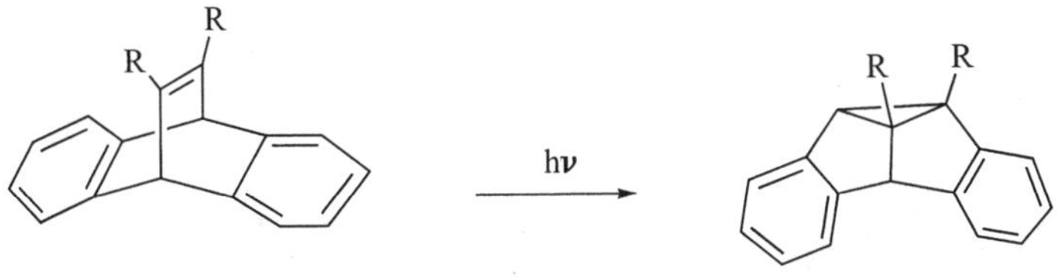

16

16a; R = CO$_2$iPr

16b; R = P(O)Ph$_2$

17

17a; *ee* = 95%

17b; *ee* = 89%

Scheme 7

16b crystallize in a chiral space group. Upon irradiation of these chiral cystals, both the derivatives yield a sole dibenzosemibullvalene product **17**. It was seen that the product from **16a** was obtained in 95% *ee*, while that from **16b** was obtained in 89% *ee*. Irradiation of the opposite enantiomorph of the crystal led to the optical antipode of the product. Reactions such as these, in which an achiral reactant is transformed into an enantiomerically enriched product without the external intervention of preexisting optical activity, have been termed absolute asymmetric synthesis. One of the drawbacks of solid-state irradiations like the above is that there would be a breakdown of the crystal structure upon prolonged irradiation. The irradiation times therefore have to be kept short, which lowers the efficiency of the process. Probably a greater drawback is that one cannot predict if and when a compound would crystallize in a chiral space group. More details on this topic are to be found in Chaps. 11 and 13.

The above drawbacks of crystalline photoreactions are circumvented in an approach called the ionic auxiliary approach which is outlined below [185–207]. Scheffer and coworkers make a salt of a substrate having a carboxylic acid with a chiral organic amine or vice versa. The chirality of the amine would necessitate that the salt crystallize in a chiral space group. The crystal formed in this fashion would also be able to withstand higher conversions due to stronger lattice forces in these crystals. An example of this type of approach is shown in Scheme 8. Treatment of the dibenzobarralene derivative **18**, with the *tert*-butyl ester of (*S*)-proline **19**, afforded salt **20**; irradiation of the crystals of this salt gave diester **21**. In this reaction, only one regioisomer **21** is formed, and this product is formed in over 95% *ee*. More examples are to be found in Chap. 12.

An inclusion compound is composed of two or more distinct molecules held together by noncovalent forces in a definable structural relationship. Chiral inclusion complexes have been used by the groups of Lahav, Leisorowitz, and Toda to carry out asymmetric photoreactions [208–245]. The chirality of the host

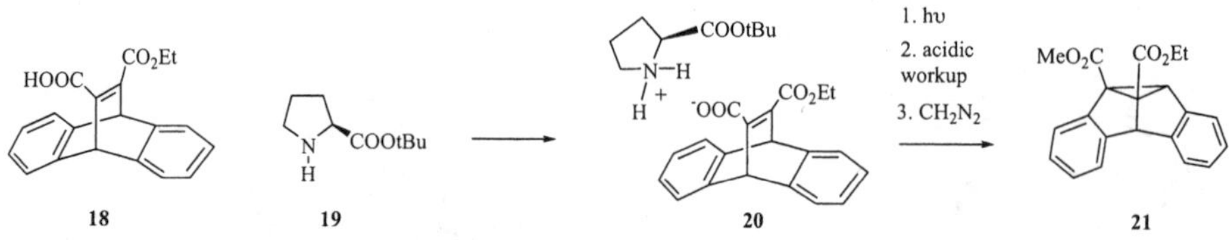

Scheme 8

provides an asymmetric environment for the reaction, yielding enantioenriched products.

Lahav, Leiserowitz and their coworkers have used inclusion complexes of deoxycholic acid to carry out asymmetric photochemical reactions. The authors reported that acetophenone forms a 2 : 5 crystalline channel inclusion complex with deoxycholic acid **22** as the host and that irradiation of the complex in the solid state leads to abstraction of a hydrogen atom by the acetophenone **23a** from C5 of the steroid followed by coupling of the resulting radical pair to produce photoproduct **24** (Scheme 9) [216,219]. The authors were able to follow the course of the reaction by x-ray crystallography. Only one diastereomer of the product was obtained in the reaction. Aoyama et al. have studied the reaction of *N,N*-dialkylpyruvamides in deoxycholic acid inclusion crystals (Scheme 10) [246]. Solid-state irradiation of the inclusion complex of **22** and **25** gave the

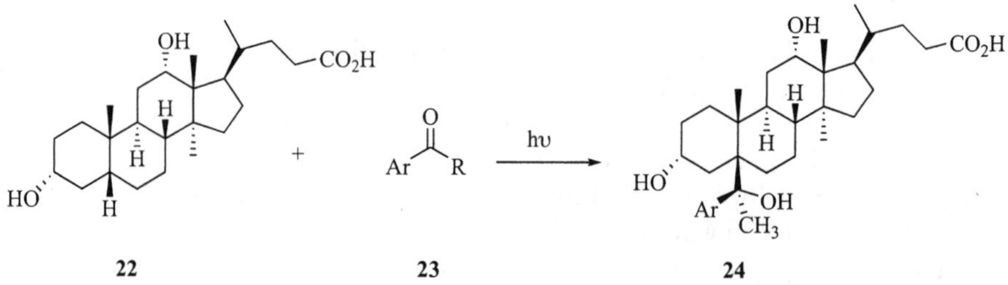

a: Ar = Ph; R = CH$_3$

b: Ar = *p*-FC$_6$H$_4$; R = CH$_3$

c: Ar = Ph; R = CH$_2$CH$_3$

Scheme 9

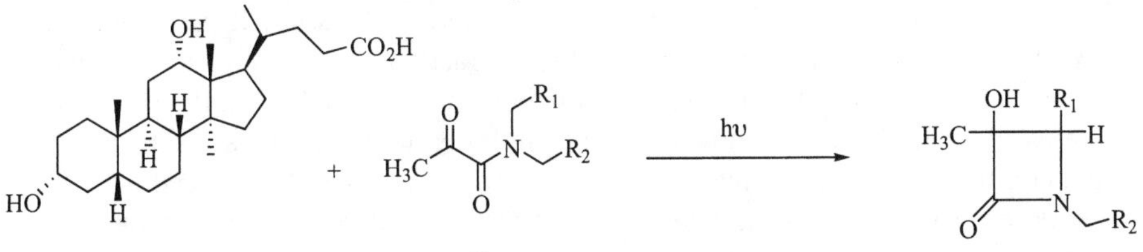

Scheme 10

corresponding enantioenriched β-lactams. Other hosts that have been used for carrying out asymmetric photoreactions include perhydrotriphenylene (PHTP), triorthothymotide (TOT), cyclodextrins (CD), and urea [26,183,247–262]. In fact, the first asymmetric transformation was carried out with resolved PHTP by Farina et al. The authors carried out the polymerization of 1,3-*trans*-pentadiene complexed with resolved PHTP. A polycrystalline sample of this complex subjected to γ-irradiation yielded a highly stereospecific 1,4-*trans* isotactic polymerization, although with low *ee*.

The inclusion complex **26**, shown in Scheme 11, has been used as a host by Toda and coworkers to carry out a number of enantioselective reactions [231]. For example, irradiation of a 1 : 1 host–guest assembly of α-tropolone methyl ether **27** and (**S,S**)-(−)-**26**, in the solid state gave (**1S,5R**)-(−)-**28** of 100% *ee*. The authors state that the high stereoselectivity is a result of the steric hindrance to disrotatory ring closure from one direction due to the structure of the host. This leads to the formation of only one enantiomer of the product. More details on this topic are available in Chap. 13.

The above discussions show that asymmetric reactions can be carried out in homogeneous and heterogeneous media. Most of the reactions performed in

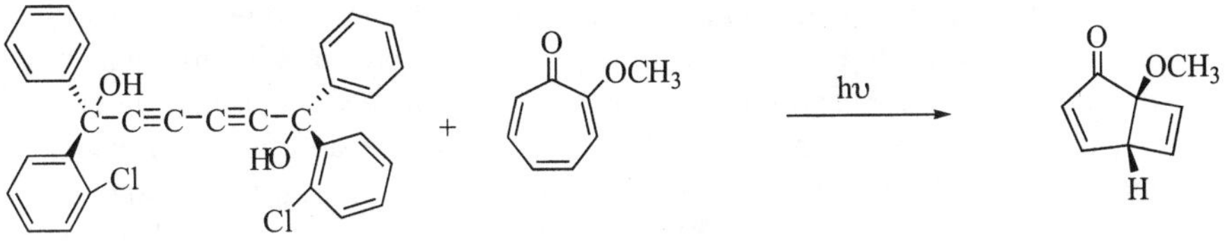

Scheme 11

solution fail to yield a substantial degree of stereoselectivity. In solution, the substrate and the chiral source are unable to interact positively within the excited-state lifetime of the substrate. These limitations are resolved when the reactions are carried out in crystals and inclusion complexes. In these media, the reactants are confined within a restricted reaction cavity and are therefore forced to react stereoselectively. Since the geometry is already fixed in a particular fashion, stereoselectivities obtained by these methods are often very high. But these media offer the disadvantage of being unpredictable. The field of crystal engineering has not advanced to a stage where one can predict a priori if a particular reactant will crystallize in a chiral space group, which is necessary for the reaction to proceed enantioselectively. In cases where the ionic auxiliary approach is used to form a chiral crystal, it has the disadvantage of being applicable only to systems having carboxylic acid or amine functionality. Inclusion crystals can only be used in those cases which form host–guest complexes with the guest molecule.

One should view the studies in asymmetric induction in photochemical reactions from the perspective of physical organic chemistry and understanding of intermolecular interactions. Organic photochemistry's success remains in the area of materials science, photomedicine, and photobiology. In these disciplines even a small asymmetric induction can be enormously significant. Therefore it would be a mistake to ignore the low *ee* and *de* often obtained in photochemical reactions. Asymmetric induction obtained in photochemical reactions, however small it might be, is conveying information that is yet to be fully deciphered.

II. ZEOLITE

A. Zeolite as a Host for Asymmetric Photoreactions

We believe that zeolites offer a number of advantages over the above host systems to carry out asymmetric photoreactions. A number of reactions can be carried out within the reaction cavities of zeolites, the only limitation being that the size of the substrate should match that of the supercage of the zeolite. Reactions within zeolites take place with more selectivity than in solution and with greater ease than in crystals or organic host–guest systems [263]. In order to use zeolites as inclusion hosts to carry out asymmetric reactions, an asymmetric environment must be inherently present within the zeolite, or an asymmetric environment must be created by inclusion of chiral compounds. Owing to the significance of chiral zeolites, their synthesis has been the focus of many research groups. Although many zeolites like ZSM–5 and ZSM–11 can, in theory, exist in chiral forms, there have not been any reports of a zeolite being isolated in a "pure and stable" chiral form. However, it was reported that zeolite beta and titanosilicate ETS–10 have unstable chiral polymorphs [264–273]. Davis and Lobo have reported the synthesis of zeolite beta-enriched in the presence of polymorph A over that of normal zeolite beta [224,275]. They showed that the utility of this enriched zeolite in carrying out asymmetric reactions by using it in the ring opening reaction of *trans*-stilbene oxide. They were able to obtain the (*R,R*)-diol with an *ee* of 5%. Meanwhile, owing to the unavailability of pure chiral zeolites, we have adopted the strategy of creating an asymmetric environment within the zeolite by adsorbing chiral organic molecules within the supercages of the zeolite. Another approach would be to modify the silanol groups of the zeolite interior with chiral molecules, which is being currently pursued in our laboratory.

B. Zeolite: A Brief Introduction to Its Structure

Zeolites are inorganic microporous and microcrystalline materials capable of complexing or adsorbing small- and medium-sized organic molecules [276]. Adsorption of organic compounds on zeolites can occur on both the external and the internal surfaces of a zeolite crystal. Internal complexation occurs by diffusion of the guest into the channels and cavities within the zeolite crystal. While internal surface areas largely exceed external surfaces, intracrystalline adsorption may occur only if the kinetic diameter of the guest is smaller than the diameter of the intracrystalline cavities. There are approximately 40 naturally occuring and over 100 synthetic forms of zeolites. The research detailed in this chapter has been carried out mainly in faujasite zeolites X and Y. These synthetic forms of faujasite zeolite have the following typical unit cell composition:

$$X \text{ type} \quad M_{86} (AlO_2)_{86} (SiO_2)_{106} \cdot 264 \, H_2O$$

$$Y \text{ type} \quad M_{56} (AlO_2)_{56} (SiO_2)_{136} \cdot 253 \, H_2O$$

where M is a monovalent cation. The faujasite framework has two main structural features. The main supercage is a result of assembly of the basic unit "sodalite cage." The sodalite cages combine to form an even larger cage, the supercage. While sodalite cages are too small to accommodate organic molecules, the spherical supercages are ~ 13 Å in diameter. Access to the supercages is afforded by four 12-ring windows 7.5 Å in diameter tetrahedrally distributed about the center of the supercage. The supercages form a three-dimensional network in which each supercage is connected tetrahedrally to four other supercages through the 12-membered ring window. A unit cell of X and Y zeolites consists of eight supercages. Charge compensating cations present in the internal structure of zeolites are known to occupy three different positions in zeolites X and Y. The first type (site I), with 16 cations per unit cell (both X and Y), is located on the hexagonal prism faces between the sodalite units (Fig. 1). The second type (site II), with 32 per unit cell (both X and Y), is located in the open hexagonal faces. The third type (site III), with 38 per unit cell in the case of the X type and only eight per unit cell in the case of the Y type, is located on the walls of the larger

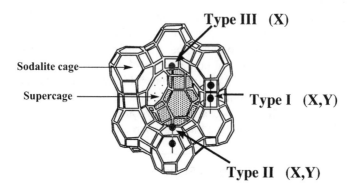

Figure 1   Structure of supercages in X and Y zeolites. Positions of cations are indicated as circles (types I, II and III).

cavity. Only cations at sites II and III are expected to be readily accessible to the organic molecule adsorbed within a supercage. These structural characteristics make zeolites ideal hosts to carry out photochemical reactions. Above-mentioned features and the commercial availability of zeolites give them advantages over other organic and inorganic host systems to carry out asymmetric photochemical reactions.

III. ASYMMETRIC INDUCTION WITHIN ZEOLITES: CHIRAL INDUCTOR AS A REAGENT DURING PHOTOREDUCTION OF CARBONYLS

The strategy of employing a chirally modified zeolite as a reaction medium requires the inclusion of two different molecules, **C** (a chiral inductor) and **R** (a reactant) within the interior spaces of an achiral zeolite. This strategy by its very nature does not allow quantitative chiral induction. When two different molecules **C** and **R** are included within a zeolite, the distribution is expected to follow the pattern shown in Fig. 2. The six possible distributions of guest molecules are (type I) cages containing a single **C**, (type II) two **C**'s, (type III) single **R**, (type IV) two **R**'s, (type V) one **C** and one **R** molecules, and (type VI) none at all. The chiral induction obtained from the photoreaction of **R** is an average of inductions that occur in cages of types III, IV (racemic product), and V (enantiomerically enriched product). In order to obtain high chiral induction, every reactant molecule (**R**) has to be placed next to a chiral inductor molecule (**C**); i.e., the ratio of type V has to be enhanced to the sum of types III and IV cages. Alterna-

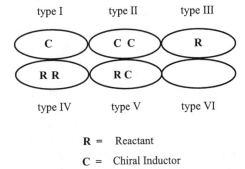

R = Reactant

C = Chiral Inductor

Figure 2 Possible distributions of reactant (R) and chiral inductor (C) molecules within zeolite supercages. Molecules are assumed to be stationed within supercages and not between supercages.

tively, one could devise reactions that are restricted to cages that contain both **C** and **R**. Such a condition eliminates the possibility of formation of the product of interest in the cages of types III and IV.

The photoreaction investigated to test the above model is the well-known electron-transfer-initiated intermolecular hydrogen abstraction reaction of carbonyl compounds [277]. Under the conditions employed, one of the guest mole-

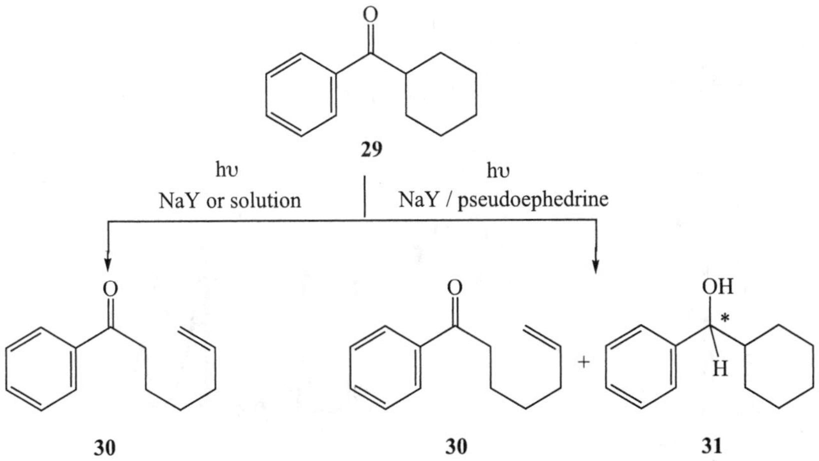

29

hυ
NaY or solution

hυ
NaY / pseudoephedrine

30

30 + **31**

Scheme 12

cules (e.g., ephedrine, norephedrine, pseudoephedrine) assumes the dual role of a chiral inductor and an electron donor. This approach would give a chirally enriched product only if the chiral amine reacts with the achiral ketone. Such a reaction would eliminate product formation from cages of type III and IV and would represent true enantioselectivity observed in the products. This approach is illustrated below with one example—photoreduction of phenyl cyclohexyl ketone (**29**) (Scheme 12) [278–282]. In isotropic media, **29** yields a γ-hydrogen abstraction Norrish–Yang cleavage product (**30**). Consistent with the solution behavior, irradiation as a hexane slurry of **29** included within NaY gave **30** as the only product. On the other hand, irradiation of **29** included within ephedrine, pseudoephedrine, or norephedrine modified NaY gave *inter*molecular hydrogen

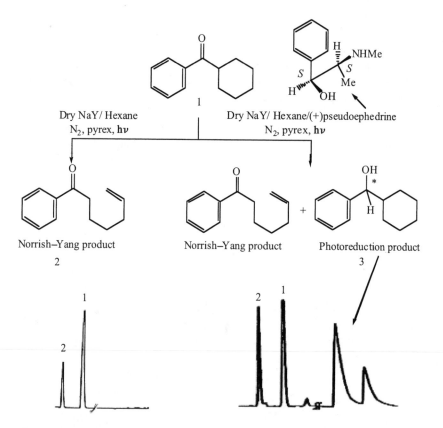

Figure 3   Photoreaction of phenyl cyclohexyl ketone within NaY in presence and absence of a chiral inductor. GC (Supelco β-dex column) traces are shown above.

abstraction product **31**, in addition to the expected product of intramolecular hydrogen abstraction **30** (Fig. 3). The alcohol **31** was the product of electron transfer from the amino group of the above chiral inductors, since it was not formed in their absence. Absence of formation of **31** when ephedrine hydrochloride or (−) diethyl tartarate (no amino group) was used as the chiral inductor further supports the above suggestion. If the role of the amine was to serve as the electron donor, the ratio of inter vs. intramolecular hydrogen abstraction products (**31** to **30**) should depend on the electron donating ability of the chiral inductor. As expected, the ratio of **31/30** was higher when secondary amine chiral inductors such as ephedrine and pseudoephedrine were the donors than when the primary amine chiral inductor norephedrine was (Fig. 4).

Thus the ketone **29** present in cages of types III and IV (Fig. 2) gave only **30**, whereas those present in cages of type V yielded both **30** and **31**. This predicted that the ratio of inter vs. intramolecular hydrogen abstraction products would depend on the ratio of the cages that contain the chiral inductor and those that do not, which in turn depended on the loading level of the chiral inductor. The GC traces of the product distribution upon irradiation of **29** included within NaY at two loading levels of pseudoephedrine are presented in Fig. 5. Clearly, the amount of intermolecular reduction product increased with the loading level of pseudoephedrine. An additional important point noticeable from Fig. 5 was that although the amount of the alcohol product **31** (with respect to **30**) was dependent on the loading level of the chiral inductor, the ratio of the optical isomers of **31** (*ee*) was almost independent of the loading level of the chiral inductor. This observation supported the view that reduction occurred only in cages containing the chiral inductor.

The ability to restrict the reduction reaction only to the cages containing both the reactant and the chiral inductor allowed us to examine, for the first time, chiral induction within a zeolite without any interference from reactions that occurred in cages lacking the chiral inductors. Thus far, chiral induction within a zeolite has been complicated by racemic reactions within cages that do not contain the chiral inductor. By restricting the photoreaction to cages containing the chiral inductor, moderate chiral induction during the photoreduction of **29** was achieved. The *ee* obtained in this study is noteworthy, as earlier attempts on achieving chiral induction during photoreduction of aryl alkyl ketones by chiral amines resulted in *ee* less than 8% (at room temperature). Furthermore, the major products in earlier studies were pinacols (diols) and not the ketone derived alcohols. The results (*ee*) obtained with various chiral inductors are summarized in Scheme 13. Of the various chiral inductors tested, best results were obtained with norephedrine (*ee* 68%). The use of (+)-norephedrine afforded the optical antipode of the photoproduct produced by the use of (−)-norephedrine, indicating that the system was well behaved. Similar to the observations with other systems, the *ee* obtained was dependent on the water content of the zeolite. When the

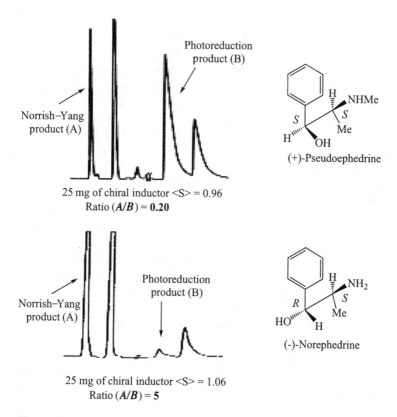

Figure 4 The ratio of the intermolecular to intramolecular hydrogen abstraction products from phenyl cyclohexyl ketone within NaY depended on the nature of the electron donor. The GC traces of the product mixture are shown above. <S> indicates the number of chiral inductor molecules per cage.

above prepared zeolite complex was intentionally made "wet" by adsorption of water, the *ee* was small relative to that obtained under dry conditions (dry: 68%; wet: < 2%).

The strategy presented above with phenyl cyclohexyl ketone has been established to be general by investigating a number of aryl alkyl ketones and diaryl ketones [281,283]. The best cases of % *ee* are summarized in Schemes 14 and 15. 2-Ethoxybenzophenone **32** gives intramolecular cyclization product **33** as the only product in solution as well as within NaY. However, in NaY in the presence of chiral amines, intermolecular reduction product **34**, in addition to **33**, was obtained (Scheme 16). More importantly, with pseudoephedrine and (1R,2R)-

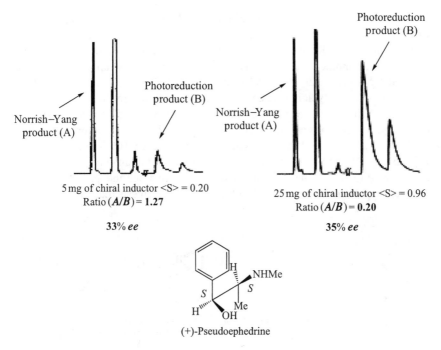

Figure 5 The dependence of the inter- and intramolecular hydrogen abstraction products from phenyl cyclohexyl ketone on the loading level of the chiral inductor, pseudoephedrine within NaY. The GC traces of the product mixture are shown above. The *ee* remained the same under the two conditions. $<S>$ indicates the number of chiral inductor molecules per cage.

diaminocyclohexane as chiral inductors, moderate % *ee* was obtained on the product **34** (39 and 51%, respectively). Several other substituted benzophenones were also investigated and the best cases of % *ee* are given in Scheme 15.

The studies with phenyl cyclohexyl ketone provide a few generalizations. Examination of Scheme 13 reveals that the probability of getting chiral induction from chiral inductors possessing two or more functional groups like amino and hydroxyl was more than with monofunctional groups. For example, norephedrine and pseudoephedrine gave more *ee* compared to α-methyl benzyl amine or bornyl amine. Studying the chiral inductors, it was also seen that the phenyl group was important in the chiral induction process. Norephedrine and other ephedrine family inductors that possess hydroxyl, phenyl, and amino groups were very successful. On the other hand, valinol, alaninol, and diethyltartarate were not effective in general.

Comparing norephedrine, ephedrine, and *N*-methyl ephedrine (Scheme 17), it can be seen that all three inductors have similar structural features and chiral configurations (1S,2R), except that norephedrine is a primary amine, ephedrine is a secondary amine, and *N*-methyl ephedrine is a tertiary amine, but the *ees* obtained from these vary from 68% to 16% to 27%, all enhancing the R(+) enantiomer in the α-cyclohexyl benzyl alcohol. Comparing ephedrine and pseudo-ephedrine (Scheme 17), which have identical molecular and structural formulae and are both secondary amines except for the chiral configuration at C_2, which is 2R for ephedrine and 2S for pseudoephedrine, gave 16%R and 35%R *ee*, respectively. Amino phenyl ethanol, which does not have any chiral center at C_2 also gave *ee* (26%R). That the same R isomer is obtained independently of the configuration at the C_2 carbon of the chiral inductor suggests that chiral informa-

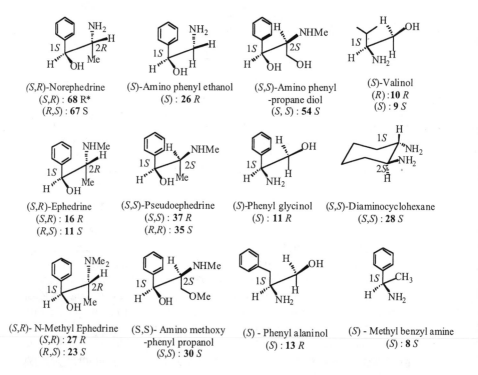

(S,R)-Norephedrine
(S,R) : **68 R***
(R,S) : **67 S**

(S)-Amino phenyl ethanol
(S) : **26 R**

(S,S)-Amino phenyl
-propane diol
(S, S) : **54 S**

(S)-Valinol
(R) : **10 R**
(S) : **9 S**

(S,R)-Ephedrine
(S,R) : **16 R**
(R,S) : **11 S**

(S,S)-Pseudoephedrine
(S,S) : **37 R**
(R,R) : **35 S**

(S)-Phenyl glycinol
(S) : **11 R**

(S,S)-Diaminocyclohexane
(S,S) : **28 S**

(S,R)- N-Methyl Ephedrine
(S,R) : **27 R**
(R,S) : **23 S**

(S,S)- Amino methoxy
-phenyl propanol
(S,S) : **30 S**

(S) - Phenyl alaninol
(S) : **13 R**

(S) - Methyl benzyl amine
(S) : **8 S**

* (S,R) represents the configuration of the chiral centers in the precursor;
R represents the configuration of the newly formed chiral center in the product.

Scheme 13

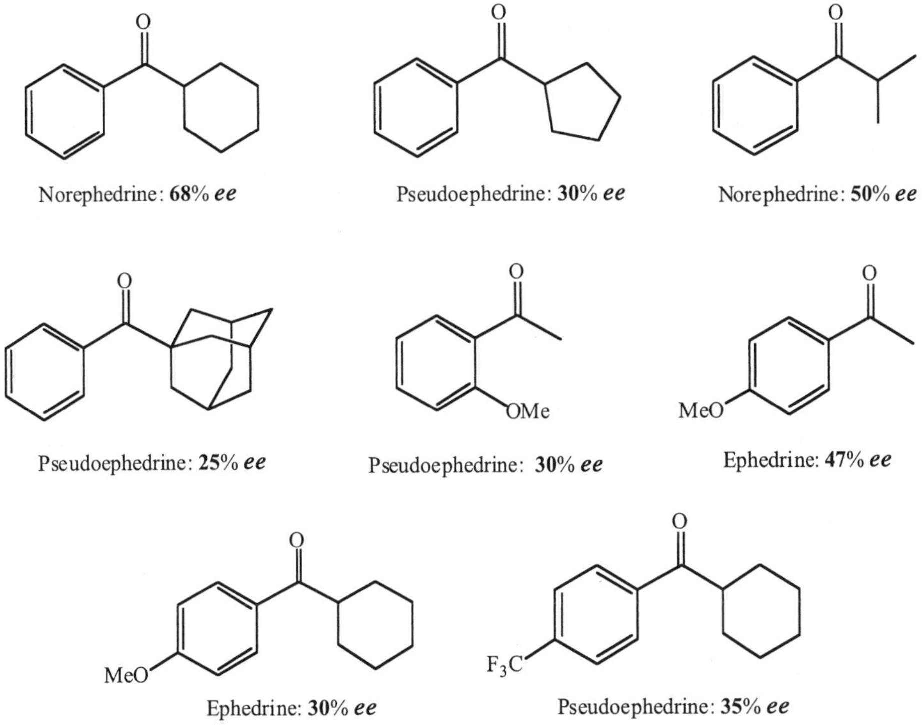

Norephedrine: **68%** *ee*   Pseudoephedrine: **30%** *ee*   Norephedrine: **50%** *ee*

Pseudoephedrine: **25%** *ee*   Pseudoephedrine: **30%** *ee*   Ephedrine: **47%** *ee*

Ephedrine: **30%** *ee*   Pseudoephedrine: **35%** *ee*

Scheme 14

tion transfer occurs mainly from substituents present at the C_1 carbon of the chiral inductor.

An insight into the role of the alkali ion in the reduction process is revealed by the results of computations (B3LYP/6–31G*; Gaussian 98). When both phenyl cyclohexyl ketone and norephedrine were allowed to bind to the Li^+ ion, the latter holds the two together by interacting with the carbonyl oxygen of the phenyl cyclohexyl ketone and the hydroxyl oxygen and amino nitrogen of norephedrine (Fig. 6). Of the various computed structures for the ternary complex, the above structure is the most stable. It is important to note that such a structure allows electron transfer from a nitrogen lone pair to the excited carbonyl. Also it reveals the lesser importance of the C_2 carbon in terms of chiral induction.

It is unlikely that the asymmetric photoreduction procedure described here will be able to compete with the available thermal methods for the reduction of carbonyl compounds. Generality of the zeolite-based method even with respect

Pseudoephedrine: **43%** *ee*   1,2 Diaminocyclohexane: **44%** *ee*

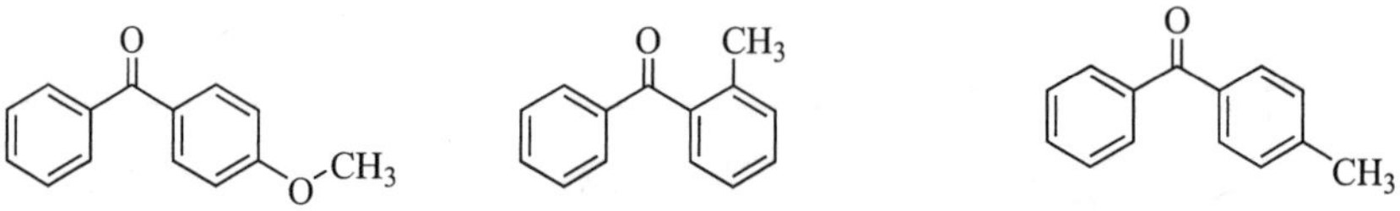

Ephedrine: **45%** *ee*   Ephedrine: **23%** *ee*   Pseudoephedrine: **22%** *ee*

Ephedrine: **34%** *ee*   Pseudoephedrine: **35%** *ee*

Scheme 15

to other carbonyl systems is not yet fully established. However, the novelty of
the method becomes obvious when one compares the results obtained within
zeolites to those in solution. That the alkali ions within zeolites are able to force
a closer interaction between the chiral inductor and the achiral reactant is novel
and worthy of further attention.

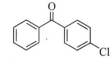

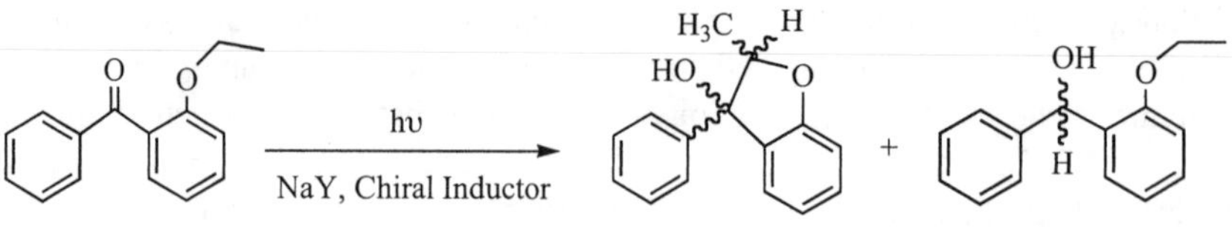

$$\text{32} \xrightarrow[\text{NaY, Chiral Inductor}]{h\upsilon} \text{33} + \text{34}$$

Scheme 16

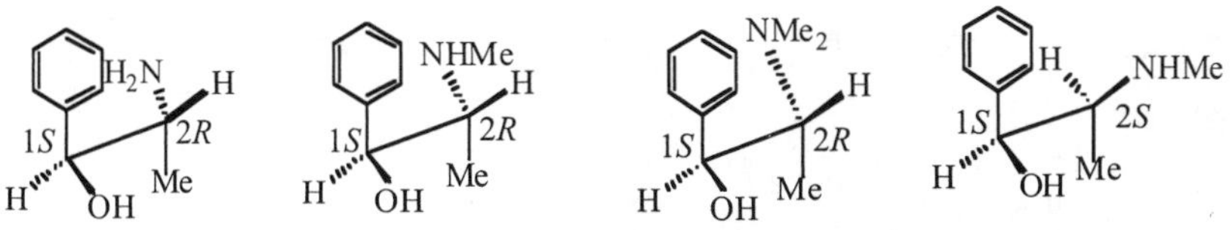

Norephedrine	Ephedrine	N-Methyl ephedrine	Pseudophedrine
68% *R*	16% *R*	27% *R*	37% *R*

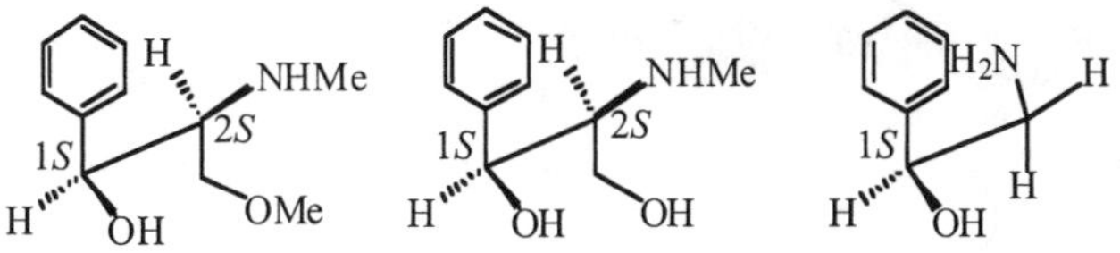

Amino methoxy phenyl propanol	Amino phenyl propane diol	Amino phenyl ethanol
30% *S*	54% *S*	26% *R*

Scheme 17

IV. DIASTEREOSELECTIVITY IN PHOTOREACTIONS WITHIN ZEOLITES

A. Background

To be sure that the chiral inductor and the reactant molecules stay together within a single cage we have explored another strategy. In this method the two components are linked via a covalent bond. This forces the chiral inductor and the reactant parts of a single molecule to stay close to each other. Because of the prior presence of a chiral center in the reactant molecule, the reactant is chiral and the products are formed as diastereomers. Elegant examples of diastereoselective photoreactions in solution are discussed in Chap. 5. We show below that chiral auxiliaries that are ineffective in solution function well within zeolites. In every one of the examples discussed in this section the zeolite is essential to obtain a significant *de*. We wish to emphasize that the examples should be examined from the perspective of the information they offer in the context of supramolecular interactions.

B. Diastereoselective Photorearrangements

Tropolone ether **35** undergoes a disrotatory 4π electrocyclic ring closure to yield the corresponding bicyclo[3.2.0] product **36** (Scheme 18) [284,285]. The chirality

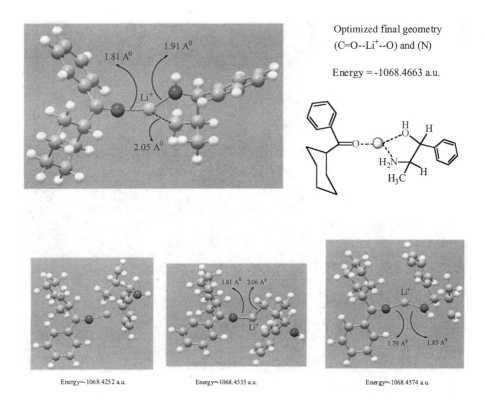

Optimized final geometry
(C=O--Li⁺--O) and (N)

Energy = -1068.4663 a.u.

Figure 6 Structures of complexes Li^+ bound to phenyl cyclohexyl ketone and norephedrine as calculated by Gaussian-98/B3LYP/6-31G*. The energy is included below each structure. Of the four structures, the structure (a) is the most stable.

is introduced into the product through the two equally allowed disrotatory processes. In spite of tropolone ether **35** having a chiral auxiliary covalently attached, irradiation in isotropic solvents like hexane, methylene chloride, and acetonitrile gave a 1 : 1 diastereomeric mixture of the product **36**. The chiral center in the tropolone derivative is unable to provide chiral discrimination in the 4π electrocyclization reaction. On the other hand, irradiation of **35** (4 mg, $<S> = 0.28$, where $<S>$ is the occupancy number, i.e., the number of molecules per supercage) in NaY led to the formation of **36** in 53% *de* [282,286–288]. This result shows the potential of zeolites to enhance the diastereoselectivity of a photoreaction. One possible mechanism for the observed diastereoselectivity in this reaction is that one molecule in the ground state is taking on the role of a chiral inductor for another molecule in the excited state. If this were true, one would expect the

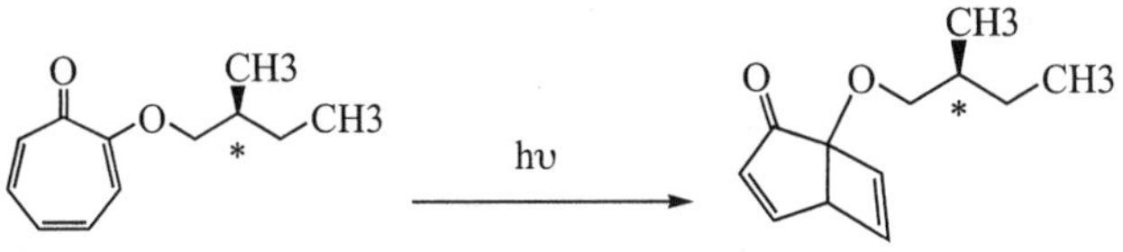

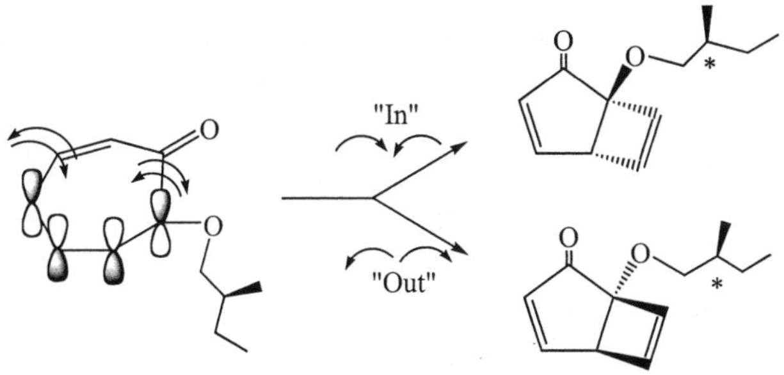

Scheme 18

diastereoselectivity to increase with an increase in the loading level of the substrate within the zeolite. But when the loading level of **35** in NaY was increased, irradiation of the NaY-**35** complex led to a decrease in the selectivity. As the loading level was increased to 0.56 and 0.83, the *de* decreased to 41% and 28%, respectively. This suggests that when a molecule of **35** in the ground state is located close to another one in the excited state, it induces the formation of diastereomer B. The observed diastereoselectivity within NaY at low loading levels must be the result of the intramolecular rather than intermolecular effect. Simple confinement of the substrate within the reaction cavity of the zeolite has enabled the transfer of chiral information to occur from the substituent to the reaction center. As illustrated below with several examples, the above approach has proven to be general in terms of both chiral auxiliary and photoreactions.

In Scheme 19, results of photocyclization of twelve tropolones covalently linked with chiral auxiliaries via amide linkages are presented [289,290]. In no case is the diastereomeric excess in solution above 15%. Neither the variation of solvent-nor the concentration of the reactant improved the selectivity. On the other hand, within zeolites the *de*s are well above 40%. The extent of *de* is dependent on the alkali ion present within Y zeolites and the best numbers are provided in the scheme.

The second set of examples consists of pyridones **37**, which upon excitation undergo 4π cyclization [291]. As with tropolones, the mode of cyclization deter-

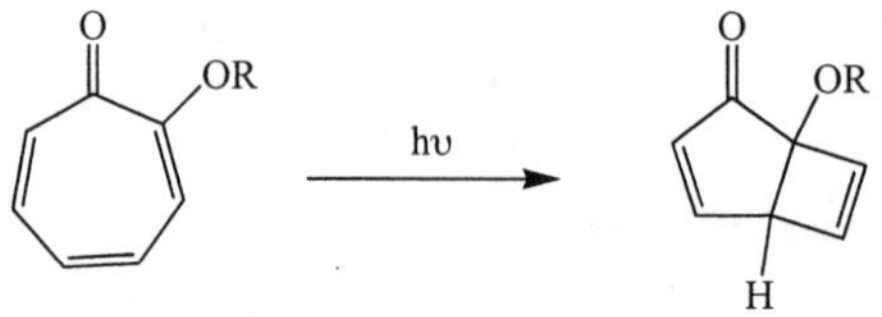

R= CH$_2$COR*

R*	% de				R*	% de			
	CH$_3$CN	NaY	KY	LiY		CH$_3$CN	NaY	KY	LiY
—NH, Me, Me, H, CH$_3$	3	44	15	17	—HN,,, COOMe, H	7	55	2	15
—NH, CH$_2$OMe, H, Me	0	54	12	24	—N,,, H, COOMe, H	2	53	15	29
—NH, Ph, H, Me	2	83	26	64	—N,,, H, COOMe, H	3	76	86	49
—NH, H, Me	2	45	28	36	N, COOMe, H	8	20	46	10
—NH, H, Me	10	88	41	80	—HN, OH	9	74	2	39
—HN, Me, OH	3	77	22	50	—NH, CH$_2$OMe, OH	14	65	50	68

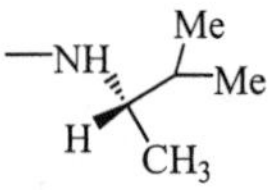

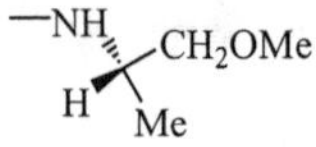

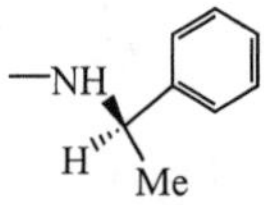

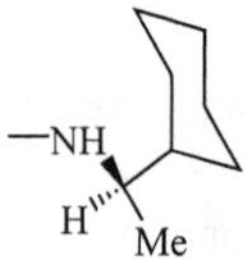

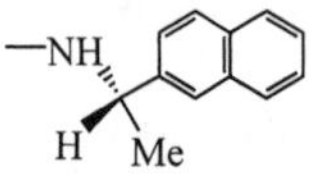

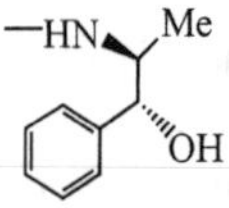

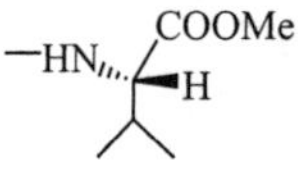

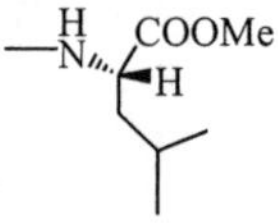

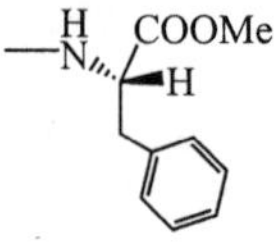

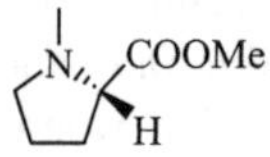

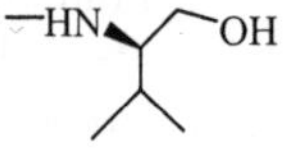

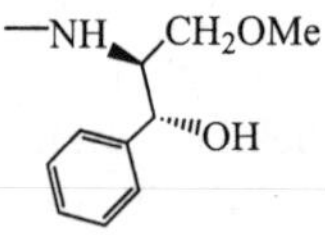

Scheme 19

mines the chirality of the product (Scheme 20). Pyridones appended with chiral auxiliary yield nearly 1 : 1 diastereomeric mixture in solution. However, within zeolites, cyclized products are obtained in respectable *de* [292].

The third set of molecules that bring out the uniqueness of zeolites in the context of asymmetric induction is of 2,4-cyclohexadienones (6,6-dimethyl-2,4-cyclohexadienones **38**, and 2,2-dimethyl-1,2-dihydronaphthalenones **39**) [279,282,287,289,293–295]. The basic chromophore in these molecules is the conjugated dienone that undergoes oxa-di-π-methane rearrangement to give a bicyclic product (Scheme 21). According to the accepted mechanism, the chirality is introduced into the system at the first step, yielding the diradical intermediate

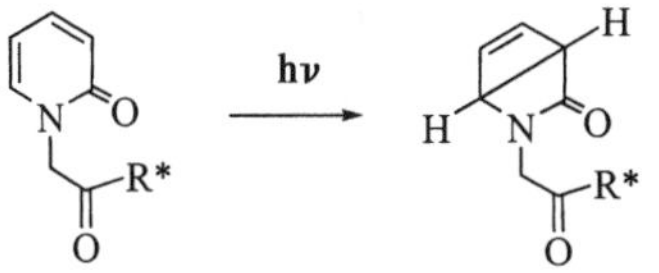

R*	CH$_3$CN	LiY	KY	MY	R*	CH$_3$CN	LiY	KY	MY
	% de					**% de**			
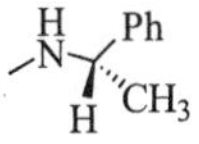 H–N–Ph, H, CH$_3$	2	28	42	–	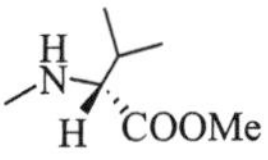 H–N–COOMe (*t*-Bu)	2	5	14	53(RbY)
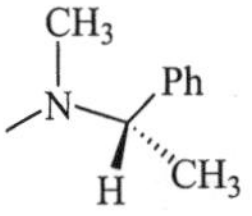 CH$_3$–N–Ph, H, CH$_3$	1	54	76	84(NaY)	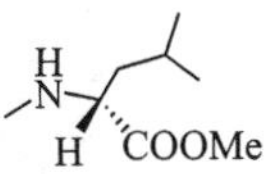 H–N–COOMe	1	28	5	–
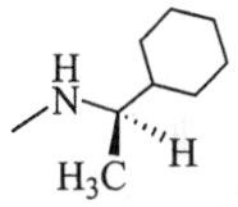 H–N–(cyclohexyl), H$_3$C, H	2	24	1	–	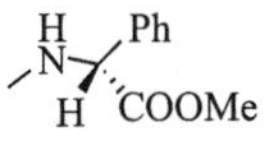 H–N–Ph, COOMe	2	33	8	–
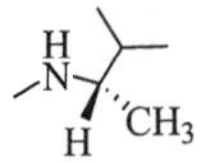 H–N–(isopropyl), H, CH$_3$	1	34	75	–	 H–N–Ph, COOMe	3	20	78	–
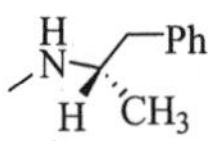 H–N–Ph, H, CH$_3$	2	8	67	–	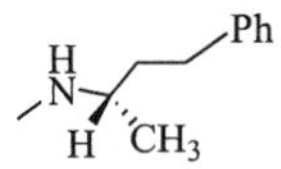 H–N–Ph, H, CH$_3$	3	25	76	88(RbY)

Scheme 20

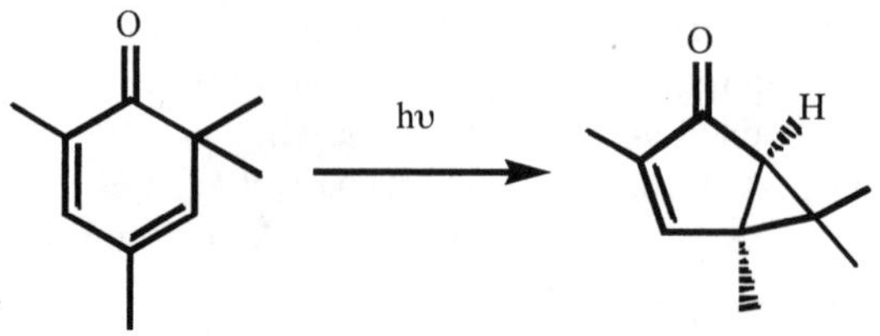

38

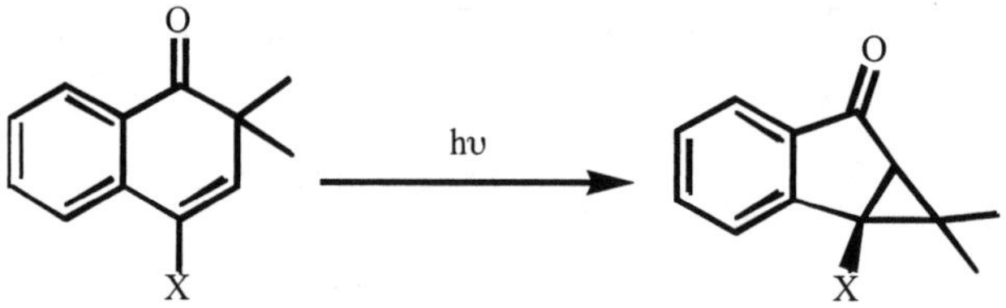

Scheme 21

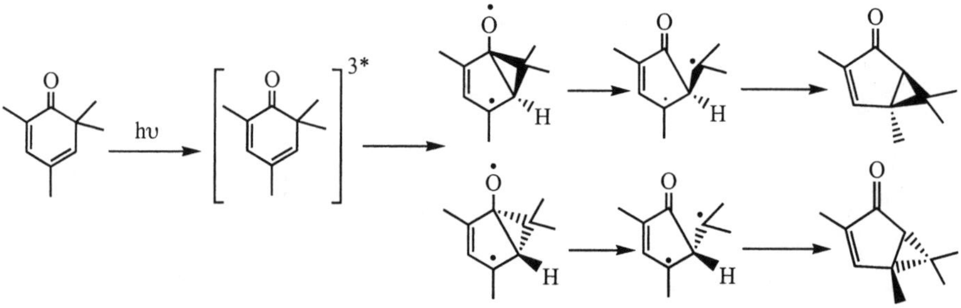

Scheme 22

(Scheme 22). Results of irradiation of a number of 6,6-dimethyl-2,4-cyclohexa-dienones and 2,2-dimethyl-1,2-dihydronaphthalenones are presented in Scheme 23 and 24. As with tropolones, in these systems the chiral auxiliaries, which have little influence in solution, are very effective within zeolites.

The final photoreaction that we discuss in the context of chiral auxiliary effect in photorearrangements is the geometric isomerization of 2,3-diphenylcy-clopropane-1-carboxylic acid derivatives **40** and 2,3-diphenyl-1-benzoylcyclo-propane derivatives **41** [279,282,287,290,296–300]. As shown in Schemes 25 and 26, the isomerization in the former occurs through the cleavage of the C_2–C_3 bond, while that in latter occurs through the cleavage of the C_1–C_2 and C_1–C_3 bonds. We have examined twenty-two 2,3-diphenylcyclopropane-1-carboxylic acid derivatives and fourteen 2,3-diphenyl-1-benzoylcyclopropane derivatives appended with a variety of chiral auxiliaries. The results in solution and zeolites are presented in Schemes 27 and 28.

Examination of the above four systems leads to the following conclusions: (a) a chiral auxiliary is more effective within a zeolite than in solution; (b) the extent of *de* depends on the structure of the chiral auxiliary and the nature of the cation present in the zeolite.

C. Role of Cations During Diastereoselective Photorearrangements

We analyze one system, the photoisomerization of 2,3-diphenylcyclopropane-1-carboxylic acid derivatives, in detail to illustrate the role of alkali ions during the asymmetric induction process within zeolites [282,296]. All other systems investigated here could be analyzed on the basis of the same model. Computational results have been used to gain an insight into the interaction among the reaction site, the chiral auxiliary, and the cation. While the computations provide information about interactions in the gas phase, the reactions that we are concerned with occur in much more complex environments, namely zeolites. In spite of this severe limitation, the computational results serve to build a preliminary model that can be used to plan further experiments.

A number of observations (Scheme 29) suggest that alkali metal ions present in zeolites play an important role in the asymmetric induction process. (a) The *de* was dependent on the nature of the alkali metal ion (e.g., % *de* in the case of **42** in LiY, NaY, KY, RbY, and CsY are 80, 28, 14, 5, and 5, respectively). (b) The *de* varied with the water content of NaY used (**42**: dry 80%, wet 8%). (c) The *de* upon irradiation of **42** adsorbed on silica gel, a surface that does not contain cations, was only 8%. (d) Diastereomeric excess in the case of **42** decreased from 80% to 10% when the Si/Al ratio of NaY zeolite was changed from 2.4 to 40. The less the aluminum on the framework of the zeolite, the less the number of alkali metal ions. The number of cations per unit cell decreases from 55 to 5

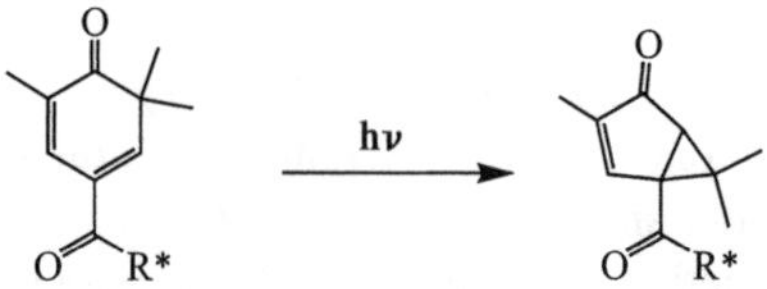

R*	% *de* CF$_3$CH$_2$OH	NaY	RbY	MY
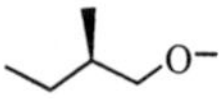	2	5	12	22 (LiY)
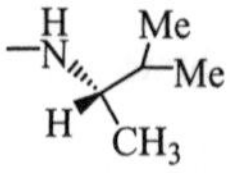	0	2	38	31 (CsY)
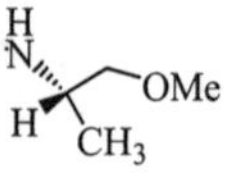	0	73	25	40 (LiY)
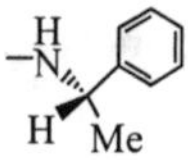	5	59	39	34 (KY)
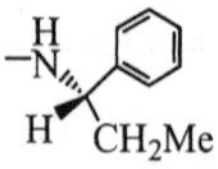	0	41	13	34 (LiY)
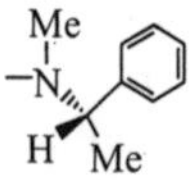	0	20	8	39 (LiY)
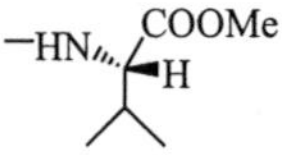	4	53	7	32
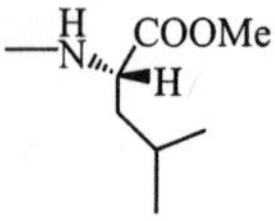	2	22	40	25
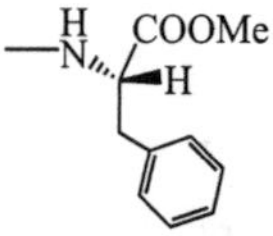	4	25	17	35 (KY)
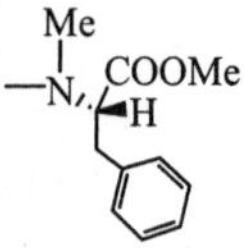	29	27	60	31 (CsY)
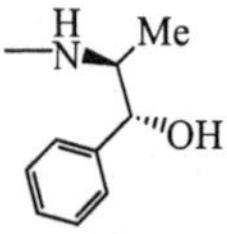	11	59	18	25 (KY)

Scheme 23

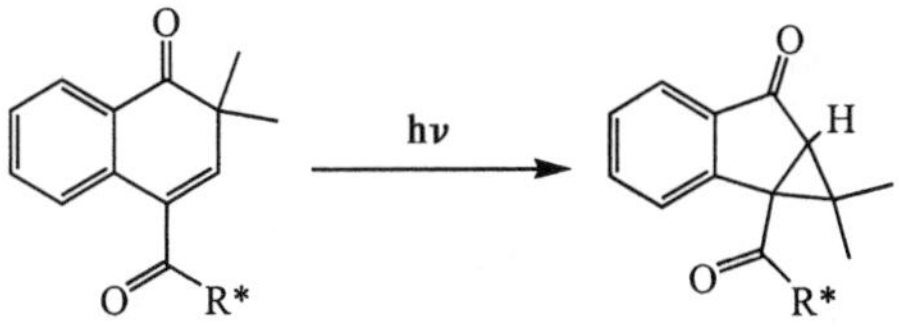

Scheme 24 reaction: 2,2-dimethyl-4-(C(=O)R*)-naphthalen-1(2H)-one →(hν)→ cyclopropane-fused indanone with C(=O)R*.

R*	% de				R*	% de			
	CH₃CN	NaY	KY	MY		CH₃CN	NaY	KY	RbY
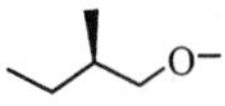 (2-methylbutyl)–O–	0	10	8	31 (LiY)	–N(H)–CH(Me)COOMe (H)	9	21	43	10
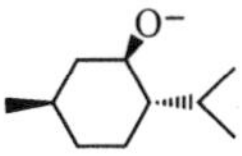 menthyl–O–	9	46	60	48 (RbY)	–N(H)–CH(iPr)COOMe (H)	0	12	58	20
 bornyl–O–	3	57	34	24 (CsY)	–N(H)–CH(iBu)COOMe (H)	2	0	45	10
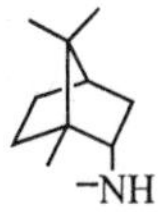 bornyl–NH–	13	48	35	34 (RbY)	–N(H)–CH(Ph)COOMe (H)	8	5	41	5
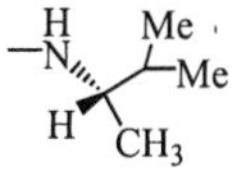 –N(H)–CH(CH₃)CH(Me)Me (H)	6	45	23	35 (CsY)	–N(H)–CH(CH₂Ph)COOMe (H)	8	11	30	7
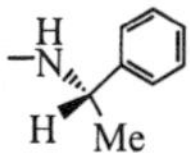 –N(H)–CH(Me)Ph (H)	0	48	81	46 (RbY)	–N(H)–CH(Me)CH(Ph)OH	9	48	12	15
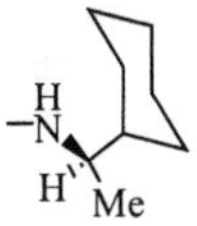 –N(H)–CH(Me)(cyclohexyl) (H)	3	35	58	45 (CsY)	–N(Me)–CH(Me)CH(Ph)OH	6	19	57	30 (LiY)
–N(Me)–CH(Me)Ph (H)	8	13	13	25 (LiY)					

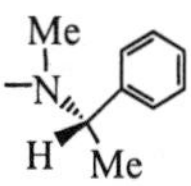

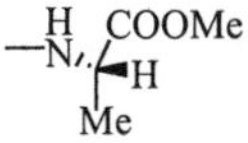

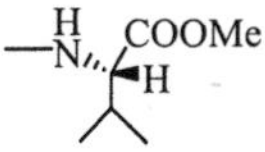

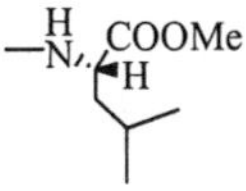

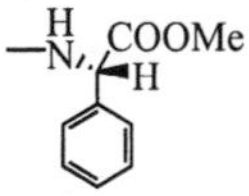

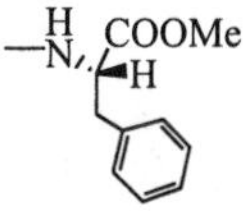

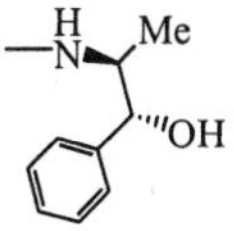

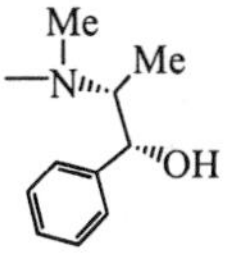

Scheme 24

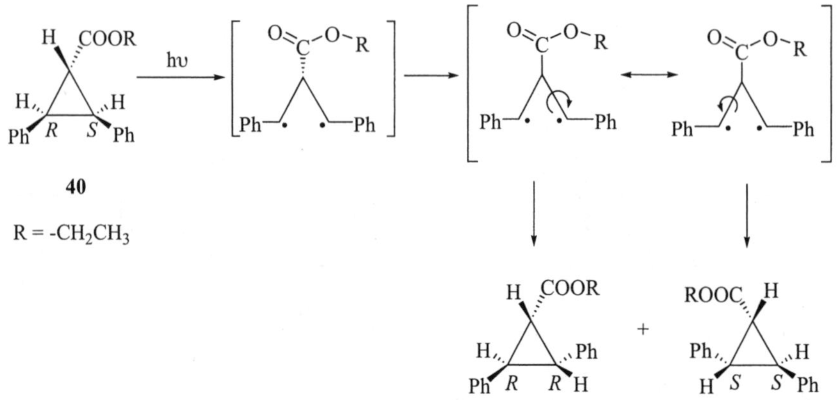

Scheme 25

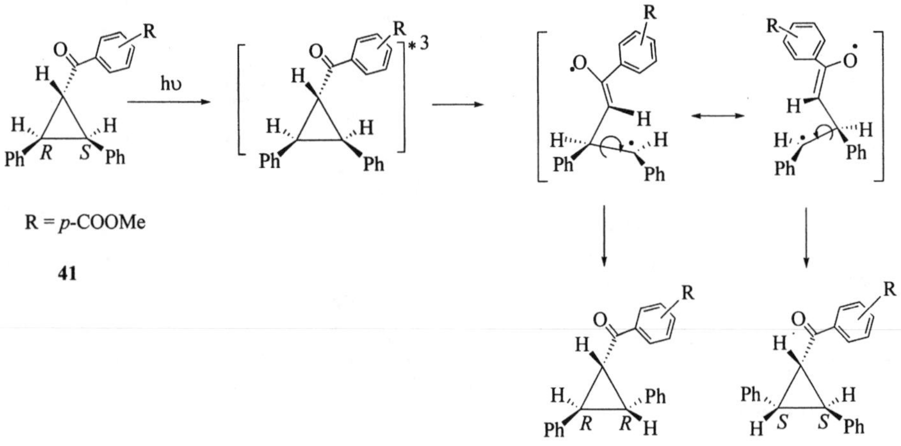

Scheme 26

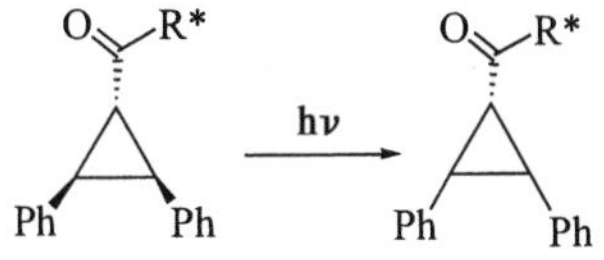

Reaction scheme: an α,β-diphenylcyclopropyl ketone (O=C–R*, with Ph and Ph substituents) undergoes photochemical ($h\nu$) conversion to a diastereomeric product.

R*	%de				R*	%de			
	CH₃CN	LiY	NaY	MY		CH₃CN	LiY	NaY	MY
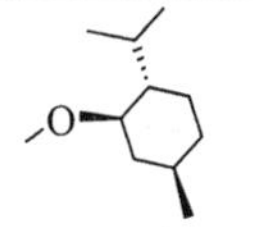 menthyloxy	4	50	55	–	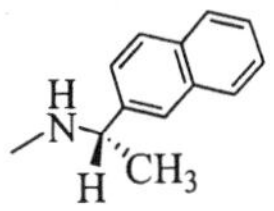 N-(1-naphthylethyl)amino (CH₃)	4	69	7	–
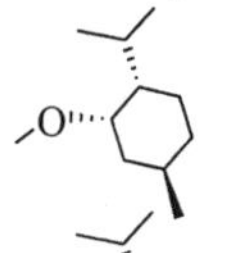 menthyloxy (epimer)	3	25	40	–	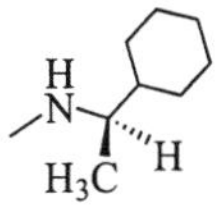 N-(cyclohexyl- CH₃)amino	2	29	24	–
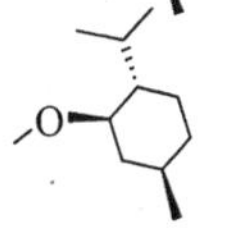 neomenthyloxy	5	12	32	–	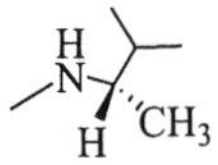 N-(isopropyl, CH₃)amino	2	7	7	–
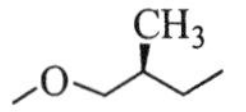 2-methylbutoxy (CH₃)	0	14	19	–	 N-(isopropyl, COOMe)amino	2	83	21	80(KY)
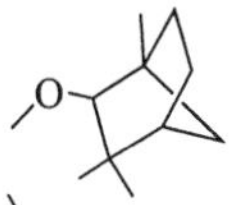 bornyloxy	4	7	16	–	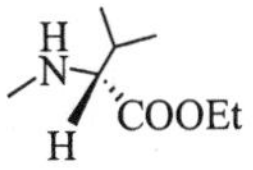 N-(isopropyl, COOEt)amino	20	39	31	61(KY)
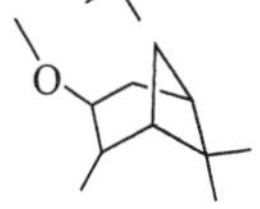 pinanyloxy	5	2	6	–	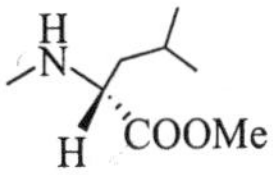 N-(isobutyl, COOMe)amino	3	26	22	46(KY)
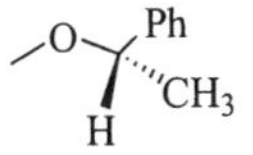 O–CH(Ph)(CH₃)	4	78	19	–	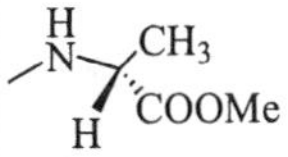 N-(CH₃, COOMe)amino	5	34	22	–
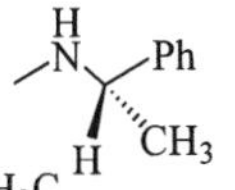 N-(Ph, CH₃)amino	2	80	28	–	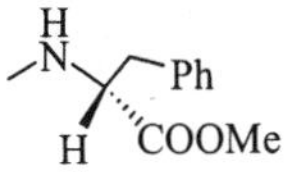 N-(Ph, COOMe)amino	2	10	32	53(KY)
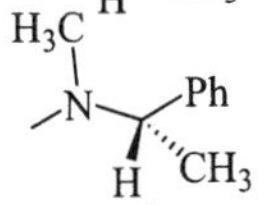 N-(Ph, CH₃)amino	3	17	30	–	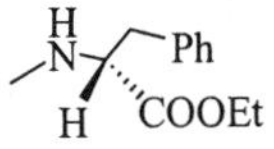 N-(Ph, COOEt)amino	1	21	40	–
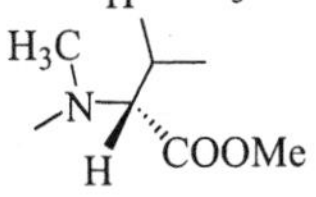 N-(isopropyl, COOMe)dimethylamino	7	3	8	–	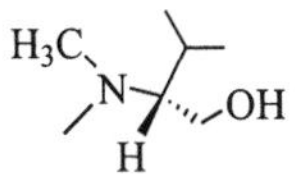 N-(isopropyl, OH)amino	8	11	9	23(RbY)
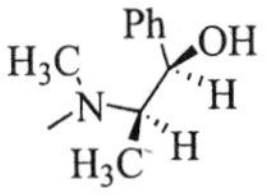 N-(Ph, OH, CH₃, H)dimethylamino	22	58	89	–	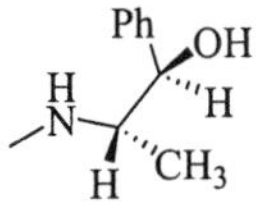 N-(Ph, OH, CH₃)amino	30	25	60	–

Scheme 27

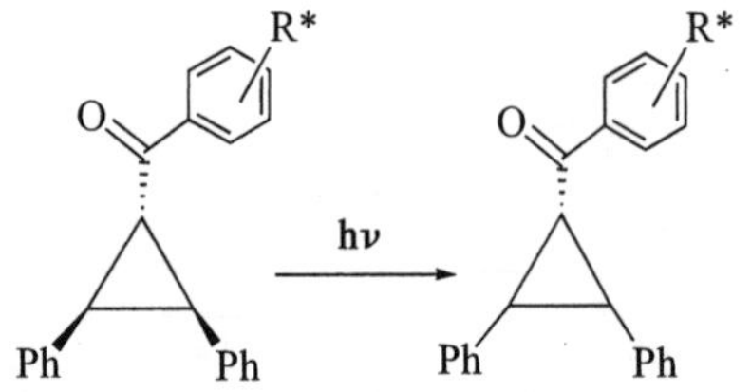

Meta derivative % de			R*	Para derivative % de		
CH₃CN	NaY	KY		CH₃CN	NaY	KY
0	71	14	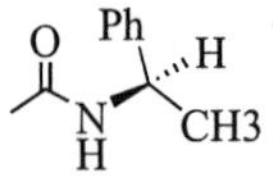 Ph, H, CH3 acetamido	2	11	1
1	30	14	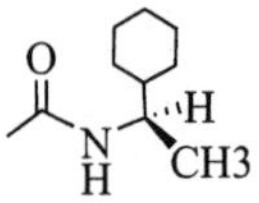 cyclohexyl, H, CH3 acetamido	2	4	3
4	41	29	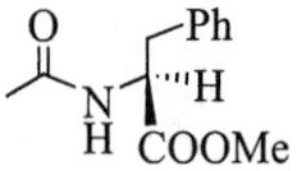 Ph, H, COOMe acetamido	1	2	5
3	38	15	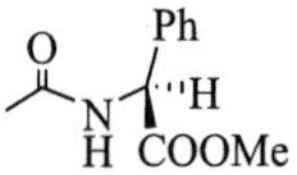 Ph, H, COOMe acetamido	1	15	7
0	25	52	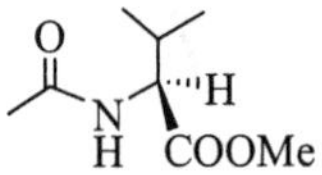 isopropyl, H, COOMe acetamido	0	1	11
0	11	14	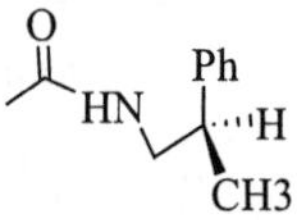 Ph, H, CH3 acetamido-HN	0	3	6
1	2	7	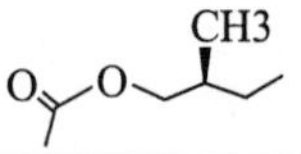 CH3 ester	2	1	0

Scheme 28

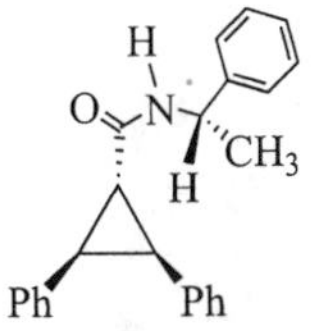

Scheme 29

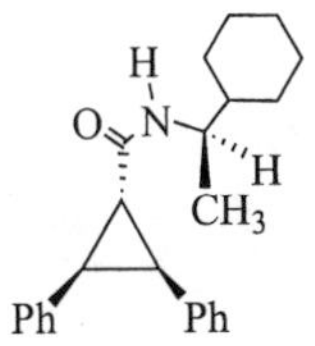

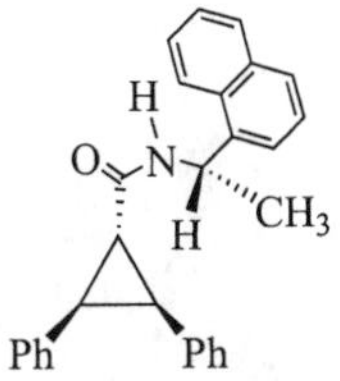

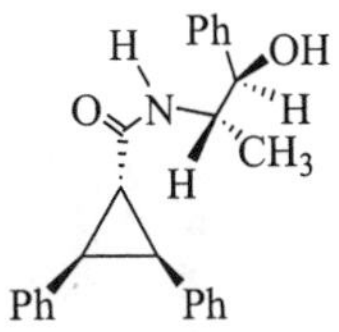

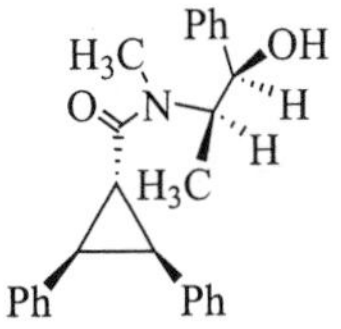

upon changing the Si/Al ratio from 2.4 to 40. The Si/Al ratio of 40 would be a silicious zeolite, and hence the results were similar to the selectivity observed in silica. Clearly the above control studies showed that the alkali metal ions present in the zeolite are critical in achieving high stereoselectivity.

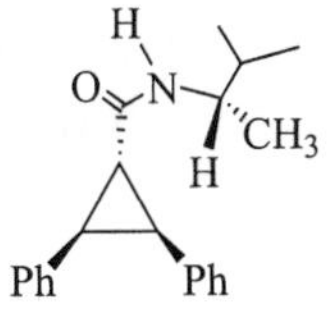

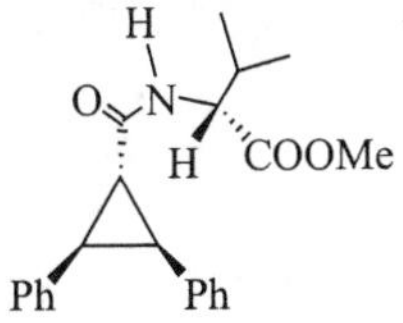

Examination of Scheme 27 (22 total chiral auxiliaries employed) indicates that aryl chiral auxiliaries (as in **42**) are more effective than alkyl chiral auxiliaries (as in **43**) within zeolites. Replacing the phenyl part in **42** by the cyclohexyl moiety as in **43** had a dramatic effect on the observed stereoselectivity. The diastereoselectivity dropped from 80% to 29% in LiY. Replacing the phenyl group by the naphthyl group (**44**; another aryl chiral auxiliary) gave 70% *de* in LiY. Aryl chiral auxiliaries, **45** and **46** based on aminoalcohols gave 60% and 89% *de*, respectively.

The important question concerning the reason for the differential behavior of the aryl and alkyl chiral auxiliaries was answered by *ab initio* computations at the Hartree-Fock level. The binding affinities and geometries (Fig. 7) of alkali metal ion bound **42** (representing the aryl chiral auxiliary) and **43** (representing the alkyl chiral auxiliary) provided insight into the role of the aryl group in enhancing the power of a chiral auxiliary within MY zeolites (M = alkali metal ion). Both the molecules interacted strongly with Li^+ and Na^+ ions with binding affinities > 70 kcal/mol. Since alkali metal ions were bound to the surface of a zeolite, the binding affinities between the cation and the guest molecules within

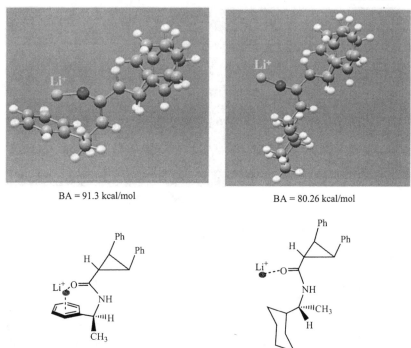

Figure 7   RHF/3-21G optimized structures of Li$^+$ bound 2β,3β-diphenylcyclopropane-1α-carboxamides of 1-phenylethylamine (a) and 1-cyclohexylethylamine (b). Binding affinities are included at the bottom of each structure.

a zeolite are expected to be smaller than the values computed above, but the trend is expected to remain the same. In the case of **42** (with the aryl chiral auxiliary) the cation interacts simultaneously with the phenyl group by cation-π interaction and the amide carbonyl oxygen by dipolar interaction (Fig. 7a). Such an interaction is expected to reduce the rotational freedom of the chiral auxiliary and thus make it rigid. On the other hand, in the case of **43** (with the cyclohexyl chiral auxiliary) the cation interacts only with the amide carbonyl oxygen via dipolar type interaction and does not interact with the chiral auxiliary part that contains the cyclohexyl group (Fig. 7b). Such an interaction would have no effect on the rotational mobility of the chiral auxiliary. A model based on differences in flexibility of the chiral auxiliary parts owing to differences in cation binding accounts for the observed variation in *de* between aryl and alkyl chiral auxiliaries within MY zeolites.

Examination of **47** and **48** showed that when the "methyl" group in **47** is replaced by a "carbomethoxy" group in **48** the *de* significantly increases. The diastereoselectivity observed in LiY zeolite with **48** was 83%, whereas in the case of **47** it was 7%. Once again, *ab initio* computations at the Hartree-Fock level provided an insight. Both **47** and **48** interacted strongly with Li$^+$ (Fig. 8) with binding affinities > 80 kcal/mol. In the case of **48** (with the carbomethoxy chiral auxiliary) the cation interacts simultaneously with both the carbonyl of the carbomethoxy group and the amide carbonyl oxygen. Such an interaction is expected to reduce the rotational freedom of the chiral auxiliary and thus make it rigid as in the case of the aryl chiral auxiliary. On the other hand, in the case

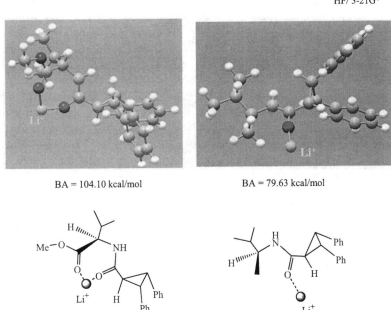

Figure 8 RHF/3-21G optimized structures of Li$^+$ bound 2β,3β-diphenylcyclopropane-1α-carboxamides of l-valine methyl ester (a) and R-(+)-3-methyl-2-butylamine (b). Binding affinities are included at the bottom of each structure.

of **47** (with the methyl group) the cation interacts only with the amide carbonyl oxygen via dipolar interaction (as in the case of **43**) and does not interact with the chiral auxiliary part that contains the methyl group. Such an interaction would have no effect on the rotational mobility of the chiral auxiliary.

The zeolite interior is clearly able to enhance the power of a chiral auxiliary. A model based on differences in flexibility of the chiral auxiliary parts imparted by cation binding would account for the observed differences in selectivity between the two classes of chiral auxiliaries (aryl vs. alkyl; carbomethoxy vs. alkyl) within MY zeolites. The prominent interaction between alkali metal ions and the aryl group could be characterized as cation-π interaction, and the interaction between the cation and the carbonyl group would be dipolar in nature. These prominent noncovalent interactions would restrict the freedom of the chiral auxiliary there by bringing about a close interaction between the chiral center and the reactant. In doing so, both the confined space and the cations present in a zeolite are used effectively by the zeolite. The cations and the confined space coerce the molecule to adopt a geometry (conformation) resulting in a stronger interaction between the chiral center and the site of reaction present within a single molecule. A model that could be employed to visualize the zeolite's ability to enhance the power of a chiral auxiliary is represented in Fig. 9. The very low diastereomeric excess for most cases in solution suggests that the conformation of the reactant molecule is likely to be such that the chiral auxiliary and the reactive site of the molecule are far apart (Fig. 9 **top**). Confining the molecule as in siliceous Y-zeolite (Si/Al = 40) would force a moderate interaction between the reactive part and the chiral auxiliary, resulting in moderate selectivity (Fig. 9 **middle**). On the other hand, within the confined space of a zeolite, the cations would force the molecule to adopt a folded conformation there by bringing the chiral environment closer to the reactant part (Fig. 9 **bottom**). The presence of cations within a zeolite, besides restricting the conformational flexibility of the chiral auxiliary, may bring it closer to the reaction center. Weak interactions between chiral auxiliary, cation, and reaction site play an important role in the overall asymmetric induction process within a zeolite.

D. Diastereoselective Norrish–Yang Photocyclization: Restriction of Rotational Isomerism Within Zeolites

In Schemes 30 to 35, products of the irradiation of several systems containing carbonyl chromophores (oxoamides, adamantyl phenyl ketones, α-mesitylacetophenones, and α-benzonorbornyl aceotophenones) are presented [282,301–303]. Primary photoreaction in every one of these cases is intramolecular γ- or δ-hydrogen abstraction. In the majority of these cases the final products of interest are cyclobutanols (Scheme 36). Of the forty compounds examined, the *de* of the cyclobutanols or cyclopentanols in solution is less than 15%. On the other hand,

Isotropic media

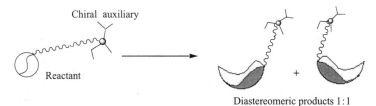

Diastereomeric products 1:1

Confined space (role of confinement)

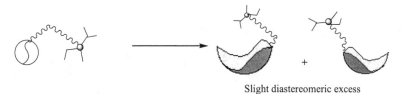

Slight diastereomeric excess

Inside zeolites (role of cations and confinement)

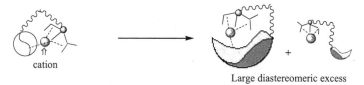

Large diastereomeric excess

Figure 9 A cartoon representation of the mechanism of asymmetric induction in solution, confined media, and confined media, with alkali ions (top to bottom).

the *de*s within zeolites reach as high as 87%. Clearly the zeolite is able to enhance the ability of the chiral auxiliary to tilt the reaction toward one diastereomer. As with the examples discussed in the previous section, the alkali ion present within a zeolite is the key to better performance of a chiral auxiliary within a zeolite. We examine one system (adamantyl phenyl ketones) in detail to illustrate how the cations favor one diastereomer over the other [302]. The conclusions drawn from this system apply to others as well.

In the 2-benzoyladamantyl system there are two prochiral hydrogens (at C_4 and C_6). The computed minimum energy conformation [RB3LYP/6–31G(d)] for 2-methyl-2-benzoyl adamantane indicates that the carbonyl group is tilted toward one of the two prochiral hydrogens (Fig. 10). The enantiomer of this structure having identical energy would be expected to be present in equal

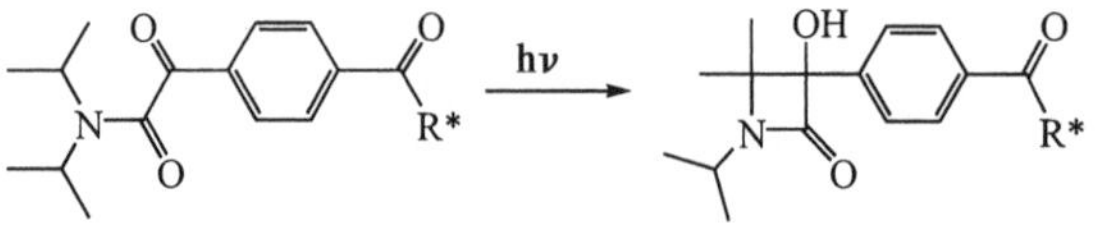

R*	% de				R*	% de			
	CH$_3$CN	NaY	KY	MY		CH$_3$CN	NaY	KY	MY
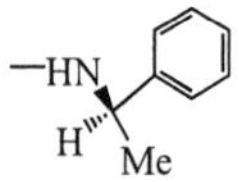 —HN–CH(Me)Ph	2	62	25	16(RbY)	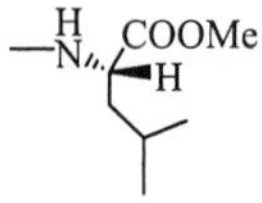 —N(H)–CH(COOMe)CH$_2$CH(Me)$_2$	1	60	11	47(CsY)
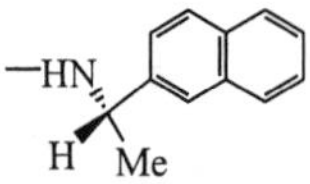 —HN–CH(Me)naphthyl	2	54	60	30(LiY)	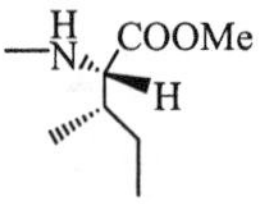 —N(H)–CH(COOMe)CH(Me)CH$_2$CH$_3$	1	20	5	30(RbY)
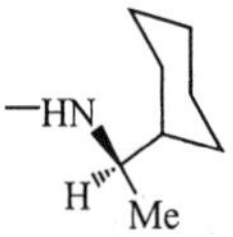 —HN–CH(Me)cyclohexyl	2	22	13	19(LiY)	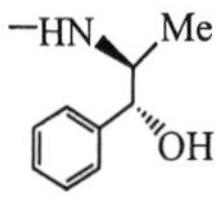 —HN–CH(Me)CH(OH)Ph	5	20	60	25(RbY)
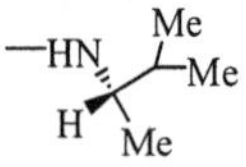 —HN–C(Me)$_2$(Me)	6	30	19	17(LiY)	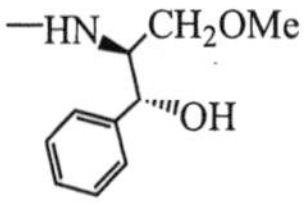 —HN–CH(CH$_2$OMe)CH(OH)Ph	5	68	74	40(LiY)
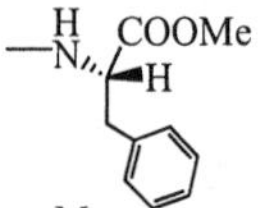 —N(H)–CH(COOMe)CH$_2$Ph	2	83	54	16(RbY)	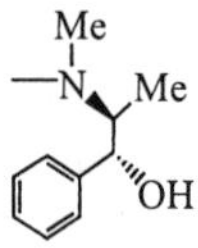 —N(Me)–CH(Me)CH(OH)Ph	5	50	9	41(LiY)
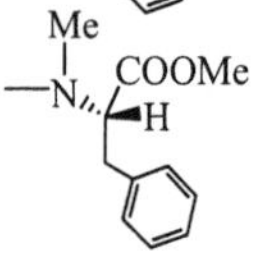 —N(Me)–CH(COOMe)CH$_2$Ph	5	43	12	20(RbY)	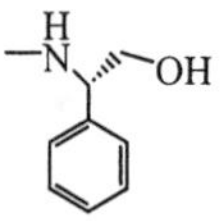 —N(H)–CH(CH$_2$OH)Ph	5	9	10	8(RbY)
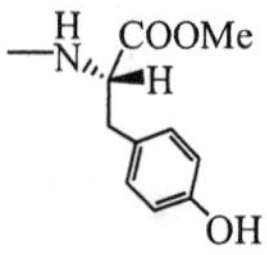 —N(H)–CH(COOMe)CH$_2$C$_6$H$_4$OH	3	40	47	12(RbY)	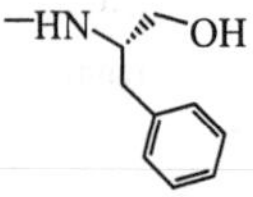 —HN–CH$_2$CH(OH)CH$_2$Ph	15	9	16	20(LiY)
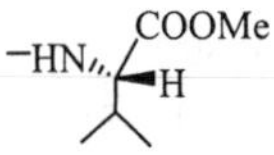 —HN–CH(COOMe)CH(Me)$_2$	6	22	17	25(CsY)					

Scheme 30

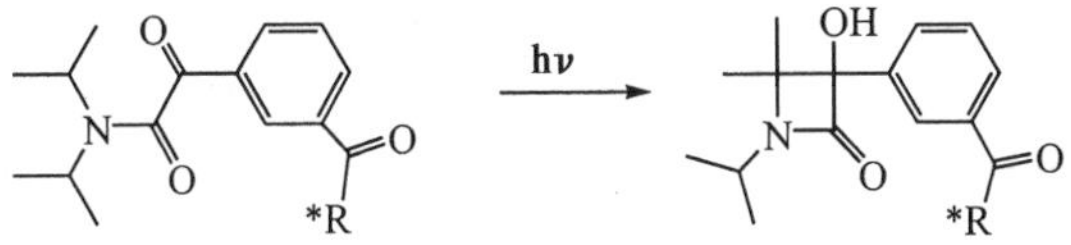

	% de				R*	% de			
R*	CH₃CN	NaY	KY	MY		CH₃CN	NaY	KY	MY

R*	CH₃CN	NaY	KY	MY
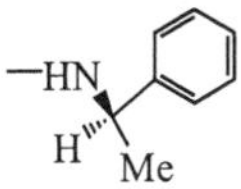 —HN⟨CH(Ph)⟩ H Me	0	45	19	13(LiY)
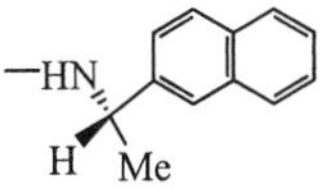 —HN⟨C(naphthyl)⟩ H Me	7	38	1	14(LiY)
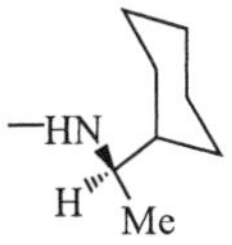 —HN⟨C(cyclohexyl)⟩ H Me	0	25	1	40(RbY)
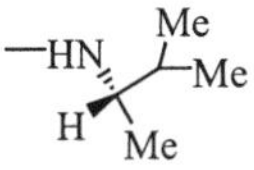 —HN⟨C(Me)(Me)⟩ H Me	2	40	19	44(CsY)

R*	CH₃CN	NaY	KY	MY
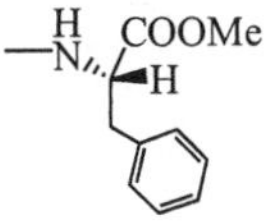 H COOMe —N,,,⟨⟩H CH₂Ph	6	35	9	10(LiY)
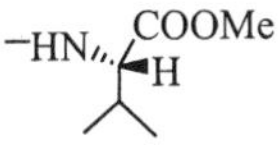 —HN,,,⟨COOMe⟩H iPr	0	49	40	35(RbY)
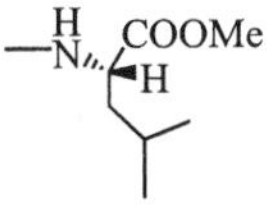 H COOMe —N,,,⟨⟩H sBu	7	83	42	—
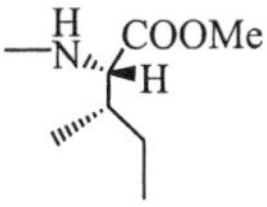 H COOMe —N,,,⟨⟩H	2	45	19	10(RbY)
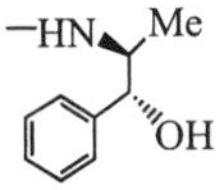 —HN⟨CH(Ph)(Me)⟩ OH	3	25	24	9(LiY)

Scheme 31

amounts in solution. Within a zeolite, the cation is expected to interact with the reactant ketone **54**. The computed structure [RB3LYP/6–31G(d)] for the Li^+-2-methyl-2-benzoyl adamantane complex shown in Fig. 10b also indicates that the carbonyl group is tilted toward one side. Based on symmetry, the two structures in which the carbonyl group is tilted toward either C4 or C6 are of equal energy. The presence of equal amounts of the two structures would yield the endo cyclobutanol as a racemic mixture, and this is what happens in the absence of a chiral influence both in solution and in zeolites. However, the presence of a chiral auxiliary changes the scenario.

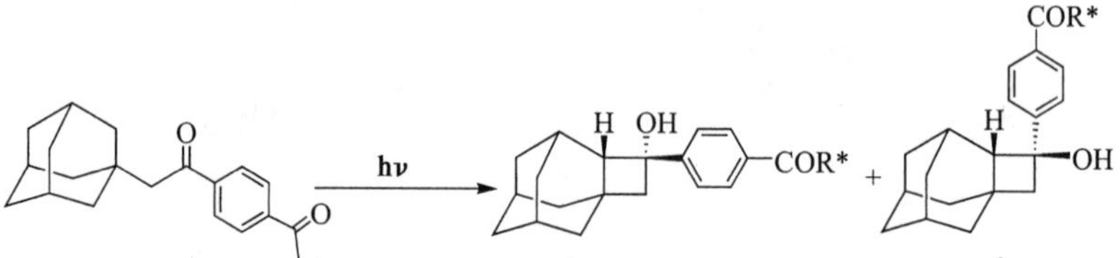

R*	% *de* on *trans*-CB				% *de* on *cis*-CB			
	CH₃CN	LiY	NaY	RbY	CH₃CN	LiY	NaY	RbY
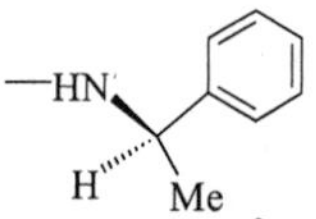 —HN, H, Me (phenyl)	3	29	62	1	–	–	–	–
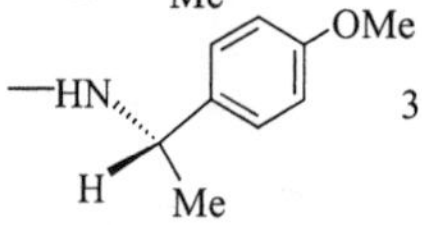 —HN, H, Me (4-OMe phenyl)	3	18	80	5	–	–	–	–
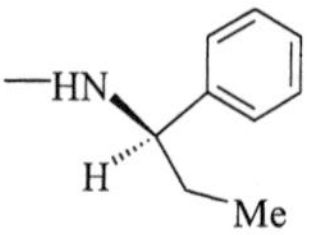 —HN, H, Me (phenyl, CH₂)	5	25	55	15	5	4	62	11
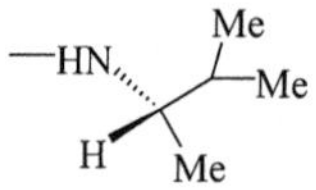 —HN, H, Me, Me, Me	1	13	10	27	1	6	27	5
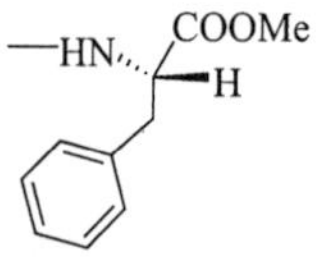 —HN, COOMe, H (benzyl)	3	30	42	5	–	–	–	–

Scheme 32

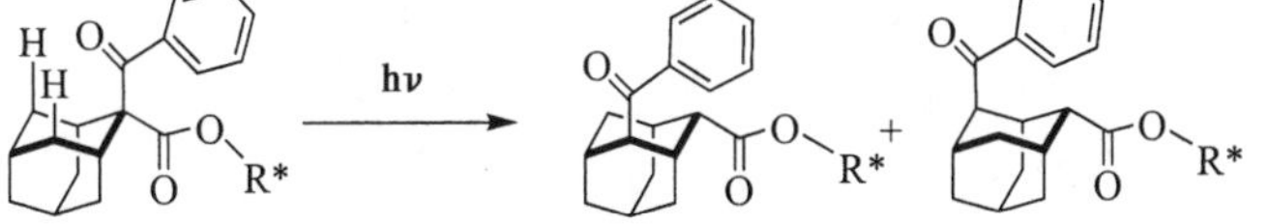

	% de					% de			
R*	CH$_3$CN	LiY	NaY	KY	R*	CH$_3$CN	LiY	NaY	KY
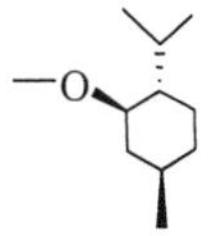 (menthyl structure)	22	79	60	31	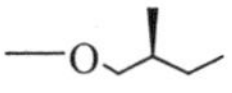 (structure)	5	26	54	30
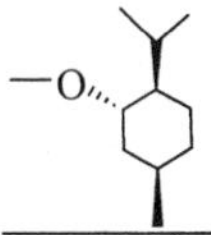 (neomenthyl structure)	14	65	79	3	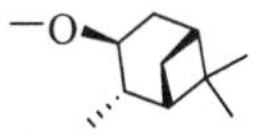 (bornyl structure)	5	52	62	4

Scheme 33

As in achiral ketone **54**, the systems with a chiral auxiliary attached to the C$_2$ carbon should have two conformations in which the carbonyl chromophore is tilted toward one or the other diastereotopic hydrogens at C$_4$ and C$_6$. The two lowest energy conformations computed [B3LYP/6–31G(d)] for the menthyl ester of 2-benzoyladamantane-2-carboxylic acid are shown in Fig. 11. Clearly the carbonyl is tilted toward either side, and the two conformers do not have the same energy. As indicated in Fig. 11, the energy difference between the two structures is 2.3 kcal/mol. This small difference allows equilibrium between the two conformers favoring low de in solution.

Computational results once again provide a clue to what is likely to be responsible for the observed enhancement of de within zeolites. When the ketones are introduced within a zeolite, the cations interact with the chromophores and thus influence the conformation of the molecule. The structures of the Li$^+$ complex of the menthyl ester of 2-benzoyladamantane-2-carboxylic acid was computed at the B3LYP/6–31G(d) level. Of the various computed structures the one in which the cation is interacting with both keto oxygen and ester oxygen is the most stable (Fig. 12). As seen in Fig. 12, the keto oxygen is tilted toward one of the two prochiral hydrogens. The two conformers in which the carbonyl is tilted toward either side do not have the same energy. Closer examination indicates that the nature of interaction between Li$^+$ and menthyl ester of 2-benzoylada-

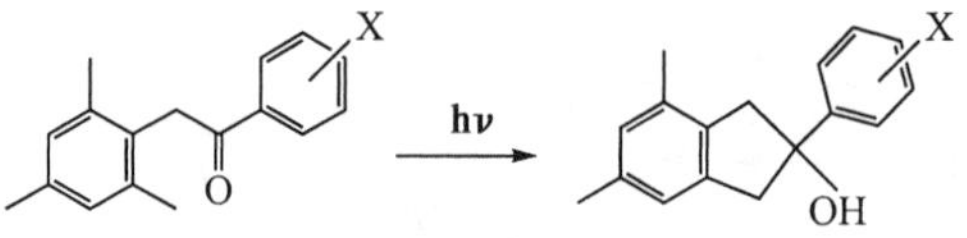

$$X = para\text{-}\overset{\overset{\textstyle O}{\|}}{C}\text{-}R^*$$
$$X = meta\text{-}\overset{\overset{\textstyle O}{\|}}{C}\text{-}R^*$$

R*	% de				R*	% de			
	CH₃CN	NaY	KY	RbY		CH₃CN	NaY	KY	RbY
—HN–CH(Me)(Ph)	1	60	42	10	—HN–CH(Me)(Ph)	7	32	45	20
—HN–CH(Me)–C(Me)₂Me	0	56	22	12	—HN–CH(Me)–C(Me)₂Me	1	26	43	54
—HN–CH(Me)–CH₂OMe	6	28	25	10	—HN–CH(Me)–CH₂OMe	1	30	8	7
—HN–CH(COOMe)(iPr)	4	11	58	62	—N(H)–CH(COOMe)(iPr)	2	74	60	23
—N(H)–CH(COOMe)(CH₂iPr)	2	7	15	45	—N(H)–CH(COOMe)(CH₂iPr)	4	57	23	5
—HN–CH(Me)–CH(Ph)(OH)	5	32	30	45	—N(Me)–CH(Me)(Ph)	1	52	50	18

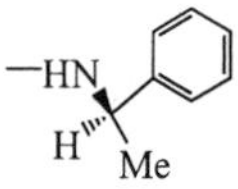

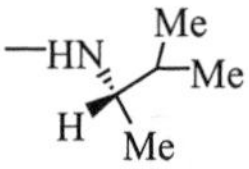

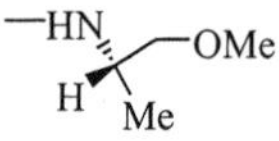

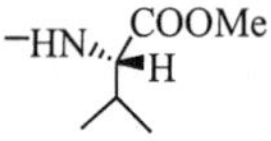

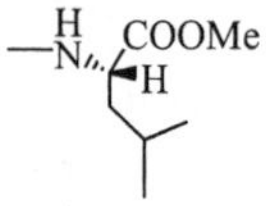

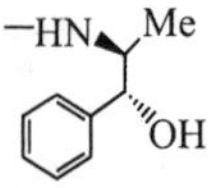

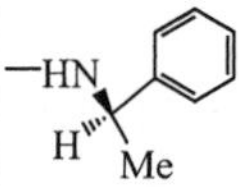

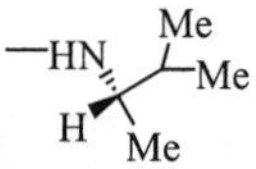

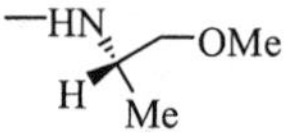

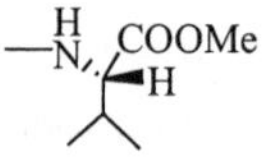

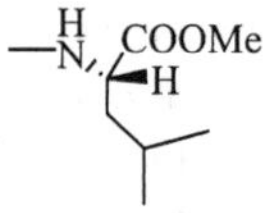

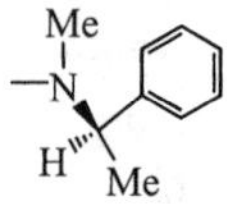

Scheme 34

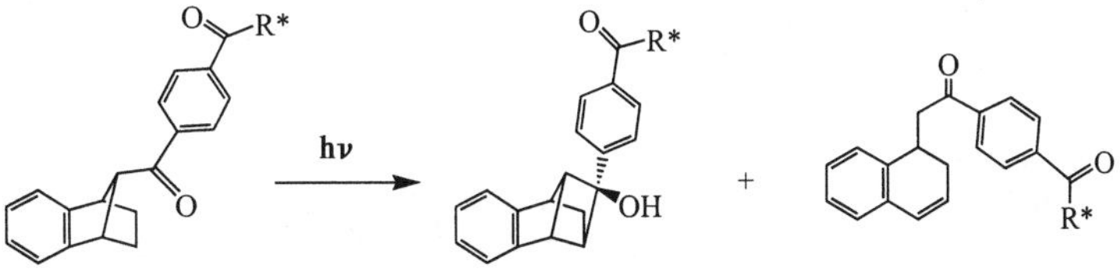

*endo*cyclobutanol + cleavage

R*	% *de* (*endo*CB)				R*	% *de* (*endo*CB)			
	CH₃CN	NaY	RbY	CsY		CH₃CN	NaY	CsY	MY
—HN⟨Me Me Me⟩ H	1	15	37	48	—HN⟨CH₂OME, OH⟩ phenyl	0	73	35	27(LiY)
—HN⟨cyclohexyl Me⟩ H	0	45	51	43	—N⟨H, OH⟩ isopropyl	1	23	64	57(RbY)
—N⟨H, COOMe, H⟩ phenyl	5	87	42	37	—N⟨Me, Me, OH⟩ phenyl	0	5	30	80(RbY)
—HN⟨OH⟩ phenyl	2	64	33	75	—HN⟨Me, OH⟩ phenyl	10	62	65	63(LiY)

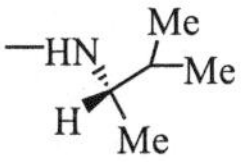

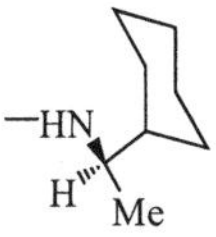

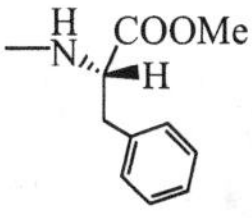

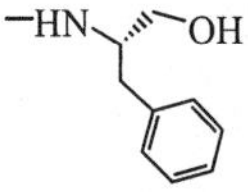

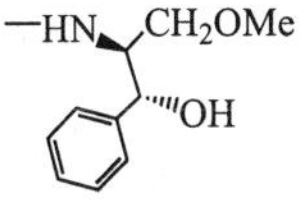

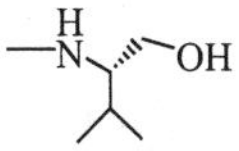

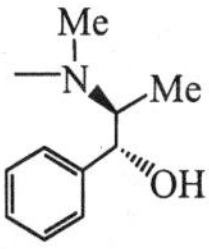

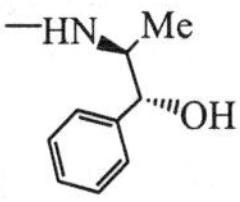

Scheme 35

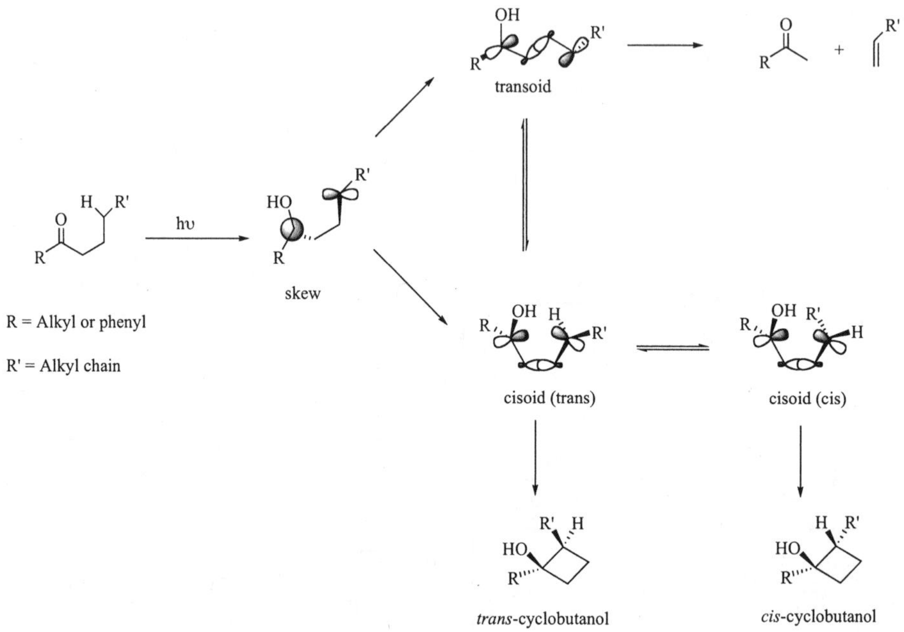

Scheme 36

mantane-2-carboxylic acid is different in the two conformers. In structure (a) the interaction is between the cation and the oxygens of the keto and ester group (CO-O-C), and in structure (b) the interaction is between cation and the oxygens of the two carbonyls of the keto and ester groups. The latter is more stable by 10.42 kcal/mol (compare with 2.3 kcal/mol in the absence of cation). In these structures the cation acts as glue to restrict the relative motions of the reactive and chiral auxiliary portions of the molecule (Fig. 12). By this process the chiral auxiliary is able to exert a stronger influence on the γ-hydrogen abstraction reaction. Once interconversion between the two conformers is restricted, the reaction will take place from the most stable of the two. Thus while the chiral auxiliary is essential to differentiate the two diastereotopic hydrogens, the cation is important to restrict rotations and freeze the molecule in a conformation that leads primarily to one cyclobutanol diastereomer. Computational results on oxoamides and α-benzonorbornyl aceotophenones suggest that a similar phenomenon is responsible for higher *de* in these systems as well.

    The examples presented above demonstrate convincingly that the confined space of a zeolite can serve as a useful medium to achieve asymmetric induction

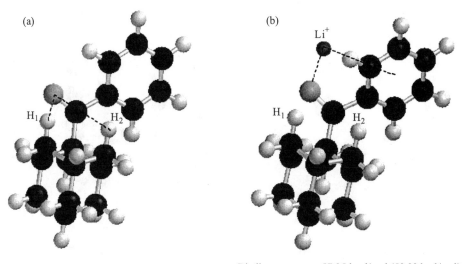

Binding energy = -57.35 kcal/mol (59.29 kcal/mol)

| C=O - H$_1$ | 2.48 A°(2.38) | Li$^+$—O=C | 1.84 A°(1.85) |
| C=O - H$_2$ | 3.46 A°(3.31) | Li$^+$---Ph | 2.01 A°(1.83) |

Figure 10 Conformation of (a) 2-methyl-2-benzoyl adamantane and its complex with Li$^+$ as computed at RB3LYP/6-31G(d) level. Note that the benzoyl carbonyl is tilted toward one of the two prochiral hydrogens. The distances between the carbonyl oxygen and the prochiral hydrogens are included. The binding energy computed at the RHF/3-21G* level is given in parentheses.

during a photoreaction. Consistently higher *de*s have been obtained within zeolites than in solution. The confined space and the cations present within zeolites are believed to be responsible for asymmetric induction.

V. ENANTIOSELECTIVITY WITHIN ZEOLITES THROUGH THE USE OF NONREACTIVE CHIRAL INDUCTORS

A. Background

The two methods described above have restrictions. In the first type, one is restricted to reactions wherein the chiral inductor also acts as a reactant. In the second type, the chiral auxiliary needs to be covalently linked to the reactant and delinked from the product. Most of the examples provided above use either ester

(a)   (b)

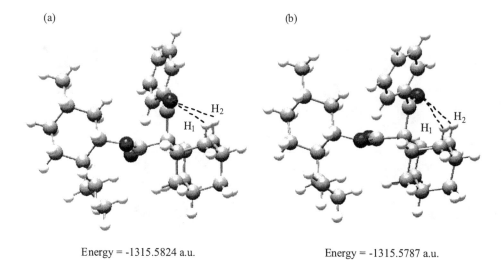

Energy = -1315.5824 a.u.   Energy = -1315.5787 a.u.

Difference in energy is 2.3 kcal/mol (3.1 kcal/mol)

C=O -H$_1$	2.49 A°(2.33)	C=O -H$_1$	3.65 A° (3.35)
C=O -H$_2$	3.48 A°(3.24)	C=O -H$_2$	2.65 A° (2.40)

Figure 11 The two conformations of menthyl ester of 2-benzoyladamantane-2-carboxylic acid as computed at the RB3LYP/6-31G(d) level. Note that the two conformations in which the carbonyl group is tilted toward the two prochiral hydrogens have different energies. The distances between the carbonyl oxygen and the prochiral hydrogens are included. The difference in energy computed at the RHF/3-21G* level is given in parentheses.

or amide functionality to attach a chiral auxiliary. The best general approach would be to obtain chiral products from an achiral reactant with the help of a chiral inductor of choice. In this approach the chiral inductor's function is restricted to being a spectator. We provide a number of examples below that illustrate that zeolites have the ability to force an interaction between an achiral reactant and a chiral inductor. In most of these examples, the enantiomeric excesses obtained are moderate at best. That the *ee* is not zero suggests that zeolite provides an environment different from that in solution, and further work could be rewarding.

The chiral inductors used in these investigations are selected based on several criteria, including availability, size, functionality, and photostability. An in-

(a) (b)

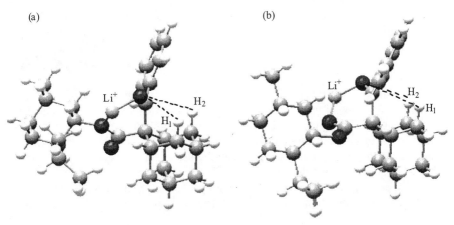

Binding energy = -69.90 kcal/mol (82.89 kcal/mol)   Binding energy = -80.32 kcal/mol (90.36 kcal/mol)

Li⁺-O=C·Ph	1.84 A°(1.81)	Li⁺-O=C·Ph	1.86 A°(1.84)
Li⁺-O–CO	1.92 A°(2.13)	Li⁺-O=C·O –	1.86 A°(1.82)
C=O--H₁	2.55 A°(2.52)	–C=O -H₁	3.79 A°(3.79)
C=O--H₂	3.51 A°(3.45)	–C=O -H₂	2.78 A°(2.79)

Figure 12   The most stable structure computed (RB3LYP/6-31G(d)) for the Li^+ complex with menthyl ester of 2-benzoyladamantane-2-carboxylic acid. In the two structures the sites of interaction of Li^+ are different. The carbonyl is tilted toward different prochiral hydrogens in the two structures. Note that the interaction energies are considerably different for the two structures. The interaction energies computed at the RHF/3-21G* level are given in parentheses.

ductor has to be small enough to fit into the interior of the zeolite and small enough to accommodate another guest molecule in the same supercage. Also, a chiral inductor should have functional groups that are capable of interacting with the zeolitic interior as well as with the functional groups of the substrate.

B.   Enantioselective Norrish-Yang Photocyclizations

In the context of chiral induction in photochemical reactions, the Norrish–Yang cyclization, which has been extensively studied in isotropic and anisotropic media, has been examined. Both cis and trans cyclobutanols exist as pairs of enantiomers (Scheme 36). These reactions are good candidates for photoreactions inside zeolites, wherein under appropriate conditions, one of the enantiomeric cyclobutanols can be formed preferentially. The results of asymmetric Norrish–Yang cyclization of several aryl alkyl ketones are briefly described below. In solution, all molecules, even in the presence of chiral inductors, gave the cyclobutanols as racemic mixtures.

Equation (1) in Scheme 37 represents the conversion of *cis*-4-*tert*-butyl-cyclohexyl ketones of general structure **49** into the corresponding cyclobutanols of structure **50**. Ephedrine proved to be the best chiral inductor, giving 25–30% enantiomeric excess [287,304,305]. The use of (+)-ephedrine afforded the optical antipode of the photoproduct produced by the use of (−)-ephedrine, indicating that the system is well behaved.

Ketone **51** undergoes Norrish–Yang cyclization, giving cyclobutanols as products (Eq. (2), Scheme 37). Unlike **49**, which gave only one cyclobutanol

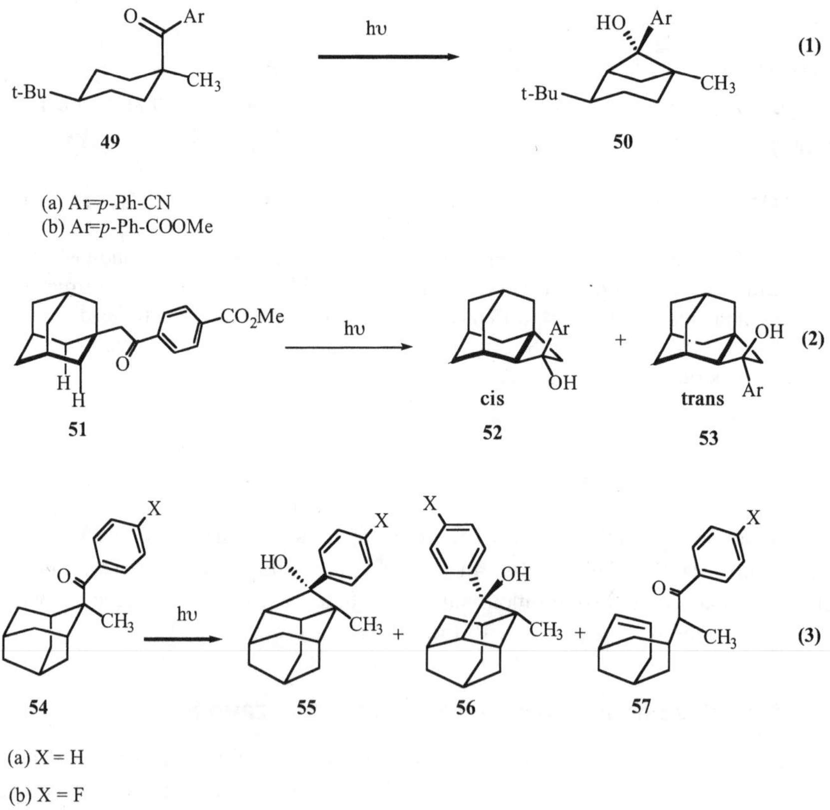

(a) Ar=*p*-Ph-CN
(b) Ar=*p*-Ph-COOMe

(a) X = H

(b) X = F

Scheme 37

diastereomer both in solution and in zeolite, ketone **51** gave both *cis*-cyclobutanol **52** and *trans*-cyclobutanol **53**. Within NaY, while **52** was formed in moderate enantioselectivity (35%), **53** was formed in low enantioselectivity (5%) [305].

The irradiation of phenyl-2-methyltricyclo[3.3.1.1]dec-2-yl methanone **54a** (Scheme 37) included in faujasite was carried out in presence of chiral inductors. The maximum enantioselectivity obtained using the chiral inductor approach was 32% *ee* both with (−)-pseudoephedrine and with (+)-2-amino-3-methoxy-1-phenyl-1-propanol [302]. Similar studies were done for 1-(4-fluorophenyl)-2-methyltricyclo[3.3.1.1]dec-2-yl methanone **54b**. The maximum enantioselectivity obtained in this case was 30% *ee* with (−)-pseudoephedrine.

Enantiomeric excesses obtained in product cyclobutanols from achiral α-oxoamide, and in cyclopentanols from α-mesityl acetophenone, are listed in Schemes 38–40. In all cases the products in solution, even in presence of chiral inductors, are racemic. Enantiomeric excess in the range of 30% is routinely obtained within NaY modified with chiral inductors. The examples presented in this section show that chiral inductors function better within a zeolite than in solution.

C. Enantioselective Photorearrangements

Photoisomerization of 1,2-diphenylcyclopropane has played a central role in photochemical asymmetric induction processes. A number of such systems have been examined within zeolites. Ethyl ester **40** undergoes photoisomerization as shown in Scheme 41. The reaction in an isotropic medium gave a racemic mixture of the corresponding trans isomer. Enantiomeric excess of 17% was achieved by the chiral inductor approach with cyclohexylethylamine [298]. Compound **41** is similar to **40**, except that the ester group is changed to a keto group. Upon excitation, **41** is converted to the chiral trans isomers (Scheme 41). Within NaY, 20% enantiomeric excess was achieved with norephedrine as the chiral inductor [299].

6,6-Dimethyl-2,4-cyclohexadienone photorearranges to bicyclic [3.1.0] product via oxa-di-π-methane rearrangement (Scheme 41). This achiral molecule photorearranges in solution to the chiral bicyclic product as a racemic mixture. This molecule within NaY zeolites in the presence of chiral inductors yields products with respectable *ee*. The best numbers are 50% (ephedrine as a chiral inductor at − 55°C) and 28% (pseudoephedrine at room temperature) [294]. These numbers are significant when one recognizes that in solution only a racemic mixture is obtained.

Photobehavior of *N*-alkylpyridones within zeolites provides support to the claim that chirally modified zeolite is a useful chiral medium (Scheme 42) [292]. Of the three examples provided, two with phenyl substitution yield cyclized products in moderate enantioselectivity.

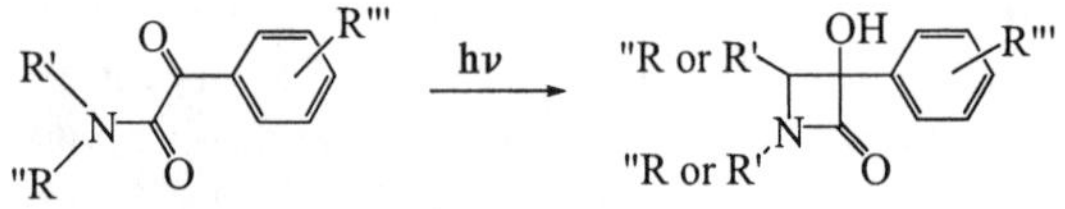

(a) R' = R"= Me   (e) R' = R"= iPr; R'''= p-COOMe
(b) R' = R"= iPr   (f) R' = R"= iPr; R'''= m-COOMe
(c) R' = cyclohexyl; R"= iPr   (g) R' = R"= iPr; R'''= o-COOMe
(d) R' = R"=cyclohexyl

	Reactant	Chiral inductor	Product	%*ee* (MY)
a)	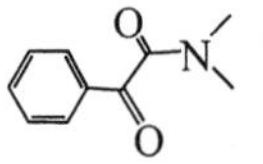	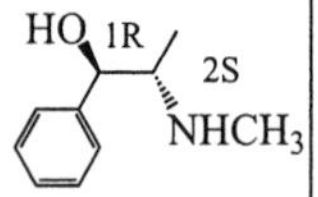	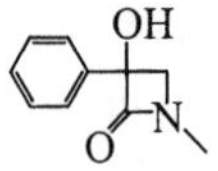	10(NaY)
b)	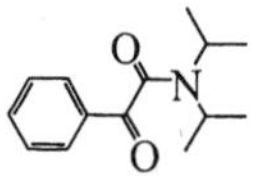	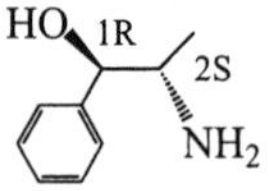	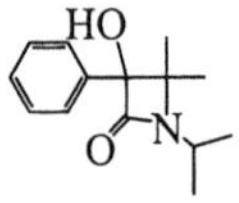	44(NaY)
c)	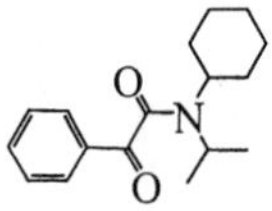	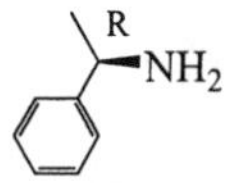	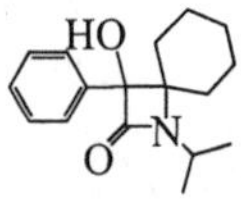	23(NaY)
d)	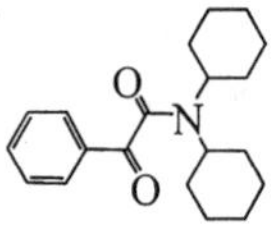	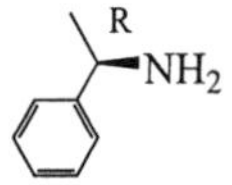	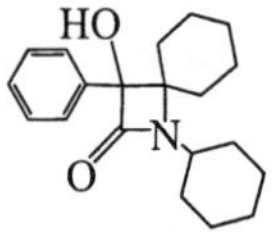	15(NaY)
e)	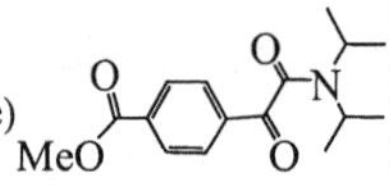	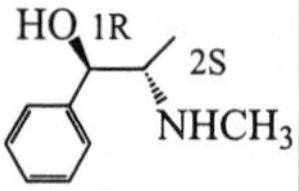	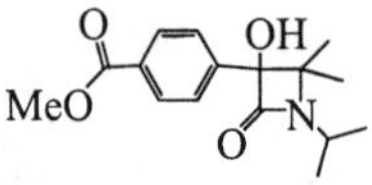	27(NaY)
f)	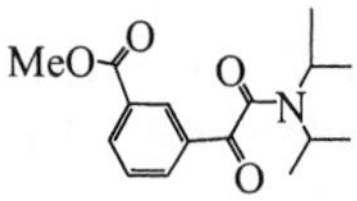	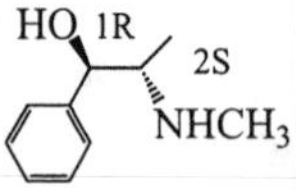	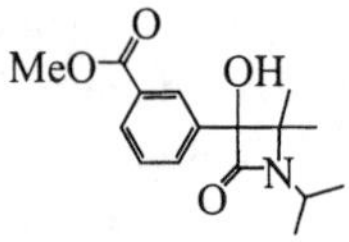	25(NaY) RT= -65°C
g)	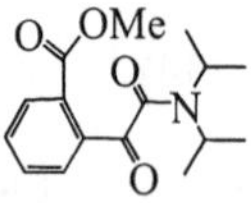	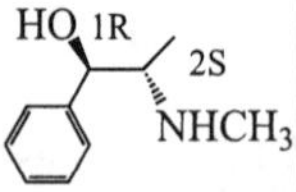	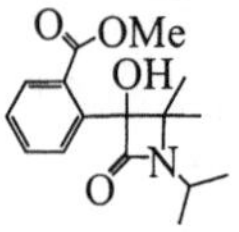	13(KY)

Scheme 38

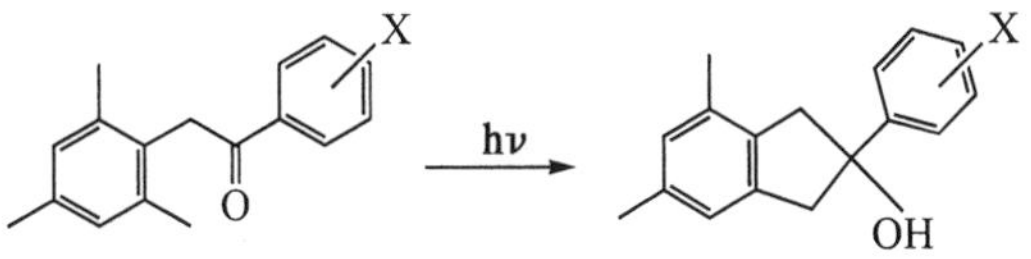

Chiral inductor	MY	%ee		
		X = H	X = p-COOMe	X = m-COOMe
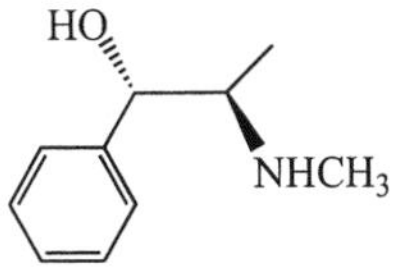 (+)ephedrine	NaY	37	33	41
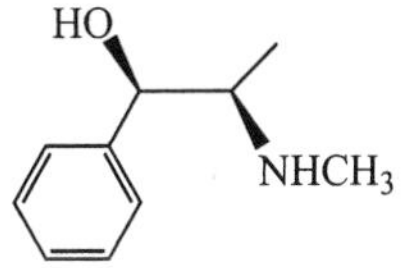 (-)pseudoephedrine	NaY	28	12	33

Scheme 39

If not for the behavior of tropolone ethers one would have come to the conclusion that the zeolite-based method has only very limited potential. Results obtained with tropolone ethers are encouraging and are discussed in detail below. Tropolone alkyl ethers upon excitation undergo 4π disrotatory ring closure to yield racemic products; depending on the mode of disrotation opposite enantiomers are obtained. The enantiomeric excess (*ee*) obtained with tropolone alkyl ethers within NaY is dependent on the chiral agent (Scheme 43), the alkoxy substituent (Scheme 44), the water content within the supercage (Scheme 45), the nature and number of cations (Scheme 45), and the temperature. The high *ee* (69%) obtained with tropolone ethyl phenyl ether is most encouraging (Scheme 45).

In principle, enantiomerically pure cyclized products from tropolone ethers can be obtained by controlling the mode of ring closure. On the basis of this premise, the observed selectivity within zeolites could be rationalized. When tropolone alkyl ether is adsorbed on a surface, cyclization inward or outward is

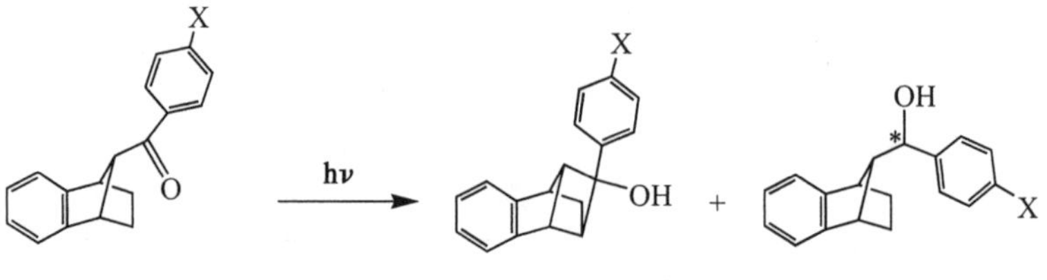

*endo*cyclobutanol reduction product

Chiral inductor	Zeolite	%*ee* *endo*cyclobutanol		%*ee* reduction product	
		X = H	X = F	X = H	X = F
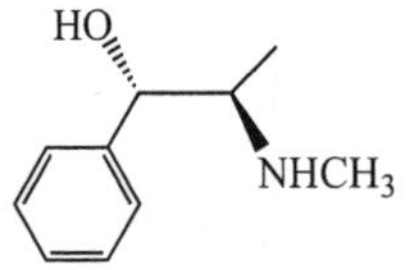 (+)Ephedrine	NaY	30	24	65	57
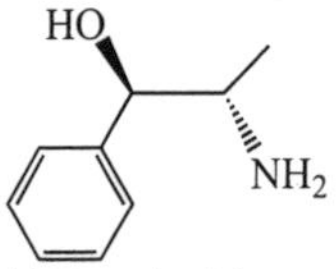 (-)Norephedrine	NaY	40	20	Not formed	

Scheme 40

expected to experience different extents of steric interaction with the surface. Since tropolone alkyl ether is not expected to show a preference for adsorption from one enantiotopic face over the other (Fig. 13), adsorption on a surface by itself is not expected to influence enantioselectivity. This is indeed true, as no *ee* is obtained when a silica surface is used as the medium. To achieve chiral induction, preferential adsorption from one enantiotopic face of the tropolone alkyl ether must be achieved, and this probably occurs on chirally modifying the surface. In Fig. 13, a cartoon representation of how a chiral inductor present on a surface may control the mode of adsorption by tropolone alkyl ether is provided. A surface that can hold the chiral inductor firmly in place is required to achieve the desired goal. Zeolitic cations are expected to interact strongly with chiral inductors and thus present them in certain geometries to the reactant molecule. The above model is preliminary and has no experimental support. At this stage

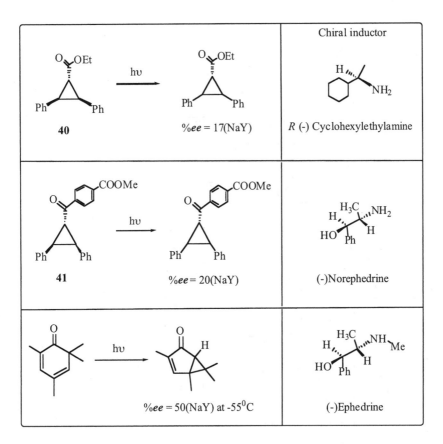

Scheme 41

it only serves as a starting point for understanding the origin of asymmetric induction within zeolites.

Based on the observation that the best *ee* is obtained with bifunctional chiral agents (ephedrine, pseudoephedrine, norephedrine, and valinol; see Scheme 43), we tentatively conclude that a multipoint interaction between the reactant molecule, the chiral inductor, and the zeolite interior is necessary to induce preferential adsorption of tropolone alkyl ether from a single enantiotopic face. The dependence of chiral induction (% *ee*) on the nature of cations (Scheme 45) suggests a crucial role of the cation present in the supercages in the chiral induction process. This is further strengthened by the results observed with wet and dry zeolites. The presence of water decreases chiral selectivity (Scheme 45). Water molecules

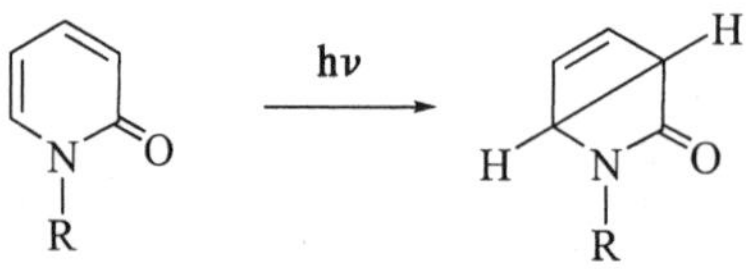

Scheme 42

$$\text{(pyridinone)} \xrightarrow{h\nu} \text{(bicyclic product)}$$

with substituents H, N–R, O as shown.

R	Chiral inductor	%ee MY
CH$_2$Ph	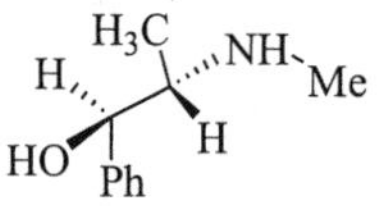 H$_3$C, H, ,,,NH–Me, HO, Ph, H (-) Ephedrine	30(NaY)
CH$_2$CH$_2$Ph	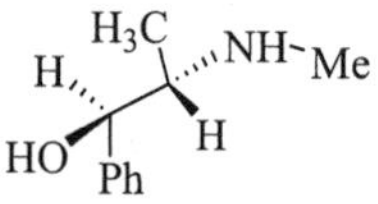 H$_3$C, H, ,,,NH–Me, HO, Ph, H (-) Ephedrine	53(KY)
CH$_2$CH$_2$CH$_2$Ph	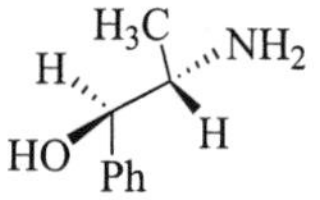 H$_3$C, H, ,,,NH$_2$, HO, Ph, H (-) Norephedrine	50(NaY)

that are expected to hydrate the cation will make the latter less effective in holding the tropolone alkyl ether on the zeolite surface. Also they disrupt the close interaction between the reactant and the chiral inductor. A simple cartoon representation for the influence of water is presented in Fig. 14.

The most promising result was obtained when (S)-tropolone 2-methylbutyl ether (chiral ether) was irradiated within an ephedrine included NaY (Fig. 15) [295,306]. In the absence of ephedrine, diastereomer A is obtained in 53% diastereomeric excess. When (−)ephedrine was used as the chiral inductor, the same isomer was enhanced to the extent of 90%. The importance of this result becomes more apparent when one recognizes that irradiation in solution of the same com-

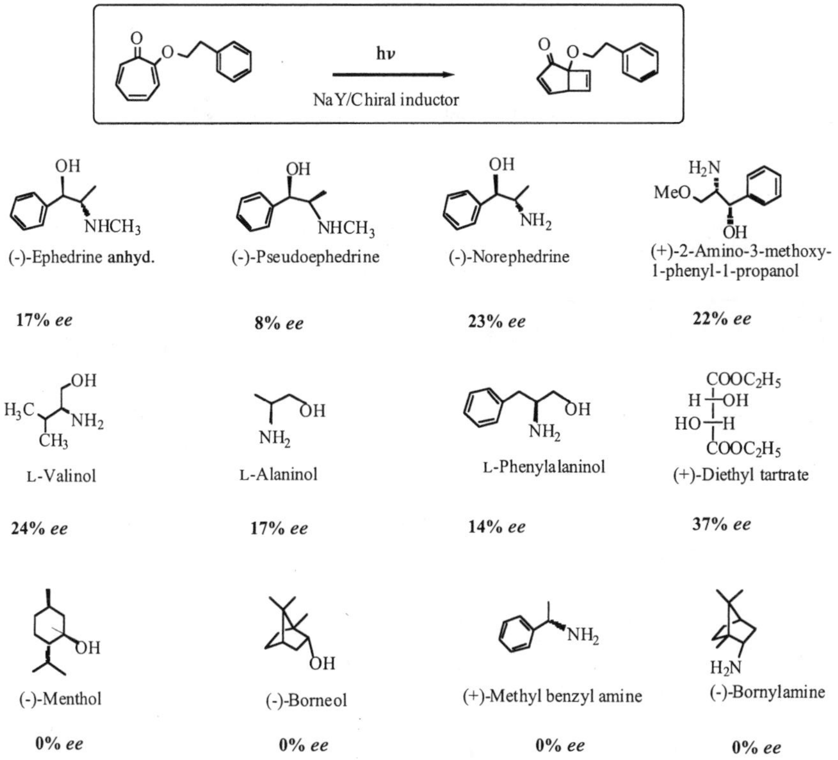
hν
NaY/Chiral inductor
OH
NHCH3
(-)-Ephedrine anhyd.
17% ee
OH
NHCH3
(-)-Pseudoephedrine
8% ee
OH
NH2
(-)-Norephedrine
23% ee
H2N
MeO
OH
(+)-2-Amino-3-methoxy-1-phenyl-1-propanol
22% ee
OH
H3C
NH2
CH3
L-Valinol
24% ee
OH
NH2
L-Alaninol
17% ee
OH
NH2
L-Phenylalaninol
14% ee
COOC2H5
H—OH
HO—H
COOC2H5
(+)-Diethyl tartrate
37% ee
OH
(-)-Menthol
0% ee
OH
(-)-Borneol
0% ee
NH2
(+)-Methyl benzyl amine
0% ee
H2N
(-)-Bornylamine
0% ee

Scheme 43

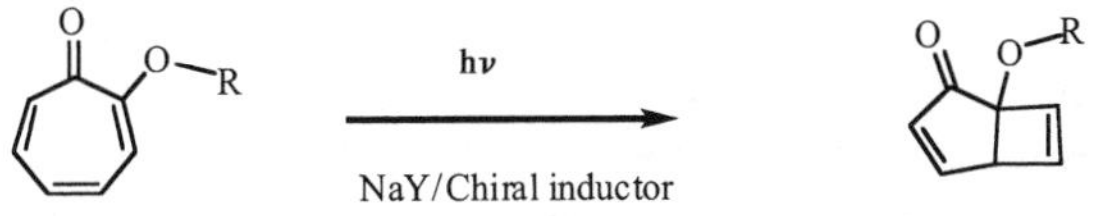
O
O
R
hν
NaY/Chiral inductor

	Chiral inductor	% *ee*	Chiral inductor	% *ee*
R = CH_3	Ephedrine	**17**	Norephedrine	**35**
R = CH_2CH_3	Ephedrine	**8**	Norephedrine	**11**
R = $CH(CH_3)_2$	Ephedrine	**0**	Norephedrine	**6**

Scheme 44

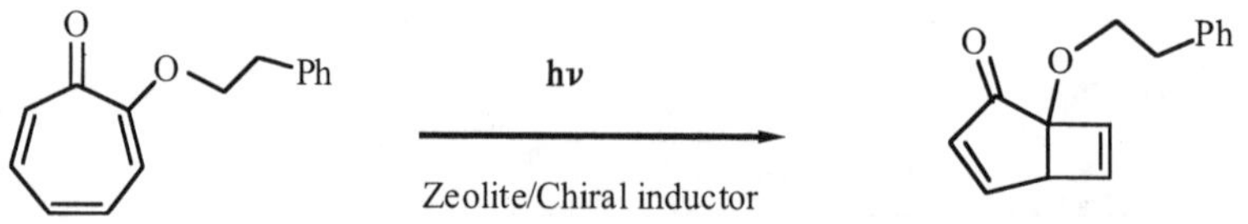

Chiral inductor	% ee	
	wet	dry
Na Y/ (+)Ephedrine	17	69
Na Y/(−) Pseudoephedrine	8	20
Na Y/(−) Norephedrine	23	38
Na X/ (+)Ephedrine	9	
Li Y/ (+)Ephedrine		22
KY/ (+)Ephedrine		11
Rb Y/ (+)Ephedrine		2

Scheme 45

pound in the presence of ephedrine gave a 1 : 1 diastereomeric mixture. It is clear that zeolite is essential to achieve the high *de*.

The results discussed above show that chirally modified zeolite could serve as a chiral medium to achieve low to moderate enantiomeric excess in photochemical reactions. Given that most examples thus far examined give less than moderate *ee*, one tends to get discouraged. At the same time there are a few examples that give respectable *ee*, which provides hope for the success of this approach.

VI. OPTIMISTIC OVERVIEW

The thrust of most studies in asymmetric photochemistry at the present time is to gain an understanding of intermolecular interactions. During the last three

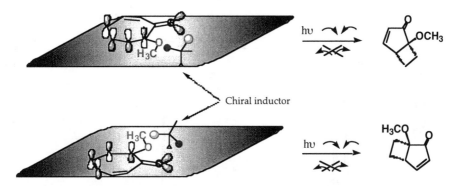

Figure 13 Adsorption of tropolone alkyl ether (TAE) on a surface. The chiral inductor may control the enantiotopic face by which TAE adsorbs. In the absence of a chiral inductor, TAE will show no preference for adsorption from either enantiotopic face. Note that the same chiral inductor interacts differently when the TAE adsorbs through different enantiotopic faces.

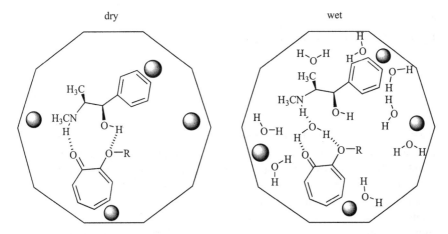

Figure 14 A cartoon representation of tropolone methyl ether and norephedrine included within a supercage under wet and dry conditions. The model helps to rationalize the difference in *ee* obtained under the two conditions. Dark circles represent the cations. Hydrogen bonding between the chiral inductor and interaction between the cation and TME are disturbed by water molecules.

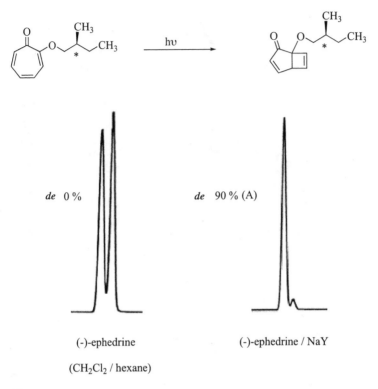

de 0 %   de 90 % (A)

(-)-ephedrine   (-)-ephedrine / NaY

(CH₂Cl₂ / hexane)

Figure 15 Results of studies on the photocyclization of (*S*)-tropolone 2-methyl butyl ether within chirally modified NaY zeolite. The % *de* (diastereomeric excess) and the isomer enhanced are shown on the HPLC traces.

decades, confined media have been utilized to achieve different types and extents of selectivities in photochemical reactions. Current studies dealing with asymmetric photochemistry extend the use of confined media in achieving ultimate selectivity, namely chiral selectivity, in photochemical reactions. Such studies have brought out the use and importance of alkali ions in achieving selectivity in photochemical reactions. In principle, one should be able to achieve similar selectivity in solution with the help of alkali ions. Unfortunately this has not been possible owing to the poor solubility of alkali salts in organic solvents. When alkali ions dissolve, they are coordinated to the solvent molecules and are not free to interact with the reactant molecules. It is in this context that zeolites have been useful. Cations present in zeolites, being only partially coordinated to the surface oxygens, are free to interact with included guest molecules. X and Y zeolites contain high concentrations (~5 M) of exchangeable cations, making them resemble open structures of alkali salts with the zeolite framework as the anion. We believe that alkali ions present in zeolites can be exploited to control the photochemical behavior of organic molecules. The cost and environmentally benign nature of the zeolite and the "greenness" of the reagent (light) used to register the chirality of the product justify further exploration of zeolite-mediated asymmetric induction processes.

ACKNOWLEDGMENTS

VR thanks the National Science Foundation for support of the research (CHE-9904187 and CHE-0212042) presented in this summary. We also thank S. Uppili, G. Sundarababu, R. J. Robbins, K. Ponchot, Y. Yoshimi, and T. Shichi for their experimental contributions.

REFERENCES

1. Rau H. Chem. Rev. 1983; 83:535–547.
2. Demuth M, Mikhail G. Synthesis 1989:145–162.
3. Buschman H, Scharf H-D, Hoffmann N, Esser P. Angew. Chem. Int. Ed. Engl. 1991; 30:477–515.
4. Inoue Y. Chem. Rev. 1992; 92:741–770.
5. Pete J-P. In:. Adv. Photochem. Neckers D. C. , Volman D. H. , Von Bunan G., Eds. Vol. 21: John Wiley, 1996:135–216.
6. Everitt SRL, Inoue Y. In:. Molecular and Supramolecular Photochemistry Ramamurthy V. , Schanze K., Eds. Vol. 3. New York: Marcel Dekker, 1999:71–130.
7. Griesbeck AG, Meierhenrich UJ. Angew. Chem. Int. Ed. Eng. 2002; 41:3147–3154.
8. Rabitz H. Science 2003; 299:525–527.
9. Nishino H, Nakamura A, Shitomi H, Onuki H, Inoue Y. J. Chem. Soc., Perkin Trans. 2 2001:1706–1713.
10. Nishino H, Nakamura A, Inoue Y. J. Chem. Soc., Perkin Trans. 2 2001:1693–1700.
11. Nakamura A, Nishino H, Inoue Y. J. Chem. Soc., Perkin Trans. 2 2001:1701–1705.
12. Zandomeneghi M, Cavazza M, Festa C, Pietra F. J. Am. Chem. Soc. 1983; 105: 1839–1843.
13. Cavazza M, Morganti G, Zandomeneghi M. J. Chem. Soc., Perkin Trans. II 1984: 891–895.
14. Shimizu Y, Kawanishi S. Chem. Commun. 1996:819.
15. Cavazza M, Zandomeneghi M. Gazetta Chimica Italiana 1987; 117:17–21.
16. Kagan H, Moradpour A, Nicoud JF, Balavoine G. Tetrahedron Lett. 1971; 27: 2479–2482.
17. Moradpour A, Kagan H, Baes H, Morren G, Martin RH. Tetrahedron 1975; 31: 2139–2143.

18. Nicoud JF, Eskenzai C, Kagan HB. J. Org. chem 1977; 42:4270–4272.
19. Zandomeneghi M, Cavazza M. J. Am. Chem. Soc 1984; 106:7261–7262.
20. Bernstein JW, Calvin M, Buchardt O. J. Am. Chem. Soc. 1972; 94:494–498.
21. Cavazza M, Zandomeneghi M, Festa C, Lupi E, Sammuri M, Pietra F. Tetrahedron Lett 1982; 23:1387–1390.
22. Bernstein JW, Calvin M, Buchadt O. J. Am. Chem. Soc. 1973; 95:527–532.
23. Blume R, Rau H, Scheneider M, Schuster O. Berichte der Bunsen-Gesellschaft 1977:33–39.
24. Flores JJ, Bonner AW, Massey AG. J. Am. Chem. Soc. 1977; 99:3622–3625.
25. Balavoine G, Moradpour A, Kagan HB. J. Am. Chem. Soc. 1974; 96:5152–5158.
26. Bernstein WJ, Calvin M, Buchardt O. Tetrahedron Lett. 1972; 22:2195–2198.
27. Nishino H, Osaka A, Hembury GA, Aoki F, Miyauchi K, Shitomi H, Onuki H, Inoue Y. J. Am. Chem. Soc. 2002; 124:11618–11627.
28. Inoue Y, Tsuneishi H, Hakushi T, Yagi K, Awazu K, Onuki H. Chem. Commun. 1996:2627–2628.
29. Hammond GS, Cole RS. J. Am. Chem. Soc. 1965; 87:3256–3257.
30. Inoue Y, Shimoyama H, Yamasaki N, Tai A. Chem. Lett. 1991:593–596.
31. Vondenhof M, Mattay J. Chem. Ber. 1990; 123:2457–2459.
32. Becker J, Weiland R, Rau H. J. Photochem. Photobio, A: Chem 1988; 41:311–330.
33. Inoue Y, Yamasaki N, Shimoyama H, Tai A. J. Org. Chem. 1993; 58:1785–1793.
34. Ouannes C, Beugelmans R, Roussi G. J. Am. Chem. Soc. 1973; 95:8472–8474.
35. Faljoni A, Zinner K, Weiss RG. Tetrahedron Lett. 1974; 13:1127–1130.
36. Horner L, Klaus J. Liebigs Ann. Chem. 1981:792–811.
37. Inoue Y, Yokoyama T, Yamasaki N, Tai A. Nature 1989; 341:225–226.
38. Saito R, Kaneda M, Wada T, Katoh A, Inoue Y. Chem. Lett. 2002:860–861.
39. Inoue Y, Yokoyama T, Yamasaki N, Tai A. J. Am. Chem. Soc. 1989; 111:6480–6482.
40. Masuyama K, Inoue T, Inoue Y. Synthesis 2001; 8:1167–1174.
41. Inoue Y, Matsushima E, Wada T. J. Am. Chem. Soc. 1998; 120:10687–10696.
42. Inoue Y, Ikeda H, Kaneda M, Sumimura T, Everitt SRL, Wada T. J. Am. Chem. Soc. 2000; 122:406–407.
43. Inoue T, Matsuyama K, Inoue Y. J. Am. Chem. Soc. 1999; 121:9877–9878.
44. Tsuneishi H, Hakushi T, Tai A, Inoue Y. J. Chem. Soc., Perkin Trans. 2 1995: 2057–2062.
45. Shi M, Inoue Y. J. Chem. Soc., Perkin Trans. 2 1998:2421–2428.
46. Inoue Y, Kunitomi Y, Takamuku S, Sakurai H. Chem. Commun. 1978:1024–1025.
47. Inoue Y, Takamuku S, Kunitomi Y, Sakurai H. J. Chem. Soc., Perkin Trans. 2 1980:1672–1677.
48. Inoue Y, Yamasaki N, Yokoyama T, Tai A. J. Org. Chem. 1992; 57:1332–1345.
49. Inoue Y, Yamasaki N, Yokoyama T, Tai A. J. Org. Chem. 1993; 58:1011–1018.
50. Inoue Y, Sugahara N, Wada T. Pure Appl. Chem. 2001; 73:475–480.
51. Tsuneishi H, Hakushi T, Inoue Y. J. Chem. Soc., Perkin Trans. 2 1996:1601–1605.
52. Inoue Y, Wada T, Asaoka S, Sato H, Pete J-P. Chem. Commun. 2000:251–259.
53. Kohomoto S, Ono Y, Masu H, Yamaguchi K, Kishikawa K, Yamamoto M. Org. Lett. 2001; 3:4153–4155.

54. Green SB, Rabinsohn Y, Rejto M. Chem. Commun. 1975:313–314.
55. Langer K, Mattay J. J. Org. Chem 1995; 60:7256–7266.
56. Asaoka S, Ooi M, Jiang P, Wada T, Inoue Y. J. Chem. Soc., Perkin Trans. 2 2000: 77–84.
57. Asaoka S, Horiguchi H, Wada T, Inoue Y. J. Chem. Soc., Perkin Trans. 2 2000: 737–747.
58. Inoue Y, Okano T, Yamasaki N, Tai A. J. Photochem. Photobiol., A: Chem. 1992; 66:61–68.
59. Dopp D, Pies M. Chem. Commun. 1987:1734–1735.
60. Demuth M, Palomer A, Sluma HD, Dey AK, Kruger C, Tsay YH. Angew. Chem. Int. Ed. Engl. 1986; 25:1117–1119.
61. Faure S, Piva-le-Blanc S, Bertrand C, Pete JP, Faure R, Piva O. J. Org. Chem. 2002; 67:1061–1070.
62. Lange GL, Humber CC, Manthorpe JM. Tetrahedron: Asymm. 2002:1355–1362.
63. Sato M, Abe Y, Takayama K, Sekiguchi K, Kaneko C, Inoue N, Furuya T, Inukai N. J. Heterocycl. Chem. 1991; 28:241–52.
64. Bonvalet C, Bouquant J, Feigenbaum A, Pete JP, Scholler D. Bull. Soc. Chim. Fr. 1994; 131:687–692.
65. Demuth M. Pure Appl. Chem. 1986; 58:1233–1238.
66. Vaen R, Runsink J, Scharf H-D. Chem. Ber. 1986; 119:3492–3497.
67. Horner L, Klaus J. Liebigs Ann. Chem. 1979:1232–1257.
68. Sato M, Takayama K, Abe Y, Furaya T, Inukai N, Kaneko C. Chem. Pharm. Bull 1990; 38:336–339.
69. Termont D, Keukeleire DD, Vanderwalle M. J. Chem. Soc., Perkins Trans. I 1977; 87:2349–2353.
70. Carreira EM, Hastings CA, Shepard MS, Yerkey LA, Millward DB. J. Am. Chem. Soc. 1994; 116:6662–6630.
71. Sato M, Takayama K, Furuya T, Inukai N, Kaneko C. Chem. Pharm. Bull. 1987; 35:3971–3974.
72. Kaneko C, Sato M, Skaki J-I. J. Heterocycl. Chem. 1990; 27:25–30.
73. Shepard MS, Carreira EM. J. Am. Chem. Soc. 1997; 119:2597–2605.
74. Winkler JD, Shao B. Tetrahedron lett. 1993; 34:3355–3358.
75. Organ GM, Froese DJ, Goddard JD, Taylor NJ, Lange GL. J. Am. Chem. Soc. 1994; 116:3312–3323.
76. Lange GL, Decicco C, Tan SL, Chamberlain G. Tetrahedron Lett. 1985; 26: 4707–4710.
77. Schreiber SL, Hoveyda AH, Wu H-J. J. Am. Chem. Soc. 1983; 105:660–661.
78. Lange GL, Lee M. Tetrahedron Lett. 1985; 26:6163–6166.
79. Lange GL, Decicco CP. Tetrahedron Lett. 1988; 29:2613–14.
80. Lange GL, Decicco C, Lee M. Tetrahedron Lett. 1987; 28:2833–6.
81. Blanc LS, Pete J-P, Piva O. Tetrahedron Lett. 1993; 34:635–638.
82. Amougay A, Pete J-P, Piva O. Tetrahedron. Lett. 1992; 33:7347–7350.
83. Faure S, Blanc SP-L, Piva O, Pete J-P. Tetrahedron Lett. 1997; 38:1045–1048.
84. Crimmins MT, Choy AL. J. Am. Chem. Soc. 1997; 119:10237–10238.
85. Chung W-S, Turro JN, Mertes J, Mattay J. J. Org. Chem. 1989; 54:4881–4887.

86. Meyers AI, Fleming SA. J. Am. Chem. Soc. 1986; 108:306–307.
87. Bruneel K, De Keukeleire D, Vandewalle M. J. Chem. Soc., Perkin Trans. I 1984: 1697–1700.
88. Gotthardt H, Lenz W. Tetrahedron Lett. 1979; 31:2879–2880.
89. Gotthardt H, Lenz W. Angew. Chem. Int. Ed. Engl. 1979; 18:868.
90. Jarosz S, Zamojski A. Tetrahedron 1982; 38:1447–1451.
91. Oppenlander T, Schonholzer P. Helv. Chim. Acta 1989; 72:1792–1800.
92. Tolbert LM, Ali MB. J. Am. Chem. Soc. 1982; 104:1742–1744.
93. Jarosz S, Zamojski A. Tetrahedron 1982; 38:1453–1456.
94. Griesbeck AG, Stadtmuller S. Chem. Ber. 1990; 123:357–362.
95. Koch H, Runsink J, Scharf H-D. Tetrahedron Lett. 1983; 24:3217–3220.
96. Nehrings A, Scharf H-D, Runsink J. Angew. Chem. Int. Ed. Engl. 1985; 24: 877–878.
97. Koch H, Scharf H-D, Runsink J, Leismann H. Chem. Ber. 1985; 118:1485–1503.
98. Runsink J, Koch H, Nehrings A, Scharf H-D, Nowack E, Hahn T. J. Chem. Soc., Perkin Trans. 2 1988:49–55.
99. Herzog H, Koch H, Scharf H-D, Runsink J. Tetrahedron 1986; 42:3547–3558.
100. Herzog H, Koch H, Scharf H-D, Runsink J. Chem. Ber. 1987; 120:1737–1740.
101. Weuthen M, Scharf H-D, Runsink J, Vaen R. Chem. Ber. 1988; 121:971–976.
102. Pelzer R, Jutten P, Scharf H-D. Chem. Ber. 1989; 122:487–491.
103. Buschmann H, Scharf H-D, Hoffmann N, Plath MW, Runsink J. J. Am. Chem. Soc. 1989; 111:5367–5373.
104. Hoffmann N, Scharf H-D, Runsink J. Tetrahedron Lett. 1989; 30:2637–2638.
105. Zagar C, Scharf H-D. Chem. Ber. 1991; 124:967–969.
106. Buschmann H, Hoffmann N, Scharf H-D. Tetrahedron Asym. 1991; 2:1429–1444.
107. Hoffmann N, Buschmann H, Raabe G, Scharf H-D. Tetrahedron 1994; 50: 11167–11186.
108. Haag D, Scharf H-D. J. Org. Chem. 1996; 61:6127–6135.
109. Adam W, Stegmann VR, Weinkotz S. J. Am. Chem. Soc. 2001; 123:2452–2453.
110. Kang T, Scheffer JR. Org. Lett. 2001; 3:3361–3364.
111. Bach T, Pelkmann C, Harms K. Tetrahedron Lett. 1999; 40:2103–2104.
112. Bach T, Schorder J, Harms K. Tetrahedron Lett. 1999; 40:9003–9004.
113. Bach T, Schroder J, Brandl T, Hecht J, Harms K. Tetrahedron 1998; 54:4507–4520.
114. Adam W, Stegmann VR. Synthesis 2001; 8:1203–1214.
115. Bach T, Jodicke K, Wibbeling B. Tetrahedron 1996; 52:10861–10878.
116. Bach T, Jodicke K, Kather K, Hecht J. Angew. Chem. Int. Ed. Engl. 1995; 20: 2271–2273.
117. Bach T, Schroder J. J. Org. Chem. 1999; 64:1265–1273.
118. Bach T, Brummerhop H. Angew. Chem. Int. Ed. Engl. 1998; 37:3400–3402.
119. Adam W, Peters K, Peters EM, Stegmann VR. J. Am. Chem. Soc. 2000; 122: 2958–2959.
120. Bach T, Jodicke K. Chem. Ber. 1993; 126:2457–2466.
121. Bach T. Tetrahedron Lett. 1994; 35:5845–5848.
122. Bach T, Jodicke K, Kather K, Frohlich R. J. Am. Chem. Soc. 1997; 119:2437–2445.
123. Pelzer R, Scharf H-D, Buschmann H, Runsink J. Chem. Ber. 1989; 122:1187–1192.

124. Brunker H-G, Adam W. J. Am. Chem. Soc. 1995; 117:3976–3982.
125. Adam W, Brunker H-G, Kumar AS, Peters E-M, Schneider U, Schnering HGv. J. Am. Chem. Soc. 1996; 118:1899–1905.
126. Gerdil R, Barchietto G, Jefford CW. J. Am. Chem. Soc. 1984; 106:8004–8005.
127. Linker T, Frohlich L. J. Am. Chem. Soc. 1995; 117:2694–2697.
128. Wyss C, Rohit B, Lehmann C, Sauer S, Giese B. Angew. Chem. Int. Ed. Engl. 1996; 35:2529–2531.
129. Griesbeck AG, Heckroth H, Schmickler H. Tetrahedron Lett. 1999; 40:3137–3140.
130. Sauer S, Schumacher A, Barbosa F, Giese B. Tetrahedron Lett. 1998; 39: 3685–3688.
131. Griesbeck AG, Kramer W, Lex J. Angew. Chem. Int. Ed. Eng. 2001; 40:577–579.
132. Bach T, Aechtner T, Neumuller B. Chem. Commun. 2001:607–608.
133. Weigel W, Schiller S, Henning H-G. Tetrahedron 1997; 53:7855–7866.
134. Lindemann U, Neuburger M, Neuburger-Zehnder M, Wulff-Molder D, Wessig P. J. Chem. Soc., Perkin Trans. 2 1999; 10:2029–2036.
135. Lindemann U, Wulff-Molder D, Weissg P. J. Photochem. Photobio., A: Chemistry 1998; 119:73–83.
136. Lindemann U, Wulff-Molder D, Wessig P. Tetrahedron Asymm. 1998; 9: 4459–4473.
137. Giese H, Wettstein P, Stahelin C, Barbosa F, Neuburger M, Zehnder M, Wessig P. Angew. Chem. Int. Ed. Eng. 1999; 38:2586–2587.
138. Wessig P, Wettstein P, Neuburger M, Zehnder M. Heliv. Chim. Acta 1994; 77: 829–837.
139. Steiner A, Wessig P, olborn K. Helv. Chim. Acta 1996; 79:1843–1862.
140. Green BS, Heller L. Science 1974; 185:525–527.
141. Addadi L, Lahav M. Pure Appl. Chem. 1979; 51:1269–1284.
142. Addadi L, Lahav M. J. Am. Chem. Soc. 1978; 100:2838–2844.
143. Addadi L, Mil JV, Lahav M. J. Am. Chem. Soc. 1982; 104:3422–3429.
144. Addadi L, Lahav M. J. Am. Chem. Soc. 1979; 101:2152–2156.
145. Addadi L, Cohen MD, Lahav M. Mol. Cryst. Liq. Cryst. 1976; 32:137–141.
146. Elgavi A, Green SB, Schmidt GMJ. J. Am. Chem. Soc. 1973; 95:2058–2059.
147. Green BS, Lahav M, Rabinovich D. Acc. Chem. Res. 1979; 12:191–197.
148. Vaida M, Shimon LJW, Mil JV, Ernst-Cabrera K, Addadi L, Leiserowitz L. J. Am. Chem. Soc. 1989; 111:1029–1034.
149. Tissot O, Gouygou M, Dallemer F, Daran J-C, Balavoine GA. Angew. Chem. Int. Ed. Eng. 2001; 40:1076–1078.
150. Mil J, Addadi L, Gati E, Lahav M. J. Am. Chem. Soc. 1982; 104:3429–3434.
151. Garcia-Garibay M, Scheffer JR, Trotter J, Wireko F. J. Am. Chem. Soc. 1989; 111: 4985–4986.
152. Garcia-Garibay M, Scheffer JR, Trotter J, Wireko F. Tetrahedron Lett. 1987; 28: 4789–4792.
153. Kaupp G, Haak M. Angew. Chem. Int. Ed. Engl. 1993; 32:694–695.
154. Wu L-C, Cheer CJ, Olovsson G, Scheffer JR, Trotter J, Wang S-L, Liao F-L. Tetrahedron Lett. 1997; 38:3135–3138.
155. Sakamoto M, Iwamoto T, Nono N, Ando M, Arei W, Mino T, Fujita T. J. Org. Chem. 2003; 68:942–946.

156. Sakamoto M, Takahashi M, Mino T, Fujita T. Tetrahedron Lett. 2001; 57: 6713–6719.
157. Sakamoto M. Chem. Eur. J. 1997; 3:684–689.
158. Sakamoto M, Takahashi M, Kamiya K, Yamaguchi K, Fujita T, Watanabe S. J. Am. Chem. Soc. 1996; 118:10664–10665.
159. Suzuki T, Fukushima T, Yamashita Y, Miyashi T. J. Am. Chem. Soc. 1994; 116: 2793–2803.
160. Sakamoto M, Takahashi M, Fujita T, Watanabe S, Iida I, Nishio T, Aoyama H. J. Org. Chem. 1993; 58:3476–3477.
161. Sakamoto M, Takahashi M, Hokari N, Fujita T, Watanabe S. J. Org. Chem 1994; 59:3131–3134.
162. Sakamoto M, Takahashi M, Fujita T, Nishio T, Iida I, Watanabe S. J. Org. Chem. 1995; 60:4682–4683.
163. Sakamoto M, Takahashi M, Fujita T, Watanabe S, Nishio T, Iida I, Aoyama H. J. Org. Chem. 1997; 62:6298–6308.
164. Sakamoto M, Takahashi M, Arai T, Shimizu M, Yamaguchi K, Mino T, Watanabe S, Fujita T. Chem. Commun. 1998:2315–2316.
165. Takahashi M, Sekine N, Fujita T, Watanabe S, Yamaguchi K, Sakamoto M. J. Am. Chem. Soc. 1998; 120:12770–12776.
166. Takahashi M, Fujita T, Watanabe S, Sakamoto M. J. Chem. Soc., Perkin Trans. 2 1998:487–491.
167. Sakamoto M, Takahashi M, Arai W, Mino T, Yamaguchi K, Watanabe S, Fujita T. Tetrahedron 2000; 56:6795–6804.
168. Sakamoto M, Hokari N, Takahashi M, Fujita T, Watanabe S, Iida I, Nishio T. J. Am. Chem. Soc. 1993; 115:818.
169. Irngartinger H, Fettel PW, Siemund V. Eur. J. Org. Chem. 1998:2079–2082.
170. Sakamoto M, Takahashi M, Shimizu M, Fujita T, Nishio T, Iida I, Yamaguchi K, Watanabe S. J. Org. Chem. 1995; 60:7088–7089.
171. Sakamoto M, Takahashi M, Moriizumi S, Yamaguchi K, Fujita T, Watanabe S. J. Am. Chem. Soc. 1996; 118:8138–8139.
172. Roughton AL, Muneer M, Demuth M, Klopp I, Kruger C. J. Am. Chem. Soc. 1993; 115:2085–2087.
173. Chung C-M, Hasegawa M. J. Am. Chem. Soc. 1991; 113:7311–7316.
174. Hasegawa M, Chung C-M, Muro N, Maekawa Y. J. Am. Chem. Soc. 1990; 112: 5676–5677.
175. Toda F, Yagi M, Soda S-I. Chem. Commun. 1987:1413–1414.
176. Chen J, Pokkuluri PR, Scheffer JR, Trotter J. Tetrahedron Lett. 1990; 31: 6803–6806.
177. Fu TY, Liu Z, Scheffer JR, Trotter J. J. Am. Chem. Soc. 1993; 115:12202–12203.
178. Evans SV, Garcia-Garibay M, Omkaram N, Scheffer JR, Trotter J, Wireko F. J. Am. Chem. Soc. 1986; 108:5648–5650.
179. Toda F, Miyamoto H. J. Chem. Soc. Perkins Trans. I 1993:1129–1132.
180. Hashizume D, Kogo H, Sekine A, Ohashi Y, Miyamoto H, Toda F. J. Chem. Soc., Perkin Trans. 2 1996:61–66.
181. Sekine A, Hori K, Ohashi Y, Yagi M, Toda F. J. Am. Chem. Soc. 1989; 111: 697–699.

182. Addadi L, Cohen MD, Lahav M. Chem. Comm. 1975:471–473.
183. Aoyama H, Hasegawa T, Omote Y. J. Am. Chem. Soc. 1979; 101:5343–5347.
184. Mil JV, Addadi L, Leiserowitz L. Chem. Comm. 1982:584–587.
185. Scheffer JR. Can. J. Chem. 2001; 79:349–357.
186. Leibovitch M, Olovsson G, Scheffer JR, Trotter J. Pure Appl. Chem. 1997; 69: 815–823.
187. Gamlin JN, Jones R, Leibovitch M, Patrick B, Scheffer JR, Trotter J. Acc. Chem. Res. 1996; 29:203–209.
188. Jayaraman S, Uppili S, Natarajan A, Joy A, Chong KCW, Netherton MR, Zenova A, Scheffer JR, Ramamurthy V. Tetrahedron Lett. 2000; 41:8231–8235.
189. Leibovitch M, Olovsson G, Scheffer JR, Trotter J. J. Am. Chem. Soc. 1998; 120: 12755–12769.
190. Cheung E, Netherton MR, Scheffer JR, Trotter J. J. Am. Chem. Soc. 1999; 121: 2919–2920.
191. Cheung E, Kang T, Netherton MR, Scheffer JR, Trotter J. J. Am. Chem. Soc. 2000; 122:11753–11754.
192. Cheung E, Netherton MR, Scheffer JR, Trotter J, Zenova A. Tetrahedron Lett. 2000; 41:9673–9677.
193. Janz KM, Scheffer JR. Tetrahedron Lett. 1999; 40:8725–8728.
194. Cheung E, Kang T, Raymond JR, Scheffer JR, Trotter J. Tetrahedron Lett. 1999; 40:8729–8732.
195. Cheung E, Rademacher K, Scheffer JR, Trotter J. Tetrahedron Lett. 1999; 40: 8733–8736.
196. Cheung E, Netherton MR, Scheffer JR, Trotter J. Tetrahedron Lett. 1999; 40: 8737–8740.
197. Patrick BO, Scheffer JR, Scheffer CS. Angew. Chem. Int. Ed. Eng. 2003; 42: 3775–3777.
198. Chen J, Garcia-Garibay M, Scheffer JR. Tetrahedron Lett. 1989; 30:6125–6128.
199. Leibovitch M, Olovsson G, Scheffer JR, Trotter J. J. Am. Chem. Soc. 1997; 119: 1462–1463.
200. Scheffer JR, Wang K. Synthesis 2001; 8:1253–1257.
201. Cheung E, Rademacher K, Scheffer JR, Trotter J. Tetrahedron 2000; 56:6739–6751.
202. Jones R, Scheffer JR, Trotter J, Yang J. Tetrahedron Lett. 1992; 33:5481–5484.
203. Cheung E, Chong KCW, Jayaraman S, Ramamurthy V, Scheffer JR, Trotter J. Org. Lett. 2000; 2:2801–2804.
204. Gudmundsdottir AD, Scheffer JR. Tetrahedron Lett. 1990; 31:6807–6810.
205. Gudmundsdottir AD, Scheffer JR. Photochem. Photobiol. 1991; 54:535–538.
206. Gudmundsdottir AD, Li W, Scheffer JR, Rettig S, Trotter J. Mol. Cryst. Liq. Cryst. 1994; 240:81–88.
207. Gudmundsdottir AD, Scheffer JR, Trotter J. Tetrahedron Lett. 1994; 35:1397–1400.
208. Tanaka K, Toda F. In:. Organic Photoreactions in the Solid State Toda F., Ed. New York: Kluwer, 2002:109–157.
209. Toda F. Acc. Chem. Res. 1995; 28:480–486.
210. Toda FI. Mol. Inclusion Mol. Recognit.—Clathrates 2 Weber E., Ed. Vol. Vol. 149: Springer-Verlag, 1988:211–238.

211. Toda F, Tanaka K, Yagi M. Tetrahedron 1987; 43:1495–1502.
212. Toda F, Miyamoto H, Matsukawa R. J. Chem. Soc., Perkin Trans. 1 1992: 1461–1462.
213. Kaftory M, Toda F, Tanaka K, Yagi M. Mol. Cryst. Liq. Cryst. 1990; 186:167–176.
214. Lahav M, Leiserowitz L, Popovitz-Biro R, Tang C-P. J. Am. Chem. Soc 1978; 100:2542–2543.
215. Popovitz-Biro R, Tang CP, Chang HC, Lahava M, Leiserowitz L. J. Am. Chem. Soc. 1985; 107:4043–4058.
216. Chang CH, Popovitz-Biro R, Lahav M, Leiserowitz L. J. Am. Chem. Soc. 1987; 109:3883–3893.
217. Friedman N, Lahav M, Leiserowitz L, Popovitz-Biro R, Tang CP, Zaretskii ZVI. Mol. Cryst. Liq. Cryst. 1976; 32:127–129.
218. Friedman N, Lahav M, Leiserowitz L, Popovitz-Biro R, Tang C-P, Zaretzkii ZVI. Chem. Comm 1975:864–865.
219. Weisinger-Lewin Y, Vaida M, Popovitz-Biro R, Chang HC, Mannig F, Frolow F, Lahav M, Leiserowitz L. Tetrahedron 1987; 43:1449–1475.
220. Toda F. Mol. Cryst. Liq. Cryst. 1988; 161:355–362.
221. Toda F, Tanaka K. Chem. Lett. 1987:2283–2284.
222. Toda F, Tanaka K, Mak TCW. Chem. Lett. 1989:1329–1330.
223. Toda F, Tanaka K. Tetrahedron Lett. 1988; 29:551–554.
224. Toda F, Miyamoto H, Koshima H, Urbanczyk-Lipkowska Z. J. Org. Chem. 1997; 62:9261–9266.
225. Tanaka K, Nagahiro R, Urbanczyk-Lipkowska Z. Org. Lett. 2001; 3:1567–1569.
226. Tanaka K, Mochizuki E, Yasui N, Kai Y, Miyahara I, Hirotsu K, Toda F. Tetrahedron 2000; 56:6853–6865.
227. Hosomi H, Ohba S, Tanaka K, Toda F. J. Am. Chem. Soc. 2000; 122:1818–1819.
228. Ohba S, Hosomi H, Tanaka K, Miyamoto H, Toda F. Bull. Chem. Soc. Jpn. 2000; 73:2075–2085.
229. Toda F, Miyamoto H, Kanemoto K. Chem. Commun. 1995:1719–1720.
230. Toda F, Tanaka K, Sekikawa A. Chem. Commun. 1987:279–280.
231. Toda F, Tanaka K. Chem. Commun. 1986:1429–1430.
232. Tanaka K, Kakinoki O, Toda F. Chem. Commun. 1992:1053–1054.
233. Toda F, Miyamoto H, Kanemoto k. J. Org. Chem. 1996; 61:6490–6491.
234. Toda F, Miyamoto H, Kikuchi S, Kuroda R, Nagami F. J. Am. Chem. Soc. 1996; 118:11315–11316.
235. Toda F, Miyamoto H, Tamashima T, Kondo M, Ohashi Y. J. Org. Chem. 1999; 64:2690–2693.
236. Toda F, Miyamoto H, Takeda K, Matsugawa R, Maruyama N. J. Org. Chem. 1993; 58:6208–6211.
237. Toda F, Miyamoto H, Kikuchi S. In:. Chem. Commun., 1995:621–622.
238. Toda G, Miyamoto H, Inoue M, Yasaka S, Matijasic I. J. Org. Chem. 2000; 65: 2728–2732.
239. Hashizume D, Ohashi Y, Tanaka K, Toda G. Bull. Chem. Soc. Jpn. 1994; 67: 2383–2387.
240. Tanaka K, Toda F. J. Chem. Soc., Perkin Trans. 1 1992:943–944.

241. Fujiwara T, Nanba N, Hamada K, Toda F, Tanaka K. J. Org. Chem. 1990; 55: 4532–4537.
242. Toda F, Tanaka K, Oda M. Tetrahedron Lett. 1988; 29:653–654.
243. Toda F, Tanaka K, Omata T, Nakamura K, Oshima T. J. Am. Chem. Soc. 1983; 105:5151–5152.
244. Tanaka K, Honke S, Urbanczyk-Lipkowska Z, Toda F. Eur. J. Org. Chem. 2000: 3171–3176.
245. Toda F, Tanaka K, Miyamoto H. In:. Understanding and Manipulating Excited-State Processes Ramamurthy V. , Schanze K. S., Eds. Vol. Vol. 8. New York: Marcel Dekker, 2001:385–425.
246. Aoyama H, Miyazaki K, Sakamoto M, Omote Y. Chem. Commun. 1983:333–334.
247. Yellin RA, Green BS. J. Am. Chem. Soc 1980; 102:1157–1158.
248. Yellin RA, Green PS, Brunie S, Knossow M, Tsoucaris G. Mol. Cryst. Liq. Cryst. 1978:275–276.
249. Yellin RA, Green BS, Knossow M, Rysanek N, Tsoucaris G. J. Inclusion. Phenom 1985; 3:317–333.
250. Brown JF, White DM. J. Am. Chem. Soc. 1960; 82:5671–5677.
251. White DM. J. Am. Chem. Soc. 1960; 82:5678–5685.
252. Farina M, Audisio G, Natta G. J. Am. Chem. Soc. 1967; 89:5071.
253. Inoue Y, Wada T, Sugahara N, Yamamoto K, Kimura K, Tong L-H, Gao X-M, Hou Z-J, Liu Y. J. Org. Chem. 2000; 65:8041–8050.
254. Takeshita H, Kumamoto M, Kouno I. Bull. Chem. Soc. Jpn. 1980; 53:1006–1009.
255. Rao VP, Turro NJ. Tetrahedron. Lett. 1989; 30:4641–4644.
256. Inoue Y, Kosaka S, Matsumoto K, Tsuneishi H, Hakushi T, Tai A, Nakagawa K, Tong L-H. J. Photochem. Photobiol. A: Chem. 1993; 71:61–64.
257. Vizvardi K, Desmet K, Luyten I, Sandra P, Hoornaert G, Eycken EVD. Org. Lett. 2001; 3:1173–1175.
258. Inoue Y, Dong F, Yamamoto K, Tong L-H, Tsuneishi H, Hakushi T, Tai A. J. Am. Chem. Soc. 1995; 117:11033–11034.
259. Weber L, Imiolczyk I, Haufe G, Rehorek D, Hennig H. Chem. Commun. 1992: 301–303.
260. Koodanjeri S, Sivaguru J, Pradhan AR, Ramamurthy V. Proc. Ind. Nat. Sci. Acad. 2002; 68:453–463.
261. Shailaja J, Karthikeyan S, Ramamurthy V. Tetrahedron Lett. 2002; 43:9335–9339.
262. Koodanjeri S, Ramamurthy V. Tetrahedron Lett. 2002; 43:9229–9232.
263. Ramamurthy V, Shailaja J, Kaanumalle LS, Sunoj RB, Chandrasekhar J. Chem. Commun. 2003:1987–1999.
264. Yilmaz A, Bu X, Kizilyalli M, Stucky GD. Chem. Mater. 2000; 12:3243–3245.
265. Stalder SM, Wilkinson AP. Chem. Mater. 1997; 9:2168–2173.
266. Bruce DA, Wilkinson AP, White MG, Bertrand JA. J. Chem. Soc., Chem. Commun. 1995:2059–2060.
267. Akporiaya DE. J. Chem. Soc., Chem. Commun. 1994:1711–1712.
268. Gray MJ, Jasper JD, Wilkinson AP, Hanson JC. Chem. Mater. 1997; 9:976–980.
269. Newsam JM, Treacy MMJ, Koetsier WT, De Gruyter CB. Proc. R. Soc. Lond. A 1988; 420:375–405.

270. Treacy MMJ, Newsam JM. Nature 1988; 332:249–251.
271. Higgins JB, LaPierre RB, Schlenker JL, Rohrman AC, Wood JD, Kerr GT, Rohrbaugh WJ. Zeolites 1988; 8:446–452.
272. Tomlinson SM, Jackson RA, Catlow CRA. J. Chem. Soc., Chem. Commun., 1990: 813–816.
273. Johnson BFG, Raynor SA, Shephard DS, Mashmeyer T, Thomas JM, Sankar G, Bromley S, Oldroyd R, Gladden L, Mantle MD. In:. Chem. Commun. (Cambridge), 1999:1167–1168.
274. Davis ME, Lobo RF. In:. Chem. Mater.. Vol. Vol. 4, 1992:756–68.
275. Davis ME. In:. Microporous Mesoporous Mater.. Vol. Vol. 21, 1998:173–182.
276. Breck DW. Zeolite Molecular Sieves. Malabar: Robert E. Krieger, 1974.
277. Cohen SG, Parola A, Parsons GHJ. Chem. Rev. 1973; 73:141–161.
278. Shailaja J, Ponchot KJ, Ramamurthy V. Org. Lett. 2000; 2:937–940.
279. Shailaja J, Sivaguru J, Uppili S, Joy A, Ramamurthy V. Microporous and Mesoporous Mater. 2001; 48:319–328.
280. Coyle JD, Carless HAJ. Chem. Soc. Rev. 1972:465–480.
281. Shailaja J. Tulane, 2002.
282. Sivaguru J, Natarajan A, Kaanumalle LS, Shailaja J, Uppili S, Joy A, Ramamurthy V. Acc. Chem. Res. 2003; 36:509–521.
283. Ponchot KJ. Tulane, 2003.
284. Dauben WG, Koch K, Smith SL, Chapman OL. J. Am. Chem. Soc. 1963; 85: 2616–2621.
285. Chapman OL, Pasto DJ. J. Am. Chem. Soc 1960; 82:3642–3648.
286. Joy A, Ramamurthy V. Chem.—Eur. J. 2000; 6:1287–1293.
287. Sivaguru J, Shailaja J, Uppili S, Ponchot K, Joy A, Arunkumar N, Ramamurthy V. In:. Organic Solid-State Reactions Toda F., Ed: Kluwer Academic, 2002:159–188.
288. Joy A. Tulane, 2000.
289. Kaanumalle LS, Ramamurthy V. Unpublished results.
290. Kaanumalle LS, Sivaguru J, Arunkumar N, Karthikeyan S, Ramamurthy V. Chem. Commun. 2003:116–117.
291. Corey EJ, Streith J. J. Am. Chem. Soc. 1964; 86:950–951.
292. Karthikeyan S, Ramamurthy V. Unpublished results.
293. Uppili S. Tulane, 2002.
294. Uppili S, Ramamurthy V. Org. Lett. 2002; 4:87–90.
295. Joy A, Uppili S, Netherton MR, Scheffer JR, Ramamurthy V. J. Am. Chem. Soc. 2000; 122:728–729.
296. Sivaguru J, Scheffer JR, Chandrasekhar J, Ramamurthy V. Chem. Commun. 2002: 830–831.
297. Sivaguru J, Shichi T, Ramamurthy V. Org. Lett. 2002; 4:4221–4224.
298. Kaanumalle LS, Sivaguru J, Sunoj RB, Lakshminarasimhan PH, Chandrasekhar J, Ramamurthy V. J. Org. Chem. 2002; 67:8711–8720.
299. Chong KCW, Sivaguru J, Shichi T, Yoshimi Y, Ramamurthy V, Scheffer JR. J. Am. Chem. Soc. 2002; 124:2858–2859.
300. Sivaguru J. Tulane, 2003.
301. Natarajan A, Wang K, Ramamurthy V, Scheffer JR, Patrick B. Org. Lett. 2002; 4: 1443–1446.

302. Natarajan A, Joy A, Kaanumalle LS, Scheffer JR, Ramamurthy V. J. Org. Chem. 2002; 67:8339–8350.
303. Natarajan A, Ramamurthy V. Unpublished results.
304. Leibovitch M, Olovsson G, Sundarababu G, Ramamurthy V, Scheffer JR, Trotter J. J. Am. Chem. Soc. 1996; 118:1219–1220.
305. Sundarababu G, Leibovitch M, Corbin DR, Scheffer JR, Ramamurthy V. Chem. Commun. 1996:2159–2160.
306. Joy A, Ramamurthy V. Chem. Eur. J. 2000; 6:1287–1293..

16
Photochemistry of Chiral Polymers

Eiji Yashima
Nagoya University
Nagoya, Japan

I. INTRODUCTION

Photoresponsive polymers have been extensively developed over the past decades [1–3], not only to mimic biological photoresponsive systems [4–9] but also for their possible applications to optoelectronics, bioelectronics, and information-storage devices [10–14]. Photoresponsive polymers can be prepared via the incorporation of photosensitive molecules, such as azobenezenes (**1**), spiropyrans (**2**), and triphenylmethanes (**3**) into the polymer main chains or the pendants, which reversibly photoisomerize through trans–cis isomerization, zwitterion formation, and ionic dissociation, respectively (Scheme 1). Upon photoirradiation, the physical and chemical properties including conformation, shape, viscosity, permeability, solubility, and functions of the photoresponsive polymers can be reversibly regulated [1]. These particularly interesting properties have been used to construct novel photoswitchable functional materials [1–3,11–13], biomaterials [4–6], and liquid crystals [10,14], in which the functions, catalytic activities of the proteins, and phase separation in liquid crystalline phases have been photoregulated. Although a number of achiral or optically inactive photoresponsive polymers have been synthesized for use in materials science, limited examples of chiral polymers, with optical activity due to the main chain chirality bearing chiral or achiral photosensitive moieties, are available [15]. Optically active photochromic molecules [16], such as chiral overcrowded alkenes [17], diarylethenes [18], and fulgides [19], are often introduced into the achiral polymer's side groups [16,20]. In these cases, the polymers are simply used as supporting materials, and therefore the chiroptical properties of the polymers are largely governed by the chiral

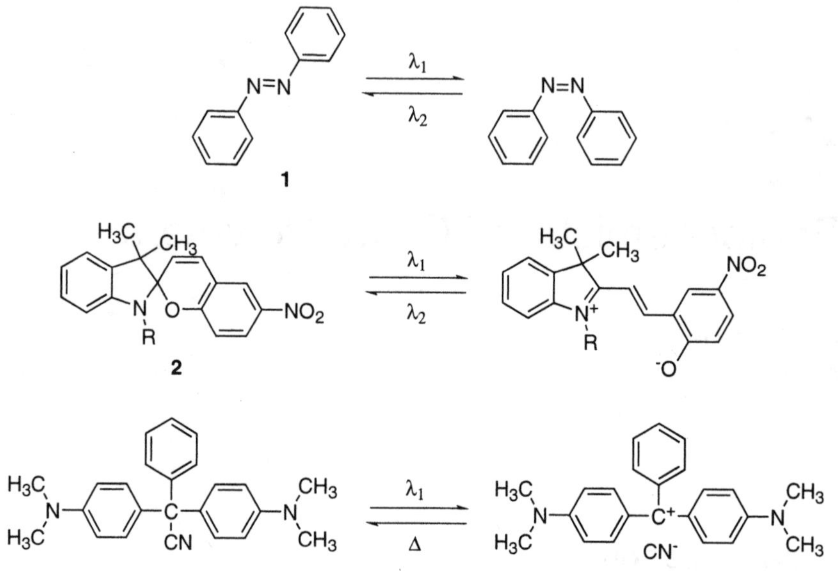

Scheme 1 Photoisomerizations of azobenzenes (**1**), spiropyrans (**2**), and triphenylmethanes (**3**).

pendant photochromic residues. This means that unexpected interesting photoresponsive phenomena cannot be anticipated in the systems.

On the other hand, most biological macromolecules such as proteins and nucleic acids are optically active and possess a one-handed helical structure (right-handed α-helix and double-helix, respectively) because of the homochirality of their components (D-sugars and L-amino acids). The chiral nature of the biological macromolecules appears to play an essential role in their sophisticated and fundamental functions such as molecular and chiral recognition ability, replication, and catalytic activity in living systems. Hence synthetic helical polymers with optical activity due to the helicity have received considerable attention not only to mimic unique structures and functions as observed in nature but also for their broad applications in chiral materials for enantiomer separation and enantioselective catalysis [21–30]. To date, a number of helical polymers have been prepared. Photoresponsive helical polymers have also been synthesized in order to realize reversible photoinduced structural changes such as the helix–helix transition in the polymer backbone [3,15,16]. Polypeptides [6–9] and polyisocyanates [31] with photochromic moieties belong to this category. These polymers exhibit a particularly interesting conformational change and exhibit a change in their optical rotatory dispersion (ORD) spectra or circular dichroism (CD) spectra in the absorption region of the polymer backbone upon photoirradiation of the chromophoric pendants.

In this chapter, we mainly discuss and deal with photoresponsive, chiral biorelated and synthetic polymers bearing a configurational and/or conformational chirality in the polymer main chains. The photocontrol of the chiral recognition ability of chiral polymers and chirality induction on achiral polymer films by circular polarized light (CPL) are also briefly reviewed.

II. PHOTORESPONSIVE POLYPEPTIDES

Polypeptides are among the most readily available optically active biopolymers. They can adopt well defined, ordered structures such as α-helix and β-sheet depending on their amino acid components and sequence. Moreover, the ordered structures of some polypeptides undergo conformational changes, for instance, the helix–helix transition and the helix–coil transition regulated by external stimuli such as change in the pH, temperature, and solvent [6–9]. This stimulated chemists to introduce photosensitive units into the polypeptide pendants to produce photoinduced conformational changes in the polypeptides. Photoresponsive polypeptides can be easily prepared by incorporating photosensitive chromophores into the polypeptide pendants through polymer reactions, that is, the modification of the polypeptide's pendants, or by the polymerization or copolymerization of the corresponding amino acid N-carboxy anhydrides. The first photoresponsive polypeptides were prepared by Goodman et al. in 1966 [32]. They observed changes in the chiroptical properties of the chromophoric azobenzene moieties incorporated into the polypeptides derived from phenylazophenyl-L-alanine and γ-benzyl L-glutamate, associated with the photoisomerization. However, the trans–cis photoisomerization of the pendant azo group induced almost no conformational variation in the polypeptide backbone. The light-induced conformational changes of the polypeptides were first accomplished by Ueno et al. [33] and later by Ciardelli and Pieroni et al. [34]. Comprehensive review articles on photoresponsive polypeptides with historical background are available elsewhere [6–9].

A. Photoresponsive Helical Polypeptides

Typical photoresponsive polypeptides exhibiting a helix–helix transition upon photoirradiation are shown in Fig. 1. These polypeptides are composed of L-aspartates as the amino acid component. The poly(L-aspartic acid esters) are the most thoroughly investigated polypeptides with respect to their ordered structures

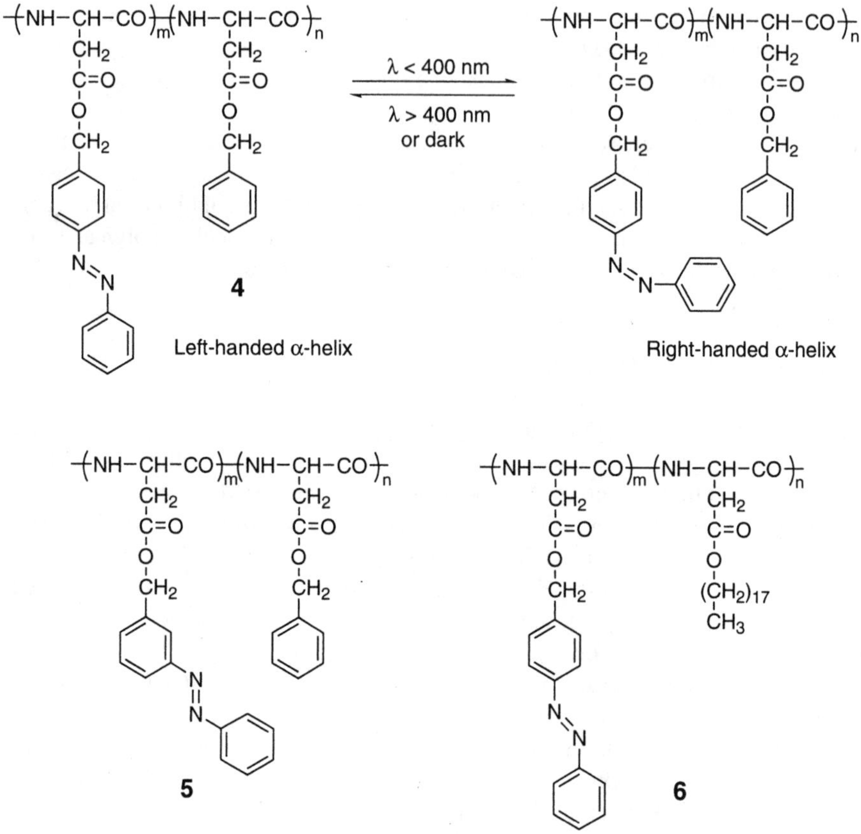

Figure 1 Photoresponsive helical polypeptides showing a helix–helix transition.

and conformational changes [6–9]. Most poly(L-aspartic acid esters) have either a right- or a left-handed α-helix in solution as well as in the solid state. However, some poly(L-aspartic acid esters), including substituted aromatic alkyl esters of poly(L-aspartic acid), are known to undergo a helix–helix transition (from right to left or the reverse) induced by changing the external conditions [35–37]. These results suggest that the difference in energy between the right- and left-handed helical poly(L-aspartic acid esters) is relatively small and there is a delicate balance between the helices, which might lead to such a helix inversion by changing the external conditions.

On the basis of these observations, Ueno et al. designed and synthesized a series of poly(L-aspartic acid ester)-containing azobenzene pendants as the photosensitive units and investigated the effect of light on the polypeptide conformations. The copolypeptides of β-benzyl L-aspartate with β-*p*- (**4**) [38] or *m*-(phenylazo)benzyl L-aspartate (**5**) showed a positive CD band at around 220 nm assigned to the left-handed α-helices in 1,2-dichloroethane. After irradiation, however, the Cotton effect sign of **4** with more than 50 mol% azo residues (59 and 81%) changed from a positive to a negative sign, indicating a reversal of the helix sense induced by the trans-to-cis photoisomerization of the azo groups. Relaxation of the polymer in the dark allowed it to go back to the original left-handed helix. On the other hand, the *m*-substituted peptide **5** did not show such a dramatic change under the same conditions; it showed a decrease in the CD intensity, probably owing to a helix-to-coil transition. The helix sense preference of poly(L-aspartic acid esters) is delicately balanced between both helices with a small energy difference as described above, so that the helix sense can be controlled by tuning the external condition with photoirradiation. The *m*-substituted **5** exhibits a preference for left- and right-handed helices in pure 1,2-dichloroethane and trimethyl phosphate, respectively, and the helix sense inversion occurred in the solvent mixtures as expected. Upon irradiation in an adequate solvent mixture, **5** exhibited a helix–helix transition accompanied by remarkable changes in their CD spectra [39]. For this particular photoregulation system, only a small amount of the azo residue (9.7%) in **5** was required.

A similar photoinduced helicity inversion was also observed for the copolypeptides of *n*-octadecyl L-aspartate with β-para-(phenylazo)benzyl L-aspartate (**6**) [40]. The key to this helicity inversion of the peptide is that the octadecyl L-aspartate residues in a polypeptide sequence have a moderate preference to take a right-handed α-helix, while the aspartate residue having the *trans*-azophenyl moieties has an opposite left-handed α-helix, which can be transformed to a right-handed one during the trans-to-cis photoisomerization. Consequently, the copolypeptides containing more than 50 mol% azo residues (68 and 89%) caused the reversal in the helix sense from a left- to a right-handed helix upon photoirradiation. The copolypeptides also showed a helix–helix transition in 1,2-dichloroethane in the presence of an increasing amount of trifluoroacetic acid and also by changing the temperature of the solution as well as of the film state. Because of the limited mobility of the polypeptide backbones in the solid films, photocontrol of the helix sense of the photoresponsive polypeptides in the solid state appears to be very difficult [41], whereas a reversible helix sense inversion of **6** [42] and poly(β-phenylpropyl L-aspartate) [37,43] concertedly occurred in the solid state along the polypeptide backbone by changing the temperature.

The photoresponsive helicity inversion process of **4** was applied to the photoregulation of permeability across a membrane from a graft copolymer containing the photoresponsive peptide (**7**) as a branch (Fig. 2) [44]. The graft polymer membrane casted from a 1,2-dichloroethane solution, followed by immersion in trimethylphosphate, exhibited a positive CD band at 215 nm, characteristic to

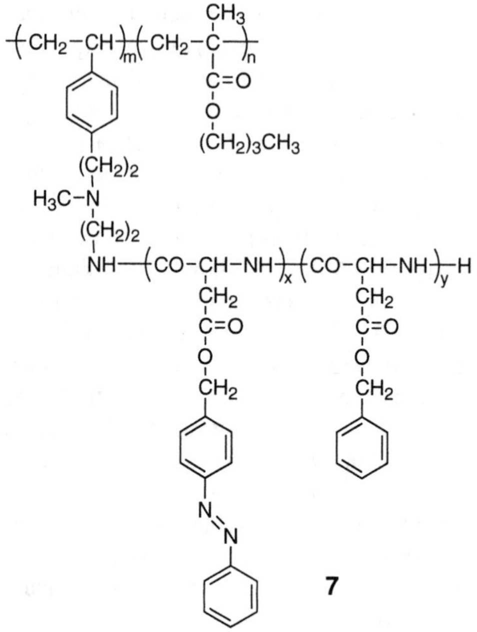

Figure 2   Graft polymer containing a photoresponsive polypeptide.

a left-handed α-helix, the sign of which was completely inverted to form a right-handed α-helix upon photoirradiation of the azobenzene groups. The helix contents before and after the UV-irradiation were estimated to be 63 and 88%, respectively. The permeation rates of the racemic mandelic acid and N-((benzyloxy)carbonyl)alanine across the membrane immersed in trimethylphosphate were raised by about six and four times with UV irradiation (trans-to-cis isomerization of the azobenzene groups), respectively, and were suppressed upon irradiation with visible light (cis-to-trans isomerization). During the alternative photoirradiation, the helix sense of the polypeptide backbones in the membrane also changed from left- to right-handed. Therefore it was concluded that the changes in permeability upon photoirradiation were induced by the inversion of the helix sense of the polypeptide backbone in the membrane. In these experiments, racemic compounds were used, and there should be a chance to permeate preferentially one of the enantiomers through the permeation process because the membrane is optically active, although there is no description regarding enantiomer enrichment across the membrane. Photoresponsive enantiomer separation through a solid membrane is attractive for the development of separation technology, but it is still a very difficult task. As for the photocontrol of the chiral recognition by photoresponsive chiral polymer films, see the following section.

B. Other Photoresponsive Polypeptides

A number of other photoresponsive polypeptides containing photosensitive units in the pendants, such as an azobenzene, stilbene, or spiropyran group, have been prepared [6–9]. Photoresponsive conformational changes in the polypeptides are a very active research area endowed with a large number of reports. Details of such conformational changes have been thoroughly reviewed [6–9]. Most of the polypeptides consist of poly(L-glutamic acid) (PLGA) (**8–10**) [45–48] or poly(L-lysine) (PLL) (**11–14**) [49–52]. These polypeptides also exhibit a photoinduced conformational change, including α-helix-to-random coil, coil-to-α-helix, and α-helix-to-β-structure transitions. Typical polypeptides showing reversible photoinduced structural changes in the peptide backbones are shown in Fig. 3. The changes in their CD spectra before and after photoirradiation were used to monitor the conformational change in the peptide backbone. In contrast, helix sense inversion was not observed in these polypeptides. This might be due to a rather large energy difference between the right- and left-handed helical structures of the PLGA and PLL derivatives as compared with that of the poly(L-aspartic acid esters). The reason that poly(L-aspartates) exhibit an inversion of helicity by changing the chemical structure, solvent, temperature, and irradiation of light has not yet been elucidated, but the poly(L-aspartates) may have a dynamic helical conformation like polyisocyanates as will be described in detail later.

III. PHOTORESPONSIVE OLIGONUCLEOTIDES

In addition to proteins and polypeptides, nucleic acids such as DNA and RNA are other important biological macromolecules with optical activity. They form right-handed double and triple stranded helices, and the control of their formation and dissociation, which is closely related to the regulation of the gene expression, by external stimuli is currently one of the most important and attractive topics. Asanuma and Komiyama et al. prepared photoresponsive DNA bearing an azobenzene moiety in the side chain of a residue and successfully photoregulated the formation and dissociation of the DNA duplex and triplex [53–57].

An azobenzene-modified oligonucleotide 5′-AAAXAAAA-3′ [**15**, **X** is the residue having an azobenzene moiety in the side chain (Fig. 4A)] was prepared and further separated into two diastereomers (**15a** and **15b**) based on the chirality of the stereogenic carbon atom of **15** by reversed-phase HPLC [53]. The melting temperature (T_m) of the duplex of each diastereomer of **15** with its complementary oligonucleotide counterpart (5′-TTTTTTTT-3′) was photoregulated by the trans-

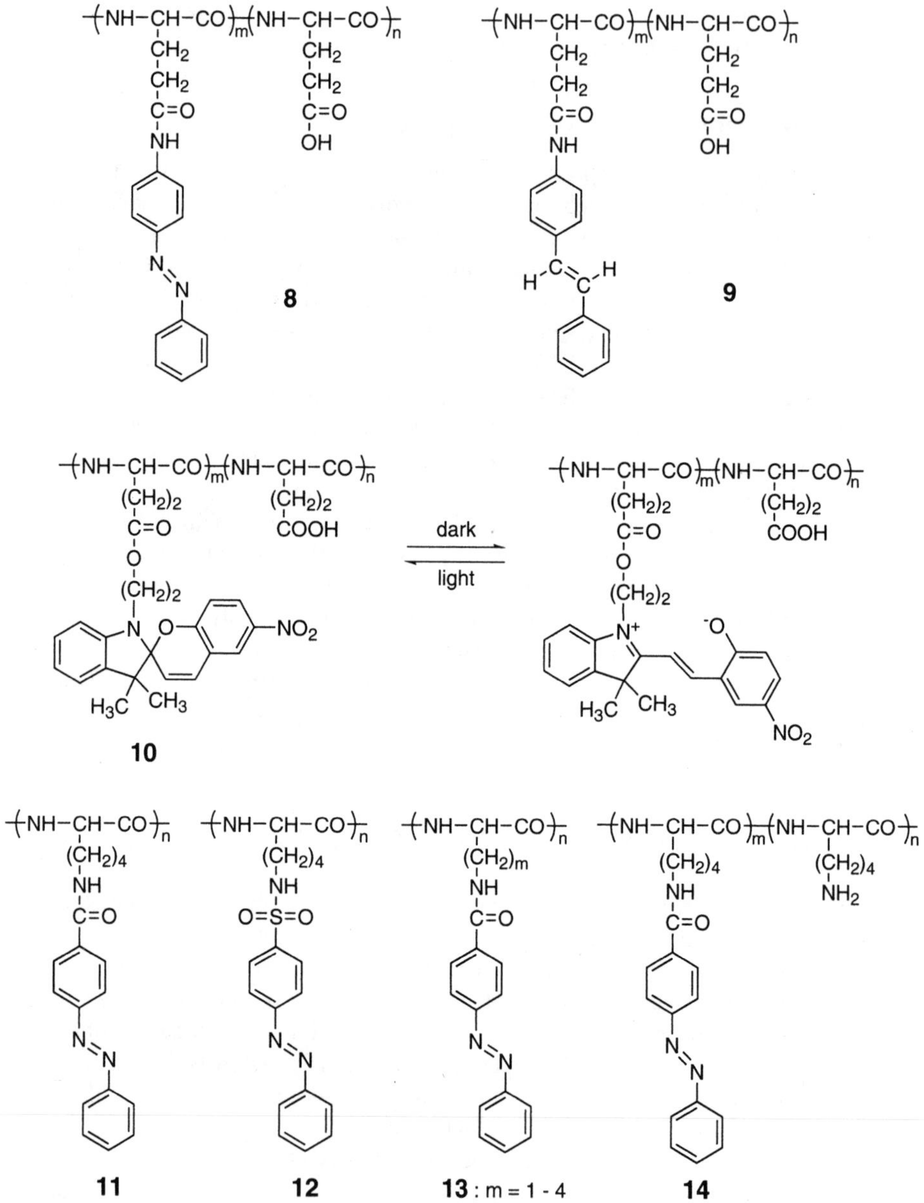

Figure 3 Photoresponsive polypeptides showing a reversible conformational change.

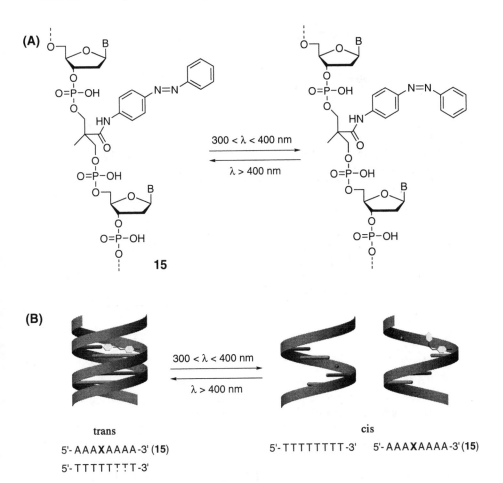

Figure 4 Isomerization of the azobenzene moiety in the side chain of the residue upon photoirradiation of oligonucleotide **15** (A) and schematic illustration of photoregulation of the formation and dissociation of a DNA duplex (B).

to-cis isomerization of the azobenzene moiety (Fig. 4B). The T_m of the duplex significantly decreased by the trans-to-cis isomerization (from 24.8 to 15.9°C for **15a**) upon photoirradiation, and the duplex was dissociated into two single-stranded oligonucleotides. The duplex was formed again by the cis-to-trans isomerization of the azobenzene moiety upon irradiation with visible light. In the same way, photocontrol of the DNA triplex formation and dissociation was achieved using an oligo(thymidine) bearing the photosensitive azobenzene-modified residue [54,55].

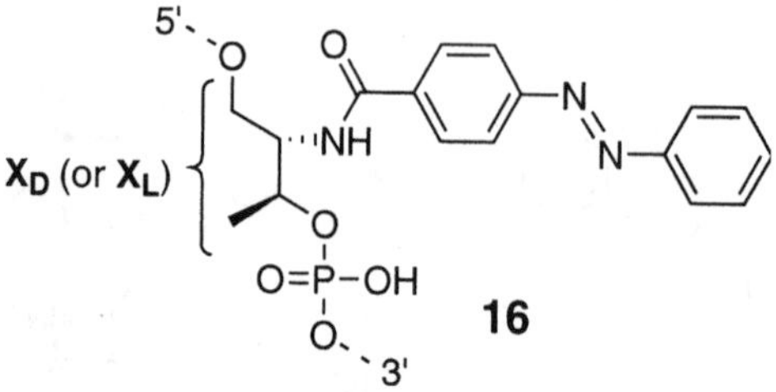

X_D (or X_L)

16

Table 1 Melting Temperature (T_m) of the Duplexes Between the Modified Oligonucleotides and 3′-CGCTCAGC-5′

16	Sequence	T_m [°C]		ΔT_m [°C]
		trans	*cis*	
L1	5′-GCGAX$_L$GTCG-3′	45.1	40.8	4.3
D1	5′-GCGAX$_D$GTCG-3′	50.9	36.6	14.3
LL2	5′-GCX$_L$GAGTX$_L$CG-3′	25.4	25.5	−0.1
DD2	5′-GCX$_D$GAGTX$_D$CG-3′	43.9	22.4	21.5
Control	5′-GCGAGTCG-3′	46.6		

Source: Adaped from Ref. 56; © 2001, Wiley-VCH.

To improve the photoregulation ability of the duplex formation, multiple azobenzene groups connected through an optically active linker were enantioselectively incorporated into oligonucleotides (**16**) [56]. Optically pure L- and D-threoninol were used as the chiral linkers. The chirality of the linker significantly affected the melting behavior. The UV-photolytic trans-to-cis isomerization of the duplex formed between the oligonucleotide having one or two D-threoninol-tethered azobenzene moieties (D1 and DD2) and its complementary counterpart caused a dramatic decrease in the T_m value (from 43.9 to 22.4°C for DD2) (Table 1). The duplex was strongly stabilized by the *trans*-azobenzene tethered by D-threoninol. The planar *trans*-azobenzene moiety intercalated between the base pairs stabilizes the duplex as a result of a π–π stacking interaction, while nonplanar *cis*-azobenzene destabilizes the duplex. This strategy was further applied to photoregulation of the transcription reaction of the T7 RNA polymerase by tethering an azobenzene to the promoter [57].

IV. PHOTORESPONSIVE SYNTHETIC CHIRAL POLYMERS

Chiral polymers with optical activity (optically active polymers) can be synthesized by the polymerization of optically active monomers or by the asymmetric polymerization of achiral and prochiral monomers with chiral catalysts or initiators. The latter involves an asymmetrical synthesis process generating a configurational and/or conformational main chain chirality in the polymers, and it has attracted great attention over the past two decades. Details of the synthesis, conformation, and function of optically active polymers have already been thoroughly reviewed [22,26]. We describe two types of photoresponsive optically active synthetic polymers whose optical activities are derived from helicity and configurational chirality in the main chain.

A. Photoresponsive Synthetic Helical Polymers

Synthetic helical polymers exhibiting optical activity due to helicity can be classified into two types with respect to the nature of the helical conformation [26,58]. One is a helical polymer stable even in solution; poly(triarylmethyl methacrylates) (PTrMA) [22], polyisocyanides [24], and polychloral [22] belong to this category. Another is a dynamic helical polymer as exemplified by polyisocyanates [59,60] and polysilanes [30]. The former helix is rigid, and therefore a helical polymer with an excess of single screw sense can be obtained by the asymmetric polymerization of the corresponding monomer with chiral initiators or catalysts. As expected, it may be difficult to vary such a rigid helical conformation by external stimuli including photoirradiation. Nevertheless, a polyisocyanide with an azobenzene pendant was found to undergo a reversible conformational change upon photoirradiation.

Polyisocyanides, typical helical polymers, form a stable 4_1 helical structure when they have a bulky side group [61]. For example, an optically active poly(phenyl isocyanide) bearing an L-menthyl pendant exhibited an intense CD in the $n-\pi^*$ transition region of the imino chromophore main chain [62]. The helical structure was stable in solution even at high temperatures. After UV-irradiation, however, the CD intensity of **17** (30 mer) decreased by 26% accompanied by a slight decrease in the UV-visible spectra [63]. The decrease in optical activity was highly dependent on the molecular weights of the polyisocyanides; the lower molecular weighted **17** (10 mer) almost lost its optical activity after photoirradiation. The origin of this unusual decay in the optical activity is not clear, but syn–anti configurational photoisomerization around the C-N double bond may be occurring, which can bring about relaxation of the helical main chain (Fig. 5).

Photosensitive azobenzene units were further incorporated into the polyisocyanide **17** by random living copolymerization of the corresponding monomers [64]. The helical polyisocyanide **18** (30 mer) contains 33% azobenzene units as pendants and caused a slight decrease in the UV-visible and CD intensities (8%) upon UV-irradiation. Visible light irradiation effected the reverse photochromic process (Fig. 5).

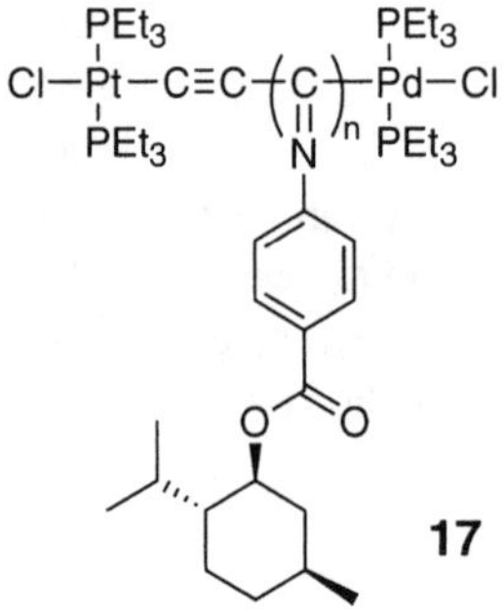

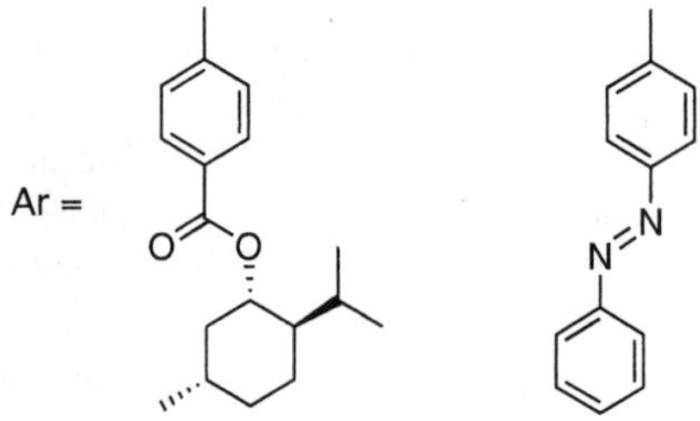

Figure 5 Photoresponsive helical polyisocyanides.

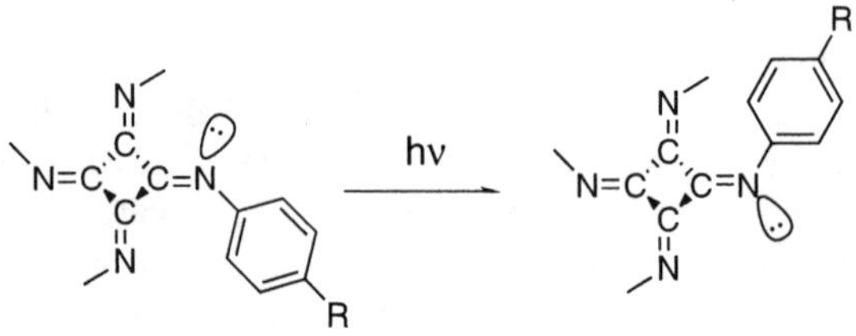

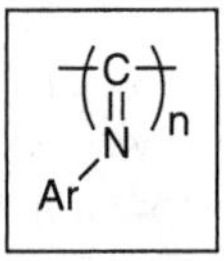

Triphenylmethyl methacrylate (TrMA) and azobenzene-modified methacrylates were randomly copolymerized in toluene at $-78°C$ with chiral catalysts to give optically active helical copolymers (**19** in Fig. 6) [65]. The optical activity (optical rotation) of the copolymers decreased with the increasing content of the azobenzene-modified methacrylates in the copolymers. The single helical conformation of PTrMA is quite stable in solution, but the copolymers of TrMA with less bulky methacrylates cannot keep their helical structure and lose their optical activity during the polymerization or after the polymerization in solution, which is highly dependent on the bulkiness of the comonomers [22]. The copolymer (**19**; $x = 2$) containing 26 mol% azobenzene units, also lost its optical activity upon irradiation within 20 min. This change is due to the helix-to-coil transition of the copolymer and can occur in the dark.

A more remarkable and dramatic change in the conformation of helical polymers upon photoirradiation has been observed for the azobenzene-modified polyisocyanates. Polyisocyanates are typical dynamic helical polymers [59,60]. Even the optically inactive poly(n-hexyl isocyanate) (**20**), which is devoid of stereogenic centers, exists as an equal mixture of right- and left-handed helical conformations. Equilibrium exists in solution between both helices separated by

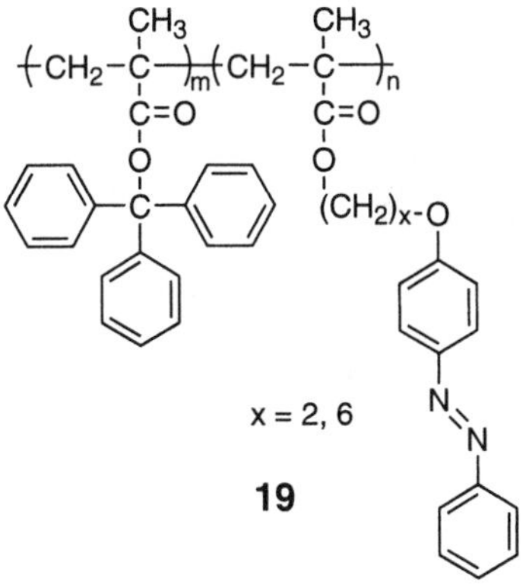

Figure 6 Photoresponsive helical polymethacrylates.

the helix reversal points that move along the polymer backbone (Fig. 7A). However, the helix inversion barriers are very small, so that optically active polyisocyanates with a prevailing homochiral helix can be obtained through the copolymerization of achiral monomers with a small amount of optically active monomers [66–68] or polymerization of achiral isocyanates with optically active initiators [69,70]. This can be considered as a typical example of chirality amplification in a polymer. This high cooperative phenomenon is called by Green et al. the sergeants and soldiers effect (Fig. 7B). Moreover, copolymers composed of a mixture of (R)- and (S)-enantiomers with a small enantiomeric excess (ee) also form a predominantly homochiral helical conformation, and only 12% ee is sufficient to give a homochiral helical polymer for poly(2,6-dimethylheptylisocyanate) [71]. The minority units obey the helical sense of the majority units in order to avoid introducing energetic helical reversals. This phenomenon is called the majority rule (Fig. 7C).

On the basis of these observations, Zentel and Mager designed and synthesized photoresponsive, optically active polyisocyanates and investigated the effect of photoirradiation on the conformation of the polymers 21–23 [31]. They discovered that the helix sense of the optically active polyisocyanates containing photosensitive azobenzene side groups could be controlled by the photoisomerization of the azobenzene moiety from the trans to cis state. Copolymers of chiral isocyanates bearing an azobenzene pendant containing one (21, 22) or two (23) stereogenic centers with an achiral isocyanate were prepared (Fig. 8). When the copolymers 21 and 22 were irradiated with UV light (365 nm), the photoisomerization of the azo moiety from the trans to cis state occurred, and the copolymers exhibited a change in the ORD and CD spectra, reflecting a shift in the population

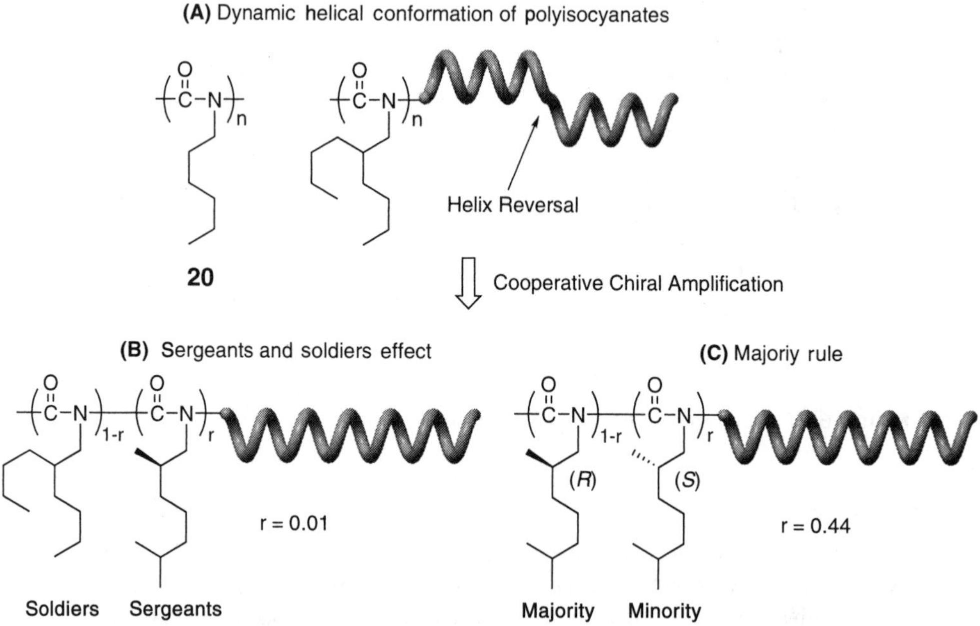

Figure 7 Characteristics of dynamic helical polyisocyanates.

of the helical segments (right- and left-handed helices). However, the preferred helical screw sense was the same in both states (Fig. 8A) [72,73]. On the other hand, the predominant helicity of the copolymer 23 containing two stereogenic centers in the photochromic side chain was switched upon the photochemical trans-cis isomerization (Fig. 8B) [74]. The CD spectral pattern of the polymer was completely inverted upon photoirradiation (helix–helix transition). By incorporating the stereogenic center in the alkyl spacer, which is closer to the main chain, a much higher helical preference and molar ellipticity in the CD was induced at lower chiral chromophore concentrations for 22 compared to 21 with a more remote chiral center [73]. The chiroptical properties of the films consisting of a chiral photochromic copolymer (22) incorporated into the poly(methyl methacrylate) (PMMA) matrix can also be photochemically switched [75]. The photochemically modified helical conformation was stable despite the thermal relaxation of the azo chromophores from the cis to the trans state, which offers the possible use of this system as an optical data storage material.

Circularly polarized light (CPL), often used as a source for the absolute asymmetric synthesis of chiral compounds [76–79], can be used as a trigger

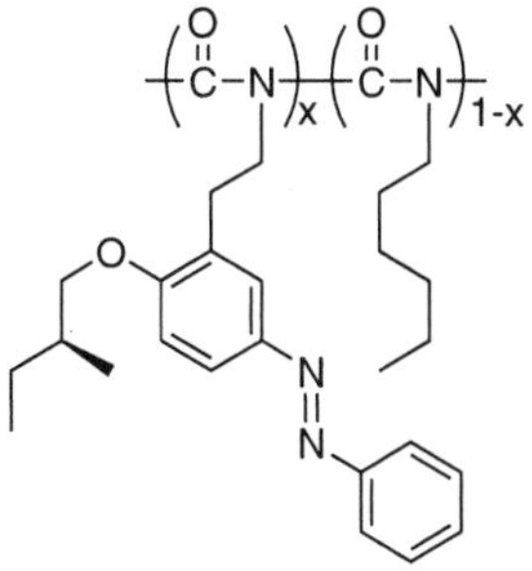

21

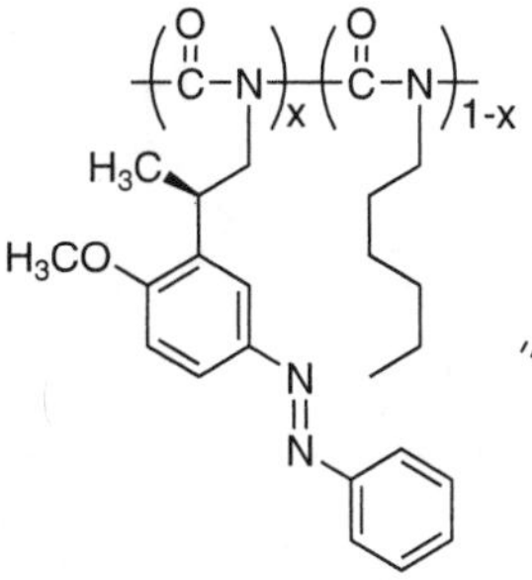

22

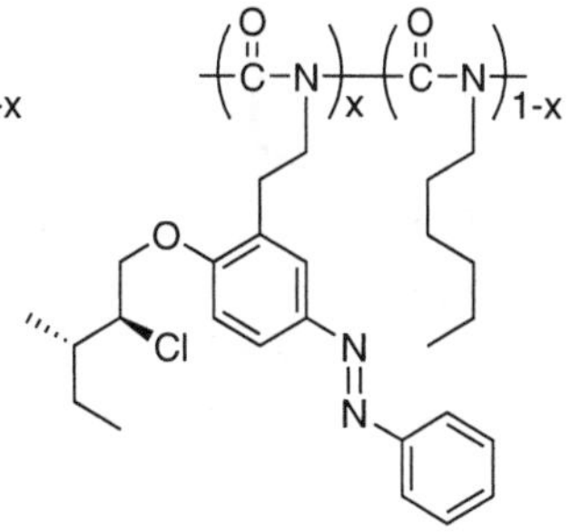

23

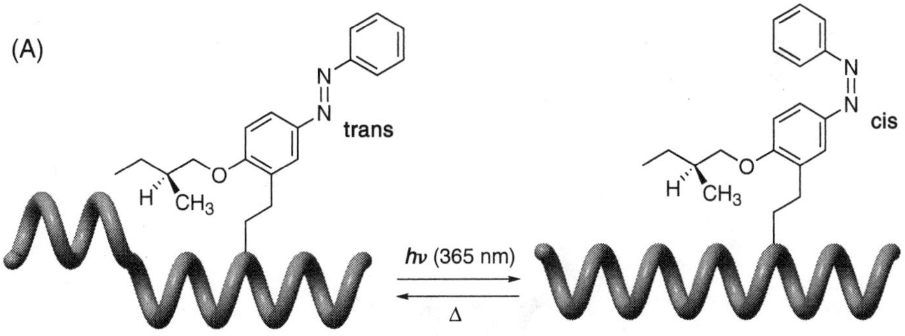

21

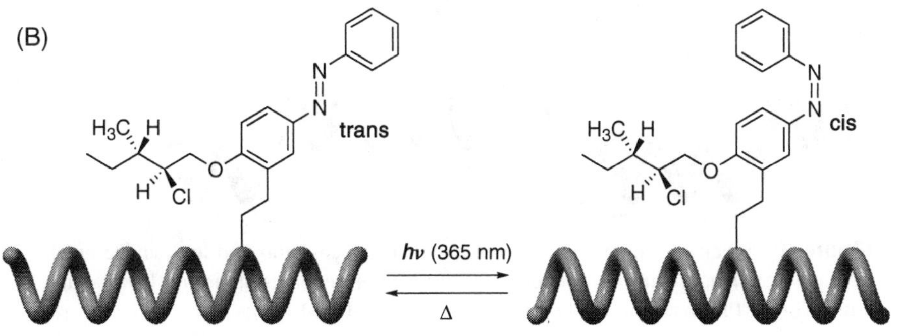

23

Figure 8 Schematic representation of the shift in the equilibrium between left- and right-handed helices (A) and the transition from left- to right-handed helix of polyisocyanates (B) by a photochemical trans–cis isomerization of the azobenzene unit.

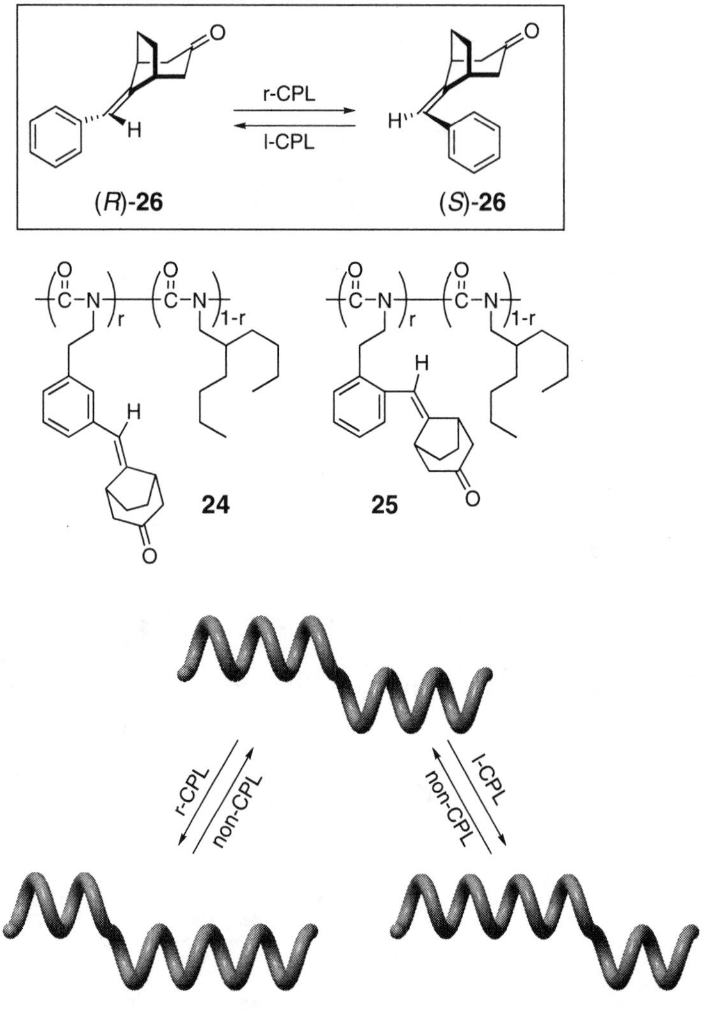

Figure 9 Schematic representation of the shift between right- and left-handed helices of a polyisocyanate with bicycloketone chromophore pendants subjected to irradiation with *r*- or *l*-CPL or noncircularly polarized light (non-CPL).

to induce the helical chirality and control the helix sense of polyisocyanates (Fig. 9). Green et al. prepared a series of polyisocyanate copolymers ($r = 1$) or terpolymers (**24, 25**) consisting of various proportions of chiral units bearing a photoresolvable pendant group and achiral units [80]. They selected the styryl-substituted, axially chiral bicyclo[3.2.1]octane-3-one (**26**) group as the chiral photoresolvable pendant group, which has the property required for reversible photoresolution [81]. When the chiral pendant group is racemic, right- and left-handed helical isocyanate units equally coexist in the main chain, and no CD signal was observed. However, irradiation of **24** and **25** with a right- or left-handed-CPL (r- or l-CPL) in the ketone's chromophore region produced a tiny enantiomeric imbalance (enantiomeric excess (ee) $< 1\%$) in the pendant group, which resulted in measurable CD signals in the polymer backbone region. The majority rule effect transfers the chirality of the pendant group of a very small ee to the polymer backbone as an excess of one helical sense, leading to the amplification of the photoresolution. The CD signal changed sign with a change in the sense of the CPL (Figs. 9 and 10). Further irradiation with unpolarized light racemizes the ketone unit, and the CD signal instantly disappears. This demonstrates that the helical sense of the polymer main chain can be reversibly switched by alternating irradiation with r- or l-CPL or returned to the racemic state by irradiation with unpolarized light. These results were theoretically analyzed based on the

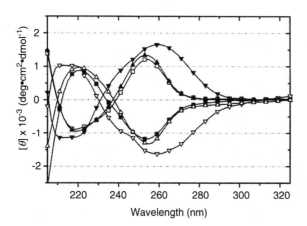

Figure 10   CD spectra of orthocopolymer **25** ($r = 1$) in methylene chloride and tetrahydrofuran (1 : 1) solution ($\blacktriangledown,\triangledown$), meta-terpolymers **24** ($r = 0.02$) in n-hexane solution ($\blacktriangle,\triangle$), and orthoterpolymer **25** ($r = 0.04$) in n-hexane solution ($\blacksquare,\square$) irradiated with CPL. The filled up-triangle, down-triangle, and square are CD spectra of the polymers irradiated with r-CPL, and the open up-triangle, down-triangle, and square are CD spectra of the polymers irradiated with l-CPL. (Reproduced from Ref. 80; © 2000, American Chemical Society.)

quenched random-field Ising model [80]. This is the first clear demonstration of the amplification of a photochemical resolution by the response of the conformation of a polymer backbone. Dynamic and cooperative characteristics of the helical polymers play a dominant role for this intriguing amplification process.

Polysilanes are also dynamic helical polymers like polyisocyanates [30]. This suggests that photoresponsive helical polymers may be designed by using polysilanes as a motif. However, polysilanes are sensitive to light, resulting in photolysis that produces low molecular weight oligomers upon photoirradiation, so that photoregulation of a helical polysilane by incorporating chromophores, such as azobenzene units in the side group, is not possible. However, Fujiki utilized the photoactivity of the polysilanes and developed an intriguing method to reconstruct a single-handed helix from an optically active helical polysilane consisting of both the helices with an excess of a screw sense by the cut-and-paste technique [82].

Poly(methyl-((S)-2-methylbutyl)silane) (27) is a flexible, chainlike, optically active helical polymer in solution and shows bisigned CD bands in the main chain region. On the basis of comprehensive spectroscopic investigations of the polymer conformation together with viscosity measurements of the polymer in solution, it was found that the polymer was composed of both tight P-helical and loose M-helical segments in the same polymer chain of 27. The segments coexist like multiblock copolymers (Fig. 11) and have different conformation-sensitive absorption bands (4.0 and 4.5 eV), which exhibit positive and negative Cotton bands, respectively (Fig. 12B). Selective photoirradiation of a segment (loose M-helix at 4.0 eV) in CCl_4 induced a selective photolysis of the segment, resulting in tight P-helix segments with the Si-Cl terminals. The Cl-terminated, reactive P-tight helix segments were further allowed to react to each other in the presence of Na in hot toluene to reconstruct a polymer 27* consisting of predominantly tight P-helix segments. The resulting polysilane mainly possesses a positive Cotton CD band at 4.5 eV associated with a weak negative Cotton CD band at 4.0 eV. This cut-and-paste technique was interesting and useful for preparing a homochiral helical polysilane from a helical polysilane with an imperfect helix sense.

B. Photoresponsive Optically Active Vinyl Polymers with Configurational Chirality

A large number of azobenzene-based amorphous and liquid crystalline polymers, particularly polyacrylates and polymethacrylates with chiral azobenzene pendants, have been prepared for the development of data storage and photonic devices [1–3,11–14]. For instance, the introduction of optically active mesogenic azobenzene residues into the side groups of the polymers produces chiral nematic and cholesteric phases, which are regulated by photoisomerization of the azobenzene units [10,14]. In most cases, however, the optical activity and chiroptical

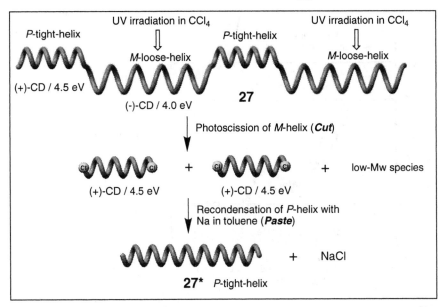

Figure 11 Schematical illustration of cut-and-paste method for the selective photolysis of M-loose-helix and reconstruction of P-screw-sense helical polymer **27***.

properties of the photoresponsive polymers are governed by the chiral pendants. Optical activity derived from conformational and configurational main chain chirality may not be able to contribute to the chiroptical functions [15], particularly for the homopolymers, even if some higher ordered stereocontrol (tacticity) is realized during the polymerization process [22]. In addition, vinyl polymers are transparent in the main chains (CD inactive), in sharp contrast to polypeptides, polyisocyanates, and polysilanes described above. Hence it is difficult to discuss the conformational and configurational chirality of the main chains using conventional spectrometers. However, when nonchromophoric chiral vinyl monomers are copolymerized with achiral vinyl monomers having a chromophoric pendant, such as an azobenzene residue, optical activity may be induced on the achiral

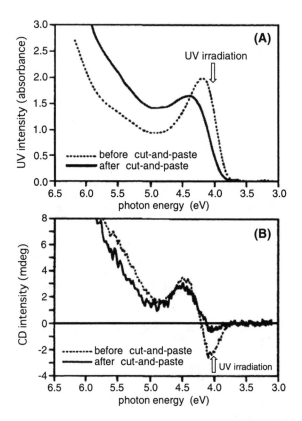

Figure 12 Change in UV (A) and CD (B) spectra of **27** and **27*** in isooctane at 20°C before and after UV-irradiation followed by reconstruction using cut-and-paste method. (Reproduced from Ref. 82; © 1994, American Chemical Society.)

chromophore region, which can be detected by CD spectroscopy [15,22]. This polymerization is called asymmetric synthesis polymerization. Even in the absence of a helical conformation, the main chain of the polymers can possess true stereogenic centers induced by chiral monomer units during the copolymerization, which contributes more or less to their optical activity. As for the homopolymers, true stereogenic centers with optical activity are induced only in a specially limited case [22].

Figure 13 shows several optically active vinyl polymers and copolymers with azobenzene or stilbene residues in the side groups. All these polymers were prepared by the conventional radical polymerization, so that the main chains should be atactic or syndiotactic and may not be able to take a helical conformation even in the presence of chiral pendants. The homopolymers (**28–31**) [83–85]

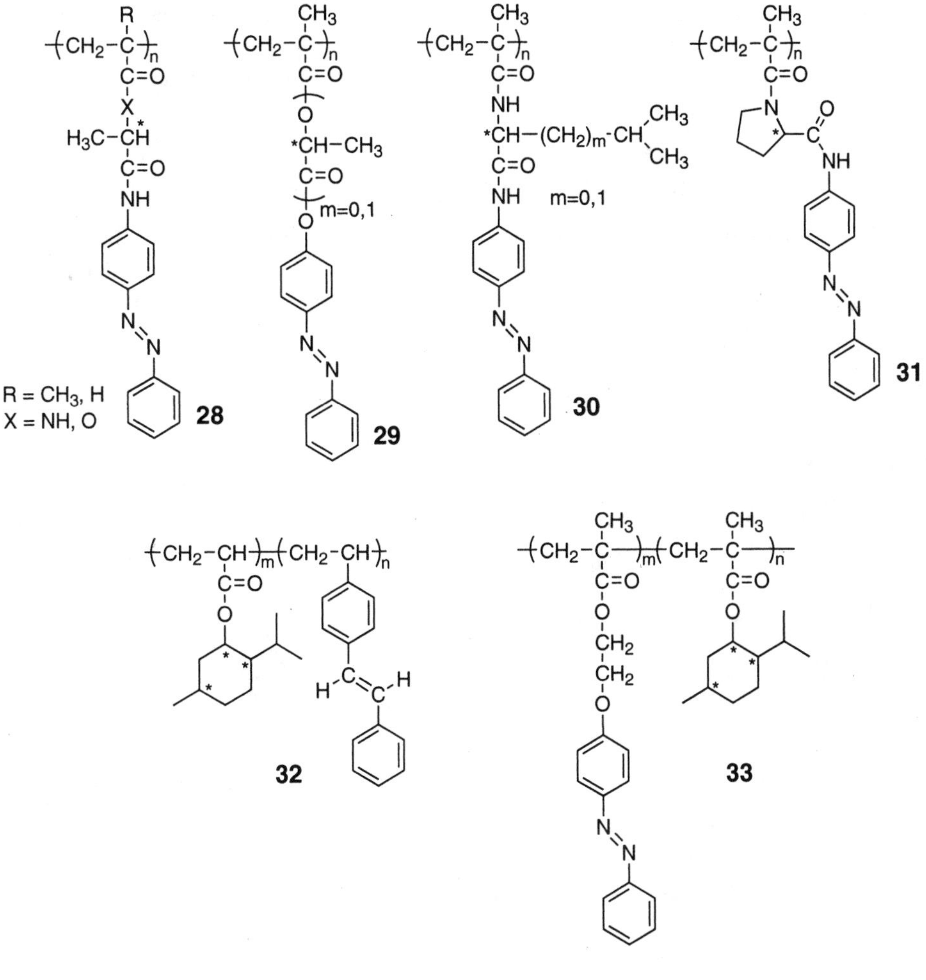

Figure 13 Photoresponsive chiral vinyl homopolymers and copolymers.

and copolymers (**32, 33**) [86,87] exhibited an induced CD in the stilbene or azobenzene chromophore regions. Photoisomerization of the chromophore units caused a series of reversible changes in their CD and absorption spectra. The photoinduced order–disorder conformational transition may be ruled out for the homopolymers, but the existence of an ordered secondary structure was pointed out for the copolymer **32** [86]. The tacticity effect on the chiroptical and photochromic properties was also investigated for the copolymer **33** [87]. A highly isotactic copolymer was prepared by anionic polymerization in order to investigate the relationship between the microstructure (tacticity) and chiroptical properties induced by the photoirradiation in synthetic chiral photoresponsive polymers. However, no appreciable effect was observed upon photoisomerization of the pendant azobenzene units.

C. Photoresponsive Optically Active Polyamides

Optical activity can also be induced on achiral azobenzene units when they are incorporated into the polymer main chains through covalent bonding to chiral subunits (34–37 in Fig. 14) [88–92]. Axially chiral amino- or carboxy-substituted biphenyl and binaphthyl and tartaric acid derivatives were used as the optically

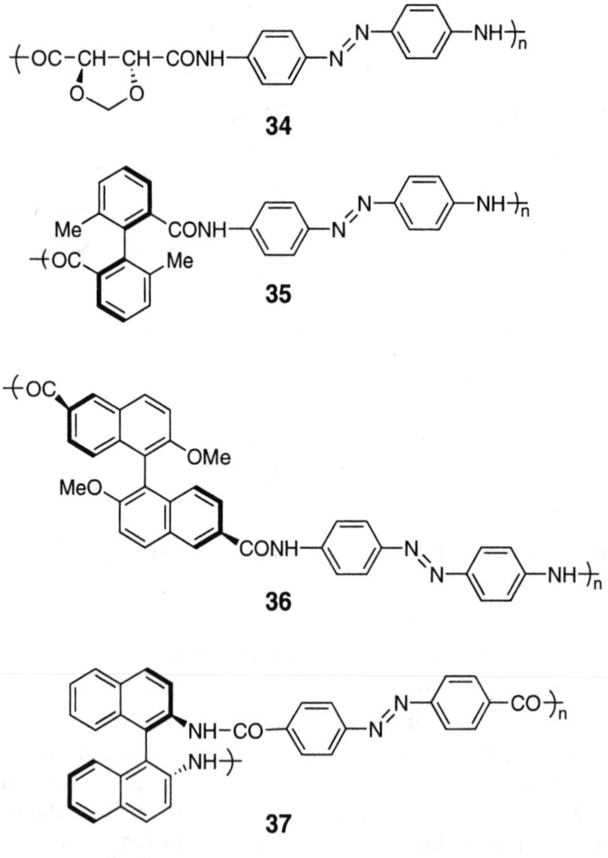

Figure 14 Photoresponsive chiral polyamides.

active components. These chiral diamines or dicarboxylic acids were polymerized with 4,4′-diamino- or 4,4′-dicarboxyazobenzene by a conventional polycondensation reaction to afford optically active polyamides. The polyamides showed a large optical rotation together with Cotton effects in the achiral azobenzene regions as well as in the chiral aromatic regions before photoisomerization. However, the optical activity of all the polyamides decreased through the trans-to-cis photoisomerization. These observations were considered to be due to the helix–coil transition of the polymer main chains during the trans-cis isomerization. Model compounds, however, also showed a similar change in their CD spectra. Optical activity induced in the achiral azobenzene units of the *trans*-polyamides may indicate the existence of a helical conformation with an excess screw sense, but the helical conformation might be dynamic in nature and may be very elusive. Since the specific rotation of **37** changed from the positive to the negative direction during the photoisomerization from the trans to the cis form, **37** may be used as a photoswitchable material.

V. CHIRALITY OR HELICITY INDUCTION ON ACHIRAL PHOTORESPONSIVE POLYMER FILMS BY CIRCULAR POLARIZED LIGHT

The irradiation of a right or left CPL to photoreactive racemic or achiral compounds may lead to partial photoresolution or asymmetric photosynthesis, respectively, as mentioned previously [79]. Although the expected *ee* values of the products are quite low, if chirality is transferred to the dynamic polymer main chains [80] or the external chiral matrix [79,93] (for example, liquid crystalline) with a large amplification, CPL can be used as a powerful tool to develop a photochemical switching material. In these cases, CPL triggers the chemical reactions, and the observed optical activity originates from the enantiomerically enriched chiral molecules attached to the polymer backbone or in liquid crystals. However, Nikolova et al. reported an interesting unprecedented chirality induction (CD and optical rotation) to achiral liquid crystalline (**38**) [94,95] and amorphous (**39**) [96] azobenzene-containing polymers by irradiation using a circular polarized Ar laser beam at 488 nm (Fig. 15). The amorphous film **39** required preirradiation of the linearly polarized (LP) light before the CPL irradiation for the chirality induction. The *r*- and *l*-CPLs induced an opposite optical activity from each other on the films. These effects were explained by photoinduced changes in the structure of the polymer films through reorientation of the azobenzene chromophores followed by circular momentum transfer from the CPL.

Similar chirality induction by CPL at 488 nm was also observed for achiral liquid crystalline (**40**) [97] and amorphous polymers (**41**) (Fig. 16) [98]. Irradiation of CPL of the opposite handedness induced mirror-image CD spectra on the

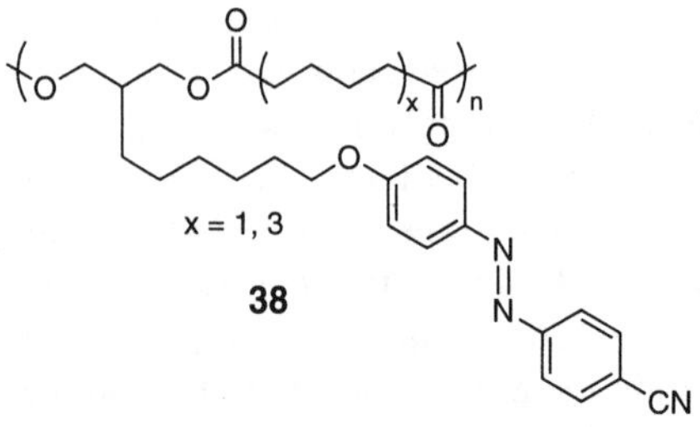

38

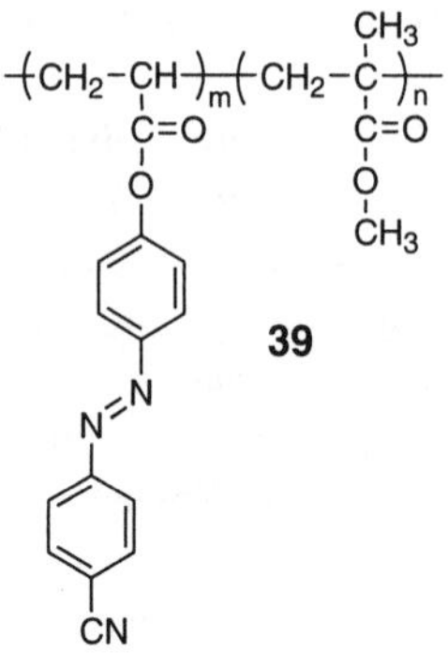

39

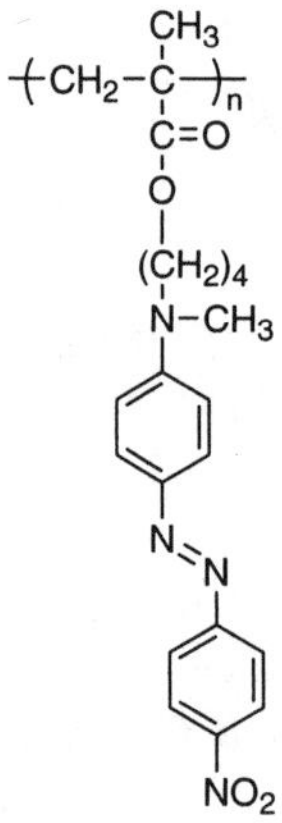

40

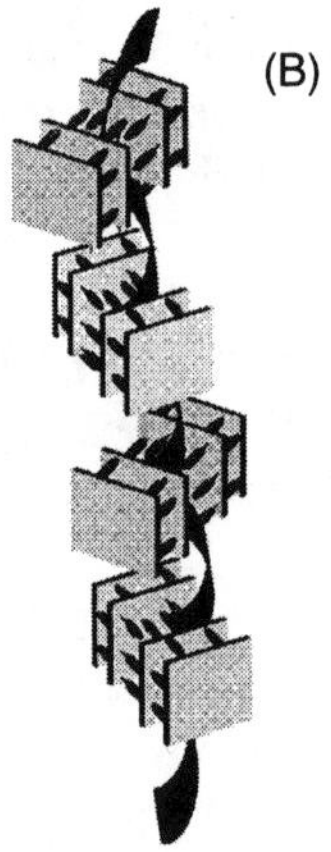

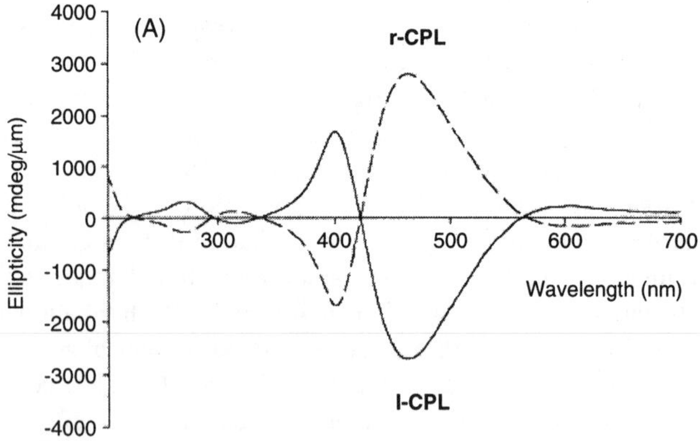

Figure 15 CD spectra of thin films (140 nm) of **40** (A) recorded after irradiation with right- or left-CPL (514 nm, 75 mW/cm^2). The schematic model for the helically chiral assembly is also shown in (B). (Reproduced from Ref. 97; © 2000, American Chemical Society.)

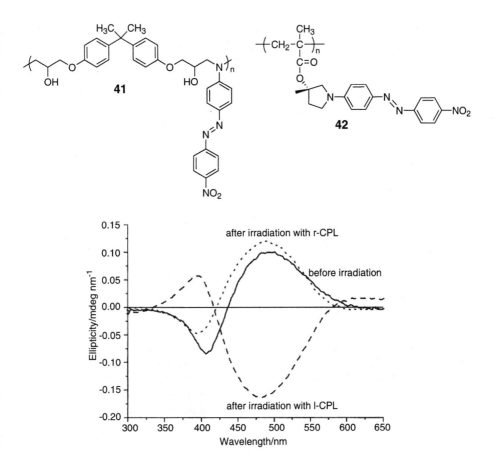

Figure 16 CD spectra of a thin film (340 nm) of **41** on glass, as prepared (full line) and after sequential irradiation of *l*-CPL (dashed line) and *r*-CPL (dotted line) at 488 nm ($I \approx 160$ mW cm^{-2}). (Reproduced from Ref. 99; © 2002, Wiley-VCH.)

40 film (Fig. 15A). Switching the chirality that is induced to the film is possible by alternating the irradiation with *r*- and *l*-CPLs. A helically arranged chiral supramolecular assembly of the pendant azobenzene units in the smectic liquid crystalline phase is proposed for the origin of the chirality induction (Fig. 15B). On the other hand, the amorphous film **41** did not show any optical activity by irradiation with *r*- or *l*-CPL at 488 nm, whereas the film exhibited an intense CD when the linear polarized portion to the incident CPL (elliptically polarized light [EPL]) was introduced [98]. These results suggest that the linear orientation of the azobenzene chromophores is a key factor for chirality induction with CPL.

A thin film of an optically active nonliquid crystalline polymer (**42**) showed a CD, the sign of which, however, completely inverted upon irradiation with *l*-CPL [99]. In this case, a preliminary ordering (orientation) of the film with LP irradiation was not necessary for the inversion of the CD signal. *r*-CPL irradiation induced a small change in the CD pattern, resulting in the restoration of the original chiral information (Fig. 16).

This method may have a great advantage because the chirality can be induced on achiral materials with chromophoric units only by the handedness of the light, although the origin and mechanism of the induced chirality by CPL irradiation with or without LP irradiation are not fully understood, and they should be elucidated before concluding the induction of chirality and after careful evaluation of the CD data, since optical activity in the solid, particularly in the solid-state CD using a commercially available instrument, is necessarily accompanied by an artifact originating from the macroscopic anisotropies of the sample, so that special care must be taken [100,101].

VI. PHOTOCONTROL OF CHIRAL RECOGNITION BY CHIRAL POLYMERS

The photocontrol of chiral recognition is one of the most challenging themes in host–guest and supramolecular chemistry. However, it is still very difficult to realize [16]. One of the pioneering works was performed by Okamoto et al. about two decades ago [102]. They prepared cellulose tris(4-phenylazophenylcarbamate) (**43**) having photoresponsive pendant groups and demonstrated that the cellulose derivative showed a different chiral recognition ability dependent on the structure of the azobenzene pendants (trans and cis form) when it was used as a chiral stationary phase (CSP) in high-performance liquid chromatography (HPLC). Polysaccharide derivatives, in particular phenylcarbamates of cellulose and amylose, have excellent chiral recognition abilities as CSPs. These can be used to resolve a wide range of aliphatic and aromatic racemic compounds without derivatization, and some polysaccharide-based CSPs have already been commercialized [103].

Figure 17 shows the resolution of the *trans*-stilbene oxide (**44**) and Tröger base (**45**) on a **43** column with different trans contents. The cellulose derivative exhibited CD bands in the azo chromophore regions, which remarkably changed upon trans-to-cis photoisomerization in solution; the CD intensity at 365 nm significantly decreased with increasing cis content. This change was reversible. The photoisomerization of the azobenzene units from the trans to the cis form in the film state was difficult, and therefore, the cis-rich **43** (trans = 30%) was first prepared in solution by the irradiation of light; then the polymer was coated on a macroporous silica gel, followed by packing into an HPLC column. The

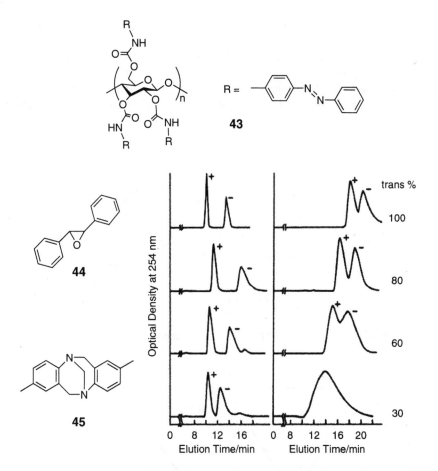

Figure 17 Resolution of **44** and **45** on a **43** column. Eluent, hexane-2-propanol (90 : 10 vol%); flow rate, 0.5 mL/min; temperature, 25°C. (Reproduced from Ref. 102; © 1986, Chemical Society of Japan.)

trans content of **43** in the HPLC column gradually increased with time, and the chiral recognition ability of **43** with different trans contents was evaluated. The trans isomer **43** completely resolved many enantiomers including **44** and **45**, while the cis-rich **43** resolved only **44** into two peaks, probably due to a disordered structure of *cis*-**43**. On the *cis*-**43** column, most of the racemic compounds were less retained than the *trans*-**43** column. This means that expeditious chromatographic resolution of enantiomers may be possible, if reversible trans–cis photoisomerization of **43** on a silica gel is possible in a UV-permeable glass column [104].

Polysaccharide phenylcarbamate derivatives have a great advantage for the easy preparation of a film (membrane), which can be used as a new device for the rapid separation of enantiomers through enantioselective adsorption [105]. Photoresponsive cellulose and amylose membranes bearing azobenzene residues

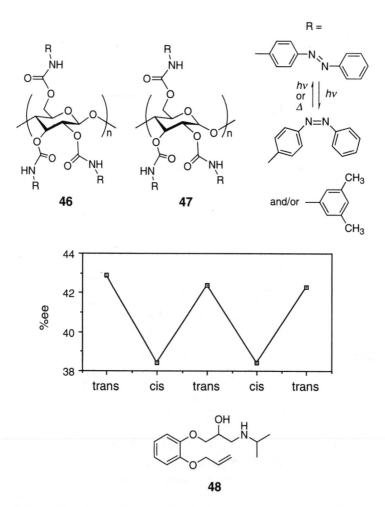

Figure 18 Changes in *ee* of **48** adsorbed on the **46** membrane during the trans–cis isomerization. The portions of trans and cis were 100 and 80%, respectively. (Reproduced from Ref. 106; © 1995, American Chemical Society.)

(**46, 47**) respond to light and/or heat and exhibited chiral recognition [106]. The trans membranes showed a higher enantioselectivity than the cis membranes. The enantioselectivity was thus for the first time reversibly controlled in an on–off fashion by photoisomerization of the pendant azobenzene residues as shown in Fig. 18. Oxprenolol (**48**) was enantioselectively adsorbed in 43 (trans) and 38% *ee* (cis) on the photoresponsive cellulose membrane.

VII. CONCLUSION

In this chapter, the photoresponsive properties and functions of chromophore-containing chiral polymers with a configurational or conformational chirality in the polymer main chains are mainly described. Photosensitive chromophores such as azobenzene derivatives are usually incorporated into polymers as pendants. Photoisomerization of the chromophores can induce more or less of a conformational change in the attached polymers. The most important feature of chiral polymers is optical activity, and therefore such a conformational change can be transformed into a change in the optical activity of the polymers. If a change that occurred in the remote side chain upon photoirradiation can be transformed into the main chain conformational change with a large amplification, such systems may provide the basis of constructing light-driven switching systems. Helical polymers, and in particular dynamic helical polymers like polyisocyanates with an excess of a screw sense, might be good candidates for this purpose. Moreover, if the main chain is also chromophoric as in conjugated polymers [107], the output signal derived from the order–disorder or helix–helix conformational changes induced by CPL or non-CPL photoirradiation may be largely enhanced when chiral conjugated polymers have a dynamic helical structure [108–111].

To date, a number of optically active polymers, particularly helical polymers with an excess one-handedness, have been prepared with much interest. However, helical poly(triaryl methacrylates) are the only examples of successful application to HPLC stationary phases [22,103]. Helical polymers combined with photoresponsive molecules will offer promising functions and applications as realized by natural macromolecules.

REFERENCES

1. Irie M. Adv Polym Sci 1990; 94:27–67.
2. Nuyken O, Scherer C, Baindl A, Brenner AR, Dahn U, Gärtner R, Röhrich S K, Kollefrath R, Matusche P. Prog Polym Sci 1997; 22:93–183.
3. Natansohn A, Rochon P. Chem Rev 2002; 102:4139–4175.
4. Kinoshita T. Prog Polym Sci 1995; 20:527–583.

5. Willner I, Rubin S. Angew Chem Int Ed Engl 1996; 35:367–385.
6. Willner I. Acc Chem Res 1997; 30:347–356.
7. Pieroni O, Fissi A, Popova G. Prog Polym Sci 1998; 23:81–123.
8. Ciardelli F, Pieroni O. In. Molecular Switches. Weinheim: Wiley-VCH, 2001 Chap. 13:399–441.
9. Pieroni O, Fissi A, Angelini N, Lenci F. Acc Chem Res 2001; 34:9–17.
10. Ichimura K. Supramol Sci 1996; 3:67–82.
11. Natansohn A, Rochon P. Adv Mater 1999; 11:1387–1391.
12. Kawata S, Kawata Y. Chem Rev 2000; 100:1777–1788.
13. Delaire JA, Nakatani K. Chem Rev 2000; 100:1817–1845.
14. Ichimura K. Chem Rev 2000; 100:1847–1873.
15. Ciardelli F, Pieroni O, Fissi A, Carlini C, Altomare A. Brit Polym J 1989; 21: 97–106.
16. Feringa BL, van Delden RA, Koumura N, Geertsema EM. Chem Rev 2000; 100: 1789–1816.
17. Feringa BL. Acc Chem Res 2001; 34:504–513.
18. Irie M. Chem Rev 2000; 100:1685–1716.
19. Yokoyama Y. Chem Rev 2000; 100:1717–1739.
20. Oosterling MLCM, Schoevaars AM, Haitjema HJ, Feringa BL. Isr J Chem 1996; 36:341–348.
21. Okamoto Y, Yashima E. Prog Polym Sci 1990; 15:263–298.
22. Okamoto Y, Nakano T. Chem Rev 1994; 94:349–372.
23. Gellman SH. Acc Chem Res 1998; 31:173–180.
24. Rowan AE, Nolte RJM. Angew Chem Int Ed 1998; 37:63–68.
25. Hill DJ, Mio MJ, Prince RB, Hughes TS, Moore JS. Chem Rev 2001; 101: 3893–4011.
26. Nakano T, Okamoto Y. Chem Rev 2001; 101:4013–4038.
27. Brunsveld L, Folmer BJB, Meijer EW, Sijbesma RP. Chem Rev 2001; 101: 4071–4097.
28. Cornelissen JJLM, Rowan AE, Nolte RJM, Sommerdijk NAJM. Chem Rev 2001; 101:4039–4070.
29. Green MM, Cheon K, Yang S, Park J, Swansburg S, Liu W. Acc Chem Res 2001; 34:672–680.
30. Fujiki M. Macromol Rapid Commun 2001; 22:539–563.
30. Fujiki M, Koe JR, Terao K, Sato T, Teramoto A, Watanabe J. Polym J 2003; 35: 297–344.
31. Mayer S, Zentel R. Prog Polym Sci 2001; 26:1973–2013.
32. Goodman M, Kossoy A. J Am Chem Soc 1966; 88:5010–5015.
33. Ueno A, Anzai J, Osa T. J Polym Sci Polym Lett Ed 1977; 15:407–410.
34. Houben JL, Pieroni O, Fissi A, Ciardelli F. Biopolymers 1978; 17:799–804.
35. Bradbury EM, Carpenter BG, Goldman H. Biopolymers 1968; 6:837–850.
36. Ueno A, Takahashi K, Anzai J, Osa T. Macromolecules 1980; 13:459–460.
37. Watanabe J, Okamoto S, Abe A. Liq Cryst 1993; 15:259–263.
38. Ueno A, Anzai J, Osa T, Kadoma Y. Bull Chem Soc Jpn 1979; 52:549–554.
39. Ueno A, Takahashi K, Anzai J, Osa T. J Am Chem Soc 1981; 103:6410–6415.

40. Ueno A, Adachi K, Nakamura J, Osa T. J Poly Sci Poly Chem 1990; 28:1161–1170.
41. Ueno A, Morikawa Y, Anzai J, Osa T. Makromol Chem Rapid Commun 1984; 5: 639–642.
42. Ueno A, Nakamura J, Adachi K, Osa T. Makromol Chem Rapid Commun 1989; 10:683–686.
43. Watanabe J, Okamoto S, Satoh K, Sakajiri K, Furuya H, Abe A. Macromolecules 1996; 29:7084–7088.
44. Aoyama M, Watanabe J, Inoue S. J Am Chem Soc 1990; 112:5542–5545.
45. Pieroni O, Houben JL, Fissi A, Costantino P, Ciadelli F. J Am Chem Soc 1980; 102:5913–5915.
46. Fissi A, Houben JL, Rosato N, Lopes S, Pieroni O, Ciadelli F. Makromol Chem Rapid Commun 1982; 3:29–33.
47. Ciardelli F, Fabbri D, Pieroni O, Fissi A. J Am Chem Soc 1989; 111:3470–3472.
48. Fissi A, Pieroni O, Angelini N, Lenci F. Macromolecules 1999; 32:7116–7121.
49. Yamamoto H, Nishida A. Macromolecules 1986; 19:943–944.
50. Fissi A, Pieroni O, Balestreri E, Amato C. Macromolecules 1996; 29:4680–4685.
51. Yamamoto H, Nishida A. Polym Int 1991; 24:145–148.
52. Fissi A, Pieroni O, Ciardelli F. Biopolymers 1987; 26:1993–2007.
53. Asanuma H, Ito T, Yoshida T, Liang X, Komiyama M. Angew Chem Int Ed Engl 1999; 38:2393–2395.
54. Asanuma H, Liang X, Yoshida T, Yamazawa A, Komiyama M. Angew Chem Int Ed Engl 2000; 39:1316–1318.
55. Liang X, Asanuma H, Komiyama M. J Am Chem Soc 2002; 124:1877–1883.
56. Asanuma H, Takarada T, Yoshida T, Tamaru D, Liang X, Komiyama M. Angew Chem Int Ed Engl 2001; 40:2671–2673.
57. Asanuma H, Tamaru D, Yamazawa A, Liu M, Komiyama M. ChemBioChem 2002: 786–789.
58. Yashima E. Anal Sci 2002; 18:3–6.
59. Green MM, Park J, Sato T, Teramoto A, Lifson S, Selinger RLB, Selinger JV. Angew Chem Int Ed 1999; 38:3138–3154.
60. Green MM, Peterson NC, Sato T, Teramoto A, Cook R, Lifson S. Science 1995; 268:1860–1866.
61. Nolte RJM. Chem Soc Rev 1994:11–19.
62. Takei F, Yanai K, Onitsuka K, Takahashi S. Chem Eur J 2000; 6:983–993.
63. Iyoda T, Ito S, Kawai T, Takei F, Onitsuka K, Takahashi S. Polym Prep Jpn 1999; 48:1819–1820.
64. Shiga K, Ishii K, Abe J, Iyoda T. Polym Prep Jpn 2001; 50:1586–1587.
65. Chen JP, Gao JP, Wang ZY. J Polym Sci Polym Chem 1997; 35:9–16.
66. Green MM, Reidy MP, Johnson RJ, Darling G, O'Leary DJ, Willson G. J Am Chem Soc 1989; 111:6452–6454.
67. Gu H, Nakamura Y, Sato T, Teramoto A, Green MM, Jha SK, Andreola C, Reidy MP. Macromolecules 1998; 31:6362–6368.
68. Jha SK, Cheon K, Green MM, Selinger JV. J Am Chem Soc 1999; 121:1665–1673.
69. Okamoto Y, Matsuda M, Nakano T, Yashima E. Polym J 1993; 25:391–396.
70. Okamoto Y, Matsuda M, Nakano T, Yashima E. J Polym Sci Part A: Polym Chem 1994; 32:309–315.

71. Green MM, Garetz BA, Munoz B, Chang H, Hoke S, Cooks RG. J Am Chem Soc 1995; 117:4181–4182.
72. Mäller M, Zentel R. Macromolecules 1994; 27:4404–4406.
73. Mayer S, Zentel R. Macromol Chem Phys 1998; 199:1675–1682.
74. Maxein G, Zentel R. Macromolecules 1995; 28:8438–8440.
75. Mayer S, Zentel R. Macromol Rapid Commun 2000; 21:927–930.
76. Inoue Y. Chem Rev 1992; 92:741–770.
77. Rau H. Chem Rev 1983; 83:535–547.
78. Avalos M, Babiano R, Cintas P, Jiménez JL, Palacios JC. Chem Rev 1998; 98: 2391–2404.
79. Feringa BL, van Delden RA. Angew Chem Int Ed 1999; 38:3418–3438.
80. Li J, Schuster GB, Cheon K, Green MM, Selinger JV. J Am Chem Soc 2000; 122: 2603–2612.
81. Zhang Y, Schuster GB. J Org Chem 1995; 60:7192–7197.
82. Fujiki M. J Am Chem Soc 1994; 116:11976–11981.
83. Altomare A, Solaro R, Angiolini L, Caretti D, Carlini C. Polymer 1995; 36: 3819–3824.
84. Angiolini L, Caretti D, Giorgini L, Salatelli E, Altomare A, Carlini C, Solaro R. Polymer 1998; 39:6621–6629.
85. Carlini C, Fissi A, Maria A, Galletti R, Sbrana G. Macromol Chem Phys 2000; 201:1161–1168.
86. Altomare A, Carlini C, Panattoni M, Solaro R. Macromolecules 1984; 17: 2207–2213.
87. Altomare A, Ciardelli F, Lima R, Solaro R. Chirality 1991; 3:292–298.
88. Kondo F, Iwaizumi T, Kimura H, Takeishi M. Kobunshi Ronbunshu 1997; 54: 908–913.
89. Kondo F, Kakimi S, Kimura H, Takeishi M. Polym Int 1998; 46:339–344.
90. Kondo F, Takahashi D, Kimura H, Takeishi M. Polymer J 1998; 30:161–162.
91. Lustig SR, Everlof GJ, Jaycox GD. Macromolecules 2001; 34:2364–2372.
92. Jaycox GD. Polymer J 2002; 34:280–290.
93. Huck NPM, Jager WF, de Lange B, Feringa BL. Science 1996; 273:1686–1688.
94. Nikolova L, Todorov T, Ivanov M, Andruzzi F, Hvilsted S, Ramanujam PS. Optl Mater 1997; 8:255–258.
95. Naydenova I, Nikolova L, Ramanujam PS, Hvilsted S. J Opt A Pure Appl Opt 1999; 1:438–441.
96. Ivanov M, Naydenova I, Todorov T, Nikolova L, Petrova T, Tomova N, Dragosti-nova V. J Mod Opt 2000; 47:861–867.
97. Iftime G, Labarthet FL, Natansohn A, Rochon P. J Am Chem Soc 2000; 122: 12646–12650.
98. Kim M-J, Shin B-G, Kim J-J, Kim D-Y. J Am Chem Soc 2002; 124:3504–3505.
99. Angiolini L, Bozio R, Giorgini L, Pedron D, Turco G, Daurŭ A. Chem Eur J 2002; 8:4241–4247.
100. Shindo Y, Ohmi Y. J Am Chem Soc 1985; 107:91–97.
101. Harada T, Kuroda R. Chem Lett 2002:326–327.
102. Okamoto Y, Sakamoto H, Hatada K, Irie M. Chem Lett 1986:983–986.

103. Okamoto Y, Yashima E. Angew Chem Int Ed 1998; 37:1020–1043.
104. Nakagama T, Yamaguchi A, Hirasawa K, Yoshida K, Uchiyama K, Hobo T. Anal Sci 2002; 18:49–53.
105. Yashima E, Noguchi J, Okamoto Y. Chem Lett 1992:1959–1962.
106. Yashima E, Noguchi J, Okamoto Y. Macromolecules 1995; 28:8368–8374.
107. McQuade DT, Pullen AE, Swager TM. Chem Rev 2000; 100:2537–2574.
108. Yashima E, Okamoto Y. In: Circular Dichroism: Principles and Applications. 2d ed. New York: John Wiley, 2000, Chap. 18:521–546.
109. Yashima E, Matsushima T, Okamoto Y. J Am Chem Soc 1997; 119:6345–6359.
110. Yashima E, Maeda K, Okamoto Y. Nature 1999; 399:449–451.
111. Yashima E, Maeda K, Sato O. J Am Chem Soc 2001; 123:8159–8160.

Index